Textbook of Medical Biochemistry

SECOND EDITION

S. RAMAKRISHNAN, M.A., Ph.D.
*Professor and Head of the Department of Biochemistry, Rajah Muthiah Medical College and Hospital,
Annamalai University, Annamalai Nagar
Formerly Professor and Head of the Department of Biochemistry, JIPMER, Pondicherry*

K.G. PRASANNAN, M.Sc., Ph.D.
*Associate Professor of Biochemistry, Faculty of Medicine, Al-Fateh University, TRIPOLI, Libya
Formerly Assistant Professor of Biochemistry, JIPMER, Pondicherry*

R. RAJAN M.B.B.S., M.D.
Formerly Assistant Professor of Biochemistry, JIPMER, Pondicherry

Orient Longman

Textbook of Medical Biochemistry
© **Orient Longman Limited 1980, 1989**

First Published 1980
Reprinted 1982, 1983, 1985, 1986
Second edition 1989
Reprinted 1990 (twice), 1991, 1994

ISBN 0 86131 917 6

Orient Longman Limited

Registered Office:

3-6-272 Himayatnagar, Hyderabad 500 029 (A.P.), India

Other Offices

Kamani Marg, Ballard Estate, Bombay 400 038
17 Chittaranjan Avenue, Calcutta 700 072
160 Anna Salai, Madras 600 002
1/24 Asaf Ali Road, New Delhi 110 022

80/1 Mahatma Gandhi Road, Bangalore 560 001
Plot No. 365, Saheed Nagar, Bhubaneshwar 751 007
S. C. Goswami Road, Panbazar, Guwahati 781 001
3-6-272 Himayatnagar, Hyderabad 500 029 (A.P.)
House No. 28/31, 15 Ashok Marg, Lucknow 226 001
1st Floor, City Centre, Ashok Govind Mitra Road, Patna 800 004

Published by
Orient Longman Ltd.,
160 Anna Salai, Madras 600 002.

Phototypeset by
Printers Plates
231, Royapettah High Road, Madras 600 01.

Printed in India by offset at
Ashwin Printing Agency,
40, Peters Road,
Royapettah,
Madras 600 014.

TEXTBOOK OF MEDICAL BIOCHEMISTRY

Foreword

I have gone through the manuscript of the book entitled *Textbook of Medical Biochemistry* written by Dr S. Ramakrishnan, Dr K.G. Prasannan and Dr R. Rajan, Department of Biochemistry, Jawaharlal Institute of Postgraduate Medical Education and Research, Pondicherry.

This book is the result of the joint efforts of three teachers in biochemistry. The authors possess considerable experience in teaching biochemistry to M.D., M.Sc and M.B.B.S. students in medical colleges. They have also got vast experience in the subject with many research publications.

The subject matter of the book covers the area of biochemistry which is within the purview of the biochemistry portion for the M.B.B.S. students. The book will also serve as a useful textbook for those who are studying for M.Sc./M.D. in biochemistry or in the biological sciences. The introduction of current views on special topics like insulin, parathyroid hormone, calcitonin, cyclic AMP, vitamin D, vitamin C, immunoglobulins, calcium and iron, biological oxidation, biosynthesis of proteins, clinical enzymology, muscular contraction, blood coagulation, atherosclerosis and modern techniques of biochemistry are some of the highlights of this textbook.

T.A. VENKITASUBRAMANIAN
Professor of Biochemistry, University of Delhi;
President, Society of Biological Chemists (India),
Indian Institute of Science, Bangalore;
Dean, Faculty of Medical Sciences, University of Delhi.

Preface to the Second Edition

This edition of **Textbook of Biochemistry** has been extensively revised and updated in response to a felt need among students and teachers of medicine. Some of the newly added topics are biostatistics, metabolism in starvation, alcoholism and aldehydism, photosynthesis, interferons, regulations of gene expression in eukaryotes, hypothalamic release factors, ELISA and HPLC. Many subject areas have been revised thoroughly, e.g., atherosclerosis, glycogen storage diseases, porphyrias, gout, body fluids, liver function tests, etc.

While we continue to extend our gratitude to all the teachers who have been mentioned in the first edition, out sincere thanks are due to the following teachers and well-wishers, Dr. P.H. Ananthanarayanan and Dr. R. Sundaresan of Jipmer, Pondichery; Dr. Sultan Sheriff and Dr. Mohammed Shameemulla, Faculty of Medicine, Arab Medical University, Benghaji; Prof. P.A. Kurup, Retd Professor, University of Kerala, Trivandrum; Dr. A. Ramiah, Professor of Biochemistry, All India Institute of Medical Sciences; Sri S. Venkatasubramanian of Bharath Graphics, Madras; Sri V. Anantaraman Charted Accountant, Cuddalore; and Shri Meganathan of RMMC & H, Annamalai Nagar.

Special thanks are due to Dr. R. Sathiamurthy, Reader in Statistics, Department of Statistics, Annamalai University, and the Lecturer in Statistics, JIPMER, Pondicherry, for their critical comments and suggestions.

We sincerely thank M/s Orient Longman for bringing out the second edition of the Textbook of Medical Biochemistry.

AUTHORS

Contents

Foreword   v
Preface to second edition   vi

Section I—The Fundamental and Physiological aspects of Biochemistry

1. The Cell   1
2. Carbohydrates   8
3. Lipids   29
4. Proteins   46
5. Nucleoproteins   65
6. Hemoglobin   84
7. Enzymes, Coenzymes, and Isozymes   97
8. Biological Oxidations and Bio-Energetics   117
9. Chemistry of Blood   135
10. Composition and Structure of Muscle and Biochemistry of Muscular Contraction   155
11. Biochemistry of Digestion and Absorption   162

Section II—Metabolism

12. Carbohydrate Metabolism   176
13. Lipid Metabolism   209
14. Protein and Amino Acid Metabolism   245
15. Biochemical Genetics   295
16. Metabolism of Nucleoproteins, Purines and Pyrimidines   319
17. Vitamins   331
18. Hormones, their Chemistry and Functions   358
19. Metabolism of Minerals   400
20. Methods of Study of Metabolisms   420
21. Body fluids   435
22. Detoxication (Bio-Transformation)   444

Section III—Energy Metabolism, B.M.R., S.D.A., Nutrition, Composition of Foods and Balanced Diet

23. Energy Metabolism, Nutrition, Composition of Foods and Balanced Diet   448

Section IV—Modern Techniques in Biochemistry

24. Modern Techniques in Biochemistry   465
    - Chromatography   465
    - Electrophoresis   473
    - Photometry   476
    - Electron microscopy   480
    - Respirometry   480
    - Ultra-centrifugation   482
    - Auto-analysers   484
    - Techniques using Isotopes   488
    - Radio immuno-assay   492

Section V—Function Tests

25. Biochemical Evaluation and Function Tests   495
    - Gastric Analysis   500
    - Renal Function Tests   503
    - Thyroid Function Tests   505
    - Para-thyroid Function Tests   506
    - Pancreas Function Tests   506
    - Adrenal Function Tests   509

Section VI—Physico-chemical aspects of Biochemistry

26. Physico-Chemical Aspects of Biochemistry
    - Hydrogen Ion Concentration, pH and Buffers   510
    - Colloidal State   517
    - Membrane phenomena   521
    - Osmosis   529
    - Adsorption   534
    - Surface Tension   535
    - Viscosity   536

Section VII—Biostatistics

27. Fundamentals of Biostatistics   539

References   561

Index   565

SECTION — I
THE FUNDAMENTAL AND PHYSIOLOGICAL ASPECTS OF BIOCHEMISTRY

1 The Cell

STRUCTURE AND COMPOSITION

The living body is made up of innumerable units or cells. The combination of a large number of cells is called a tissue. The tissues form organs and a combination of organs makes the whole living organism. The cells from a living organism have to perform diversified functions and are accordingly specialised for the division of labour.

The cell theory was propounded by Schleiden and Schwann in 1839. Virchow's demonstration in 1859 that cells divided to give rise to daughter cells gave a new interest to the field of cytology. The invention of the microscope and further improvements in microscopy helped to view the cell closely and to study its finer details and highly complicated entities. After the invention of the electron microscope, the complexities of the cellular organelles have become more and more familiar to the modern scientist.

Porter *et al* have given the composite picture of the cell as a highly organised structure consisting of convoluted and inter-connected elements which vary from organism to organism, tissue to tissue and cell type to cell type.

The prokaryotic cells are the first cells to arise in biological evolution. They are small and simple, have only a single cell membrane and do not contain the nucleus, mitochondria and endoplasmic reticulum. They have only one chromosome containing a single molecule of DNA (double-helical). Their reproduction is asexual. Examples of prokaryotic cells are found in blue green algae, spirochetes, eubacteria and mycoplasma.

On the other hand, eukaryotic cells are much larger and much more complex and often 1,000 to 10,000 times larger in size than prokaryotic cells. All animal and plant cells, protozoa and fungi are examples of eukaryotic cells. The latter contain a nucleus enclosed in a nuclear membrane. Several chromosomes are present in the nucleus which undergoes mitosis during cell multiplication. Other membrane-bound organelles like mitochondria, Golgi, endoplasmic reticulum, lysosomes, etc., are also present in these cells, and the enzymes that are needed for metabolic sequences are kept in different compartments of the cells.

The cells from various organs differ in their origin, size, morphological features and functions. All cells have some features like cellular components and sub-cellular organelles in common. A cell from the liver tissue is usually considered a typical cell and is used for general study. Such a cell has an average diameter of 20 μ (2×10^5 Å) and a volume of 5000 μ^3.

The structure of a typical cell

The principal structural units of a typical cell are shown diagrammatically in Fig. 1.1. The chief organelles present in a cell are the cell membrane, the nucleus, mitochondria, endoplasmic reticulum, ribosomes, lysosomes and Golgi bodies. These are described below:

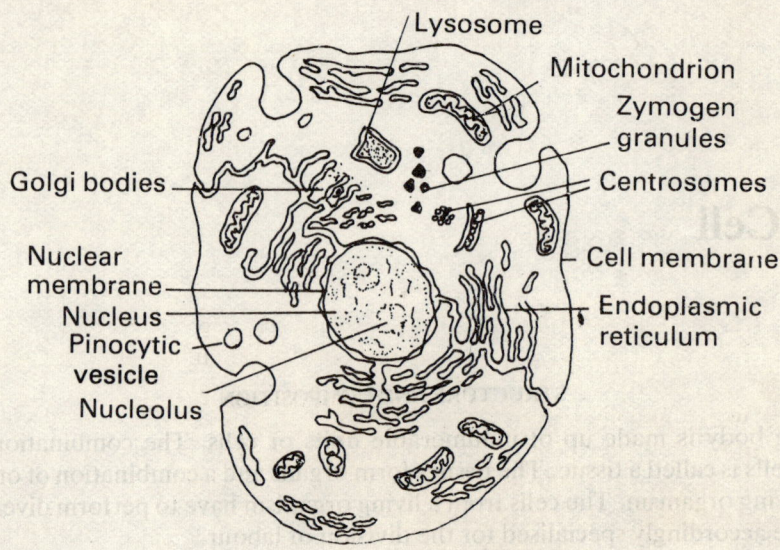

Fig. 1.1 Structure and components of a typical cell

1) *Cell membrane:* The cell is enveloped by a thin membrane called the cell membrane or plasma membrane with an average thickness of 75 Å. Electron microscopy has revealed the cell membrane as a rigid and organised structure consisting of a lipid bilayer primarily of phospholipids and penetrated by protein molecules forming a mosaic-like pattern. The hydrophobic ends (fatty acid tails) of the phospholipids are oriented towards the centre of bilayer while their hydrophilic ends (polar heads) point outwards (Fig. 1.2.) The membrane phospholipids act as a 'solvent' for the membrane proteins. In other words, the presence of phospholipids makes the proteins amphipathic, the latter having hydrophilic regions protruding inside and outside the membrane but connected by a hydrophobic core of the bilayer. This is important for an ideal environment for maintaining protein function. Membrane proteins are usually glycoproteins. Those that traverse the hydrophobic core of the lipid bilayer are called *integral proteins* and those which are closer to the outside and are weakly bound to the phospholipids are called *peripheral proteins*. Examples of integral proteins are rhodopsin of the retinal cells, immunoglobulins of the leukocytes and hormone receptor proteins of the various cell membranes. Ankyrin of the erythrocytes and several secretory proteins, peptide hormones and mitochondrial ATP synthetase can be considered peripheral proteins. There are also trans-membrane proteins that are peripheral proteins which extend along the membrane and bind themselves to the integral proteins and thereby acting as ion channels. The band III protein of the erythrocytes binds the integral protein ankyrin with another protein spectrin and stabilises the biconcave shape of these cells.

The fluid mosaic model of the cell membrane of Singer and Nicholson is widely accepted as most suitable to explain the structure of cell membranes (Fig. 1.2).

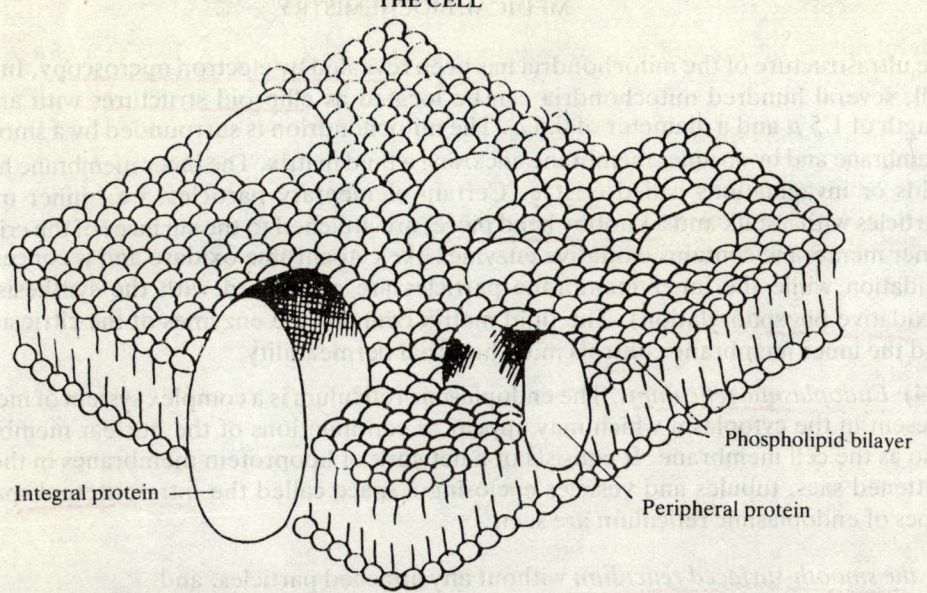

Fig. 1.2. Fluid mosaic model of a cell membrane (after Singer and Nicholson)

Among the enzymes present on the cell membrane are adenylate cyclase, ATPase, phospholipase, etc. The membrane has small pores, approximately 8Å wide, which are more permeable to water and hydrated K^+ than to Na^+. These are concerned with selective permeability and active transport of ions and molecules. The cell membrane is the site for some enzymatic reactions. Cells have the capacity to internalise extra-cellular materials involving the formation of endocytotic vesicles, and this type of transport process is called endocytosis. Endocytotic vesicles are formed by the invagination of plasma membrane which engulfs some fluid and then either pinches off or seals the neck. thus helping in the cellular uptake of materials. This process, often called pinocytosis, is receptor mediated and is a very common mode of internalisation of macro-molecules by certain cells. An example of this process is found in the disposal of LDL (*see* Lipid Metabolism).

2) *Nucleus:* The nucleus of the cell is easily seen under the microscope after suitable staining. It is a spherical body approximately 5μ in diameter and $16\mu^3$ in volume. Each cell has usually one nucleus. The nucleus is enveloped by the nuclear membrane which is 75Å thick and consists of three layers. The thickness however differs in different portions. The nuclear membrane is not always continuous and may have openings, or connections with a network of membranes of the endoplasmic reticulum. The macro-molecules synthesised in the nucleus make their way into the cytoplasm through these openings. The nucleus consists of thick protoplasm called the nucleoplasm, which contains various enzymes, free amino acids, lipids, Ca^{++} and Mg^{++} ions, RNA and DNA. Inside the nucleoplasm are the nucleoli, spherical structures rich in RNA (nuclear RNA) which comprises 10-20 per cent of the total RNA of the cell. Almost the entire DNA of the cell, the determinant of genetic traits, occurs in the nucleus distributed as chromatin. Enzymes like DNA polymerases are present in the nucleus. During cell division, the chromatin gets highly organised into chromosomes, which are linear thread-like structures with bead-shaped bodies. The number of chromosomes in a human somatic cell is constant (23 pairs). During cell division (mitosis), the same number of chromosomes is passed on to a daughter cell.

3) *Mitochondria:* The mitochondria are the largest organelles in the cytoplasm and can be seen in a light microscope, after staining with Janus green, as small granules. In recent years,

the ultrastructure of the mitochondria has been revealed by electron microscopy. In a normal cell, several hundred mitochondria can be located as ellipsoid structures with an average length of 1.5 μ and a diameter of 0.5 μ. The mitochondrion is surrounded by a smooth outer membrane and by an inner membrane enclosing a fluid matrix. The inner membrane has several folds or invaginations called cristae. Certain elementary particles, i.e., inner membrane particles with a stalk and a knobby head piece, are attached to the surfaces of the cristae. The inner membrane contains oxidative enzymes like cytochrome oxidase and is concerned with oxidation while the inner membrane particles are concerned with the synthesis of ATP (oxidative phosphorylation). The fluid matrix contains the enzymes of the citric acid cycle, and the inner membrane controls mitochondrial permeability.

4) *Endoplasmic reticulum*: The endoplasmic reticulum is a complex system of membranes present in the cytoplasm which may appear as continuations of the nuclear membrane and also as the cell membrane. It consists of a network of lipoprotein membranes in the form of flattened sacs, tubules and vesicles enclosing a space called the intra-cisternal space. Two types of endoplasmic reticulum are seen:

a) *the smooth-surfaced reticulum* without any attached particles, and
b) *the rough-surfaced reticulum* lined with sub-microscopic spherical particles called ribosomes, which are essentially molecules of ribo-nucleoproteins.

The endoplasmic reticulum has four important functions:
a) biosynthetic process concerned with the production of proteins, lipids and polysaccharides;
b) transport of these products to their sites of storage or utilisation;
c) regulation and control of utilisation of newly arrived molecules; and
d) acting as a precursor of other membranes like the nuclear membrane, cell membrane and Golgi bodies.

An important enzyme present is glucose-6-phosphatase.

5) *Ribosomes*: The ribosomes are present on the rough surfaced reticulum and have a particle weight of 4×10^6 daltons and a sedimentation constant of 75 S to 80 S. They contain about 50 per cent RNA and about 50 per cent protein. The number of ribosomal particles is much more in the cells designed for high protein synthesis and secretion, like the cells of the exocrine pancreas, than in most other cells.

6) *Golgi bodies*: The Golgi bodies are closely related to the endoplasmic reticulum in structure and function. However, they consist of a double membrane with smooth surfaces containing more lipid than the reticulum. The proteins produced by the ribosomes are stored in the form of secretory granules or zymogen granules in the vacuoles of the Golgi bodies. Some vacuoles carry secretory granules synthesised by the endoplasmic reticulum. Lysosomal enzymes are also packaged and become primary lysosomes which, after fusing with phagosomes, become secondary lysosomes. Pro-insulin produced from pre-proinsulin in the endoplasmic reticulum is transferred to the Golgi bodies where it is converted to insulin and packaged before secretion. The chief enzyme present in the Golgi bodies of the liver cell is alkaline phosphatase.

7) *Lysosomes*: Lysosomes are present in the cytoplasm and may resemble mitochondria in shape and size. They are bound by a lipoprotein membrane and appear as bags filled with digestive enzymes such as ribonuclease, glycosidases, acid phosphatase and cathepsins. When the lysosomal membrane ruptures, these enzymes are released, break up the large molecules and bring about the dissolution (lysis) of the cell.

8) *Centrosomes:* These are apparent only when a cell is preparing to divide. They form the poles of a spindle apparatus involved in chromosomal replication.

9) *Cytosol:* The cytosol refers to the aqueous phase in which all the cytoplasmic organelles are suspended. No characteristic structure is assigned to this. The soluble enzymes, like those concerned with the glycolytic pathway of glucose break-down, are present in the cytosol along with other soluble proteins. Organic nutrient substances like glucose, amino acids and lipids, and products of metabolism like urea and creatinine, and ions like chloride, bicarbonate, Na^+, K^+, Mg^{++}, Ca^{++} and phosphate are present in the cytosol.

Cell fractionation

There are no standard methods that can be applied uniformly to the fractionation of all types of cells. Various methods have to be modified according to the type of tissue to be studied. Much progress has been made in recent years in fractionation and partition of cell constituents by cellular disintegration and differential centrifugation in suitable media like phosphate buffer, isotonic NaCl and sucrose. Cells can be broken in aqueous media by the application of shearing forces, with the aid of various homogenisers, colloid mills, Waring blenders, etc. They can also be disrupted in non-aqueous media in which fragmentation is done either in a dry state or in an inert organic solvent. The aim is to minimise the transfer of soluble substances from cytoplasm to nucleus or vice versa. Therefore, rapid freezing and removal of water by lyophilisation (dehydration under high vacuum) is done, followed by grinding. Other methods make use of freezing and thawing, sonic and ultrasonic vibrations, osmotic shock, etc.

The homogenate technique: Tissue homogenates are usually obtained by grinding pieces of tissue in an isotonic medium (an isotonic medium is one with an osmolality equal to that of blood serum; e.g., 0.25 M sucrose, 0.9% NaCl, etc.).

Suspension medium: The ideal medium should approximate in composition to the soluble phase of the cytoplasm so that the cell particulates may remain morphologically, structurally and functionally intact. Such an ideal medium, however, has not yet been devised and many modifications have become necessary.

The sucrose medium is usually preferred as it can maintain the ATP during homogenisation. Na^+, K^+ and Ca^{++} ions are usually avoided for the homogenate technique with tissue slices. Homogenisation is done either using a glass mortar and pestle or the Potter-Elvehjem tissue homogeniser with a teflon pestle, the operation being carried out in a cold room. The tissue is ground with the medium nine times its weight in order to obtain a 10 per cent homogenate. From the crude homogenate, the cell debris and the connective tissue fragments are removed by filtration through muslin cloth.

Isolation of particular components of liver tissue
(Schneider and Hogeboom, 1950)

The typical tissue for this purpose is fresh liver obtained from a rat. The homogenate is prepared as described above, subjected to differential centrifugation and separated into various fractions, *viz.*, nuclear, mitochondrial, sub-microscopic particles and a soluble fraction (supernatant). The centrifugation is carried out at low temperatures using a refrigerated ultra-centrifuge with a multi-speed attachment. Fractional centrifugation helps in separation by sedimentation of particles differing in density, shape or size at different rates in a field of centrifugal force. Special unbreakable lusteroid (cellulose acetate) tubes are used for this high speed centrifugation.

The centrifugation at 700 g of a homogenate prepared in 0.88 M sucrose for 10 min. sediments the nuclei along with some unbroken cells. The sediment is washed by 0.88 M sucrose with the aid of a homogeniser pestle and re-centrifuged. The final sediment is re-suspended in 0.88 M sucrose and is called *nuclear fraction*. Almost the entire DNA of the homogenate is recovered in this fraction. However, 10 per cent of the mitochondria is lost along with the nuclei owing to their clumping properties.

The supernatant obtained in the first step is centrifuged at 7000 g for 20 min. Three layers are obtained after this operation:

1) a small deposit of brown material at the bottom,
2) an intermediate layer of firmly packed material of tan colour, and
3) a bulky layer at the top consisting of poorly sedimented material.

The top layer is removed by a teat pipette and called *supernatant* or *soluble fraction*. It contains all the glycolytic enzymes and nucleoside phosphorylase.

The material remaining in the tube is re-suspended in 0.88 M surose and again subjected to centrifugation at 7,000 g. The supernatant obtained is transferred to another lusteroid tube. The residue is mostly due to mitochondria. This exhibits a strong birefringence property. The supernatant on centrifugation at 53,000 g sediments a part of the microsomal fraction. Microsome is a term that does not represent any definite particle or unit but refers to a collection of small fragments of endoplasmic reticulum mixed with ribosomes. It is an operational term.

The supernatant obtained is re-centrifuged at 105,000 g when the ribosomes get sedimented as residue and the supernatant consisting of all the soluble enzymes is separated.

A summary of the fractionation procedure is given in the following table:

TABLE 1

A SUMMARY OF THE FRACTIONATION PROCEDURE

Centrifugal force	Fractions
700 g	Nuclei, cell membrane
7000 g	Mitochondria
53,000 g	A part of microsomes
105,000 g	Ribosomes (rest of the microsomes)
Unsedimented	Soluble fraction (supernatant)

The isolation of chromosomes is accomplished by grinding the tissue with extremely fine sand (the grains being smaller than the nuclei) to break the nuclear membrane and thus release the nuclear contents. Differential centrifugation is then done as described earlier.

TABLE 2

BIOCHEMICAL CONSTITUENTS AND THE FUNCTIONS OF VARIOUS CELLULAR FRACTIONS

Cytological structure (*) = Marker	Biochemical constituents	Functions
NUCLEUS (* NAD pyrophosphorylase, DNA polymerases)	DNA, RNA, DNA replicase, DNA_+ dependant RNA polymerase NAD pyrophosphorylase, Histones.	Site of coenzyme formation. Controls some enzymes in cytoplasm *via* the coenzymes. Transmission of hereditary characteristics.

(Contd.)

Cytological structure (*) = Marker	Biochemical constituents	Functions
MITOCHONDRIA (* Succinate Dehydrogenase, Glutamate Dehydrogenase)	Citric acid cycle enzymes; Fatty acid oxidase system, Electron transport chain. ALA synthease, Coproporphyrinogen to Heme converting enzymes Monoamine oxidase. Diamine oxidase, Ketogenic enzymes in liver.	Oxidative phosphorylation, aerobic respiration, detoxication of ammonia *via* urea cycle.
LYSOSOME (* Acid phosphatase)	Acid DNAse, Acid RNAse, Acid-phosphatase, Cathepsin. Collagenase, ß Glucuronidase, ß Galactosidase, ß-Mannosidase, Arylsulfatase.	Intra-cellular digestion, autolysis of the cell.
ENDOPLASMIC RETICULUM & RIBOSOMES (* Glucose-6-phosphatase)	Glucose-6-phosphatase, RNA, NADH & NADHP linked mono oxygenases.	Synthesis of proteins, complex carbohydrates, lipids. Transport & control over utilisation, metabolism of drugs. Precursor of other membranes.
GOLGI BODIES (* Uridine Diphosphatase)	Alkaline phosphatase.	Concentration and packaging of proteins, precursors of zymogen granules.
SUPERNATANT (* Transketolase, Phosphofructokinase, Glucose-6-Phosphate Dehydrogenase)	Glycolytic enzymes, enzymes concerned with conversion of Phosphoenolpyruvate to glycogen. HMP shunt, NADPH-linked dehydrogenases. Fatty acid biosynthesis, Triacyl glycerols, phospholipids & cholesterol synthetic enzymes.	Glycolysis, gluconeogenesis, generation of NADPH for reductive lipogenesis.
PLASMA MEMBRANE (* Phospholipase A)	Glycolipids.	Reception & secretion of hormones, transport

2 Carbohydrates

The subject of carbohydrates is important in the study of biochemistry as many members belonging to this class of organic compounds are very useful to the living systems in some way or the other. Glucose is the 'sugar of life' and is required for energy. Fructose also contributes to meeting the energy needs of the body. Galactose forms part of the lactose of breast milk and some complex lipids like cerebrosides. Ribose and deoxyribose are important sugars contained in nucleic acids. Starch is the major constituent of our food while sucrose and lactose also have dietary value. Glycogen is a reserve carbohydrate stored in the liver and the muscles. Glycosaminoglycans (muco-polysaccharides) are complex carbohydrates containing amino-sugars and uronic acids and are found in cartilage, tendons and bones in combination with certain proteins like collagen and elastin. Cellulose is a polymer of glucose and is abundantly found in the plant kingdom. It is not easily digestible for man.

CARBOHYDRATES AND SUGARS

These are a class of organic compounds most of which are naturally occurring. They usually contain carbon, hydrogen and oxygen; hydrogen and oxygen are present in the ratio in which they occur in water, i.e., 2:1. Carbohydrates have the empirical formula $C_x(H_2O)_n$ (hence the name carbohydrate, implying hydrate of carbon, which is not strictly correct). Acetic acid, formaldehyde, lactic acid and inositol are not carbohydrates but have the ratio of hydrogen and oxygen as 2:1, while rhamnose is a carbohydrate with the formula $C_6H_{12}O_5$. Some carbohydrates have nitrogen atom in their molecule, e.g., glucosamine.

Sugar is a carbohydrate, sweet to taste, dissolves in water and chars on heating. Glucose (grape sugar), fructose (fruit sugar), sucrose (cane sugar), lactose (milk sugar) and maltose (malt sugar) are some examples of sugars. All sugars are carbohydrates but not all carbohydrates are sugars. Glycogen and inulin, for example, are carbohydrates but not sugars.

Chemically, all carbohydrates contain alcoholic groups with either free aldehydic or ketonic groups or potential aldehydic or ketonic groups. They are therefore poly-hydroxy-aldehydes or ketones.

Classification of carbohydrates

Carbohydrates are classified into 1) monosaccharides, 2) disaccharides and oligosaccharides, and 3) polysaccharides.

1) *Monosaccharides* are simple sugars, non-hydrolysable and have usually six carbon atoms or less in a molecule. If the sugar is a hydroxy-aldehyde, it is called aldose, e.g., glucose is an aldohexose. If the sugar is a hydroxy-ketone, it is called ketose, e.g., fructose is a ketohexose. If the number of carbon atoms is 5, the sugar is a pentose, if 4 it is a tetrose. The simplest sugar is a triose. All sugars are optically active as they possess asymmetric carbon atom or atoms. Typical examples of some physiologically important monosaccharides are:

a) trioses: glyceraldehyde and dihydroxy acetone
b) tetroses: erythrose
c) pentoses: ribose, deoxyribose, xylose, xylulose
d) hexoses: glucose, fructose, galactose and mannose

2) *Disaccharides and Oligosaccharides:* A disaccharide represents an anhydride derived from two molecules of the same or different monosaccharides by the elimination of one molecule of water. It is easily hydrolysed into monosaccharides by boiling with dilute acids or treatment with enzymes. Some examples of disaccharides are lactose, maltose and sucrose which have the molecular formula $C_{12}H_{22}O_{11}$. These are also sweet to taste and are sugars. Disaccharides are optically active as well. There can be reducing or non-reducing disaccharides depending upon whether they possess the free or potential reducing groups. Lactose is a reducing disaccharide while sucrose is a non-reducing one.

Oligosaccharides may be trisaccharides, tetra- or pentasaccharides, etc. containing upto about 10 monosaccharide units, e.g., raffinose, gentianose.

3) *Polysaccharides:* These are carbohydrates of high molecular weight and contain many monosaccharide units. They are mostly insoluble in water and tasteless. On hydrolysis with enzymes or by boiling with dilute acids, they are converted into simple sugars. If the polysaccharide is made up of only one type of sugar, it is referred to as a homoglycan, e.g., glycogen, starch, inulin. They have the general formula $(C_6H_{10}O_5)n$. If different sugars or sugar derivatives go to make up a polysaccharide, the latter is called a heteroglycan, e.g., heparin, hyaluronic acid, chondroitin sulphate.

MONOSACCHARIDES

Glucose

Glucose being a typical representative of the monosaccharides and also the main sugar of the body, its structure and reactions are discussed first. Many reactions of glucose can be taken as being representative of all sugars.

Structure of glucose: The molecular formula of glucose is $C_6H_{12}O_6$. On acetylation, glucose gives a penta-acetyl derivative indicating the presence of 5–OH groups. As glucose is a stable substance, no 2–OH groups are attached to the same carbon atom.

The reducing properties of glucose, its oxidation to gluconic acid with the same number of C atoms and reaction with HCN all indicate the presence of –CHO group.

Kiliani's reaction, i.e., reaction of glucose with HCN, hydrolysis of the cyan hydrin and reduction with HI, gives normal heptylic acid which is a straight chain compound. Hence according to this reaction glucose should be a straight chain compound.

By distributing the 5–OH groups and giving one – CHO group to glucose with the formula $(C_6H_{12}O_6)$, the structure is

$$C - C - C - C - C - CHO$$
$$|||||$$
$$OHOHOHOHOH$$

After satisfying the other valencies of carbon with the remaining hydrogen atoms, the structure becomes

$$CH_2 - CH(OH) - CH(OH) - CH(OH) - CH(OH) - CHO$$
$$|$$
$$OH$$

The above structure is however unable to explain the following:
1) muta-rotation of glucose,
2) no compound formation with $NaHSO_3$,
3) formation of two isomeric methyl glucosides with methyl alcohol and dry HCl gas, and
4) no restoration of colour of Schiff's reagent.

To explain the above, Fisher gave the ring structure to glucose which has both –CHO and –OH groups which can interact to form hemiacetal. Both a six-membered ring (amylene oxide, pyranose) and a five-membered ring (butylene oxide, furanose) are possible, and experimental evidence has proved the existence of two types of rings. However, as the six-membered ring is more stable, glucose is predominantly in the pyranose form. The pyranose form can have the H of the interacting –CHO on the left and –OH on the right in which case it is called alpha-glucose. If the –OH is on the left and H on the right, it is called beta-glucose. These two are called anomers.

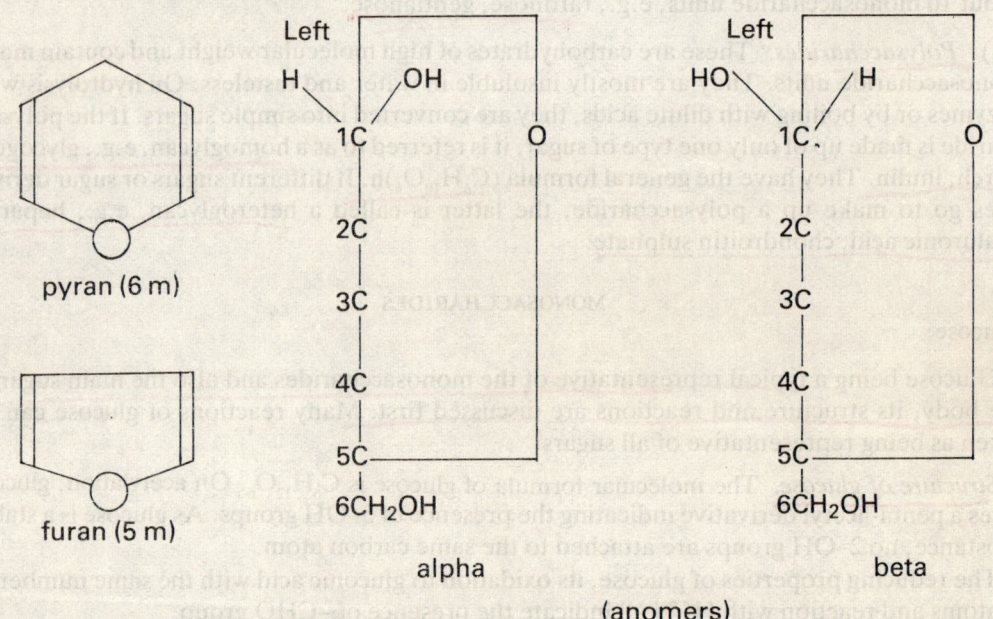

Configuration: According to the accepted convention, originated from Emil Fischer, configuration, namely D and L forms, is given to sugars. These have nothing to do with the direction of optical rotation but depend on the spatial distribution of H and OH in the carbon atom adjacent to the –CH_2OH group. Fischer fixed glyceraldehyde as the arbitrary standard; D-glyceraldehyde having H on the left and –OH on the right. This is referred to as the D configuration. If –OH is on the left and –H on the right, it is said to have L configuration. The direction of optical rotations, namely, dextro and levo, should be denoted by + and –.

All monosaccharides derived from D glyceraldehyde belong to the D series of sugars. Both alpha- and beta-glucoses have the D configuration and have the following structure:

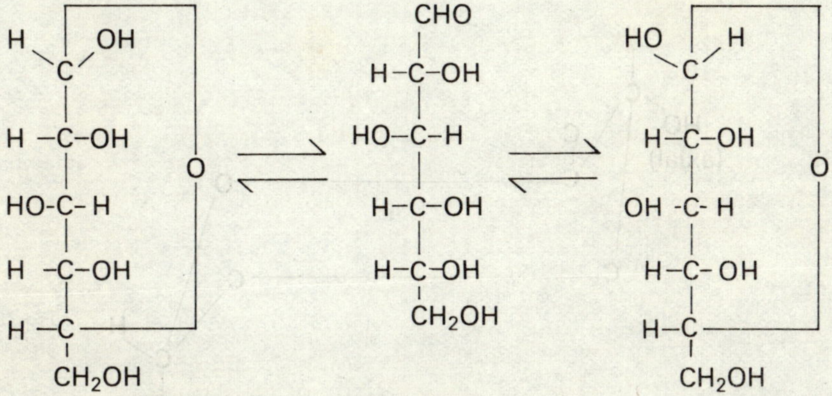

```
    H   OH                    CHO                    HO    H
     \ /                       |                       \ /
      C                     H–C–OH                      C
      |                        |                        |
   H–C–OH                   HO–C–H                   H–C–OH
      |          ⇌             |            ⇌          |
   HO–C–H   O               H–C–OH                  OH–C–H    O
      |                        |                        |
   H–C–OH                   H–C–OH                   H–C–OH
      |                        |                        |
    H–C                     CH₂OH                    H–C
      |                                                 |
    CH₂OH                                             CH₂OH

  alpha D-glucose            aldehyde              beta D-glucose
                               form
                            (free CHO)
```

The alpha D glucose and beta D glucose structures are said to have the potential aldehydic form since from the ring (which does not have the free aldehydic form) an aldehyde group can be made available on cleavage. There is an equilibrium between the ring structures and the aldehydic form. It is because of the ring structures which do not possess free –CHO group that a few reactions of –CHO are not answered by glucose.

As the ring structures of Fischer are difficult to write, Haworth gave a perspective formula for sugars, incorporating the ring structures, as given below.

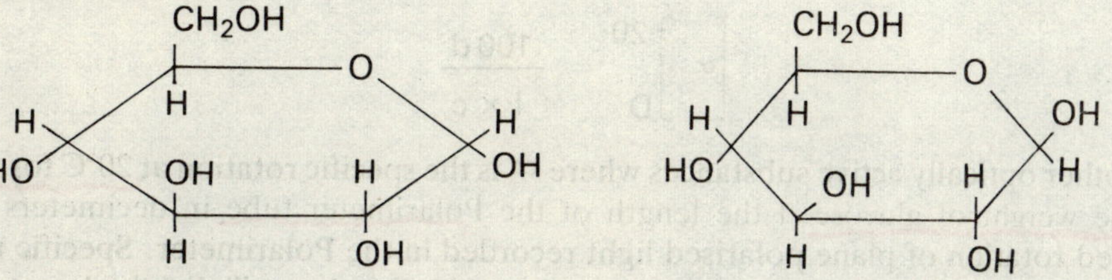

```
        CH₂OH                           CH₂OH
         |    O                          |    O
      H  |       H                    H  |       OH
       \ |      /                      \ |      /
    HO   OH  H   OH                 HO   OH  H   H
         |                                |
         H   OH                           H   OH
      Alpha D-glucose                 Beta D-glucose
```

Conformation: A six-membered ring can have either a boat form or a chair form.

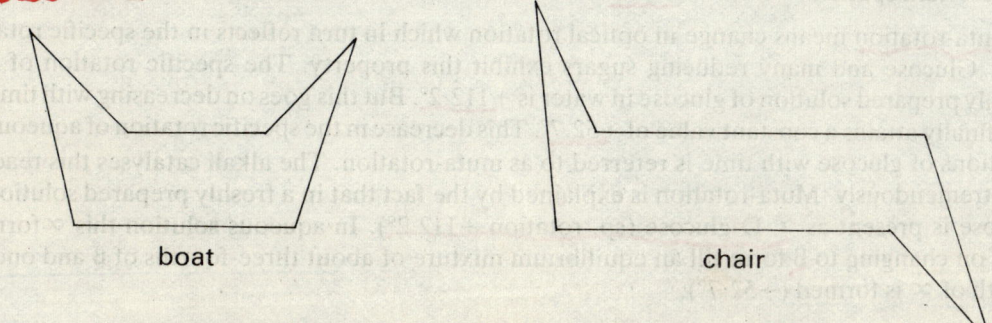

boat   chair

From energy considerations, the six-membered glucose ring is said to exist as a chair form. As a corollary, the covalent bonds of the carbon atoms can be either directed upwards and downwards (polar bonds) or laterally (axial or equatorial). In the figure, of the 2–OH groups forming part of –CH OH, there is difference in their conformational arrangement, one being axial and the other polar.

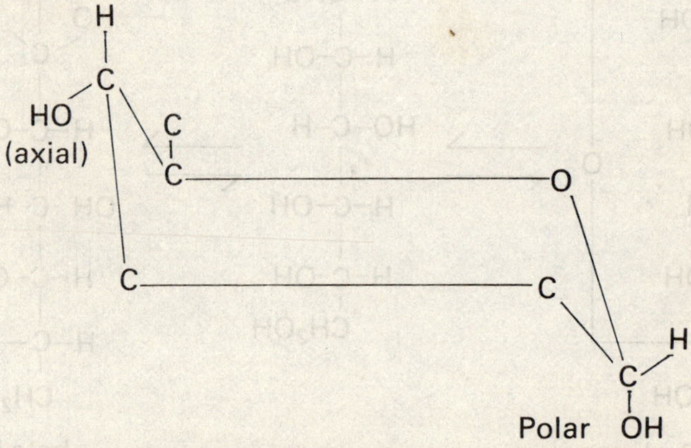

Note: The letters (d) and (l) used in some books to denote dextro and laevo rotations should not be confused with D and L. While D-glyceraldehyde and D-bromo lactic acid are dextro-rotatory (+), D-glyceric acid and D-lactic acid are laevo rotatory (−) showing that D is not concerned with (+) or (−). Fischer's fixing of D-glyceraldehyde as that with H on the left and OH on the right has since been confirmed by X-ray studies on the absolute configuration of L-tartaric acid in 1954.

Reactions of glucose: Reactions of glucose are typical reactions of all monosaccharides. The special reactions of some monosaccharides will be discussed separately.

Optical activity, specific rotation and muta-rotation: Glucose is optically active and is dextro-rotatory (hence called dextrose). The specific rotation is calculated from the equation:

$$\left[\alpha\right]_D^{20} = \frac{100\,d}{l \times c}$$

as for other optically active substances where α is the specific rotation at 20°C for the D line, c is the weight of glucose, l the length of the Polarimeter tube in decimeters and d the observed rotation of plane polarised light recorded in the Polarimeter. Specific rotation of α D-glucose is + 112.2° while for ß D-glucose it is + 18.7° (values will slightly change with temperature). In an aged solution of glucose the equilibrium is established between the two and the resultant specific rotation is + 52.7°.

Muta-rotation means change in optical rotation which in turn reflects in the specific rotation. Glucose and many reducing sugars exhibit this property. The specific rotation of a freshly prepared solution of glucose in water is +112.2°. But this goes on decreasing with time and finally attains a constant value of +52.7°. This decrease in the specific rotation of aqueous solutions of glucose with time is referred to as muta-rotation. The alkali catalyses this reaction tremendously. Muta-rotation is explained by the fact that in a freshly prepared solution glucose is present as α D-glucose (sp. rotation +112.2°). In aqueous solution this α form goes on changing to ß form till an equilibrium mixture of about three-fourths of ß and one-fourth of α is formed (+52.7°).

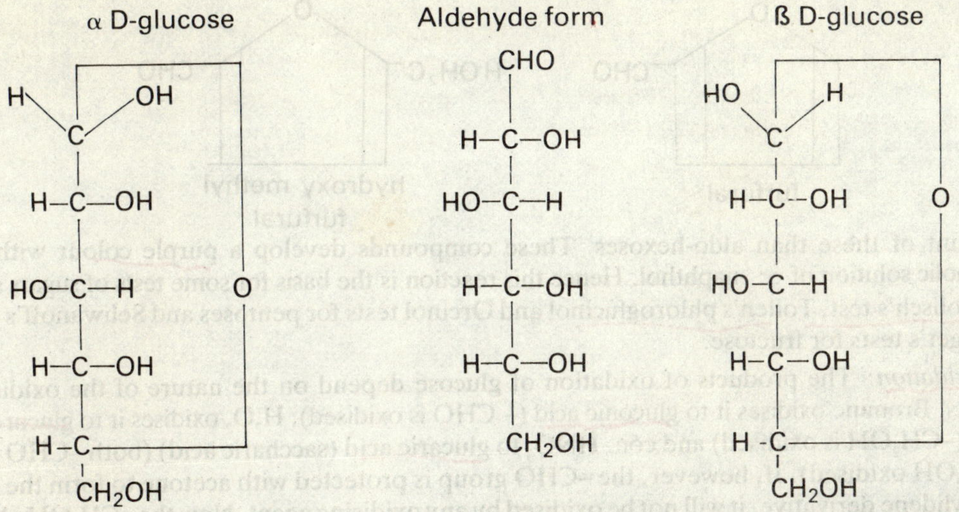

CARBOHYDRATES

α D-glucose

```
    H   OH
     \ /
      C
      |
   H—C—OH
      |
   HO—C—H        O
      |
   H—C—OH
      |
   H—C
      |
    CH₂OH
```

$[\alpha]_D^{20} = +112.2°$

Aldehyde form

```
    CHO
     |
   H—C—OH
     |
  HO—C—H
     |
   H—C—OH
     |
   H—C—OH
     |
    CH₂OH
```

β D-glucose

```
    HO   H
     \ /
      C
      |
   H—C—OH
      |
   HO—C—H        O
      |
   H—C—OH
      |
   H—C
      |
    CH₂OH
```

$[\alpha]_D^{20} = +18.7°$

Alkalies: 1) Very dilute alkali: The immediate reaction with very dilute alkali is the conversion of α to β anomer. But with dilute alkali (.05 N NaOH) and *on standing*, glucose undergoes the Lobry de Bruyn Alberda Van Ekenstein transformation. It is partly isomerised to D-fructose and D-mannose, and hence all the three sugars will be present in equilibrium. The same transformation takes place with fructose or mannose. This is due to the formation of an enediol (2 carbon atoms bound by double bond and carrying hydroxyl groups) from these sugars with dilute alkali. The enediol subsequently isomerises to the other two members as these sugars differ only in the configuration of two carbon atoms; the –CHO group carbon and its neighbour.

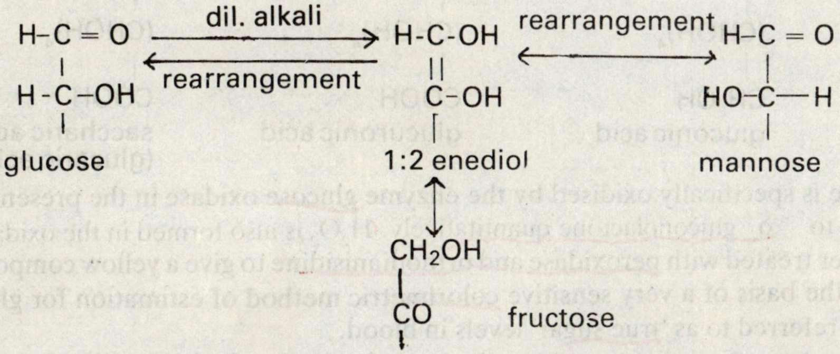

```
   H-C=O       dil. alkali      H-C-OH      rearrangement     H-C=O
    |        ⇌                    ||        ⇌                  |
   H-C-OH      rearrangement     C-OH                        HO-C—H
    |                             |                            |
  glucose                      1:2 enediol                   mannose
                                  |
                                CH₂OH
                                  |
                                  CO       fructose
                                  |
```

2) Strong alkalies (Moore's test): When boiled with strong alkalies sugars are caramelised, and yellow to brown resinous products are formed. (This is one reason why Benedict's reagent containing Na_2CO_3 is preferred to Fehling's solution containing NaOH.)

Acids: Weak acids have no significant action on sugars. Treatment with strong sulphuric acid and heating breaks the sugars into furfural or its derivatives. Keto-hexoses give larger

furfural   hydroxy methyl furfural

amount of these than aldo-hexoses. These compounds develop a purple colour with an alcoholic solution of α-naphthol. Hence this reaction is the basis for some tests of sugars such as Molisch's test, Tollen's phloroglucinol and Orcinol tests for pentoses and Seliwanoff's and Foulger's tests for fructose.

Oxidation: The products of oxidation of glucose depend on the nature of the oxidising agents. Bromine oxidises it to gluconic acid (– CHO is oxidised), H_2O_2 oxidises it to glucuronic acid (–CH_2OH is oxidised) and con. HNO_3 to glucaric acid (saccharic acid) (both –CHO and –CH_2OH oxidised). If, however, the –CHO group is protected with acetone to form the isopropylidene derivative, it will not be oxidised by any oxidising agent. Now the –CH_2OH alone can be oxidised and the isopropylidene group subsequently can be hydrolysed to give back the –CHO group. This is a convenient method of preparing glucuronic acid.

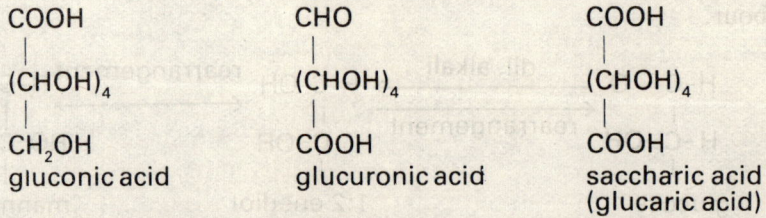

Formulae of acids obtained on oxidation

COOH	CHO	COOH
(CHOH)$_4$	(CHOH)$_4$	(CHOH)$_4$
CH$_2$OH	COOH	COOH
gluconic acid	glucuronic acid	saccharic acid (glucaric acid)

Glucose is specifically oxidised by the enzyme glucose oxidase in the presence of molecular oxygen to δ gluconolactone quantitatively. H_2O_2 is also formed in the oxidation. This H_2O_2 is further treated with peroxidase and orthodianisidine to give a yellow compound. This reaction is the basis of a very sensitive colorimetric method of estimation for glucose and gives what is referred to as 'true sugar' levels in blood.

Reduction: On reduction with sodium amalgam, the alcohol sorbitol is formed.

Reducing reactions of glucose: Glucose reduces alkaline copper reagent (Benedict's qualitative reagent and also Fehling's solution) to the red precipitate of cuprous oxide. The cupric ions of the sparingly soluble cupric hydroxide are kept in solution by the citrate of the Benedict's reagent or tartrate of Fehling's solution offering enough cupric ions in solution. The sugar is converted to enediol with alkali and this group is a powerful reducer.

glucose + alkali ⟶ enediol
enediol + Cu^{++} ⟶ Cu$^+$ + sugar acid
(cupric)   (cuprous)
Cu$^+$ + OH$^-$ ⟶ CuOH ⟶ Cu$_2$O
                                     (cuprous oxide) (red)

Benedict's reagent contains copper sulphate, sodium carbonate (functions as mild alkali) and sodium citrate while Fehling's solution contains copper sulphate, sodium hydroxide and Rochelle salt (sodium potassium tartrate). Benedict's reagent is preferable to Fehling's solution because it is stable, and traces of sugars which might be destroyed by the strong alkali of Fehling's are left intact. Fehling's reagent has to be obtained by mixing Fehling solution 1 and Fehling solution 2, which means extra work. Benedict's *quantitative* reagent contains, in addition, potassium thiocyanate and potassium ferrocyanide. On boiling glucose with this reagent, a *white* precipitate of cuprous thiocyanate is obtained. This reagent is used for the estimation of glucose in urine.

Glucose and other monosaccharides reduce Barfoed's reagent (cupric acetate in acetic acid) to red cuprous oxide on boiling. Due to the acid medium, enediol formation which happens with Benedict's reagent does not happen, and reduction is not as marked in acid pH as in alkaline medium. Reducing monosaccharides give a precipitate within about 30 seconds to one minute while reducing disaccharides give the precipitate after boiling for about 7-12 minutes. Hence this test is employed to differentiate reducing monosaccharides from reducing disaccharides.

Glucose reduces Tollen's reagent (ammoniacal silver nitrate) Ag(NH$_3$)$_2$OH to bright metallic silver.

R CHO + 2 Ag(NH$_3$)$_2$ OH → R COO NH$_4$ + 2 Ag + 3 NH$_3$ + H$_2$O

Molisch's Test: This is a general test for all carbohydrates, sugars and sugar-protein complexes (muco and glycoproteins). Glucose with alcoholic alpha-naphthol and a few drops of con. H$_2$SO$_4$ added through the sides of the test tube forms a purple ring at the junction of the liquids, due to the formation of furfural from glucose with con. H$_2$SO$_4$ and the condensation of furfural with alpha-napthol.

Anthrone test: Carbohydrate with anthrone in the presence of con. H$_2$SO$_4$ gives a green

or blue green colour. The carbohydrate is decomposed by acid to a furfural derivative which reacts with anthrone. This forms the basis for a useful colorimetric estimation of sugars.

Osazones: With phenyl hydrazine (usually as phenyl hydrazine hydrochloride) glacial acetic acid and sodium acetate, glucose, on boiling, gives a yellow crystalline substance, glucosazone. This being insoluble separates even when the solutions are being heated, within a few minutes. The crystals have needle-like (corn leaf-like) shape under the microscope. This test, therefore, helps in purification and characterisation of glucose and to differentiate it from lactose when there is a doubt whether the sugar in the urine is glucose or lactose.

$$\begin{array}{c} H \\ H-C = \boxed{O\ H_2}\ N\ N\ C_6H_5 \\ | \quad\quad + \\ H-C-OH \\ | \\ R \end{array} \xrightarrow{H_2O} \begin{array}{c} H \\ | \\ H-C = N-N-C_6H_5 \\ | \\ H-C-OH \\ | \\ R \end{array}$$

phenyl hydrazone

$$\begin{array}{c} H \\ H-C = N-N-C_6H_5 \\ | \\ H-C-O\!+\!H + H_2N \\ | \\ R \end{array} -NH-C_6H_5 \xrightarrow[C_6H_5NH_2]{NH_3} \begin{array}{c} H \\ H-C = \ N-N-C_6H_5 \\ | \\ C = O \\ | \\ R \end{array}$$

$$\begin{array}{c} H \\ H-C = N-N-C_6H_5 \\ | \\ C = \boxed{O + H_2}\ N-NH-C_6H_5 \\ | \\ R \end{array} \xrightarrow{H_2O} \begin{array}{c} H \\ HC = N-N-C_6H_5 \\ | \\ C = N-NH-C_6H_5 \\ | \\ R \end{array}$$

Use of acetic acid and sodium acetate is to offer a buffer for suitable reactions. The precise mechanism of the reaction is explained by Amadori rearrangement.

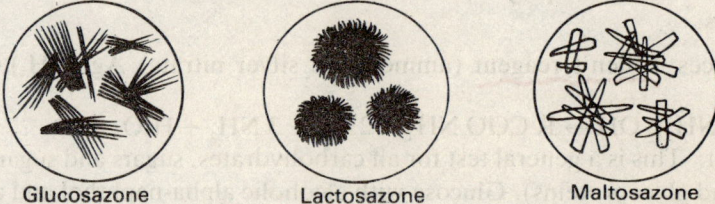

Glucosazone   Lactosazone   Maltosazone

Fermentation: Fermentation is the breakdown of complex organic substances into simpler ones by the activity of living cells through the agency of enzymes. Glucose is fermented to ethyl alcohol and CO_2 by yeast through the enzyme zymase. As alcohol is produced in this type of fermentation, this is called alcoholic fermentation.

Fructose

This is also a reducing monosaccharide and is a ketohexose. Like glucose this also possesses ring structures pyranose and furanose.

$$\begin{array}{c} CH_2OH \\ | \\ CO \\ | \\ HO-C-H \\ | \\ H-C-OH \\ | \\ H-C-OH \\ | \\ CH_2OH \end{array} \quad\quad \begin{array}{c} CH_2OH \\ | \\ HO-C\text{---------}\! \\ | \quad\quad\quad\quad | \\ HO-C-H \quad\quad | \\ | \quad\quad\quad\quad O \\ H-C-OH \quad\ | \\ | \quad\quad\quad\quad | \\ H-C\text{-----------}\! \\ | \\ CH_2OH \end{array} \quad\quad \begin{array}{c} CH_2OH \\ | \\ HO-C\text{----------}\! \\ | \quad\quad\quad\quad\ | \\ HO-C-H \quad\quad | \\ | \quad\quad\quad\quad\ O \\ H-C-OH \quad\ | \\ | \quad\quad\quad\quad\ | \\ H-C-OH \quad\ | \\ | \quad\quad\quad\quad\ | \\ CH_2\text{------------}\! \end{array}$$

furanose   pyranose

Of the two ring forms, the pyranose form is stabler and is levo rotatory. Hence fructose exists as D-fructo pyranose and is called levulose. $[\alpha]_D^{20}$ for ß-D fructose is $-133.5°$. It is the

CARBOHYDRATES

sweetest of all sugars. The configuration of D-fructose is identical with that of glucose in respect of the 4 carbon atoms other than the ones constituting $-CO-CH_2OH$ group. In many reactions it resembles glucose. It answers Molisch's test, Benedict's, Barfoed's, Tollen's reactions, undergoes Van Ekenstein transformation with very dilute alkalies and is fermented to alcohol.

The reducing reactions of fructose are not due to the isolated $-C = O$ group, as ketonic group does not have the powerful reduction reactions of $-CHO$ group. It is due to the presence of $-CO-CH_2OH$ group forming with alkali enediol which is the actual functional reducing group of fructose.

Fructose forms the same osazone as glucose and glucosazone and fructosazone are identical as they have the same crystalline shape and structure. Osazone is formed in a few minutes even in hot conditions.

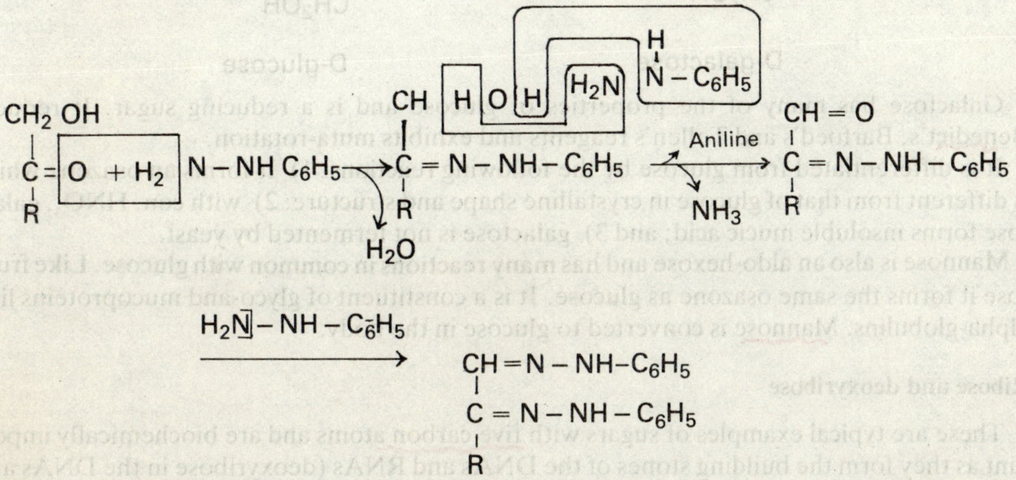

$$\begin{array}{c}
CH_2OH \\
| \\
C=O \\
| \\
R
\end{array} + H_2N-NH\,C_6H_5 \longrightarrow \begin{array}{c} CH \\ | \\ C=N-NH-C_6H_5 \\ | \\ R \end{array} \xrightarrow[NH_3]{Aniline} \begin{array}{c} CH=O \\ | \\ C=N-NH-C_6H_5 \\ | \\ R \end{array}$$

$$\xrightarrow{H_2N-NH-C_6H_5} \begin{array}{c} CH=N-NH-C_6H_5 \\ | \\ C=N-NH-C_6H_5 \\ | \\ R \end{array}$$

The formation of osazone affects only the groups $-CHOH-CHO$ in glucose and $-CO-CH_2OH$ in fructose to form, in both cases, the same substituents in those two carbons. Since the configuration of glucose and fructose is the same in respect of the other carbon atoms, the osazones formed by the two sugars are identical. In fact it could be said that fructose also forms glucosazone. Muta-rotation is answered by fructose, but unlike glucose. it is a conversion of a six-membered pyranose ring to the five-membered furanose ring of fructose till an equilibrium is obtained. At equilibrium, fructose solution has $[\propto]$ of $-92°$. The oxidation of fructose yields mesotartaric acid and glycollic acid.

Differentiation between glucose and fructose

Seliwanoff's test: When fructose is boiled with Seliwanoff's reagent (resorcinol in HCl), a cherry-red colour is formed. Glucose does not have such a reaction.

Foulger's test: On boiling with Foulger's reagent (urea in H_2SO_4 with stannous chloride), fructose gives a blue colour while glucose gives a green or amethyst colour.

Paper chromatography can also be used.

Galactose

This is an aldo-hexose-like glucose but differs in the orientation of $-H$ and $-OH$ around only one carbon atom when compared with glucose. Such isomers are called epimers. In liver, galactose is converted to glucose with the help of enzymes. This process is referred to as epimerisation.

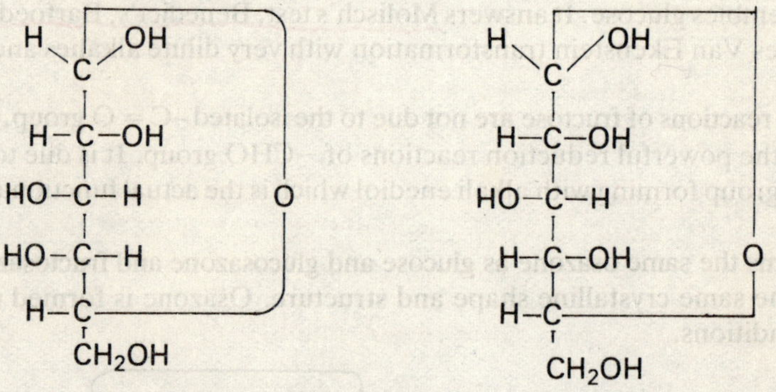

```
    H   OH                          H   OH
     \ /                             \ /
      C                               C
   H—C—OH                          H—C—OH
  HO—C—H      O                   HO—C—H      O
  HO—C—H                           H—C—OH
   H—C                              H—C
     CH₂OH                            CH₂OH

   D-galactose                      D-glucose
```

Galactose has many of the properties of glucose and is a reducing sugar. It reduces Benedict's, Barfoed's and Tollen's reagents and exhibits muta-rotation.

It is differentiated from glucose by the following reactions: 1) It forms an osazone which is different from that of glucose in crystalline shape and structure: 2) with con. HNO_3, galactose forms insoluble mucic acid; and 3) galactose is not fermented by yeast.

Mannose is also an aldo-hexose and has many reactions in common with glucose. Like fructose it forms the same osazone as glucose. It is a constituent of glyco-and mucoproteins like alpha-globulins. Mannose is converted to glucose in the body.

Ribose and deoxyribose

These are typical examples of sugars with five carbon atoms and are biochemically important as they form the building stones of the DNAs and RNAs (deoxyribose in the DNAs and ribose in RNAs). Ribose is also present in many coenzymes and cofactors like ATP, GTP, CoA, NAD^+, $NADP^+$ and flavoproteins.

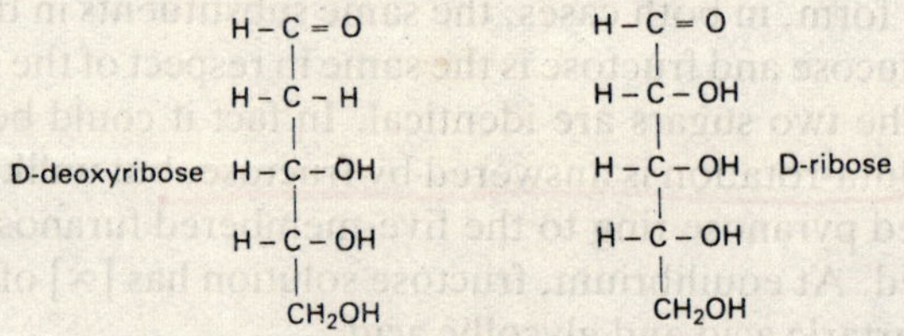

```
      H—C=O                         H—C=O
      H—C—H                         H—C—OH
D-deoxyribose  H—C—OH          H—C—OH   D-ribose
      H—C—OH                        H—C—OH
       CH₂OH                         CH₂OH
```

Being an aldo-pentose, ribose reduces Benedict's, Barfoed's and Tollen's reagents. It also forms a characteristic osazone, and answers Molisch's test. Deoxy sugar cannot form osazone, is unstable and easily forms resins. Unlike ribose, it restores the colour of Schiff's reagent (fuchsin-sulphurous acid) indicating the presence of considerable free –CHO form in solution.

Bial's orcinol test: Both ribose and deoxyribose answer Bial's orcinol-hydrochloric acid test. The reagent contains a little $FeCl_3$ also. It is heated to boiling point and several drops of sugar solution are added immediately after removing it from the flame. Pentoses give a green colour (also certain uronic acids like glucuronic and galacturonic acids which, on heating with acid, give pentoses as well as trioses). Keto-hexoses and methyl-pentoses produce orange-coloured solutions which separate as dark green precipitate on standing.

Tollen's phloroglucinol-HCl test: Pentoses when boiled with a solution of phloroglucinol in HCl develop a cherry-red colour. Galactose also gives this test.

Test for deoxy sugar: When a pine splinter is placed in con. HCl containing deoxy sugar, the splinter develops a green colour. Treating the deoxy sugar in glacial acetic acid containing a ferrous salt and H_2SO_4 produces a blue colour (Kiliani's test).

Sugar amines, uronic acids and sugar phosphates

Sugars containing the amino groups are called amino sugars. Sugar amines are present in many muco-polysaccharides. Glucosamine is present in hyaluronic acid, heparin, blood group substances and galactosamine (also called chondrosamine) in chondroitin sulphates while mannosamine is present in sialic acids. Several antibiotics, such as erythromycin and carbomycin, contain amino-sugars which contribute to their physiological activity.

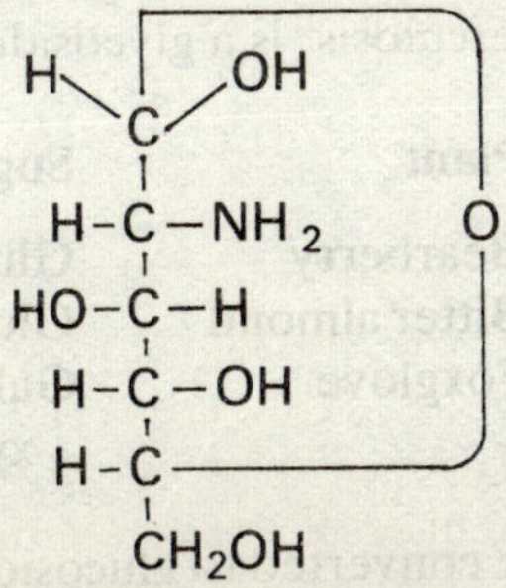

$$\begin{array}{c} H-C-OH \\ H-C-NH_2 \\ HO-C-H \\ H-C-OH \\ H-C \\ CH_2OH \end{array}$$

A typical uronic acid is glucuronic acid. Any uronic acid should contain the reducing group like –CHO and the acid group. Glucuronic acid is an important uronic acid and is present in many muco-polysaccharides. This is also produced in an alternate metabolic pathway of glucose in the body called the 'uronic acid pathway'. Glucuronic acid also conjugates with substances like bilirubin, phenols, secondary alcohols, sterols, benzoic and phenyl acetic acids and helps in the detoxication and disposal of these substances from the body. Common uronic acids are tested by Tollen's naphthoresorcinol test. The sugar is dissolved in 1N HCl containing a small amount of naphthoresorcinol and heated for about five minutes. The mixture is cooled and extracted with ether. The ether layer becomes purple, violet or pink if the test is positive.

Sugar phosphates are intermediates in the metabolism of many sugars. Examples are glucose-6-phosphate, glucose-1-phosphate, fructose-6-phosphate (Neuberg ester), fructose 1,6-

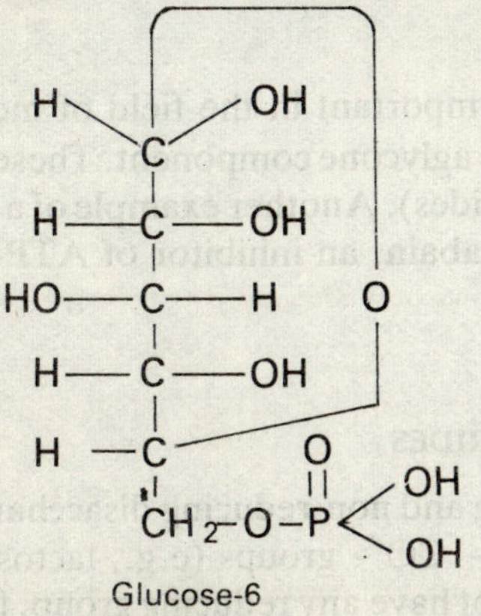

Glucose-6 Phosphate (Robinson ester)

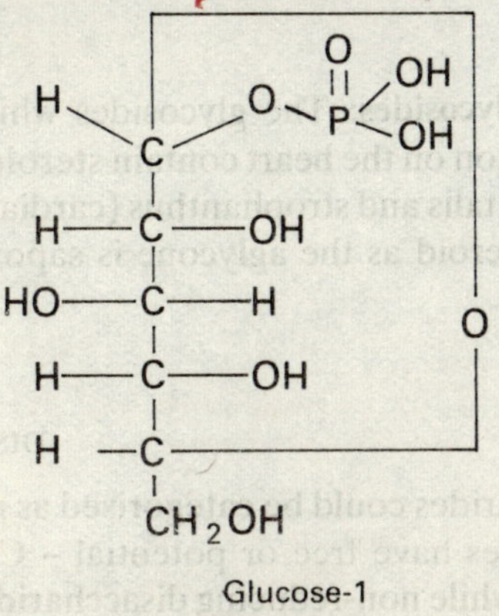

Glucose-1 Phosphate (Cori ester)

diphosphate (Harden Young ester) and 3-phospho-glyceraldehyde. The phosphorylation of sugars is an important step in the metabolism of sugars.

Glycosides

These are mostly naturally occurring derivatives of sugars and are present in plants. They are acetals derived from a combination of various hydroxy compounds with sugars, and are designated as glucosides, if glucose is the sugar, mannosides if mannose is the sugar and so on. All glucosides are glycosides but all glycosides need not be glucosides.

When a sugar is combined with a non-sugar, the latter is described as an aglycone. (Sugar-O-aglycone). The glycosides are hydrolysed, by mineral acids, to the sugar and the aglycone. Enzymes emulsin and maltase which may also be present in plants hydrolyse glycosides: emulsin hydrolyses beta glycoside while maltase hydrolyses alpha glycoside. Streptomycin, an antibiotic used for treating tuberculosis, is a glycoside.

Examples of glycosides:	Plant	Sugar	Aglycone
Arbutin (glucoside)	Bearberry	Glucose	Hydroquinone
Amygdalin (glucoside)	Bitter almond	Glucose	Mandelonitrile
Digitonin (Galactoside and xyloside)	Foxglove	Galactose and xylose	Digitogenin

In the laboratory, glucose can be converted to glucoside, viz. methyl glucoside by treating a solution of glucose in boiling methyl alcohol with 0.5 per cent hydrogen chloride as a catalyst.

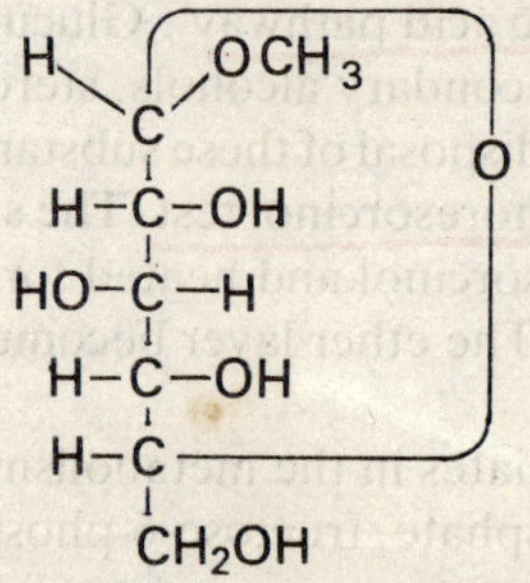

$$\begin{array}{c} HOCH_3 \\ C \\ H-C-OH \\ HO-C-H \\ H-C-OH \\ H-C \\ CH_2OH \end{array}$$

Steroidal glycosides: The glycosides which are important in the field of medicine because of their action on the heart contain steroids as the aglycone component. These include derivatives of digitalis and strophanthus (cardiac glycosides). Another example of a glycoside which contains steroid as the aglycone is saponin. Ouabain, an inhibitor of ATP-ase is a steroid glycoside.

DISACCHARIDES

Disaccharides could be categorised as reducing and non-reducing disaccharides. Reducing disaccharides have free or potential – CHO or – CO – groups (e.g., lactose, maltose and turanose) while non-reducing disaccharides do not have any reducing group, free or potential (e.g., sucrose and trehalose).

Lactose

$C_{12}H_{22}O_{11}$ (milk sugar) which occurs in the milk of mammals (about five per cent) is a white crystalline solid, soluble in water and is optically active. It gives lactic acid by lactic fermentation.

Lactose is a reducing sugar. It reduces Benedict's reagent. Barfoed's reagent is reduced slowly (7-12 minutes of boiling). Lactose reacts to Molisch's test and exhibits muta-rotation. It forms osazone, somewhat slowly, which separates on cooling. The shape of lactosazone crystals is like powder-puff, uncombed hair or a hedgehog.

Each molecule of lactose is made up of one molecule of glucose and one molecule of galactose with the potential aldehydic group of glucose available for reducing properties.

ß D galactosyl – 1,4 – D glucose

It can be hydrolysed into glucose and galactose by acid and also by lactase. Lactose gives rise to mucic acid on prolonged boiling with HNO_3 from the galactose derived from it by hydrolysis. Lactose cannot be fermented. It is hydrolysed by emulsin.

Sucrose

Cane sugar $C_{12}H_{22}O_{11}$ which occurs in many plants, e.g., sugarcane, is a white crystalline solid, soluble in water, sweet to taste and is dextro-rotatory.

Sucrose reacts to Molisch's test. Con. H_2SO_4 chars it at ordinary temperature.

It is a non-reducing sugar and has no free or potential aldehydic or ketonic group in the molecule. As such, it does not reduce Fehling's solution, Benedict's solution and Barfoed's reagent. It does not exhibit muta-rotation and does not form an osazone.

Inversion and invert sugar: Sucrose is dextro-rotatory (+66.5), but by treating it with dilute acid or sucrase (invertase) it is hydrolysed to equal amounts of glucose (+) and fructose (−). The resulting mixture is levo-rotatory as the magnitude of levo-rotation (−92) of fructose is greater than the magnitude of dextro-rotation of glucose (+52.7). Since the dextro-rotatory sucrose is converted to a levo-rotatory mixture with inversion of optical rotation, this process is called inversion. The mixture of glucose and fructose is called invert sugar. Hudson has shown that on hydrolysis sucrose at first gives only D-fructo furanose (+) which takes some time to isomerise to D-fructo pyranose (−). Hence sucrose proper contains D-fructo furanose.

Sucrose + H_2O $\xrightarrow{\text{Enzyme or dilute acid}}$ D–gluco pyranose + D fructo furanose
(dextro)   (+ 52.7)   (+)
+ 66.5 $\downarrow$ isomerization
                                                                            D fructo pyranose
                                                                            (− 92)

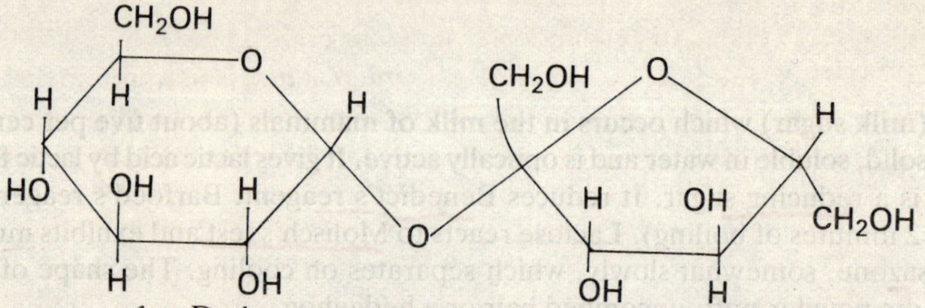

1 ∝ D-gluco pyranosyl 2β-D-fructo furanoside.

It is fermentable. As sucrose is hydrolysed to reducing sugars, the reducing reactions of sugars will be answered by the hydrolysate of sucrose after necessary neutralisation.

Tests for sucrose: Molisch's test is positive, and sucrose chars with con. H_2SO_4. It does not reduce Benedict's solution. But the hydrolysate reacts to all typical reactions of reducing sugars and forms osazone. Sucrose also gives positive reactions with Seliwanoff's and Foulger's reagents.

Maltose

Malt sugar ($C_{12}H_{22}O_{11}$) results from digestion or hydrolysis of starch with diastase (amylase). It is made up of glucose only and is fermentable. It is a reducing sugar answering all the typical reactions of reducing sugars and forms an osazone. Maltosazone also appears after heating for considerable time and separates only after cooling. Crystals of maltosazone are sunflower-like. On hydrolysis, maltose gives only glucose.

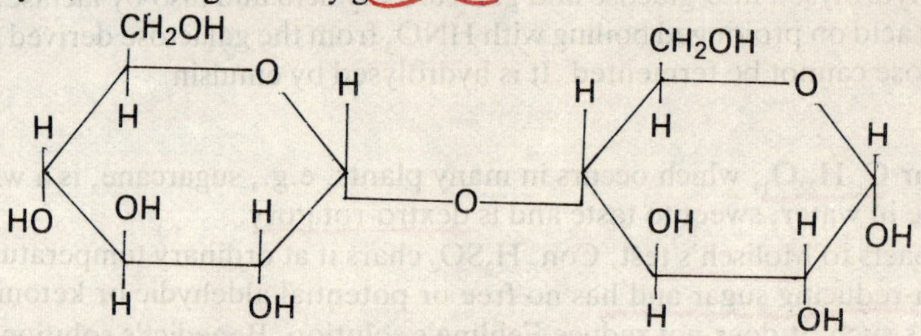

4 ∝ D glucopyranosyl ∝ D glucopyranoside

Reducing disaccharides react positively to Fearon's test (methyl amine and NaOH).

The body cannot utilise unhydrolysed disaccharides like sucrose, lactose, etc. If they are injected they will be simply excreted. Hence to derive the nutritional value of these sugars they must be taken orally as only then they are converted to utilisable sugars by the enzymes of the gastro-intestinal tract.

POLYSACCHARIDES (GLYCANS)

These are classified into homoglycans and heteroglycans. Starch and glycogen are homoglycans as they are made of only glucose and are called glucans or glucosans. On the other hand, muco-polysaccharides like heparin, hyaluronic acid and chondroitin sulphates are heteroglycans as they are made up of different monosaccharide units, like uronic acids, acetylated sugar amines with or without $-SO_2OH$ groups.

Starch

Amylum ($C_6H_{10}O_5 \cdot H_2O)_n$: Widely distributed in the vegetable kingdom (cereals, potatoes), starch grains are associated with about 14-19 per cent water of which about 10 per cent is bound chemically. The starch grain consists of a central core of water-soluble amylose (10–20 per cent) and an outer insoluble portion of amylopectin (80–90 per cent). Hence it is only slightly soluble in cold water but dissolves in hot water giving a viscous solution. The formation of the pasty solution is due to amylopectin. Amylose gives a blue colour with iodine solution while amylopectin, violet colour. The starch-iodine reaction is so sensitive that it can be used to detect traces of iodine or of starch. The colour disappears when the solution is heated and reappears on cooling. This is because, for the development of colour, large molecular aggregates (micelles) of amylose and amylopectin molecules are required to form the micelle-I_2 complex which is coloured. These micelles are destroyed on heating and reformed on cooling.

Starch is hydrolysed to glucose by acids on heating and is therefore a glucan and a homoglycan.

$$(C_6H_{10}O_5)n + n(H_2O) \rightarrow n(C_6H_{12}O_6)$$
$$\text{D-glucose}$$

Enzyme hydrolysis: By the action of the enzyme diastase (amylase) abundantly present in germinating seeds, in saliva and in the pancreatic juice of animals, starch is converted in steps to soluble starch (blue colour with iodine) amylodextrin (violet with iodine), erythrodextrin (red with iodine), achrodextrin (no colour) and maltose. There are two amylases, α and β, which have different modes of action. While ß amylase is an exo-amylase α is an endo-amylase.

DIFFERENCES BETWEEN AMYLOSE AND AMYLOPECTIN

Amylose	Amylopectin
Low molecular weight	High molecular weight
About 300 glucose units	About 1000 glucose units
Soluble in water	Sparingly soluble
Blue colour with dilute iodine solution	Violet colour
Not branched	Branched

Structures of amylose and amylopectin:

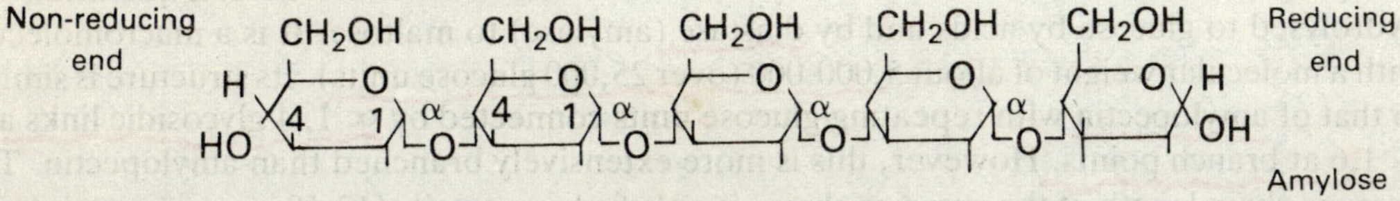

Amylose (containing glucose units linked by 1,4 α glycosidic links) has one reducing and one non-reducing end.

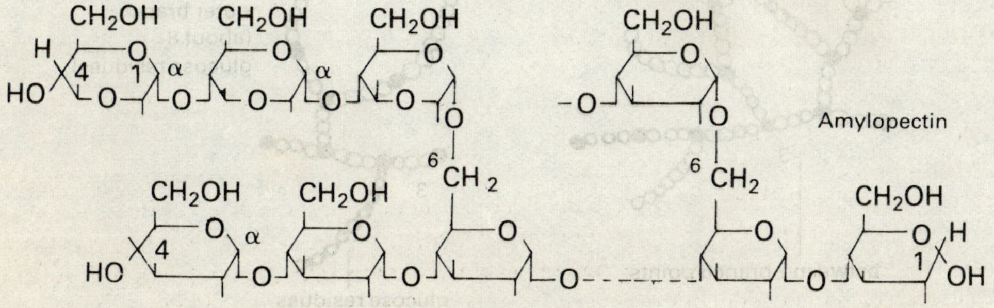

Amylopectin is made of glucose units with 1,4 α glycosidic links and 1,6 at branch points. Starch is tasteless. If heated by itself to about 200°C, a gum-like product, dextrin, is formed. The microscopic form of the granules is characteristic of the source of starch.

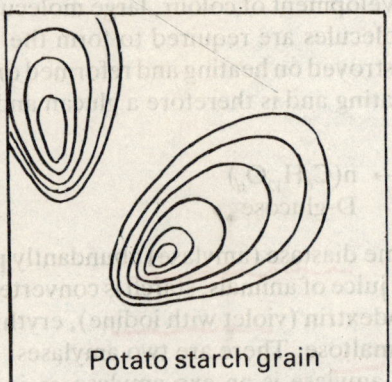
Potato starch grain

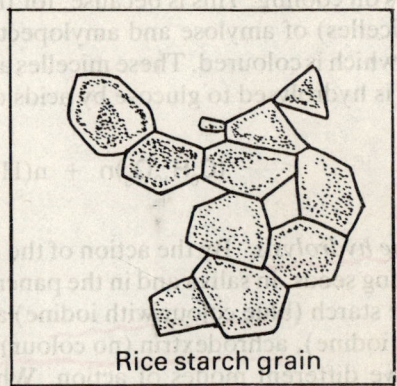
Rice starch grain

Starch in the plant kingdom and glycogen in the animal kingdom with close structural resemblance and similar functions (reserve carbohydrate) indicate unity in diversity.

Glycogen $(C_6H_{10}O_5)_n$

This is the polysaccharide of the animal body and is often called animal starch. It is present in muscle, liver and to a small extent in the brain and some other tissues.

Glycogen is the reserve carbohydrate of the body and the liver glycogen is converted to D-glucose during fasting. Muscle glycogen is not as labile as liver glycogen but on occasions like intense physical exercise, muscle glycogen is converted to lactic acid. Thus glycogen is a source of energy for the body.

It is a white, amorphous, tasteless powder soluble in hot water. The solution is dextro-rotatory and does not have reducing properties. It forms a red brown colour with iodine solution.

Glycogen shows considerable similarity to starch in chemical behaviour. It is quantitatively hydrolysed to glucose by acids and by diastase (amylase) to maltose. It is a macromolecule with a molecular weight of about 5,000,000 (over 25,000 glucose units). Its structure is similar to that of amylopectin with repeating glucose units connected by α 1, 4 glycosidic links and α 1,6 at branch points. However, this is more extensively branched than amylopectin. The average chain length of the exterior chains is only 8 glucose units (13-18 in amylopectin) and in the main chain there is a branch point to every three units on an average (every 5-6 units in amylopectin).

GLYCOGEN

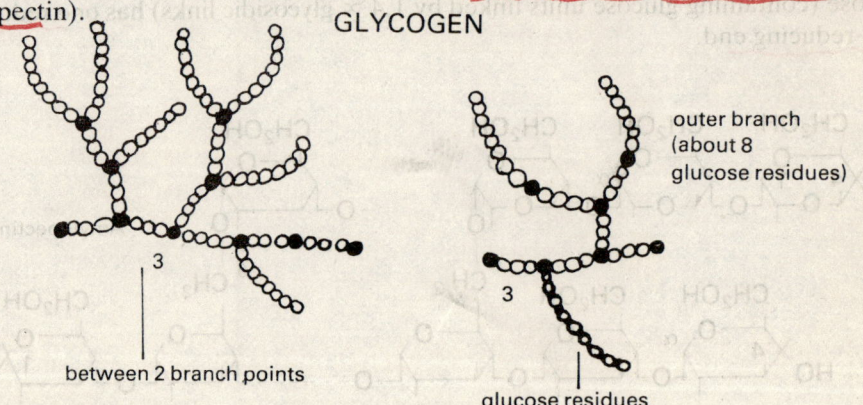

between 2 branch points

outer branch (about 8 glucose residues)

glucose residues

Glycogen from the liver of a fed animal is extracted as follows: The liver is digested with about 30 per cent aqueous KOH. The tissue is degraded and the glycogen goes into solution. On adding excess alcohol, glycogen gets precipitated as white powder as it is insoluble in alcohol.

Dextrins: These are glucans and are formed from starch either on heating it under controlled conditions or in the course of hydrolysis (refer Starch).

Dextrins give violet to red colour with iodine. Complete hydrolysis of dextrins with acid yields glucose.

Inulin: This is a fructosan as it gives only fructose on hydrolysis. It is not metabolised in the body and hence is used for an important kidney function test (Inulin clearance test).

Heteroglycan

Agar, gum arabic, gum acacia, pectins, acid muco-polysaccharides, blood group polysaccharides and bacterial polysaccharides are examples of heteroglycans.

Agar is a sulphuric acid ester of a complex galactose polysaccharide ordinarily combined with metal ions like Na^+, Ca^{2+}, etc. Gums on hydrolysis yield galactose, arabinose or xylose.

Glycosamino glycans (Acid muco-polysaccharides)

Glycosamino glycans are heteroglycans and generally contain repeating units of amino sugars and sugar uronic acids. Hence the name 'glycosamino glycans'. The amino groups are acetylated. In addition, sulphuric acid groups may also be present. It is the sulphuric acid groups that give significant acidity to these classes of compounds. Carboxylic acid groups also contribute to the acidity but to a limited extent — hence the other name 'acid muco-polysaccharides'.

Glycosamino glycans are present in animal tissues as structural units. They are generally in combination, in part, with protein as mucoproteins or mucoids. They are also called proteoglycans and some of these contain polysaccharides to the extent of 95 per cent and protein to the extent of only five per cent. They form the ground substance of connective tissue.

PROTEO GLYCAN

Schematic figure of proteoglycan aggregate in **CONNECTIVE TISSUE**

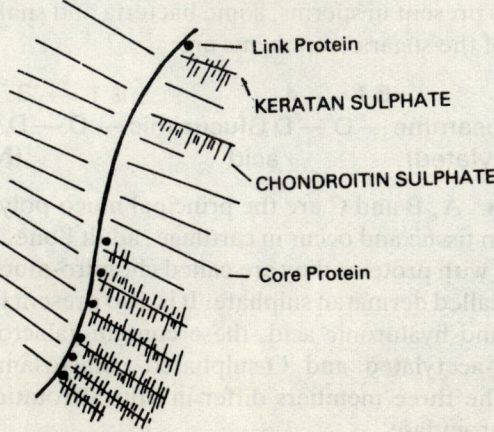

The various glycosamino glycans are:

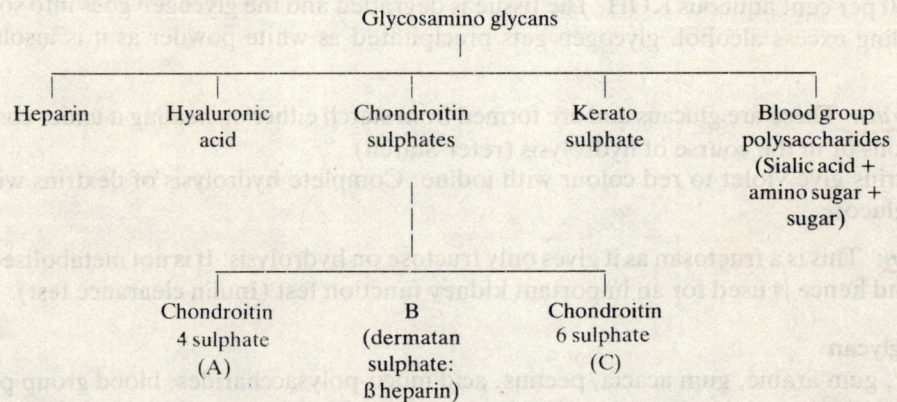

Heparin: Alpha heparin is present in liver, lungs, spleen and blood. It is an anti-coagulant, and is made of repeating units of D-glucuronic acid and D-glucosamine whose $-NH_2$ is sulphated. Some of the $-OH$ groups are also sulphated. This is a strongly acidic substance due to the sulphuric acid groups.

The repeating units of sugar derivatives in heparin are

$$\text{—D glucuronic acid (O-sulphated)} \xrightarrow{\alpha\ 1\to 4} O \text{—D glucosamine (N and O sulphated)} \xrightarrow{\alpha\ 1\to 4} O \text{—D glucuronic acid (O-sulphated)}$$

Hyaluronic acid occurs in vitreous humour, synovial fluid, skin and umbilical cord. It is an integral part of the gel-like ground substance of connective and other tissues, and hyalin functioning as lubricant and shock absorber in joints, and as tissue barrier resisting penetration by bacteria. It is made up of repeating units of D-glucuronic acid and N-acetyl D-glucosamine. This is one important heteroglycan containing no sulphuric acid group. Total hydrolysis gives equal quantities of D-glucosamine, D-glucuronic acid and acetic acid. Hyaluronic acid is depolymerised by testicular hyaluronidase followed by mineral acid to hyalo-biuronic acid which is a combination of D-glucuronic acid and D-glucosamine. The enzyme 'hyaluronidase' (the spreading factor) is present in sperms, some bacteria and snake venoms.

The repeating units of the sugar components are:

$$\text{—D glucosamine (N-acetylated)} \xrightarrow{\beta\ 1\to 4} O \text{— D Glucuronic acid} \xrightarrow{\beta\ 1\to 3} O \text{— D glucosamine (N-acetylated)}$$

Chondroitin sulphates: A, B and C are the principal muco-polysaccharides in the ground substance of mammalian tissue and occur in cartilage, adult bone, tendons, cornea, skin and heart valves. Occurring with protein, they are called chondro-mucoid and osseo-mucoid. As B is present in skin it is called dermatan sulphate. It is also present in the heart valves and tendons. Unlike heparin and hyaluronic acid, these contain galactosamine (acetylated). The repeating units are N-acetylated and O-sulphated galactosamine and glucuronic acid (iduronic acid in B). The three members differ in optical rotation and in their behaviour towards testicular hyaluronidase.

CARBOHYDRATES

The repeating units of sugar components in chondroitin sulphate A are:

$$\text{—D galactosamine} \xrightarrow{\beta 1 \to 4} O \text{—D glucuronic} \xrightarrow{\beta 1 \to 3} O \text{—D galactosamine—}$$
(N-acetylated and   acid   (N-acetylated and
O-sulphated)   O-sulphated)

While the sulphuric acid group is in the fourth carbon of galactosamine in chondroitin sulphate A, it is in the sixth carbon in C which is the only structural difference between the two. Presently they are referred to as chondroitin-4-sulphate and chondroitin-6-sulphate, respectively.

Chondroitin sulphate B is also known as dermatan sulphate and ß heparin and has a primary structure different from that of A and C.

The primary structure of chondroitin sulphate B is:

$$\text{—L iduronic acid} \xrightarrow{\alpha 1 \to 3} O \text{—N acetyl galactosamine} \xrightarrow{\beta 1 \to 4} O \text{—L iduronic—}$$
(O-sulphated)   acid

Keratosulphate occurs in the cornea and is made of repeating units of N-acetylated glucosamine, which is also sulphated and galactose (rare case where the uronic acid is absent).
Keratosulphate consists of

$$\text{— D galactose} \xrightarrow{\beta 1 \to 4} O \text{— D glucosamine} \xrightarrow{\beta 1 \to 3} O \text{— D galactose — O —}$$
(N-acetylated
O-sulphated)

Sialic acids are widely distributed in tissues particularly in mucins and blood group substances. They are also present in vertebrate tissues. Sialic acids are structurally related to neuraminic acids which have a skeleton of nine carbon atoms with an amino group. A typical sialic acid is N-acetyl neuraminic acid.

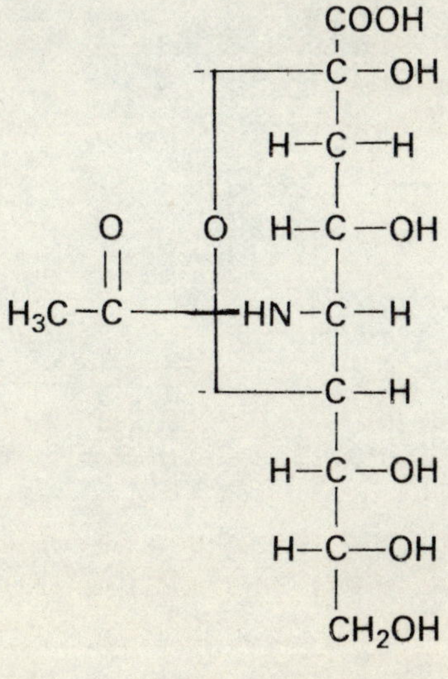

```
                      COOH
                       |
                    —C—OH
                       |
                     H—C—H
                       |
         O     O    H—C—OH
         ||    |      |
   H₃C—C———HN—C—H
                       |
                    —C—H
                       |
                    H—C—OH
                       |
                    H—C—OH
                       |
                      CH₂OH
```

Enzymes in the liver of rats and bovine sub-maxillary glands can accomplish the biosynthesis of N-acetyl neuraminic acid. N-acetyl neuraminic acid undergoes cleavage to pyruvic acid and N-acetyl D-mannosamine in a reaction catalysed by the enzyme N-acetyl neuraminic acid aldolase.

The mucin of saliva contains N-acetyl neuraminyl N-acetyl galactosamine with protein. It cannot be considered a *muco-polysaccharide* as it is only a *disaccharide*. Numerous molecules of this disaccharide are associated with a protein molecule. The *blood group* polysaccharides also contain sialic acids, amino sugars and simple sugars (L-fucose, a methyl pentose is present to about 19 per cent in some). With proteins they form A, B, O, Rh factors of erythrocytes which constitute the antigens of RBCs. These antigens of the RBCs react with the antibodies of serum causing agglutination. The O group antigen of RBC is not agglutinated by different types of antibodies of serum and the person with O group blood is referred to as the universal donor (O group blood can be given to any person safely).

Mucoproteins and glycoproteins

If the carbohydrate associated with protein is greater than four per cent, the complex protein is called a mucoprotein. If less, it is termed glycoprotein. In human plasma, alpha$_1$ and alpha$_2$ globulins are glycoproteins and contain acetyl hexosamine, mannose, galactose, fucose, and sialic acids. Some hormones like pituitary gonadotropin and thyrotropin are glycoproteins. Oroso-mucoid contains 30-50 per cent carbohydrate and is a mucoprotein. Glycophorin is an integral membrane mucoprotein of the erythrocyte. It contains 60 per cent carbohydrate and 40 per cent polypeptide.

Mucous secretions contain mucoproteins which are made up of specific proteins with acetyl hexosamines, hexoses like mannose and galactose and L-fucose (methyl pentose).

3 Lipids

Lipids are substances which, according to Bloor, possess the following characteristics:
1) They are insoluble in water but soluble in fat solvents such as ether, benzene, chloroform, carbon tetrachloride, alcohol, etc. Some lipids are soluble only in hot alcohol.
2) They possess some relationship with fatty acids such as esters, either actual or potential.
3) They are utilised by living organisms.

The various types of oils, fats, sterols, waxes and phospholipids belong to this class. Lipids are present both in the plant and animal kingdoms. In animal tissues, the distribution of lipids is variable within wide limits both in quality and quantity, thus, peri-renal fatty tissue, subcutaneous tissue, mesentery and yellow bone marrow contain lipid to the extent of 90 per cent which is mainly due to neutral fat. In the non-depot tissue this value amounts to 10 per cent while in embryonic tissue it is only 1-2 per cent. Brain and spinal cord contain 7.5-30 per cent of lipid which is mostly due to cholesterol, glycolipids and phospholipids.

Classification of lipids

Lipids vary in their chemical composition so widely and differently that their classification has been rendered difficult. The classification by Bloor is widely used.

I. *Simple lipids:* Esters of fatty acids with alcohols
   1) Neutral fats: Triacyl glycerols (Triglycerides). Esters of various fatty acids with glycerol, e.g. tripalmitin, oleodistearin.
   2) Waxes: Esters of fatty acids with monohydroxy aliphatic alcohols other than glycerol.
      a) True waxes: Esters of palmitic, stearic, oleic or other fatty acids with cetyl alcohol $CH_3(CH_2)_{14}CH_2OH$ or other straight chain alcohols.
      b) Cholesterol esters: Esters of fatty acids with cholesterol.
      c) Vitamin A esters.
      d) Vitamin D esters.

II. *Compound lipids:* Esters of fatty acids with alcohols and containing other groups in addition to alcohols and acids.
   1) Phospholipids (Phosphatides): Esters containing phosphoric acid and a nitrogenous base, an amino acid or some other moiety like inositol.
      a) Lecithin (Phosphatidyl choline)
      b) Cephalin (Phosphatidyl ethanolamine)
      c) Phosphatidyl serine
      d) Acetal phospholipid (Plasmalogen)
      e) Inositol phospholipids (Lipositol)
      f) Sphingomyelins
   2) Phosphatidic acids: Phospholipids without the organic base. Cardiolipins are polyphosphatidic acids.

3) **Glycolipids:** Esters containing a carbohydrate and a nitrogenous base, but no phosphoric acid and glycerol.
   a) Cerebrosides
   b) Gangliosides
   c) Cerebroside sulphatides
4) **Sulpholipids:** Esters containing sulphuric acid.
5) **Lipoproteins:** Lipids attached to proteins.

III. *Derived lipids:* Derivatives obtained by the hydrolysis of I and II and which still possess the physical characteristics of lipids.
   1) Saturated and unsaturated fatty acids
   2) Monoacyl glycerols and diacyl glycerols
   3) Alcohols
      a) Straight chain alcohols: Water-insoluble alcohols of higher molecular weight like cetyl alcohol,
         $CH_3(CH_2)_{14}CH_2OH$.
      b) Sterols.
      c) Vitamin A alcohol (Retinol) and other carotenols containing the ß-ionone ring.

IV. *Miscellaneous lipids:*
   1) Aliphatic hydrocarbons like iso-octadecane found in liver fat and pentacosane found in bees' wax.
   2) Carotenoids: Alpha, beta and gamma carotenes, cryptoxanthin and other pigments which are fat-soluble.
   3) Squalene: An unsaturated straight chain hydrocarbon present in shark and mammalian liver and human sebum.
   4) Vitamins D, E and K (fat-soluble vitamins).
   5) Various steroid compounds including the steroid hormones and bile acids.

FATTY ACIDS

Most lipid structures have either a fatty acid as a component or possess some ability to combine with it. Most of the fatty acids occurring in common lipids are aliphatic monocarboxylic acids, with an even number of carbon atoms ranging from C_4 to C_{24}. Some of the fatty acids are also dicarboxylic, but not found so often. The fatty acids are classified as follows:

I. *Saturated fatty acids:* ($C_nH_{2n+1}COOH$)
   Butyric C_3H_7COOH present in butter fat.
   Palmitic $C_{15}H_{31}COOH$ * present in coconut oil, animal fats.
   Stearic $C_{17}H_{35}COOH$ * present in plant and animal fats.
   (* more common fatty acids)

II. *Unsaturated fatty acids:*
   1) Fatty acids containing one double bond (monoenoic series or crotonic acid series).
      $C_nH_{2n-1}COOH$
      Palmitoleic $C_{15}H_{29}COOH$ present in butter fat, fish oils and animal fats.
      Oleic $C_{17}H_{33}COOH$ present in plant oils and animal fats.
   2) Fatty acids containing two double bonds (dienoic series or linoleic acid series)
      $C_nH_{2n-3}COOH$
      Linoleic acid (9, 12 octa deca dienoic), $C_{17}H_{31}COOH$ present in plant and animal fats.
      $CH_3-(CH_2)_4-CH=CH-CH_2-CH=CH-(CH_2)_7-COOH$
   3) Fatty acids containing three double bonds (trienoic series or linolenic series)

C_nH_{2n-5}COOH
Linolenic acid (9, 12, 15-octa deca trienoic)

$C_{17}H_{29}$COOH present in linseed oil, fresh liver oil, mammalian fats and plant oils.

$$CH_3-CH_2-CH=CH-CH_2-CH=CH-CH_2$$
$$-CH=CH(CH_2)_7-COOH$$

4) Fatty acids with four or more double bonds (polyunsaturated or polyenoic fatty acids)
Arachidonic acid (5, 8, 11, 14-eicosatetrenoic) C_nH_{2n-7}COOH
$C_{19}H_{31}$COOH present in animal phospholipids and liver fat.
$$CH_3-(CH_2)_4-(CH=CH-CH_2)_4(CH_2)_2-COOH$$

III. *Hydroxy fatty acids:*
 1) Saturated hydroxy acids
   Beta-hydroxybutyric acid C_3H_6(OH)COOH, an intermediate in fatty acid metabolism and one of the ketone bodies.
 2) Unsaturated hydroxy acids
   Ricinoleic
   $C_{17}H_{32}$(OH) COOH present in castor oil.

IV. *Cyclic fatty acids:*
   Hydno= carpic $C_{15}H_{27}$ COOH present in chaulmoogra oil.

V. *Eicosanoids:* Compounds derived from eicosapolyenoic fatty acids. These include prostaglandins, prostacyclins, leukotrienes and thromboxanes.

Prostaglandins

A group of compounds known as prostaglandins, originally isolated from seminal plasma and now shown to be present in many tissues all over the body, are related to the arachidonic acid. They are C_{20} monocarboxylic fatty acids containing two or more double bonds and hydroxyl groups and a five membered ring. PGE_1, 11 α, 15 α dihydroxy 9 keto prost – 13 enoic acid.

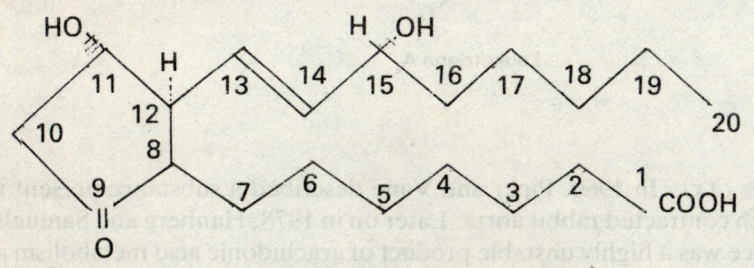

(For Prostacyclins refer to the chapter on Hormones.)

Prostacyclin PGI_2

Leukotrienes: The leukotrienes are a family of biologically active molecules formed from the arachidonic acid in leukocytes, mastocytoma cells, macrophages and other cells in response to immunological and non-immunological stimuli. The name 'leuko - trienes' was derived from 'leuko' indicating leukocytes and 'trienes' suggesting the presence of three conjugated double bonds in the molecule. They were first recognised as slow reacting substances (SRS) as they cause slow and prolonged contraction of guinea pig ileum and anaphylaxis.

Leukotrienes are of different types, A, B, C, D, E and F, depending on the substituent group. The suffix shows the number of double bonds, e.g., LTA5; LTB4; LTD4 and LTF4; LTA5 has five double bonds.

LTC4, D4 and E4 are potent stimulators of the airway smooth muscle and hence have been proposed to have a role in asthma. LTC4 and D4 induce leakage and increase vascular permeability. They also induce transient constriction of the arterioles. LTC4, D4, E4 and F4 induce contraction of the ileum, and their effect on stomach explains the anaphylactic contractions. The constriction of the coronary arterioles increases blood pressure.

Biological studies indicate that the LTs containing cysteinyl residues are important in immediate hypersensitivity reactions like human asthma and anaphylaxis through their potent effect as broncho-constrictors and stimulators of vascular leakage. It is conceivable that some of the therapeutic effects of the steroids which are not shared by the aspirin type (non-steroidal anti-inflammatory drugs) are due to the inhibition of LT formation.

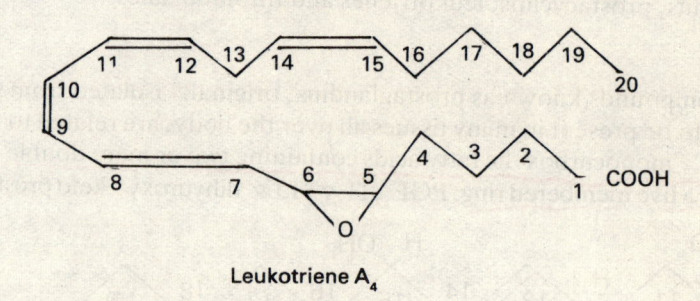

Leukotriene A_4

Thromboxanes (Tx): In 1960, Piper and Vane described a substance present in platelet-rich plasma which contracted rabbit aorta. Later on in 1978, Hanberg and Samuelsson found that this substance was a highly unstable product of arachidonic acid metabolism and named it as Thromboxane A_2 (TxA2).

The name 'thromboxane' indicates that the substances were first isolated from the thrombocytes and that they contain an oxane ring. Two types of Thromboxanes A(TxA) are reported: TxA1 from 8, 11, 14 eicosa trienoic acid, and TxA2 from 5, 8, 11, 14 eicosa tetrenoic acid (arachidonic acid). Thromboxanes B1 and B2 are the stable end products of TxA1 and TxA2, respectively. TxA is the functionally potent form and has a half life of 30 seconds. In humans TxA2 is the existing form.

Functions of TxA2: TxA2 has a major role to play in changing the shape, in the release, reaction and irreversible secondary phase of the aggregation of platelets and is therefore a powerful aggregator of platelets. It is also a potent vaso-constrictor capable of causing broncho-constriction. TxA2 is reported to have a role in the regulation of cell-mediated immune response at the inductive phase.

The platelet is widely recognised as circulating muscle. TxA2 acts as physiological ionophore, decreases the cAMP level in the platelets and extrudes Ca^{2+} from the storage vesicles into the cytoplasm, thus facilitating platelet aggregation.

TxA2 has been reported to be released in Cardiac Immediate Hypersensitivity Reactions (CIHSR), K^+ depletion in the kidneys, pre-eclampsia, diabetes mellitus, ischemic heart disease and malignancies. Smoking increases TxA2.

Tx synthetase inhibitors, anti-oxidants, dietary manipulation with fatty acids like eicosa polyenoic acids are used to keep down TxA2 in anti-thrombotic therapy.

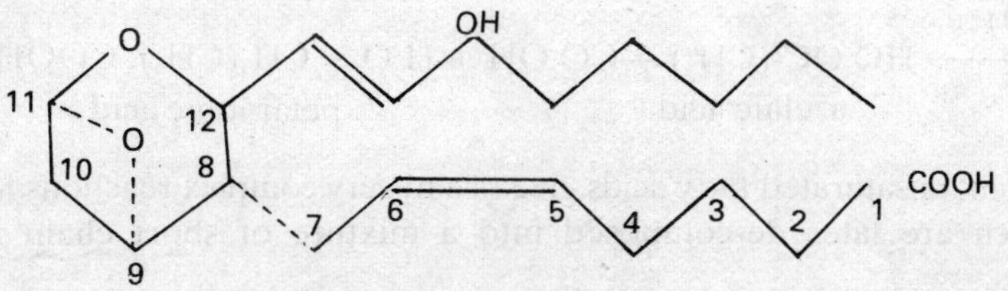

Thromboxane A_2

Among the various fatty acids classified earlier, the most common fatty acids present in plant and animal fats are the lauric, myristic, palmitic, stearic, oleic, linoleic, linolenic and arachidonic acids. Oleic acid is the most abundant fatty acid present at more than 50 per cent in many of the oils and fats and is found in all fats including the phospholipids. Animal fats contain more of stearic acid than oleic acid. Liver fat from cold-blooded animals like shark and cod contains more of unsaturated fatty acids than that of the warm-blooded species. The saturated fatty acids mostly found in animal fats are the stearic and palmitic acids.

Properties of fatty acids

1) Fatty acids from butyric (C_4) to caproic acid (C_{10}) are liquids at ordinary temperature, while the higher members above C_{10} are solids.

2) The melting point of the fatty acids increases with chain length.

3) Solubility in water decreases with chain length. Butyric acid is soluble in water in all proportions. Caproic acid is moderately soluble in water. The higher fatty acids are all insoluble in water.

4) The short chain fatty acids are steam-volatile, the volatility decreasing with chain length.

5) Most of the fatty acids are soluble in hot alcohol.

6) Fatty acids form salts with Na^+ and K^+, which are soaps. Soaps are excellent emulsifying and cleansing agents. The sodium salts of oleic, palmitic and stearic acids which form the commercial soaps are soluble in water and alcohol, but insoluble in ether and benzene.

7) A fatty acid can be reduced to the corresponding alcohol by hydrogen at 200 atmospheres and 320°C in the presence of nickel as catalyst. This alcohol can be converted to a sulphuric acid derivative on treatment with fuming H_2SO_4. The sodium salts of such derivatives are used as synthetic detergents.

8) <u>Unsaturated fatty acids can be oxidised</u> at the double bond by alkaline $KMnO_4$ to give dihydroxy fatty acids which are later broken down to simpler fatty acids. Thus,

$$CH_3(CH_2)_7 CH = CH-(CH_2)_7 COOH \xrightarrow{+(O)}_{+HOH}$$
Oleic acid

$$CH_3(CH_2)_7\text{-}CH\text{-}CH\text{-}(CH_2)_7 COOH$$
$$\quad\quad\quad\quad\;\; |\;\;\; |$$
$$\quad\quad\quad\quad OH\; OH$$
Dihydroxy stearic acid

$$\xrightarrow{3(O)} HOOC\text{-}(CH_2)_7\text{—}COOH + H_2O + CH_3(CH_2)_7 COOH$$
$$\quad\quad\quad\quad\text{azelaic acid} \quad\quad\quad\quad\quad\quad \text{pelargonic acid}$$

9) With oxygen, unsaturated fatty acids give rise to very complex reactions forming unstable peroxides, which are later de-composed into a mixture of <u>short chain fatty acids</u> and <u>aldehydes</u>. Thus.

$$CH_3 - (CH_2)_7 CH = CH - (CH_2)_7 COOH \xrightarrow{O_2}$$

$$CH_3\text{—}(CH_2)_7 CH\text{—}CH\text{—}(CH_2)_7 COOH$$
$$\quad\quad\quad\quad\quad\quad |\quad\;\; |$$
$$\quad\quad\quad\quad\quad\quad O\text{—}O$$

10) <u>Ozone</u> also decomposes unsaturated fatty acids in a similar way, first forming an ozonide.

11) <u>Beta-oxidation of a fatty</u> acid can be carried out using $KMnO_4$ in alkaline solution or with H_2O_2.

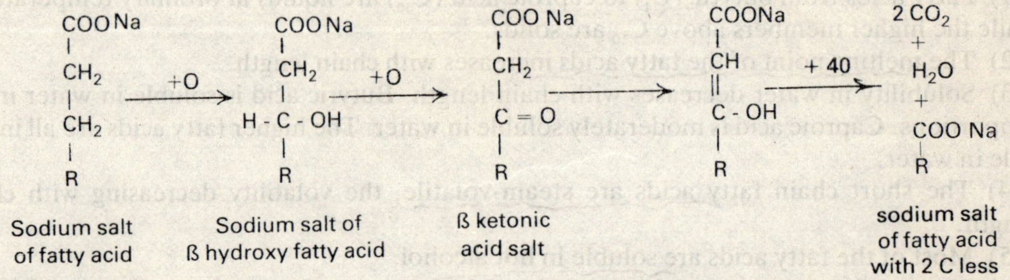

```
 COONa          COONa          COONa          COONa        2 CO2
  |              |              |              |            +
  CH2    +O      CH2     +O     CH2            CH    +40     H2O
  |      →       |       →      |       →      ||            +
  CH2            H-C-OH          C=O            C-OH         COONa
  |              |              |              |            |
  R              R              R              R            R
Sodium salt   Sodium salt of   ß ketonic                   sodium salt
of fatty acid ß hydroxy        acid salt                   of fatty acid
              fatty acid                                    with 2 C less
```

12) <u>Hydrogenation of unsaturated fatty acids using nickel as catalyst yields saturated fatty acids.</u>

$$CH_3(CH_2)_7 CH = CH — (CH_2)_7—COOH \;+2H \longrightarrow CH_3(CH_2)_{16} COOH$$
oleic acid $\quad\quad\quad\quad\quad\quad\quad\quad\quad\quad\quad\quad\quad\quad\quad\quad$ stearic acid

13) Iodine and bromine are taken up by unsaturated fatty acids giving halogenated derivatives.

Fats and oils

Neutral fats or triacyl glycerols are present as fats and oils and are abundantly distributed in the plant and animal kingdoms. Vegetable seeds and nuts contain oil as a reserve food material. Coconut, groundnut, linseed and olive oils and cotton seed oil belong to this category of vegetable oils. Lard (pork fat), butter, fish liver oil and tallow (fat from ox or sheep) are the commonly known animal fats.

Neutral fat is the triacyl glycerol formed from one molecule of glycerol by esterification with three molecules of the same or different fatty acids (triglyceride). Thus we have simple and mixed triacyl glycerol. Usally the fatty acids present are the palmitic, oleic and stearic acids in natural fats, with smaller amounts of other fatty acids.

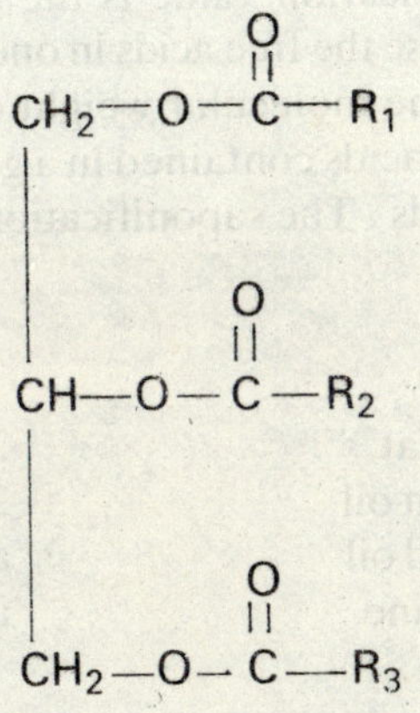

$$\begin{array}{c} O \\ \parallel \\ CH_2-O-C-R_1 \\ \\ O \\ \parallel \\ CH-O-C-R_2 \\ \\ O \\ \parallel \\ CH_2-O-C-R_3 \end{array}$$

The triacyl glycerols which are useful to man as food and which are absorbed and metabolised well are considered edible fat and edible oils. Thus groundnut oil is an edible oil while castor oil is not. Oils are generally liquids at the ordinary temperature while fats are either semisolids or solids at the same conditions. Chemically both are triacyl glycerols. The melting point and consistency of a fat depend upon the nature of the fatty acids and their structure. The presence of more oleic acid decreases the melting point of the fat, while stearic acid raises the melting point.

Tests for purity of fat or oil

These tests are useful in detecting the presence of adulterants. The common tests which are applied are determination of melting point, iodine value, saponification value, Reichert Meissl number and acid number.

1) Melting point: The melting point of a fat composed of even chain saturated fatty acids is directly proportional to the chain length. The presence of even a small amount of unsaturated fatty acids may lower the melting point significantly. The melting point is of little importance in the case of vegetable oils which are liquids at ordinary temperature.

2) Iodine value (iodine number): Iodine value is defined as the number of grams of iodine absorbed by 100 g of fat or oil. It is a measure of the extent of unsaturation present in the fats and is important in the identification of the fat or oil as well as in determining whether it is free from **adulteration**. The iodine numbers of a few oils are given below for comparison.

Coconut oil	8.4
Olive oil	79-83
Cottonseed oil	103-111
Linseed oil	175-202

If a given sample of olive oil has an iodine value above 83, it indicates adulteration with cottonseed oil or some other oil with a higher iodine value. On the other hand, if a commercial sample of linseed oil has an iodine value lower than 175, it might be adulterated.

Oils having high iodine value, e.g., linseed oil, dry up quickly and are called drying oils, while coconut oil is a non-drying oil.

3) Saponification value: Saponification value is the number of mg of KOH required to hydrolyse the glycerides and neutralise the free acids in one gram of oil or fat. The saponification value is inversely proportional to the molecular weight of oil or fat. Saponification is also a measure of the chain length of fatty acids contained in a given fat. This value is high in oils and fats containing short chain fatty acids. The saponification values of some of the common fats and oils are given below:

Butter fat	233-240
Coconut oil	250-264
Mustard oil	about 174
Margarine	about 195

4) Reichert-Meissl number: The Reichert-Meissl number is the number of ml of 0.1 N alkali (KOH) required to neutralise the soluble volatile fatty acids distilled from five g of fat. This number is of special value in the characterisation of butter fat and for testing its purity. Other edible fats do not contain as much of volatile fatty acids as butter fat. The volatile fatty acids which come out during the steam-distillation of fat are butyric acid, caproic acid and, to some extent, caprylic acid which are soluble in water. The RM value of butter is 26 – 33 while that of coconut oil is 6-8. It is less than one for other edible oils. The admixture of certain fats may be used to prepare synthetic butter which may simulate butter in most of the constants except RM value and hence can be detected.

5) Acid value (or acid number): Refined oils should not contain free fatty acids. The presence of free fatty acids in any fat or oil sample indicates that it is not pure. Free fatty acids may also arise from decomposition of fat during storage, and bacterial or chemical action. The acid number measures the free fatty acids present in a sample of fat or oil. It is defined as the number of mg of KOH required to neutralise the free fatty acids present in one gram of fat or oil and is measured by titration with standard KOH using phenolphthalein as indicator. The edibility of a fat is inversely proportional to the acid number.

Apart from the above indices, there are several other commercially useful ones which are used for testing the purity of oils and fats. They are the acetyl number, Polenske number, thiocyanogen value, ester number, refractive index, specific gravity, etc.

Analytical methods for separation of fats and fatty acids

1) Paper and column chromatography methods are used for analytical separation of the fatty acids in chloroform (99 per cent) +n-butanol (1 per cent), using filter paper soaked in grease, with water as the mobile phase (reversed phase chromatography) or columns of silica gel or Hyflo-supercel.

2) Thin layer chromatography is a modern technique which uses glass plates layered with silica gel G, alumina or Kieselguhr for separation of the fatty acids.

3) Gas chromatography can be used to separate the various fatty acids as their methyl esters, (for details refer to chapter on Modern Techniques).

4) High performance liquid chromatography (HPLC) is a very recent technique employed for the separation of fatty acids.

Rancidity of fats and oils

When certain fats and oils are exposed to heat, light, air, moisture and bacterial action, they develop an offensive odour and taste which is called rancidity. Rancid fats and oils contain certain short chain dicarboxylic acids, aldehydes and ketones which are formed of unsaturated fatty acids. Rancid fats are unsuitable for human consumption and have undesirable physiological effects if consumed. Rancidity is enhanced by light, moisture and the enzyme lipase (from bacteria). Unsaturated fat is readily spoilt by rancidity as compared to saturated fat, because it is the double bond of the molecule which is attacked first. Oxygen is necessary for rancidity. In the initial stage, oxygen is absorbed by the fat or oil which is used for the formation of a peroxide. This peroxide breaks down to short chain acids like azelaic acid and other derivatives like aldehydes, all of which have an offensive odour.

$$\begin{array}{c} \text{H} \quad \text{H} \\ | \quad\quad | \\ R - C = C - R \\ \downarrow \\ O - O \\ | \quad\quad | \\ R - C - C - R \\ | \quad\quad | \\ \text{H} \quad \text{H} \\ \text{Peroxide} \\ \downarrow \\ \text{R CHO} \quad\quad \text{R COOH} \\ \text{Decomposition products} \end{array}$$

Anti-oxidants: Compounds which prevent oxidation and rancidity of fat or oil are called anti-oxidants. Hydroquinone, gallic acid, tocopherols (vitamin E), o- and p-diphenols, alpha-naphthol, etc. are the common anti-oxidants used for protecting oils and fats from becoming rancid. Anti-oxidants like butylated hydroxyanisole (BHA) and butylated hydroxytoluene (BHT) are often used as food additives with a view to prevent spoiling of food by the onset of rancidity.

PHOSPHOLIPIDS

Phospholipids are conjugated lipids containing phosphorus present as esterified phosphoric acid. The first phospholipid was isolated from egg yolk and was named lecithin (Greek *lecithos* = egg yolk). Phospholipids are distributed in various animal tissues and plant seeds. The brain lipids consist of a large percentage of phospholipids. Most of the brain phospholipids are in the white matter. Soya beans contain vegetable lecithins.

Classification of phospholipids

I. Phosphoglycerides or glycerophosphatides: These are phospholipids which yield glycerol on hydrolysis.
  a) Phosphatidyl choline (lecithin)
  b) Phosphatidyl ethanolamine (cephalin).
  c) Phosphatidyl serine (cephalin).
  d) Acetal phospholipid (Plasmalogen).
  e) Cardiolipins (Polyphosphatidic acid)

II. Phospho-inositides or inositol phospholipids: This type of phospholipids yields inositol on hydrolysis.
  a) Mono-phospho-inositide
  b) Diphospho-inositide

III. Phospho-sphingosides (Sphingomyelins): This type of phospholipids does not contain glycerol, but a nitrogen containing alcohol called spingosine, along with fatty acid, phosphoric acid and choline.

Phosphatidyl choline or lecithin

Lecithins are derivatives of phosphatidic acid. The naturally occurring lecithins contain L-phosphatidic acid.

$$\begin{array}{l} CH_2O-\overset{O}{\overset{\|}{C}}-R_1 \\ \quad\quad\;\; O \\ \quad\quad\;\; \| \\ CHO-C-R_2 \\ \quad\quad\;\; O \\ \quad\quad\;\; \| \\ CH_2O-P-OH \\ \quad\quad\;\; | \\ \quad\quad\;\; OH \end{array}$$

alpha-phosphatidic acid

$$\begin{array}{l} CH_2O-\overset{O}{\overset{\|}{C}}-R_1 \\ \quad\quad\;\; O \\ \quad\quad\;\; \| \\ CHO-C-R_2 \\ \quad\quad\;\; O \\ \quad\quad\;\; \| \\ CH_2O-P-O-CH_2-CH_2-N^+\begin{array}{l} CH_3 \\ CH_3 \\ CH_3 \end{array} \\ \quad\quad\;\; | \\ \quad\quad\;\; OH \end{array}$$

alpha-lecithin

Lecithin is present in egg yolk, brain tissue and a wide variety of animal fats and plant oils. It is a waxy white substance when pure but may turn brown on exposure to air and light due to partial oxidation. Lecithins are soluble in all fat solvents except acetone. They have an affinity for water and form a milky emulsion with it. They are excellent emulsifying agents.

Lecithins can be hydrolysed by dilute acids, alkalies as well as enzymes called lecithinases or phospholipases. Dilute H_2SO_4 splits off a molecule of choline leaving behind phosphatidic acid. When boiled with alkalies, choline is split off first, followed by the hydrolysis of the phosphatidic acid to give the soaps of fatty acids and the salt of glycerophosphoric acid.

The phospholipases are of different types. The phospholipase A1 present in rice hulls, *Penicillium notatum, Aspergillus oryzae*, etc., attacks the ester bond in position 1 of phospholipids. Phospholipase A2, which is present in the pancreatic juice, and in snake venom, attacks the ester bond in position 2 of the phospholipids giving rise to free fatty acids and lysophospholipid. The lysophospholipid is attacked by lysophospholipase (phospholipase B) splitting one more acyl group and forming glycerophosphoryl base. Phospholipase C present in mammalian brain and *Clostridium welchii* splits off choline phosphate and gives 1,2-di-acyl glycerol. Thus it is a glycerophosphatase acting on the phosphoric acid ester linkage.

LIPIDS

Phospholipase D present in various plants, cabbage, cottonseed, etc. hydrolyses the nitrogenous residue giving rise to phosphatidic acid.

Cobra venom and rattle snake venom contain phospholipase A2 which converts lecithin to lysolecithin, a potent hemolytic agent. Lysolecithin is also formed when the enzyme lecithin-cholesterol acyl transferase (LCAT) found in plasma and liver acts on lecithin in vivo, catalysing the transfer of its acyl residue in position 2 to cholesterol to form cholesterol ester.

Role of lecithins: Lecithins are widely distributed in the cells of the body and the cell membranes and maintain their integrity. Owing to their highly polar structure, lecithins tend to associate in an orderly fashion. Hence they are helpful in the organisation of the protoplasm into its orderly and characteristic structure. They have the power of lowering the surface tension of water and this is more marked if they are bound to proteins. Dipalmityl lecithin is a major constituent of 'lung surfactant' which prevents the adherence of the inner surfaces of the alveoli of the lungs (thereby preventing the collapse of the alveoli) by its surface tension lowering effect. The absence of this in the alveolar membrane of some premature infants causes the respiratory distress syndrome in them. Lecithins are also particularly important in the metabolism of fat by the liver. The choline content of these substances may be of use in many ways. (Refer also chapter on Lipid Metabolism.)

Cephalins

Like lecithins, cephalins are distributed widely in animal tissues, the brain and spinal cord being rich sources. They are structurally identical with lecithins except that choline is replaced by ethanolamine or serine. Cephalins have properties similar to lecithins. They are soluble in ether but insoluble in acetone and hot alcohol. They are soluble in methanol and glacial acetic acid and with water give colloidal systems. Cephalins are a part of thromboplastin enzyme and by themselves have some thromboplastic activity.

$$
\begin{array}{cc}
\text{CH}_2-\text{O}-\overset{\text{O}}{\underset{\|}{\text{C}}}-\text{R}_1 & \text{CH}_2-\text{O}-\overset{\text{O}}{\underset{\|}{\text{C}}}-\text{R}_1 \\
\text{CHO}-\overset{\text{O}}{\underset{\|}{\text{C}}}-\text{R}_2 & \text{CHO}-\overset{\text{O}}{\underset{\|}{\text{C}}}-\text{R}_2 \\
\text{CH}_2\text{O}-\overset{\text{O}}{\underset{\|}{\text{P}}}-\text{O}-\text{CH}_2-\text{CH}_2-\text{NH}_2 & \text{CH}_2\text{O}-\overset{\text{O}}{\underset{\|}{\text{P}}}-\text{O}-\text{CH}_2-\overset{\text{NH}_2}{\underset{|}{\text{CH}}}-\text{COOH} \\
\text{OH} & \text{OH} \\
\text{Phosphatidyl ethanolamine} & \text{Phosphatidyl serine}
\end{array}
$$

The phospholipases which act on lecithin have a similar action on cephalin.

Plasmalogens (acetal phospholipids)

Plasmalogens are insoluble in ether, benzene and hexane but are soluble in methanol, chloroform and hot ethanol. On hydrolysis, plasmalogen yields glycerol, choline or ethanolamine, a molecule of fatty acid and a molecule of fatty acid aldehyde in addition to phosphoric acid. The aldehyde of the fatty acid is not present as such in the molecule but is only an artefact formed during the hydrolysis of the alpha, beta unsaturated ether. Plasmalogens have been found to be present in the brain, heart muscle and liver.

$$CH_2-O-CH=CH-R$$
$$CHO-\overset{O}{\underset{\|}{C}}-R_1$$
$$CH_2-O-\underset{\underset{OH}{|}}{\overset{O}{\underset{\|}{P}}}-O-CH_2-CH_2-NH_2$$

Cardiolipins

Cardiolipins have been obtained from the lipid extract of the heart tissue. They occur to the extent of 16 per cent of the total phospholipids in mitochondrial membrane and are considered necessary for biological function. They are widely used for the serological test for syphilis (VDRL) as an antigen. Cardiolipins are insoluble in water but soluble in alcohol and acetone. Chemically, they are made up of condensed phosphatidic acid molecules and are phosphatidyl glycerols.

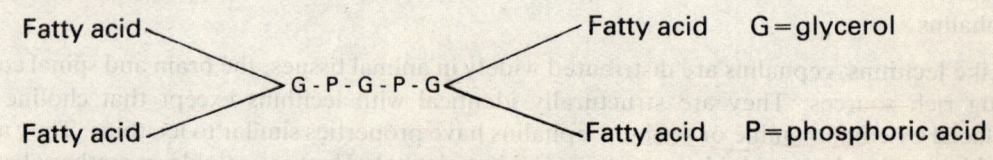

```
Fatty acid \           / Fatty acid      G = glycerol
            G-P-G-P-G
Fatty acid /           \ Fatty acid      P = phosphoric acid
```

The simplest cardiolipin molecule has three glycerol, two phosphoric acid and four fatty acid molecules in it.

Phospho-inositides

Inositol-containing phospholipids are present abundantly in the brain (25 per cent of the total cephalin) and moderatley in soya bean and in the tubercle bacilli. Lipositol from soya beans contains 16 per cent of inositol. Lipositol is a white creamy powder soluble in petroleum ether, benzene, ether and chloroform. A trace of moisture seems to be necessary for its solubility in the above-mentioned solvents. Lipositol is insoluble in methanol, acetone and ethanol but forms a colloidal emulsion with water.

Phosphatidyl inositol

The function of phosphatidyl inositol is mediation in the action of some hormones and to play a similar role as other phospholipids.

Phospho-sphingosides

The phospho-sphingosides or sphingomyelins are characterised by their insolubility in ether, and the absence of glycerol in them. They contain sphingosine, a complex amino alcohol, a fatty acid, phosphoric acid and choline. The sphingomyelins occur abundantly in the nervous tissue and brain, and in less amounts in the adrenals, liver, egg yolk and blood phospholipids. The levels of sphingomyelin are increased markedly in the liver, spleen, brain and other organs in Niemann-Pick's disease.

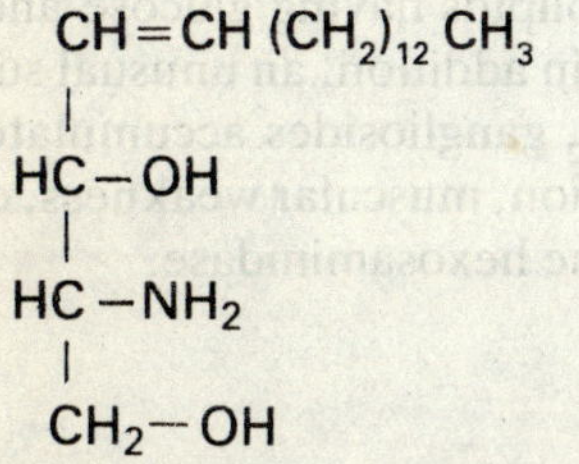

$$
\begin{array}{l}
CH=CH\ (CH_2)_{12}\ CH_3 \\
| \\
HC-OH \\
| \\
HC-NH_2 \\
| \\
CH_2-OH
\end{array}
\qquad
\begin{array}{l}
CH=CH\ (CH_2)_{12}\ CH_3 \\
| \\
HC-OH \qquad\qquad O \\
| \qquad\qquad\qquad\ \ \ \| \\
HC-NH-C-R \\
| \\
\qquad\qquad O \qquad\qquad\qquad\qquad CH_3 \\
\qquad\qquad \| \qquad\qquad\qquad\qquad\ \ / \\
CH_2O-P-O-CH_2-CH_2-N^+-CH_3 \\
\qquad\qquad | \qquad\qquad\qquad\qquad\ \ \backslash \\
\qquad\qquad OH \qquad\qquad\qquad\qquad CH_3 \\
\qquad\qquad\qquad\qquad\qquad\qquad\qquad\quad OH^-
\end{array}
$$

Sphingosine 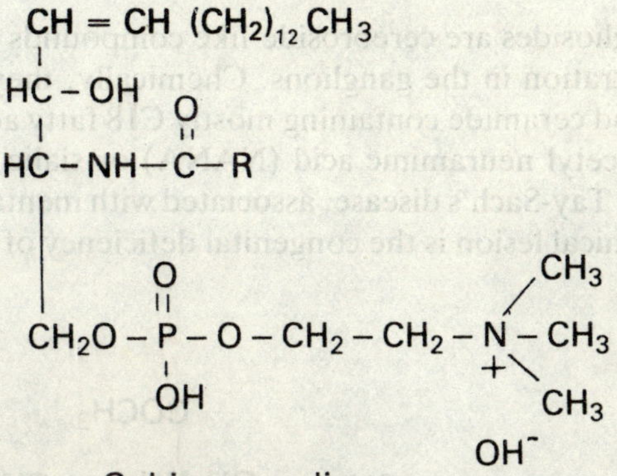   Sphingomyelin

GLYCOLIPIDS OR CEREBROSIDES

Another class of conjugated lipids present in the brain and spinal cord but which do not contain phosphoric acid are called cerebrosides. They contain a carbohydrate moiety, which is usually galactose, and the nitrogenous base sphingosine instead of glycerol and have a fatty acid attached to the amino group of sphingosine. Cerebrosides occur along with phospholipids and are present in high concentration in the white matter of brain as well as in the myelin sheaths. They are present in less concentration in the grey matter.

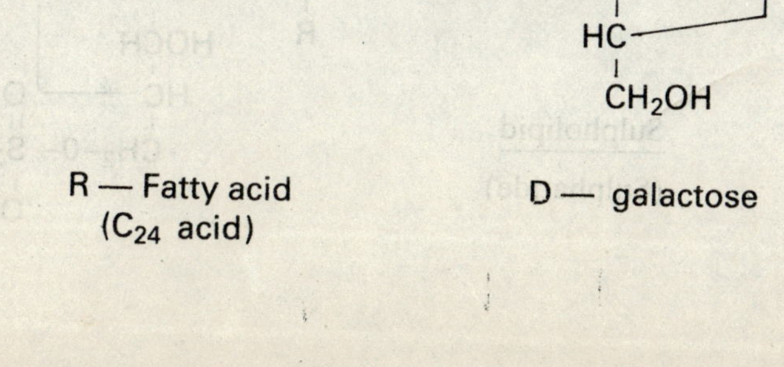

$$
\begin{array}{l}
\qquad\qquad\qquad\qquad\qquad OH \\
\qquad\qquad\qquad\qquad\qquad\ | \\
CH_3-(CH_2)_{12}-CH=CH-CH-CH-CH_2O-C-H \\
\qquad\qquad\qquad\qquad\qquad\qquad\ | \qquad\qquad\ | \\
\qquad\qquad\qquad\qquad\qquad\quad NH \qquad\ HC-OH \\
\qquad\qquad\qquad\qquad\qquad\qquad\ | \qquad\qquad\ | \\
\qquad\qquad\qquad\qquad\qquad\quad CO \qquad\ HO-CH \\
\qquad\qquad\qquad\qquad\qquad\qquad\ | \qquad\qquad\ | \qquad\quad O \\
\qquad\qquad\qquad\qquad\qquad\quad\ R \qquad\quad HO-CH \\
\qquad\qquad\qquad\qquad\qquad\qquad\qquad\qquad\quad | \\
\qquad\qquad\qquad\qquad\qquad\qquad\qquad\qquad\ HC \\
\qquad\qquad\qquad\qquad\qquad\qquad\qquad\qquad\quad | \\
\qquad\qquad\qquad\qquad\qquad\qquad\qquad\qquad\ CH_2OH
\end{array}
$$

R — Fatty acid   D — galactose
(C$_{24}$ acid)

Four types of cerebrosides have been characterised, depending upon the type of fatty acid contained in them. Thus we have phrenosin (cerebron), kerasin, nervon and oxynervon according to the fatty acid, cerebronic, lignoceric, nervonic or oxynervonic acids, respectively. All these are C_{24} fatty acids. The carbohydrate present is normally galactose, but in certain pathological conditions like Gaucher's disease, the galactose is replaced by glucose. These glucose-containing cerebrosides accumulate in the liver and spleen.

Cerebrosides are all insoluble in water, petroleum ether and ether, but are soluble in the chloroform-methanol mixture and acetone.

Gangliosides

Gangliosides are cerebroside-like compounds present in the brain, occurring also in high concentration in the ganglions. Chemically, they are glycolipids having glucose and galactose, and ceramide containing mostly C18 fatty acids, and, in addition, an unusual sugar called N-acetyl neuraminic acid (NANA) or sialic acid. GM_2 gangliosides accumulate in the brain in Tay-Sach's disease, associated with mental retardation, muscular weakness, etc. The biochemical lesion is the congenital deficiency of the enzyme hexosaminidase.

```
                    COCH3
                     |
        OH      OH  NH       OH  OH
        |       |   |        |   |
    HOOC-C-CH2-CH-CH-CH-CH-CH-CH2-OH
        |_____O_____|
                                            NANA
```

Sulpholipids

Sulphur-containing lipids are present in the brain, liver and kidney, the white matter of brain being rich in this material. They consist of sphingosine, galactose, fatty acid and sulphuric acid and have the following formula:

```
CH3-(CH2)12-CH=CH-CH-CH-CH2-O
                |   |        |
                OH  NH       HC——
                    |        |
                    CO       HCOH
                    |        |
                    R        HOCH      O
                             |
                             HOCH
                             |
                             HC————
                             |       O
                             CH2-O-  ‖
                                     S-OH
        Sulpholipid                  |
        (Sulphatide)                 OH
```

Fractionation

A simple scheme for the fractionation of a mixture of major phospholipids and glycolipids is as follows:

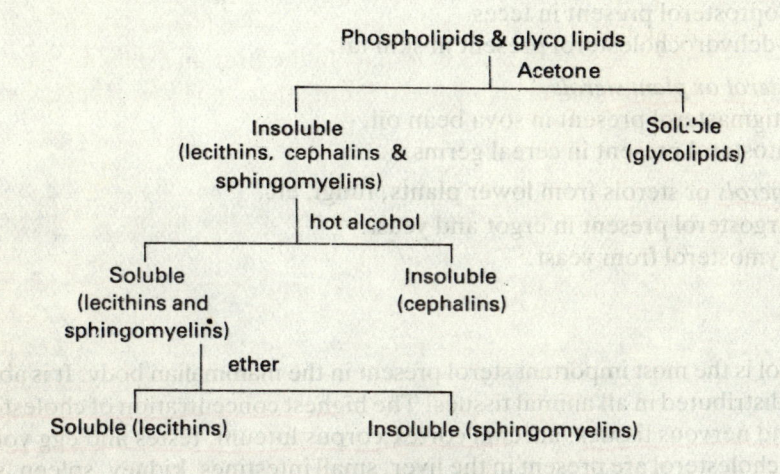

```
              Phospholipids & glyco lipids
                       |
                    Acetone
              ┌────────┴────────┐
          Insoluble           Soluble
     (lecithins, cephalins &  (glycolipids)
        sphingomyelins)
              |
          hot alcohol
         ┌────┴────┐
      Soluble    Insoluble
   (lecithins and (cephalins)
   sphingomyelins)
        |
      ether
    ┌───┴───┐
 Soluble    Insoluble (sphingomyelins)
 (lecithins)
```

STEROLS

Sterols are complex monohydroxy alcohols having a steriod ring system. The steroids have a carbon skeleton consisting of a fused ring system known as cyclopentano perhydrophenanthrene. Sterols are derived from steroid hydrocarbons and have hydroxyl group at position 3 of ring A (see below). They are waxy solids at ordinary temperature and have high melting points as compared with other alcohols. Hence they are named sterols (meaning solid alcohols). Cholesterol is a typical animal sterol.

Steroid skeleton
(Cyclopentanoperhydro-
phenanthrene)

Cholesterol

There are many steroid compounds which are physiologically important and widely distributed in the plant kingdom as well as in the animal kingdom. In the plant kingdom, we have ergosterol, sitosterol, sapogenins of digitalis, saponins and the aglycone part of cardiac glycosides, while in the animal kingdom, apart from cholesterol, there are many compounds like the bile acids, sex hormones, provitamin D_3 and adrenal cortical hormones belonging to the steroid family.

Classification of sterols

1) *Zoosterols or animal sterols.*
   e.g., Cholesterol, present in all animal cells.
   Coprosterol present in feces.
   7-dehydrocholesterol present in skin fat.

2) *Phytosterol or plant sterols*
   e.g., Stigmasterol present in soya bean oil.
   Sitosterol present in cereal germs.

3) *Mycosterols* or sterols from lower plants, fungi, etc.
   e.g., Ergosterol present in ergot and yeast.
   Zymosterol from yeast.

Cholesterol

Cholesterol is the most important sterol present in the mammalian body. It is absent in plants but widely distributed in all animal tissues. The highest concentration of cholesterol occurs in the brain and nervous tissues, adrenal cortex, corpus luteum, testes and egg yolk. Moderate amounts of cholesterol are present in the liver, small intestines, kidney, spleen, skin, etc. The muscles and skeleton contain the least amount of cholesterol. Among foods of animal origin, milk has the minimum cholesterol content.

Structure: Cholesterol has a steroid ring system consisting of 19 carbon atoms and a side chain hydrocarbon consisting of eight carbon atoms. It has a double bond at carbons 5–6 and a secondary alcoholic hydroxyl group attached to C_3. The C_{10} and C_{13} carry two angular methyl groups which are numbered respectively as C_{19} and C_{18}. The side chain is attached to C_{17}.

Properties: Cholesterol was first isolated from bile. The name cholesterol means 'solid alcohol of bile'. Gallstone or biliary calculus consists mostly of cholesterol as its component. Cholesterol is insoluble in water, but soluble in solvents such as ether, chloroform, benzene, hot alcohol, etc. and crystallises out from these solutions as rhombic plates with characteristic notches at the corners (see below). It is an unsaturated compound containing one double bond and can take up two atoms or bromine per molecule.

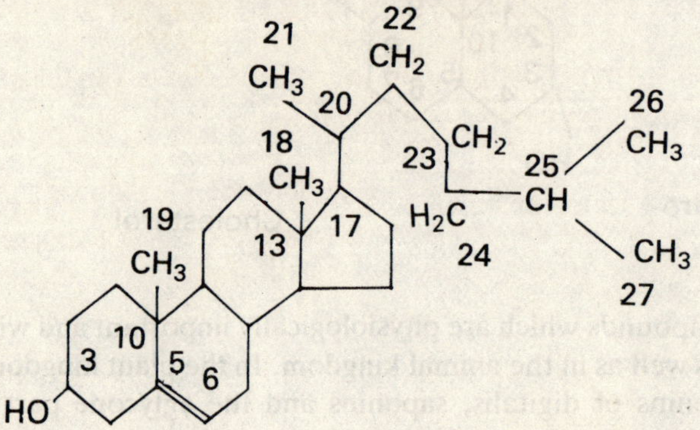

Cholesterol $C_{27}H_{45}OH$

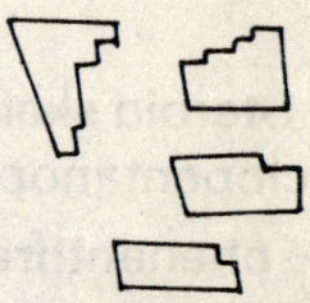

Cholesterol crystals.

Reactions of cholesterol: 1) A number of colour reactions are given by cholesterol and some of these are used for qualitative and quantitative tests. In the Salkowsky's test, concentrated sulphuric acid is added carefully to a chloroformic solution of cholesterol along the sides of the test tube. The heavier acid which forms the lower layer assumes a yellowish colour with a green fluorescence whereas the upper chloroform layer becomes bluish red first, gradually turning violetish-red.

2) In the Liebermann-Burchard reaction, acetic anhydride and concentrated sulphuric acid are added to a chloroformic solution of cholesterol. A blue colour formed first turns to emerald green which persists for more than an hour. The same reaction is used for the quantitative determination of cholesterol in blood and tissues by the colorimetric method in clinical laboratories. In the presence of concentrated sulphuric acid, cholesterol loses a molecule of water at the C_3 and gets oxidised to 3,5 cholestadiene and then polyenes. The latter acts as a chromophore and forms cholestapolyene sulphonic acid which is green.

3) Cholesterol gives a purple colour with a reagent containing ferric chloride, glacial acetic acid and concentrated H_2SO_4 (Zak's method of estimation). The chromophore in this is polyene cation (cholestapolyene carbonium ion).

4) It also forms a precipitate with digitonin because of its alcoholic group. This reaction is useful for the estimation of free and ester cholesterol in blood.

Role of cholesterol in the body: Cholesterol is a poor conductor of heat and hence is an insulator. It is also a poor conductor of electricity and has a high dielectric value. As an abundant constituent of the brain, nerves and spinal cord, it seems to function as an insulating covering for electrical impulse-generating and transmitting structures. Cholesterol is a precursor of bile acids, cortical hormones, sex hormones and vitamin D_3.

Other animal sterols

Among the other sterols are included 7-dehydro-cholesterol (provitamin D_3) and the fecal sterol, coprosterol. 7-dehydro-cholesterol is formed from cholesterol in the skin fat and is the precursor of vitamin D_3 or cholecalciferol which is biosynthesised in the presence of the sun's rays, by ultraviolet irradiation.

Coprosterol of feces is formed from the cholesterol entering the intestines and getting reduced by intestinal bacteria. Coprosterol has no double bond unlike cholesterol and is not absorbed from the intestines.

Bile acids are produced by the oxidation of cholesterol in liver and are discharged into the duodenum in bile. Bile acids have more –OH groups than cholesterol, and in addition also possess a –COOH group and have a high affinity for water. They act as excellent emulsifying agents and are responsible for keeping cholesterol in solution in bile.

4 Proteins

The name protein was given by Mulder in 1839 to a large group of nitrogenous substances of biological origin which had a fundamental role in life. It is derived from the Greek word 'proteios' meaning 'primary' or 'holding first place'.

Proteins are the building blocks for cellular and sub-cellular structures and are present in all living cells.

Structural proteins are those which are required for the support and protection of tissues. Bone and cartilage have the structural protein collagen, while nails and hair contain keratin.

Proteins associated with cellular and other physiological functions are called functional proteins. All enzymes (biocatalysts) and many hormones which regulate biochemical reactions are functional proteins. The hemoglobin of erythrocytes is the protein which functions as a carrier of oxygen from the lungs to the peripheral tissues. Myoglobin acts as the carrier for oxygen in the muscles and helps to keep some oxygen stored there for use in an emergency.

Protein is almost the only source of nitrogen in the body. There is a daily excretion of about 15g of nitrogen mostly as urea by the normal adult, due to the dynamic flux of the body proteins. To make good this loss, protein is a 'must' in the diet. In addition to supplying amino acids required for protein synthesis in the body, protein meets the requirement for sulphur as well. Protein could also be tapped for energy purposes when the necessity arises.

Definition

Proteins are complex organic molecules of large molecular weight ranging from a few thousand to a million or above. They contain the elements, carbon, hydrogen, oxygen, nitrogen and, in many cases, sulphur. These biopolymers are built up of fundamental units of amino acids through peptide bonds. Hence, the properties and behaviour of proteins are dependent on those of the component amino acids.

AMINO ACIDS

Amino acids are a class of organic compounds which contain the two functional groups, the basic amino group ($-NH_2$) and the carboxylic acid group ($-COOH$). If the amino group is attached to the carbon atom carrying the $-COOH$ group, it is called an alpha amino acid. Most of the amino acids in the body are alpha amino acids (which have the general formula):

$$R-CH-COOH$$
$$|$$
$$NH_2$$

There could be β, γ, δ and ω amino acids. The amino acids proline and hydroxy proline contain the secondary amino group $>NH$.

Classification of amino acids

According to one method of classification, amino acids are classified into three types: the acidic amino acids, the basic amino acids and the neutral amino acids.

Acidic amino acids contain more than one carboxylic acid group per molecule.

PROTEINS

1) E.g., aspartic acid HOOC–CH–CH$_2$–COOH:
$$\underset{NH_2}{|}$$

2) Glutamic acid HOOC–CH$_2$–CH$_2$–CH–COOH
$$\underset{NH_2}{|}$$

Basic amino acids possess more than one amino group per molecule.

E.g., lysine NH$_2$–CH$_2$–CH$_2$CH$_2$CH$_2$CH–COOH
$$\underset{NH_2}{|}$$

Arginine HN=C–NH–CH$_2$CH$_2$CH$_2$–CH–COOH
$$\underset{NH_2}{|} \qquad\qquad\qquad \underset{NH_2}{|}$$

Neutral amino acids are those which contain equal number of amino groups and carboxylic acid groups.

E.g., alanine CH$_3$–CH–COOH   2) glycine H$_2$N–CH$_2$–COOH
$$\underset{NH_2}{|}$$

According to another method of classification, amino acids are classified into seven groups:

1) Amino acids containing aliphatic side chains:

Glycine H$_2$ H–CH$_2$–COOH

Alanine CH$_3$–CH–COOH
$$\underset{NH_2}{|}$$

Valine $\begin{matrix}H_3C\\H_3C\end{matrix}\!\!> $CH–CH–COOH
$$\underset{NH_2}{|}$$

Leucine $\begin{matrix}H_3C\\H_3C\end{matrix}\!\!> $CH–CH$_2$–CH–COOH
$$\underset{NH_2}{|}$$

Isoleucine $\begin{matrix}C_2H_5\\H_3C\end{matrix}\!\!> $CH–CH–COOH
$$\underset{NH_2}{|}$$

2) Amino acids containing hydroxyl groups:

Serine HO–CH$_2$–CH–COOH
$$\underset{NH_2}{|}$$

Threonine H$_3$C–CH–CH–COOH
$$\underset{OH}{|}\ \underset{NH_2}{|}$$

3) *Sulphur-containing amino acids:*

Cysteine HS–CH$_2$–CH(NH$_2$)–COOH

Cystine S–CH$_2$–CH(NH$_2$)–COOH
         S–CH$_2$–CH(NH$_2$)–COOH

Methionine H$_3$C–S–CH$_2$–CH$_2$–CH(NH$_2$)–COOH

(Homocysteine and taurine also contain sulphur but they are not present in proteins.)

4) *Acidic amino acids:*
Aspartic acid HOOC–CH$_2$–CH(NH$_2$)–COOH
Glutamic acid (for formula, refer back)

5) *Basic amino acids:*
Lysine H$_2$N–CH$_2$–CH$_2$–CH$_2$–CH$_2$–CH(NH$_2$)–COOH
Arginine (for formula, refer back)

Histidine (imidazole ring)–CH$_2$–CH(NH$_2$)–COOH

(Histidine contains the imidazole ring.)

6) *Aromatic amino acids*

Phenyl Alanine C$_6$H$_5$–CH$_2$–CH(NH$_2$)–COOH

Tyrosine HO–C$_6$H$_4$–CH$_2$–CH(NH$_2$)–COOH

Tryptophan (indole)–CH$_2$–CH(NH$_2$)–COOH

(Tryptophan contains the indole nucleus.)

7) Secondary amino acids (also called imino acids):

Proline — (pyrrolidine ring with NH)–COOH

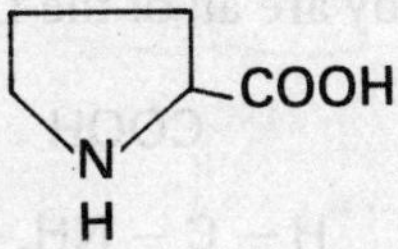

Hydroxy proline — HO-substituted pyrrolidine ring with NH)–COOH

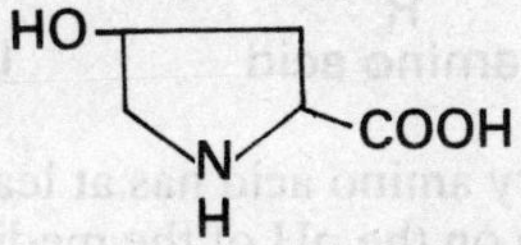

(These contain reduced pyrrole nucleus.)

The amino acids, numbering about 20, occur in the common proteins with variation in quantity and sequence. In some proteins a few of the above may be absent. There are some amino acids of biological importance which do not occur in proteins.

E.g.,

Taurine $H_2N - CH_2 - CH_2 - SO_3H$

γ-amino butyric acid $H_2N - CH_2 - CH_2 - CH_2 - COOH$

Ornithine $H_2N - CH_2 - CH_2 - CH_2 - CH - COOH$
                                                  |
                                                  NH_2

Citrulline $H_2N - CO - NH - CH_2 - CH_2 - CH_2 - CH - COOH$
                                                        |
                                                        NH_2

Thyroxine HO–⟨ring⟩–O–⟨ring⟩–$CH_2 - CH - COOH$
                                      |
                                      NH_2

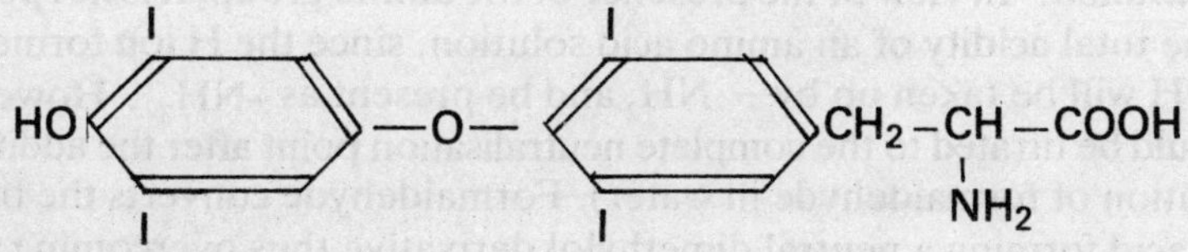

Tri iodo Thyronine HO–⟨ring⟩–O–⟨ring⟩–$CH_2 - CH - COOH$
                                              |
                                              NH_2

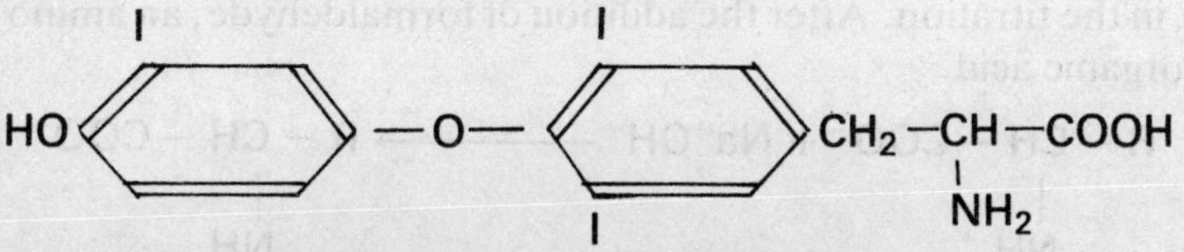

The last two occur in the special protein thyroglobulin of the thyroid gland.

Reactions of amino acids

Amino acids are colourless, crystalline substances soluble in water and insoluble in organic solvents. Aspartic and glutamic acids, cystine and the aromatic amino acids have limited solubility in water. All amino acids are soluble in dilute acids and alkalies.

Except glycine all other alpha amino acids are optically active as they possess an asymmetric carbon atom in the molecule.

$$R - \overset{*}{C}H - COOH$$
$$\qquad\quad |$$
$$\qquad\; NH_2$$

(*C is the asymmetric carbon.)

As in the case of the sugars, amino acids can have either the D configuration or the L configuration (It is to be noted that the D or the L form is concerned with the arrangement of the amino group, and the H relating to a particular carbon atom and not concerned with optical rotation.) The amino acids of the body are all of the L configuration.

$$\begin{array}{cc} \text{COOH} & \text{COOH} \\ | & | \\ H_2N-C-H & H-C-NH_2 \\ | & | \\ R & R \\ \text{L amino acid} & \text{D amino acid} \end{array}$$

Ionisation: As every amino acid has at least two ionisable groups, it can exist in different ionic forms depending on the pH of the medium.

In aqueous solution a neutral amino acid is in the zwitter ion form which is dipolar. It is, therefore, an amphoteric electrolyte.

$$\begin{array}{c} R-CH-COO^- \\ | \\ NH_3^+ \end{array}$$

In strongly acid pH, it is in cationic form while in strongly alkaline pH, it is in anionic form.

$$\begin{array}{cc} R-CH-COOH & R-CH-COO^- \\ | & | \\ NH_3^+ & NH_2 \\ \text{(cationic form)} & \text{(anionic form)} \end{array}$$

Each amino acid has an iso-electric pH, at which it will be in the zwitter ion form.

Formol titration: In view of the presence of the amino group, it is not possible to titrate and estimate the total acidity of an amino acid solution, since the H ion formed by the ionisation of $-COOH$ will be taken up by $-NH_2$ and be present as $-NH_3^+$. However, an amino acid solution could be titrated to the complete neutralisation point after the addition of excess of formalin (solution of formaldehyde in water). Formaldehyde converts the basic $-NH_2$ group of the amino acid forming a neutral dimethylol derivative thus overcoming the interference by $-NH_2$ in the titration. After the addition of formaldehyde, an amino acid behaves as any ordinary organic acid.

$$\begin{array}{c} R-CH-COO^- \\ | \\ NH_3^+ \end{array} + Na^+OH^- \longrightarrow \begin{array}{c} R-CH-COO^- \\ | \\ NH_2 \end{array} + Na^+ + H_2O$$

(This basic NH_2 group should be modified to neutral group.)

$$\begin{array}{c} R-CH-COO^- \\ | \\ NH_2 \end{array} + 2\,HCHO \longrightarrow \begin{array}{c} R-CH-COO^- \\ | \\ N(CH_2OH)_2 \end{array}$$

neutral dimethylol derivative

(It may be noted that formaldehyde does not react with the zwitter ion form.)

The method of titration of amino acid after adding formaldehyde is called Sorensen's formol titration.

For example, if 10 ml of 0.1 N solution of glycine is directly titrated against 0.1 N sodium hydroxide, the alkali consumed at the end point may be less than one ml. If adequate formal is added to 10 ml of 0.1 N solution of glycine and titrated, the volume of 0.1 N NaOH would be 10 ml. Thus, formal releases the H^+ ions bound to NH_2 during titration against alkali.

Reactions of the amino group

Amino acids form salts with acids like the hydrochloric acid.

$$R-CH(COOH)-NH_2 + HCl \rightarrow R-CH(COOH)-NH_3^+ Cl^-$$

With nitrous acid the $-NH_2$ group is converted to $-OH$ with liberation of nitrogen.

$$R-CH(COOH)-NH_2 \xrightarrow{HNO_2} R-CH(COOH)-OH + N_2 + H_2O$$

With aldehydes amino acids form Schiff's bases.

$$-N|H_2 + O| = HC- \rightarrow -N = HC-$$

Ninhydrin: A delicate colour reaction for amino acids is the ninhydrin reaction. Most amino acids give a blue or purple colour on warming with a solution of ninhydrin in acetone. Ninhydrin oxidises the amino acids releasing ammonia. The reduced ninhydrin reacts with the liberated ammonia and another molecule of ninhydrin to form the coloured complex. The keto acids formed from ∝ amino acids are immediately and quantitatively decomposed to their corresponding aldehydes liberting CO_2 which is not the case with ß and other amino acids. (Ninhydrin is tri keto hydrindene hydrate.)

(Ninhydrin; Tri keto hydrindene hydrate) + Amino acid ⟶ Keto acid + Ammonia + (Reduced ninhydrin; hydrindantin)

$$\xrightarrow{-3H_2O}$$

$$\xrightarrow{+NH_4}$$

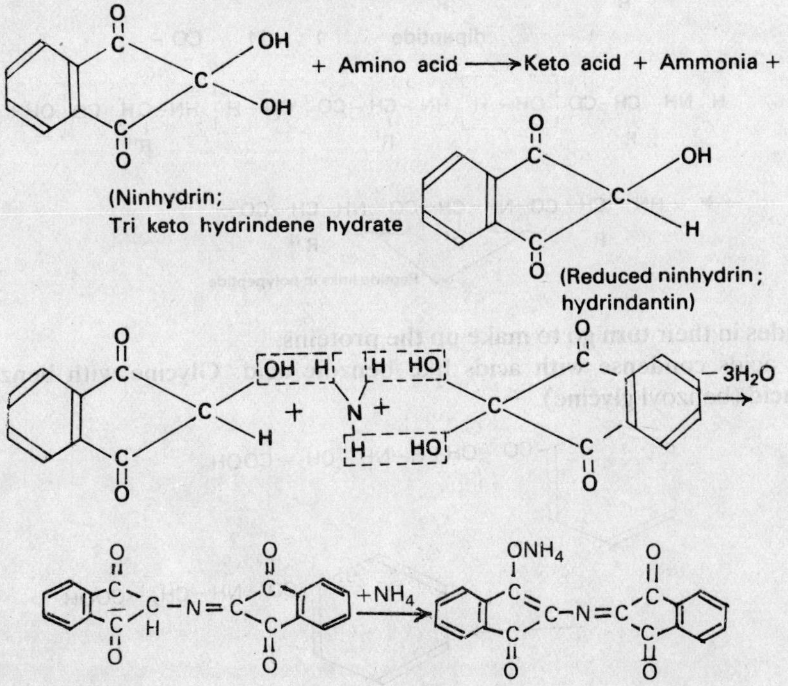

(*Exception*: Proline and hydroxy proline produce yellow colour with ninhydrin.)

Dinitroflurobenzene: Amino acids react with 2, 4-dinitrofluorobenzene to form yellow substances called DNP amino acids (dinitro-phenyl amino acids). This reagent is a valuable reagent introduced by Sanger in studying the amino acid sequence of proteins.

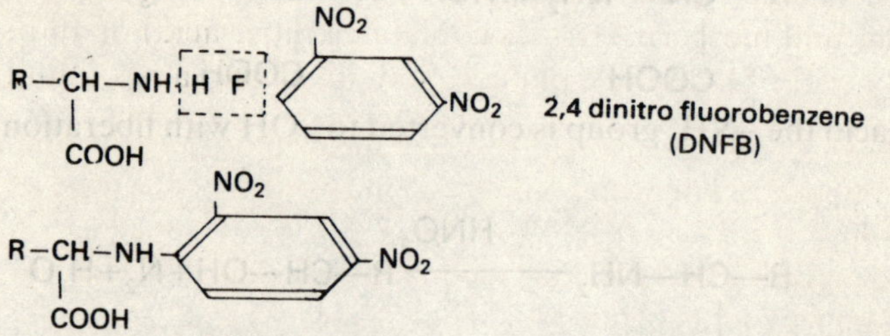

R—CH—NH—H F + O_2N-C$_6$H$_3$-NO$_2$ 2,4 dinitro fluorobenzene (DNFB)
     |
    COOH

R—CH—NH—C$_6$H$_3$(NO$_2$)$_2$
     |
    COOH
(DNP amino acid)

Metal complexes: Alpha amino acids form five-membered ring complexes with metal ions like copper and cobalt.

Peptide bond: Amino acids under certain experimental conditions could condense with one another with the elimination of water and form large molecules called polypeptides. The carboxyl group of an amino acid condenses with the $-NH_2$ group of another amino acid giving rise to $-CO-NH-$ (peptide) link. If two amino acids condense, the resulting product is a dipeptide.

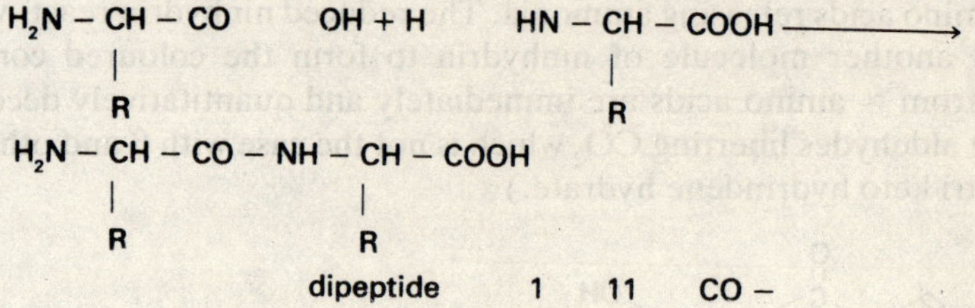

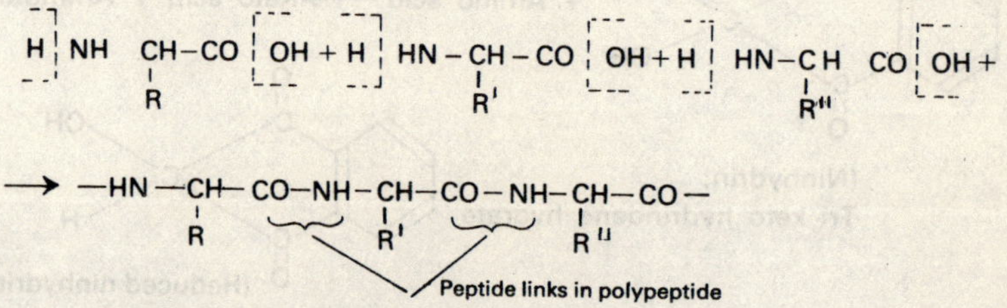

$H_2N - CH - CO$ $OH + H$ $HN - CH - COOH$ →
         | |
         R   R

$H_2N - CH - CO - NH - CH - COOH$
         | |
         R   R
         dipeptide

H NH CH—CO OH+H HN—CH—CO OH+H HN—CH CO OH+
       | | |
       R   R'   R''

→ —HN—CH—CO—NH—CH—CO—NH—CH—CO—
       | | |
       R   R'   R''
       ⏟   ⏟   ⏟
       Peptide links in polypeptide

Polypeptides in their turn go to make up the proteins.

Amino acids condense with acids like benzoic acid. Glycine with benzoic acid gives hippuric acid (benzoyl glycine).

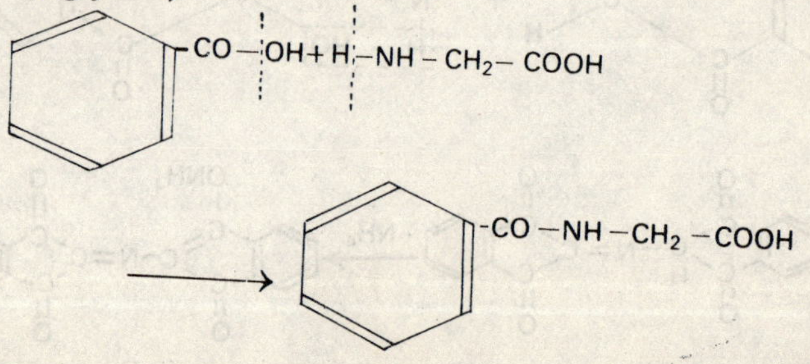

C$_6$H$_5$—CO—OH + H—NH—CH$_2$—COOH

→ C$_6$H$_5$—CO—NH—CH$_2$—COOH

Reactions of –COOH group: Amino acids form salts with bases, and esters with alcohols. On heating with soda-lime, decarboxylation takes place and amines are produced. In the body, decarboxylation of a few amino acids takes place with the help of enzymes.

Special reactions of some amino acids

Millon's test for tyrosine: When a solution of tyrosine is heated with Millon's reagent (a solution of mercurous and mercuric nitrates containing nitric acid) or 10 per cent mercuric sulphate in 10 per cent H_2SO_4 and one per cent sodium nitrite, a red colour is produced.

The Hopkins–Cole or glyoxylic acid reaction for tryptophan: A mixture of glyoxylic acid solution and tryptophan, when layered over concentrated H_2SO_4, gives a violet ring at the interface.

Tryptophan also gives a violet colour with mercuric sulphate in 10 per cent sulphuric acid, one in 500 formalin and concentrated H_2SO_4.

Sakaguchi reaction: Arginine with alcoholic alpha naphthol and sodium hypochlorite gives a deep red colour.

Lead sulphide test for sulphur–containing amino acids: The sulphur–containing amino acids are boiled with 40 per cent NaOH. The sodium sulphide formed is treated with lead acetate after acidifying the solution with acetic acid. A black precipitate of lead sulphide is formed.

Folin's reaction: With sodium ß naphthoquinone sulphonate in alkali (Folin's reagent) amino acids give a red colour.

Proline gives a blue colour with isatin.

Separation of amino acids from a mixture

The older methods like esterification of amino acids and fractional distillation of esters, selective precipitation of some amino acids using acids like HCl, picric acid or silver ions and electrophoresis (ionophoresis) of amino acids under high voltage have now been relegated to the background.

At present amino acid mixtures are separated by paper chromatography or thin layer chromatography using suitable solvents. In view of different Rf (ratio of fronts), the amino acids separate themselves in the paper or plate. The amino acids can be located by using a ninhydrin spray.

A very useful method for the separation of amino acid mixtures is by ion exchange. Suitable anion and cation exchange resins take up the amino acids in the ionic form by the ion exchange process. The amino acids held by electrostatic attraction are eluted with buffers of different pH. By this technique Moore and Stein separated the amino acids obtained by hydrolysis of as small a quantity as 5–10 mg of proteins and have constructed an automatic amino acid analyser for the separation of amino acids.

The microbiological assay method, using organisms which selectively utilise specific amino acids, and gas liquid chromatography after converting amino acids into volatile derivatives have also been used.

Assay of amino acids

There are various methods for the estimation of amino acids.
In van Slyke's method, a known volume of amino acid solution is treated with nitrous acid, and the volume of nitrogen gas liberated is measured after absorbing the oxides of nitrogen in alkaline $KMnO_4$. From the volume of nitrogen at NTP, the amount of amino acid

can be known since one molecular weight of amino acid containing one amino group gives, with HNO_2, 11,200 ml of nitrogen at NTP.

The microbiological assay uses organisms like *Lactobacillus arabinosis* whose growth is proportional to the amount of amino acids like tryptophan. The growth can be assessed either from the turbidity of the solution or by titrating the lactic acid produced due to the growth of the organism.

If the amino acid is a hydrochloride (say lysine hydrochloride), the chloride content can be estimated by Volhard's method. By assuming the molecular weight of the hydrochloride the amino acid could be estimated.

A good method of estimation of amino acids is by paper chromatography. A known volume of the solution of amino acids is subjected to paper chromatography and the amino acids made to react with ninhydrin. Known weights of amino acids are treated similarly. The intensities of colour are assessed using a densitometer and the amino acids are thus estimated. Alternately, the coloured spots are cut out, eluted with aqueous alcohol and the intensities of colour are measured in a photo-electric colorimeter.

Ion exchange chromatography is very useful in the estimation of amino acids and this technique is used in the amino acid auto-analyser.

HPLC is a very recent technique which is useful for the separation and estimation of amino acids.

Some workers have used the counter-current distribution technique. The solution containing the amino acids is extracted repeatedly with an immiscible solvent when the amino acids partition between the two solvents.

Formol titration can also be used for the estimation of amino acids and, being simple, could be adopted in many laboratories.

The total amino acids in blood are estimated by Folin's method using sodium β-naphthoquinone sulphonate.

PEPTIDES

In the body, peptides are formed during the initial digestion of the proteins and inside the cells by synthesis from amino acids.

Carnosine and anserine are dipeptides found in the voluntary muscles. An important tripeptide is glutathione which contains glutamic acid, cysteine and glycine.

(γ glutamyl cysteinyl glycine)

$$CH_2-\overset{O}{\underset{\|}{C}}-\underset{H}{\overset{|}{N}}-CH-\overset{O}{\underset{\|}{C}}-\underset{H}{\overset{|}{N}}-CH_2-COOH$$

with side chains: CH_2, $H-C-NH_2$, $COOH$ (glutamic acid); CH_2, SH (cysteine); glycine

Other peptides of biological importance containing many amino acids are glucagon, oxytocin, vasopressin, MSH, angiotensin and adreno-corticotropic hormone (ACTH).

Peptides have a positive biuret reaction, i.e., with dilute copper sulphate and excess of NaOH, they develop a pink or purple colour. Dipeptides are exceptions. Peptides could also be separated by paper chromatography and identified with ninhydrin. This was done by Ingram and his technique was called 'finger print technique' in the identification of different hemoglobins.

PROTEINS

Amino acids go to make up the polypeptides which in their turn give the finished products of proteins. Therefore, all proteins contain amino acids linked to one another by peptide bonds.

Total hydrolysis of proteins yields amino acids. Most proteins contain the common 20-22 amino acids. The size of protein molecules is quite large and hence proteins exist as colloids in solution.

Classification of proteins

According to one classification, proteins are classified into 1) catalytic proteins or enzymes, e.g., hexokinase, 2) structural proteins, e.g., keratin, 3) contractile proteins, e.g., myosin, 4) transport proteins, e.g., hemoglobin, 5) genetic proteins, e.g., nucleoproteins, 6) hormonal proteins, e.g., insulin, and 7) immune proteins, e.g., immunoglobulins.

A second and more popular classification divides proteins into three major groups, viz., simple proteins, compound or conjugated proteins and derived proteins. These three groups are in turn subdivided.

Simple proteins: These contain only amino acids. The following are the different types of simple proteins.

1) Albumins: These are soluble in water and dilute salt solutions, coagulated by heat and precipitated by the full saturation of their aqueous solutions with solid ammonium sulphate. This indicates that they are profusely hydrated. Serum albumin, lactalbumin and ovalbumin are examples.

2) Globulins: These are insoluble in water but soluble in dilute neutral salt solutions. These are coagulated by heat, and precipitated by half saturation with ammonium sulphate solution. These are less hydrated than albumins. Examples are ovoglobulin and serum globulins.

3) Glutelins: These are soluble in dilute acids and alkalies, insoluble in water and neutral solvents, and are coagulated by heat, e.g., glutelin from wheat.

4) Prolamines: Soluble in about 70 per cent alcohol, insoluble in water, absolute alcohol and other neutral solvents. They are rich in proline. Zein of corn and gliadin of wheat are good examples.

5) Histones: These are soluble in water and very dilute acids, insoluble in dilute NH_4OH and not coagulated by heat. They are strongly basic and occur in nucleoproteins.

6) Globins: They are referred to as examples of histones but are classified separately since they are not basic like histones and are not precipitated by ammonium hydroxide, e.g., globins of hemoglobin.

7) Protamines are the simplest of proteins and may be regarded as large polypeptides, rich in basic amino acids. They are soluble in water or ammonium hydroxide and not coagulated by heat. They resemble histones but are soluble in NH_4OH. Protamines occur in sperm cells.

8) Albuminoids (Sclero proteins): These are insoluble in all neutral solvents, dilute acids and alkalies, and are present in supporting tissues like bone, cartilage, horn, hair, etc.(e.g., keratin, collagen). Gelatin is derived from collagen.

Conjugated proteins (compound proteins): These are composed of simple proteins in association with some non-protein substance as the prosthetic factor.

1) Nucleoproteins are combinations of simple basic proteins, i.e., protamines or histones

with nucleic acids (DNA or RNA). These are soluble in dilute NaCl. The chromatin of cell nuclei and the viruses are nucleoproteins.

2) **Glycoproteins and mucoproteins:** These are simple proteins combined with carbohydrates (glycosamino glycans) and on hydrolysis give amino acids, amino sugars and uronic acids. The glycosamino glycans may be hyaluronic acid, chondroitin sulphates and heparin. Glycoproteins contain less than four per cent carbohydrate. Some of the plasma proteins like fibrinogen, transferrin, ceruloplasmin and immunoglobulins are glycoproteins containing different amounts of carbohydrates. Mucoproteins contain more than four per cent. Most of the mucoproteins are soluble in water, e.g., serum mucoproteins, ovo mucoid of egg white and mucin of saliva. They are important constituents of the ground substance of connective tissue and are present as tendomucoid, osseomucoid and chondroproteins in tendons, bones and cartilage, respectively.

3) **Phosphoproteins:** These contain phosphoric acid. Casein of milk and vitellin of egg yolk are phosphoproteins. The phosphoric acid is combined with serine of the proteins.

4) **Chromoproteins** are simple proteins united with coloured prosthetic groups (chromophoric or colour-producing), e.g., hemoglobin, flavoproteins, cytochromes, visual purple of the retina and catalase.

5) **Lipoproteins** are proteins conjugated to a lipid-like lecithin, cephalin, neutral fat, fatty acid or cholesterol. Lipoproteins are different from proteolipids in that the latter are soluble in organic solvents and insoluble in water. Lipoproteins occur in the blood and cell membranes.

6) **Metalloproteins** are proteins containing metallic elements such as iron, cobalt, manganese, zinc, e.g., ceruloplasmin (copper), siderophilin (iron). Iron-containing heme proteins which are classified as chromoproteins are also metalloproteins.

Derived proteins: These are divided into two broad types:

I) **Primary deriatives**

1) Denatured and coagulated proteins are grouped under this category. On heating, alone or with acids, alkalies, etc. proteins undergo intra-molecular rearrangement and consequently their solubility is affected. Under these conditions, proteins are said to be denatured. If denaturation becomes irreversible, the protein is said to be a coagulum (for more details, refer to the section on 'Denaturation of proteins').

2) Proteans are insoluble products formed from proteins by the action of water, very dilute acids and enzymes, e.g., fibrin.

3) Metaproteins are formed by the action of fairly concentrated acids and alkalies on proteins. These are soluble in very dilute acids and alkalies but insoluble in neutral solvents, e.g., acid metaproteins, alkali metaproteins.

II) **Secondary derived proteins**

These are obtained by progressive hydrolysis of proteins and are grouped by taking into account their molecular complexity. Each group consists of many different substances.

1) Proteoses are products of hydrolysis of proteins, soluble in water, not coagulated by heat but precipitated by saturation with ammonium sulphate.

2) Peptones are products of hydrolysis of proteins but much simpler than the proteoses, soluble in water, not coagulated by heat and not precipitated by saturation with ammonium sulphate.

3) Peptides are composed of relatively few amino acids united through peptide bonds. These are water soluble and not coagulated by heat. They cannot be salted out of solution and

PROTEINS

are often precipitated by phosphotungstic acid.

It is believed now that proteoses and peptones are mixtures of simple amino acids and peptide chains of various lengths. Protein chemists are abandoning these terms which are however commonly used for materials employed in the preparation of bacteriological media.

Reactions of proteins: physico-chemical properties

1) Molecular weights: In view of their molecular complexity resulting in high molecular weight, proteins in solution exist as colloids. There are practical difficulties in the determination of molecular weights. A physical method of osmotic pressure measurement has been employed to determine molecular weights in some cases. For proteins like hemoglobin, the iron content is estimated precisely and, assuming that one molecule of Hb contains one iron atom, the minimal molecular weight is determined. Svedberg measured sedimentation velocity and sedimentation equilibrium of proteins in an ultracentrifuge and calculated the molecular weights of many proteins. The gel filtration (molecular exclusion) technique is also used for determination of molecular weights.

Svedberg's experiments showed the interesting fact that proteins in solution may dissociate by a change of pH, salt concentrations or protein concentration.

MOLECULAR WEIGHTS OF SOME PROTEINS

Protein	Molecular weight
Myoglobin	16,900
Insulin	5,733
Hemoglobin (man)	63,000
Serum albumin (man)	69,000
Serum globulin	176,000
Ovalbumin	44,000
Lactoglobulin	41,500

2) Hydrolysis of proteins: Proteins are hydrolysed ultimately to amino acids. However, various intermediate products are formed especially when enzymes are used for hydrolysis. The intermediate products are metaproteins, proteoses, peptones and polypeptides. Acids and alkalies could also be used for hydrolysis. A protein could be hydrolysed to amino acids by digesting with 6 NHCl for 24 hours at 100°C. Hydrolysis is preferably carried out in a sealed evacuated tube (to exclude oxygen, as otherwise the side chain groups might be oxidised). In acid hydrolysis, tryptophan and large amounts of serine and threonine are destroyed. Glutamic acid may get dehydrated. Hydrolysis by alkali is used to identify tryptophan which is not destroyed, but serine, threonine, arginine, and cystine are lost. Enzyme hydrolysis does not destroy the amino acids but is slow. Proteolytic enzymes, trypsin and chymotrypsin, catalyse the hydrolysis of certain peptide bonds quite rapidly.

3) Iso-electric point (iso-ionic point) of proteins: Proteins are colloidal electrolytes which exist either as an anion or as a cation or as a zwitter ion. This is because protein is made up of amino acids. One end of the protein molecule has the free amino group while the other end has the free $-COOH$ group. By convention, the amino acid with free $-NH_2$ is written on the left.

$$H_2N - AA_1 - CO - NH - AA_2 - CO - NH - AA_3 \ldots COOH$$

(N terminal)   (Carboxy terminal)

Thus, at the extremities there are ionisable groups. Again in the body of the protein, if acidic and basic amino acids are present, there will be –COOH groups (of acidic amino acids) and –NH$_2$ of basic amino acids. In acid solution the –NH$_2$ groups are combined with H$^+$ ion and present as –NH$^+_3$. Hence, proteins in acid solution will be cationic. In alkaline pH, the –COOH groups ionise and are present as –COO$^-$ Hence, proteins in alkaline solution are anionic (of amino acids). When the protein has an equal number of dipolar groups, it will be a zwitter ion and the pH is called the iso–electric (iso–ionic) pH for the protein. While proteins in acid and alkaline solutions exhibit electrophoresis, at iso–electric pH there is no movement of the protein in an electrical field.

In addition to primary ionisation of – COOH and – NH$_2$ groups, there is contribution to the electrical charge on proteins by Helmholtz-Guoy electrical double layer and ζ (Zeta) potential, as proteins are colloidal and as traces of ions are adsorbed. Again, as proteins are colloidal electrolytes, they exhibit Donnan membrane phenomena.

ISO-ELECTRIC pH OF A FEW PROTEINS (APPROXIMATE)

Ovalbumin	4.6
Hb	6.7
Serum albumin	4.7
Lactoglobulin	5.19
Urease	5.0
Casein	4.6

4) Precipitation of proteins from solution: Proteins are precipitated from a solution by a variety of methods.

Being colloidal electrolytes, they can be brought to iso–electric pH by the addition of suitable ions in the form of electrolytes. Serum proteins are precipitated by ions provided by sulpho salicylic acid, trichloracetic acid, phosphotungstic acid, phosphomolybdic acid and tungstic acid. Albumin is precipitated by full saturation and globulin by half saturation with ammonium sulphate. Serum proteins, being hydrophilic colloids, can be precipitated by the addition of alcohol. Heavy metal ions like Pb^{++}, Hg^{++}, Fe^{++} cause the precipitation of proteins. Heat coagulable proteins like albumin and globulins are precipitated by heat.

5) Denaturation, flocculation and coagulation: If albumin is precipitated from a solution by alcohol, the precipitate is soluble in water, if tested immediately. But if the precipitate is left in contact with alcohol for about half an hour, it is no longer soluble in water. A change has taken place and albumin is said to be 'denatured'. Denaturation affects solubility. This is said to be due to an intra-molecular rearrangement. Denatured protein is not digested in the manner in which native protein is digested. In general, denatured proteins are insoluble in water at their iso-electric point *while they are soluble in acid and alkali*. The precipitate formed from a protein solution at the iso-electric point is referred to as a flocculum which is denatured protein (flocculum is soluble in acid and alkali).

1) A delay in the 1) redissolving of the denatured protein, 2) application of heat, 3) treatment with strong acids, alkalies, salts of heavy metals, alcohol, acetone, ether, urea, 4)exposure to light, 5) mechanical agitation, and 6) X-ray and ultra-violet irradiation, result in forming a coagulated protein which *does not dissolve in acid or alkali*. The change of flocculum to coagulum is irreversible. While a solution of proteins can be coagulated at any pH, coagulation occurs at the lowest temperature at the iso-electric point. At any other pH, coagulation is delayed or requires higher temperature. Water is necessary for denaturation and coagulation by heat. (Heating egg white even to 100^{0}C does not affect its solubility.) Coagulation is a property not only of albumins and globulins which are heat coagulable proteins but also of

other proteins which are not heat coagulable. During denaturation there is no hydrolysis but only intra-molecular structural changes with destruction of H bonds. The viscosity is changed and the surface tension lowered; hemoglobins and enzymes lose their species specificity and potency.

Other reactions of proteins

Biuret reaction: Proteins in solution answer the biuret reaction, i.e., they develop a pink or a purple colour with a solution of dilute copper sulphate and excess of NaOH. The colour is due to the presence of at least two peptide bonds. Insoluble proteins like keratin and dipeptides do not answer this reaction.

Ninhydrin reaction: Proteins, on warming with ninhydrin solution, develop a purple colour due to the amino acids present in protein.

Xanthoproteic reaction: When a native protein is treated with con. HNO_3, a white precipitate (meta protein) is formed and this turns yellow on heating. The yellow colour is due to the nitration of the benzene rings of aromatic amino acids of the protein. (xantho —yellow).

Millon's test for tyrosine: Proteins answer this test as they contain tyrosine (refer to Reactions of amino acids).

Hopkins–Cole or glyoxylic acid reaction for tryptophan: Proteins containing tryptophan answer this test (refer to Reactions of amino acids).

Nitroprusside test: Proteins containing free –SH groups (of cysteine) give a reddish colour with sodium nitroprusside in ammoniacal solution. The lead acetate test is answered by proteins (refer to Reactions of amino acids).

Molish's test: This test is answered by glycoproteins and mucoproteins as they contain carbohydrate.

Neumann's test: This is answered by phosphoproteins. When casein, for example, is boiled with con. HNO_3 and con. H_2SO_4, phosphorus present in it is oxidised to phosphoric acid. With ammonium molybdate it gives the yellow precipitate of ammonium phosphomolybdate in the cold.

Isolation of proteins

Various methods have been employed for the isolation of proteins. These methods also help to show whether a sample of protein is homogeneous or not.

Electrophoresis on paper and gels like starch, agar and polyacrylamide separates the proteins. Serum proteins are separated into five fractions: albumin, alpha$_1$ globulin, alpha$_2$ globulin, beta globulin and gamma globulin by paper or agar electrophoresis using a barbiturate buffer. Polyacrylamide 'disc' electrophoresis separates serum proteins into many other fractions like prealbumin, ceruloplasmin, transferrin, haptoglobins, etc. Recently immuno-electrophoresis has shown the presence of about 30 different protein fractions in blood serum.

Salt fractionation using ammonium sulphate or sodium sulphate has been employed for the separation of proteins. Serum albumins and globulins are isolated by using salt fractionation with ammonium sulphate. Cohn fractionated plasma proteins by using different concentrations of alcohol and then chilling to low temperature.

Paper chromatography is of little value in protein separation as proteins get adsorbed firmly to the paper and their movements are sluggish. Instead, ion exchange chromatography using modified cellulose resins has been used to separate proteins.

Fractional centrifugation by the ultra-centrifuge is also useful in the separation of proteins. The proteins and enzymes of cells are separated by differential centrifugation.

Recently a technique known as gel filtration or molecular sieving is being employed for the isolation and fractionation of proteins. In this, Sephadex gels with different porosity and cross-links are used. The molecules and particles (in the case of colloidal proteins) are 'sieved' as it were in the gels. This technique is very popular nowadays in the isolation of proteins.

Very recentlty, the techniques of iso-electric focussing and HPLC have also been employed for the separation of proteins.

Structure of proteins

The elucidation of the structure of proteins is difficult and time consuming and requires a lot of patience. However, the structures of some proteins of low molecular weight like insulin, ribonuclease, ACTH and myoglobin have been elucidated. Nobel prizes have been awarded to scientists working in this field. — Sanger for his work on insulin and Moore, Stein and Anfinsen for their contribution to the structure or ribonuclease.

The structure of proteins is very complex. To appreciate the function of proteins, Linderstrom-Lang suggested in 1953 that the structure of proteins should be considered at three different levels, 'primary', 'secondary' and 'tertiary'. A fourth level, 'quaternary', has been added subsequently.

1) The 'primary structure' reveals the sequence of amino acids linked through peptide bonds to form the 'backbone' of the protein; 2) the secondary structure refers to the twisting of the peptide chain into a helical form which is kept in position mostly by hydrogen bonds; 3) the helical form may fold itself into globular, ellipsoidal or any other conformation and which is called the tertiary structure – the main linkage bringing about this is disulphide (– S – S –) though other weak forces like van der Waal's attractive and repulsive forces and hydrophobic forces may also be involved; 4) joining into layers or clusters of sub-units having primary, secondary and tertiary structures results in the quaternary structure in the case of a few proteins.

Primary Structure: The first contribution to the knowledge of the primary structure of proteins came in 1955 through the work of Sanger on insulin, the protein-hormone. Sanger used the valuable reagent dinitrofluorobenzene (DNFB) which reacts with the free amino group of the N-terminal amino acid of the protein.

O_2N-⟨NO_2⟩-$F + H$ $HN-C(R)(H)-C-NH-AA_2-CO-NH-AA_3...COOH$

O_2N-⟨NO_2⟩-$N(H)-C(R)(H)-C(=O)-NH-AA_2-CO-NH-AA_3...COOH$

When the dinitrophenyl (DNP) derivative of protein is hydrolysed, the DNP derivative of the N-terminal amino acid is released while the other amino acids are free.

PROTEINS

$O_2N-C_6H_3(NO_2)-NH-CH(R)-CO-NH-AA_2-CO-NH-AA_3\ldots$ hydrolysis $\rightarrow$

$O_2N-C_6H_3(NO_2)-NH-CH(R)-COOH + AA_2 + AA_3$ etc.

The DNP amino acid is identified by chromatography. Using this technique, Sanger deciphered the structure of insulin. It has two peptide chains A (21 amino acids) and B (30 amino acids) linked by two disulphide (–S–S–) groups. They are broken into A and B by performic acid. The positions of the disulphide chains are 7-7 and 20-19. Chain A has glycine as N-terminal and asparagine as C-terminal while chain B has phenyl alanine as the N-terminal amino acid and alanine as C-terminal amino acid. There is an intra-chain (–S–S–) in chain A between the sixth and eleventh amino acids.

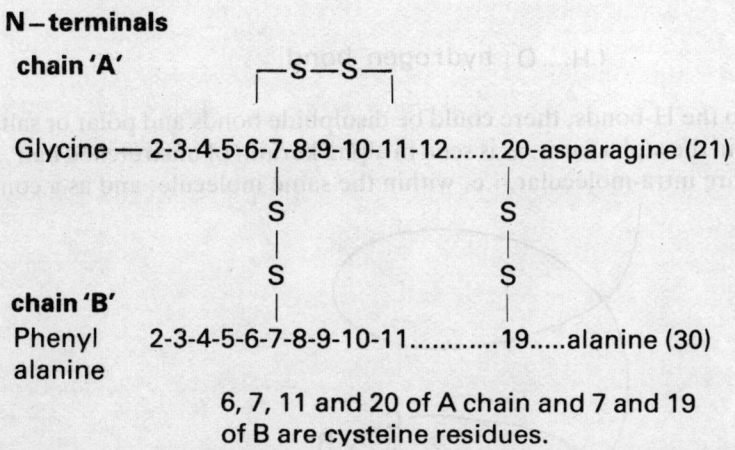

N–terminals

chain 'A'
```
          ┌─S—S─┐
Glycine — 2-3-4-5-6-7-8-9-10-11-12 ....... 20-asparagine (21)
                |                    |
                S                    S
                |                    |
                S                    S
chain 'B'       |                    |
Phenyl    2-3-4-5-6-7-8-9-10-11 ............ 19 ..... alanine (30)
alanine
```
6, 7, 11 and 20 of A chain and 7 and 19 of B are cysteine residues.

In addition to DNFB, other reagents like phenyl isothiocyanate (Edman's reagent) and Leucine amino peptidase were used to study the N–terminal, while hydrazine and carboxy-peptidase have been used for studying the C-terminal amino acid.

Glucagon was shown to have 29 amino acids. There are no disulphide bridges. The sequence of amino acids in oxytocin, vasopressin, MSH and ACTH was known by these techniques which are outstanding triumphs in biochemistry.

Moore and Stein demonstrated the complete sequence of amino acids in ribonuclease which is a much larger molecule than insulin and contains 124 amino acids.

One large protein whose primary structure was investigated was the tobacco mosaic virus. This consists of 158 amino acids with a molecular weight of 18,000.

Secondary structure: Techniques like X-ray diffraction, streaming birefringence (double refraction of flow) and electron photo-micrography, have thrown much light on the secondary, tertiary and quaternary structures of proteins.

Fibrous proteins (alpha and beta forms): X-ray diffraction studies have shown that there are two general patterns consistently associated with certain proteins like keratin, silk fibroin, and myosin. The two general forms are the alpha and the beta of which ß is the simplest and is found in silk fibroin. The ß form consists of nearly straight polypeptide chains held together by hydrogen bonds but in this the H-bonds are inter-molecular (between different polypeptide molecules), the hydrogen atoms of one chain linked to oxygen atoms of the adjoining chain.

Beta sheet

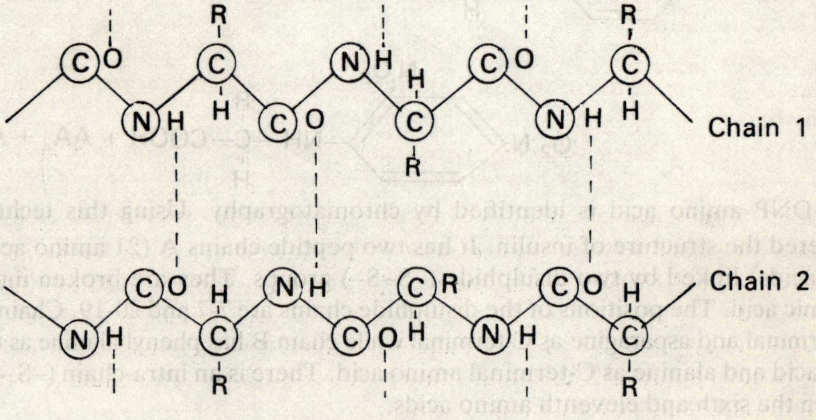

Chain 1

Chain 2

(H....O) hydrogen bond

In addition to the H-bonds, there could be disulphide bonds and polar or salt linkages. On the other hand in the alpha form, as is seen in alpha keratin of unstretched hair, the hydrogen bonds formed are intra-molecular, i.e. within the same molecule, and as a consequence the

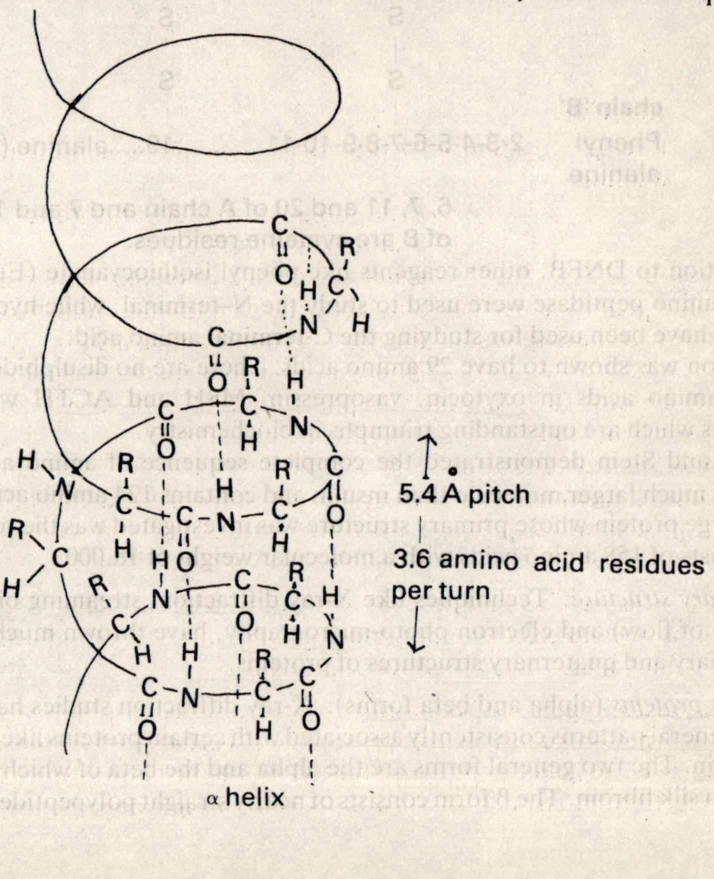

5.4 Å pitch

3.6 amino acid residues per turn

α helix

entire chain takes the shape of a helix which can be likened to a spiral staircase of which amino acids form the steps. The alpha helix contains 3.6 amino acid residues per turn. It takes five turns or 18 amino acids till an amino acid residue is situated exactly above the starting point. The distance between the turns is 5.4 Å units.

The keratin of unstretched hair has the alpha helix structure while that of stretched hair has the beta sheet structure.

Proteins having the alpha helical or the beta sheet forms are called fibrous proteins which are all elastic. The secondary structure alters on stretching and on recovery. The stretched form is the beta form. Both forms are single dimensional.

Collagen, though a fibrous protein is not strictly similar to alpha or beta as it contains considerable proline and hydroxy proline amino acids which do not particpate in H-bonding. G.N. Ramachandran has shown that collagen forms a third pattern and has three polypeptides twisted round one another and is like a long thin rod.

Tertiary structure(the globular proteins): The more soluble proteins like native albumins, globulins, myoglobin, insulin and hemoglobins are considered to be globular, or ellipsoidal which is two-dimensional. Whereas the axial ratio (length:width) of the fibrous proteins is greater than 10, it is less than 10 for globular proteins (usually 3:1 or 4:1). The globular or any other conformation is due to the intricate folding of the molecule with helical forms in some regions. The intricate folding is due to additional hydrogen bonds, disulphide linkages and polar or salt linkages between the amino and the carboxyl groups (already there would be hydrogen bonds in the formation of helical regions). The disulphide linkage is found in cysteine while polar groups are formed from, say, the free amino group of lysine and the carboxyl group of aspartic and glutamic acids.

```
     OC         NH                          NH        CO
       \       /                              \      /
        NH----OC                       HC—CH₂—S—S—CH₂—CH
       /       \                            /          \
     RCH       HCR                         CO          HN
                                          /  disulphide link \
  hydrogen bond                                              CO
                    \NH                    +          −     /
                    CH—(CH₂)₃—CH₂—NH₃   OOC—CH₂—CH₂—CH
                   /                                         \
                  CO     lysine           glutamic acid      NH
                 /                                             \
                            polar (salt) linkage
```

In addition there are the van der Waal's attractive and repulsive forces and hydrophobic forces. van der Waals's attractive forces are weak forces binding molecules together while hydrophobic forces are brought about by the coming together to a certain extent of the hydrocarbon chains of amino acids and pushing out of water molecules from the site. The various forces mentioned above give the particular conformation which is the tertiary structure. The integrity of the tertiary structure is very important for the functioning of enzymes and many other proteins. The bridge linkages in the globular proteins are, however, labile and could be easily disrupted and re-oriented.

Myoglobin from the sperm whale is a typical globular protein with a molecular weight of 17,000. It consists of 153 amino acid residues with a known sequence. The molecule has some alpha helical configuration and is highly folded. In addition to the contribution by bridge links, myoglobin is also stabilised to a large extent by the heme group which slips into a pocket as it were in the molecule of myoglobin.

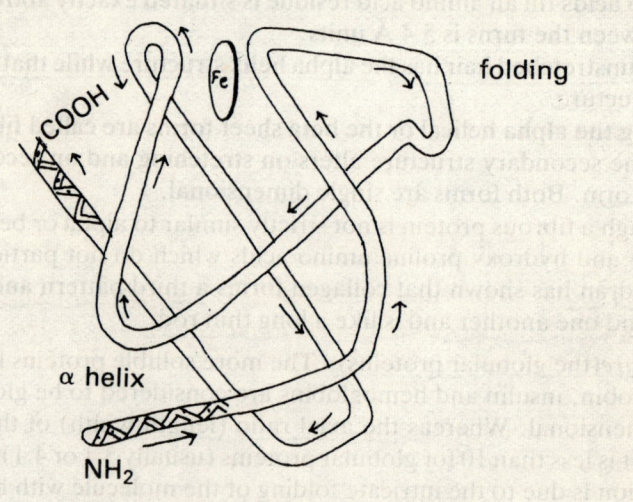

Schematic diagram of myoglobin

Quaternary structure: Some proteins may display a fourth level of organisation in which, units with primary, secondary and tertiary structures with a particular conformation may continue to form something like a polymer. The sub-units may be similar as in the case of hemoglobin or dissiminar. In hemoglobin there are four separate sub-units in two identical pairs. It is also known that the quaternary structure of hemoglobin is altered markedly during oxygenation when it passes from the tense form (T) to the relaxed form (R) and returned to normal during deoxygenation. The enzyme phosphorylase contains four identical sub-units. This quaternary structure is also essential for the potency of many enzyme proteins.

Very few proteins have a super-quaternary structure by polymerisation, e.g., α_2 macroglobulin.

Detection of proteins

Proteins are detected by the following reactions:
1) Biuret reaction
2) Xanthoproteic test
3) Millon's test (tyrosine)
4) Hopkin-Cole reaction (tryptophan)
5) Sakaguchi's reaction (arginine)
6) Ninhydrin reaction.
7) Sulphur test for S-containing amino acids.
8) Neumann's test for phosphoprotein.
9) Molish's test for glycoproteins.
  (For details refer to 'Reactions of proteins'.)

5 Nucleoproteins

Nucleoproteins occur in all living cells of animal and plant tissues, viruses, bacteria and bacteriophages and have a fundamental role in life processes. A large part of the substance is present in the chromatin of the nucleus while the rest is in the cytoplasm associated with ribosomes.

Nucleoproteins are conjugated proteins of high molecular weight and consist of the basic proteins, protamins or histones linked up with nucleic acids. It was Altmann in 1889 who gave the name 'nucleic acids' for 'nuclein' given by Miescher. The credit of isolating the nucleic acids first goes to Miescher who obtained them from pus cells by digesting them for weeks with dilute HCl.

The work of Chargaff, A.R. Todd, H.G.Khorana, Wilkins, Watson, Crick, Kornberg, Ochoa, Nirenberg, Jacob, Monod and Sanger has thrown considerable light on the biochemistry and function of nucleic acids which have been shown to constitute the deoxy ribonucleic acid (DNA) and ribonucleic acids (RNAs). DNA which is almost exclusively present in the nucleus constitutes the main substance of genes and as such has paramount importance in storing, expressing and transmitting genetic information. Growth, reproduction, tissue differentiation and hereditary characters of organisms depend on DNA. The function of RNAs which are present in the cytoplasm is the biosynthesis of proteins in accordance with the information dictated by DNA.

The building stones of nucleic acids are the nucleotides. A nucleotide is a combination of a nitrogen base (purine or pyrimidine), sugar, ribose or deoxy sugar, deoxy ribose and phosphoric acid. Nucleotide = Nitrogen base — sugar or deoxy sugar — Phosphoric acid.

We shall first consider the nitrogen bases present in the nucleic acids.

THE PYRIMIDINES

Pyrimidine is the name given to the heterocyclic nucleus having the ring structure (see below) with two nitrogen atoms in 1:3 position. The nitrogen atoms give the basicity.

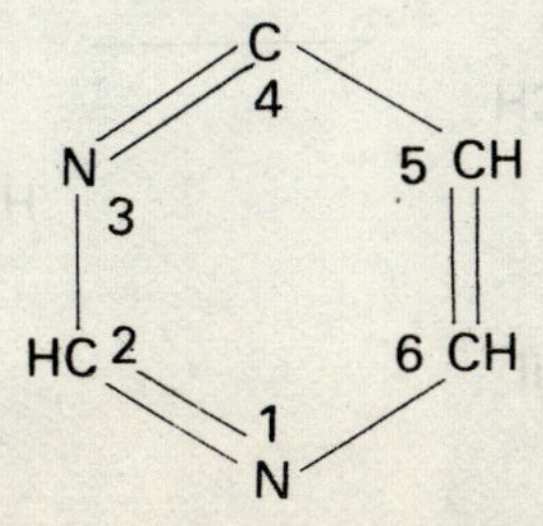

The main pyrimidine bases in nucleic acids are cytosine, uracil and thymine. While cytosine and thymine are present in DNA, cytosine and uracil occur in RNAs. Recently it has been shown that certain types of RNAs contain a very small amount of thymine also.

Cytosine

This is a 2-oxy 4-amino pyrimidine. It can exist in lactam and lactim forms. (If the group is –HN–CO–, it is called the lactam type (keto) while if the same isomerises to

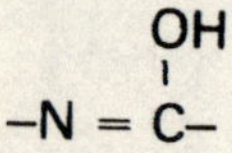

$$-N=C-\overset{\displaystyle OH}{|}-$$

it is the lactim form.

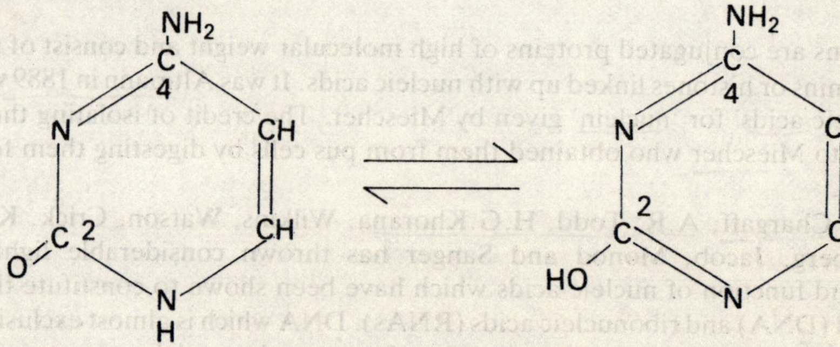

lactam form of cytosine ⇌ lactim form

Uracil

Uracil is 2, 4-dioxy pyrimidine. It also has lactam and lactim forms.

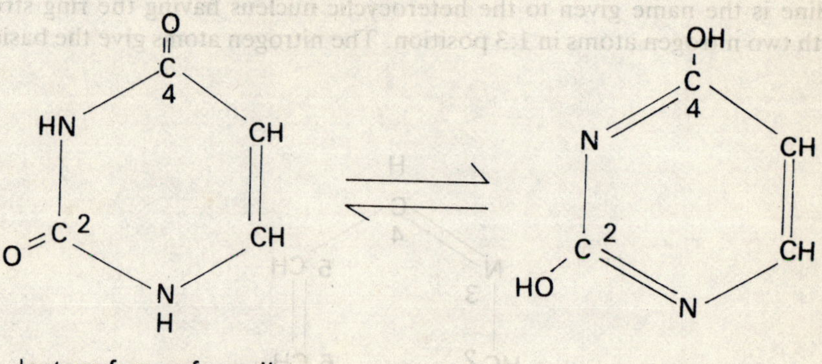

lactam form of uracil ⇌ lactim form

NUCLEOPROTEINS

Thymine: This is 5-methyl uracil and has lactim and lactam forms.

lactam form of thymine ⇌ **lactim form**

At the physiological pH, the lactam (keto) forms are predominant.

In addition to the three major pyrimidine bases, there occur in small quantities bases like 5 hydroxy methyl cytosine, methylated derivatives and reduced uracil compounds.

THE PURINES

The purine ring is more complex than the pyrimidine ring. It can be considered the product of fusion of a pyrimidine ring with an imidazole ring.

Pyrimidine   **Imidazole**   **Purine**

Adenine and guanine are two principal purines found in both DNA and RNAs. Adenine is 6-amino purine.

Adenine

Guanine is 2-amino 6-oxy purine and can be present as lactim and lactam forms.

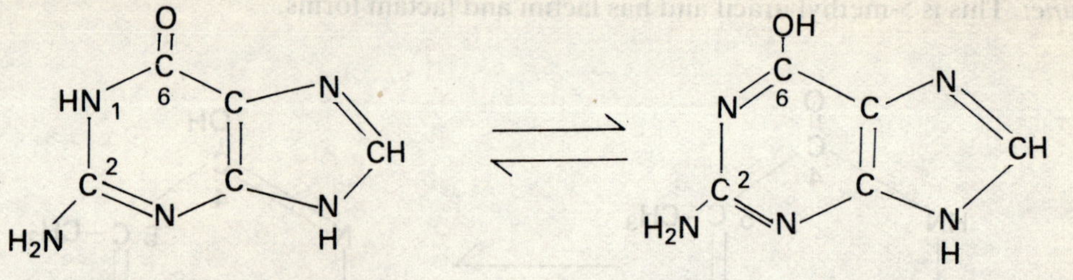

MEDICAL BIOCHEMISTRY

Guanine (lactam) ⇌ **Lactim form**

In addition, small amounts of methylated purines have been shown to be present in nucleic acids.

SUGARS

While ribose, the 5-membered aldose is present in RNAs (hence the name), deoxy ribose which is also a 5-membered aldose but has one oxygen less than ribose, is present in DNA (hence the name).

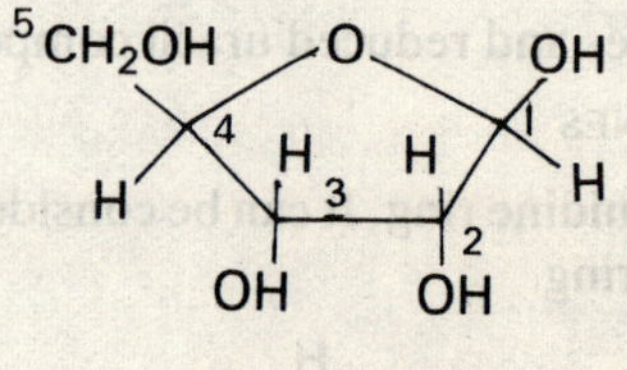

ßD-ribose 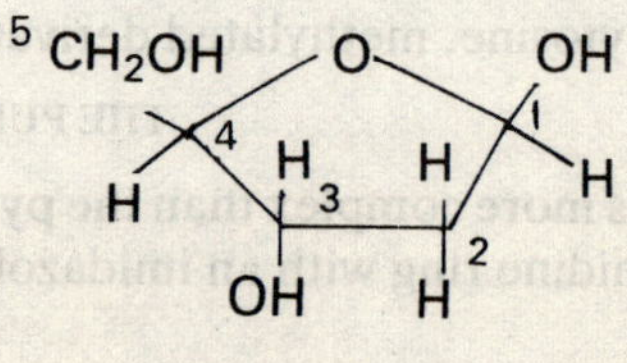   **ß-D-2 deoxy ribose**

Both the sugars occur in nucleic acids in furanose form with ß-configuration about carbon 1.

NUCLEOSIDES

The combination of pyrimidine or purine with sugar or deoxy sugar gives the nucleoside. A number of nucleosides are possible.

Base + sugar	=	Nucleoside
Adenine + ribose	=	Adenosine
Guanine + ribose	=	Guanosine
Cytosine + ribose	=	Cytidine
Uracil + ribose	=	Uridine
Thymine + deoxyribose	=	Thymidine
Thymine + ribose	=	Ribothymidine

If in place of ribose, deoxy ribose is present, the prefix 'deoxy' should be added before the name of the nucleoside, in all cases except thymidine.

If pyrimidine is linked to sugar or deoxy sugar, the N of the first position of pyrimidine is linked to carbon 1' of sugar or deoxy sugar.

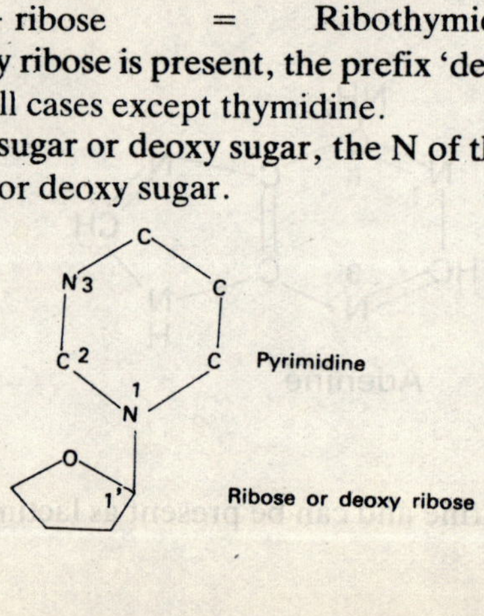

Pyrimidine

Ribose or deoxy ribose

NUCLEOPROTEINS

Thus the structure of uridine will be

It may be remembered that uridine will be present only in RNAs, but absent in DNA. Though the above is the usual type of nucleoside, relatively small amounts of what is called pseudouridine (ψ) are also present in RNA in which the carbon-5 of pyrimidine is linked to C 1' of sugar.

If purine is the N-base, the nitrogen in position 9 of the base is attached to carbon 1' of the sugar or deoxy sugar to give the purine nucleoside:

The structure of (say) adenosine will be

For deoxy adenosine, in place of –OH in position 2' there will be hydrogen.

NUCLEOTIDES

A nucleoside combined with phosphoric acid is a nucleotide. As a nucleoside is a combination of nitrogen base–sugar or deoxy sugar, a nucleotide will be a combination of nitrogen base, sugar (deoxy sugar) and phosphoric acid.

Nitrogen base – sugar (or deoxy sugar) – phosphoric acid = Nucleotide.

$$\begin{aligned}
\text{Nucleoside} + H_3PO_4 &\rightarrow \text{Nucleotide} \\
\text{Adenosine} + \text{phosphoric acid} &\rightarrow \text{Adenylic acid} \\
\text{Guanosine} + \text{phosphoric acid} &\rightarrow \text{Guanylic acid} \\
\text{Cytidine} + \text{phosphoric acid} &\rightarrow \text{Cytidylic acid} \\
\text{Uridine} + \text{phosphoric acid} &\rightarrow \text{Uridylic acid} \\
\text{Thymidine} + \text{phosphoric acid} &\rightarrow \text{Thymidylic acid}
\end{aligned}$$

If deoxy ribose is present instead of ribose, the prefix 'deoxy' should be used with the names as deoxy adenylic acid, deoxy cytidylic acid. However, uridylic acid occurs in RNA with only ribose, and no deoxy uridylic acid is possible.

The phosphoric acid esterified could be with the hydroxyl of 3'C or 5'carbon of ribose or deoxy ribose. So there can be 3' nucleotides, 5' nucleotides, 3' deoxy nucleotides, 5' deoxy nucleotides and even both hydroxyls involved in linking the nucleotides.

The structure of 3' adenylic acid would be,

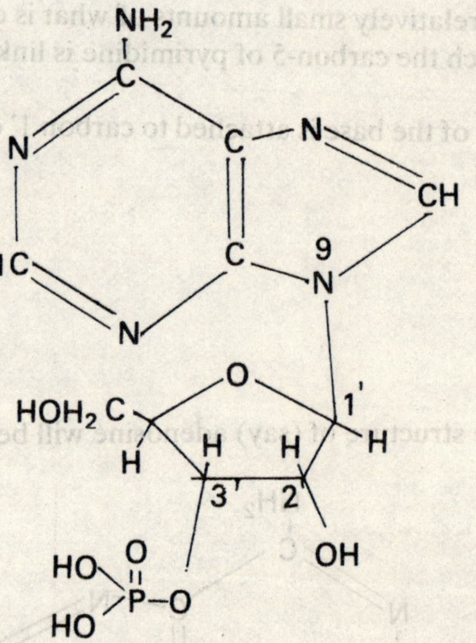

The structure of 5' adenylic acid would be,

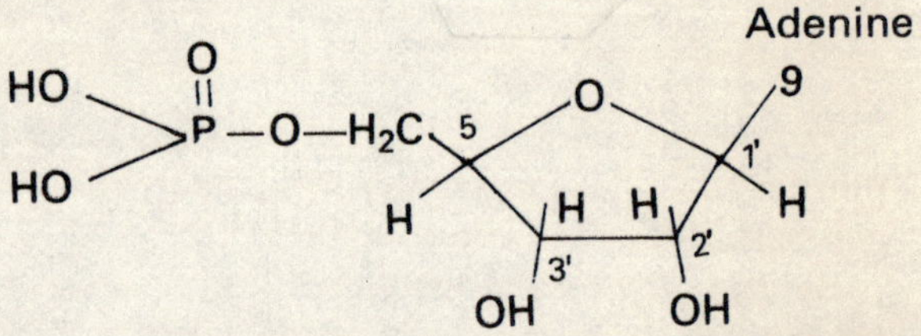

The adenylic acid is also called adenosine monophosphate. Where only one phosphoric acid group is present, it is called a mononucleotide. In nucleic acids, we have only mononucleotides. But in the body, there are dinucleotides and trinucleotides. In a dinucleotide, the **phosphate linked to sugar (or deoxy) will bear a second phosphate as in adenosine diphosphate (ADP).**

Adenine–ribose–phosphate–phosphate

NUCLEOPROTEINS

If a third phosphate is also present, it is **Adenosine triphosphate (ATP)**, one of the most important substances in bioenergetics

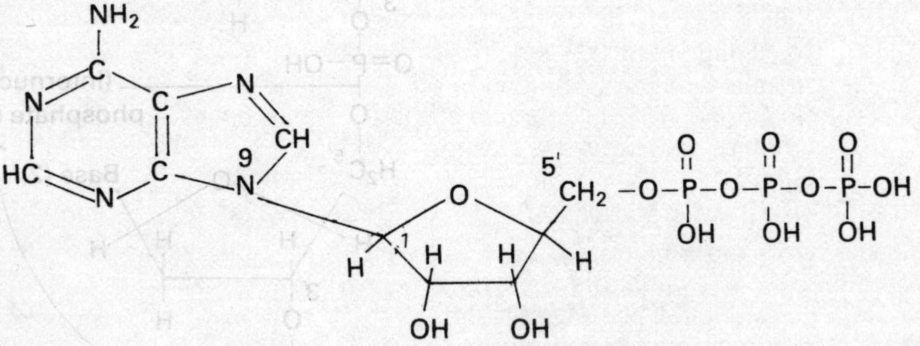

Likewise, GTP, CTP, and UTP are available in the body as trinucleotides.

NUCLEIC ACIDS

The term nucleic acids stands for both DNA and RNAs. Strictly speaking, it is a misnomer as it indicates that these are acids present only in the nucleus. It is known that RNA is present chiefly in the cytoplasm. In addition to nuclear DNA there is also mitochondrial DNA.

DNA (deoxyribonucleic acid)

DNA is present in every cell and is the genetic material. It is conveniently prepared from viruses, thymus gland, spleen, fish or invertebrate spermatozoa. In one method, the tissue is homogenised at neutral pH and centrifuged at a low speed to get the nuclear fraction. The material is extracted with 2 M NaCl which breaks the nucleoproteins into DNA and protein. DNA is in solution. This is precipitated with ethanol. To destroy any RNA, the sample can be treated with a ribonuclease which destroys only the RNA.

X-ray studies and other physical measurements like light scattering, hypochromism, optical rotation, viscosity and density have indicated that DNA has a highly complex structure. The molecules are long and thread-like, their lengths about 250 times more their breadth.

Primary structure of DNA: The chemical analysis of DNA has shown that it contains the purines (adenine and guanine) and the pyrimidines (thymine and cytosine) as the major nitrogen bases, deoxy ribose and phosphoric acid.

By using the enzymes pancreatic and micrococcal DNAse, splenic and snake venom diesterases and also treating with acids, it has been established that DNA is a polynucleotide. The macromolecule is built up of a large number of mononucleotides. The number of pairs of nucleotides may be as many as 150 million or more and the molecular weight may be as high as two billion. It has, however, been shown recently that a gene (DNA) could consist of only 77 nucleotide pairs but much longer genes certainly exist. The individual mononucleotides have been shown to be linked to one another by phosphate bridges. The phosphate linked to 3' carbon of the deoxy sugar of one nucleotide will be linked to the 5' carbon of the deoxy sugar of its immediate neighbour as shown here.

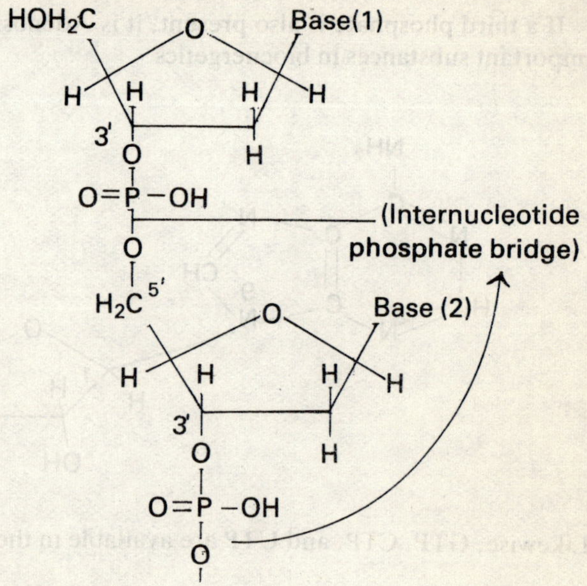

```
         HOH₂C   O    Base(1)
        H  H        H  H
           3'     H
           O
           |
       O=P—OH                ___ (Internucleotide
           |                        phosphate bridge)
           O
           |
         H₂C  5'  O    Base (2)
        H   H       H  H
           3'     H
           O
           |
       O=P—OH
           |
           O
           |
```

Thus a molecule of DNA can be schematically represented thus:

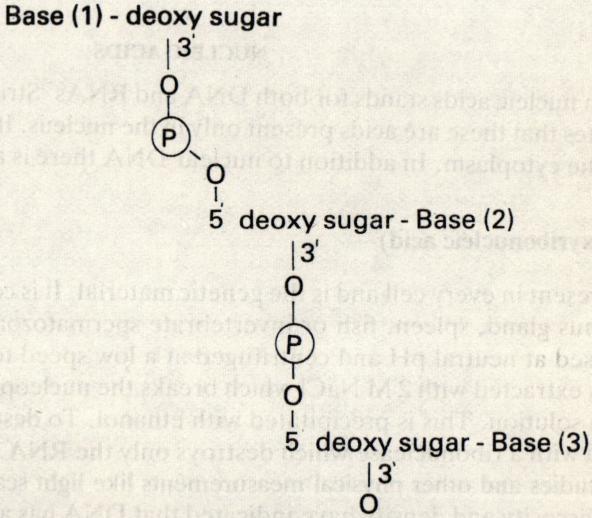

```
Base (1) - deoxy sugar
     |3'
     O
     |
     P
     |
     O
     |5
     deoxy sugar - Base (2)
     |3'
     O
     |
     P
     |
     O
     |5
     deoxy sugar - Base (3)
     |3'
     O
     |
```

The long molecule of this type of arrangement will have a definite sequence of the nitrogen bases. It is this sequence that spells out the genetic message. The sequence is specific for the species and in the species it decides the hereditary characters.

Secondary structure of DNA: Chargaff made a fundamentally important observation. The molar content of adenine is equal to that of thymine while the molar content of guanine is equal to that of cytosine, i.e., A = T and G = C. This is helpful in the interpretation of the secondary structure of DNA. X-ray studies of Watson, Crick and Wilkins revealed that the DNA molecule consists of two right-handed helical chains coiled round the same axis and held together in a double helix. The two strands run in opposite directions, i.e., they are anti-parallel, the terminal phosphate groups being at the opposite ends of the double helix. The chains are held together by inter-chain hydrogen bonds and to some extent the hydrophobic forces providing general interaction. The H-bonds exist between a single base from one chain and another single base from the opposite chain so that the two bases lie side by side and are in the same plane. There are two H-bonds between adenine and thymine and three H-bonds between guanine and cytosine.

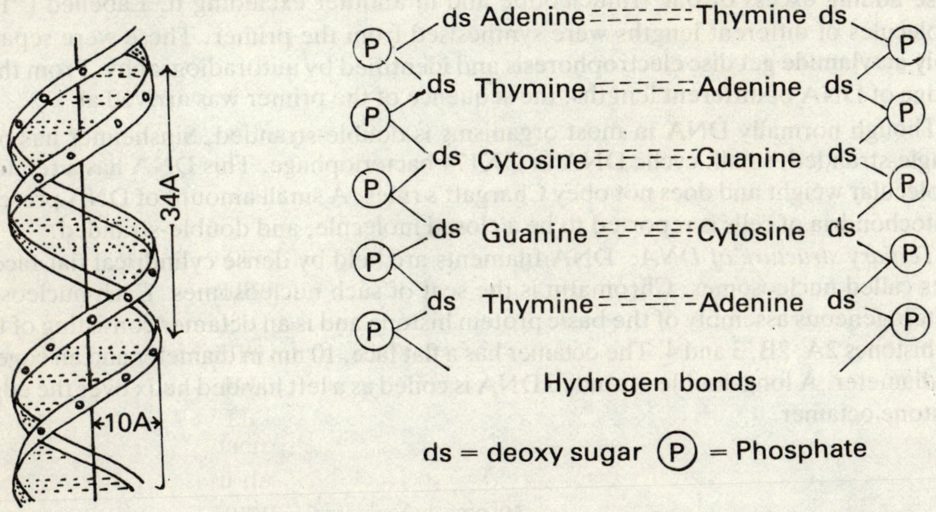

```
ds Adenine ======= Thymine ds
        (P)                        (P)
   ds Thymine ===== Adenine ds
 (P)                              (P)
   ds Cytosine ======= Guanine ds
 (P)                              (P)
   ds Guanine ======= Cytosine ds
 (P)                              (P)
   ds Thymine ===== Adenine ds
 (P)                              (P)
           Hydrogen bonds

ds = deoxy sugar   (P) = Phosphate
```

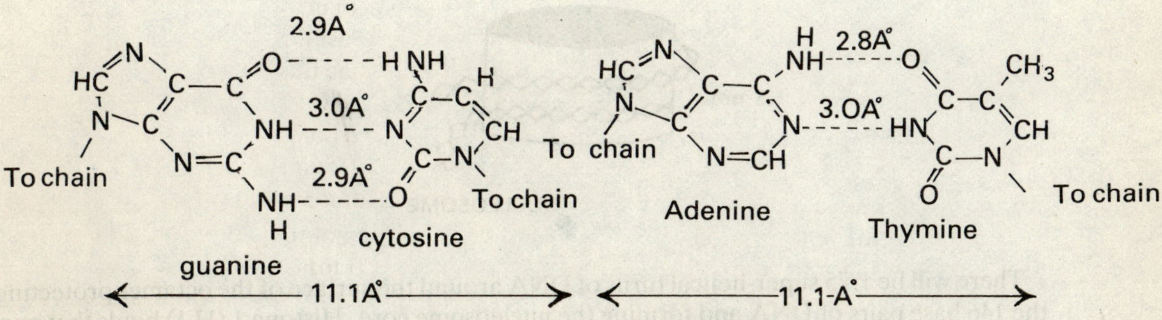

(Left figure: double helix showing 34 Å pitch and 10 Å width)

Guanine—Cytosine pairing with H-bond distances 2.9 Å, 3.0 Å, 2.9 Å; overall 11.1 Å.
Adenine—Thymine pairing with H-bond distances 2.8 Å, 3.0 Å; overall 11.1 Å.

The distance for H-bond formation has been measured by Pauling and is in the neighbourhood of 3 Å:

This double helical structure substantiates Chargaff's rules. As hydrogen bonds have to be formed between adenine and thymine, there should be as many adenines as there are thymines. Likewise to have H-bonds formed between cytosine and guanine, there should be an equal number of the two. Each one of the pair is called the complementary base for the other, e.g., adenine is the complementary base for thymine. When the new DNA is synthesised it must have this type of pairing between the complementary bases.

The elucidation of the complete sequence of nitrogen bases in DNA has been one of the difficult problems in modern biochemistry.

Kornberg has developed a technique by which the frequencies of the nearest neighbours in respect of bases in DNA could be known. This is called 'nearest neighbour analysis'. In this, a study is made on how many adenines, for instance, are nearest to how many thymines or how many other nitrogen bases. The 'nearest neighbour' frequency is specific for the DNA of different species. Recently, Khorana has synthesised a very small molecule of DNA (gene) — the one corresponding to alanine transfer RNA and containing 77 nucleotides. This is the first total synthesis of DNA with information on the complete sequence of nitrogen bases in the biochemistry of DNA.

Very recently, Sanger invented an elegant technique for the sequence studies of DNA, for which he got the Nobel Prize. He uses what is called the plus - minus technique, i.e., in one case adding excess of one trinucleotide and in another excluding it. Labelled (^{32}P), DNA molecules of different lengths were synthesised from the primer. These were separated by poly acrylamide get disc electrophoresis and identified by autoradiography. From the knowledge of DNA of different lengths, the sequence of the primer was arrived at.

Though normally DNA in most organisms is double-stranded, Sinsheimer has observed single-stranded, small, cyclic DNA in $\emptyset$X174 bacteriophage. This DNA has a relatively low molecular weight and does not obey Chargaff's rules. A small amount of DNA present in the mitochondria of cells is reported to be a closed molecule, and double-stranded.

Tertiary structure of DNA: DNA filaments are held by dense cylindrical flat-faced particles called nucleosomes. Chromatin is the seat of such nucleosomes. Each nucleosome is a heterogeneous assembly of the basic protein histone and is an octamer consisting of two each of histones 2A, 2B, 3 and 4. The octamer has a flat face, 10 nm in diameter and an edge, 5.5 nm in diameter. A long double-stranded DNA is coiled as a left handed helix over the edge of the histone octamer.

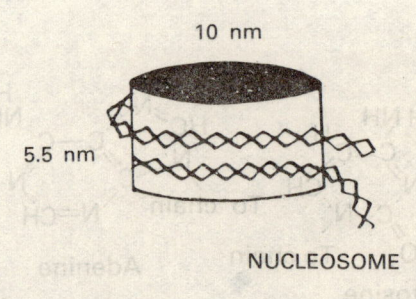

NUCLEOSOME

There will be 1.75 super-helical turns of DNA around the surface of the octamer protecting the 146 base pairs of DNA and forming the nucleosome core. Histone 1 (H 1) binds that portion of DNA which enters and leaves the adjacent nucleosome cores.

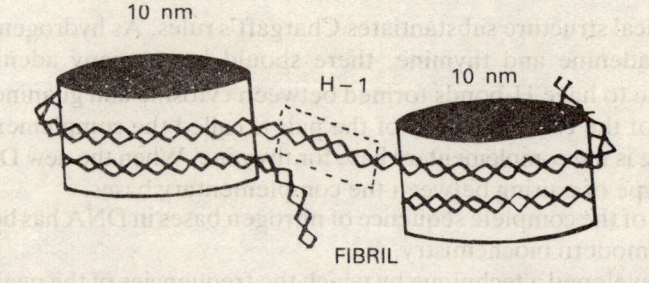

FIBRIL

In addition to histones there are also non-histone proteins most of which are acidic and larger than histones. Histones can bind non-specifically to the strongly anionic DNA by forming salt bridges.

The nucleosomes with the coiled DNA are arranged in such a way that they form the 10 nm fibrils of chromatin. The 10 nm fibrils are associated to form the 30 nm chromatin fiber. Each turn of the supercoil consisting of 6-7 nucleosomes would be relatively flat and the faces of the nucleosomes of successive turns would be nearly parallel to each other.

NUCLEOPROTEINS 75

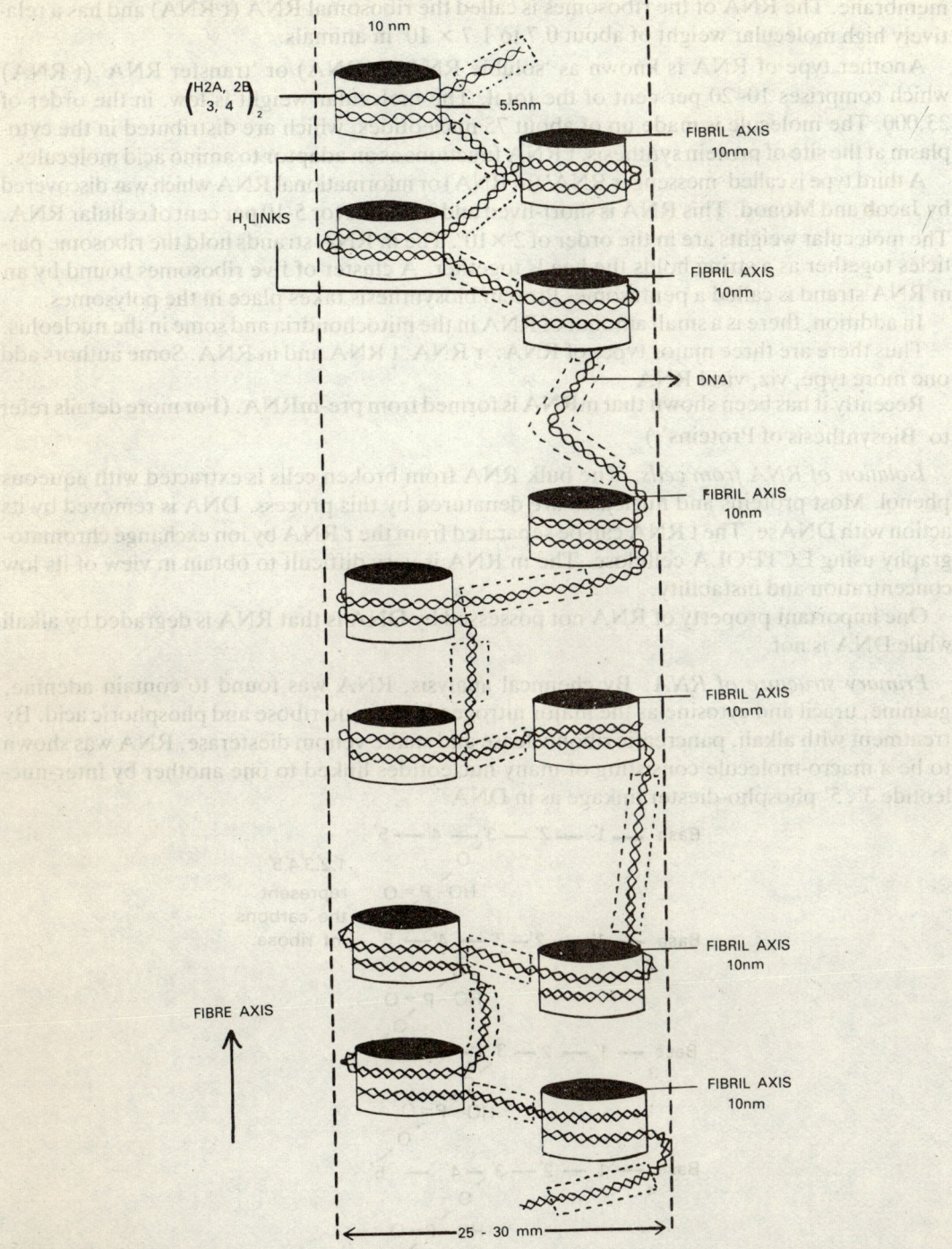

$(H2A, 2B, 3, 4)_2$

10 nm
5.5nm
FIBRIL AXIS 10nm
H_1 LINKS
FIBRIL AXIS 10nm
FIBRIL AXIS 10nm
DNA
FIBRIL AXIS 10nm
FIBRIL AXIS 10nm
FIBRIL AXIS 10nm
FIBRIL AXIS 10nm

FIBRE AXIS ↑

← 25 - 30 mm →

RNA (ribonucleic acid)

RNA is found in the cytoplasm. The largest fraction, i.e., 50-80 per cent of the total RNA is associated with small particles called ribosomes about 100-Å in diameter. They contain about 50 per cent protein and 50 per cent RNA. The ribosomes, may be free in the cytoplasm, arranged in clusters called polysomes, or they may be attached to the endoplasmic reticular membrane. The RNA of the ribosomes is called the ribosomal RNA (r RNA) and has a relatively high molecular weight of about 0.7 to 1.7×10^6 in animals.

Another type of RNA is known as 'soluble RNA' (s RNA) or 'transfer RNA' (t RNA) which comprises 10–20 per cent of the total. The molecular weight is low, in the order of 23,000. The molecule is made up of about 75 nucleotides, which are distributed in the cytoplasm at the site of protein synthesis. t RNA functions as an adaptor to amino acid molecules.

A third type is called 'messenger RNA' (m RNA) or informational RNA which was discovered by Jacob and Monod. This RNA is short-lived and accounts for 5-10 per cent of cellular RNA. The molecular weights are in the order of 2×10^6. The m RNA strands hold the ribosome particles together as a string holds the beads together. A cluster of five ribosomes bound by an m RNA strand is called a pentasome. Protein biosynthesis takes place in the polysomes.

In addition, there is a small amount of RNA in the mitochondria and some in the nucleolus.

Thus there are three major types of RNA: r RNA, t RNA and m RNA. Some authors add one more type, viz, viral RNA.

Recently it has been shown that mRNA is formed from pre-mRNA. (For more details refer to 'Biosynthesis of Proteins'.)

Isolation of RNA from cells: The bulk RNA from broken cells is extracted with aqueous phenol. Most proteins and nucleases are denatured by this process. DNA is removed by its action with DNAse. The t RNA can be separated from the r RNA by ion exchange chromatography using ECTEOLA cellulose. The m RNA is very difficult to obtain in view of its low concentration and instability.

One important property of RNA not possessed by DNA is that RNA is degraded by alkali while DNA is not.

Primary structure of RNA: By chemical analysis, RNA was found to contain adenine, guanine, uracil and cytosine as the major nitrogen bases, and ribose and phosphoric acid. By treatment with alkali, pancreatic ribonuclease and snake venom diesterase, RNA was shown to be a macro-molecule consisting of many nucleotides linked to one another by inter-nucleotide 3', 5' phospho-diester linkage as in DNA.

```
Base —— 1' —— 2' —— 3' —— 4' —— 5'
 1                        O
                    HO— P = O       1'.2'.3'.4'.5'
                          O         represent
                                    the carbons
Base —— 1' —— 2' —— 3' —— 4' —— 5'  of ribose.
 2                        O
                    HO— P = O
                          O
Base —— 1' —— 2' —— 3' —— 4' —— 5'
 3                        O
                    HO— P = O
                          O
Base —— 1' —— 2' —— 3' —— 4' —— 5'
 4                        O
                    HO— P = O
```

Secondary structure of RNA: The size and shape of RNA varies with the origin and functions. The m RNA, t RNA, r RNA and viral RNA have different sizes. Study of the secondary structure by X-ray diffraction patterns, hypochromism, optical rotation and viscosity showed RNA to be less organised than DNA and single-stranded. The single-stranded RNA is coiled to take the 'clover' leaf structure. Though there is a single strand, in certain regions there is

NUCLEOPROTEINS

a helical portion and hydrogen bonds between possible bases. The 3' terminus of all RNAs is the same, i.e., cytosine-cytosine-adenine. The amount of methylated bases is relatively higher in RNA. In addition to uridine, there is also pseudo-uridine. The structure of alanine tRNA is schematically represented below:

Structure of typical tRNA

Rare occurrence of thymine at 23rd position; amino acid attaches to 3' adenosine.

In addition to the clover leaf structure, other types of secondary structures for different t-RNA molecules are also suggested. A few of them are:

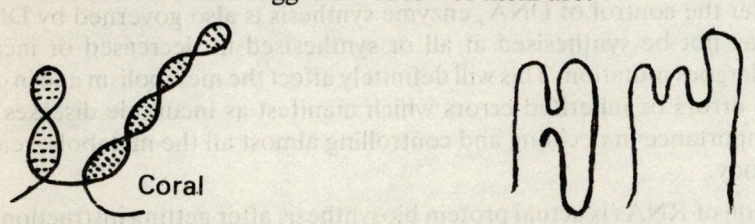

Coral

Sequence studies on RNA: The complete sequence of alanine tRNA was determined by Holley and co-workers. Holley got the Nobel Prize for the same as this type of work is a 'first' and very significant. The nucleotide sequence of yeast tRNA for tyrosine was later reported by Madison and others.

RNA was subjected to different degrees of hydrolysis by pancreatic ribonuclease and Takadiastase ribonuclease TI; the fractions were identified by chromatography. Individual fractions were also degraded by alkaline phosphatase, snake venom diesterase and finally alkali. Only on the basis of the sequence studies on alanine tRNA by Holley and co-workers, Khorana was able to synthesise the DNA (gene) corresponding to this alanine t RNA.

RNA protein complex: The rRNA is in combination with proteins in ribosomes. The proteins are rich in arginine and lysine.

In both DNA-protein and RNA-protein complexes, there exist ionic bonds between the positively charged protein and the negatively charged nucleic acid.

$$DNA^- {}^+protein: \quad RNA^- {}^+protein.$$

In the cells, the nucleic acids occur as nucleoproteins only. The nucleoproteins when hydrolysed with acid and enzymes give the various components shown below:

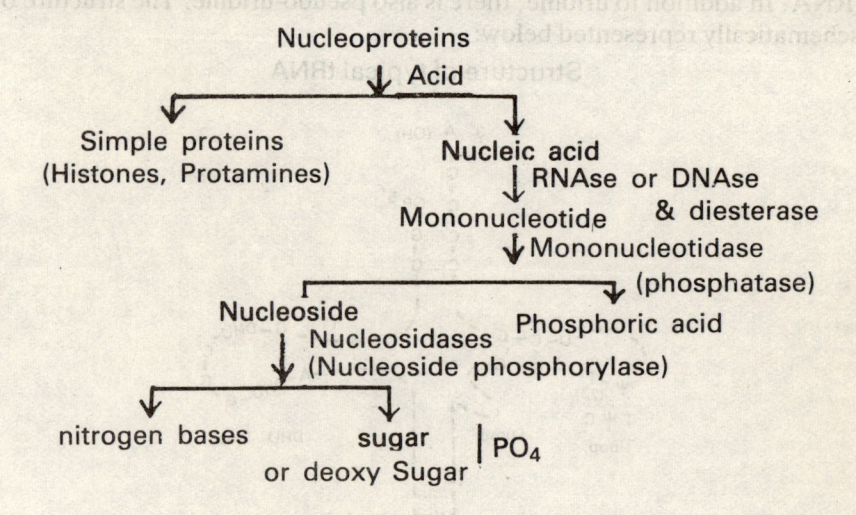

```
                        Nucleoproteins
                              ↓ Acid
          ↙                                    ↘
   Simple proteins                          Nucleic acid
 (Histones, Protamines)                      ↓ RNAse or DNAse
                                Mononucleotide  & diesterase
                                     ↓ Mononucleotidase
                   ┌─────────────────────────────────┐
                   │                              ↘ (phosphatase)
              Nucleoside                       Phosphoric acid
                   ↓ Nucleosidases
                   (Nucleoside phosphorylase)
          ↙                         ↘
    nitrogen bases              sugar      │ PO4
                             or deoxy Sugar
```

BIOLOGICAL IMPORTANCE OF NUCLEIC ACIDS

DNA, as already pointed out, has fundamental importance in the living systems as it is the genetic material spelling out the genetic message. It can safely be said that DNA is the chief for all biochemical operations in the living body. This is because all the biochemical reactions have to be catalysed at some stage or other by enzymes which are proteins. As the synthesis of all proteins is under the control of DNA, enzyme synthesis is also governed by DNA. A required enzyme may not be synthesised at all or synthesised in decreased or increased amounts if DNA undergoes mutation. This will definitely affect the metabolism and in certain cases lead to inborn errors or inherited errors which manifest as incurable diseases. Thus DNA is of utmost importance in deciding and controlling almost all the metabolic reactions taking place in the body.

RNAs: The function of RNAs is actual protein biosynthesis after getting instructions from DNA. The m RNA gets the information from DNA. The tRNAs carry the amino acids as adaptor molecules for polypeptide synthesis while the rRNA binds the m RNA strand to the ribosomes which are the sites of protein synthesis. No protein synthesis (and also enzyme) is possible in the absence of the three types of RNAs.

BIOLOGICALLY IMPORTANT NUCLEOTIDES

Apart from those nucleotides which are an integral part of the DNA and RNA molecule, there are many biologically important nucleotides present in tissues known to have other diverse biochemical functions. In this class are the adenosine nucleotides ATP, ADP, AMP, cyclic AMP, guanosine nucleotides GTP, GDP, GMP, cyclic GMP, the uridine nucleotides UTP, UDP, UMP, UDPG, coenzymes like NAD^+, $NADP^+$, FAD, FMN, etc. These nucleotides are involved in energy exchange in cells, biological oxidation and reduction, and in various steps in the metabolic pathways.

Adenosine-containing nucleotides

1) ATP: ATP may be called the 'storage battery' of the tissues. It is synthesised during the oxidative phosphorylation process and broken down when energy is needed for endergonic processes. In the ATP molecule, two of the three phosphate residues on hydrolysis release considerable energy (7.4–8.3 kcals) ATP donates phosphate for a variety of phospho-transferase reactions, e.g., the hexokinase reaction for the phosphorylation of glucose. Here ATP is converted to ADP during the phosphate-transfer. Many synthetase reactions require energy which is obtained in the splitting of ATP to AMP and PPi, e.g., argininosuccinate synthetase reaction in the urea cycle. ATP is also required in the synthesis of creatine phosphate from creatine, fatty acids from acetyl CoA, peptides and proteins from amino acids, the formation of glucose from pyruvate as well as oxaloacetate (gluconeogenesis) and other processes. ATP is an important source of energy for muscle contraction, transmission of nerve impulses, transport of nutrients across cell membranes and motility of the spermatozoa.

In vivo: ATP is converted to ADP, AMP or 3'-5' cyclic AMP, which have important roles to play in biochemical processes.

2) AMP: Adenylic acid or adenosine monophosphate (AMP) obtained by the action of RNAse and phosphodiesterase on RNA or hydrolysis of ATP or ADP is the same as the adenylic acid found in muscle. AMP was the first naturally occurring mononucleotide discovered by Embden in 1927. Embden observed that the muscle adenylic acid, however, was not identical with the yeast adenylic acid.

AMP is an activator of several enzymes in the tissues. In the glycolytic pathway, the enzyme phospho fructokinase is inhibited by ATP but the inhibition is reversed by AMP, the deciding factor for the reaction being the ratio between ATP and AMP. In resting muscle, AMP is formed from ADP according to the adenylate kinase reaction:

$$2\ ADP = ATP + AMP.$$

AMP can act as an inhibitor of certain enzymes, like fructose 1, 6-diphosphatase (FDPase) and adenylo succinate synthetase.

3', 5'-cyclic AMP: This compound, discovered by Sutherland, has a hormone-like action and is also an activator of enzymes, like the phosphorylases.

3', 5'-cyclic AMP is derived from ATP in a reaction catalysed by the enzyme adenylate cyclase and is ultimately destroyed in the tissues by the enzyme phospho-diesterase by conversion to 5'-AMP. Apart from activation of phosphorylase in the muscle, and liver, cyclic AMP converts glycogen synthetase a to b. Thus it increases glycogenolysis and decreases glycogenesis. It is involved in many other processes including gastric secretion, CNS function and release of various hormones from the corresponding endocrine glands. Cyclic AMP inhibits gastric secretion. In the adipose tissue, 3', 5' cyclic AMP activates a proteinkinase which in turn converts an inactive lipase to the active form and triggers lipolysis. The formation of cyclic AMP from ATP through the enzyme adenylate cyclase in the cells of a target tissue is effected by certain hormones, and cyclic AMP is the factor which may actually bring about the final effect due to the hormone. Thus cyclic AMP is called the 'second messenger', the hormone being the 'first messenger'. The actions of many of the hormones can be mimicked by cyclic AMP. Thus it has been demonstrated in experiments that the effect of epinephrine on contraction of the heart can be produced by 3', 5'-cyclic AMP.

ATP $\xrightarrow[\text{Mg}^{2+}]{\text{adenylate cyclase}}$ 3',5' cyclic AMP + PP

Cyclic AMP activates a protein kinase which catalyses phosphorylation and activation of the enzyme RNA polymerase. The activated RNA polymerase stimulates RNA production (mRNA) which takes part in protein synthesis. Cyclic AMP also stimulates the phosphorylation of histone of ribonucleoprotein and causes the induction of RNA synthesis. Thus a hormone which initiates production of cyclic AMP via adenylate cyclase reaction ultimately brings about protein synthesis.

The cAMP with catabolite activator protein (CAP) binds to the promoter site of DNA and causes the direct expression of genetic information by starting transcription through the formation of mRNA.

Apart from hormones, there are certain ions which stimulate the production of cyclic AMP. Thus Ca^{++} ions cause the activation of adenylate cyclase in the calcitonin-secreting cells of thyroid and the resultant cyclic AMP releases the calcium-lowering hormone, calcitonin, thus helping homeostasis.

In certain cases, cyclic AMP increases the permeability of the membrane for calcium and this is the mode of action of parathyroid hormone. Thus in many metabolic events cyclic AMP has a prominent role.

ADP: Adenosine diphosphate plays an important role as a primary phosphate acceptor in oxidative phosphorylation and photo-phosphorylation in addition to its effect on the control of cellular respiration, muscle contraction, etc.

ADP is known to be an activator of the enzyme glutamate dehydrogenase. Glutamate dehydrogenase reaction which is inhibited by GTP non-competitively, provides a negative feedback in the utilisation of amino acids as an energy source. This inhibition is reversed by ADP by which more substrate is added to the citric acid cycle as alpha ketoglutarate.

Adenosine 3', 5'-diphosphate, diphosphoadenosine (usually abbreviated as DPA) is different from ADP as the two phosphates are linked to two different carbon atoms of the ribose moiety and there is no high energy bond available. The biological importance of DPA has been recognised as a co-factor for the reactions involved in the process of bioluminescence in sea pansy (e.g., Renilla reniformis).

4) Phospho - adenosine phosphosulphate (PAPS): 3'-phosphoadenosine-5'-phosphosulphate (PAPS) or 'activated' sulphate plays a part in the reactions in the pathways of sulphate metabolism, as the co-factor for the various sulphatase enzymes. The sulphatases are enzymes which catalyse the introduction of a sulphate group to various molecules in the biosynthesis of heparin, chondroitin sulphates, keratosulphate and sulpholipids and also in the conjugation of phenols, indole and skatole to form ethereal sulphates.

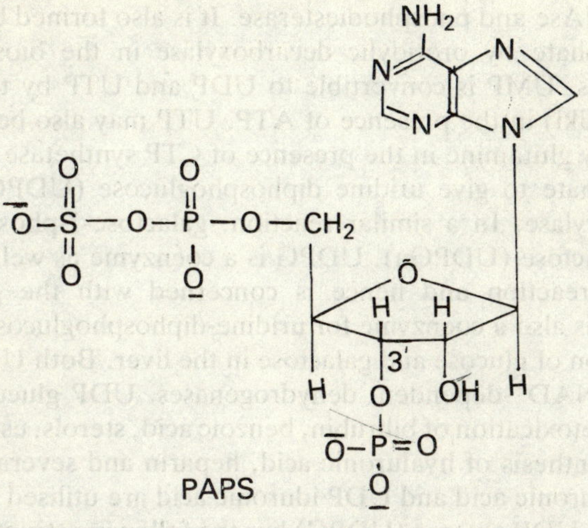

PAPS

5) Adenosine diphosphoglucose: ADP-glucose (ADPG) which carries a glucose moiety attached to the terminal phosphate group of ADP has been enzymatically synthesised and has been shown to play a role in the biosynthesis of starch in plants.

Coenzyme A: Coenzyme A is an important adenosyl nucleotide also containing pantothenic acid, beta alanine, beta-mercaptyl ethylamine. Discovered by Lipmann and shown to be an important coenzyme for a large number of reactions in the pathways in carbohydrate, lipid and protein metabolism. It has a very reactive sulphhydryl group attached to one end of the molecule. In combination with the acetyl group, it is available as acetyl CoA which links lipid metabolism with carbohydrate and protein metabolisms.

Pyridine coenzymes or nicotinamide nucleotides

Both nicotinamide adenine dinucleotide (NAD^+) and nicotinamide adenine dinucleotide phosphate ($NADP^+$) act as coezymes for a number of dehydrogenases. NAD^+ is the coenzyme for lactate dehydrogenase, malate dehydrogenase and alcohol dehydrogenase. $NADP^+$ is the coenzyme for glucose-6-phosphate dehydrogenase, malic enzyme, and isocitrate dehydrogenase (cytosol). In all cases, NAD^+ and $NADP^+$ act as H-acceptors getting themselves reduced to NADH and NADPH. NAD^+ has a chemical structure consisting of nicotinamide-ribose-phosphate-phosphate-ribose-adenine. In $NADP^+$, the third phosphate is attached to the 2' C of the ribose moiety of adenosine. ATP is required for the formation of these coenzymes. NAD^+ and $NADP^+$ are present in most natural sources, in plants and animal cells.

Riboflavin nucleotides: Flavin mononucleotide (FMN) and flavin adenine-dinucleotide FAD has, in addition, an AMP part attached to the phosphate of FMN. FMN is the co-factor for cytochrome C reductase, and dehydrogenases like L-amino acid dehydrogenase, while FAD is the co-factor for D-amino acid dehydrogenase, xanthine dehydrogenase and acyl CoA dehydrogenase.

Uridine-containing nucleotides: Uridine monophosphate (UMP) is obtained by the hydrolysis of RNA by RNAse and phosphodiesterase. It is also formed by the decarboxylation of orotidine 5′-phosphate by orotidylic decarboxylase in the biosynthetic pathway for pyrimidine nucleotides. UMP is convertible to UDP and UTP by the enzyme nucleoside diphosphokinase (nudiki) in the presence of ATP. UTP may also be converted to cytidine triphosphate (CTP) by glutamine in the presence of CTP synthetase and ATP. UTP reacts with glucose-1-phosphate to give uridine diphosphoglucose (UDPG) in the presence of UDPG pyrophosphorylase. In a similar reaction, galactose-1-phosphate and UTP form uridine diphosphogalactose (UDPGa). UDPG is a coenzyme as well as a substrate for the glycogen synthetase reaction and hence is concerned with the glycogen and glucose metabolisms. UDPG is also a coenzyme for uridine-diphosphoglucose-galactose epimerase for the inter-conversion of glucose and galactose in the liver. Both UDPG and UDPGa can be oxidised to UDP-NAD^+-dependent dehydrogenases. UDP glucuronic acid is used for the conjugation and detoxication of bilirubin, benzoic acid, sterols, estrogens and drugs. It is also used in the biosynthesis of hyaluronic acid, heparin and several other muco-polysaccharides. UDP galacturonic acid and UDP iduronic acid are utilised for the biosynthesis of chondroitin sulphates. UDP glucose (UDPG) has the following structure :

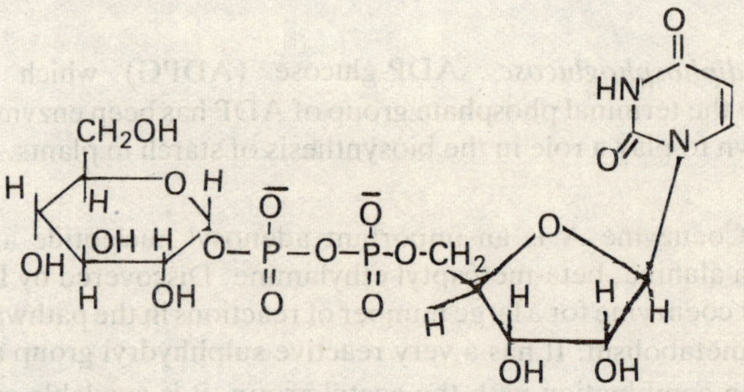

Cytidine-containing nucleotides: These include CMP, CDP, deoxy CDP derivatives of glucose, ribitol, choline, glycerol, sialic acid, etc. CDP-choline, CDP-glycerol and CDP-ethanolamine are involved in the biosynthesis of phospholipids. CMP acetyl neuraminic acid is an important precursor of cell wall polysaccharides in bacteria. CMP-sialic acid is present in salivary glands and may be concerned with the biosynthesis of salivary mucin.

Guanosine-containing nucleotides: The guanine analogue of ATP, i.e., GTP, is involved in metabolism. The oxidation of succinyl CoA in the citric acid cycle involves phosphorylation of GDP to GTP. Sugar derivatives of GDP like GDP glucose, GDP-mannose, GDP-galactose, GDP-mannuronic acid are all known. GTP is required for protein biosynthesis.

Cyclic GMP: 3′, 5′ cyclic GMP is also an important intra-cellular signal of extra-cellular events. It is formed from GTP by the action of the enzyme guanylate cyclase on GTP. Its formation does not appear to be under any hormonal control. In many cases, cyclic GMP acts in a manner antagonistic to that of **cyclic AMP**.

Guanylate cyclase is inhibited by **ATP**, but activated by Mn^{++} ions. A unique role of cyclic GMP on the excitability of the rods of the retina along with Ca^{++} ions has been found. The exposure of rhodopsin to light results in the activation of the phospho-diesterase that splits cyclic GMP to GMP.

Hypoxanthine nucleotides: Hypoxanthine mononucleotide (IMP), usually called inosinic acid, is derived from adenylic acid by deamination in the muscle tissue. Inosinic acid may interact with ATP to give ITP which may in turn be converted to IDP. Both ITP and IDP participate in phosphorylation reactions. Inosinic acid is an intermediate in the synthesis of adenylic acid and guanylic acid. ITP promotes the conversion of oxaloacetate to phosphoenol-pyruvate in the PEP carboxykinase reaction for gluco-neogenesis.

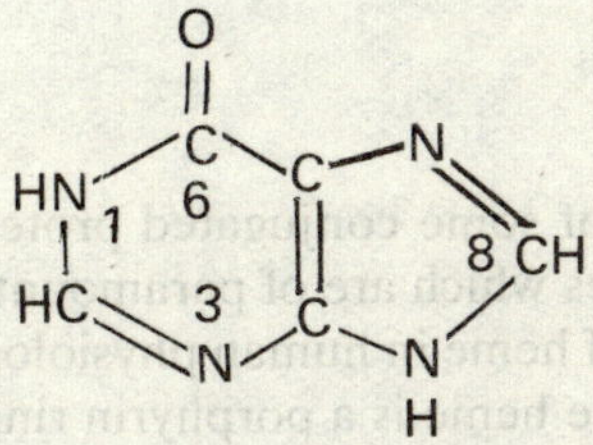

BIOLOGICALLY IMPORTANT NUCLEOSIDES

Active methionine (S-adenosyl methionine) of Cantoni is the adenine nucleoside of methionine and is a powerful methyl donor. For example, it is required for the synthesis of choline, creatine, epinephrine etc.

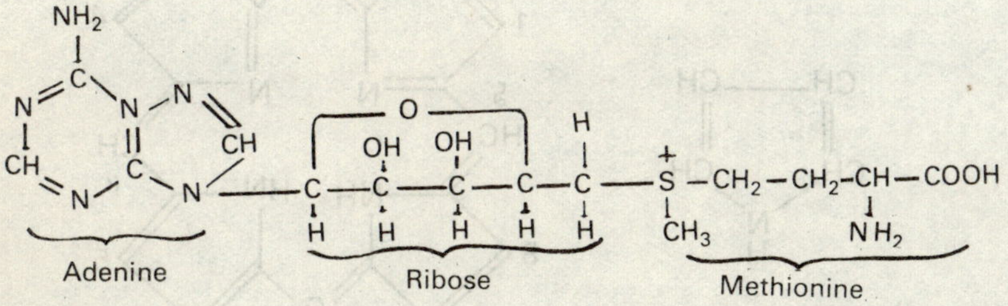

A few other examples are cobamide coenzymes containing adenosyl moiety in place of CN, adenosine, guanosine, cytidine, thymidine, uridine, etc.

IDENTIFICATION OF NUCLEIC ACIDS

RNAs have been identified histologically by the orcinol colour reaction for pentoses while DNA is recognised by the colour reaction with Feulgen's reagent (a fuchsin dye in sulphurous acid). Another method uses the intense absorption of ultraviolet light by the nitrogen bases of the nucleic acids. The purine and pyrimidine bases of nucleic acids exhibit maximum absorption in the region of 260 mμ. Hence with the help of ultraviolet spectrophotometer, nucleic acids can be estimated.

The DNA and RNA contents of a tissue may be determined by warming the sample with dilute alkali. Only the RNA is degraded to nucleotides. On subsequent treatment with acid, the nucleotides formed from RNA are soluble in acid while the DNA is unhydrolysed and is precipitated by acid. The sugar or phosphate content of the acid soluble fraction gives a measure of RNA while that of the acid insoluble fraction measures DNA.

6 Hemoglobin

Heme is the prosthetic group of some conjugated proteins in the body, like hemoglobin, myoglobin, and the cytochromes which are of paramount importance in respiration. Hence it is needless to stress the role of heme in human physiology and biochemistry. It is also present in the enzyme catalase. The heme is a porphyrin ring containing iron in ferrous condition.

The porphyrin ring is derived from porphin which incorporates 4 pyrrole rings connected by methine (–CH=) bridges. These are large flat heterocyclic ring structures with conjugated double bonds.

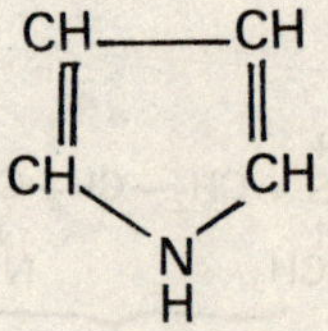

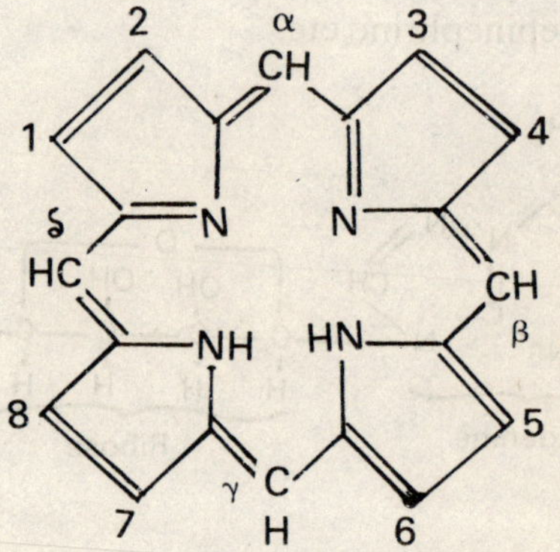

Porphin

When the porphin which has eight available positions accommodates substituents, it becomes porphyrin.

$$\text{pyrrole} \xrightarrow{(-CH=) \text{ bridge}} \text{porphin} \xrightarrow{} \text{porphyrin (substituents)} \xrightarrow{Fe} \text{heme}$$

Porphyrinogens are similar to porphyrins but have methylene (–CH$_2$–) bridges connecting the pyrroles.

Depending upon the type of arrangement of substituents, two main types of porphyrin derivatives are possible. They are Type I and Type III. In the body, important substances like heme are Type III though some Type I and III compounds, viz. coproporphyrin I and III, are excreted in urine and feces.

HEMOGLOBIN

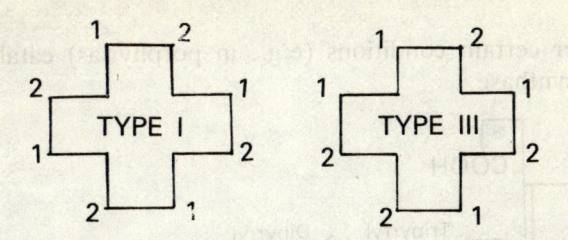

Schematic diagram to show Types I and III

The porphyrins, by nature are basic substances as they contain tertiary nitrogen atoms. However, presence of –COOH in the side chain can alter their acid-base behaviour.

Biosynthesis of porphyrins

By using labelled glycine, it has been shown that glycine is utilised for the biosynthesis of porphyrins in respect of its $-NH_2$ and $-CH_2$ groups while its $-COOH$ group is not utilised.

Glycine condenses with succinyl-CoA (active succinate) in a reaction requiring pyridoxal phosphate. The hypothetical product alpha amino beta keto adipic acid loses CO_2 spontaneously to form δ-amino levulinic acid (δ-ALA) thereby eliminating the glycine carboxyl carbon atom. Synthesis of delta ALA takes place in the mitochondria.

Two molecules of delta-ALA condense to form **porphobilinogen**, catalysed by the enzyme aminolevulinic acid dehydratase. This enzyme is cytosolic and contains zinc. It is very sensitive to inhibition by Pb and is affected during lead poisoning.

Porphobilinogen

Four molecules of porphobilinogen undergo deamination and simultaneous polymerisation catalysed by the enzyme uroporphyrinogen III cosynthase to form uroporphyrinogen III. Though uroporphyrinogen III is the normal product, another side product uroporphyrinogen I is also formed under certain conditions (e.g., in porphyrias) catalysed by the enzyme uroporphyrinogen I synthase.

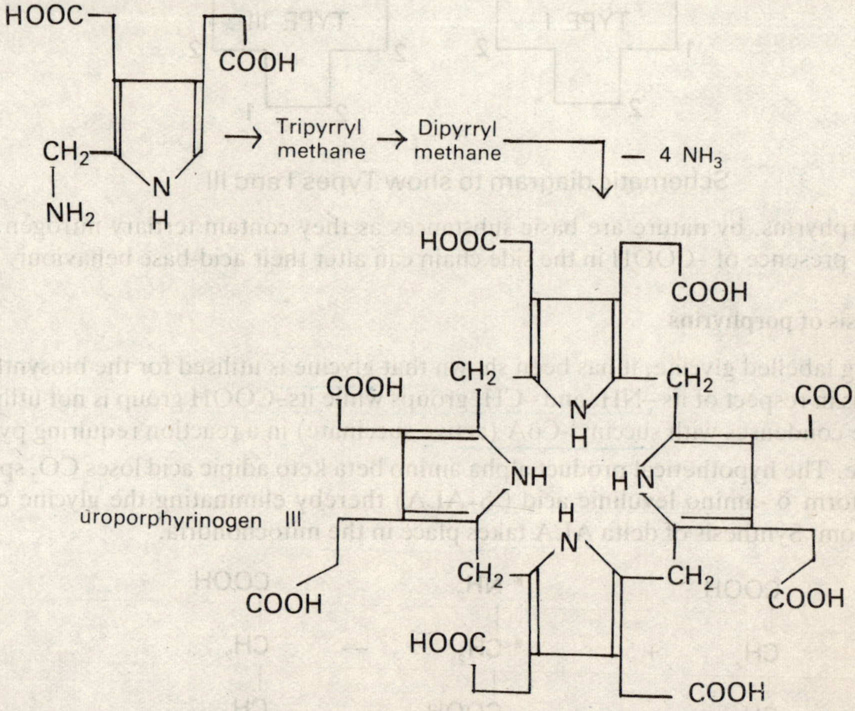

[Structure: pyrrole with CH$_2$NH$_2$, HOOC and COOH groups] → Tripyrryl methane → Dipyrryl methane → − 4 NH$_3$ ↓

uroporphyrinogen III

Uroporphyrinogen III undergoes decarboxylation in the acetate groups which change to methyl groups to form proto-porphyrinogen III, catalysed by uroporphyrinogen decarboxylase. The protoporphyrinogen III is further oxidatively decarboxylated in the propionic acid residues to form protoporphyrin III by the enzyme protoporphyrinogen oxidase.

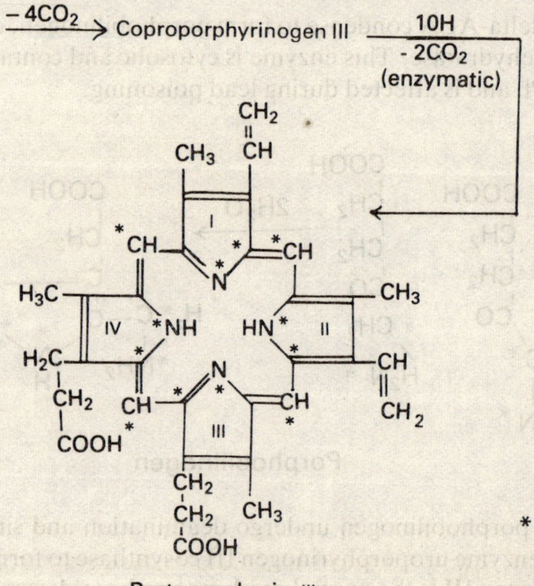

$$\xrightarrow{-4CO_2} \text{Coproporphyrinogen III} \xrightarrow[\substack{-2CO_2 \\ \text{(enzymatic)}}]{10H}$$

Protoporphyrin III

* from glycine, the others are from active succinate

Protoporphyrin as could be seen from the diagram, has four methyl, two vinyl and two propionic acid side chains. This, with ferrous iron, forms heme catalysed by enzyme heme synthetase or ferrochelatase. In this process, two hydrogens of the nitrogen atoms of 2 pyrrole are replaced by the ferrous (divalent) iron which is also linked to the two other N atoms of the porphyrin by chelation.

Heme

Uroporphyrin and coproporphyrin are different only in the nature of side chains.

Side chain in the 8 positions of porphin

	1	2	3	4	5	6	7	8
Uroporphyrin I	A	P	A	P	A	P	A	P
Uroporphyrin III	A	P	A	P	A	P	P	A
Coproporphyrin I	M	P	M	P	M	P	M	P
Coproporphyrin III	M	P	M	P	M	P	P	M
Proto-porphyrin III	M	V	M	V	M	P	P	M

$A = -CH_2 - COOH$; $P = -CH_2 - CH_2 - COOH$.
$M = -CH_3$ $V =$ Vinyl

Porphyrins in the body

In normal individuals, Type III isomers are relatively abundant. Porphyrins are coloured compounds while porphyrinogens are not. The porphyrins in strong mineral acids or organic solvents emit red fluorescence under ultraviolet light. This characteristic property is used for the detection of porphyrins, say in urine.

Porphyrins are found in urine, feces, bile, blood, and bone marrow. In blood most of the porphyrins are in the red blood cells in the form of proto-porphyrins (about $20\mu g$) while the concentration of coproporphyrins is negligible (about $1\mu g$). Normal urine contains about 166 $\pm 45\mu g$ of coproporphyrins excreted per day (slightly less for females) of which about 30 percent is Type III. Fecal co-proporphyrins are 300 to $1100\mu g$ per day of which about 70 to 90 per cent is Type I. Excretion of uroporphyrins is insignificant (about $20\mu g$) each in urine and in feces). Porphobilinogen amounts to 1-1.5 mg/24 hr. Isotopic tracer experiments have shown that the excreted porphyrins are not derived from Hb but are only the by-products of porphyrin synthesis.

It is the porphyrinogens which are the intermediate compounds formed during the biosynthesis of protoporphyrin. The porphyrins which are excreted are the oxidation products of these porphyrinogens.

The colour of porphyrins is due to the Resonance of the molecule comprising conjugated double bonds. They have specific absorption maxima, a property which is used in the spectroscopic analysis of Hb and its derivatives.

As pyridoxal phosphate is required in the synthesis of heme, its deficiency will lead to anemia similar to that of iron deficiency anemia (hypochromic, microcytic) in spite of the availability of enough iron in the body.

Heme ($C_{34}H_{32}O_4N_4Fe^{++}$) and related compounds

Heme and its derivatives are the co-ordination complexes of porphyrins with iron. Heme is the prosthetic group of hemoglobin. It contains iron in ferrous form. While the iron atom displaces two protons from the two NH groups when it enters the ring, it produces an equivalent of two negative charges on the two N atoms. It appears that these negative charges become distributed between all the N atoms due to Resonance. Hence the four valencies of Fe are equivalent.

HEMOGLOBIN (Hb)

The combination of heme with globin through different linkages like salt linkages (polar) and van der Waal's forces gives the red pigment, hemoglobin, which does the vital function of transport of oxygen and to a minor degree CO_2 during respiration. The oxygen that is inhaled should be taken to each cell to be used for cellular respiration by mitochondria in the utilisation of nutrients like glucose. Heme alone without globin cannot combine with oxygen reversibly.

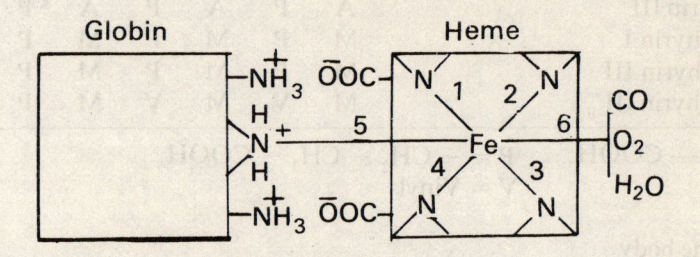

The organisation of Hb molecule from the peptide chain and heme has been extensively studied by X-ray analysis. The globin is made up of four polypeptide chains of which two are alpha and two beta in the normal adult hemoglobin (Hb A). Alpha chain has 141 aminoacids while beta 146, each with a definite sequence. If an aminoacid is displaced or the sequence of amino acid is disturbed, abnormal hemoglobins (like Hb S) will result, and it is a 'molecular disease' (Pauling).

Human hemoglobin contains 0.34 percent iron which corresponds to a molecular weight of 16,400. But osmotic pressure measurements give a molecular weight of about 65,000 for hemoglobin (suggesting four iron atoms per molecule of Hb). According to this, there should be four heme with four globin peptide chains each combined with a heme and all held together in a definite arrangement or conformation by H-bonds, salt linkages, van der Waal's forces, etc. The heme groups are buried in pockets formed by the polypeptide chains. X-ray analysis has shown that the Fe of heme is linked to the imidazole nitrogen of the histidine in positions 58 and 87 in the case of alpha, while they are 92 and 63 in the case of beta and gamma. With these two extra linkages, Fe gets a co-ordination valency of 6.

HEMOGLOBIN

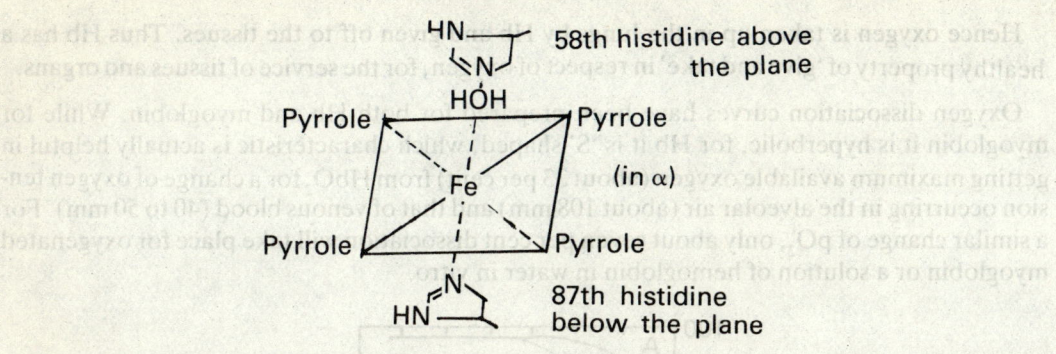

```
         HN─┐    58th histidine above
           ╲N╱         the plane
          HOH
 Pyrrole ┌────┐ Pyrrole
         │ Fe │        (in α)
 Pyrrole └────┘ Pyrrole
           N
         ╱N╲
        HN─┘    87th histidine
                 below the plane
```

In the formation of oxy Hb, the O-N bond being weaker, it breaks in such a way that Fe is bound only to the 87th histidine.

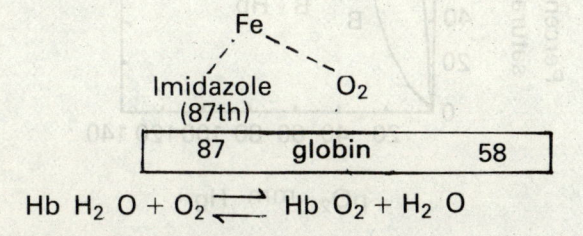

```
              Fe
             ╱   ╲
      Imidazole   O₂
        (87th)
      ┌──────────────────────┐
      │ 87      globin    58 │
      └──────────────────────┘
```

$$Hb\ H_2O + O_2 \rightleftharpoons Hb\ O_2 + H_2O$$

It is possible to dissociate the two alpha chains from the two beta chains by treatment with mild acid (decrease of pH). But on raising the pH to neutral value, the alpha and beta chains will combine to form the original globin. It is also possible to completely denature the globin while it is present in the molecule, or to detach and denature the same. Ferrous can be oxidised to ferric and new groups can be introduced by chemical means. By these a number of derivatives of Hb are obtained.

Oxyhemoglobin

The loose complex of Hb with O_2 is called oxyhemoglobin. But ferrous ion is not oxidised though oxygen is accommodated. The reaction is freely reversible. The molecule has greater acidic nature as well and this property is very important in the chloride-bicarbonate shift. X-ray diffraction studies have shown that the HbO_2 molecule is spheroid, 64 Å long, 55 Å wide and 50 Å high.

The oxygenation of Hb to HbO_2 is a typical reversible reaction and obeys the Law of Mass Action. Hence the amount of oxyhemoglobin is directly proportional to oxygen tension (partial pressure of oxygen). At high oxygen tension, as in the lungs, the equilibrium is shifted to the right with the formation of more of HbO_2. At a tension of 100 mm of Hg or more, Hb is completely saturated. Under these conditions, about 1.34 ml of oxygen is combined with each gm of Hb. For a concentration of about 14.5 gm/100 ml of blood, the total oxygen which would be carried as oxyhemoglobin would be 14.5 × 1.34 = 19.43 ml/100 ml. To this, an amount of 0.393 ml of physically dissolved oxygen is to be added making a total of about 20 volumes per cent. But in tissues where oxygen pressure is relatively less, the equilibrium is shifted to the left, favouring dissociation of HbO_2.

$$Hb + O_2 \underset{tissues}{\overset{lungs}{\rightleftharpoons}} HbO_2$$

Hence oxygen is taken up in the lungs by Hb and given off to the tissues. Thus Hb has a healthy property of 'give and take' in respect of oxygen, for the service of tissues and organs.

Oxygen dissociation curves have been prepared for both Hb and myoglobin. While for myoglobin it is hyperbolic, for Hb it is 'S' shaped, which characteristic is actually helpful in getting maximum available oxygen (about 33 per cent) from HbO_2 for a change of oxygen tension occurring in the alveolar air (about 108 mm) and that of venous blood (40 to 50 mm). For a similar change of pO_2, only about seven per cent dissociation will take place for oxygenated myoglobin or a solution of hemoglobin in water in vitro.

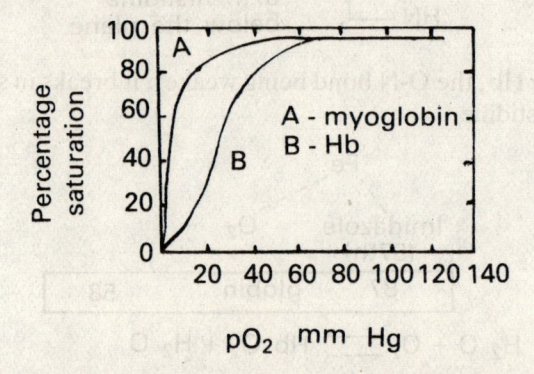

The difference between the two types of curves for myoglobin and Hb (though myoglobin has almost the same tertiary structure as Hb) is said to be due to the fact that the oxygenation of the four heme units in Hb is not independent. The equilibrium constant for the combination of oxygen with any one of the heme units is influenced by the state of oxygenation of the other units. Under physiological conditions the combination of Hb with O_2 takes place in four stages, these being simultaneous.

$$Hb_4 + O_2 \longrightarrow Hb_4O_2$$

$$Hb_4 + 2\,O_2 \longrightarrow Hb_4O_4$$

$$Hb_4 + 3\,O_2 \longrightarrow Hb_4O_6$$

$$Hb_4 + 4\,O_2 \longrightarrow Hb_4O_8$$

X-ray studies have also shown that when HbO_2 loses O_2, the two beta chains move apart by a distance of about 7 Å which is about 10 per cent of the diameter of the whole molecule (myoglobin does not exhibit this effect).

Hb can also combine to a limited extent with CO_2 to form carbamino Hb and by this process a little of CO_2 is carried by Hb towards the lungs.

$$Hb\,NH_2 + CO_2 \rightarrow Hb-NH-COOH$$

Changes occurring during binding of oxygen to T-form of hemoglobin

A molecule of oxygen can get into the heme pockets in the alpha subunits of hemoglobin. However, this is not easy in the heme pockets of the beta subunits as they are blocked by the Val residue.

When hemoglobin is in T-form (i.e. taut, tense or deoxygenated form), the Fe^{2+} is attached to the proximal Hist which is not in the plane of the porphyrin ring. When oxygen enters and binds to heme, the Fe^{2+} moves back into the plane of porphyrin ring along with the Hist. The Tyrosyl residues get wedged into a pocket between the F and H helices and this results in a number of salt bridges being broken within and between the subunits, thereby the T-form gets snapped and changes to the R-form (relaxed or oxygenated form). In the R-form, the hemoglobin molecule has a far greater affinity for oxygen and this is called 'cooperativity' or 'co-operative effect'. The block by the Val residue is removed and the binding of one or two oxygen molecules to the alpha subunits makes the binding of subsequent molecules of oxygen easier. In the peripheral tissues with less of oxygen tension, the oxygen starts getting released and again the Fe gets displaced above the plane of the porphyrin ring, displacing the F-helix simultaneously and reopening the space between the F and H-helices. This allows the tyrosyl residue to occupy the space. The reformation of the salt bridges leads to the snapping back of the hemoglobin to the T-form. When one or two molecules of oxygen are released, the same co-operative effect helps to push out the other oxygen molecules as well.

When oxygen is bound to the T-form, the breaking of the salt bridges generate protons (H^+ ions) which are released from the nitrogen atoms of the Hist of the beta chain. These H^+ ions are used up for the formation of carbonic acid by combination with HCO_3^- ions for releasing CO_2 into alveolar blood. On the other hand, H^+ ions from the peripheral tissues protonate the Hist residues of the beta chain and re-form the salt bridges and force the release of oxygen from the R-form of hemoglobin (oxygenated form). Thus an increase in protons causes oxygen release and an increase in oxygen causes proton release. This explains the Bohr effect.

Role of 2,3 - Diphosphoglycerate binding on hemoglobin function

2, 3-diphosphoglycerate (DPG) is formed from 1, 3-DPG, an intermediate of the glycolytic cycle in erythrocytes. It is known to play a role in the binding of hemoglobin to help release or unload oxygen in the peripheral tissues. One molecule of DPG is bound per hemoglobin tetramer, in its central cavity, in the T-form, when the space between the H-helices of the beta chains is just wide enough to accommodate it. The Val, Lys and Hist residues of the beta chains make salt bridges with DPG and thus the T-form gets stabilised.

DPG has the ability to reduce the oxygen affinity of hemoglobin. Binding of one molecule of DPG is equivalent to the unloading of four molecules of O_2 as per the following equation:

$$Hb\text{-}DPG + 4 O_2 \rightarrow Hb(O_2)_4 + DPG$$

DPG binding to fetal hemoglobin is very weak as compared to that with adult hemoglobin, because the amino acid His residue with which DPG is normally bound is replaced by Ser of the gamma chain of fetal hemoglobin. It is not possible to form salt bridges efficiently, and stabilise the T-form in such cases. Thus fetal hemoglobin has a greater affinity to oxygen than adult hemoglobin.

Abnormal hemoglobins

Alterations in the primary structure on the protein chains of hemoglobins, especially in the alpha and beta subunits, brought about by mutations, affect the biological function. We have thus a wide variety of abnormal hemoglobins identified.

Adult and fetal hemoglobins

The normal or adult hemoglobin HbA has four protein chains or subunits, two of them being designated as alpha and two of them as beta. The alpha chains have 141 amino acids while the beta chains have 146 amino acids. While the normal adult hemoglobin HbA is alpha$_2$ beta$_2$, there is a minor adult hemoglobin called HbA$_2$ which is alpha$_2$ delta$_2$ and which constitutes two per cent of the total hemoglobin. The fetus has another type of hemoglobin in the early embryonic stage with the subunit structure alpha$_2$ epsilon$_2$. This is followed later by the true fetal hemoglobin HbF which has the subunit structure alpha$_2$ gamma$_2$. The fetal hemoglobin gradually starts disappearing in the newborn infant and is replaced by adult hemoglobin.

Hemoglobin S

Hemoglobin S or sickle-cell hemoglobin, found in patients with sickle-cell anemia, has glut acid replaced in the beta chain by Val. As the polar glut acid is replaced by a non-polar amino-acid Val, it gives rise to a sticky patch on the outside of the beta chain. In the deoxygenated form, this sticky patch binds to a complementary patch and that enhances the polymerisation of deoxyhemoglobin S forming a fibrous insoluble material (tactoid). This distorts the shape of the erythrocytes to sickle shape. The formation of the polymer is not possible in the oxygenated state as the complements that help polymerisation are masked in that state. Thus, it is only the T-form of hemoglobin that is subjected to sickling. When the sickle cells penetrate the splenic sinusoids, they are liable to undergo lysis resulting in anemia. Sickle cell anemia is common in Central and Equatorial Africa where malaria is a major cause of death.

Spectroscopic examination of Hb and its derivatives

Hb (or reduced hemoglobin) has a characteristic absorption band – a single, broad, faint band with the mid-point at 559 λ in the yellow green region of the spectrum seen through a spectroscope. The solution is cherry red in colour.

Oxyhemoglobin (HbO$_2$) shows two bands in the green region at 578 λ and 542 λ. The band closer to the D line is narrow while the other one is broader. The solution is bright red in colour.

Oxygention of Hb is achieved quite simply; by just shaking Hb vigorously in air.

Reduced hemoglobin is obtained by treating HbO$_2$ with a reducing agent like sodium hydrosulphite. Fe is in the ferrous state in HbO$_2$

CARBOXY HEMOGLOBIN (HbCO)

Hb combines with carbon monoxide irreversibly (under normal conditions) to carboxy hemoglobin. Two bands are seen in the green region of the spectrum, (570 and 542) very much like the oxyhemoglobin bands, but shifted to slightly shorter wavelengths. Fe is in the ferrous state. The affinity of Hb for CO is about 200 times that for oxygen.

Carboxy hemoglobin HbCO and HbO$_2$ can be distinguished by the following means:

1. On treatment with sodium hydrosulphite, there is no change in the absorption bands of HbCO, while the two bands of HbO$_2$ are changed to one broad band as oxyhemoglobin is reduced to Hb.

2. Hartridge's Reversion Spectroscope, if used, could differentiate HbCO from HbO$_2$.

METHEMOGLOBIN (HEMATIN PLUS NATIVE GLOBIN)

Treatment of oxyhemoglobin with potassium ferricyanide oxidises ferrous to ferric, and methemoglobin is formed. The red colour of the solution turns brown. There is a prominent band in the red region, the centre of the band being 634 λ. There are also two faints bands in the green region. Other oxidising agents like ozone, $KMnO_4$, $KClO_3$ also would yield methemoglobin.

Methemoglobin can be reduced by sodium hydrosulphite to Hb. In blood, the little methemoglobin that is formed (less than 1 per cent) is reduced by glutathione and an enzyme methemolgobin reductase (a diaphorase) to hemoglobin.

HEMATIN

Acid and alkali hematins contain ferric iron and hence the hematin has one + ve charge to take an anion.

Acid hematin (hematin salt of acid): When blood is heated gently with glacial acid, acid hematin (hematin salt of acid) is formed due to the detachment of globin and the oxidation of ferrous to ferric ion. The ether extract containing acid hematin gives a prominent band in the red region with the centre at 640 λ .

Alkali hematin: When blood is boiled with con. ammonia, cooled and a little alcohol added, alkali hematin is formed. It has a faint broad band adjacent to the D-line with the centre at 600 λ . Formula of alkali hematin is $C_{34}H_{32}O_4N_4Fe^{+++}OH^-$ (hydroxy hemin). The globin gets detached.

Hemin ($C_{34}H_{32}O_4N_4Fe^{+++}Cl^-$) is the chloride of hematin. When Hb (of the blood) is heated gently with acetic acid and sodium chloride or Nippe's fluid (KCl with KBr and Kl in acetic acid), the globin is detached,

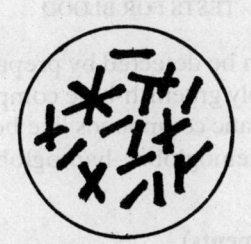

Hemin crystals

and the chloride ion gets attached to hematin. Hemin crystals have a characterisitic crystalline shape and can be identified under the microscope. These crystals can be prepared even from old samples of blood or blood stains in clothes and hence this test has medico-legal significance.

HEMATOPORPHYRINS (ACID HEMATOPORPHYRIN)

With con. acids like con. H_2SO_4, not only the globin but the iron of heme also gets detached. The unsaturated vinyl group of proto-porphyrin becomes saturated. The solution becomes purple. Two prominet bands are seen with centres at 606 λ and 558 λ , respectively.

HEMOCHROMOGEN

This is a very important derivative of hemoglobin. The importance lies in the fact that it absorbs light even in very great dilutions. A few ml of 1 in 100 dilution of blood are heated with a few drops of 5 per cent sodium hydroxide. The solution turns yellow forming alkali hematin. A pinch of sodium hydrosulphite is then added. The solution turns pink due to the formation of hemochromogen (heme + denatured globin). On spectroscopic examination, two bands are seen in the green portion of the spectrum. The band closer to the D-line is called the alpha band with the centre at 555 λ and the other band is called the beta band which is fainter, with the centre at 525 λ . In high dilutions (say 1 in 1000), only the alpha band will

be seen. This test is important in testing for the presence of traces of blood in clinical medicine and in forensic medicine in the detection of suspected blood stains.
The relation between various derivatives of hemoglobin is shown below:

Heme	Ferrous iron plus proto-porphyrin
Hematin	Ferric iron plus proto-porphyrin
Hemin	Chloride of hematin
Hemoglobin	Heme plus native globin
Methemoglobin	Hematin plus native globin
Hematoporphyrin	Proto-porphyrin with vinyl group saturated (no iron; globin detached and denatured).
Hemochromogen	Heme plus denatured globin
Acid hematin	Acid salt of hematin } globin detached and denatured
Alkali hematin	Alkali salt of hematin }

TESTS FOR BLOOD

As described previously, blood can be detected by preparing derivatives of Hb like hemochromogen and hemin crystals. Deeply greenish-blue compounds are formed in the presence of blood when H_2O_2 reacts with aromatic compounds like benzidine, o. tolidine, guaiacol, etc. This is due to the catalytic effect of hemoglobin–haptoglobin complex which has peroxidase activity.

Degradation products of Hb (Bile pigments)

It is estimated that there are about 750 gm of Hb in the total circulating blood of a man weighing 70 Kg and that about 6.25 gm are produced and destroyed every day. The degradation of Hb yields various bile pigments like bilirubin, biliverdin etc. The bile pigments consist of an open chain of four pyrrole rings joined by methylene or methine groups. They are derived from the proto-porphyrin IX by the oxidative breakdown of the alpha methine bridge to form biliverdin in the reticulo-endothelial cells of liver, spleen and bone marrow. Biliverdin is the first bile pigment formed. Biliverdin is easily reduced to bilirubin, the major pigment in human bile, which is transported to the liver in the plasma in combination with albumin. It is conjugated with glucuronic acid to form bilirubin diglucuronide in the liver, and to a smaller extent, bilirubin monoglucuronide.

(It appears that precursors of biliverdin, as they are formed, contain an open ring of porphyrin with attached globin.)

The conjugated bilirubin is discharged into the intestines from the liver through the bile. There, it is reduced by bacterial enzymes to urobilinogen identical with stercobilinogen. Some of the urobilinogen is oxidised to urobilin, and stercobilinogen to stercobilin. The urobilinogen not oxidised in the intestines is returned to the liver (entero-hepatic circulation) and oxidised to bilirubin which is re-excreted into bile. Normal urine therefore contains very little urobilinogen.

The urobilin formed normally in the intestine is identical with stercobilin. The stercobilin gets excreted through the feces and gives the brown colour to it.

HEMOGLOBIN

Circulation and excretion of bile pigments

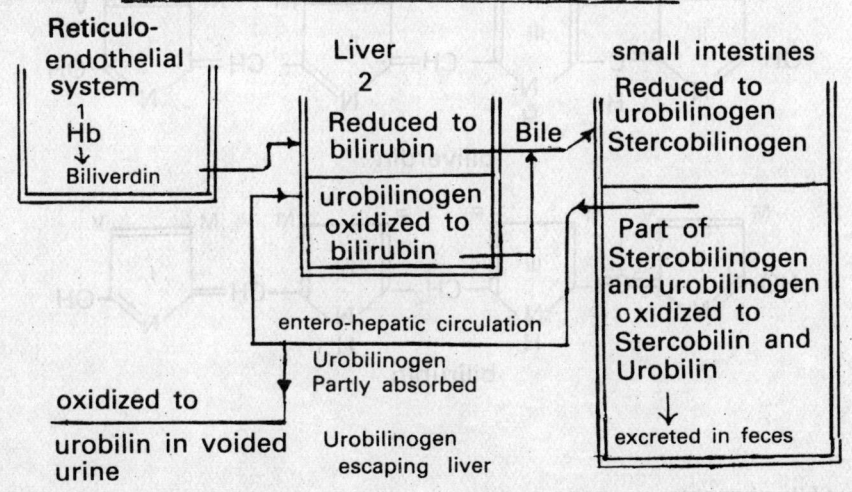

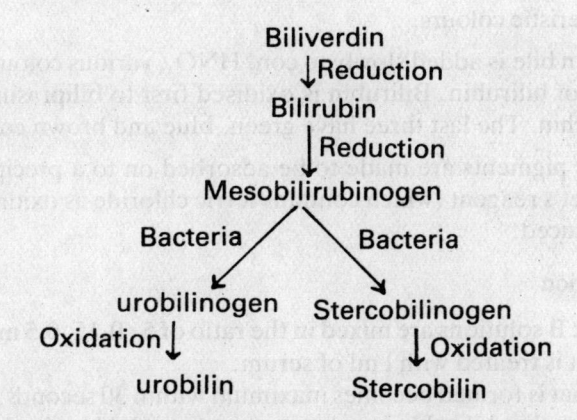

```
Reticulo-           Liver                small intestines
endothelial          2                   Reduced to
system           Reduced to    Bile      urobilinogen
  1              bilirubin  ────────→    Stercobilinogen
  Hb         ┌→ ──────────────────
  ↓          │  urobilinogen              ←─────────────
  Biliverdin │  oxidized to              Part of
             │  bilirubin                Stercobilinogen
             │                           and urobilinogen
             │  entero-hepatic circulation  oxidized to
             │           │                 Stercobilin and
             │           ↓                 Urobilin
                   Urobilinogen              ↓
  oxidized to      Partly absorbed        excreted in feces
  ─────────
  urobilin in voided    Urobilinogen
  urine                 escaping liver
```

Biliverdin
↓ Reduction
Bilirubin
↓ Reduction
Mesobilirubinogen
  Bacteria ↙   ↘ Bacteria
urobilinogen   Stercobilinogen
 ↓ Oxidation   ↓ Oxidation
 urobilin   Stercobilin

Structure of bile pigments

```
          V       H       M
      M  ┌─┐      C      ┌─┐  V
         │I│     α       │II│
          N               N
         HC               CH
          N               N
      M  │IV│            │III│ M
         └─┘     CH      └─┘
          P              P
```

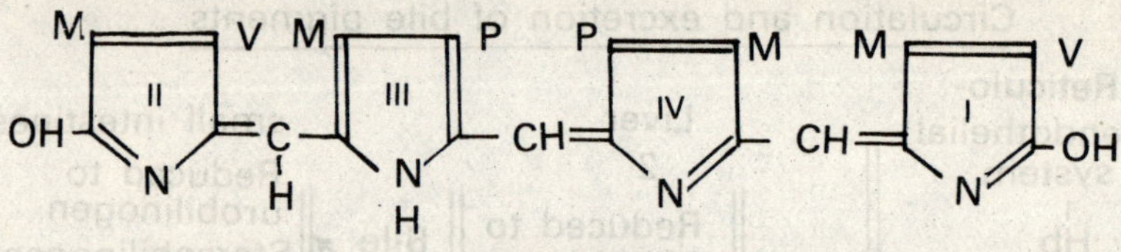

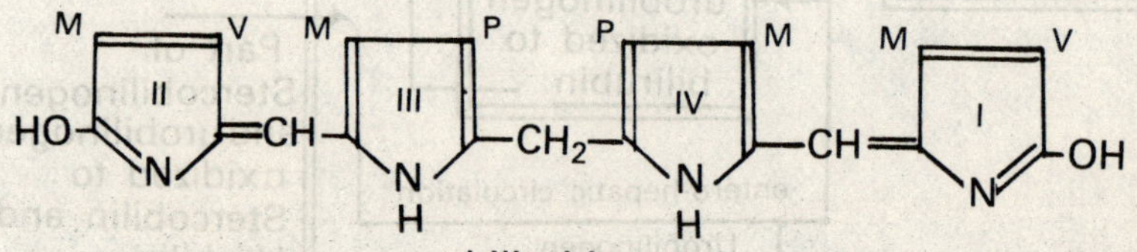

biliverdin

bilirubin

Reactions of bile pigments

When bile pigments are oxidised, different products of oxidation are obtained. These have got their own characteristic colours.

Gmelin's test: When bile is added slowly to con. HNO_3, various coloured rings are formed due to the oxidation of bilirubin. Bilirubin is oxidised first to biliprasin, then to biliverdin, bilicyanin and bilifuschin. The last three have green, blue and brown colours, respectively.

Fouchet's test: Bile pigments are made to be adsorbed on to a precipitate of $BaSO_4$. If a drop or two of Fouchet's reagent (which contains ferric chloride as oxidising agent) is added, a green colour is produced.

Van den Bergh's reaction

Van den Bergh A & B solutions are mixed in the ratio of 5 : 0.15. 0.5 ml of freshly prepared mixture of the reagent is treated with 1 ml of serum.

If the pink colour that is formed becomes maximum within 30 seconds, it is a direct positive reaction. This indicates that bilirubin is present as water soluble bilirubin glucuronide. The direct Van den Bergh reaction suggests obstructive jaundice (intra-hepatic or post-hepatic).

If no colour develops, 2.5 ml of methanol is added. If a pink colour develops, the reaction is indirect positive. This is a reaction for free bilirubin which is water insoluble but is made to dissolve in methanol for the test. An indirect positive Van den Bergh reaction is indicative of hemolytic jaundice or early hepato-cellular jaundice without obstruction.

If there is no colour after addition of methanol, the Van den Bergh reaction is said to be negative. If the bilirubin levels are normal, 0.1 - 0.8 mg/100 ml of serum, the reaction will be negative. Van den Bergh's solution A is sulphanilic acid in concentrated HCl. Solution B is 0.5% sodium nitrite.

The Van den Bergh test is very sensitive and is performed routinely in clinical biochemical laboratories for the diagnosis and differential diagnosis of obstructive and hemolytic jaundice.

7 Enzymes, Coenzymes and Isozymes

The biochemical reactions taking place in living cells at body temperature proceed at a sufficiently rapid rate in a regular order. Such reactions would have been extremely slow, had they not been catalysed by biocatalysts known as enzymes, which are present in every living cell. There are thousands of enzymes present in the living body meant for catalysing myriads of reactions. These reactions consist of processes involved in energy production and energy consumption, the gross effect appearing as the life process. Enzymes are all recognised as heat-labile proteins, highly specific in their action, some of them containing a non-protein prosthetic group also. Enzymes are synthesised in a living cell, but they can also act independently of the cell. Unlike inorganic catalysts, enzyme catalysts undergo a certain amount of destruction during their activity. Thus fresh synthesis of enzymes by the cell becomes necessary again and again.

The intra-cellular enzymes on dialysis can be separated into a protein portion called apoenzyme and a dialysable portion referred to as coenzyme. The active enzyme which comprises both apoenzyme and coenzyme is called holoenzyme. The term 'prosthetic group' is used for a non-dialysable part of a complex enzyme, the non-dialysable part functioning as a co-factor. In the case of certain digestive enzymes, e.g. pepsin of gastric juice, there is no coenzyme portion and the entire enzyme is a simple protein.

ENZYMES

Classification of enzymes

The name of any enzyme could be called by adding the suffix 'ase' to the substrates, e.g., protein and protease, amylum and amylase, maltose and maltase, etc. The nomenclature of enzymes now adopted is with reference to the class and type to which they belong and the reactions they catalyse according to the classification by the Enzyme Commission of the International Union of Biochemistry (IUB) system. According to this system, there are six classes of enzymes catalysing six major types of reactions. A code number or EC (Enzyme Commission) number is given to each enzyme with four numbers or premises, the first number denoting the class, the second number the sub-class, the third number the sub-sub-class and the fourth number the particular enzyme. For example, if the first number is 1, it indicates an oxido-reductase. If the second number is 1, it means the group which is oxidised is $> CH(OH)$. (The second number will be 3 if the group oxidised is $-CH_2-CH_2-$.) The third number stands for the coenzyme or cofactor required for oxidation. If NAD^+ is the coenzyme, the third number is 1, while it is 3 if oxygen is the co-factor. The fourth number just gives the serial number of the enzyme after taking into account the first three numbers. According to this, Lactate dehydrogenase is EC 1.1.1.27. The first number 1 shows that it is an oxido-reductase. The second number 1 shows that the group oxidised is $>CH(OH)$. The third number 1 shows that NAD^+ is the co-factor. The fourth number 27 is just the serial number of lactate dehydrogenase among those which have the first three premises 1.1.1.

(A few other examples are cytochrome oxidase EC 1.9.3.1. Succinate dehydrogenase. EC 1.3.99.1)

The main groups or classes to which enzymes are classified with their first numbers are as follows:

1) Oxido-reductases: These include dehydrogenases and oxidases which are concerned with biological oxidation. (EC 1).
    a) Dehydrogenases remove H from a substrate in the presence of the H-acceptor, e.g., lactate dehydrogenase.
    b) Oxidases activate molecular O_2. e.g., cytochrome oxidase.

2) Transferases: These include enzymes which catalyse the transfer of certain groups from one compound to another. (EC 2).
    a) Methyl groups: e.g., Transmethylase.
    b) Aldehyde or Ketonic groups: e.g., Transketolase.
    c) Acyl groups: e.g, Acyl transferase.
    d) Sugar groups: e.g., Sucrose 1-fructosyl transferase.
    e) Amino and Keto groups: e.g., Amino transferase or transaminase.

3) Hydrolases (Enzymes which add water to the substrate and hydrolyse or decompose it to give products. (EC 3).
    a) Lipases: e.g. Glycerol ester hydrolase.
    b) Phosphatases: e.g., Glucose-6-phosphatase.
    c) Choline esterases: e.g., Choline esterases which split acetyl choline.
    d) Peptide hydrolases: which split peptides to give rise to aminoacids.
    e) Nucleases, nucleotidases, nucleosidases acting on polynucleotides, mononucleotides and nucleosides, respectively. Nucleosidases while hydrolysing the nucleosides also phosphorylate the sugars.
    f) Carbohydrases which split disaccharides and polyaccharides. These include
        i) Amylases which split starch and glycogen.
        ii) Disaccharidases like invertase, lactase and maltase.
    g) Enzymes attacking C-N linkages.
        i) Urease which converts urea to ammonia.
        ii) Asparaginase.
        iii) Glutaminase.
        iv) Arginase.

4) Lyases: Enzymes that catalyse removal of groups from substrates giving rise to compounds with double bonds as intermediate products. (EC 4.)
    a) Decarboxylases, e.g., Oxaloacetate decarboxylase.
    b) Carbonic anhydrase.
    c) Cysteine desulfurase.

5) Isomerases: Enzymes which catalyse the interconversion of 1) optical isomers and 2) geometrical isomers. (EC 5.)
    a) Racemases or Epimerases: e.g., Alanine racemase
    b) Cis-trans isomerase, e.g., All trans retinene-isomerase.
    c) Intramolecular transferases, mutases, e.g., phosphoglyceromutase.

6) Ligases or synthetases: Enzymes catalysing the linking together of two compounds, e.g., Pyruvate carboxylase (EC 6).

Specificity of enzyme action

Enzyme action is specific, and this is because the enzyme attacks a compound at a definite type of linkage. This specific action of enzyme on the substrate has been compared to the action of a key on a lock by Emil Fischer. Thus, it is found that peptidases do not attack starches and amylases do not split fats. Urease acts only on urea. Maltase splits maltose and alpha glucosides. Emulsin on the other hand splits only beta glucosides but not the alpha. L-aminoacid dehydrogenase acts on L-amino acids only and not D-amino acids.

There are enzymes which have absolute specificity, i.e. an enzyme will act only on one substance, urease acts only on urea. Other examples are arginase, sucrase.

Some enzymes do not have absolute specificity but have relative specificity. The enzyme will act on chemically related substances, e.g., pepsin and trypsin will hydrolyse many different proteins.

L-amino acid dehydrogenase and D-amino acid dehydrogenase are enzymes which have configurational specificity. So also the alpha and beta dehydrogenases.

Properties of enzymes:

Enzymes are proteins and hence possess physical and chemical properties characteristic of proteins—solubility in water or salt solutions, formation of colloidal systems. Enzymes are globular proteins with molecular weights of 13,000-500,000. They act as amphoteric colloidal electrolytes and can be precipitated at their iso-electric points and with optimum concentration of salts. They show electrophoretic mobilities. Enzymes become inactive when heated to above 75°C (due to denaturation). They are also inactivated by the alteration of pH and undue dilution. When enzymes are injected into the bloodstream, they produce antibodies.

Extraction of enzymes from biological material

This requires disruption of the cell, which can be achieved by grinding with sand, chopping, high pressure, autolysis, and desiccation followed by pulverisation. The solvents used for extraction are saline, glycerol, alcohols (aqueous alcohols), buffers, dilute acids, dilute alkalies, etc. The extracts are then purified by dialysis, adsorption on suitable adsorbents and finally their elution. Gel filtration, electrophoresis and affinity chromatographic techniques are also employed. The polyacrylamide gel electrophoresis, one-dimensional and two-dimensional, has been used to separate even minor components of samples for the purification of enzymes. Several enzymes have now been obtained in pure crystalline state, e.g., urease, trypsin, chymotrypsin, catalase.

Factors influencing enzyme action

The rate or velocity of enzyme reactions is influenced by various factors. Contact between the enzyme and the substrate is very important for enzyme action. This is usually effected by good admixture, emulsification (in the case of fats), etc. The main factors influencing enzyme action are as follows:

1) *Concentration of enzyme:* Within fairly wide limits, the velocity of an enzyme reaction is directly proportional to the enzyme concentration. Hence, one method of speeding up an enzyme reaction is to increase the concentration of the enzyme preparation. If it is estimation of serum enzyme, the volume of fresh serum may be increased. The effect of enzyme concentration is of course subject to the availability of the substrate.

2) *pH (Hydrogen ion concentration):* The pH influences the velocity of enzyme reactions considerably. Each enzyme has an optimum pH, at which the reaction proceeds most rapidly. On either side of this optimum pH, the velocity of reaction will be lower and at a certain pH, the enzyme becomes completely inactive.

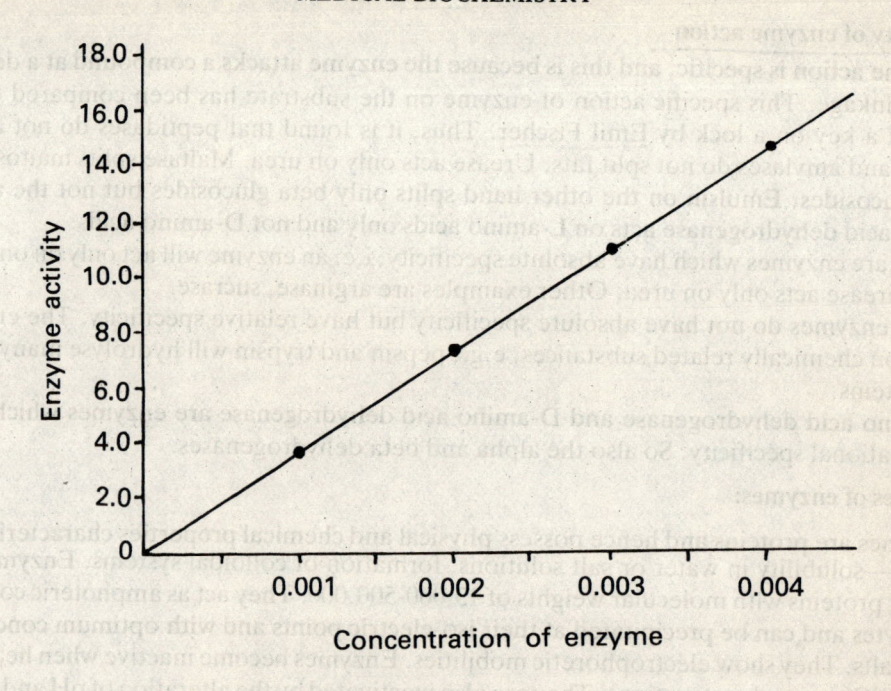

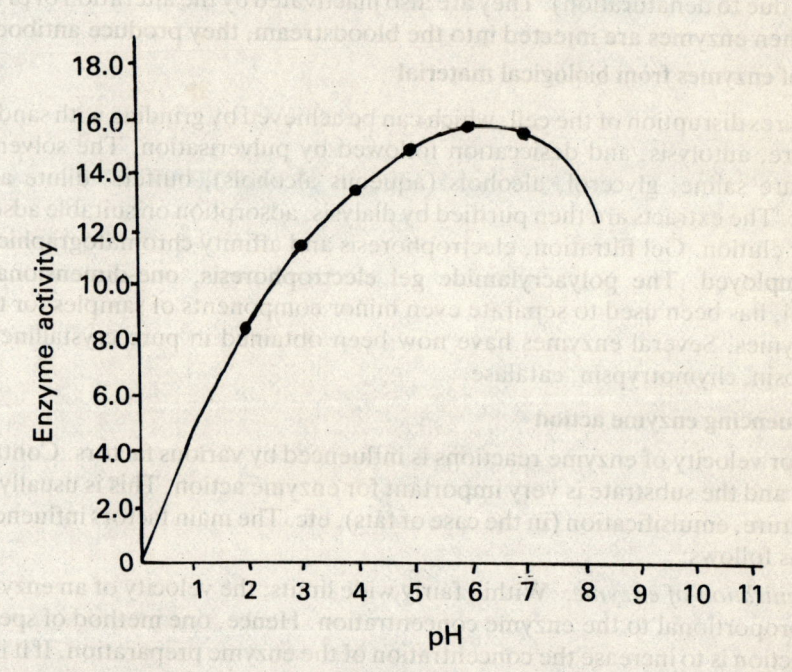

Some of the digestive enzymes like pepsin act at a high acidic pH of 1.5 while trypsin acts at an alkaline pH of 8.0. Alkaline phosphatases act at an alkaline pH of 9.5 while acid phosphatases act at an acid pH of 4.8.

The effect of pH on enzyme action is due to various factors like primary ionisation of the enzyme, influencing the formation of enzyme-substrate complex, co-factors etc. Hydration and ionic strength could also be affected.

ENZYMES COENZYMES AND ISOZYMES

3) *Temperature:* The velocity of an enzyme reaction increases with increase in temperature. The van't Hoff's temperature coefficient Q_{10} is defined as the increase in the velocity of an enzyme reaction for an increase in temperature of 10°C. It is usually double. Above 50°C, the enzyme activity slows down and it ceases at 70–80° when the enzyme gets denatured and destroyed. For most animal enzymes, the optimum temperature lies between 40–50°C.

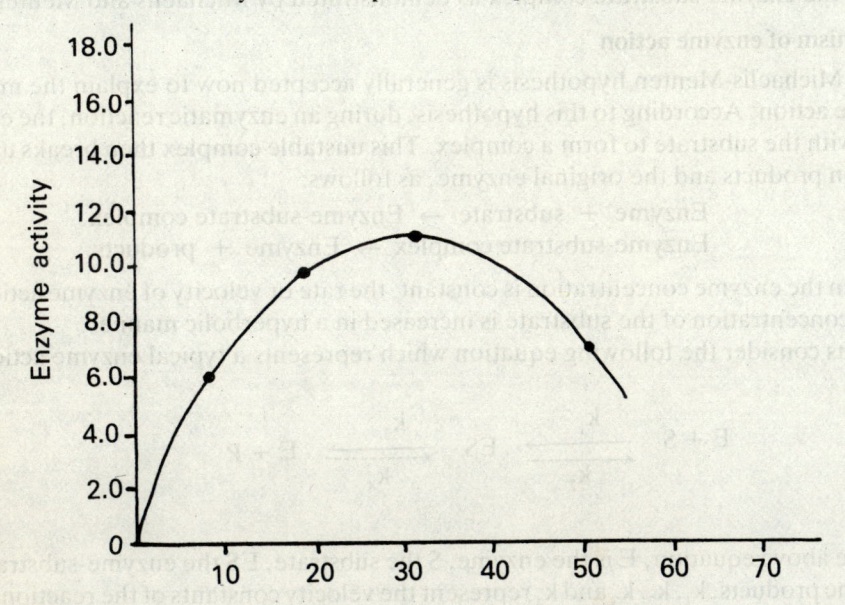

4) *Concentration of substrate* The velocity of an enzyme reaction increases as the concentration of the substrate increases, first in a linear fashion, but, later on, the curve flattens and becomes a plateau.

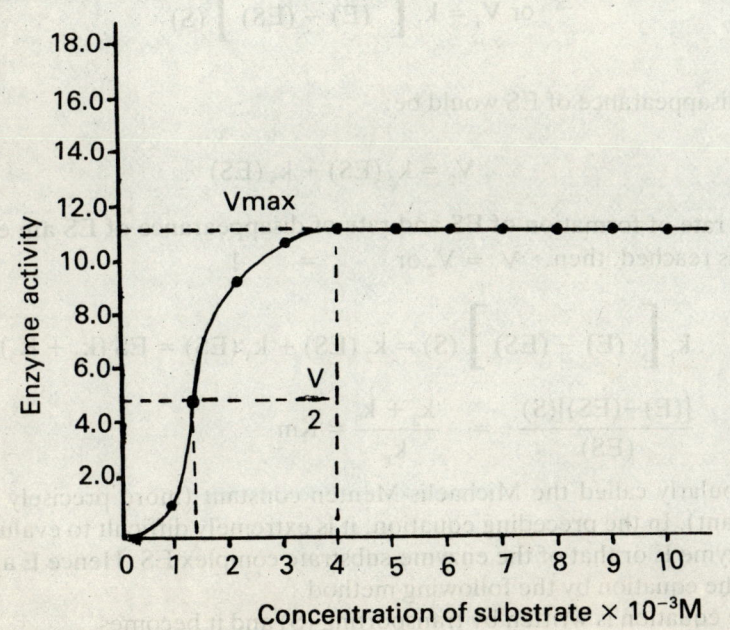

So, just by increasing the concentration of the substrate indefinitely, the velocity of the reaction cannot be increased beyond a certain stage. This is due to the mutual dependence of enzyme and substrate concentrations in the overall reaction. The rate of the reaction depends not on the individual concentrations of either the enzyme or the substrate but the concentration of the enzyme-substrate complex as demonstrated by Michaelis and Menten (1913).

Mechanism of enzyme action

The Michaelis-Menten hypothesis is generally accepted now to explain the mechanism of enzyme action. According to this hypothesis, during an enzymatic reaction, the enzyme combines with the substrate to form a complex. This unstable complex then breaks up to give the reaction products and the original enzyme, as follows:

$$\text{Enzyme + substrate} \rightarrow \text{Enzyme-substrate complex.}$$
$$\text{Enzyme-substrate complex} \rightarrow \text{Enzyme + products.}$$

When the enzyme concentration is constant, the rate or velocity of enzyme action increases as the concentration of the substrate is increased in a hyperbolic manner.

Let us consider the following equation which represents a typical enzyme action:

$$E + S \underset{k_2}{\overset{k_1}{\rightleftarrows}} ES \underset{k_4}{\overset{k_3}{\rightleftarrows}} E + P$$

In the above equation, E is the enzyme, S the substrate, ES the enzyme-substrate complex and P the products, k_1, k_2, k_3 and k_4 represent the velocity constants of the reactions (K_4 is negligible). The rate of formation of ES is proportional to the concentration of uncombined enzyme, i.e., (E)-(ES) and the concentration of the substrate (S) as the reaction, though bimolecular, exhibits pseudo, first-order kinetics.

$$\text{or } V_1 = k_1 \Big[(E) - (ES) \Big] (S)$$

The rate of disappearance of ES would be:

$$V_2 = k_2 (ES) + k_3 (ES)$$

When the rate of formation of ES and rate of disappearance of ES are equal, i.e., when equilibrium is reached, then, $V_1 = V_2$ or $\quad = \quad 1$

$$k_1 \Big[(E) - (ES) \Big] (S) = k_2 (ES) + k_3 (ES) = ES (k_2 + K_3)$$

$$\frac{[(E)-(ES)](S)}{(ES)} = \frac{k_2 + k_3}{k_1} = Km$$

Km is popularly called the Michaelis-Menten constant (more precisely Brig's and Haldane's constant). In the preceding equation, it is extremely difficult to evaluate the concentration of enzyme E or that of the enzyme-substrate complex ES. Hence E and ES are eliminated from the equation by the following method.

The above equation is written by transporting (S) and it becomes,

ENZYMES COENZYMES AND ISOZYMES

$$\frac{(E)}{(ES)} - \frac{(ES)}{(ES)} = \frac{Km}{(S)}$$

or

$$\frac{(E)}{(ES)} - 1 = \frac{Km}{(S)}$$

and

$$\frac{(E)}{(ES)} = \frac{Km}{(S)} + 1 = \frac{Km + (S)}{(S)}$$

Inverting the equation, it becomes,

$$\frac{(ES)}{(E)} = \frac{(S)}{Km + (S)}$$

or

$$(ES) = \frac{(E)(S)}{Km + (S)}$$

If the determined velocity of the overall enzyme reaction is V_3, it may be written also as $V_3 = k_3 (ES)$ and $V_3 = V$

Substituting for ES,

$$V_3 = V = k_3 \frac{(E)(S)}{Km + (S)}$$

At maximum velocity, all the enzyme is present in combination with substrate and $(E) = (ES)$.

The maximum velocity, $Vmax = k_3 (E)$

Therefore,

$$\frac{V}{Vmax} = \frac{(S)}{Km + (S)}$$

or

$$V = \frac{Vmax \times (S)}{Km + (S)}$$

$$Km + (S) = \frac{Vmax}{V} \times (S)$$

$$Km = \left\{ \frac{Vmax}{V} \times (S) \right\} - (S) = (S) \left(\frac{Vmax}{V} - 1 \right)$$

If $V = \frac{1}{2} Vmax$, $Km = (S) \left\{ \frac{Vmax}{\frac{1}{2} Vmax} - 1 \right\}$

$Km = (S) \cdot 1 = (S)$

Thus, if the concentration of the substrate is determined at half the maximum velocity, the Michaelis-Menten constant is obtained. Km value for most enzymes ranges between 10^{-2} and 10^{-5} or sometimes even higher. Investigation of Km can be done by using the Michaelis-Menten curve or the double reciprocal curve of Lineweaver and Burk, i.e., graph drawn between

$1/V$ and $1/S$. Km can be calculated from the slope or the intercept by the equations. Intercept in X-axis is $-\dfrac{1}{Km}$; Slope $= \dfrac{Km}{V}$,

In the figure, intercept in X axis is -2.5×10^3;

$-\dfrac{1}{Km} = -2.5 \times 10^3; \dfrac{1}{Km} = 2.5 \times 10^3; Km = \dfrac{1}{2.5 \times 10^3} = 4 \times 10^{-4} M$

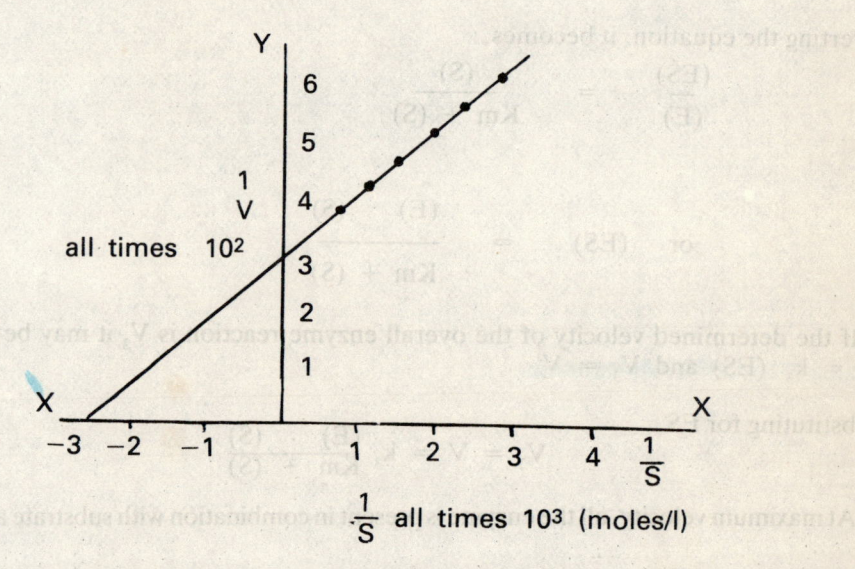

[Lineweaver-Burk plot: Y-axis labeled $\dfrac{1}{V}$ all times 10^2 with values 1–6; X-axis labeled $\dfrac{1}{S}$ all times 10^3 (moles/l) with values −3 to 4]

5) *Time:* The time required for the completion of an enzyme reaction increases with decrease in temperature from the optimum. Under optimum pH and optimum temperatures, the time required for enzymatic reaction is less. Time can be shortened or lengthened by altering temperature and pH.

6) *Products of reaction:* The addition of the reaction products to a reaction system retards the rate of enzyme action. Probably the products form a loose complex with the enzyme which makes the active centres less available. Mass action may also influence the reaction. Sometimes a reverse reaction may take place. Some of the products of a reaction may inhibit the pathway by feedback inhibition, e.g., CTP in the case of aspartate transcarbamoylase.

7) *Physical agents like light, etc.:* Light rays can inhibit or accelerate certain enzyme reactions. The activity of salivary amylase is increased by red and blue light. On the other hand, it is inhibited by ultraviolet rays.

8) *Enzyme activators (modifiers):* i) Inorganic modifiers (Positive modifiers): Metals like Cu, Co, Mn and Mg activate certain enzymes. Some enzymes have metal atoms inseparable from them, e.g., molybdenum in xanthine dehydrogenase and iron of cytochromes and catalase. In other cases, the presence of metals promotes the reaction. Ca and Mg promote reactions of enzymes requiring ATP. Metals activate enzyme action by various mechanisms as direct participation, combination with substrate, formation of enzyme-substrate-metal complex and effecting a conformational change in the enzyme.

H_2S activates some enzymes. H^+ activates pepsinogen while NaCl activates amylase.

ii) **Organic positive modifiers:** Glutathione and cysteine activate some enzymes. Chymotrypsinogen is activated by trypsin.

iii) **Allosteric activation.**

The activity of an enzyme is sometimes enhanced by certain metabolic intermediates which bind themselves to the allosteric sites of the enzymes and improve the availability of active sites. These metabolic intermediates are called allosteric effectors. The activation of pyruvate carboxylase by acetyl CoA, phosphofructokinase by AMP, carbamoyl phosphate synthetase by N-acetyl glutamate, b form of glycogen synthetase by glucose-6-phosphate and tryptophan oxygenase by L-tryptophan are examples of the allosteric activation of enzymes.

Pyruvate is carboxylated in the mitochondria to produce oxaloacetate, for gluco-neogenesis. The activity of pyruvate carboxylase is controlled by its allosteric effector, viz., acetyl CoA. The levels of intra-mitochondrial acetyl CoA resulting from the oxidation of pyruvate in the absence of malate cause saturation of the allosteric sites. During starvation, the increased availability of acetyl CoA from fatty acid breakdown activates pyruvate carboxylase and helps gluco-neogenesis.

Phosphofructokinase is allosterically activated by AMP and also ADP. However, ATP is an inhibitor for this enzyme. Both AMP and ADP bind to specific allosteric sites of phosphofructokinase. When these effectors bind themselves to allosteric sites of the enzyme, they change its conformation, resulting in an increase in the affinity of the enzyme to the substrate.

N-acetyl glutamate acts as a positive allosteric effector for carbamoyl phosphate synthetase and initiates the Krebs-Henseleit urea cycle.

Glycogen synthetase exists in two forms a and b. The phosphorylated form, i.e., synthetase b, is dependent on glucose-6-phosphate for its activity. Here glucose-6- phosphate acts as its allosteric effector. In the absence of glucose-6-phosphate, the dephosphorylated form or synthetase 'a' functions, as it does not need this effector.

Thus metabolic intermediates can bring about allosteric activation of certain enzymes and enhance the metabolic pathways.

9) *Enzyme inhibition:* i) **Irreversible negative modifiers:** Metals like Pb, Ag and Hg inhibit the action of many enzymes. In many cases, these combine with the active – SH groups of enzymes. Arsenic and cyanide are enzyme poisons.

Iodo acetamide, EDTA, p-chloro mercury benzoate, fluoride and di-isopropyl flourophosphate also bring about inhibition of many enzymes. The inhibition is irreversible and referred to as non-competitive inhibition (e.g., enolase by fluoride; hexokinase by p-chloro mercury benzoate). The negative modifiers normally combine with the enzyme at various sites and alter the nature of the enzyme chemically.

ii) **Allosteric inhibition:** In some cases, the inhibitors occupy the allosteric sites and bring about a conformational change in the enzyme whose active sites are thereby affected. The substrate molecules cannot therefore bind to the active site. This type of inhibition is called allosteric inhibition, e.g., ATP for phosphofructokinase and phosphorylase b, estrogens and thyroxine for glutamate dehydrogenase and 5'AMP for fructose, 1-6 diphosphatase.

iii) **Feedback inhibitors (reversible):** This is a type of allosteric inhibition. The products formed occupy the allosteric sites in some cases and inhibit the enzyme action e.g., a) CTP in the case of aspartate trans-carbamoylase; b) ATP on glutamate dehydrogenase and c) alpha keto glutarate on isocitrate dehydrogenase. If the products are removed, the rate of enzyme reaction will be increased.

iv) **Competitive (reversible) inhibition:** Certain inhibitors being structurally similar, compete with the substrate for the enzyme, and are called competitive inhibitors. These inhibitors occupy the available active sites, and decrease the rate of formation of E-S complex. Thus, they inhibit the enzyme action, e.g., inhibition of succinate dehydrogenase by malonic acid (structural analogue of succinic acid).

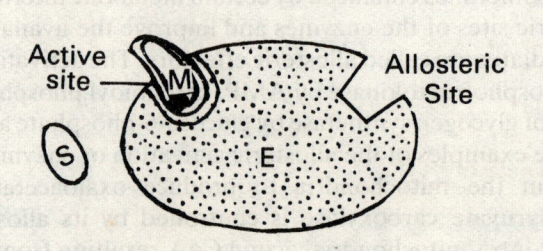

E: Enzyme
S: Substrate (succinic acid)
M: Malonic acid

Competitive inhibition

The effect of aminopterin on folic acid enzymes, sulphonamides on PABA required for folic acid synthesising enzymes, and fluorocitrate on aconitase, is competitive inhibition. Such inhibition is reversible, and by increasing the concentration of substrate, the inhibition by structural analogues can be corrected.

Enzyme substrate complex formation and active site of enzymes

It is established by various methods that there is a definite enzyme-substrate complex formation during enzyme action. It can be shown visually in peroxidase-involved reaction in which the colour changes from brown to green and then to red, the change of colour indicating the formation of the E-S complex. Spectrophotometry has proved E-S complex formation from absorption peaks. Kinetic evidence has been given by Michaelis and Menten, Brigs and Haldane, and Lineweaver and Burk. Use of radioactive label like ^{32}P, as glucose 1,6 diphosphate and di-isopropyl fluoro phosphate has also established E-S complex formation. Recently iso-electric focussing has also proved the formation of E-S complex.

Emil Fischer compared the action of enzyme on substrate to the lock and key interaction. Later on, biochemists obtained evidence that a specific region of the enzyme molecule functions as the 'active site' which may also be called 'catalytic site' or 'substrate site'. According to a recent hypothesis by Koshland, a conformational change in the enzyme molecule, which is induced by the presence of substrate, alters the alignment of the amino acid residues at the active site by altering the geometry of the protein. Koshland's 'induced fit' hypothesis has gained a lot of experimental support in recent years. Some of the amino acid residues of the enzyme molecule have groups protruding into space, e.g.; the hydroxyl groups of serine and –SH groups of cysteine, the epsilon amino group of lysine. The experimental evidence for this induced fit model is the conformational change which can be demonstrated in the case of enzymes like creatine kinase, carboxy peptidase and phospho-glucomutase.

Koshland has differentiated the amino acid residues of enzymes into different types. Any small-sized molecule, acting as an activator, substrate or inhibitor, coming into contact with an amino acid residue may bring about conformational change. Such an amino acid residue is called 'contact residue'. It may be involved in the specificity of an enzyme as well as its 'catalytic activity' to some extent. The catalytic residue may include an amino acid which is involved in the formation of co-valent bonds with the substrate, thus helping catalysis. These residues may also be situated at sites far from one another in the primary structure of the enzyme protein.

ENZYMES COENZYMES AND ISOZYMES

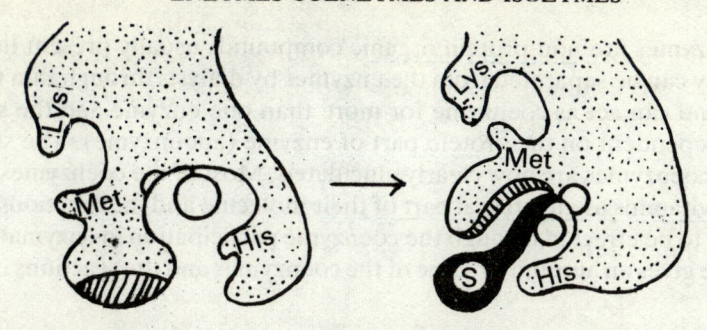

Induced fit by a conformational change in protein structure.

Different amino acids in enzyme molecules have been shown to act as active centres.

Enzyme	Amino acid
Phosphoglucomutase	Serine
Chymotrypsin	Serine and histidine
Choline esterase	Serine
Papain	Tyrosine
Aldolase	Lysine
Pancreatic ribonuclease	Histidine

Again a three–pronged attack is suggested between the enzyme and the substrate by Ogsten.

An indirect proof for this three - pronged attack is the action of aconitase on citric acid. The **symmetrical molecule of citric acid is made unsymmetrical by a three-pronged contact.** Though the two $-CH_2COOH$ groups are identical in citric acid, it is only the $-CH_2COOH$ from oxalo acetate (and not the one from acetyl CoA) that takes part in the elimination process.

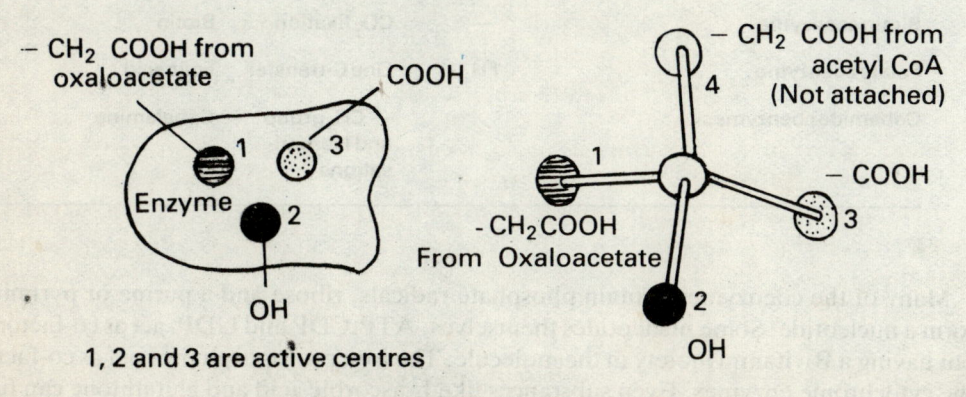

1, 2 and 3 are active centres

3-pronged attachment of a substrate to active sites of enzyme

Coenzymes: Coenzymes are non-protein organic compounds usually present in association with enzymes. They can be separated from the enzymes by dialysis through thin membranes. The same compound can act as coenzyme for more than one enzyme and the specificity of enzyme action is dependent on the protein part of enzyme (apoenzyme). The structure and functions of many coenzymes are now clearly elucidated. Most of the coenzymes have one or the other of the B-vitamins as an integral part of their molecule and the functions of these vitamins are believed to be effected through the coenzyme participation in enzymatic reactions. The following table gives an account of some of the coenzymes and the reactions in which they take part:

TABLE 2

COENZYMES RELATES TO B-VITAMINS

Name	Abbreviation	Groups transferred	B-vitamin component
Nicotinamide adenine dinucleotide	NAD$^+$	H$^+$ + 2e	Nicotinamide
Nicotinamide adenine dinucleotide phosphate	NADP$^+$	H$^+$ + 2e	Nicotinamide
Flavin mononucleotide	FMN	2H$^+$ + 2e	Riboflavin
Flavin adenine dinuecleotide	FAD	2H$^+$ + 2e	Riboflavin
Coenzyme A	CoA	Acetyl group, and acyl group	Pantothenic acid
Thiamine pyrophosphate as lipothiamide pyrophosphate	TPP	Ketol group, C$_2$ aldehyde group	Thiamine
Pyridoxal phosphate	—	Amino and Keto groups	Pyridoxine, pyridoxal Pyridoxamine
Biotin coenzyme	—	CO$_2$ fixation	Biotin
Folate coenzyme	FH$_4$	One C-transfer	Folic acid
Cobamide coenzymes		— CH$_3$ group and isomerisations	Cobalamine

Many of the coenzymes contain phosphate radicals, ribose and a purine or pyrimidine to form a nucleotide. Some nucleotides themselves, ATP, CDP and UDP, act as co-factors without having a B-vitamin moiety in the molecule. The iron protoporphyrins act as co-factors for the cytochrome enzymes. Even substances like L-ascorbic acid and glutathione can function as co-factors of certain enzymes.

Nicotinamide nucleotides

Nicotinamide is present in two of the coenzymes NAD^+ and $NADP^+$ associated with the dehydrogenases. The IUB has recommended these terms in place of DPN and TPN,

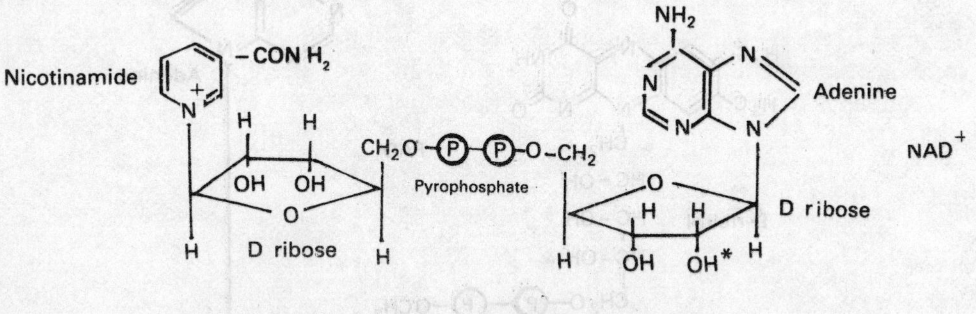

*The position of the third phosphate in $NADP^+$

Both NAD^+ and $NADP^+$ function as H-acceptors in various dehydrogenase-catalysed reactions involved in the oxidation of primary and secondary alcoholic groups. NAD^+ usually functions in conjunction with the enzymes of the respiratory chain but $NADP^+$ transfer H in other processes concerned with the biosynthesis of many compounds like lipids, pentoses, steroid hormones, etc. NAD^+ is the co-enzyme for lactate dehydrogenase, malate dehydrogenase, etc. while $NADP^+$ is required for glucose-6-phosphate dehydrogenase and malic enzyme.

When NAD^+ and $NADP^+$ act as H-acceptors, it is known that only one hydrogen and two electrons get added to the coenzyme, while the other hydrogen atom is converted to a proton (H^+) which escapes into the medium. From the substrate, 2H atoms are removed resulting in oxidation ($2H^+$ 2e).

$$\text{Thus} \quad \begin{array}{c} CH_3 \\ | \\ CHOH + NAD^+ \\ | \\ COOH \\ \text{Lactic acid} \end{array} \quad \xrightleftharpoons[]{\text{Lactate dehydrogenase}} \quad \begin{array}{c} CH_3 \\ | \\ CO + NADH + H^+ \\ | \\ COOH \\ \text{Pyruvic acid} \end{array}$$

Two types of dehydrogenases are known, the alpha and the beta, depending on the position of H added to the pyridine ring of NAD^+. For example, alcohol dehydrogenase and lactate dehydrogenase are alpha dehydrogenases while beta hydroxy steroid dehydrogenases are beta dehydrogenases.

Flavin nucleotides

In the respiratory chain, the transfer of H^+ is also effected by flavoproteins along with the nicotinamide coenzymes. Flavoproteins are yellow enzymes which contain the iso-alloxazine ring as an integral part (prosthetic group). Vitamin B_2 is a constituent of flavin mononucleotide (FMN) and flavin adenine dinucleotide (FAD), FAD includes FMN as a portion of its molecule.

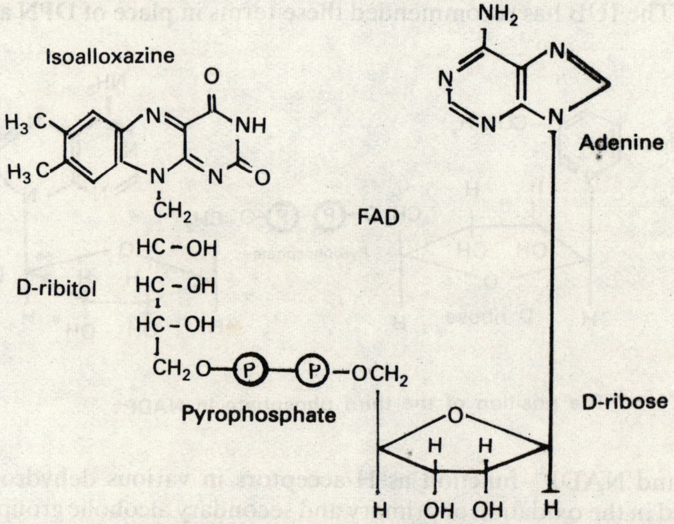

FMN is the coenzyme for various enzymes like glycolic acid oxidase, L-amino acid dehydrogenase, etc., while FAD is required for succinate dehydrogenase, xanthine dehydrogenase and acyl CoA dehydrogenase. When FMN or FAD accepts H, two atoms of H are removed from the substrate and added to the coenzyme.

$$FAD + Substrate \rightarrow FADH_2 + Oxidised\ substrate$$

Many of the flavoprotein enzymes also require a metal which functions in the oxidation-reduction systems. Thus succinate dehydrogenase requires Fe and xanthine dehydrogenase contains Mo.

Cytochromes (co-factors)

Cytochromes do not contain B-vitamins, but they function efficiently as co-factors in the oxidative chain. Cytochromes, chemically, are iron-porphyrin complexes and are related to hemoglobin. There are three principal cytochromes, a, b and c. Cytochrome c is the one which is present in large amounts in cells. The Fe atom of cytochromes undergoes reversible oxidation and reduction by the loss or gain of electrons. Once reduced, it can be re-oxidised by molecular oxygen in the presence of cytochrome aa_3.

Coenzyme A

Coenzyme A participates in the transfer of acyl groups like the acetyl groups and thus takes part in the oxidation of fatty acids and compounds containing the –COOH group. Acetic, malonic and succinic acids have to be converted to their CoA derivatives for participation in metabolic reactions. The coenzyme A molecule contains pantothenic acid, one of the B-vitamins. The structure of coenzyme A is given in the next page.

The functional part of coenzyme A is the terminal -SH group. When the carboxylic group reacts with reduced CoASH, the acids are converted to acyl CoA. Thus

$$Palmitic\ acid + Co\text{-}enzyme\ A \xrightarrow[ATP]{Thiokinase} Palmityl\ CoA$$

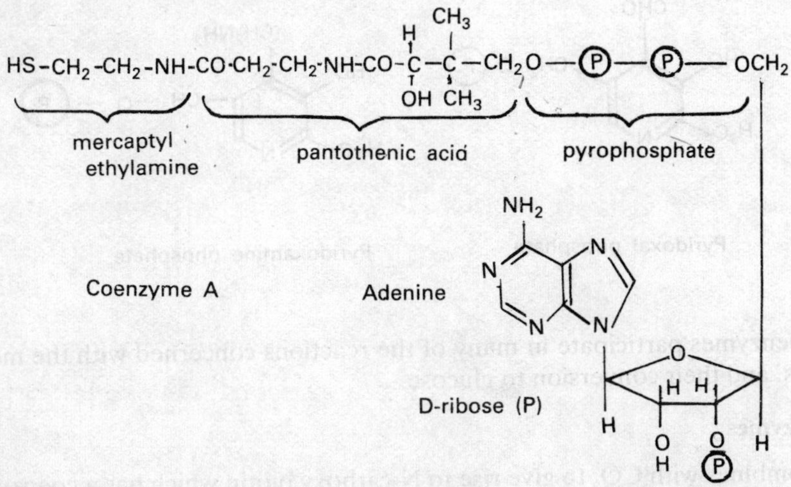

Coenzyme A structure

$$\underbrace{HS-CH_2-CH_2-NH-CO-CH_2-CH_2-NH-CO}_{\text{mercaptyl ethylamine}}\underbrace{-\overset{H}{\underset{OH}{C}}-\overset{CH_3}{\underset{CH_3}{C}}-CH_2O}_{\text{pantothenic acid}}\underbrace{-\text{P}-\text{P}-}_{\text{pyrophosphate}}OCH_2-\text{Adenine, D-ribose (P)}$$

Coenzyme A combines with acetic acid and ATP to form 'active acetate' or acetyl CoA. Acetyl CoA is readily formed from pyruvic acid during oxidation by pyruvate dehydrogenase complex. Coenzyme A is a very important factor in many of the reactions in carbohydrate, lipid and, protein metabolisms. (For further details refer to pantothenic acid under vitamins.)

Thiamine pyrophosphate

Vitamin B_1 or thiamine, functions in the body as thiamine diphosphate or thiamine pyrophosphate. As lipothiamide pyrophosphae (LTPP), it is a coenzyme for the decarboxylation of alpha-keto acids, in the pyruvate dehydrogenase complex as well as the alpha keto glutarate system. It is also a coenzyme for the transketolase reaction in the HMP pathway of glucose metabolism. Its structure is given below:

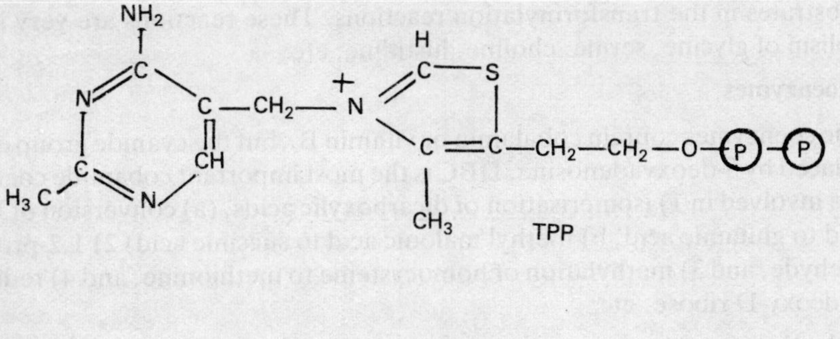

TPP

Pyridoxal phosphate

Vitamin B_6 or pyridoxine and its derivatives, pyridoxal and pyridoxamine, function as coenzymes in their phosphorylated form. They function as co-transaminases, co-decarboxylases and co-racemases.

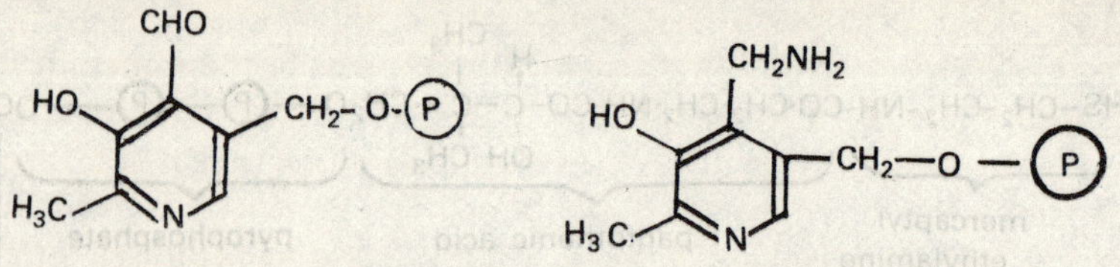

MEDICAL BIOCHEMISTRY

Pyridoxal phosphate **Pyridoxamine phosphate**

These coenzymes participate in many of the reactions concerned with the metabolism of amino acids, and their conversion to glucose.

Biotin coenzymes

Biotin combines with CO_2 to give rise to N-carboxy biotin which has a coenzyme function of transferring this CO_2 to the substrates and helping the biosynthesis of a molecule with one more C-atom. This is one of the first steps in the biosynthesis of fatty acids from acetyl CoA.

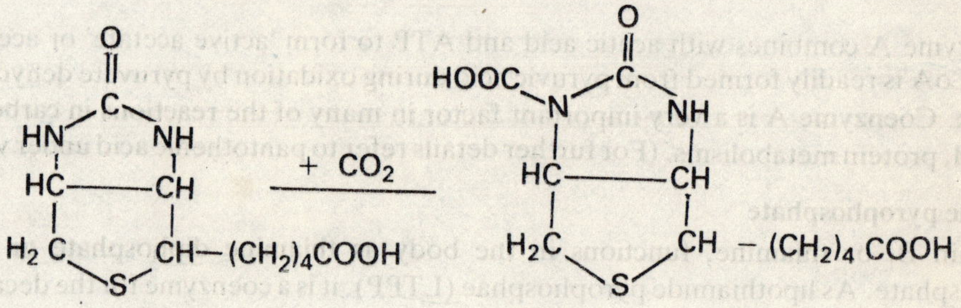

Biotin + CO_2 → **N-Carboxy biotin**

Folate coenzymes

Formate and 1-C metabolism require folic acid, one of the B-vitamins. The formyl group combines with tetra-hydrofolic acid to form $f^{10}FH_4$. The coenzyme transfers the formyl group to other substrates in the transformylation reactions. These reactions are very important in the metabolism of glycine, serine, choline, histidine, etc.

Cobamide coenzymes

Cobamide coenzymes contain cobalamin or vitamin B_{12} but the cyanide group on the cobalt atom is replaced by 5-deoxyadenosine. DBC is the most important cobamide coenzyme. This coenzyme is involved in 1) isomerisation of dicarboxylic acids, (a) conversion of Beta methyl aspartic acid to glutamic acid, b) methyl malonic acid to succinic acid 2) 1,2-propanediol to propionaldehyde, and 3) methylation of homocysteine to methionine, and 4) reduction of D-ribose to 2-deoxy-D ribose, etc.

Uridine nucleotides (co-factors)

Uridine diphosphate (UDP) and uridine triphosphate (UTP) act as co-factors in the synthesis of glycogen from glucose-1-phosphate as well as in the synthesis of many sugars.

Cytidine phosphates (co-factors)

CTP and CDP have similar functions in the synthesis of phospholipids.

Isozymes (Isoenzymes)

Isozymes (Isoenzymes) are the physically distinct forms of the same enzyme but catalyse the same chemical reaction or reactions. For example, lactate dehydrogenase is an enzyme catalysing the oxidation of lactic acid to pyruvic acid. In blood serum there are as many as five physically distinct isozymes of this enzyme, which are known as lactate dehydrogenase 1, LDH 1, LDH 2, LDH 3, LDH 4 and LDH 5. All these isozymes though different physically, catalyse the same reaction of oxidation of lactic acid to pyruvic acid. The different forms can be separated by electrophoresis. Tissues might also have isozymes. Another enzyme of clinical importance, alkaline phosphatase of serum, is also made up of many isozymes produced by different tissues, kidneys, liver, small intestines and bone.

The difference in electrophoretic mobilities is due to different electric charges on the isozymes. This is because of the difference in the content of acidic and basic amino acids. LDH 1 has the highest negative charge and moves fastest during electrophoresis. It contains a higher proportion of aspartic and glutamic acids than the other forms. LDH 5 is the slowest moving fraction.

Though the same chemical reaction is catalysed, the different isozymes may catalyse the same reaction at different rates. The rate of oxidation of Beta hydroxy butyrate is greater by LDH 1 and LDH 2 when compared with the rate of oxidation by LDH 4 and LDH 5. The isozymes may have different physical properties also. LDH 4 and LDH 5 are easily destroyed by heat while LDH 1 and LDH 2 are not, if heated upto about 60°. The isozymes have different pH optima and Km values, and have immunological differences as well.

It has been shown that the existence of different isozymes for an enzyme is due to the difference in the quaternary structure of the enzyme protein. Two sub units, H and M, have been shown to constitute the five isozymes of LDH, as five combination are possible, viz. HHHH, HHHM, HHMM, HMMM, MMMM. Such existence has been proved experimentally by breaking and remaking the isozymes. Breaking of (1) or (5) *separately* and remaking can give only (1) or (5) as the case may be. But if a *mixture* of (1) and (5) is broken and reconstructed, 2, 3 and 4 isozymes could be formed.

Clinical enzymology and isozymology

(Serum enzymes and isozymes in diagnosis of diseases)

Clinical enzymology is a branch of biochemistry dealing with the diagnostic value of enzyme estimation in serum and tissues in diseases. Rona was the first to introduce serum lipase estimation in pancreatic disease. Subsequently estimations of many serum enzymes have become popular in the clinical biochemical laboratory. In fact, estimations of alkaline phosphatase and alanine amino transferase have become almost a daily routine.

The enzymes present in circulating plasma are of two types: 1) the plasma specific or functional plasma enzymes, and 2) the plasma non-specific or non-functional plasma enzymes. The former type of functional plasma enzymes have their substrates in the plasma itself. These enzymes, though synthesised in the liver, are found in circulating plasma at higher concentrations than in tissues. Examples of functional plasma enzymes are lipoprotein lipase, pseudo-choline esterase, and enzymes associated with the clotting of blood.

There are many other enzymes circulating in plasma but do not have any substrate in plasma and have no role. These are usually present in small amounts but their levels are altered in pathological conditions. In many cases there is increase in their concentration and only in a few cases is there a decrease. Examples of plasma non-specific enzymes are alkaline phosphatase, acid phosphatase, transaminases, viz., glutamate oxalo acetate transaminase (G.O.T., presently known as AST, Aspartate amino transferase), glutamate pyruvate transaminase (G.P.T., presently known as ALT, alanine amino transferase), creatine phospho kinase (presently known as creatine kinase), lipase, amylase, etc.

The increase in the serum levels of an enzyme may be due to the excessive production of the enzyme by an acute inflamed organ which releases the same into the bloodstream. This happens during acute pancreatitis when the enzymes of exocrine secretions of pancreas, namely amylase and lipase, are released in large amounts into the bloodstream. In respect of many enzymes which are extra-cellular, the presence of such enzymes in serum is the result of routine normal destruction of tissue cells (autolysis), erythrocytes and leukocytes. In necrotic conditions associated with significant destruction of tissue cells, for example in myocardial infarction, the intra-cellular enzyme, AST of the tissue overflows into the blood and increases the normal concentration. A rare case of decrease is that of amylase in liver diseases.

Whatever may be the mechanism, a rise or fall in serum enzyme levels has been of immense value for a clinician. Some serum enzymes which are altered in diseased conditions are listed in Table 3.

The overall increase or decrease of an enzyme in serum depends on various factors like the number of cells damaged, the type and degree of damage, concentration of the enzyme in the affected cell, the location of the enzyme in the cell, its membrane permeability, catabolism and excretion.

In any particular disease, before arriving at a diagnosis, it is worthwhile to estimate three or four enzymes in serum rather than a single enzyme. Thus in heart disease, it is advisable to estimate AST, creatine kinase and hydroxy butyrate dehydrogenase (HBD which reflects the levels of LDH 1 and LDH 2) while in liver diseases, the enzymes alkaline phosphatase, ALT, isocitrate dehydrogenase, LDH 4 and LDH 5 isozymes could be investigated. Likewise in pancreatic diseases serum amylase and lipase estimations (also urine amylase) will be of value. In primary myopathies, aldolase and creatine kinase (CK) might be investigated.

While interpreting the values of the serum enzyme levels, it has to be taken into account that some enzymes are altered even in physiological conditions. Amylase is absent in the newborn. CK is low at rest but increased after exercise; alkaline phosphatase is raised in the newborn and during pregnancy. Acid phosphatase, aldolase, and AST are raised in children and infants.

In addition to variation of serum enzymes and urine enzyme (e.g.amylase), there are reports about variation of enzymes in cerebrospinal fluid and in the tissues and about the usefulness of their studies in diseases.

The study of isozymes has ushered in the new field of clinical isozymology. Knowledge of isozymes has helped in the differential diagnosis of certain diseases. For instance, the simple estimation of LDH in serum will not reveal precisely whether the disease is that of heart (myocardial infarction) or of the liver (infective hepatitis). But, if LDH 1 and LDH 2 are increased and not LDH 4 and LDH 5, it can be stated that it is the case of heart disease as heart tissue contains higher amounts of LDH 1 and LDH 2 (H-form). On the other hand, if LDH 4 and LDH 5 are increased, it is a case of liver involvement as these two isozymes (M-form) are present in larger amounts in liver. Thus the study of isozymes of LDH is helpful in the differential diagnosis of heart and liver diseases. Another enzyme which has isozymes is creatine kinase (CK). Total enzyme activity is increased in acute or chronic disorders of muscle and myocardial infarction as it is abundant in the skeletal and cardiac muscles.

Two sub units of the enzyme are the M unit (Muscle) and the B unit (Brain). There are three different dimers: 1) CK (BB), 2) CK (MB), and 3) CK (MM).

Of the above, the estimation of the CK-MB form has attained much importance in the differential diagnosis of muscle and heart diseases.

Table 3

Name of the enzyme	Clinical condition	Observation	Remarks
A.S.T. (G.O.T.)	Myocardial infarction. Also liver and pancreatic diseases.	Peak value 24 hrs. Raised activity for about 5 days	Normal in angina pectoris.
A.L.T. (G.P.T.)	Infective hepatitis. Toxic hepatitis.		Not increased in cirrhosis of liver.
Alkaline phosphatase	Bone diseases like i) Rickets, Paget's disease; ii) Obstructive jaundice–quite high, iii) Infective hepatitis and cirrhosis-slightly raised, iv) hyper-parathyroidism.		Normal in hemolytic jaundice.
Acid phosphatase	Increased in metastasising prostatic cancer, bone diseases like Paget's disease.		It is the tartrate-labile acid phosphatase which increases in prostatic cancer.
Lactate dehydrogenase	Raised i) myocardial infarction, ii) muscular dystrophy, iii) acute infective hepatitis, iv) carcinomas, v) pernicious anemia.		Normal in cirrhosis and post-hepatic jaundice.
Isocitrate dehydrogenase	Inflammatory, malignant and toxic liver diseases		Normal in cirrhosis and obstructive jaundice.
Creatine kinase	Myocardial infarction and progressive muscular atrophy.		Normal in poliomyelitis.
Aldolase	Muscular dystrophy, Myocardial infarction, Acute infective hepatitis.		Normal in poliomyelitis and cirrhosis.
Pseudo-choline esterase	*Decrease* in liver diseases (damage of liver cells), uremia, poisoning by organo phosphorus compound (also congenital).		Normal in post-hepatic Jaundice.
Amylase	Remarkable increase in acute pancreatitis also increased in intestinal obstruction, acute peritonitis perforated peptic ulcer (may increase in carcinoma of the pancreas), mumps (returns to normal in 2-6 days). Decreased in liver diseases.		Not significant in chronic pancreatitis.
Lipase	Increased in acute pancreatitis and other conditions mentioned against amylase except mumps (comes to normal after some days).		Not increased in chronic pancreatitis.

In normal individuals, the CK-MB isozyme accounts for less than two per cent of the total CK of plasma. By contrast, CK-MB accounts for 4.5 – 20 per cent of the total plasma enzyme of patients with a recent myocardial infarction. CKMB is not elevated to significant levels in skeletal muscle diseases. Likewise the study of the isozymes of alkaline phosphatase of serum will throw light as to whether it is a case of bone disease or other diseases like obstructive jaundice, but a simple estimation of alkaline phosphatase may not surely reveal whether the increase in the serum enzyme level is due to bone involvement or some other disease. Another interesting example is given by $NADP^+$ dependent iso-citrate dehydrogenase which exists as four isozymes ICD_1, ICD_2, ICD_3, and ICD_4 in blood. Though both the heart and the liver tissues have the enzyme, there is no increase of it in blood in myocardial infarction, while there is its increase in liver diseases. This is because, the liver contains the fast, stable form of the isozyme, i.e., ICD_1 while the heart contains the slow and unstable form. Hence iso-citrate dehydrogenase is specifically increased in blood in inflammatory, malignant and toxic liver diseases. Analysis of isozymes is done by disc gel electrophoresis.

Clinical enzymology and isozymology are thus invaluable in ascertaining some pathological conditions, when even sophisticated techniques like electrocardiography have not helped in the early diagnosis.

8 Biological Oxidations and Bio-Energetics

BIOLOGICAL OXIDATIONS

Biological oxidations, viz, oxidations taking place in living systems are vital for the existence and functions of cells, as in these reactions energy is produced and conserved as ATP molecules. If one has to respire air from birth to death, it is only for the purpose of biological oxidations in the cells. The oxygen of the air is carried to each and every cell of the body. Hence the term 'cellular' or 'internal respiration' has been used by some for biological oxidations. Another term used is 'tissue oxidation'.

During biological oxidations, the reacting chemical systems move from a higher energy level to a lower one and hence there is liberation of energy. The oxidations are therefore exergonic. The energy liberated as heat is converted to chemical energy, to the extent possible, by the formation of ATP which is one of the important energy-rich compounds. The formation of ATP from ADP and Pi is termed **phosphorylation**, and as it is concurrent with oxidation, the process is called **oxidative phosphorylation**. The oxidation is tightly coupled with phosphorylation. The energy of oxidations trapped in ATP is used for synthetic reactions, muscular contraction, nerve conduction, active transport and other processes that require energy (endergonic processes).

Oxidation is defined as addition of oxygen, removal of hydrogen or removal of electrons. Most biological oxidations are only biological dehyrogenations in which hydrogen is removed from the metabolites. In a few cases, however, as in the conversion of tryptophan to formyl kynurenine, oxygen is added. Cytochromes are oxidised by removal of electrons from the ferrous iron contained in them.

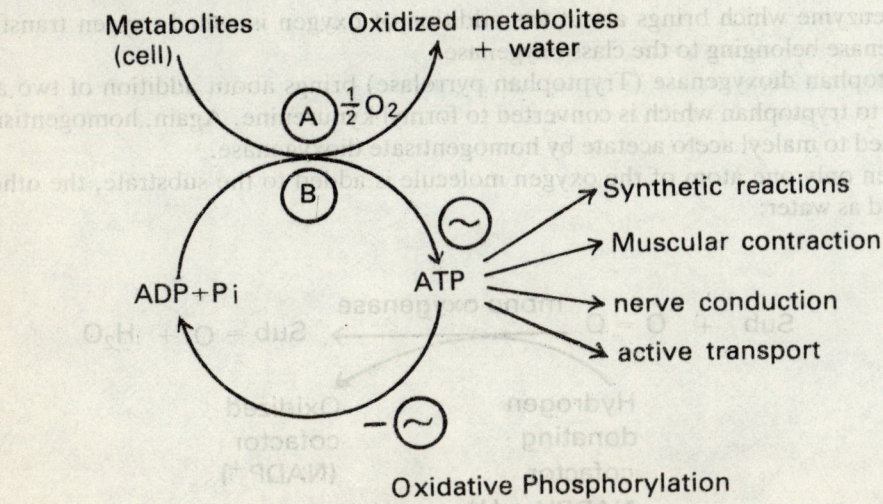

Oxidative Phosphorylation
(A) Oxidation (B) Phosphorylation

$$A + O_2 \rightarrow AO_2 \text{ (addition of oxygen)}$$
$$AH_2 + \tfrac{1}{2}O_2 \rightarrow A + H_2O \text{ (removal of hydrogen)}$$
$$A^{++} \xrightarrow{-e} A^{+++} \text{ (removal of electron,}$$

Reduced state   Oxidised state

Biological oxidations are of many types and are brought about with different enzymes and coenzymes and electron carriers. Actual utilisation of respiratory oxygen takes place in the mitochondria of the cells, with the help of enzymes, coenzymes and electron carriers. A few oxidations however take place without the assistance of coenzymes and electron carriers.

1) Different enzymes associated with oxidation are oxidases. oxygenases, aerobic dehydrogenases. anaerobic dehydrogenases. cytochrome oxidase (cytochrome a_3).

2) The coenzymes involved are NAD^+, $NADP^+$, FMN, FAD, electron transferring flavoprotein (ETF), coenzyme Q, lipoate.

3) Electron carriers: Heme proteins—cytochrome b, c_1, c, a, Another substance required is non-heme iron (NHI).

The main participants are metabolites like pyruvate, alpha keto glutarate. beta hydroxy acyl CoA, glutamate, malate, iso-citrate, acyl CoA, succinate, glycerol 3-phosphate and molecular oxygen, ADP and P_1.

Products of oxidation are ATP (in most cases in which oxidation is tightly coupled with phosphorylation) H_2O or H_2O_2 and reduced metabolites.

Biological oxidations are classified into different types by Dixon.

I. Addition of oxygen, oxygenases: It may be that both the oxygen atoms of the molecule could be added or only one atom added.

a) Where both oxygen atoms added:

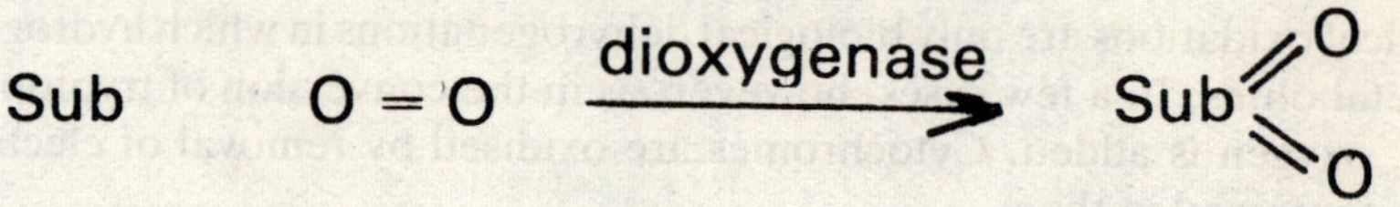

$$\text{Sub} \quad O=O \xrightarrow{\text{dioxygenase}} \text{Sub}\!\begin{array}{c}\diagup O \\ \diagdown O\end{array}$$

The enzyme which brings about the addition of oxygen is called oxygen transferase or dioxygenase belonging to the class oxygenase.

Tryptophan dioxygenase (Tryptophan pyrrolase) brings about addition of two atoms of oxygen to tryptophan which is converted to formyl kynurenine. Again, homogentisic acid is converted to maleyl aceto acetate by homogentisate dioxygenase.

b) When only one atom of the oxygen molecule is added to the substrate, the other being removed as water:

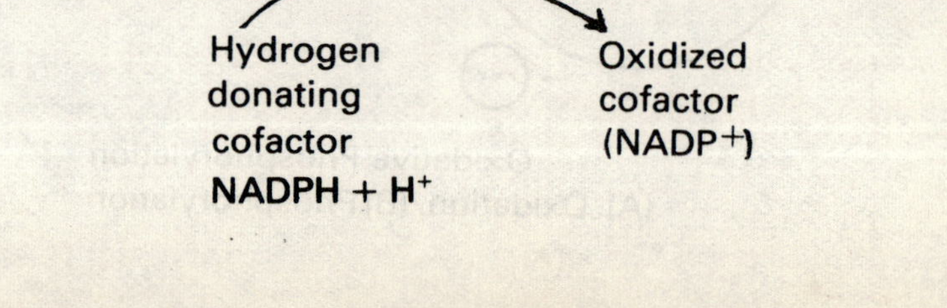

$$\text{Sub} + O=O \xrightarrow{\text{mono oxygenase}} \text{Sub}=O + H_2O$$

Hydrogen donating cofactor
NADPH + H$^+$

Oxidized cofactor ($NADP^+$)

The enzyme is called mono-oxygenase (mixed function oxidase or hydroxylase).

As an example, phenyl alanine is oxidised to tyrosine by phenyl alanine oxidase which is a mono-oxygenase. Hydroxylation of steroids takes place with such oxygenases. Tyrosinase is another example.

II. Hydroperoxidases: These are enzymes which use hydrogen peroxide as a substrate. Examples are catalase and peroxidase.

III. Oxidases: The molecular oxygen is activated by the enzyme and the hydrogen from the metabolite is directly transferred to the activated oxygen:

$$MH_2 + \tfrac{1}{2}O_2 \xrightarrow{\text{oxidase}} M + H_2O$$

The enzymes concerned are called oxidases. Cytochrome oxidase is a typical example. Molecular oxygen is directly utilised by cytochrome oxidase. Uricase is an oxidase present in subprimate mammals oxidising uric acid to allantoin. Phenolase and mono amine oxidase are other examples. (Uricase and mono-amine oxidases form H_2O_2).

IV. Aerobic dehydrogenases: These enzymes catalyse the removal of hydrogen from the metabolite. The hydrogens can be given to molecular oxygen or artificial acceptors like methylene blue. The subtle difference between oxidases and aerobic dehydrogenases is that oxidases cannot use artificial hydrogen acceptors like methylene blue. Again, in the case of aerobic dehyrogenases, hydrogen peroxide is the product of oxidation which is decomposed by catalase to water and oxygen:

$$MH_2 + O_2 \xrightarrow{\text{aerobic dehydrogenase}} M + H_2O_2$$
$$\downarrow \text{catalase}$$
$$H_2O + O_2$$

Examples of aerobic dehydrogenases are L-amino acid dehydrogenase, D-amino acid dehydrogenase, glucose oxidase, xanthine dehydrogenase. All these are flavoprotein enzymes.

V. Anaerobic dehydrogenases: In biological oxidations brought about by anaerobic dehydrogenases, the hydrogen from the metabolite is accepted by coenzyme systems. Subsequently, the electron carriers complete the oxidation.

There are two types of dehydrogenases – one type depends initially on the coenzyme NAD^+ (niacinamide adenine dinucleotide) and the other type does not depend on NAD^+ but depends on Fp (Flavin nucleotide). Even among the NAD^+ dependent dehydrogenases, there are two sub-groups, the alpha dehydrogenases and the beta dehydrogenases.

In the NAD^+ dependent dehydrogenases, the first hydrogen acceptor is NAD^+ (also known as DPN^+ i.e., diphosphopyridine nucleotide; coenzyme I) and in some cases $NADP^+$, Niacinamide adenine dinucleotide phosphate (also known as TPN^+ i.e., triphosphopyridine nucleotide, coenzyme II).

NAD^+ contains the following units:
Niacinamide-Ribose-Phosphate-Phosphate Ribose-Adenine
$NADP^+$ contains one extra phosphate on the ribose attached to adenine, i.e.,
Niacinamide-Ribose-Phosphate-Phosphate-Ribose-Adenine
$$|$$
Phosphate

The active unit in NAD^+ and $NADP^+$ is the niacinamide part which can reversibly take up hydrogen and lose it.

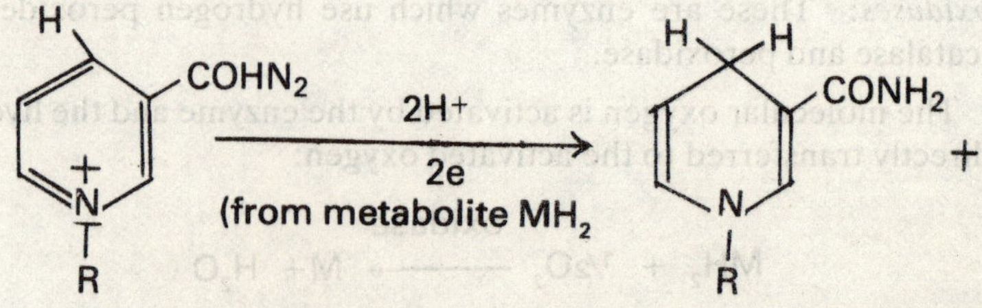

$$\underset{R}{\underset{|}{\overset{+}{N}}}\text{-COHN}_2 \quad \xrightarrow[\text{(from metabolite MH}_2\text{)}]{2H^+ \; 2e} \quad \underset{R}{\underset{|}{N}}\text{-CONH}_2 + H^+$$

Niacinamide unit

The general expression for the anaerobic dehydrogenase reactions is,

$$MH_2 + \text{acceptor} \xrightarrow{\text{anaerobic dehydrogenase}} M + \text{Reduced acceptor}$$

$$MH_2 + NAD^+ \longrightarrow M + NADH + H^+$$

$NAD^+/NADH$ is a reduction oxidation system having a standard electrode potential of $-0.32V$ (E_o).

The reaction does not stop at this stage. The NADH (reduced NAD^+) is oxidised by an enzyme NADH dehydrogenase which is a flavoprotein. The enzyme contains FMN (flavin mono-nucleotide) as co-factor and non-heme iron. (The role of flavin in biological oxidation was brought to light by the work of Warburg and Christian in 1932 when they isolated a yellow ferment of yeast. It was even called Warburg's yellow enzyme, the yellow colour being due to riboflavin. In 1967, it was shown that two flavoproteins were involved and they were Fp_1 and Fp_2. The redox potential of one is similar to that of $NAD^+/NADH$, while that of the other is relatively higher. FMN has the following structural unit:

<u>Iso-alloxazine — ribitol — phosphate</u>
<u>Riboflavin</u>

It may be observed from future discussion that FAD, i.e., flavin adenine dinucleotide also is used as a hydrogen carrier during biological oxidation in certain other cases.

The effective structural unit in FMN and FAD that takes up hydrogen is the iso-alloxazine ring.

$$\begin{matrix}FMN\\FAD\end{matrix} + NADH + H^+ \longrightarrow NAD^+ + FMNH_2 \; / FADH_2$$

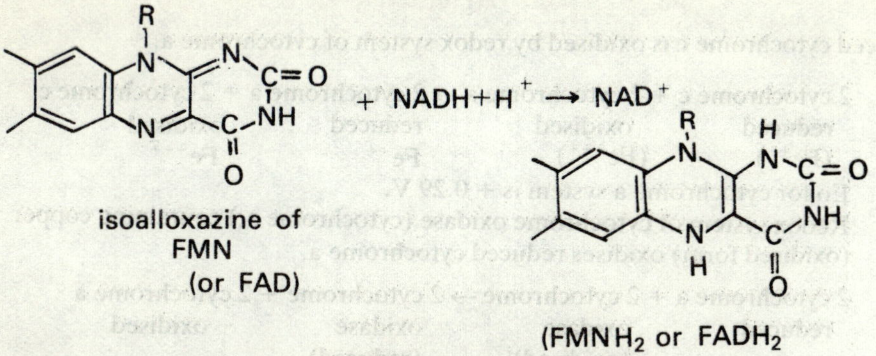

BIOLOGICAL OXIDATIONS AND BIO-ENERGETICS

isoalloxazine of FMN (or FAD) + NADH + H$^+$ → NAD$^+$ + (FMNH$_2$ or FADH$_2$)

FMN/FMNH$_2$ is the second redox system having Eo = 0.

The reduced flavin coenzyme is re-oxidised by coenzyme Q (ubiquinone) which accepts hydrogen.

$$\left.\begin{array}{c}\text{FMN H}_2\\ \text{FAD H}_2\end{array}\right\} + \text{CoQ} \longrightarrow \begin{array}{c}\text{FMN}\\ \text{FAD}\end{array} + \text{CoQH}_2$$

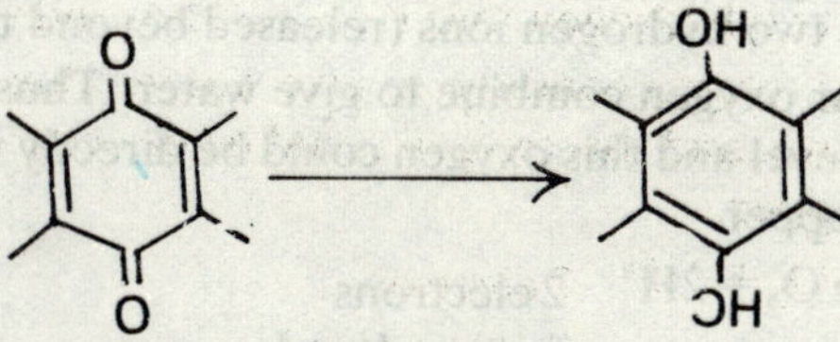

Thus CoQ/CoQH$_2$ constitutes a third redox system.

Hydrogen is transferred upto the stage of CoQ but beyond that the electrons are removed. This removal of electrons is done with the help of cytochromes b, c$_1$, c, a and a$_3$. Cytochromes are heme proteins and contain iron which could be either in **ferrous form (ferrocytochrome, i.e., reduced form)** or **ferric form (ferricytochrome. i.e., oxidised form)**.

CoQH$_2$ is oxidised by ferricytochrome b to CoQ while ferricytochrome b is reduced to ferrocytochrome b.

$$\text{CoQH}_2 + \text{2 molecules of cytochrome b} \xrightarrow{-e} \text{CoQ + 2 cytochrome b reduced}$$
(oxidised) Fe^{+++} → (Fe^{++}) + 2H$^+$

Eo for cytochrome b system is slightly above zero. Cytochrome b has non-heme iron (NHI).

Cytochrome c$_1$, c and a continue the relay of electrons. Reduced cytochrome b is oxidised by the redox cytochrome c$_1$.

2 cytochrome b + 2 cytochrome c$_1$ ⟶ 2 cytochrome b + 2 cytochrome
reduced   oxidised   oxidised   c$_1$ reduced
(Fe^{++})   (Fe^{+++})   (Fe^{+++})   (Fe^{++})

Reduced cytochrome c$_1$ is oxidised by redox system of cytochrome c.

2 cytochrome c$_1$ + 2 cytochrome c ⟶ 2 cytochrome c + 2 cytochrome
reduced   oxidised   reduced   c$_1$ oxidised
(Fe^{++})   (Fe^{+++})   (Fe^{++})   (Fe^{+++})

Eo for cytochrome c system is + 0.26 V.

Reduced cytochrome c is oxidised by redox system of cytochrome a.

$$2 \text{ cytochrome c} + 2 \text{ cytochrome a} \rightarrow 2 \text{ cytochrome a} + 2 \text{ cytochrome c}$$
$$\text{reduced} \quad\quad \text{oxidised} \quad\quad\quad \text{reduced} \quad\quad \text{oxidised}$$
$$(Fe^{++}) \quad\quad (Fe^{+++}) \quad\quad\quad Fe^{++} \quad\quad Fe^{+++}$$

Eo for cytochrome a system is + 0.29 V.

Redox system of cytochrome oxidase (cytochrome a_3) containing copper (oxidised form) oxidises reduced cytochrome a.

$$2 \text{ cytochrome a} + 2 \text{ cytochrome} \rightarrow 2 \text{ cytochrome} + 2 \text{ cytochrome a}$$
$$\text{reduced} \quad\quad \text{oxidase} \quad\quad\quad \text{oxidase} \quad\quad \text{oxidised}$$
$$\quad\quad\quad\quad (\text{oxidised}) \quad\quad (\text{reduced})$$
$$(Fe^{++}) \quad\quad (Fe^{+++}) \quad\quad\quad (Fe^{++}) \quad\quad (Fe^{+++})$$

Redox potential for cytochrome a_3 system is +0.53V.

Recently, it has been shown that both cytochrome a and a_3 are bound to the same protein and can be put as aa_3.

Finally, the reduced cytochrome oxidase is oxidised by molecular oxygen to the oxidised form. At this stage the two hydrogen ions (released beyond the CoQ system), the two electrons and the molecular oxygen combine to give water. Thus there is utilisation of respired oxygen at the cellular level and this oxygen could be directly utilised by cytochrome oxidase only, which contains copper.

$$\begin{array}{lll} 2\text{ cytochrome} + \frac{1}{2}O_2 + 2H^+ & 2 \text{ electrons} & 2\text{ cytochrome} + H_2O \\ \text{oxidase} & \text{from reduced} & \text{oxidase} \\ \text{reduced} & \text{cytochrome} & \text{oxidised} \\ (Fe^{++}) & \text{oxidase} & (Fe^{+++}) \end{array}$$

Eo for this system is +0.8 V.

Super oxide ion can form at this step.

Thus the hydrogen atoms from the metabolite are removed with the help of 'middle men', i.e., coenzyme systems, while electrons are removed by electron carriers and handed over to molecular oxygen. This sequence of reactions is said to constitute the respiratory chain.

Oxidations take place from the metabolite with Eo of about -0.43 V, and end up with molecular oxygen with Eo of $+0.8$ V. The various participants (redox systems) in the respiratory chain are given below:

Substrate $\rightarrow NAD^+ \rightarrow F_p \rightarrow CoQ \rightarrow$ cytochrome b $\rightarrow$ cytochrome $c_1 \rightarrow c \rightarrow$ cytochrome $aa_3 \rightarrow O_2$.

Metabolite								Molecular Oxygen
MH_2	NAD^+	FPH_2	C_oQ	Cyt b Fe^{3+}	Cyt C_1 Fe^{3+}	Cyt C Fe^{3+}	Cyt a Fe^{3+}	Cyt a_3 Fe^{3+}	H_2O
	2e	2e	2e	2e	2e	2e	2e	2e	
M	NADH $+H^+$	FP	C_oQH_2	Cyt b Fe^{2+}	Cyt C_1 Fe^{2+}	Cyt C Fe^{2+}	Cyt a Fe^{2+}	Cyt a_3 Fe^{1+}	$\frac{1}{2}O_2 + 2H^+$
E - 0.43 V	-0.32 V	0		Slightly above 0	+0.26		+0.29	+0.53	
				2H (Medium)					

The sequence of the redox systems is such that the Eo increases from negative value of about -0.43 V to O and reaches the positive value of $+0.82$. Hence this is an exergonic process. The energy transfers involved in the oxidation-reduction systems are measured by differences in the electrical potential of the various systems. The system with a higher potential will oxidise one with a lower potential with a consequent liberation of energy for vital process. This is the sequence of redox systems in most biological oxdiations.

Respiratory chain

The respiratory chain is organised on the inner membrane of the mitochondria, which is often designated as the 'initra-cellular power house' of the cell. The beta-oxidation of fatty acids, the oxidation of intermediates of glucose metabolism, and the C-skeleton of amino acids, all taking place through the citric acid cycle (terminal pathway of metabolism) produce energy. The energy derived from such oxidations within the tissues is captured in the form of ATP which is achieved in the mitochondria with the help of a series of enzyme systems along with the co-factors. The respiratory chain is concerned with the transport of reducing equivalents (H^+ and electrons) from the substrate to molecular O_2. It is also linked with the enzymes of beta oxidation and citric acid cycle which are present in the matrix and from which NADH and $FADH_2$ (also $FMNH_2$) are supplied for oxidation. In the respiratory chain, the components are arranged in the order of their increasing redox potentials and the flow of electrons is from the most electronegative substrate to the most electro-positive O_2.

Not all substrates are linked to the respiratory chain through NAD^+-linked dehydrogenases. Some are linked directly to flavo-protein dehydrogenases which are then linked to cytochromes.

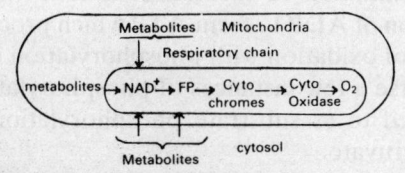

```
                Metabolites    Mitochondria
                     ↓     Respiratory chain
   metabolites → NAD⁺ → FP → Cyto → Cyto → O₂
                           chromes  Oxidase
                     ↑
                Metabolites      cytosol
```

Betahydroxy acyl-CoA, betahydroxy butyrate, malate, isocitrate, glutamate, 3-phosphoglyceraldehyde are all NAD^+ dependent. The hydrogen atoms are directly taken up by NAD^+. Pyruvate and alpha ketoglutarate are oxidised by lipoate, Fp combination and subsequently NAD^+ followed by flavin system.

On the other hand succinate and alpha glycerophosphate do not depend on NAD^+. They are oxidised by flavoprotein system (Fp) which contains non-heme iron and shunted on to CoQ system. While acyl-CoA also does not require NAD^+, it requires in addition to flavoprotein (Fp) system another flavoprotein system, i.e., 'electron transferring flavoprotein' (ETF) and the relay is continued by CoQ. All these are shown in the following figure.

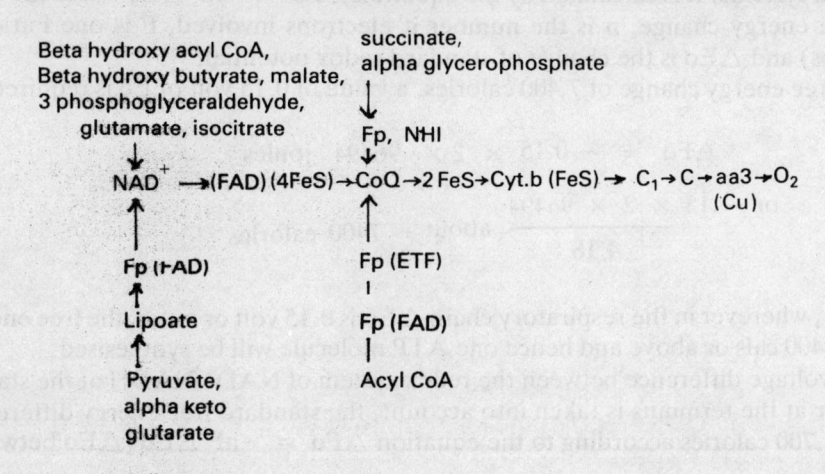

```
Beta hydroxy acyl CoA,          succinate,
Beta hydroxy butyrate, malate,  alpha glycerophosphate
3 phosphoglyceraldehyde,              ↓
glutamate, isocitrate
       ↓                          Fp, NHI
                                     ↓
    NAD⁺ →(FAD)(4FeS)→ CoQ → 2 FeS → Cyt.b (FeS) → C₁ → C → aa3 → O₂
       ↑                                                        (Cu)
    Fp (FAD)                      Fp (ETF)
       ↑                             |
    Lipoate                       Fp (FAD)
       ↑                             |
    Pyruvate,                     Acyl CoA
    alpha keto
    glutarate
```

Coenzyme Q (ubiquinone) is one of the components of mitochondrial lipids and plays an important role in the respiratory chain. It is present in mitochondria in larger amounts than the other components. It is lipid soluble and hence very much mobile in the mitochondrial membrane, since it is not attached to any protein and is in the free state. It can be easily reduced to quinol (CoQ H_2). From CoQ H_2 onwards, only electrons are transferred through the cytochromes and the $2H^+$, released at the CoQ stage, join, with a pair of electrons and molecular O_2 at the last stage to form H_2O.

While the iso-citrate in the mitochondria requires NAD^+, the iso-citrate in the cytosol depends on $NADP^+$. Malic enzyme and glucose-6-phosphate dehydrogenase are $NADP^+$ dependent. The site of biological oxidation is the mitochondria of cells as already indicated. Various participating enzymes are present in the inner mitochondrial wall (see chapter 1). The integrity of the double wall of cristae and mitochondria should be maintained for a proper oxidative role and the normal permeability also should be preserved. When there is swelling of the mitochondria as brought about by old age or by thyroxine, oxidations will be affected.

Valuable byproduct of oxidation (ATP): Oxidative phosphorylation

The importance of the oxidation of substances is that most biological oxidations are coupled with phosphorylation of ADP to form ATP which process is referred to as oxidative phosphorylation. Coupling of oxidation with phosphorylation is done in the interests of the organism. There are of course a few instances of phosphorylation of ADP to ATP without oxidation. They are referred to as substrate phosphorylation, e.g., conversion of phosphoenol pyruvate to enol pyruvate.

Oxidative phosphorylation is a valuable side-effect in oxidation by anaerobic dehydrogenases. Whether at any step of oxidation by anaerobic dehydrogenases assisted by coenzymes and electron carriers, ATP would be formed depends on the quantum of free energy change occurring when hydrogen or electrons move from one redox system to another. The heat of formation of ATP from ADP and Pi is about 7,400 cals. Hence, if the free energy change when hydrogen or electrons pass between two redox systems is about 7,400 cals or more, a molecule of ATP will be formed and conserved for future use. If the free energy change is less than about 7,400 cals, it is dissipated as heat.

The free energy change depends on the standard oxidation reduction potential Eo for the individual systems. It is calculated by the equation $\triangle Fo = -nF \triangle Eo$ where $\triangle Fo$ is the standard free energy change, n is the number if electrons involved, F is one Faraday (96494 Coulombs) and $\triangle Eo$ is the change of standard redox potential.

For a free energy change of 7,400 calories, a value of 0.15 volt of Eo is required as per the

$$\triangle Fo = -0.15 \times 2 \times 96494 \text{ joules}$$

$$\text{or} -\frac{.15 \times 2 \times 96494}{4.18} \text{ about} - 7400 \text{ calories}$$

Hence, wherever in the respiratory chain $\triangle Eo$ is 0.15 volt or more, the free energy change will be 7,400 cals or above and hence one ATP molecule will be synthesised.

If the voltage difference between the redox system of $NAD^+/NADH$ at the start and oxygen/water at the terminus is taken into account, the standard free energy difference will be about 51,700 calories according to the equation $\triangle Fo = -nF \triangle Eo$ ($\triangle Eo$ between the two

about 51,700 calories according to the equation $\star Fo = -nF \star Eo$ ($\star Eo$ between the two systems is 1.12 volts). 51,700 calories would be equivalent to (51,700/7,400) about seven molecules of ATP. However, this entire energy is not available for ATP formation as in certain steps energy less than 7,400 calories is liberated. In such cases, the liberated energy is simply dissipated as heat. Hence the number of ATP molecules produced in many oxidations is 3. This is represented as phosphorus oxygen ratio (P:O) or phosphorus-2 electron ratio (P: 2 e), where P stands for ATP molecules synthesised. So, in most cases, P:O is 3:1, i.e., for each atom of oxygen ($\frac{1}{2} O_2$) utilised, 3 ATP molecules are produced. The percentage of energy captured as ATP molecules is about 43. The actual site of phosphorylation in the respiratory chain is between Fp_1 and Fp_2, (1 ATP), cytochrome b and c_1, (1 ATP) and cytochrome aa_3 and oxygen (1ATP) as shown below:

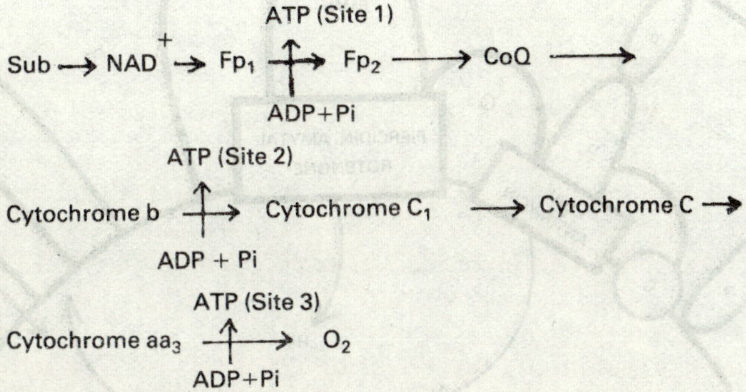

$$Sub \rightarrow NAD^+ \rightarrow Fp_1 \xrightarrow[ADP+Pi]{ATP\ (Site\ 1)} Fp_2 \rightarrow CoQ \rightarrow$$

$$Cytochrome\ b \xrightarrow[ADP+Pi]{ATP\ (Site\ 2)} Cytochrome\ C_1 \rightarrow Cytochrome\ C \rightarrow$$

$$Cytochrome\ aa_3 \xrightarrow[ADP+Pi]{ATP\ (Site\ 3)} O_2$$

However, in oxidations in which there is a bypass and the hydrogen from the metabolite is directly taken over by Fp, as in the case of the oxidations of succinate and acyl CoA, the P:O ratio is 2:1 and only two ATPs are produced. Actual ATP synthesis takes place in the inner membrane particles which have in their head portion the enzyme F_1 ATPase.

Usually, plenty of ADP is available for phosphorylation to ATP and hence the oxidation is tightly coupled with phosphorylation.

Cellular respiration is influenced by the following factors: 1) availability of ADP and substrate, 2) availability of substrate only, 3) capacity of the respiratory chain itself, 4) availability of ADP only, and 5) availability of oxygen.

In most cells at rest, cellular respiration is controlled by the availability of ADP. When work is performed, the energy required is obtained by the hydrolysis of ATP (group transfer potential) to ADP. In view of increased availability of ADP, respiration is increased, and this, in turn, replenishes the store of ATP.

Inhibitors of the respiratory chain: The inhibitors of the respiratory chain act at sites I, II and III of the chain.

Site I (NAD dehydrogenase step, i.e., $NAD^+ \rightarrow Fp$ step) is inhibited by barbiturates (amobarbial), piercidin A (antibiotic), rotenone (fish poison), some steroid drugs and mercurials.

Site II i.e., cyt b$\rightarrow$cyt c_1 step) is inhibited by BAL (dimercaprol) and antimycin A.

Site III (Cytochrome oxidase, i.e., at cytochrome a_3 step) is inhibited by **CN**, carbon monoxide, H_2S and azide.

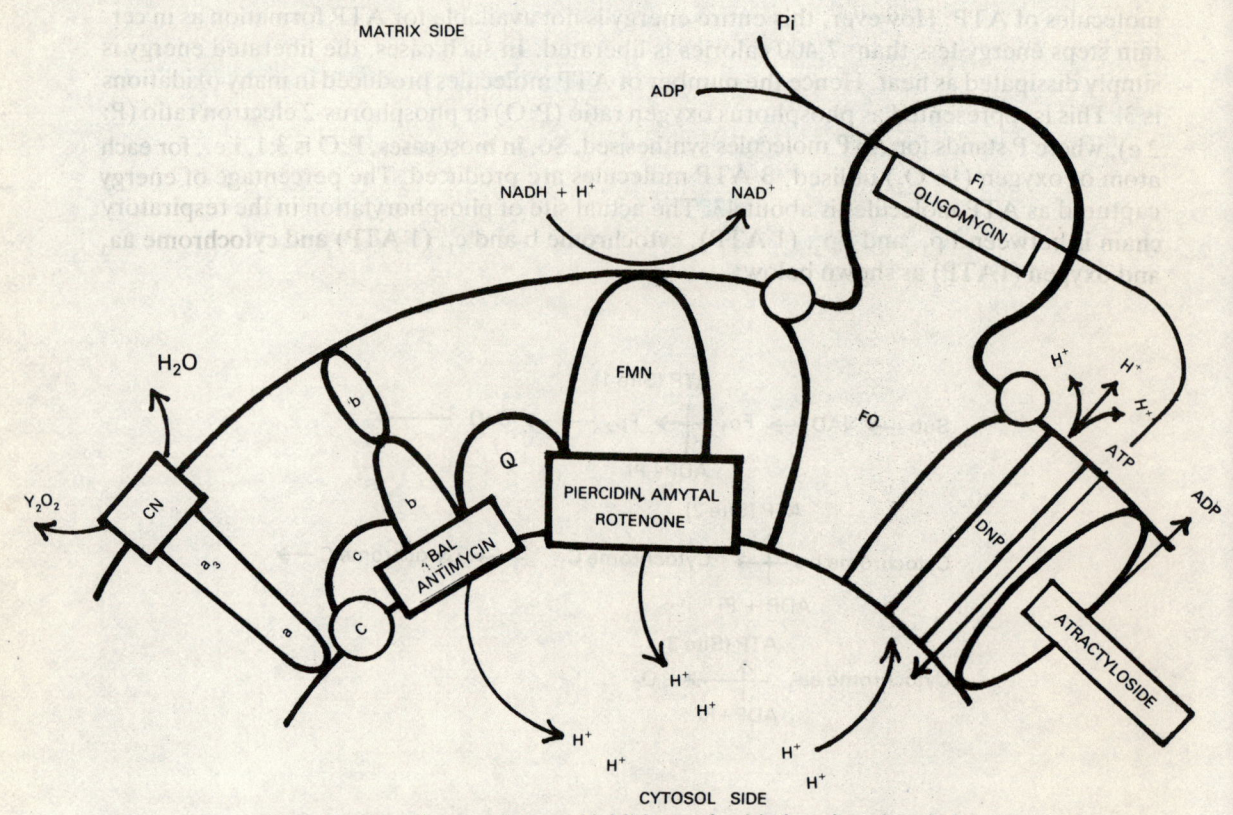

Site of action of uncouplers and inhibitors of oxidative phosphorylation.

Uncoupling of oxidative phosphorylation

While normally biological oxidation is coupled with phosphorylation, in certain conditions the two processes can be uncoupled. Chemical agents like dinitrophenol, azide and thyroxine could cause uncoupling of oxidative phosphorylation. Swelling of the mitochondria could also be responsible.

Uncoupling of oxidative phosphorylation results in energy being dissipated as heat without being captured as ATP, while the respiration (oxidation) proceeds well, because of the availability of ADP. ADP stimulates the respiration of mitochondria. This ADP cannot be phosphorylated to ATP for want of energy through electrochemical potential difference.

All respiratory inhibitors also inhibit ATP production. Some compounds like atractyloside, oligomycin, valinomycin etc, inhibit ATP production by different mechanisms. Atractyloside is an inhibitor of ATP transport and thereby its translocation. Oligomycin interferes with the phosphorylation of ADP while valinomycin acting as an ionophore eliminates the pH gradient, by allowing K^+ ions to enter the lipid membranes.

Mechanism of oxidative phosphorylation

Mitchel has put forward the chemiosmotic hypothesis according to which oxidative phosphorylation takes place by the translocation of protons (H^+ ions) to the exterior of the coupling membrane. The latter is normally impermeable to protons which accumulate during oxidation outside the membrane, creating an electro-chemical potential difference. Two potentials result : one is the chemical potential due to the difference in pH and the other is the membrane potential. The electro-chemical potential difference drives the vectorial membrane-bound enzyme ATPase, which though called ATPase, is actually ATP synthetase. This enzyme is called vectorial as there is a geometric direction for the reaction. The vectorial ATPase causes the formation of ATP from ADP and Pi.

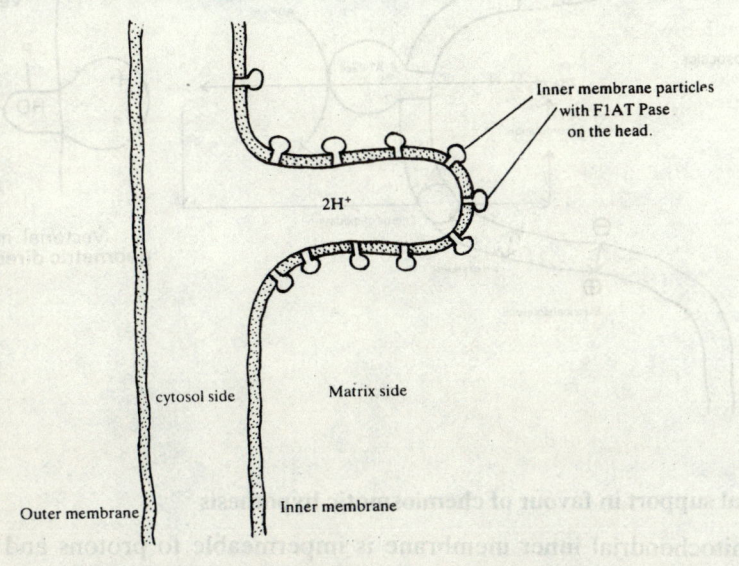

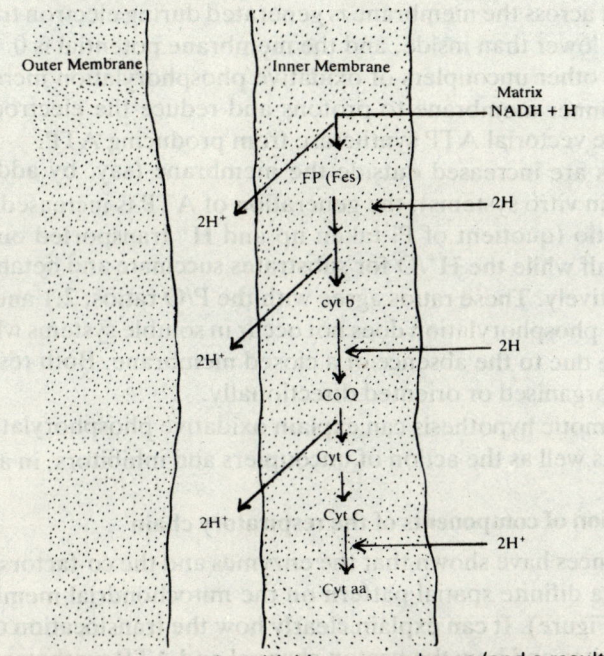

Accumulation of H$^+$ on the Cytosol side of mitochondrial inner membrane (and a resultant accumulation of $-$ OH on the matrix side)

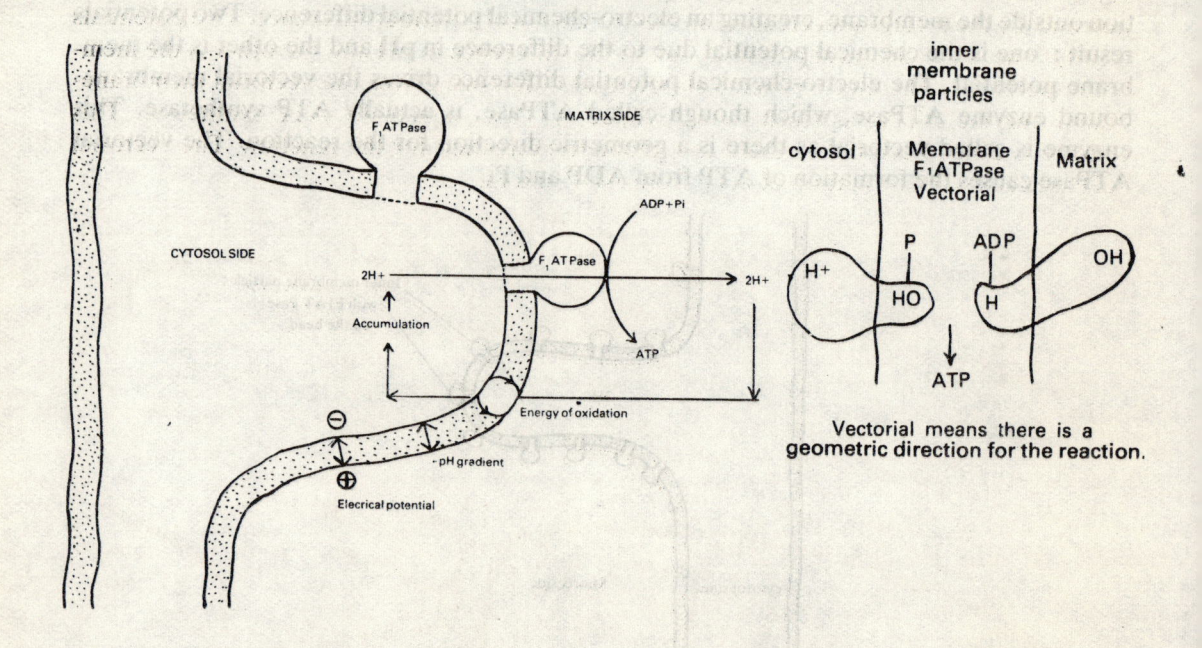

Experimental support in favour of chemiosmotic hypothesis

1) The mitochondrial inner membrane is impermeable to protons and other ions, and hence proton translocation to the outside of the coupling membrane is possible. Actually a proton gradient across the membrane is generated during electron transport. The pH outside is also 1.4 units lower than inside, and the membrane potential is 0.14v.

2) DNP and other uncouplers of oxidative phosphorylation increase the permeability of mitochondrial inner membrane to protons and reduce the electrochemical potential. This may prevent the vectorial ATP synthetase from producing ATP.

3) If protons are increased outside the membrane (say, by adding acid to the external medium in the in vitro systems), the generation of ATP is increased.

4) The Pi ratio (quotient of Pi taken up and H^+ transported out respectively) for ATP synthetase is half while the H^+/O for substrates succinate and betahydroxybutyrate are four and six, respectively. These ratios agree with the P/O ratios, 2:1 and 3:1, respectively.

5) Oxidative phosphorylation does not occur in soluble systems where there is no vectorial ATP synthetase due to the absence of a closed membrane. Both respiratory chain and ATP synthetase are organised or oriented directionally.

The chemiosmotic hypothesis can explain oxidative phosphorylation, respiratory control, ion transport, as well as the action of uncouplers and inhibitors, in a satisfactory manner.

Spatial orientation of components of the respiratory chain

Recent evidences have shown that the enzymes and the co-factors of the respiratory chain are oriented in a difinite spatial pattern on the mitochondrial membrane to give maximum efficiency (see Figure). It can explain clearly how the translocation of protons takes place at three different sites and how the proton channel and ATP synthetase system function simultaneously during respiration of the mitochondria.

The components of the respiratory chain are not the same as far as their solubility is concerned. Cytochrome c is water soluble while cytochrome b is lipid soluble. NAD^+ has affinity towards aqueous environment. It is the lipid soluble components like CoQ and the mitochondrial phospholipids that help to position the other components along with the proteins, in their proper orientation and conformation, to maximise the productive interactions and bring about maximum efficiency in this energy producing 'power house' of the cells.

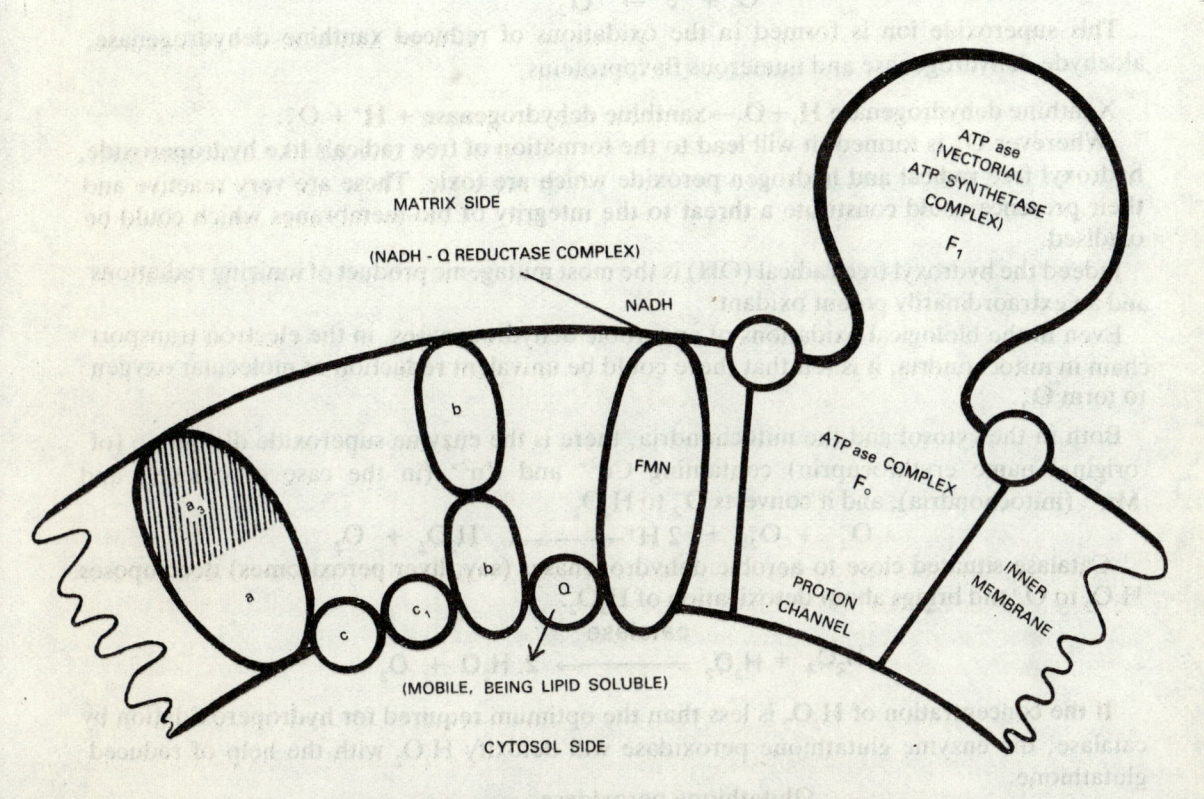

Fig. Spatial orientation of components of respiratory chain

Accessory factors for oxidation

In addition to the various intermediates whose precise role has been established in biological oxidations, a non-specific role has been assigned to glutathione and Vitamin C in oxidations. These substances are generally required to keep the enzymes in active form, i.e. as enzyme SH.

Cellular respiration can be demonstrated experimentally by Warburg's monometric technique (see 'Modern Techniques') and by Thunberg's technique using methylene blue.

Cytosolic oxidations; superoxide-ion

In oxidations by xanthine dehydrogenase, and aldehyde dehydrogenase in the cytosol, a divalent oxidation does not take place at one stroke as given below:

Xanthine dehydrogenase $H_2 + O_2 \rightarrow$ xanthine dehydrogenase $+ H_2O_2$.

This is because molecular oxygen is para-magnetic and contains two unpaired electrons with parallel spins. These unpaired electrons reside in separate orbitals, as two electrons cannot occupy the same orbital unless their spins are opposed. Reduction of oxygen by direct insertion of a pair of electrons into its partially filled orbitals is not possible without inversion of one electronic spin, and such an inversion of spin is a slow change. Hence, electrons are added to molecular oxygen as single electrons successively.

When the oxygen molecule takes up one electron it becomes superoxide ion.

$$O_2 + e \rightarrow O_2^-$$

This superoxide ion is formed in the oxidations of reduced xanthine dehydrogenase, aldehyde dehydrogenase and numerous flavoproteins.

Xanthine dehydrogenase $H_2 + O_2 \rightarrow$ xanthine dehydrogenase $+ H^+ + O_2^-$.

Wherever O_2^- is formed, it will lead to the formation of free radicals like hydroperoxide, hydroxyl free radical and hydrogen peroxide which are toxic. These are very reactive and their presence could constitute a threat to the integrity of bio-membranes which could be oxidised.

Indeed the hydroxyl free radical (OH) is the most mutagenic product of ionizing radiations and an extraordinarily potent oxidant.

Even in the biological oxidations of anaerobic dehydrogenases, in the electron transport chain in mitochondria, it is felt that there could be univalent reduction of molecular oxygen to form O_2^-.

Both in the cytosol and the mitochondria, there is the enzyme superoxide dismutase (of original name erythrocuprin) containing Cu^{2+} and Zn^{2+} (in the case of cytosol) and Mn^{2+} (mitochondria), and it converts O_2^- to H_2O_2

$$O_2^- + O_2^- + 2H^+ \dashrightarrow H_2O_2 + O_2$$

Catalase situated close to aerobic dehydrogenases (say, liver peroxisomes) decomposes H_2O_2 to O_2 and brings about detoxication of H_2O_2.

$$H_2O_2 + H_2O_2 \xrightarrow{\text{catalase}} 2H_2O + O_2$$

If the concentration of H_2O_2 is less than the optimum required for hydroperoxidation by catalase, the enzyme glutathione peroxidase will detoxify H_2O_2 with the help of reduced glutathione.

$$2\,GSH + H_2O_2 \xrightarrow{\text{Glutathione peroxidase}} G-S-S-G + 2H_2O$$

Thus, the superoxide ion formed, is converted to H_2O_2 by superoxide dismutase, and finally to oxygen by catalase. It could also be oxidised to oxygen by ferricytochrome.

$$\underset{\text{Cyto. C}}{O_2^- + Fe^{3+}} \longrightarrow \underset{\text{Cyto. C}}{Fe^{2+} + O_2}$$

In the cytosolic hydroxylations of sterols and drugs brought about by cytochrome P_{450} and cytochrome b_5, superoxide ion has been shown to be an intermediate. The hydrogens for these hydroxylations are donated by NADH (or NADPH) routed through FP_1 and FP_2, and cytochrome P_{450} or b_5.

$$\text{Cytochrome } P_{450} - \text{Substrate} - Fe^{2+} O_2 \xrightarrow{-e}$$
$$\text{Cytochrome } P_{450} - \text{Substrate} - Fe^{2+} O_2^-$$
$$\Big\downarrow \begin{array}{l} 2H^+ \\ + 2e \end{array}$$
$$\text{Cytochrome } P_{450}\, Fe^{3+} + \text{Substrate} - OH + H_2O$$

BIOENERGETICS

Bioenergetics aims at the study of the energy aspects of biological reactions. The various reactions could be broadly classified into *exergonic* reactions and *endergonic* reactions. Exergonic reactions are spontaneous reactions while endergonic reactions are not spontaneous and require energy for driving them. Many reactions in the body are endergonic. Synthesis, and transport of substances against electrical and/or concentration gradient (active transport) are endergonic. Even though endergonic, they do take place as they are coupled with some other reactions to make the coupled reaction exergonic.

Any reaction will be spontaneous and exergonic, if it is accompanied by a decrease of free energy. Free energy is the actual available energy, i.e., the chemical energy capable of causing chemical reactions. The absolute free energy of a compound is not known. It is the difference in free energy between the reactants and products that is measured. It is represented as ΔF or ΔG.

Free energy is related to heat content (H) and entropy (S) by the following equation:

$$\Delta F = \Delta H - T\Delta S$$

Enthalpy (H) which is a measure of the heat content represents the entire chemical energy. Entropy (S) represents the disorder or state of randomness in the molecule. Thus, the free or available energy is not the entire chemical energy contained, but the chemical energy less the entropy which is influenced by temperature (T).

ΔF_o is referred to as the standard free energy change and it is related to ΔF by the equation:

$$\Delta F = \Delta F_o + RT \ln \frac{(products)}{(reactants)}$$

ΔF_o the standard free energy change is the change of free energy when the concentrations (activities) of the participating substances are unity.

If a reaction as given below is to take place,

$$A + B \rightarrow C + D$$

the free energies of A and B should be higher than those of C and D so that the reaction is accompanied by a decrease of free energy. Only then, the reaction will take place spontaneously as in exergonic reaction. To generalise, exergonic reactions take place with decrease of free energy. On the other hand, if the free energies of A and B are such that they are less than those of the products, the overall reaction will mean an increase of free energy. Such a reaction cannot take place as it is like an uphill movement. This is what is encountered in endergonic reactions in which ΔF is positive. But, as indicated already, an endergonic reaction can be clubbed with an energy-giving reaction, such as hydrolysis of ATP, so that the coupled reaction is accompanied by a decrease of free energy. Thus the endergonic reaction is made into an exergonic one. This is what happens in the biosynthesis of many substances.

For a system in equilibrium, the change of free energy is zero.

A typical reaction which is exergonic is,

$$ATP + H_2O \rightleftharpoons ADP + Pi$$

$$\Delta F = -7.4 \text{ K cals/mol.}$$

An example of an endergonic reaction is the combination of glucose and fructose to form sucrose which is not possible as there would be increase of free energy.

$$\text{Glucose + Fructose} \rightarrow \text{Sucrose} + H_2O$$
$$\Delta F = +K \text{ cals}$$

The same reaction with UDP glucose is exergonic and takes place as follows:

$$\text{UDP glucose + fructose} \rightarrow \text{sucrose} + UDP + Pi$$
$$\Delta F = -K \text{ cals}$$

It has, however, to be understood that even in the case of exergonic reactions accompanied by a decrease of free energy, a certain amount of activation energy has to be supplied to the reactants. Only after acquiring it, will the substances react. The extent of activation energy needed to start the reaction can, however, be diminished by the influence of catalysts. Hence, the role of a catalyst and also a biocatalyst (enzyme) is to decrease the activation energy required for starting a spontaneous reaction accompanied with decrease of free energy. Many exergonic reactions will not take place if activation energy is not supplied.

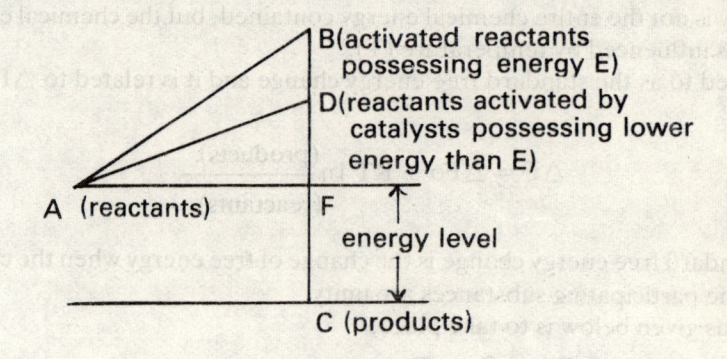

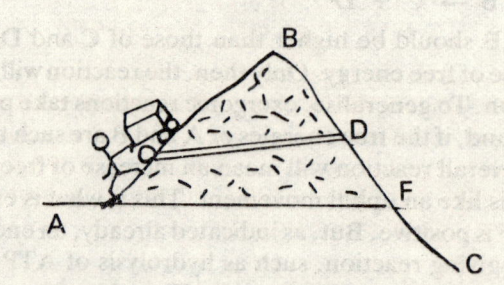

The trolley has to reach point B to run downhill of its own accord. This corresponds to activation energy (E) to be supplied to the reactants. However, the trolley can be taken with less energy through an alternate route to point D from where it will move down easily and reach the same destination as from B. This corresponds to the role of catalysts. The figure also shows that unless some initial energy is spent, the trolley will not ever move downhill.

There are different methods for the determination of ΔF_o. One of the simple methods is from the knowledge of the equilibrium constant K and using the equation,

$\Delta F_o = -RT \ln K$

where K is the equilibrium constant.

F_o can be determined from the equation relating it with electrode potential change in the reaction,

$$\Delta F_o = -nF \Delta E_o$$

where n is the number of electrons involved, F, a Faraday (96494 coulombs, and ΔE_o the standard redox potential in volts. The product obtained in joules is divided by 4.18 to give calories.

A third method is from the equation,

$$\Delta F = \Delta H - T\Delta S$$

for which knowledge of enthalpy and entropy changes is required. Another method is to use the data of standard free energies of formation of the substances taking part in a reaction and arriving at ΔF. The standard free energy of formation of elements is zero. Thus, for the oxidation of glucose represented by the equation,

$$C_6H_{12}O_5 + 6 O_2 \rightarrow 6 CO_2 + 6 H_2O$$

The standard free energies of formation in kilocals/mol for glucose, CO_2 and water at 25°c and atmospheric pressure are 219.2, 92.3 and 56.7, respectively. (It is zero for oxygen). ΔF_o for the reaction can be worked out as –675 K cals/mol of glucose; using the data available, i.e., $219.2 + 0 \rightarrow (6 \times 56.7) + 6 \times 92.3$

From a knowledge of ΔF_o (standard state), ΔF for any particular concentrations of the participants is calculated using the equation,

$$\Delta F = \Delta F_o + \frac{RT}{nF} \ln \frac{(Products)}{(Reactants)}$$

Depending on the concentration of the products and reactants, the value for ΔF can be changed from −ve to +ve. Hence, the concentrations of substances are very important in guiding the course of a reaction.

For example, the conversion of, say, glucose-1-phosphate and glucose-6-phosphate depends on the relative concentrations of the two substances.

Glucose-1 (P) $\rightleftharpoons$ Glucose-6-(P)

K for the reaction is 19 at 38°C

$\Delta F_o = -RT \ln K$
$= -RT \, 2.303 \log_{10} 19$
$= -1.987 \times 311 \times 2.303 \times 1.278 = -1800$ cals/mol.

If glucose-1-(P) is .01 M while glucose-6-(P) is .001 M,

$$\Delta F = \Delta F_o + RT \ln \frac{gl\ 6\ (P)}{gl\ 1\ (P)}$$

Substituting values for $\triangle F_o$ and concentrations, $\triangle F$ becomes equal to -3200 cal/mol. Hence, the forward reaction, i.e., conversion of glucose-1-(P) to glucose-6-(P) takes place with ease as there is decrease of free energy. If, on the other hand, the concentration of glucose-1 (P) is very low, i.e., 0001 M and glucose-6-(P) 01 M, $\triangle F$ becomes $+1040$ cal/mol. The forward reaction cannot take place as $\triangle F$ is positive. Instead, glucose-6-(P) will get converted to 1-(P). Thus the course of the reaction is decided by concentrations of substances taking part in a reaction.

It was shown that many endergonic reactions in the body are made to take place by supplying them with energy from energy-rich substances. These energy-rich substances transfer some of their groups to the substances taking part in the reaction and the potential arising out of such a transfer is termed 'the group transfer potential'. The twin substances of importance in Bioenergetics are ATP (adenosine triphosphate) and creatine phosphate. ATP donates energy for many biological reactions and makes the otherwise endergonic reactions exergonic. That is why in most reactions ATP will figure on the left side of the equation as its 'group transfer potential', will give the equivalent energy of about 7.4 K cals/mol. Other high energy compounds are 1:3 diphosphoglycerate, enol phosphate, arginine phosphate, carbamoyl phosphate, acetyl phosphate, active sulphate, S-adenosyl methionine, acyl CoA compounds, acyl adenylate and amino acyl adenylate. Methionine has to be converted to active methionine which takes part in a number of methylations. Fatty acids cannot be metabolised through ß oxidation if they are not initally converted to fatty acyl CoA. Likewise amino acids have to be converted first to amino acyl adenylate before they take part in protein synthesis.

In the list of high energy substances given above, 1:3 diphosphoglycerate and phosphoenol pyruvate have higher free energy than ATP itself. Hence they could combine with ADP (Adenosine diphosphate) to give ATP.

$$1:3 \text{ diphosphoglycerate} + \text{ADP} \longrightarrow \text{ATP} + 3 \text{ phosphoglycerate}$$
(higher energy than ATP)
$$\text{phosphoenol pyruvate} + \text{ADP} \longrightarrow \text{ATP} + \text{enolpyruvate}$$
(higher energy than ATP)

The above reactions, in which ATP is formed without oxidation at the substrate level are good examples of substrate phosphorylation. These are the few instances where ATP is on the right side of the equation.

ATP and creatine phosphate have about equal energy and hence the course of the reaction is decided by relative concentrations of ATP and CP.

$$\text{CP} + \text{ADP} \rightleftharpoons \text{ATP} + \text{C} \quad \text{(Lohmann's reaction)}$$

When a large amount of ATP is produced (oxidation of food), the equilibrium of the reaction is shifted to the left towards formation of CP. On the other hand, if ATP store is depleted and ADP is available in large amounts, the forward reaction, i.e., reaction of CP with ADP will take place, say, in fasting.

Thus, only from bioenergetics, i.e., the knowledge of free-energy change, one can explain why certain reactions occur freely in the body and how the course of reactions can be controlled and conducted at will.

9 Chemistry of Blood

Blood is considered a tissue consisting of red blood corpuscles (erythrocytes), white corpuscles (leukocytes), platelets and the liquid plasma. It is a carrier for gases, oxygen, carbon dioxide, metabolites, products of digestion, hormones, enzymes and clotting factors. It has also many buffer systems. The packed cell volume or hematocrit is about 45 per cent.

The specific gravity of blood is 1.054 – 1.060 for whole blood and 1.024 – 1.028 for plasma. Its viscosity is about 4.5 times that of water. The pH of blood is 7.35 – 7.45 relatively less for venous blood compared to arterial blood.

A 70 kg individual has a blood volume of about six litres (85 ml/kg), i.e., about one-twelfth of the body weight and about three litres (45 ml/kg) of plasma (i.e., about half that of whole blood). Blood volume and plasma volumes can be determined by using dyes like Evans blue (a blue dye) or radio-active substances like ^{131}I labelled human serum albumin or radio chromium.

Blood has many diverse functions:

1) Respiration—transport of oxygen from the lungs to the tissues and CO_2 from the tissues to the lungs.
2) Nutrition—transport of digested and absorbed nutrients.
3) Excretion—transport of the metabolic wastes to the excretory organs.
4) Maintenance of acid-base equilibrium by buffering action.
5) Regulation of fluid and electrolyte balance.
6) Maintenance of body temperature and osmotic pressure.
7) Defence against infections.
8) Transport of metabolites and hormones from the site of production to target organs and enzymes, chiefly the plasma-specific enzymes.

The substances present in the blood could be divided into three major types: the proteins like albumins and enzymes, neutral molecules like glucose and cholesterol and ionic species (electrolytes) like sodium, potassium and bicarbonate. Regarding nitrogen-containing substances alone, they are of two types viz., the proteins (6 – 8g/100 ml) and the non-protein nitrogen substances (15 – 35 mg/100 ml).

PLASMA PROTEINS

A list of plasma proteins and their normal levels in health is given in the next page. The difference between plasma and serum is that fibrinogen is present in plasma and absent in serum.

g per cent of serum unless otherwise specified:

Total	6 – 8	
Albumins	3.2 – 5.5	
Globulins	1.5 – 3	
Alpha$_1$ globulin	0.06 – 0.39	
Alpha$_2$ globulin	0.28 – 0.74	
Beta globulin	0.69 – 1.25	
Gamma globulins	0.8 – 2	
IgG	0.8 – 1.8	(75 per cent immunoglobulin)
IgA	0.15 – 0.4	(20 per cent immunoglobulin)
IgM	0.08 – 0.18	(5 per cent immunoglobulin)
IgD	0.003	
Fibrinogen	0.2 – 0.4	(plasma)

Fractionation and identification of plasma proteins

Various techniques have been employed for the fractionation of plasma proteins. Howe used a solution of sodium sulphate of different concentrations to separate plasma proteins into albumins, globulins and fibrinogen. While globulins are precipitated by half saturation using 27 per cent solution, albumins will be precipitated only on full saturation. Fibrinogen is a type of globulin precipitated on half saturation with NaCl. Cohn has resorted to alcohol treatment and low temperature to separate serum proteins. Ultracentrifugation, paper and gel (agar) electrophoresis, polyacrylamide gel disc electrophoresis, immuno-diffusion, and immuno-electrophoresis and iso-electric focussing are the modern techniques used to separate and characterise serum proteins. Such identifications, and a knowledge of the ratio of albumin and globulin are important as they may be of help in the diagnosis of diseases. The albumin-globulin ratio, A/G, is normally 1.2 : 1. But in chronic conditions it is reversed, as the globulin concentration is increased with a decrease of albumin. In certain kidney diseases (nephrotic syndrome) there is a generalised decrease of all the proteins, but alpha$_2$ globulin alone is increased. In acute infections, alpha$_1$ is increased, and in diseases referred to immunoglobulinopathies, there could be increase or decrease of immunoglobulins.

The normal electrophoretic pattern of human serum in paper or agar electrophoresis with veronal buffer shows five fractions, albumin, alpha$_1$ globulin, alpha$_2$ globulin, beta globulin and gamma globulin. (For diagram see 'Electrophoresis' under Modern Techniques.) In disc electrophoresis with polyacrylamide gel and tris buffer, serum proteins can be resolved into pre-albumin, post-albumin, ceruloplasmin, transferrin and haptoglobins also.

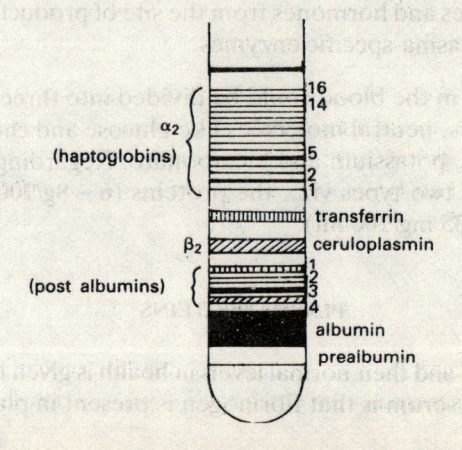

CHEMISTRY OF BLOOD

The separation of the proteins during electrophoresis is due to electrical charge on these colloidal proteins and to some extent on the molecular diameter. The latter is given below in respect of some proteins in comparision with Na^+ and Cl^- ion and glucose molecule.

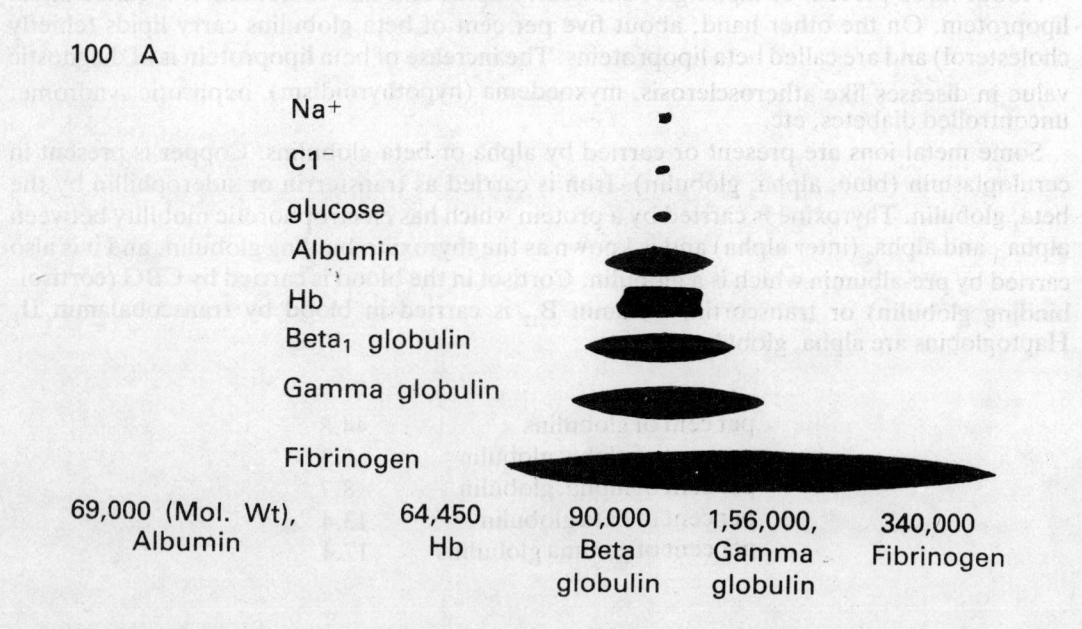

```
100   A
      Na+
      Cl-
      glucose
      Albumin
      Hb
      Beta₁ globulin
      Gamma globulin
      Fibrinogen
```

| 69,000 (Mol. Wt), Albumin | 64,450 Hb | 90,000 Beta globulin | 1,56,000 Gamma globulin | 340,000 Fibrinogen |

Albumins: They are present to the extent of 55.2 per cent in serum being synthesised in liver. Their half life is 17 – 20 days.

One of the functions of albumin is its major contribution to oncotic pressure of colloidal proteins. Albumin exerts greater oncotic pressure than globulin as its molecular diameter is less than that of globulin. If serum albumin is decreased to levels below about 2 g per cent edema can result, as in nephrotic syndrome. In chronic conditions like cirrhosis of liver, there is hypo-albuminemia (decreased albumin in blood). Though the levels of serum albumin (g per cent) are investigated routinely in a clinical biochemistry laboratory, the more useful investigation will be assessing the total circulating albumin in the body rather than the simple percentage of albumin in serum.

Howe's albumin (fractionated by Howe's method) has higher values than electrophoretic albumin, as the albumin in the former method is contaminated with unprecipitated globulins. Even the electrophoretic albumin is not absolutely homogeneous. However, one component constituting about two-thirds is homogeneous and is called mercaptalbumin as it contains one free – SH (mercaptyl group) per mol.

Albumin carries in blood, calcium, bilirubin, free fatty acids, cortisol and certain dyes.

Globulins: Globulins are very complex in structure and functions. Routine paper and agar electrophoresis separtes them into alpha₁, alpha₂, beta and gamma. Globulins are bigger molecules when compared with albumin, and hence for the same concentration by weight there will be fewer globulin molecules than albumin. Hence, the contribution of globulins to oncotic pressure (a colligative property) is less than that of albumin. Again, in some diseases of the kidney, albumin is lost to a much greater extent than globulins since filterability is decided by molecular size. Albumin which is a smaller molecule is more easily filtered than the globulins (Mol.wt.90,000 and above).

$Alpha_1$ and $alpha_2$ globulins contain carbohydrates and hence are called glycoproteins.

About three percent of alpha globulins carry lipids and this combination is called alpha lipoprotein. On the other hand, about five per cent of beta globulins carry lipids (chiefly cholesterol) and are called beta lipoproteins. The increase of beta lipoprotein is of diagnostic value in diseases like atherosclerosis, myxoedema (hypothyroidism), nephrotic syndrome, uncontrolled diabetes, etc.

Some metal ions are present or carried by alpha or beta globulins. Copper is present in ceruloplasmin (blue, $alpha_2$ globulin). Iron is carried as transferrin or siderophillin by the $beta_1$ globulin. Thyroxine is carried by a protein which has electrophoretic mobility between $alpha_1$ and $alpha_2$ (inter alpha) and is known as the thyroxine-binding globulin, and it is also carried by pre-albumin which is a globulin. Cortisol in the blood is carried by CBG (cortisol-binding globulin) or transcortin. Vitamin B_{12} is carried in blood by transcobalamin II. Haptoglobins are $alpha_2$ globulins.

per cent of globulins	44.8
per cent of $alpha_1$ globulin	5.3
per cent of $alpha_2$ globulin	8.7
per cent of beta globulin	13.4
per cent of gamma globulins	17.4

The $alpha_1$ globulin is increased in acute infections like pneumonia, typhoid, cholera, acute meningitis, etc., while $alpha_2$ is increased in diseases like nephrotic syndrome, diabetes mellitus, rheumatoid arthritis, pulmonary tuberculosis and leprosy. $Alpha_1$ and $alpha_2$ globulins are produced by the liver. Beta globulins are increased in diabetes mellitus, sarcoidosis and infectious mononucleosis.

Gamma globulins: These were originally considered to be proteins functioning as antibodies. Recently the name immunoglobulin has been given to proteins which have antibody function because there are certain gamma globulins which do not have antibody function, and antibody activity could be associated with certain beta and even alpha globulins. Unlike other proteins, these are not synthesised in the liver but are produced in the plasma cells, lymphoid tissue and reticulum cells. The organs are thymus, spleen, lungs and bone marrow. Gamma globulins have an approximate mol.wt. of 1,56,000.

The gamma globulins in the blood are increased in infective hepatitis, cirrhosis of liver, collagen diseases like lupus erythematosis and chronic infections like TB, kala azar, malaria and lymphogranuloma venereum (polyclonal gammopathy). In multiple myeloma, Waldenstrom's macroglobulinemia and Heavy Chain disease, abnormal immunoglobulins are produced (monoclonal gammopathy). In all the above, the A/G ratio is reversed.

Immunoglobulins: These serum proteins are the 'soldiers of the body' in that they perform the function of defence from invaders. Such an activity is referred to as antibody function. Immunoglobulins are therefore called antiobodies. The infecting agents or substances are the antigens. The antibodies react with the antigens and destroy the latter or make them powerless. Immunoglobulins have been defined by the WHO as follows:

1) They should be proteins.
2) They should be of animal origin.
3) They should have a common structure.
4) They should function as antibodies. However, related proteins like myeloma proteins and Bence Jones proteins which do not have antibody function can also be considered as immunoglobulins.

The structure of each molecule of immunoglobulin is made up of what are known as light chains and heavy chains. As the name itself indicates, the light chains have less molecular weight and less number of amino acids in the polypeptide chain than heavy chains. In a molecule of immunoglobulin there are two light chains and two heavy chains connected by disulphide linkages. In addition there are intra-chain disulphide linkages. The molecule also possesses carbohydrate moiety attached to heavy chain.

In addition to higher molecular weight, heavy chains are also governed by a specific sequence of amino acids in selected regions. This applies to light chains also. One half of the light and heavy chains containing the N terminus is called the variable portion while the other half is the constant region in respect of amino acid composition and sequence. Recently 'hypervariable' portions responsible for antibody action have been reported.

Immunoglobulins - Classes: Immunoglobulins are represented by 'Ig'. Five different heavy chains are known so far. They are gamma, alpha, μ, δ and ε. Depending on the heavy chain, five classes of immunoglobulins are known. They are IgG, IgA, IgM, IgD and IgE, respectively, i.e., Ig G has two heavy chains which are gamma while Ig A has two heavy chains which are alpha. G stands for gamma, A for alpha. M for μ, D for δ, and E for ε.

Types: The light chains are of two types: one is kappa 'k' and the other lambda 'λ'. For each class of immunoglobulin, two types are possible, decided by the nature of the light chain. Thus IgG can have either the kappa light chain or the lambda light chain. Similarly other immunoglobulins also have two types. Thus five classes can have two types each, and thus a total of ten different immunoglobulins have been found.

If the IgG contains the kappa light chain, it belongs to the K type (after Dr Korngold), while if it has the lambda light chain, it belongs to the L type (after Dr Lipari).

Table 1

Class	Type	Heavy chain	Light chain
Ig G	K	γ	k
Ig G	L	γ	λ
Ig A	K	α	k
Ig A	L	α	λ
Ig M	K	μ	k
Ig M	L	μ	λ
Ig D	K	δ	k
Ig D	L	δ	λ
Ig E	K	ε	k
Ig E	L	ε	λ

Typical structure of Ig G: This has two light chains (Mol.wt.22,000 with a total of 214 amino acids) and two heavy chains (Mol. wt. 55,000 and a total of 446 amino acids) united by di-sulphide links, and the whole structure has the shape of the letter Y. There is also a hinge portion. The molecule is about 200 Å in length and 40 Å in breadth.

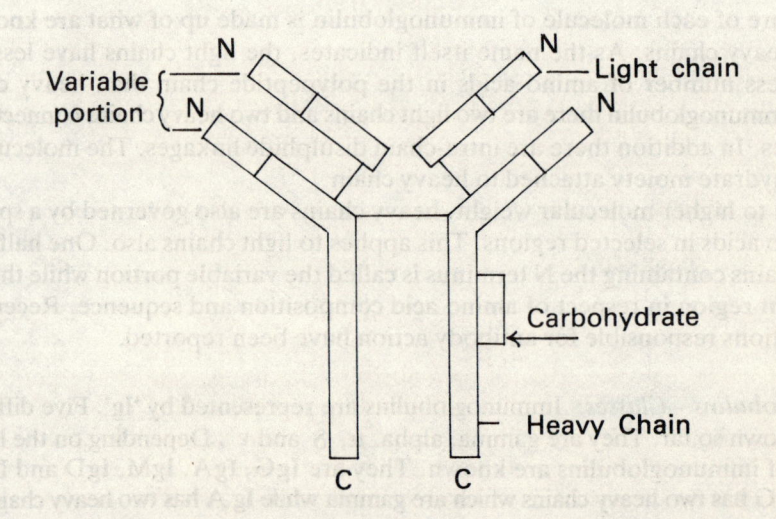

```
         N           N
Variable {   N          N  } ── Light chain
portion  {                    
                              
                              ── Carbohydrate
                              
                              ── Heavy Chain
         C           C
```

The antigen combining sites are in the N terminus of the variable portions of the light and heavy chains. Recently, it has been found that it is not even the variable portion but the hyper variable portion (specific locations in the variable portion) that acts as the antigen-combining site.

Only Ig G can cross the placenta. Of the various immunoglobulins, it is present to a much greater extent in serum and has a low carbohydrate content. The molecular weight of Ig M is quite high (900,000) while others have only 150,000 to 196,000.

The relative amounts present in serum are:

$$\text{IgG (75 per cent) at 1.24 g per cent}$$
$$\text{IgA (20 per cent) at 0.039 g per cent}$$
$$\text{IgM (5 per cent) at 0.12 g per cent}$$
$$\text{IgD (1 per cent) at 0.003 g per cent}$$

Like various other substances in the body, immunoglobulins also undergo flux. The biological half life period Ig G is about 23 days. It appears that prior to destruction, the antibody has to form a complex with the antigen, and this combination undergoes phagocytosis in the reticulo-endothelial elements.

The study of the structure of immunoglobulins has been done mostly by using myeloma proteins which have all properties in common with Ig except antibody function. (They are altered Ig molecules.) These accumulate in multiple myeloma (a fatal disease) with local tumours particularly in the bones as well as with disseminated infiltration of the bone marrow. The elucidation of the structure of Ig has been made easier by a selective cleavage of the immunoglobulin molecules by the enzymes pepsin and papain.

SPECIAL PROTEINS

Ceruloplasmin

This has copper as an integral part of a glycoprotein with 7-10 per cent carbohydrate and an alpha, globulin, and is a ferro-oxidase (mol. wt. about 160,000). It contains 0.34 per cent copper (6 – 8 atoms of copper per molecule – half as cuprous and the other half as cupric). The normal serum contains about 30 mg per cent of ceruloplasmin. There is deficiency of ceruloplasmin in Wilson's disease.

Ceruloplasmin promotes hematopoiesis and is needed for the oxidation of ferrous to ferric. It has the enzyme effect and oxidises certain amines like epinephrine. It has oxidase activity in vitro also.

Thyroxine-binding globulin

This is a glycoprotein (molecular weight 50,000) containing sialic acid and binds to T_3 and T_4 with 100 times more affinity than thyroxine-binding pre-albumin (TBPA). The amount of thyroxine is 1.5 mg/7g of total protein. It has an electrophoretic mobility between $alpha_1$ and $alpha_2$ – inter-alpha. TBG is synthesised in liver and its production is increased by estrogens. This happens during pregnancy and after use of birth control pills, due to the increased circulating estrogens.

Pre-albumin

Pre-albumin moves ahead of albumin in electrophoresis and is a globulin. There are different fractions. The molecular weight is 66,000. The tryptophan-rich pre-albumin is important as it carries thyroxine. It is therefore called thyroxine binding pre-albumin. It also binds retinol-binding protein (RBP) with retinol. About one third of TBPA of human serum circulates as the TBPA-RBP complex.

Transferrin

This is a beta globulin and contains iron. It has 5.5 per cent carbohydrates.

Haptoglobins

The class of haptoglobins bind free hemoglobin and protect the kidneys from being damaged by filtration of hemoglobin in hemolytic condition. There are different types: 1.1(Hp^1), 2.1 and 2.2 (Hp^2). Type Hp^1 binds one molecule of hemoglobin per molecule while Hp^2 binds different numbers of hemoglobin molecules. Haptoglobins have two polypeptides – alpha and beta chains.

Transcortin (cortisol-binding globulin)

This is a mucoprotein with a molecular weight of 52,000. This binds cortisol and has a high affinity of 20 μg cortisol/100 ml which amounts to 34 mg CBG/100 ml. Though it has remarkable affinity, its saturation capacity is poor. On the other hand, albumin has low affinity for cortisol but high saturation capacity. CBG is also synthesised in liver and its synthesis is increased by estrogens as in the case of TBG

Transcobalamin II

Vitamin B_{12} binds to Transcobalamin II and circulates in blood.

Interferons

Interferons are a class of cellular glycoproteins (mol.wt. 26,000 – 38,000) produced early in the viral infection of cells. Other inducers are micro-organisms, inactivated viruses, fungal extracts, synthetic polynucleotides, polycarboxylates and smaller molecules such as cyclo-heximide. Interferons are non-toxic and potent, and inhibit the multiplication of viruses of both RNA and DNA types and the oncogenic viruses at the intracellular level. According to their antigenic types, they are classified into three main classes: IFN α, IFN β and IFN γ. They are not virus-specific but only cell-specific. Different viruses infecting the same cell leads to the production of the same interferon.

Interferon production is inhibited at low temperatures. Stress, serotonin, epinephrine and nor-epinephrine inhibit IFN formation.

Interferon limits or prevents viral multiplication at the place of entry and in the target organs. Circulating interferon produced within a few hours of onset of viremia will gain entry into body tissues, thereby protecting the target organs from viral infection.

Special mention may be made on the prevention of viral oncogenesis, i.e., cancer of viral etiology. Interferon inhibits ornithine decarboxylase and also inhibits increased production of polyamines in many types of cancer. IFN induces the enzyme 2', 5' oligo-A-synthetase which is needed for the polymerisation of ATP to 2',5' oligo adenylate. This substance activates one RNAse-L (latent endonuclease) which cleaves the viral RNA and prevents viral multiplication. It can also inactivate the Initiation Factor$_2$ required for protein synthesis for viral coat and activate phospho-diesterase which inhibits peptide elongation, through removal of cAMP.

Because of the above-mentioned effects, Interferon is anti-viral and anti-tumorous. It is of great significance in the treatment of renal cell cancer, multiple myeloma, lymphomas, breast cancer, leukemias of viral origin, enteric viral infection, connective tissue disorders and chronic neurological diseases. However, exogenous IFN is found to be not as active as endogenous IFN because of its short half life. Moreover, high doses are required for getting the desired effect.

ACID-BASE BALANCE AND THE MAINTENANCE OF pH OF BLOOD

Homeostasis is the maintenance of a constant internal environment in the body. The regulation of the pH of blood is one of the finest homeostatic mechanisms in the system. Acidic as well as basic factors from our diet and from diverse metabolic processes are continually being discharged into the systemic circulation. Nevertheless the pH of blood is maintained within a remarkably constant range of 7.35 – 7.45 due to several lines of defence that efficiently operate in the body.

Buffers constitute the foremost factors that bear the brunt of fluctuations in the acid-base balance, provided they are within limits. The lungs and the respiratory centre are almost concerned in a minute-to-minute role in acid-base homeostasis, either by way of elimination of carbon dioxide or retention of carbon dioxide according to the acid-base status of the individual.

The kidneys, apart from filtration, have got an equally important part to play in the secretion of hydrogen ions and the excretion of more of acid phosphates or less, as the need may be. Also, the renal production of ammonia is influenced by the acid-base status of the person. The kidneys are also concerned with re-absorption of sodium bicarbonate and its excretion, dictated by the needs.

The essential principles of buffer action is the transformation of a strongly dissociated acid into a weakly dissociated one or of a strongly dissociated base into a weakly dissociated base. To illustrate such an action, lactic acid which tends to accumulate during strenuous muscular exercise is effectively buffered by the sodium bicarbonate of plasma to form carbonic acid and sodium lactate. The reserve of bicarbonate still left over in the plasma after such a neutralisation is referred to as the alkali reserve. Hence, the alkali reserve represents the reserve of sodium bicarbonate left over after the neutralisation of most of the non-volatile acids more strongly dissociable than carbonic acid. The carbon dioxide combining power of plasma is a measure of the alkali reserve since the major form of carbon dioxide carriage is as sodium bicarbonate of plasma. The normal CO_2 combining power of plasma is 50–70 volumes per cent (22–30 mEq/L). Values much below 50 point to metabolic acidosis while values much above 70 denote metabolic alkalosis.

NORMS IN RESPECT OF BLOOD CONSTITUENTS
[Per 100 ml of blood-serum (unless otherwise stated)]

pH (arterial blood)	... 7.35–7.45
pH (venous blood)	... 7.33–7.43
Sugar (blood), Folin & Wu	... 80–120 mg
Sugar (blood), Nelson and Somogyi (values close to true sugar)	... 60–90 mg
Urea (blood)	... 14–38 mg
Proteins (total)	... 6.5–7.5 g
Albumins	... 4–5.5 g
Globulins	... 2–3 g
Non-protein nitrogen (blood)	... 10–30 mg
Electrolytes	
Sodium	... 136–149 mEq/L
Potassium	... 3.8–5.2 mEq/L
Chloride (as sodium chloride)	... 100–107 mEq/L
Cholesterol	... 140–250 mg
Bilirubin	... 0.1–0.8 mg
Uric acid	... 3–5 mg
Creatinine	... 0.6–1.4 mg
Calcium (total)	... 9.5–10.5 mg
Phosphate (inorganic)	... 3–4.5 mg (adults)
	... 4–6 mg (children)
Iron	... 65–175 micro g
Fibrinogen (Plasma)	... 200–400 mg
Plasma bicarbonate (inclusive of dissolved CO_2)	... 50–70 volumes (22–30 mEq/Litre)
Protein Bound Iodine	... 3–8 micro g
Enzymes	
Alkaline phosphatase	... 4–11 K.A. Units
Acid phosphatase (total)	... 1–5 K.A. Units
Acid phosphatase (tartrate labile-prostatic)	... Less than 0.8 K.A. Units
A.S.T.(G.O.T.)	... 2–20 m.I.U./ml
A.L.T. (G.P.T.)	... 2–15 m.I.U./ml
Amylase	.. 80–180 Somogyi Units

However, when strongly dissociated acids tend to accumulate, as in severe muscular exercise, starvation or diabetes mellitus, though the plasma bicarbonate is the first line of defence which serves to neutralise such strong acids, the pH of blood cannot be maintained at the normal level unless the carbonic acid formed, as a result of the neutralisation, is also disposed of. Though carbonic acid is weakly dissociable, a quantitative accumulation lowers the pH of the blood and can be buffered only by hemoglobin.

Like other proteins, plasma proteins, by virtue of their amphoteric property act like acids in the alkaline medium of the blood, whereby the carboxyl group exerts a firm hold on the predominant extra-cellular cation, namely, sodium ion. However, the buffer action of plasma proteins accounts for only one sixth of the buffer action of hemoglobin.

Phosphates in blood, because of their low concentration, are not of the same importance as the other buffers, though they are effective urinary buffers.

The pH of a buffer system is represented by the Henderson-Hasselbalch's equation as follows:

$$pH = pK + \log_{10} \frac{(\text{conc. of salt})}{(\text{conc. of acid})}$$

Acid here refers to the weakly dissociated acid, while salt denotes the strongly dissociated salt of the weakly dissociated acid. Applying the equation, pH of blood, i.e.,

$$7.4 = 6.1 + \log_{10} \frac{(BHCO_3)}{(H_2CO_3)}$$

6.1 being the pK of carbonic acid. On simplification

$$1.3 = \log_{10} \frac{(BHCO_3)}{(H_2CO_3)}$$

Taking the antilog of both sides, $BHCO_3/H_2CO_3 = 20/1$ (at pH 7.4). If this ratio is altered, the pH accordingly fluctuates and returns to normal if the ratio is maintained constant.

The buffers of the blood are extra-cellular as well as intra-cellular which are in the erythrocytes. The extra-cellular buffers are: 1) $H_2CO_3/NaHCO_3$, 2) NaH_2PO_4/Na_2HPO, and 3) H Protein/Na Protein. The intra-cellular pairs are 1) $H_2CO_3/KHCO_3$, 2) HHb/KHb, 3) $HHbO_2$ $KHbO_2$, and 4) KH_2PO_4/K_2HPO_4. All these buffer pairs are in equilibrium with one another. The most important extra-cellular buffer is the bicarbonate pair and the intra-cellular buffer is the hemoglobin buffer. The sodium or potassium is not confined exclusively to the plasma or erythrocytes, respectively, but each ion predominates in the respective compartment.

Role of hemoglobin in acid-base balance

Hemoglobin is the chief participant in acid-base homeostasis. The starting point is oxygenation of hemoglobin in the lungs due to the high partial oxygen tension, and the subsequent deoxygenation of oxyhemoglobin in the tissues due to the low partial oxygen tension but without any change in the pH or erythrocytes. Because of the increased CO_2 and decreased O_2 tension, HbO_2 dissociates in the erythrocytes of tissue capillaries. CO_2 diffuses into the erythrocytes and is rapidly hydrated to H_2CO_3 under the influence of carbonic anhydrase. H_2CO_3 tends to lower the pH while dissociation of HbO_2 to Hb tends to raise the pH within the erythrocyte. Consequently the protons formed from the dissociation of H_2CO_3 are accepted by the groups of Hb. The net result of these two events is to maintain the pH unchanged. Now K^+

ions get neutralised by HCO_3^- previously neutralised by HbO_2. These red cell HCO_3^- ions leave the erythrocytes of capillary into venous blood. This set of transformations is termed the 'isohydric shift'.

The iso-hydric change enables the conversion of the invading carbon dioxide to bicarbonate so that hemoglobin is the most important buffer against any pH change brought about by this.

More factors than one contribute to the disposal of carbon dioxide by hemoglobin. Carbonic anhydrase inside the RBCs is the chief enzyme concerned. Conditions are ideal for rapid diffusion of oxygen and carbon dioxide. The partial tension of the gases and other factors, like the thinness of the erythrocyte membrane and its osmotic behaviour and the small calibre of the capillaries and temperature facilitate what is known as Hamburger's phenomenon or the chloride-bicarbonate shift. The Gibbs-Donnan equilibrium partly accounts for the differential distribution of electrolytes between the plasma and the erythrocytes. The sequence of events that occur within the erythrocytes in the tissue capillaries and the pulmonary capillaries are diagrammatically represented on the next page.

In the tissues: The oxygen tension in the tissues being low compared to that in the lungs favours the dissociation of the strongly acidic oxyhemoglobin into the weakly acidic reduced hemoglobin and oxygen which readily diffuses into the tissues. This reaction facilitates the formation of $KHCO_3$ from H_2CO_3, i.e., salt from acid.

The potassium compound of reduced hemoglobin combines with carbonic acid, formed by the hydration of carbon dioxide and catalysed by carbonic anydrase, the CO_2 having entered the cells due to the higher CO_2 tension obtaining outside the cells. Consequently, there is an increase in the concentration of bicarbonate within the cells. As a result of the inequality in the concentration of bicarbonate within and outside the cell, there has to be a shift of bicarbonate ions from the cells into the plasma, which is the primary feature, necessitating in turn a secondary chloride shift from the plasma into the cells to preserve electrical neutrality. An equal number of chloride ions migrate for the number of bicarbonate ions that shift, in accordance with Donnan membrane equilibrium. Carbon dioxide also combines with hemoglobin at the amino group to form carbamino-hemoglobin under the low oxygen tension obtaining in the tissues so that carbon dioxide is in the bound state more in the tissues.

$$Hb\ NH_2 + CO_2 \rightleftharpoons Hb\ NH\ COOH$$

Within the erythrocytes in the pulmonary capillaries, due to the prevailing high oxygen tension, one encounters diametrically opposite state of affairs to what is seen in the capillaries in the tissues. Being at a higher partial tension in the alveolar air, oxygen readily diffuses into the cells and oxygenates the hemoglobin to oxyhemoglobin and oxygenates the potassium compound of hemoglobin to the potassium compound of oxyhemoglobin. Potassium bicarbonate present in the cells reacts with the strongly acidic oxyhemoglobin forming carbonic acid and potassium oxyhemoglobinate with the result that the bicarbonate concentration within the cell falls. The salt $KHCO_3$ is converted to acid H_2CO_3 in view of the acidic nature of $H\ HbO_2$. To replenish the depletion of HCO_3 there is a migration of bicarbonate ions from the plasma into the cells, and again to preserve electrical neutrality, an equal number of chloride ions shift from the cells to the plasma as a secondary shift. This time they move in the opposite direction to what is seen in the tissue capillaries, again in accordance with Donnan membrane equilibrium.

CHLORIDE SHIFT

Erythrocytes in lungs

Alveolar air (Difference in partial tension)

$H\ Hb + O_2 \rightarrow H\ Hb\ O_2$ (Strongly acidic)

$K\ Hb + O_2 \rightarrow K\ Hb\ O_2$

$KHCO_3 + H\ HbO_2 \rightarrow K\ HbO_2 + H_2CO_3$

$H_2CO_3 \xrightarrow{CA} H_2O + CO_2$

$Hb\ NH\ COOH \rightarrow Hb\ NH_2 + CO_2$

Expired air (Difference in tension)

Cl^- HCO_3^-

Plasma $Na^+ + HCO_3^-$

Erythrocytes in tissues

Tissues (Difference in tension)

$CO_2 + H_2O \xrightarrow{CA} H_2CO_3$

$CO_2 + Hb\ NH_2 \rightarrow Hb\ NH\ COOH$

$K\ Hb\ O_2 \rightarrow K\ Hb + O_2$

$H\ Hb\ O_2 \rightarrow H\ Hb + O_2$

$H_2CO_3 + K\ Hb \rightarrow H\ Hb + KHCO_3$

$KHCO_3 \rightarrow K^+ + HCO_3^-$

HCO_3^- Cl^-

$K^+ + Cl^-$

H_2O — Increase in size due to water intake

Tissue fluid

Plasma (difference in partial tension)

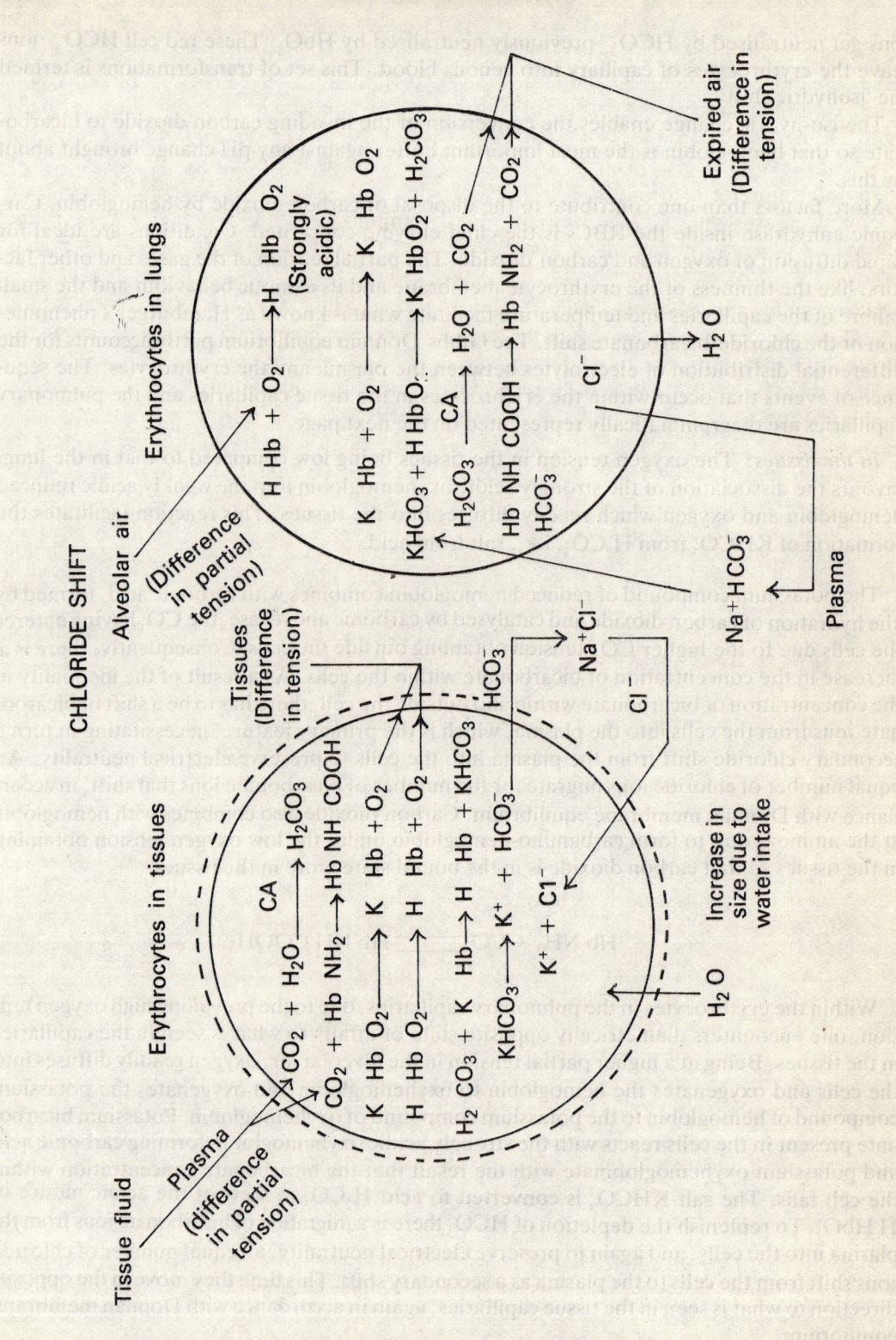

$Na^+ Cl^-$

Carbonic acid formed earlier is split up into carbon dioxide and water, catalysed by carbonic anhydrase. Carbamino hemoglobin under the prevailing oxygen tension readily splits up into carbon dioxide and hemoglobin. Hence, carbon dioxide is more in the free form in the lungs. The carbon dioxide formed readily diffuses out into the expired air because the partial pressure of carbon dioxide is higher in the cell than in the alveolar air.

The total number of osmotically active particles in the cells increases on the uptake of CO_2 and decreases when CO_2 is given off. This is compensated for by a flow of water into the cells as CO_2 is taken up, and a flow from the cells as CO_2 is given off in the lungs. This results in the erythrocytes on the venous side being slightly larger than those on the arterial side.

Role of kidneys in acid-base balance: Though carbon dioxide is eliminated chiefly via the expired air and constitutes the major pathway in acid-base homeostasis, there are non-volatile acids resulting from metabolic processes like phosphoric acid and sulphuric acid which are eliminated by the kidneys in the form of salts. Sodium partially buffers such acids, but the distal convoluted tubules of the kidneys are concerned in the acidification of urine, which helps conserve the cation reserves of the body. Though the glomerular filtrate is almost as alkaline as blood, a tubular ionic exchange is believed to be responsible for the transformation of the filtrate into an acidic urine. The hydrogen ion from the dissociation of carbonic acid formed by carbonic anhydrase, which is secreted by the tubular epithelial cells of the kidneys, gets exchanged for one of the basic sodium ions of the dibasic monohydrogen phosphate of the glomerular filtrate. This results in the formation of the monobasic sodium dihydrogen phosphate, which accounts for the titrable acidity of the urine. The other basic sodium ion combines with the bicarbonate ion to be reabsorbed as sodium bicarbonate by the tubules, thereby helping in the conversation of alkali reserve for the body.

$$Na_2 HPO_4 + H_2CO_3 \rightarrow Na H_2PO_4 + NaHCO_3$$
$$\text{i.e. } HPO_4^{--} + H_2CO_3 \rightarrow H_2PO_4^- + HCO_3^-$$

A second role of the kidney is the formation of urinary ammonia by hydrolytic deamination, chiefly of glutamine which is catalysed by the glutaminase secreted by the tubules. Ammonia formation by the kidneys serves to neutralise acidity under physiological as well as pathological conditions. Normally, one excretes 0.5 – 1g of ammonia per day almost wholly of renal origin. Moreover ammonia produced by the kidneys helps in the conservation of the fixed cations for the body, e.g, in diabetic ketosis the action of sodium bicarbonate of plasma results in the formation of carbonic acid and the sodium salts of aceto-acetic acid and beta hydroxybutyric acid. The ammonium ion gets exchanged for the basic ions of the salts resulting in the formation of the corresponding ammonium salts which are excreted in the urine while sodium bicarbonate which is also formed is reabsorbed by the tubules. When there is an alkalosis, the kidneys excrete less of acid salts unlike in acidosis where the kidneys excrete more of acid salts. In acidosis, the kidneys produce more of ammonia while it is suppressed in alkalosis as a compensatory phenomenon. Kidney diseases may, result in defective glomerular filtration of acid metabolites, or in an inadequate acidification by the distal convoluted tubules and diminished ammonia fomation. Consequently, a renal type of metabolic acidosis may supervene.

Types of acid-base imbalance and diagnostic biochemical parameters: Any alteration produced in the ratio between carbonic acid and bicarbonate results in an acid-base imbalance. Acidosis and alkalosis can either be respiratory or metabolic in origin and they may be compensated or uncompensated.

Since pCO_2 which denotes the partial pressure of carbon dioxide is related to the H_2CO_3 content, the respiratory system influences the carbonic acid content. Hence, a rise in CO_2 content is brought about in respiratory diseases causing retention of CO_2. Respiratory acidosis can occur in pneumonia, emphysema, bronchial asthma, congestive cardiac failure or in morphia-poisoning which is known to depress the respiratory centre. Inadequate pulmonary ventilation due to a defect in the respirator can also result in a respiratory type of acidosis. These conditions can be compensated for by a concomitant rise in the bicarbonate level due to the increased reabsorption of bicarbonate by the renal tubules. The kidneys on the other hand, excrete more of bicarbonate to compensate an alkalosis, though, when the latter is associated with a potassium deficit, the kidneys defer rather to the necessity of conserving elecrolytes than to maintaining acid-base homeostasis. In this case an acidic urine may be excreted and thus it is quite reconciliable with a high plasma bicarbonate.

Respiratory alkalosis is met with due to hyperventilation, on account of hypoxia at high altitudes causing excessive elimination of carbon dioxide and therefore lowering of the H_2CO_3 content. The hyperventilation may be voluntary or forced. Other causes are hysteria, anxiety states, encephalitis and early stages of salicylate poisoning as well as hepatic coma. Overzealous resuscitation measures can inadvertently convert metabolic acidosis into respiratory alkalosis. The pH of blood is helpful in the diagnosis of uncompensated states of acid-base imbalance.

The CO_2 content of plasma which is a measure of both bicarbonate and carbonic acid is higher than normal in compensated respiratory acidosis and lower than normal in compensated respiratory alkalosis.

Metabolic types of acid-base imbalance are characterised by alterations in the bicarbonate content of blood. An uncompensated type of metabolic acidosis is one where the bicarbonate content of plasma goes down without a commensurate fall in the carbonic acid content, as in diabetic ketosis or the ketosis of starvation. Compensation is brought about by stimulation of the respiratory centre due to the direct effect of excess of CO_2 and also via the chemo-receptors, i.e., the carotid and aortic bodies. On the other hand, metabolic alkalosis as in pyloric obstruction can be compensated by suppression of the respiratory centre resulting in the retention of carbon dioxide. The kidneys compensate metabolic acidosis usually by excreting more of BH_2PO_4 and more of ammonia, while they compensate alkalosis by excreting less of BH_2PO_4, more of $NaHCO_3$ and less of ammonia. Renal disease may cause a retention type of metabolic acidosis. Excessive loss of alkaline intestinal fluids because of diarrhoea or colitis can result in metabolic acidosis.

In metabolic alkalosis, the plasma bicarbonate level goes up. There may be a concomitant increase in carbonic acid or it may not be there when it is uncompensated. Apart from pyloric obstruction in general, a high intestinal obstruction, over ingestion of alkali in the treatment of pe tic ulcer or prolonged gastric suction can cause alkalosis by loss of gastric hydrochloric acid. A hypochloremic alkalosis is encountered due to loss of HCl and is often associated with potassium deficit. More chloride is lost than sodium and compensated by bicarbonate ions.

Merits and limitations of the various biochemical investigations relating to the assessment of acid-base imbalance.

The pH of blood enables the recognition of an uncompensated type of acid-base imbalance, respiratory or metabolic in origin. On the other hand, determination of the CO_2 combining power of plasma, which is a measure of the alkali reserve, can help in the diagnosis of the metabolic types of acidosis and alkalosis. But in the respiratory types, the alkali reserve determination is not helpful. On the other hand, the plasma bicarbonate may even go up in respiratory acidosis due to the compensatory role of the kidneys which reabsorb more of bicarbonate. The CO_2 combining power for example in metabolic acidosis, enables proper corrective therapy to be instituted. However, the CO_2 combining power does not give any idea of the ratio between carbonic acid and bicarbonate and so of the pH of blood as well.

The carbon dioxide content refers to the carbonic acid and bicarbonate present in blood which is collected under liquid paraffin to prevent escape of gases. It is expressed as volumes per 100 ml under standard conditions of temperature and pressure. The carbon-dioxide liberated by acidification and by imposition of vacuum is measured. Venous blood obviously has a higher carbon dioxide content than arterial blood.

Usually the CO_2 combining power and CO_2 content are the same though not identical for practical purposes unless the CO_2 tension in the alveolar air of the patient is less than 40 mm. In this case the CO_2 combining power or capacity cannot be equated with the carbon dioxide content. Under such circumstances, the necessity arises for the equilibration of the plasma of the test case with normal alveolar air, with CO_2 tension of 40 mm of Hg.

Even the determination of the CO_2 content has got its own limitations. This is illustrated by salicylate poisoning where, in the early stages, due to the toxic effect of the drug on the respiratory centre, there is a stimulation and a consequent hyperventilation resulting in a respiratory type of alkalosis. If the CO_2 content alone is determined and not the pH of blood simultaneously, there is a likelihood of the respiratory alkalosis being mistaken for metabolic acidosis. The pH of blood would be increased reflecting the real state of affairs, but subsequently a really metabolic type of acidosis supervenes due to the toxic effect of salicylates on the kidneys. Again, if one banks upon the CO_2 content for diagnostic purposes, there is a likelihood of the metabolic acidosis that has really set in being mistaken for a continuation of the respiratory alkalosis, for the correction of which therapy would have been already instituted. But now in the face of the metabolic acidosis these measures would aggravate the condition. Hence it is the pH of blood which enables a correct diagnosis. It will be lowered pointing to the acidotic state of affairs.

In short, the pH of blood, alkali reserve, as well as the CO_2 content, when determined, have to be correlated judiciously. Only then a correct conclusion can be arrived at regarding the acid-base status of the individual and the exact nature of the corrective therapy can be instituted.

BIOCHEMISTRY OF BLOOD CLOTTING

Clotting or coagulation of blood is the major mechanism to arrest the escape of blood. The simplest scheme of biochemistry of blood coagulation is the one proposed by Marawitz and Howell. According to them, the soluble protein fibrinogen in plasma is converted to insoluble fibrin by an enzyme thrombin. Thrombin itself is not present as such in normal plasma but is liberated from its inactive precursor prothrombin. It has since been found that clotting of blood is much more complicated and as many as eleven factors participate. They are as follows:

Factor		Name
I	...	Fibrinogen
II	...	Prothrombin
III	...	—
IV	...	Ca^{2+} ions
V	...	Labile factor pro-accelerin accelerator (Ac-) globulin
VI	...	—

VII	...	Proconvertin, serum prothrombin conversion accelerator (SPCA)
VIII	...	Anti-hemophilic factor, anti-hemophilic globulin (AHG)
IX	...	Plasma thromboplastin component (PTC), Christmas factor
X	...	Stuart-Prower factor
XI	...	Plasma thromboplastin antecedent (PTA)
XII	...	Hageman factor
XIII	...	Laki-Lorand factor (LLF)

There are no factors corresponding to numbers III and VI.

The four distinct biochemical processes in clotting of blood are:
1. Activation of Stuart-Prower factor, i.e., X to Xa
2. Conversion of prothrombin to thrombin
3. Conversion of fibrinogen to fibrin
4. Reaction of fibrin with Ca^{2+} ions to form calcium fibrin (physiological clot)

1) Activation of Staurt-Prower factor, i.e., X to Xa

It may be noted that the initial process is the activation of factor X (Stuart-Prower) to Xa. X is a serine protease and contains gamma carboxy glutamate (Gla) residues to bind to Ca^{2+} Its activation is made by two pathways – the extrinsic and the intrinsic.

Initiation of the clot formation in response to *tissue injury* is carried out by the extrinsic pathway. The initiation of the pure red thrombus in an area of restricted blood flow or in response to an abnormal vessel wall *without tissue injury* is carried out by the intrinsic pathway. The intrinsic and extrinsic pathways converge and activate Factor X.

Activation of X to Xa in the extrinsic pathway: Factor X is activated to Xa by the cleavage of the Arg-Isoleucine bond by two serine proteases, one in the extrinsic pathway (VIIa) and the other in the intrinsic pathway (IXa). Factor VII (proconvertin, serum prothrombin conversion accelerator or SPCA) operates exclusive in the extrinsic pathway. VII is a Gla-containing glycoprotein synthesised in the liver. It is a Zymogen and activated to VIIa by cleavage, by thrombin or factor Xa. VIIa is accelerated by the tissue factor. The tissue factor is high in the placenta, lung and the brain.

$$VII \xrightarrow{\text{Thrombin or } Xa} VIIa$$

$$X \xrightarrow{\text{VIIa, Tissue factor}} Xa$$

The extrinsic pathway reacts very fast in response to tissue injury.

Activation of X to Xa in the intrinsic pathway: An 'activating surface', perhaps collagen *in vivo* is to be exposed first to Pre-Kallikrein, High Molecular weight Kininogen. (HMK), Factors XII and XI. This exposure constitutes the contact system. XII is hydrolysed and activated to XIIa by Kallikrein.

The activating surface augments this proteolysis (XIIa so generated attacks Pre-Kallikrein to genrate more Kallikrein). XIIa activates XI to XIa. Factor XIa activates IX, a Gla-containing zymogen to IXa. IXa in presence of Ca^{2+} and acid phospholipids slowly activates X to Xa.

The role of IXa is accelerated about 500-fold by VIIIa which is obtained from VIII by thrombin. IXa with VIIIa activates X to Xa.

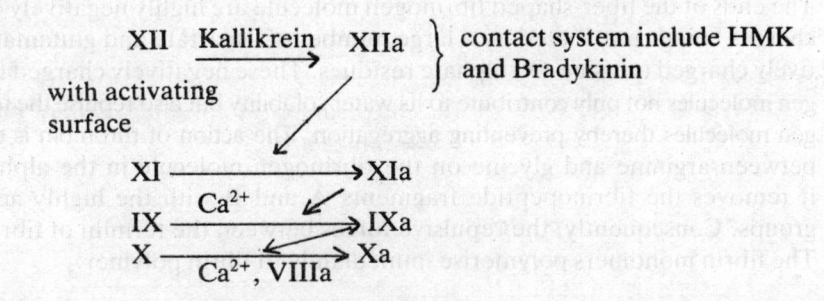

$$XII \xrightarrow[\text{with activating surface}]{\text{Kallikrein}} XIIa$$

contact system include HMK and Bradykinin

$$XI \longrightarrow XIa$$
$$IX \xrightarrow{Ca^{2+}} IXa$$
$$X \xrightarrow{Ca^{2+}, VIIIa} Xa$$

The intrinsic systme is slow due to the cascade mechanism.

Conversion of X to Xa is an autocatalytic process and hence there is tremendous amplification. This is necessary as the amount of X is only 0.01 mg/ml of plasma compared to 3 mg/ml of fibrinogen.

2) Conversion of prothrombin to thrombin

Thrombin is not found in normal plasma and that is the reason why blood remains fluid in the blood vessels. Thrombin is formed during the clotting process from prothrombin which is present in normal plasma. Prothrombin is a glycoprotein containing 4–5 per cent carbohydrate. It is synthesised in liver. Vitamin K is required for the hepatic synthesis of prothrombin. Prothrombin has Gla residues to bind Ca^{2+}. Vitamin K is needed for the formation of Gla.

Xa which is a serine protease cleaves the prothrombin to thrombin, assisted by Ca^{2+} and Factor Va (pro-accelerin).

The activation of prothrombin occurs on the platelets and requires the phospholipids present on the internal side of the platelet plasma membrane. The phospholipids on the internal side are exposed as a result of collagen induced platelet disruption. These phospholipids bind Ca^{2+} which binds to the Gla region of prothrombin. The bridging of the phospholipids via Ca^{2+} to the Gla residues of prothrombin accelerates the activation of prothrombin 50–100-fold. At the same time, Va acts as a receptor for Xa. Xa binds to prothrombin and converts the latter to thrombin. Va brings about 350-fold acceleration of the conversion. Thrombin is also a serine protease.

Binding of Factors Xa, Va, Ca^{++} and Prothrombin
To the platelet Plasma Membrane

3) Conversion of fibrinogen to fibrin monomer

Fibrinogen is present in the normal plasma to the extent of 0.2 – 0.4 g/dl. It is the least soluble of the plasma proteins. It consists of six polypeptide chains and is synthesised in the liver. The ends of the fiber-shaped fibrinogen molecule are highly negatively charged, the negative charges being contributed by a large number of aspartate and glutamate residues and negatively charged tyrosine-O-sulphate residues. These negatively charged termini of the fibrinogen molecules not only contribute to its water solubility but also repulse the termini of other fibrinogen molecules thereby preventing aggregation. The action of thrombin is on the peptide linkage between arginine and glycine on the fibrinogen molecule in the alpha and beta peptides; it removes the fibrinopeptide fragments A and B with the highly acidic electro-negative groups. Consequently, the repulsive forces between the termini of fibrin monomers vanish. The fibrin monomers polymerise immediately to fibrin polymer.

$$\text{Fibrinogen} \xrightarrow{\text{Thrombin}} \text{Fibrin monomer} + \text{Fibrino-peptides A \& B}$$

$$n \text{ Fibrin monomers} \longrightarrow (\text{Fibrin})_n \text{ polymer}$$

The monomers are also cross-linked by Factor XIIIa, i.e., the activated Laki-Lorand factor which is a transglutaminase as it brings about covalent bonding between the glutamine and lysine of different monomers. Ca^{2+} is also needed for this process. Thus the polymer is stabilised as fibrin polymer clot and traps the red cells, platelets and other components to form the red thrombus or the white thrombus (platelet plug).

The deficiency of a factor can cause inherited disease. The most common deficiency is that of Factor VIII; this produces a disease known as hemophilia A, which is sex-linked. The males are all bleeders while the females are the carriers. Deficiency of Factor VIII has played a major role in the health history of the royal families of Europe and Russia. Deficiency of Factor IX (PTC) – Christmas factor – had been found as an inherited defect in a patient by name of Christmas who had hemophilia B (Christmas disease). It also affects only males and is transmitted by females as a sex-linked recessive trait. Congenital deficiency of Factor X can occur in both the sexes and is characterised by prolonged clotting time. In the deficiency of Factor V, as in congenital para-hemophilia, either the clotting time is prolonged or the blood may not clot at all.

Plasma prothrombin deficiency (hypo-prothrombinemia) occurs usually in liver disorders, obstructive jaundice and Vitamin K deficiency. Vitamin K is necessary for the synthesis of Gla residues of proteins like prothrombin, Factors VII, X and IX.

Dicoumarol which acts as an anti-vitamin to vitamin K interferes with the hepatic synthesis of prothrombin (and other Gla containing factors) and can be counteracted only by ingestion of vitamin K. Vitamin K and dicoumarol are structural analogues and their action involves competitive antagonism.

The synthesis of fibrinogen is also affected in certain liver diseases. Low fibrinogen levels in plasma occur in certain types of leukemia, carcinoma and scurvy. In the congenital condition afibrinogenemia, fibrinogen is not synthesised in liver and therefore hemorrhagic tendencies ensue. These are usually fatal in later years.

Chemical nature of some factors

The anti-hemophilic globulin is now characterised as a heat-stable $beta_2$ globulin. The Christmas factor (PTC) is also a $beta_2$ globulin – a mucoprotein containing 19 per cent carbohydrate. The Stuart-Prower factor is an alpha globulin. The Ac globulin (V) migrates between beta and gamma globulins. Factor X is a zymogen of a serine protease and contains Gla residues. Prothrombin is a glycoprotein containing 4 – 5 per cent carbohydrate. Electrophoretically, it moves as an $alpha_2$ globulin. The amino terminal region of prothrombin contains upto 14 Gla residues and has a serine-dependent active protease site. Thrombin is a serine protease that consists of two polypeptide chains and hydrolyses four Arg-Gly peptide bonds in fibrinogen. Thrombin also converts Factor XIII to XIIIa. Fibrinogen is present to the extent of 0.2 – 0.4 g/100ml. It moves electrophoretically as a globulin. It is the least soluble of the plasma proteins and can be easily precipitated from plasma by 0.25 per cent saturation with ammonium sulphate.

Anti-thrombin III

The activity of thrombin is carefully controlled in the body to avoid uncalled-for formation of blood clots. One of the mechanisms is through the action of anti-thrombin III.

Anti-thrombin inhibits the activity of thrombin, IXa, Xa, XIa and XIIa. Antithrombin III is greatly activated by heparin. The anti-coagulant action of heparin is by the activation of anti-thrombin.

Individuals with inherited deficiency of antithrombin III are prone to develop frequent and severe widespread clots.

Anticoagulants

If blood is collected in the presence of decalcifying agents like fluoride, citrate, or oxalate, Ca^{2+} ions are not available and hence these agents act as anticoagulants. Ethylene diamine tetra acetate (EDTA) acts as an anticoagulant by binding Ca^{2+} ions forming a chelation complex. Other substances like Amberlite IR 100 (an ion exchange resin) can also be used to preserve whole blood by removal of Ca^{2+} ions-Heparin activates antithrombin and functions as an anticoagulant.

Fibrinolysis

The blood clotting system is normally in a dynamic and steady state, i.e., fibrin clots are constantly being laid down and subsequently dissolved. Plasmin is a serine protease capable of digesting fibrinogen, fibrin and Factors V and VIII. It cleaves the arginine-lysine linkages of the protein. Plasmin exists in plasma as plasminogen which is inactive.

Plasminogen is activated to plasmin by the plasminogen activator which is also a serine protease. The activator itself needs to be activated through fibrin and is present in an inactive form in most body tissues.

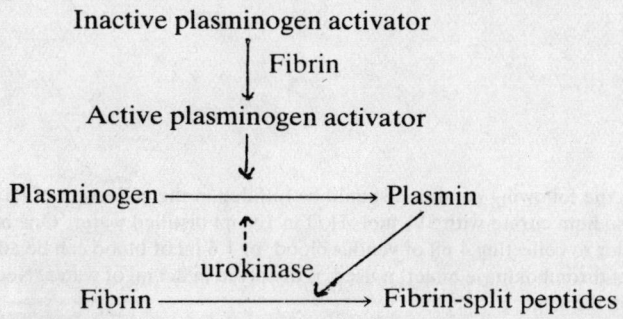

```
          Inactive plasminogen activator
                    │ Fibrin
                    ↓
          Active plasminogen activator
                    │
                    ↓
Plasminogen ─────────────→ Plasmin
                    ↑           ╲
                 urokinase       ╲
Fibrin ─────────────────────→ Fibrin-split peptides
```

Prothrombin time and prothrombin activity

'Prothrombin time' is an indirect and inverse measure of the amount of prothrombin present in blood plasma. When prothrombin level is decreased, prothrombin time is increased.

Determination of prothrombin time is usually done by Quick's one stage method which is as follows: Blood is collected in citrated* tubes and centrifuged to get the plasma. 0.1 ml of plasma is then mixed with 0.1 ml of a thromboplastin extract in a small test tube, kept in a water-bath at 37°C and to this 0.1 ml of 0.02 M $CaCl_2$ is added and a stop watch started. The tube is kept in the water-bath and gently tilted from time to time till the incipient web of fibrin is seen. The period of time required for the formation of this web is noted. In normal blood plasma, prothrombin time is usually about 11 – 13 seconds. The prothrombin time varies according to the thromboplastic material used. The results can also be expressed as a percentage of normal and is called 'percentage coagulability'.

---

*For this test, the following condition should be fulfilled in the collection of blood: A solution is prepared from 3.80 g tribasic sodium citrate with 5½ mol. H_2O in 100 ml distilled water. One ml citrate solution is drawn into a 5 ml syringe prior to collecting 4 ml of venous blood, or 1.6 ml of blood can be added to 0.4 ml of citrate solution. The commercial thrombokinase tablet, if used, is dissolved in 2.5 ml of water. Necessary Ca^{2+} has to be added, if it does not contain the same.

10. Composition and Structure of Muscle and Biochemistry of Muscular Contraction

Muscle tissue constitutes about 40 per cent of the total body weight in man. This specialised tissue is morphologically and biochemically designed for the production of chemical energy as well as its conversion to mechanical work. There are three types of muscle tissue in the body:
1) striated (voluntary) or skeletal muscle,
2) non-striated (involuntary) or smooth muscle, and
3) cardiac muscle.

Striated muscles are of two types, red and white muscles, which exhibit certain functional differences. Muscle tissues like those of the heart have a large number of mitochondria in their cells, a higher content of myoglobin and a more efficient cytochrome system than skeletal muscles, with the result that they are capable of continuous activity for very long periods without getting easily fatigued. Oxygen utilisation, oxidative phosphorylation and resynthesis of ATP are very efficient in the red muscle like the heart muscles and the flight muscles of birds. On the other hand, the white muscle which responds easily to stimulus is capable of intense activity only for short periods. The white muscles have increased stores of high energy phosphates, and the energy during intense contraction is obtained by glycolytic reactions under anaerobic conditions or oxygen debt. The anaerobic glycolysis leads to production of lactate which has to be oxidised to CO_2 and water aerobically during the resting period and the high energy phosphate stores restored.

STRUCTURE OF A TYPICAL MUSCLE

The cells of both skeletal and smooth muscles contain myofibrils which are the contractile units. These myofibrils are oriented along the fibre axis and are bathed by the fluid, sarcoplasm. In the sarcoplasm are found the nucleus and the sarcosomes. Mitochondria, lysosomes and other granules are also present in the sarcoplasm. The sarcoplasmic reticulum or the sarco-tubular system is very active in the muscle and is intimately connected with many phases of muscular activity.

Myofibrils

These are the contractile units of the striated muscle as well as smooth muscle. They have an elongated cylindrical shape with a diameter of $0.5-2.0\,\mu$ and are cross-striated alternately by A bands and I bands (see figure).

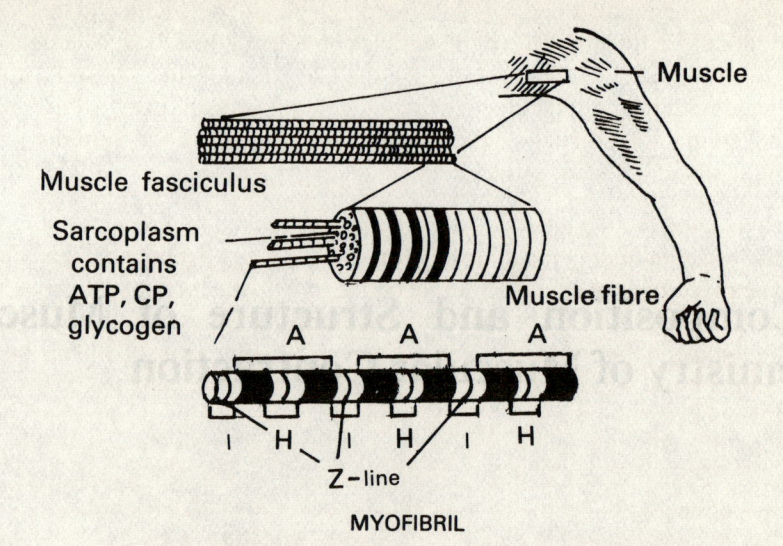

Microscopic appearance of a muscle sarcomere.

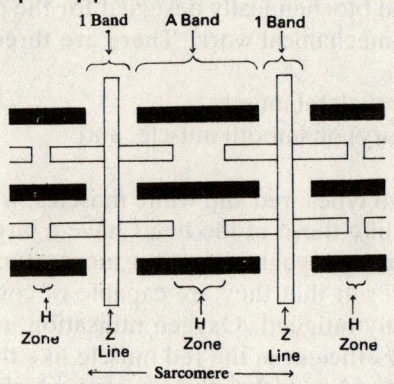

Arrangements of filaments in striated muscle.

The anisotropic A bands are strongly birefringent due to the presence of very high amounts of the protein myosin present in them. The I bands are narrow, isotropic and weakly birefringent. The A bands have thick filaments with a diameter of 110–140 Å. The I band which contains thin filaments with a diameter of 40 Å is continuous with the A band and has a Z line in the middle. Both the A band and I band present together between two Z lines comprise the complete functional unit of the myofibril which is called a sarcomere. A sarcomere in the resting muscle has a length of 2.5 μ of which 1.5 μ is due to the A band and 1.0 μ is due to the I band The length is however altered during muscular contraction.

Sarcolemma: Sarcolemma is the membrane which acts as a covering for the muscle fibres. It can be considered as having been formed of reticular fibres attached at the inner side to the Z line of the myofibrils. As in the case of most other cell membranes, sarcolemma is also polarised and depolarised alternately during the different phases of muscle contraction. The sarcolemma membrane is permeable to various ions and molecules, and an active 'sodium pump' operates on it facilitating the process of migration of Na^+.

Sarco-tubular system

The sarcoplasmic reticulum or the sarcotubular system consists of a specialised form of endoplasmic reticulum terminating in the vesicles near the Z line. These vesicles or sacs have the capacity to store Ca^{++}. During excitation, the Ca^{++} leaves the sacs and activates the filaments concerned with the contractile process. During relaxation, the Ca^{++} is picked up by the sarcotubular system and returned to the storage sacs. An active 'calcium pump' operates on the sarcotubular system for this purpose.

Muscle proteins

The muscle fibrils are mainly made up of proteins characterised by their elasticity or contractile power. The myofibrils of skeletal muscle contain 90 – 95 per cent proteins of which 50 – 55 per cent is due to myosin, 20 – 25 per cent due to actin, 10 – 15 per cent due to tropomyosin and the rest due to actinin, troponin and the nucleoproteins. Myosin is mostly present in the thick filaments of the A band while actin is mostly found in the thin filaments of the I band.

The sarcoplasm contains a mixture of soluble proteins chiefly albumins (myogen) and a globulin called globulin X. These soluble proteins also include a number of glycolytic enzymes, myoglobin and some enzymes like creatine kinase, phosphorylase, adenylate deaminase, myokinase, etc. Myoglobin or muscle hemoglobin has a molecular weight about one fourth that of hemoglobin. It has more oxygen-carrying capacity than hemoglobin.

Myosin: Myosin, the chief constituent of the thick filament, has been purified by repeated extraction with 0.6 M KCl. It has a molecular weight of 450,000 – 500,000 obtained by the light scattering method as well as by the equilibrium ultra-centrifugation methods. The myosin molecule is rod-shaped with a globular head at one end.

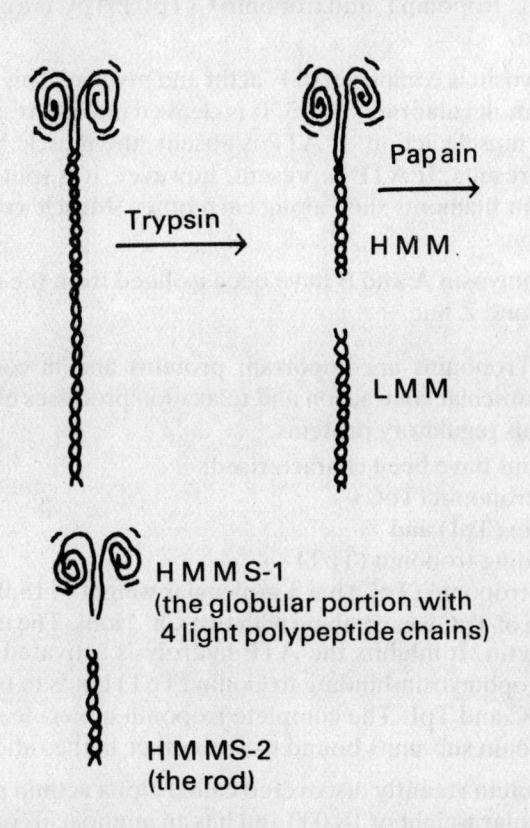

HMMS-1 (the globular portion with 4 light polypeptide chains)

HMMS-2 (the rod)

HMMS₁ (the globular portion with 4 light polypeptide chains)

The head consists of two globular sub-units which may be separated by proteolytic degradation with papain. Four light chain polypeptides bound to the two globular heads modulate the ATPase activity. On treatment with a proteolytic enzyme trypsin, myosin can be separated into two components, light meromyosin (LMM) made up of a fully coiled alpha-helix with a molecular weight of 130,000 and heavy meromyosin (HMM) with a molecular weight of 320,000 – 350,000 which includes the globular ends of the original molecule and four light chains. The ATPase activity as well as the sites for combination with actin are known to reside in the HMM portion.

Myosin contains ATPase which hydrolyses ATP to ADP and Pi. It can also act on UTP, GTP and ITP and convert them to UDP, GDP and IDP, respectively. Myosin ATPase is activated by Ca^{++} and inhibited by Mg^{++}. The ATPase activity is present on the globular head of the myosin molecule which also has the binding site for actin. In the thick filaments, the myosin molecules are stacked in such a way that their globular heads are exposed on the surface. The SH groups on the myosin molecule are also essential for the ATPase activity.

Actin: Actin, the protein of thin filaments, exists in two forms, a globular form or G-actin and a fibrous form or F-actin. Actin has now been purified and characterised as a globulin. G-actin which has a molecular weight of 43,000 is the monomer of F-actin whose molecular weight can be very high. The monomeric form G-actin, at physiological ionic strength and in the presence of Mg^{++} ions, may be polymerised to form the fibrous F-actin. Four functional proteins are present also in striated muscle in much lower concentrations and they are: tropomyosin, troponin T, troponin I, and troponin C(TpC). TpC is a calcium-binding protein analogous to calmodulin.

Actomyosin: Actomyosin is composed of F-actin and myosin. This protein is formed from actin and myosin in the molecular ratio of 1:3. It is cleaved into the respective proteins during the relaxation phase of muscle action. If ATP is absent, the muscle becomes very rigid and muscle rigor or fatigue results. If ATP is present, however, it is split up by myosin ATPase and the actin and myosin filaments slide along each other. Muscle contraction is effected in the presence of Ca^{++} ions.

Tropomyosin: Tropomyosin A and B have been isolated from the muscle. These proteins are found in the I band and Z line.

Troponin complex: Troponins are important proteins and in combination with tropomyosin B, regulate the muscular contraction and relaxation processes of the muscle. Troponins and tropomyosin are thus regulatory proteins.

Three types of troponins have been characterised:
1) Calcium-binding troponin (TpC)
2) Inhibitory troponin (TpI) and
3) Tropomyosin-binding troponin (TpT)

The calcium-binding troponin (TpC) has a molecular weight of 18,000 and is highly soluble in water. Each molecule of TpC can combine with two Ca^{++} ions. The inhibitory troponin (TpI) has a binding site for actin. It inhibits the ATP hydrolysis activated by actin by its binding effect with actin. The tropomyosin-binding troponin (TpT) binds to tropomyosin. It has also an ability to bind to TpC and TpI. The complete troponin molecule may be looked upon as consisting of three troponin sub-units bound to each other in the ratio of 1:1:1.

Actinin: Another protein recently discovered called alpha actinin present in the myofibrilar proteins has a molecular weight of 18,000 and has an aminoacid composition distinct from that of actin. It has the ability to increase the viscosity of F-actin sol. It also increases the actomyosin ATPase activity and enhances the tension of the actomyosin thread. At 37°C, it binds to the F-actin strand at the Z-line end. But at a lower temperature (0°), it is able to bind to every 10-11 G-actin monomer of the F-actin stand as has been shown by laboratrory studies.

MECHANISM OF MUSCULAR CONTRACTION

Contraction takes place by the inter-digitation of two kinds of protein filaments oriented parallel to the long axis of the muscle. The protein filaments are made of actin and myosin, respectively.

Force is produced by attaching and detaching cross-bridges (inter-molecular hydrogen bonds) between the two fibre systems causing the thin actin filaments to draw towards the centre of the myosin filaments. Relaxation is the disengagement of cross-bridges and return of the filaments to their original relative positions.

Ca^{++} is essential for muscular contraction and it plays two roles in this process.
1) Ca^{++} ions are required for the activity of the surface membranes and invaginations of sarcolemma in the inward transmission of impulses.
2) In the actual contractile process.

Ca^{2+} are maintained in the sarcoplasmic reticulum when a muscle is in the resting state accomplished by an active transport system for Ca^{2+}. This system, with the energy from ATP, lowers the concentration of Ca^{2+} in the sarcoplasm (cytoplasm) of the resting muscle while increasing the Ca^{2+} concentration within the sarcoplasmic reticulum bound there by a protein called calsequesterin. Interaction of actin and myosin to form actomyosin in the resting muscle is inhibited by troponin complex and tropomyosin when Ca^{2+} are absent.

The muscle contracts in response to a stimulus delivered down its motor nerve. When the nerve impulse arrives at the junction between the nerve ending and the muscle (the end plate), the outer membrane of the muscle fibre is depolarised. The wave of depolarisation moves along the length of the muscle fibre in either direction. It is also transmitted to the interior of the muscle fibre by a system in close proximity to a plexus of channels of sarcoplasmic reticulum. This releases Calcium, Ca^{2+} from the sarcoplasmic reticulum. The released Ca^{2+} binds to the binding sites of troponin C (TpC) lying on the actin and causes a change in the conformation of troponin C. The changed conformation of troponin C permits tropomyosin to move out of its normal position. Such a change permits the formation of cross bridges between actin and myosin. In the resting muscle, the position of tropomyosin is such that it will prevent the formation of cross-bridges (Hydrogen bonds) between actin and myosin. The formation of cross-bridges pulls the actin along getting the energy from the hydrolysis of ATP by myosin. During contraction, the actin and myosin filaments do not shorten but the sarcomere itself shortens by virtue of an inward sliding movement of the actin filaments between the myosin filaments. 'A' bands remain on the same length in view of the two arrays of filaments. The actin filaments are drawn deeply to interdigitate with the myosin filaments thus effecting the shortening of the sarcomere and the resulting contraction.

The schematic sequence of events is given below.

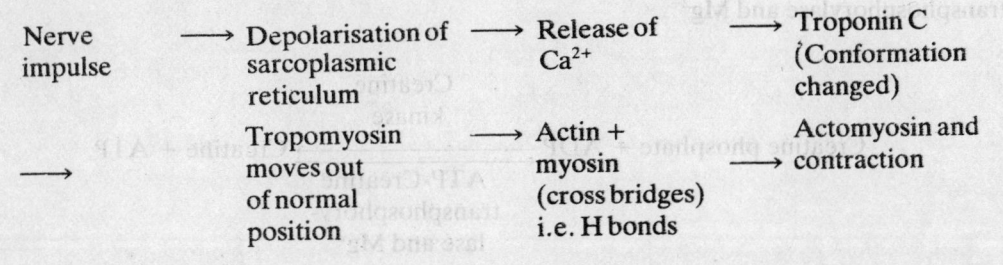

Nerve impulse ⟶ Depolarisation of sarcoplasmic reticulum ⟶ Release of Ca^{2+} ⟶ Troponin C (Conformation changed)

⟶ Tropomyosin moves out of normal position ⟶ Actin + myosin (cross bridges) i.e. H bonds ⟶ Actomyosin and contraction

During relaxation, the Ca^{2+} ions are recaptured by the active calcium pump operating in the sarcoplasmic reticulum. Decrease of Ca^{2+} could not change the normal conformation of troponin C. So tropomyosin comes back to its normal position and inhibits the interaction of the myosin head with the active sites of actin. Hence, disengagement of actin and myosin takes place and the muscle relaxes.

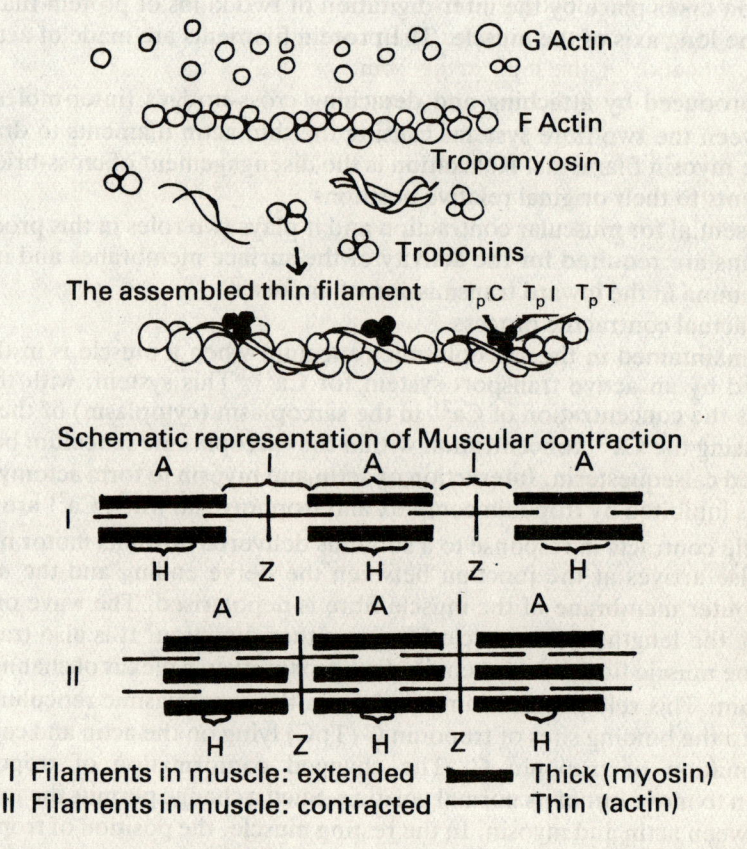

G Actin

F Actin

Tropomyosin

Troponins

The assembled thin filament T_pC T_pI T_pT

Schematic representation of Muscular contraction

I Filaments in muscle: extended ▬▬ Thick (myosin)
II Filaments in muscle: contracted ── Thin (actin)

Source of energy for muscular contraction

The immediate source of energy for muscular contraction is ATP but the ultimate source is carbohydrate or fat. The amount of ATP present in muscle at any time is sufficient for only 0.5 second of activity and hence there must be other sources for replenishing the used up ATP if it is to continue activity. Another high energy phosphate compound, creatine phosphate is stored in the muscle in greater quantities than ATP. This compound can transfer its high energy phosphate group to ADP rapidly, the reaction being catalysed by the enzyme creatine kinase (Lohmann's reaction). The reverse reaction is catalysed by ATP – Creatine transphosphorylase and Mg^{2+}.

$$\text{Creatine phosphate + ADP} \underset{\text{ATP-Creatine transphosphorylase and } Mg^{2+}}{\overset{\text{Creatine kinase}}{\rightleftharpoons}} \text{Creatine + ATP}$$

During continuous muscular activity, there is an increased breakdown of creatine phosphate under anaerobic conditions. The glycogen of the muscle is also broken down and simultaneously large amounts of lactic acid are formed. The glycolysis provides some energy for the process of muscular contraction. When the glycogen stores and creatine phosphate stores get exhausted, the muscle gets fatigued. This happens very often with the white muscles where the oxidation of lactic acid leading to the regeneration of creatine phosphate is slower than the breakdown of high energy phosphates. In the highly active red muscles (e.g., flight muscles of birds), because of the high myoglobin and cytochrome content, active respiration serves as the chief source of energy for rephosphorylation of ADP to ATP via oxidative phosphorylation.

Certain feedback control mechanisms regulate the rate of respiration, ATP consumption and ATP production. During the resting state, the ATP/ADP ratio in muscle is quite high and consequently the electron transport rate is very low. In this state, the activity of phosphofructokinase, the pacemaker enzyme for glycolysis, is low because of inhibition by negative modulator ATP. When the muscle is stimulated to activity, ATP is broken down, and ADP is available as a phosphate acceptor for the enzyme for electron transport. Some of ADP is converted to AMP by the enzyme adenylate kinase.

$$2\ ADP \rightleftharpoons ATP + AMP$$

ADP and AMP act as positive modulators of the enzyme phospho-fructokinase and thus stimulate glycolysis.

It has been found out recently that the resting muscle uses fatty acids and acetoacetate mostly as fuels for rephosphorylation of ADP rather than glucose. However, during maximal activity, glucose and glycogen are the fuels used. The cardiac muscle has a preference for fatty acids as the fuel. Fat from adipose tissue gets mobilised as FFA, which are carried to the muscles by plasma loosely bound to albumin and are oxidised. Alternately, ketone bodies produced by fatty acid oxidation are also efficiently oxidised in muscles.

In the white muscle, during a period of exertion, the whole process of contraction goes on under an oxygen debt. Extra oxygen has to be consumed later in order to remove the large excess of lactic acid and replenish ATP and phosphocreatine. The amount of extra oxygen consumed is proportional to the extent to which the oxygen debt has been incurred. The lactic acid that is produced diffuses out of the muscle and is carried by blood to the liver where it is partly reconverted to glucose and glycogen and partly oxidised in tissues to CO_2 and water by the citric acid cyle. The increased formation of lactic acid in the muscle during anaerobic glycolysis may tend to produce pH changes. However, anserine, a component present in white muscles (but not in red muscles) acts as a buffer against it.

11. Biochemistry of Digestion and Absorption

DIGESTION

Digestion serves to transform large organic molecules present in foodstuffs to smaller and preferably water-soluble molecules capable of being absorbed by the gastro-intestinal tract so as to be utilised by the organism.

It is carried out by enzymes of the hydrolases class which are secreted into the digestive tract from the exocrine glands. These consist of 1) proteolytic enzymes that cleave proteins to amino acids; 2) amylolytic enzymes that cleave starch and other carbohydrates to simple monosaccharide sugars; 3) lipolytic enzymes that split triacyl glycerols to fatty acids, monoacyl glycerol and glycerol; and 4) enzymes that act on nucleoproteins and nucleic acids.

Salivary digestion

The carbohydrate-splitting enzyme of saliva is an alpha-amylase which acts on starch, glycogen and dextrin, hydrolysing them to the disaccharide maltose. However, as the time of exposure of starchy food with saliva is very short, mostly only dextrins are produced in the mouth. Salivary amylase or ptyalin acts on alpha-1, 4-glycosidic linkages, but not on alpha-1, 6-linkages. Cl^- and Br^- ions activate this enzyme effectively.

Gastric digestion

As the food swallowed enters the stomach, chemical changes ensue, involving the digestion of proteins brought about by the gastric juice. The digestion of proteins in the stomach is only partial as pepsin is an endo-peptidase and the acid chyme left behind consists of peptones and polypeptides.

Hydrochloric acid: That the gastric mucosa can produce and secrete a strong mineral acid like HCl is unique from a biological standpoint. The chloride of blood plasma is the source of gastric HCl. It has been confirmed by isotopic experiments that radio-active chloride injected into the general circulation appears in the gastric juice within two minutes. At the time of secretion, the hydrochloric acid formed is isotonic with blood plasma and has a pH of 0.87. It is present as free HCl, but soon it gets mixed with the other secretions and a part of the HCl gets combined with mucin and other proteins.

A recent theory regarding the mechanism of HCl formation has been formulated by Hollander. According to him, parietal cells have a high concentration of the enzyme carbonic anhydrase, which catalyses the reaction.

$$H_2O + CO_2 \rightleftharpoons H_2CO_3 \rightleftharpoons H^+ + HCO_3^-$$

The CO_2 for this is derived from metabolic processes like oxidations and decarboxylations. The CO_2 diffuses from the interstitial space into the parietal cell, where it takes part in this reaction which leads to the formation of H^+ and HCO_3^- ions in the interior of the cells.

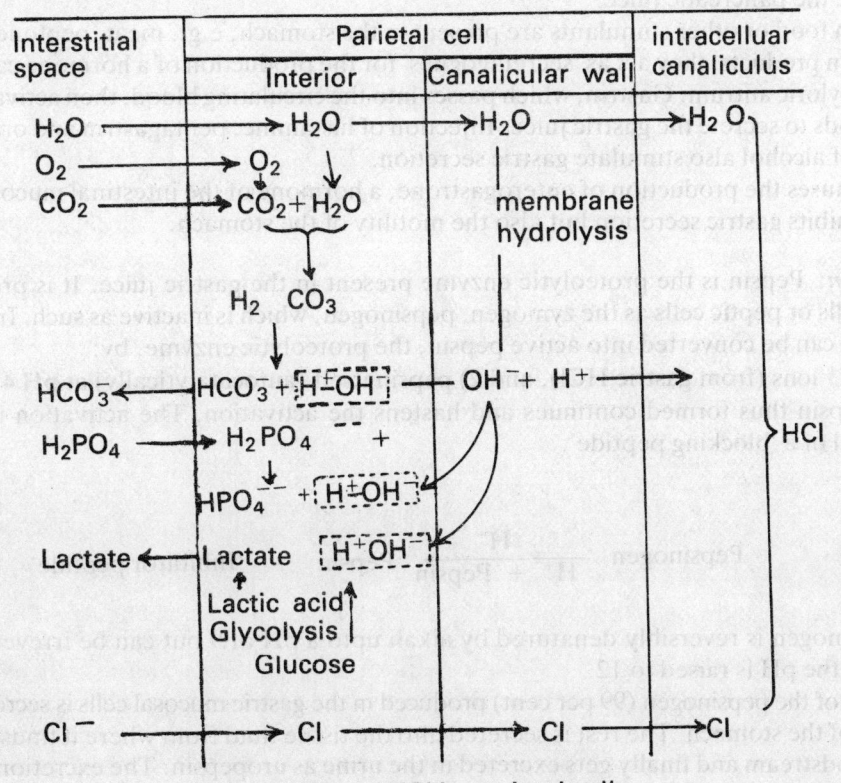

Interstitial space	Parietal cell		intracelluar canaliculus
	Interior	Canalicular wall	
$H_2O \longrightarrow$	$H_2O \longrightarrow$	$\longrightarrow H_2O \longrightarrow$	$\longrightarrow H_2O$
$O_2 \longrightarrow$	$O_2 \downarrow$		
$CO_2 \longrightarrow$	$CO_2 + H_2O$	membrane hydrolysis	
	H_2CO_3		
$HCO_3^- \longleftarrow$	$HCO_3^- + [H^+OH^-] \longrightarrow$	$OH^- + H^+ \longrightarrow$	
$H_2PO_4 \longrightarrow$	$H_2PO_4{}^- +$		HCl
	$HPO_4{}^{2-} + [H^+OH^-]$		
	$[H^+OH^-]$		
Lactate $\longleftarrow$	Lactate		
	↑ Lactic acid ↑		
	Glycolysis		
	Glucose		
$Cl^- \longrightarrow$	$Cl \longrightarrow$	$\longrightarrow Cl \longrightarrow$	$\longrightarrow Cl$

Formation of HCl in parietal cell

The membrane of the intra-cellular canaliculus (canalicular wall) is permeable only to H_2O, H^+ and Cl^-. On the other hand, the membrane of the parietal cell is permeable to Na^+ HCO_3^-, Cl^-, lactate and phosphate ions, as well as H_2O, CO_2 and oxygen from the interstitial fluid. The membrane hydrolysis of water (taking place at the canalicular wall) results in its dissociation to H^+, OH^- ions. The H^+ ions pass into the gastric juice because of selective permeability, while the OH^- ions are retained. As the H^+ ions travel, an equivalent amount of Cl^- ions also move along with it, thus forming HCl which is finally found in the secretion.

The OH^- ions which are retained in the parietal cell now react with the H^+ ion obtained by the splitting of carbonic acid, and form water. The HCO_3^- ions move out into the interstitial fluid in exchange for Cl^- ions. These HCO_3^- ions which pass into the blood in increasing amounts following gastric secretion result in the 'alkaline tide' often seen following a meal.

Biochemical functions of HCl in the stomach: 1) HCl is essential for the activation of pepsinogen by providing H^+ ions. 2) It provides a suitable pH for the activity of pepsin (pH 1.5 - 2.5). 3) It helps the swelling and denaturation of proteins, so that the enzyme pepsin may readily act on them. 4) HCl of gastric juice inhibits gastric fermentation, destroys pathogenic bacteria and thus prevents infection by virtue of its antiseptic action. 5) Iron contained in foodstuffs is present in combination with proteins. HCl of gastric juice converts this bound iron to Fe^{+++} ion which are later reduced to Fe^{++} ions by the ascorbic acid and

— SH compounds. Thus HCl facilitates the absorption of iron from foods. 6) The HCl of gastric juice stimulates the secretion of the hormone secretin in the duodenum. This hormone is carried by circulating blood to the acinar cells of the pancreas which are then stimulated to produce the pancreatic juice.

When food or other stimulants are present in the stomach, e.g., meat, peptones and other digestion products, they act as 'secretagogues' for the production of a hormone called gastrin by the pyloric antrum. Gastrin, which passes into the circulating blood, then activates the gastric glands to secrete the gastric juice. Injection of histamine, pentagastrin and oral administration of alcohol also stimulate gastric secretion.

Fat causes the production of enterogastrone, a hormone of the intestinal mucosa. This not only inhibits gastric secretion but also the motility of the stomach.

Pepsin: Pepsin is the proteolytic enzyme present in the gastric juice. It is present in the chief cells or peptic cells as the zymogen, pepsinogen, which is inactive as such. Inactive pepsinogen can be converted into active pepsin, the proteolytic enzyme, by:

1) H^+ ions (from gastric HCl), and 2) pepsin itself (autocatalytically) at pH 4.6 or below. The pepsin thus formed continues and hastens the activation. The activation involves the removal of a 'blocking peptide'.

$$\text{Pepsinogen} \xrightarrow[H^+ + \text{Pepsin}]{H^+} \text{Pepsin} + \text{'Inhibitor peptide'}$$

Pepsinogen is reversibly denatured by alkali upto a pH of 9 but can be irreversibly denatured if the pH is raised to 12.

Most of the pepsinogen (99 per cent) produced in the gastric mucosal cells is secreted into the lumen of the stomach. The rest is secreted into the tissue fluid from where it finds its way into the bloodstream and finally gets excreted in the urine as uropepsin. The excretion of uropepsin is proportional to the rate of secretion of pepsinogen into the stomach and its determination is, therefore, of clinical importance as a test for gastric function.

Pepsin acts on almost all native proteins, except silk fibroin, keratin, mucoids, mucoproteins, histones and protamines. The optimum activity of this enzyme is exhibited at a pH of 1.5 – 2.5. Once the food leaves the stomach and enters the duodenum, peptic activity stops because the duodenal pH is either neutral, slightly alkaline or alkaline.

Pepsin is an endopeptidase (acting on peptide linkages in the interior of the molecule) which attacks a variety of peptide bonds preferably those involving the amino groups of the aromatic amino acids, tryptophan, phenyl alanine and tyrosine. It exerts a more pronounced effect on denatured proteins which possess free – SH groups. Gelatin, though not containing any of the aromatic amino acids, is acted upon by pepsin. As pepsin acts only on certain specific linkages, the digestion of proteins in the stomach converts them only to peptones and polypeptides with a small amount of free amino acids.

Gastric mucus: The gastric mucus consists of a mixture of glycoproteins and mucoproteins secreted by the 'neck glands' as well as the surface epithelium. The cells of the neck glands produce the 'glandular mucoprotein' which is the main carrier of the anti-pernicious anemia factor or the 'intrinsic factor of Castle.' The intrinsic factor of Castle is essential for the absorption of vitamin B_{12}. It is a glycoprotein with a molecular weight of about 5,000 and contains three per cent glucosamine.

Gastric lipase: The amount of lipase in the gastric juice is very small. The pH of the stomach is always acidic and does not favour the activity of gastric lipase which needs a pH of 7.8 for optimum activity. Hence, the digestion of fats in the stomach is almost negligible.

Intestinal digestion: Digestion in the intestine consists of the action of three different secretions on food, i.e., pancreatic juice, bile and intestinal juice.

PANCREATIC JUICE

Pancreatic juice is secreted by the acinar cells of the pancreas (exocrine pancreas) and is carried to the duodenum by the pancreatic duct. The pancreatic duct joins with the bile duct to form the Ampulla of Vater and the pancreatic juice mixes with the bile during its arrival into the duodenum at the same point. The total volume of pancreatic juice produced per day is man is about 500 – 650 ml. It has a specific gravity of 1.007 and contains 98.7 per cent water and 1.3 per cent solids. The pH of the pancreatic juice is on the alkaline side and ranges between 7 – 8, due to the presence of $NaHCO_3$. The source of this HCO_3^- is the CO_2 and HCO_3^- of blood. This has been confirmed by Ball and co-workers, who administered $NaHCO_3$ containing ^{14}C intravenously into dogs and found the ^{14}C in the HCO_3^- of the pancreatic juice in a short time at levels four to five times those of blood serum. The HCO_3^- concentration of the pancreatic juice is normally five times that of blood serum.

Carbonic anhydrase is present in pancreatic tissue and HCO_3^- is formed as follows:

$$CO_2 + H_2O \rightleftharpoons H_2CO_3 \rightleftharpoons H^+ + HCO_3^-$$

Substances and drugs that inhibit carbonic anhydrase also decrease the HCO_3^- content of the pancreatic juice.

The inorganic ions present in the pancreatic juice are Na^+, K^+, Ca^{2+}, Cl^-, HPO_4^{--} and HCO_3^- and have the following composition:

Na^+	148*		Cl^-	81	
K^+	7	161	HPO_4^{--}	1	161
Ca^{++}	6		HCO_3^-	79	

Inorganic ionic composition of pancreatic juice.

*The ionic composition given is in gram equivalents per litre.

The organic constituents of the pancreatic juice consist of several powerful enzymes, among which are proteinases, lipases, amylase and nucleases. Trypsin, chymotrypsin and carboxypeptidase are the important proteinases. Trypsin and chymotrypsin are elaborated as their zymogens 'trypsinogen' and 'chymotrypsinogen. Two types of chymotrypsinogen, chymotrypsinogen A and chymotrypsinogen B, have been recognised.

The activation of trypsinogen to trypsin is effected by the enzyme enterokinase, secreted by the intestinal mucosa. The trypsin that is formed then activates auto-catalytically more trypsinogen and converts it to trypsin. Trypsin thus formed also catalyses the conversion of chymotrypsinogen to chymotrypsin.

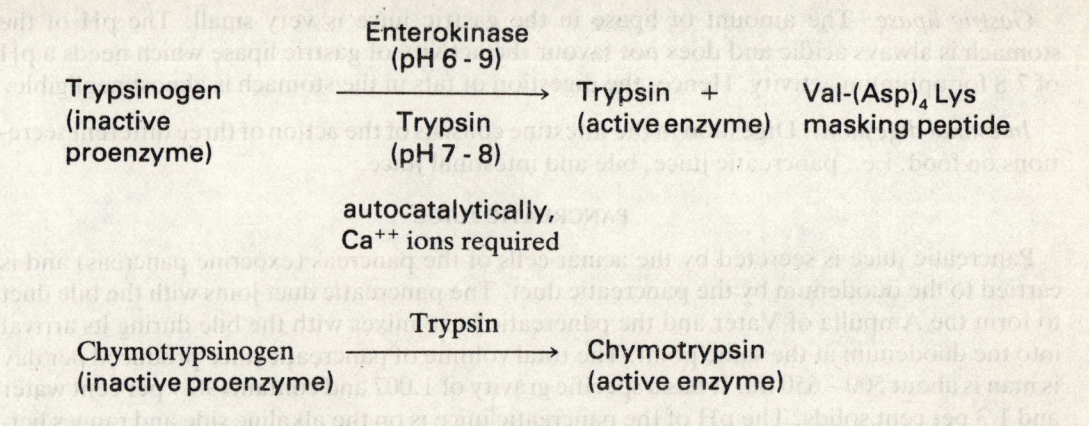

```
                    Enterokinase
                     (pH 6 - 9)
Trypsinogen    ─────────────────▶  Trypsin    +   Val-(Asp)₄ Lys
(inactive            Trypsin         (active enzyme)   masking peptide
proenzyme)          (pH 7 - 8)

              autocatalytically,
              Ca⁺⁺ ions required

                    Trypsin
Chymotrypsinogen  ─────────▶   Chymotrypsin
(inactive proenzyme)            (active enzyme)
```

The active site of typsin involves the 'histidine' region of the molecule and the 'serine' region. The same is true of chymotrypsin. 40 per cent of the positions of amino acids are the same in both trypsin and chymotrypsin.

Proteolytic enzymes of pancreatic juice: 1) Trypsin: This acts as an endopeptidase with an optimum activity at a pH of 8.0 – 9.0. It attacks practically all proteins, including basic proteins like histones and protamines which are not attacked by pepsin. It attacks denatured and partially digested proteins more rapidly than native proteins. The specific peptide linkages attacked by trypsin involve the carboxylic group of basic amino acids like lysine and arginine, which carries a side group with a positive charge — NH_3^+.

2) Chymotrypsin: This is also an endo-peptidase which acts optimally at a pH of 8.0 – 9.0. It has affinity for peptide linkages involving the carboxylic groups of amino acids like tyrosine, tryptophan and phenyl alanine, but certain proteins which contain methionine, asparagine, histidine and leucine are also attacked at the linkage adjacent to these amino acids.

3) Collagenase : Collagenase of the pancreatic juice attacks the collagen present in food (meat diet) and digests it. The excessive production of collagen in pancreatitis results in the necrosis of the pancreatic tissue.

4) Elastase and mucoproteases : These are enzymes present in the pancreatic juice and digest elastin and mucoprotein, respectively.

5) Nucleases (polynucleotidases): The nucleases present in the pancreatic juice are ribonuclease and deoxyribonuclease that act on RNA and DNA, respectively, to their corresponding oligonucleotides at an optimum pH of 7.0.

6) Carboxy-peptidases: The carboxy peptidases of the pancreatic juice are of two types, A and B. They act on the peptide-linkage at one end of the protein or polypeptide chain carrying a free – COOH group end, thus splitting the end amino acid.

The carboxy-peptidases are secreted by the acinar cells of the pancreas as their zymogens or procarboxy-peptidases and are activated by trypsin. Carboxy-peptidases contain Zn as a trace element firmly bound to the molecule. Carboxy-peptidase A has a preference for peptide linkages adjacent to the amino acids tyrosine, tryptophan or phenyl alanine. Carboxy-peptidase B, on the other hand, has a preference for peptide-linkages adjacent to lysine or arginine.

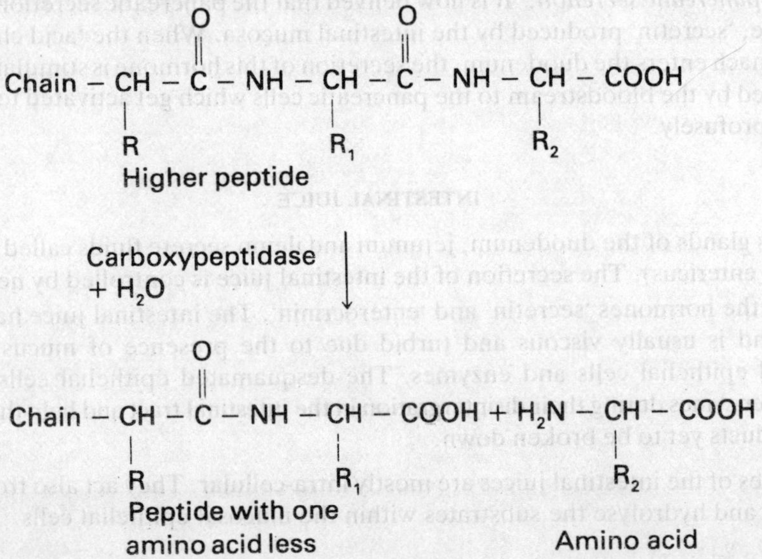

BIOCHEMISTRY OF DIGESTION AND ABSORPTION

$$\text{Chain} - \underset{\underset{R}{|}}{CH} - \overset{\overset{O}{\|}}{C} - NH - \underset{\underset{R_1}{|}}{CH} - \overset{\overset{O}{\|}}{C} - NH - \underset{\underset{R_2}{|}}{CH} - COOH$$

Higher peptide

$$\downarrow \text{Carboxypeptidase} + H_2O$$

$$\text{Chain} - \underset{\underset{R}{|}}{CH} - \overset{\overset{O}{\|}}{C} - NH - \underset{\underset{R_1}{|}}{CH} - COOH + H_2N - \underset{\underset{R_2}{|}}{CH} - COOH$$

Peptide with one amino acid less   Amino acid

Carbohydrate-splitting enzymes of pancreatic juice: The carbohydrate-splitting enzyme of the pancreatic juice is the pancreatic amylase which is a very potent alpha-amylase. Pancreatic amylase acts on starches, dextrins and glycogen at a pH range of 6.3 – 7.2, i.e., slightly alkaline, neutral or slightly acidic pH ranges. It is activated by the presence of chloride ions, which are already present in the duodenum, brought from the stomach (gastric HCl). Pancreatic amylase acts only on alpha-1, 4 glycosidic linkages and the products are mostly maltose with some oligosaccharides containing alpha-1, 6 glycosidic linkages.

The pancreatic juice does not contain maltase. However, lactase and invertase which are present convert lactose and sucrose, respectively, to their monosaccharide components.

Lipolytic enzyme (pancreatic lipase): Pancreatic lipase is secreted by acinar tissue. At the alkaline pH of the pancreatic juice, the enzyme is very active, especially in presence of Ca^{++} ions, soaps, peptides and bile salts. The optimum pH for lipase activity is 7.0–8.8.

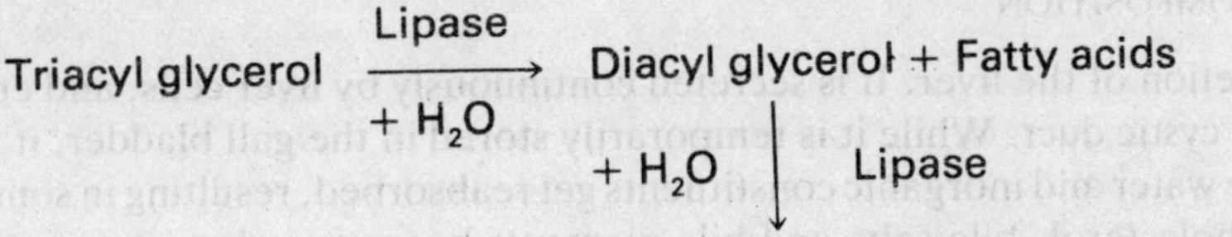

$$\text{Triacyl glycerol} \xrightarrow[+ H_2O]{\text{Lipase}} \text{Diacyl glycerol + Fatty acids}$$

$$\downarrow + H_2O \quad \text{Lipase}$$

$$\text{Monoacyl glycerol + fatty acid}$$

The digestion of fat needs initial emulsification with the help of peptides, bile salts and the sodium salts of fatty acid (soaps) at an alkaline pH. As the fats are insoluble in water, this process of emulsification increases the surface area of the substrate exposed to the lipase enzyme. The digestion of fats in the intestine is never complete.

Certain esterases like cholesterol-esterase and phospholipases are also present in the pancreatic juice and act on cholesterol esters and phospholipids, respectively, and also aid in the absorption of fats.

Control of pancreatic secretion: It is now belived that the pancreatic secretion is controlled by a hormone, 'secretin' produced by the intestinal mucosa. When the 'acid chyme' coming from the stomach enters the duodenum, the secretion of this hormone is stimulated. This hormone is carried by the bloodstream to the pancreatic cells which get activated to secrete pancreatic juice profusely.

INTESTINAL JUICE

The mucus glands of the duodenum, jejunum and ileum secrete fluids called the intestinal juice (succus entericus). The secretion of the intestinal juice is controlled by nervous stimuli as well as by the hormones 'secretin' and 'enterocrinin'. The intestinal juice has an alkaline pH of 8.3 and is usually viscous and turbid due to the presence of mucus, leukocytes, desquamated epithelial cells and enzymes. The desquamated epithelial cells liberate the intra-cellular enzymes during their disintergration in the intestinal tract and help the digestion of the food products yet to be broken down.

The enzymes of the intestinal juices are mostly intra-cellular. They act also from inside the mucosal cells and hydrolyse the substrates within the mucosal epithelial cells. They include the following:

1) *Phosphatases, mononucleotidases and nucleosidases (nucleoside phosphorylases)* that act on organic phosphate esters, mononucleotides and nucleosides, respectively.

2) *Peptidases:* a) *Aminopeptidases* that act on peptide linkages near the end of the chain carrying a free amino group. They require Mn^{++} or Mg^{++} for activity.

b) *Di and tripeptidases* which act on dipeptides and tripeptides converting them into amino acids. They require Co^{++} or Mn^{++} for their activity.

3) *Carbohydrases:* a) *Oligo-1, 6-glycosidases* (isomaltase) acting on disaccharides containing alpha-1, 6-glycosidic linkage, obtained from the digestion on amylopectin by amylases earlier.

b) *Maltase* acting on maltose converting it into glucose.

c) *Lactase* acting on lactose.

d) *Sucrase* or invertase acting on sucrose. The carbohydrases of the intestinal juice convert the disaccharides completely into monosaccharides.

BILE AND ITS COMPOSITION

Bile is a secretion of the liver. It is secreted continuously by liver cells, and enters the gall bladder via the cystic duct. While it is temporarily stored in the gall bladder, it gets concentrated when the water and inorganic constituents get reabsorbed, resulting in some of the constituents like cholesterol, bile salts and bile pigments becoming about ten times greater in quantity than that in the liver bile. The gall bladder bile is also more viscous due to the addition of mucus secreted by the walls of the gall bladder. On stimulation by certain 'cholagogues', the gall bladder contracts and simultaneously empties the bile into the duodenum. Fatty substances, cream, egg yolk and certain digestion products like peptones act as cholagogues. A hormone produced by the intestinal mucosa called 'cholecystokinin' also stimulates the contraction of the gall bladder and the discharge of bile. Bile salts themselves have a similar effect on the production of more bile by liver. This effect of bile salts in enhancing the hepatic secretion of bile is called *choleretic effect*. On the other hand, *cholagogue effect* refers to the stimulation of contraction of the gall bladder and the release of bile into the duodenum.

A healthy adult secretes 500 – 1000 ml of bile daily. Bile is a clear olive green fluid and has a bitter metallic taste. It has a pH of 7.8–8.6 at the time of secretion. It contains 2.6 per cent of solids consisting mainly of cholesterol, bile salts, bile pigments, lecithins, proteins and inorganic ions like Na^+, K^+, Ca^{++}, Fe^{++}, Cl^-, HCO^-_3 etc., small quantities of urea, fat, fatty acids and various excretory products from the liver including poisonous drugs after detoxication. The composition of bile differs depending on whether it is obtained from the gall bladder or the liver as given in the table below.

Table I
COMPOSITION OF GALL BLADDER BILE AND LIVER BILE

Constituents	Gall bladder bile g per cent	Liver bile g per cent
Water	82.0 - 89.0	96.5 - 97.5
Solids	11.0 - 18.0	2.5 - 3.5
Bile salts	5.7 - 10.8	0.9 - 1.8
Mucin	1.0 - 4.0	0.1 - 0.9
Bile pigments	1.5 - 3.0	0.4 - 0.5
Cholesterol and lipids	0.5 - 4.7	0.2 - 0.4
Inorganic salts	0.6 - 1.1	0.7 - 0.8

Bile is one of the channels for the excretion of cholesterol. The cholesterol produced in the liver enters the systemic circulation as plasma lipoprotein cholesterol, but some amount of liver cholesterol remains in solution in bile, and in this form it is discharged into the duodenum. Cholesterol is kept in solution in bile by virtue of the presence of bile salts which are also formed from cholesterol in the liver. Most of the cholesterol of liver is oxidised to bile acids in the liver and is continuously excreted through this pathway.

In certain special circumstances, biliary calculi are formed in the gall-bladder and the bile duct. These calculi are formed from the bile by precipitation as insoluble particles. The biliary calculi or gall bladder stones (gall stones) are generally hard stones, round or ovoid in shape, and may consist mostly of cholesterol and calcium as well as small quantities of bile pigments and proteins. The mechanism of gall stone formation is not clear. Usually it is initiated by an infection or injury to the gall bladder, where a clot or an association of pathogenic organisms may be formed, around which cholesterol may get deposited. The $Ca\ CO_3$ which is insoluble at the alkaline pH of bile may be deposited over it and may turn it to a hard mass. More and more cholesterol now gets deposited over it, layer upon layer, and the stone may gradually grow in size and form concretions big enough to block the bile duct and cause obstruction to bile flow.

Bile acids and bile salts: The bile salts are the sodium salts of glycocholic and taurocholic acids, the bile acids which are combinations of cholic acids with glycine and taurine. The bile acids of human bile are related to the parent acids: a) cholic acid (3, 7, 12-trihydroxycholanic acid which occurs to the extent 25 – 60 per cent of the total bile acids in human bile), b) chenodeoxycholic acid (3, 7-dihydroxy cholanic acid) occurring to the extent of 30 – 50 per cent of total bile acids in human bile, and c) deoxycholic acid (3, 12-dihydroxy cholanic acid) occurring to the extent of 5 – 25 per cent.

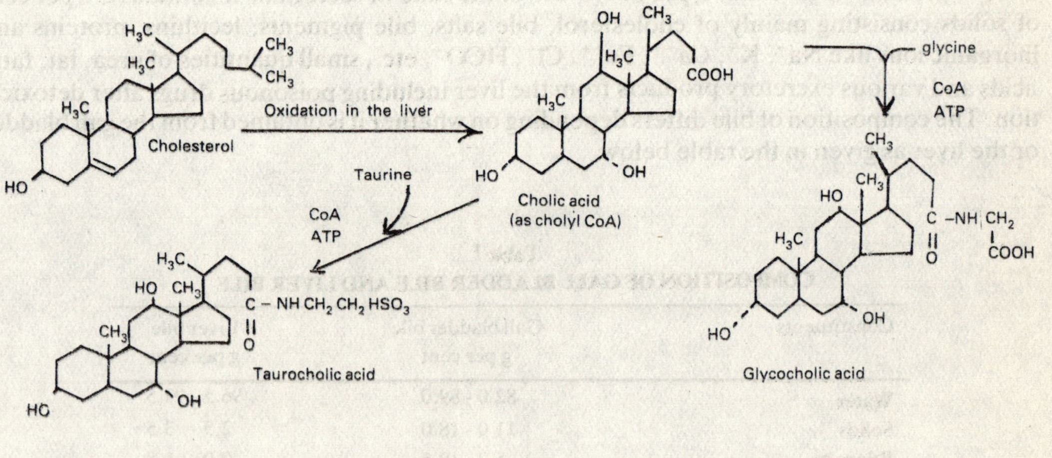

In the formation of glycocholic and taurocholic acids, coenzyme A and ATP are required for this conjugation and cholyl CoA is an intermediate.

Entero-hepatic circulation of bile salts: The bile salts that are excreted by the liver and which enter the intestine via the common bile duct are once again reabsorbed and carried by portal circulation into the liver and re-excreted into the bile. Bile salts re-enter the liver along with the absorption of fats with which they combine to form 'choleic acids'. However, small quantities of bile salts are always excreted in feces but not in urine.

Functions of bile salts: Bile salts are the most useful among the constituents of the bile. They have the following functions:

1) They lower the surface tension and thus aid in the emulsification of fats. Due to the increased surface area exposed, the enzyme lipase can act on this emulsion more efficiently.

2) Bile salts activate the pancreatic lipase in a non-specific manner.

3) Bile salts stimulate the peristalsis of intestines and the bowel movements.

4) By their choleretic action, bile salts help to stimulate the production of more bile by the liver.

5) Cholesterol is held in solution in the bile due to the hydrotropic action of bile salts.

6) Bile salts help the absorption of fat, fatty acids, carotenoid pigments, cholesterol, fat-soluble vitamins and phospholipids. In obstructive jaundice, there is a decrease in the absorption of all these substances due to lack of bile salts in the intestine.

ABSORPTION

Absorption of digestion products from the intestine

Absorption is the process which refers to the passage of products of digestion in the gastro-intestinal tract into the blood and lymph. There is hardly any absorption of nutrients from the mouth, oesophagus and stomach. The absorption of small quantities of alcohol, drugs, water, amino acids and inorganic salts from the stomach takes place to a very limited extent. The major site of absorption is, however, the small intestine. Fats and products of digestion of fats are absorbed via the intestinal lymphatics and transported to the thoracic duct and subclavian vein, from where they enter the systemic circulation directly. Other substances which are absorbed into the portal blood directly enter the liver where they undergo metabolic changes before being passed into the systemic circulation. The large intestines are mainly meant for absorption of water and inorganic salts. Whatever is unabsorbed is passed out of the large intestines into the rectum and excreted as feces.

Absorption of sugars

Only monosaccharides are absorbed from the intestines under normal circumstances. They are absorbed into the portal blood. The absorption of sugars takes place at different rates from the small intestines indicating that the absorption is an active process rather than a mere process of diffusion. The absorption coefficients of the monosaccharides (expressed as mg of sugar absorbed in a 100 g rat per hour) are in the following order:

galactose > glucose > fructose > mannose > xylose > arabinose.

If the absorption coefficient of glucose is taken as 100, that of the other monosaccharides is as follows:

D-galactose	110
D-glucose	100
D-fructose	43
D-mannose	19
Xylose	15
Arabinose	9

Galactose and glucose with a higher rate of absorption are said to be 'actively' absorbed. They are transported across the mucosal cell-membrane by an energy-dependent process. On the other hand, mannose, xylose and arabinose are said to be absorbed passively, i.e., by simple diffusion.

The 5-carbon sugars are not actively transported since they do not conform to the stereo-configurational requirements for active transport. The requirements are: 1) the hydroxyl group must be present at carbon 2; 2) a pyranose ring must be there as in glucose of galactose; and 3) one or more carbons must be attached to carbon 5. The hydroxyl group at carbon 1, 3, 4 and 6 is not essential for active transport. Phosphorylation which occurs at carbons 1 and 6 is not essential for active transport. Hence, monosaccharides with necessary specific structural requirements can be actively transported in the absence of phosphorylation.

It is also known that in the case of absorption of glucose, the absorption is independent of the concentration in the intestine, being the same from a 25 per cent, 50 per cent or 80 per cent solution.

Recently, it has been found that for the transport of glucose there should be compulsory co-transport of Na^+ so that glucose and Na^+ move together. For effecting such a movement, the sodium pump drives out Na^+ from inside the cell in exchange of K^+. This establishes a concentration gradient with a higher concentration of Na^+ outside.

When Na^+ moves into the cell down the concentration gradient, glucose and galactose, (amino acids, iodide, thiamine, etc.) move along with Na^+, without the requirement of a separate coupling of the sugar with ATP. The 'mobile carrier' is common for glucose and Na^+. The energy for the sodium pump proper is provided by ATP generated during intra-cellular metabolic reactions. The enzyme intimately related to the active exchange of Na^+ and K^+, is the membrane-bound ATPase which is inhibited by glycosides like ouabain.

Quastel (1960) confirmed by experiments that the absorption of sugars is influenced greatly by Na^+, K^+ and Mg^{++} ions. In adrenal cortical insufficiency, there is a decreased absorption of sugars and there may be dependence on the decreased concentration of Na^+ in the body fluids in such a condition.

Many substances and agents are known to depress the absorption of sugars from the intestine. Cyanide, azide, malonate and fluoroacetate inhibit respiration while dinitrophenol uncouples oxidation from phosphorylation and inhibits the absorption of sugars from the intestine. The presence of other sugars inhibits the absorption of a particular sugar. Thus, if galactose is present alone, it is absorbed very fast. But if glucose is also present, the absorption of galactose is depressed.

The absorption of sugars is enhanced by B-vitamins (especially thiamine, pyridoxine and pantothenic acid), thyroxine and adrenal corticoids. Increased levels of cations in the body fluids also enhance the absorption of sugars.

Amino acids are absorbed more or less in the same way as glucose. L-amino acids are absorbed by active transport from the mucosal to the serosal side. D-isomers are absorbed by passive diffusion. The transport of many L-isomers by different carriers is sodium-dependent. There is one for neutral amino acids, another for phenyl alanine and methionine and a third for proline and hydroxy proline. There is, in addition, a gamma glutamyl cycle in which the amino acid is absorbed as glutamyl derivative after reacting with glutathione and transported with a carrier. This mode of absorption also requires ATP.

Absorption of lipids

Practically all the absorption of fats and lipids takes place in the small intestine. The *rate* of absorption of different types of lipids is however different. Fats like lard (pork fat) are absorbed from the intestine almost completely (98 per cent) while certain oils like castor oil are not absorbed at all.

It is now believed that approximately 40 per cent of fed triacyl glycerols are hydrolysed to glycerol and fatty acids, 3–10 per cent absorbed as triacyl glycerols and about 50 per cent partially hydrolysed mainly to 2-monoacyl glycerols and free fatty acids. Pancreatic lipase is virtually specific for the hydrolysis of primary ester linkages, i.e., at positions 1 and 3 of triacyl glycerols. Hence, the major end products are the 2-monoacyl glycerols. Some diacyl glycerols also are presumed to exist.

According to another view, only less than one fourth (i.e., about 22 per cent) of the triacyl glycerol is completely hydrolysed to glycerol and fatty acids, about 72 per cent is converted to 2-monoacyl glycerol and about six percent to 1 monoacyl glycerol formed by the isomerisation of 2-monoacyl glycerol.

The various lipids form micro-emulsion or micellar solution, with bile salts, a little lecithin and cholesterol furnished by the bile. The micelles which are water-soluble enter the mucosal cell. The exact mechanism is not known. Bile salts, however, are not absorbed with the lipids but are left behind in the intestinal lumen. **They are required** only for the emulsification and solubilisation of lipids. As enzyme hydrolysis occurs only at the interface, the higher the degree of emulsification, the smaller the individual lipid droplets and the larger the total available area, the greater will be the digestion. (But emulsification by bile salts is apparently not entirely essential for digestion, as, even, when bile is totally excluded, fat is hydrolysed and salts of fatty acids have been shown to be formed.) Phospholipids are hydrolysed by phospholipase A_2, and cholesterol esters are hydrolysed to cholesterol and absorbed from micelles.

BIOCHEMISTRY OF DIGESTION AND ABSORPTION

In the mucosal cells, long chain fatty acids with 14 carbons or more are converted to acyl CoA derivatives which acylate the monoacyl glycerols to diacyl glycerols and finally to triacyl glycerols. Absorbed lysophospholipids and cholesterol are also acylated. The triacyl glycerols combine with small amounts of phospholipids, cholesterol and proteins to form chylomicrons of composition TAG 88 per cent phospholipids eight per cent, cholesterol three per cent and protein one per cent. These chylomicrons pass into the lacteals, then into the lymphatics, into the thoracic duct and finally into the subclavian vein.

Short chain fatty acids with 12 or less number of carbon atoms are transported via portal circulation. Glycerol readily enters the mucosal cell and is transported via the portal circulation.

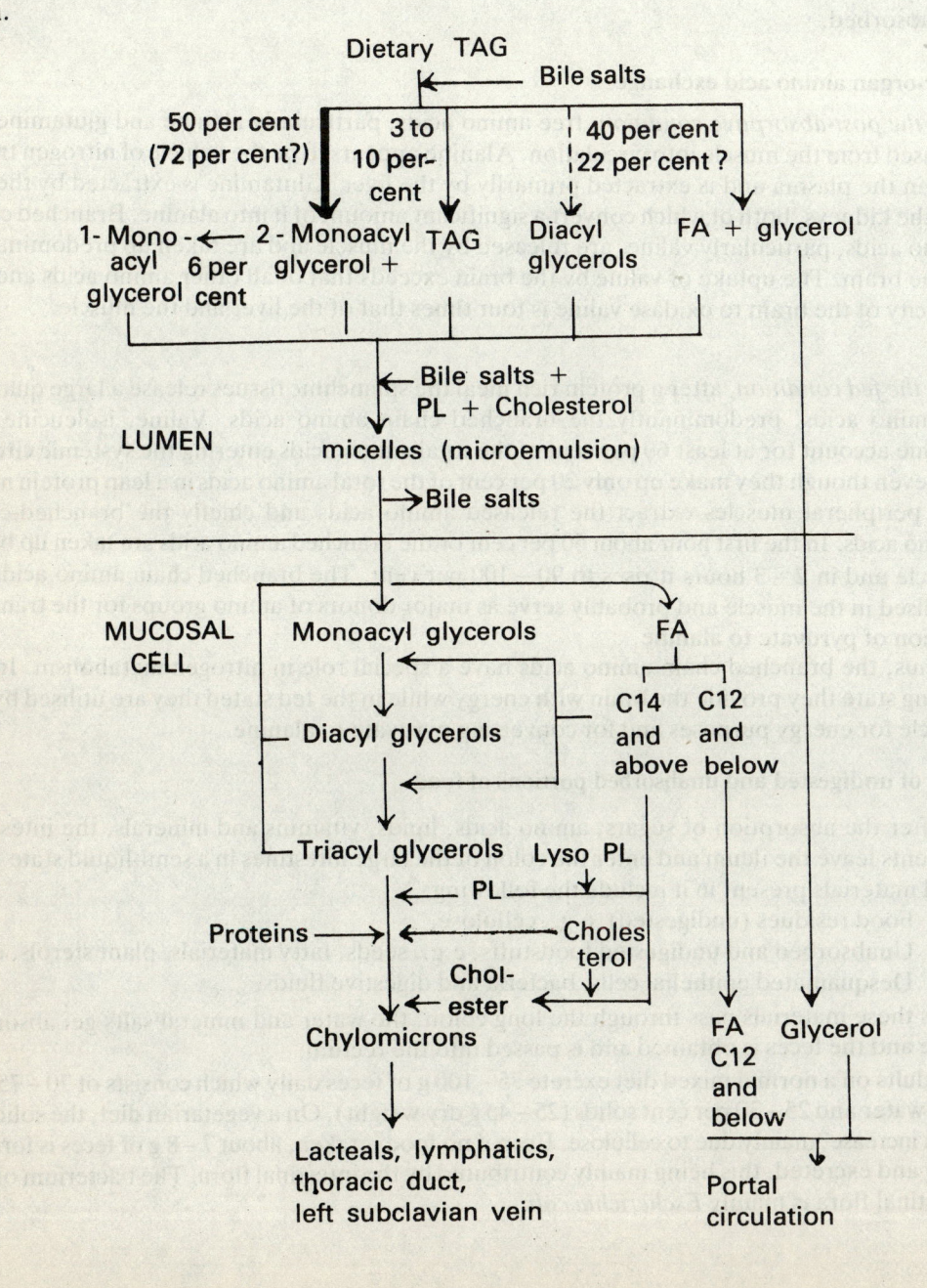

```
                          Dietary TAG
                         ←—— Bile salts
        50 per cent    3 to        40 per cent
        (72 per cent?) 10 per      22 per cent ?
                       cent
           ↓             ↓      ↓         ↓            ↓
        1- Mono- ← 2- Monoacyl  TAG    Diacyl      FA + glycerol
        acyl    6 per glycerol +        glycerols
        glycerol cent
              └─────────┬──────────┴─────────┘
                        ↓
                   ← Bile salts +
                     PL + Cholesterol
    LUMEN         micelles (microemulsion)
                        ↓
                   → Bile salts
    ─────────────────────────────────────────────────
    
    MUCOSAL        Monoacyl glycerols        FA
    CELL                 ↑ ←————————————┬─────┴─────┐
                                       C14        C12
                         ↓             and        and
                   Diacyl glycerols    above      below
                         ↑ ←———————————┘           │
                         ↓                         │
                   Triacyl glycerols   Lyso PL     │
                         ↓ ←—— PL ———————┘         │
    Proteins  →          ← ———— Choles-            │
                              terol               │
                         ← ——— Chol-              │
                              ester              │
                   Chylomicrons              FA    Glycerol
                         ↓                   C12
                                             and
                                             below
                   Lacteals, lymphatics,       ↓
                   thoracic duct,           Portal
                   left subclavian vein     circulation
```

Absorption of phospholipids and cholesterol

Most of the dietary phospholipids are absorbed after partial hydrolysis since the enzymes phospholipases are present in the intestinal tract. Intact phospholipids can also be absorbed. Phospholipid re-synthesis occurs in the cells lining the mucosa and the same is transported into the lacteals.

Absorption of cholesterol is facilitated by the presence of unsaturated fatty acids in the intestine. Saturated fatty acids inhibit absorption of cholesterol. Bile salts are necessary for absorption. Cholesterol combines with them to form choleic acid complexes. Plant sterols are not absorbed.

Inter-organ amino acid exchange

In the post-absorptive condition, free amino acids, particularly alanine and glutamine are released from the muscle into circulation. Alanine appears to be the vehicle of nitrogen transport in the plasma and is extracted primarily by the liver. Glutamine is extracted by the gut and the kidneys, both of which convert a significant amount of it into alanine. Branched chain amino acids, particularly valine, are released by the muscle and are taken up predominantly by the brain. The uptake of valine by the brain exceeds that of all other amino acids and the capacity of the brain to oxidase valine is four times that of the liver and the muscle.

In the fed condition, after a protein rich meal the splanchnic tissues release a large quantity of amino acids, predominantly the branched chain amino acids. Valine, isoleucine and leucine account for at least 60 per cent of the total amino acids entering the systemic circulation even though they make up only 20 per cent of the total amino acids in a lean protein meal. The peripheral muscles extract the released amino acids and chiefly the branched-chain amino acids. In the first hour about 50 per cent or the branched amino acids are taken up by the muscle and in 2 – 3 hours it rises to 90 – 100 per cent. The branched chain amino acids are oxidised in the muscle and probably serve as major donors of amino groups for the transformation of pyruvate to alanine.

Thus, the branched chain amino acids have a special role in nitrogen metabolism. In the fasting state they provide the brain with energy while in the fed stated they are utilised by the muscle for energy purposes and for converting pyruvate to alanine.

Fate of undigested and unabsorbed portions of food

After the absorption of sugars, amino acids, lipids, vitamins and minerals, the intestinal contents leave the ileum and enter the colon of the large intestines in a semi-liquid state. The solid materials present in it include the following:
1) Food residues (undigested), e.g., cellulose.
2) Unabsorbed and undigested foodstuffs, e.g., seeds, fatty materials, plant sterols, etc.
3) Desquamated epithelial cells, bacteria and digestive fluids.

As these materials pass through the long colon, the water and mineral salts get absorbed more and the feces is obtained and is passed into the rectum.

Adults on a normal mixed diet excrete 75 – 100 g of feces daily which consists of 70 – 75 per cent water and 25 – 30 per cent solids (25 – 45 g dry weight). On a vegetarian diet, the solids of feces increase, mainly due to cellulose. Even if no food is taken, about 7 – 8 g of feces is formed daily and excreted, this being mainly contributed by the intestinal flora. The bacterium of the intestinal flora is mainly *Escherichia coli*.

BIOCHEMISTRY OF DIGESTION AND ABSORPTION 175

The excretion of fecal N is about 0.5-1.5 g/day. About half of this is of bacterial origin, while the rest is due to unabsorbed intestinal secretions and digestive fluids, mucus, desquamated cells and food residues.

The lipid content of feces is about 5 – 25 per cent of dry weight of feces which includes neutral fat, fatty acids, soaps and sterols.

Feces have usually a dark brown colour due to the presence of pigments originating from the bile pigments and consist of stercobilin, bilifuschin and urobilin.

SECTION — II
METABOLISM

12 Carbohydrate Metabolism

METABOLISM

Life, to a biochemist, seems to be a series of reactions taking place in the living systems, both plants or animals. These reactions indicate that there is unity in diversity between plants and animals or micro-organisms and mammals.

Various biochemical reactions constitute the favourite expression of the biochemist, i.e, metabolism. The chemistry of metabolism is, to them, the chemistry of life itself.

Metabolism includes two types of reactions — reactions involving the synthesis of molecules, tissue formation and maintenance of the structures, in which case, it is referred to as anabolism, and the degradation of larger molecules referred to as catabolism. It should be understood that catabolism is not an end in itself; it takes place either for providing building units for subsequent anabolism or for the production of energy.

Differentiation of metabolism into energy metabolism and intermediary metabolism (specfic chemical reactions within the organism) is not considered important nowadays as there is hardly any reaction which is not governed by bioenergetics.

Carbohydrate metabolism

Carbohydrates are the major source of energy which are assimilable, though fats yield more energy, weight for weight. Both carbohydrates and fats are protein-sparers.

As a result of digestion by the salivary amylase and eventually the intestinal disaccharidases, carbohydrates, chiefly composed of starch, are hydrolytically split into monosaccharides. The monosaccharides, glucose, fructose and galactose are absorbed by the small intestinal mucosal cells into the blood of the portal venous system.

Rarely, a congenital disaccharidase deficiency may result in a disaccharide intolerance and consequently a disacchariduria. The metabolism of carbohydrates which is primarily intended to meet our energy requirements is constituted by important biochemical transformations, namely, glycogenesis, glycogenolysis, glycolysis, the tricarboxylic acid cycle, gluconeogenesis, the hexose monophosphate shunt and the uronic acid pathway.

Glycogenesis

Glycogen, the reserve carbohydrate in the animal kingdom, is synthesised chiefly in the liver and muscle. On a carbohydrate-rich diet, the liver may contain as much as five per cent glycogen. The total glycogen content of the body is about 300 g, being almost equally distributed between liver and muscle (total glucose in the body is 20–30 g).

In starvation, it is the liver glycogen that gets depleted quickly and not muscle glycogen which has got a specialised function to perform during muscular activity. Muscle glycogen is not directly available for blood sugar regulation unlike liver glycogen since glucose-6-phosphatase is not present in muscle.

Phosphorylation plays a significant role in glycogenesis as in most other transformations pertaining to carbohydrate metabolism. The modern concept of the mechanism of glycogenesis is one of participation of uridine triphosphate (UTP). The initial step is the phosphorylation of glucose to glucose-6-phosphate catalysed by hexokinase and glucokinase in non-ruminants, in the presence of ATP activated by magnesium ions. Glucose-6-phosphate is next converted into glucose-1-phosphate by phospho-glucomutase in the presence of manganese and magnesium ions. Glucose-1-phosphate reacts with UTP to form UDP glucose which is the nucleotide-containing activated glucose, the activation being catalysed by UDPG pyrophosphorylase. The next step involves the role of glycogen synthetase (UDP glucose glycogen transglucosylase), a glycosyl transferase catalysing the formation of an alpha 1, 4-glycosidic bond between carbon 1 of the activated glucose of UDPG and carbon 4 of the terminal glucose residue of the glycogen primer with the formation of UDP. As a result, glucose residues are added on to a pre-existing glycogen primer at the non-reducing outer end of the molecule so that the glycogen tree gets elongated to 6–11 glucose residues, as successive 1, 4-linkages occur. At the branching point, the branching enzyme (amylo 1, 4 → 1, 6 transglucosidase) operates and is concerned with the transfer of a part of 1, 4-chain to a neighbouring chain to form a 1, 6 glycosidic linkage. A minimum of 6–8 glucose units should be linked by alpha 1, 4 bonds for the formation of a branch.

Glycogenesis, like every other aspect of carbohydrate metabolism, is essentially hormone-controlled. Many hormones exert their action via cyclic AMP, a very important intra-cellular intermediate.

Glycogenesis in liver and muscle

Glucose $\xrightarrow[\text{ATP, Mg}^{2+}]{\text{Hexokinase, (Glucokinase in non-ruminants)}}$ Glucose 6-phosphate + ADP + H_2O

Glucose 6-phosphate $\xrightarrow[\text{Mn}^{2+}, \text{Mg}^{2+}]{\text{Phosphoglucomutase}}$ Glucose 1-phosphate

Glucose-1-phosphate $\xrightarrow[\text{UDPG Pyrophosphorylase}]{\text{UTP}}$ Uridine diphosphoglucose + PP. (UDPG)

UDPG + Glycogen Primer $\xrightarrow{\text{Glycogen Synthase (Glucosyl transferase)}}$ (1,4 Glucosyl units)

(1,4 Glucosyl units) $\xrightarrow{\text{Branching enzyme amylo 1,4 — 1,6 —transglucosidase}}$ Glycogen (1, 4 and 1, 6 Glucosyl units)

Glycogen synthase is present in two forms, the active forms i.e., glycogen synthase *a* (dephosphorylated form) and the inactive form i.e., glycogen synthase *b* (phosphorylated form). The active form glycogen synthase *a* can be phosphorylated by ATP in the presence of the enzyme glycogen synthase kinase 3, 4 and 5 to give the inactive form. A cyclic AMP-dependent protein kinase has also been known to carry out this inactivation. Apart from this, two calcium calmodulin dependent protein kinases (one is Ca^{2+} Calmodulin phosphorylase kinase) also play a role in the same way to inactive glycogen synthase. Glucose-6-phosphate acts as an allosteric activator of glycogen synthase *b*. Insulin promotes the activation of glycogen synthase *b*, by promoting protein phosphatase and dephosphorylation and conversion to glycogen synthase *a*.

CONTROL OF GLYCOGENESIS

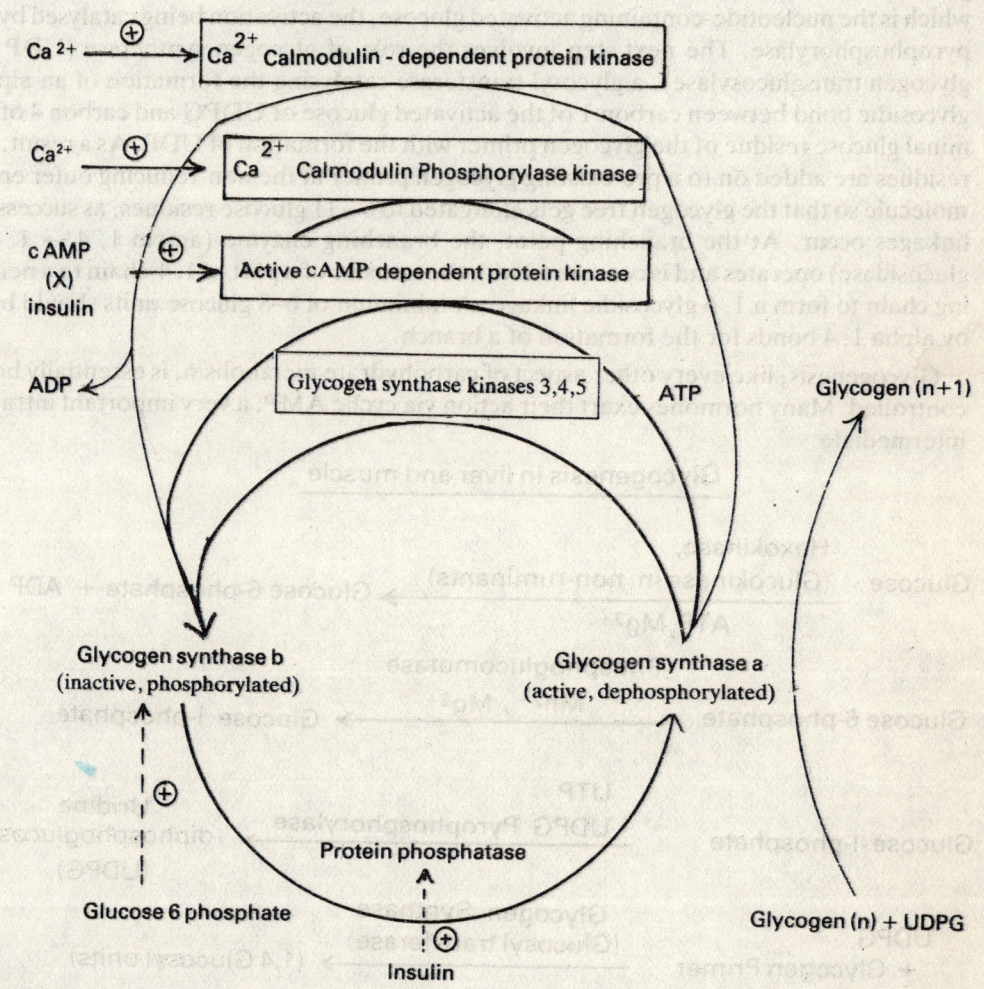

Ca^{2+} $\xrightarrow{\oplus}$ Ca^{2+} Calmodulin - dependent protein kinase

Ca^{2+} $\xrightarrow{\oplus}$ Ca^{2+} Calmodulin Phosphorylase kinase

cAMP (X) insulin $\xrightarrow{\oplus}$ Active cAMP dependent protein kinase

ADP ← Glycogen synthase kinases 3,4,5 ← ATP → Glycogen (n+1)

Glycogen synthase b (inactive, phosphorylated) ← → Glycogen synthase a (active, dephosphorylated)

Glucose 6 phosphate $\xrightarrow{\oplus}$ Protein phosphatase $\xleftarrow{\oplus}$ Insulin

Glycogen (n) + UDPG

Glycogenolysis

Glycogen undergoes breakdown to glucose in liver and lactic acid in muscle. In aerobic environment however, lactate is oxidised to pyruvate which is oxidised eventually to CO_2 and water in the tricarboxylic acid cycle. The breakdown of glycogen is termed as glycogenolysis. Glycogenolysis starts with phosphorolysis catalysed by the enzyme phosphorylase requiring also inorganic Pi to yield glucose-1 phosphate. The enzyme cleaves only the outer 1,4-glycosidic linkages of glycogen.

$$\text{Glycogen (C}_6)_n + \text{Pi} \rightarrow \text{Glycogen (C}_6)_{n-1} + \text{glucose-1-phosphate}$$

When some of the glucose residues are removed by phosphorolysis, another enzyme $\alpha 1,4 \rightarrow \alpha\text{-}1,4$-glucan transferase tranfers some three-sugar unit bits from one branch to the other exposing the 1 – 6 branch point. Later, a third enzyme amylo 1. 6-glucosidase (also called debranching enzyme) helps to cleave off the 1, 6-branching linkages and the remaining molecules are once again acted on by phosphorylase.

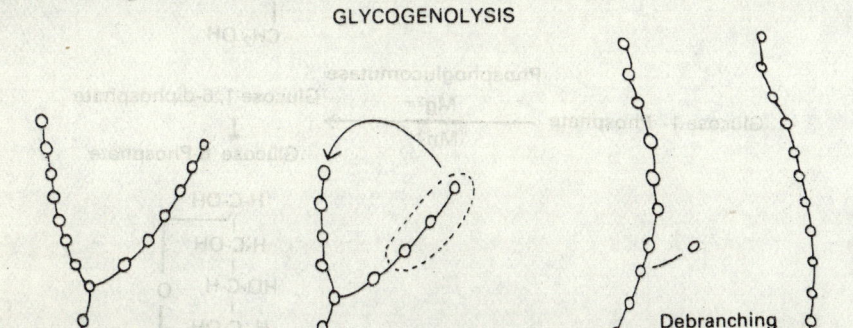

GLYCOGENOLYSIS

Phosphorylase → Glucan transferase → Debranching enzyme →

Glycogenolysis is brought about by the conversion of the inactive form of phosphorylase into the active form. Epinephrine has got a profound impact on glycogenolysis in the liver and more in the muscle, but glucose is the end product in the liver while it is lactic acid in the muscle. The mode of action of the hormone is by way of activation of adenylate cyclase to form cyclic AMP. Glucagon simulates epinephrine in its mode of action but it does not act on the skeletal muscle though it can act on the cardiac muscle. Phosphorylase activity is, therefore, the rate-limiting step in glycogenolysis. From glucose-1-phosphate, glucose-6-phosphate is formed due to phospho-glucomutase, in the presence of manganese and magnesium ions.

In the liver, kidney and intestines (but not in the muscle) glucose-6-phosphatase catalyses the hydrolytic dephosphorylation of glucose 6-phosphate into glucose which diffuses from the cell into the extra-cellular spaces. This is indicated by a rise in the blood sugar level. Even under physiological emotional states like fear or anger, hyperglycemia and glycosuria are met with due to sympathetic stimulation and elaboration of epinephrine.

In the muscle, phosphorylase exists in two forms, the active form being phosphorylase *a* (phosphorylated form) and the inactive form existing as phosphorylase *b* (dephosphorylated form). Phosphorylase *a* can be inactivated to the *b* form by an enzyme protein phosphatase I. Re-activation of phosphorylase *b* is effected by ATP in the presence of the activating enzyme phosphorylase kinase. The latter enzyme itself is present in inactive and active forms. The inactive phosphorylase kinase *b* is phosphorylated to the active *a* form by an active cAMP-dependent protein kinase or Ca^{++} which activates the calmodulin-component of phosphorylase kinase. Epinephrine can thus enhance glycogenolysis through the activation of the phosphorylase system through cAMP. Glucagon is effective in this respect only in the liver and not in the muscle.

Phosphorylase *a* of the muscle is active in the presence or the absence of AMP while phosphorylase *b* is active only in the presence of AMP.

Glycogenolysis in liver

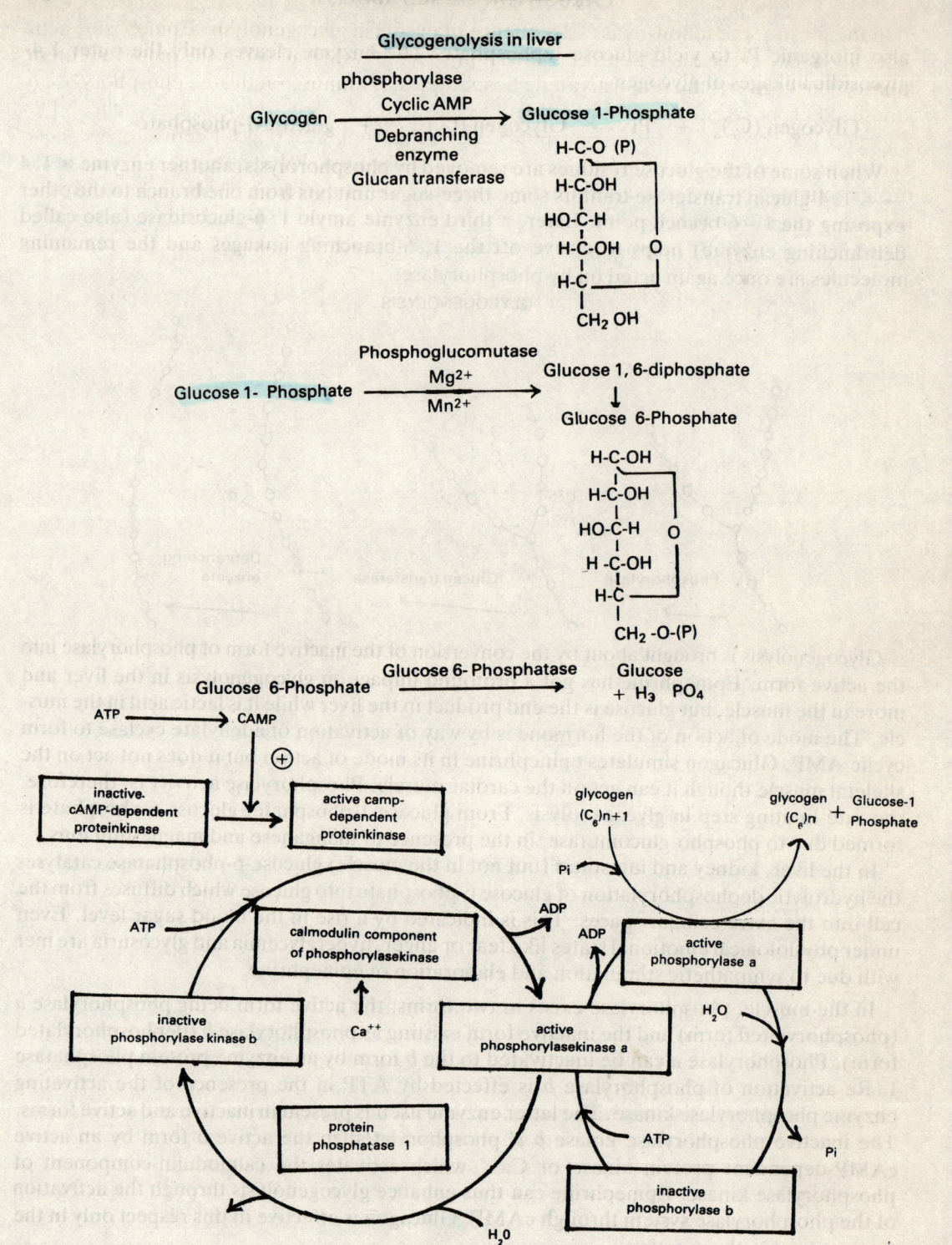

$$\text{Glycogen} \xrightarrow[\substack{\text{Debranching}\\ \text{enzyme}\\ \text{Glucan transferase}}]{\substack{\text{phosphorylase}\\ \text{Cyclic AMP}}} \text{Glucose 1-Phosphate}$$

```
H-C-O (P)
H-C-OH
HO-C-H
H-C-OH    O
H-C
CH₂ OH
```

$$\text{Glucose 1- Phosphate} \xrightarrow[\substack{Mg^{2+}\\ Mn^{2+}}]{\text{Phosphoglucomutase}} \text{Glucose 1, 6-diphosphate} \downarrow \text{Glucose 6-Phosphate}$$

```
H-C-OH
H-C-OH
HO-C-H
H-C-OH    O
H-C
CH₂ -O-(P)
```

$$\text{Glucose 6-Phosphate} \xrightarrow{\text{Glucose 6-Phosphatase}} \text{Glucose} + H_3 PO_4$$

fig. Regulation and control of glycogenolysis in muscle.

CARBOHYDRATE METABOLISM

In the muscle, the following is the sequence of events in **glycogenolysis**: Epinephrine activates adenylate cyclase and more of cAMP is produced. cAMP activates the inactive protein kinase. This converts the non-activated phosphorylase b kinase to activated phosphorylase b kinase. Phosphorylase b is now stimulated by the active phosphorylase b kinase and ATP. Phosphorylase b is converted to phosphorylase a. Thus it is seen that for activation of phosphorylase b, the enzyme phosphorylase b kinase is required. Phosphorylase b kinase in its turn is activated by two processes, i.e., by release of Ca^{2+} or through cAMP dependent protein kinase as shown below.

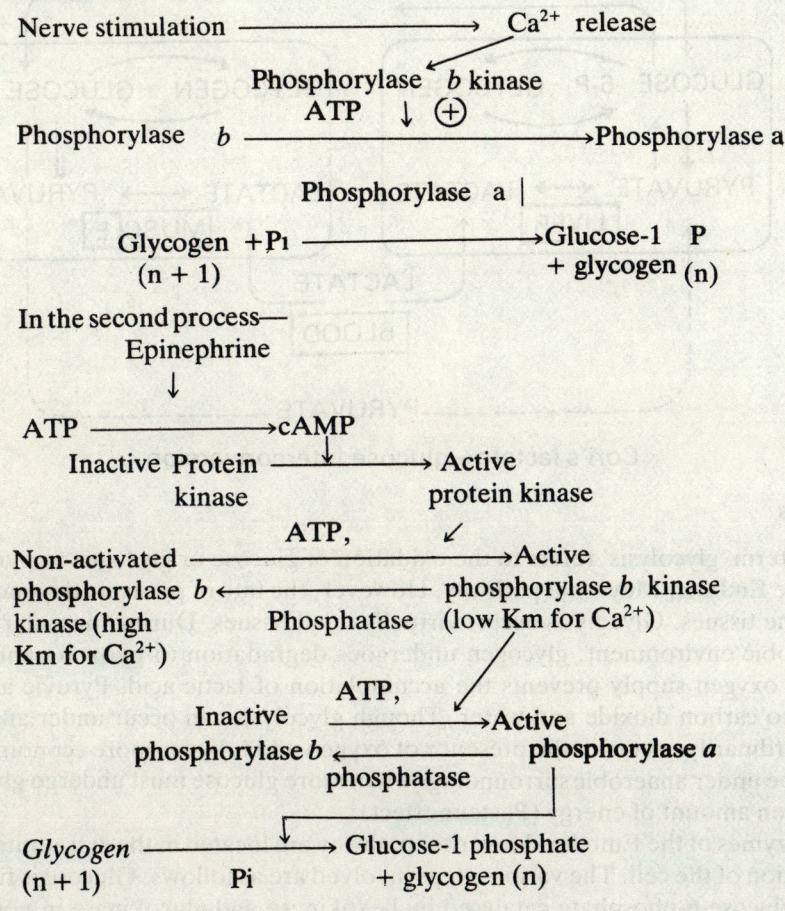

$$\text{Nerve stimulation} \longrightarrow Ca^{2+} \text{ release}$$

$$\text{Phosphorylase } b \xrightarrow[\text{ATP}]{\text{Phosphorylase } b \text{ kinase } \oplus} \text{Phosphorylase } a$$

$$\text{Glycogen} + P_i \xrightarrow{\text{Phosphorylase } a} \text{Glucose-1 P} + \text{glycogen } (n)$$
$$(n+1)$$

In the second process—

$$\text{Epinephrine} \downarrow$$

$$\text{ATP} \longrightarrow \text{cAMP}$$
$$\text{Inactive Protein kinase} \longrightarrow \text{Active protein kinase}$$

$$\text{Non-activated phosphorylase } b \text{ kinase (high Km for } Ca^{2+}) \underset{\text{Phosphatase}}{\overset{\text{ATP,}}{\rightleftarrows}} \text{Active phosphorylase } b \text{ kinase (low Km for } Ca^{2+})$$

$$\text{Inactive phosphorylase } b \underset{\text{phosphatase}}{\overset{\text{ATP,}}{\rightleftarrows}} \text{Active phosphorylase } a$$

$$\text{Glycogen} \longrightarrow \text{Glucose-1 phosphate} + \text{glycogen } (n)$$
$$(n+1) \quad P_i$$

IN LIVER

Phosphorylase b of the liver is structurally distinct from that of the muscle. Phosphorylase b kinase, highly purified, is not activated on incubation with cAMP protein kinase, but is stimulated by a Ca^{2+} dependent process. Stimulation of glycogenolysis by epinephrine through α_1 receptors is not accompanied by an increase of cAMP but an increase of calcium. mobilised from the mitochondria into the cytosol which stimulates Ca^{2+} Calmodulin-sensitive phosphorylase kinase. Interestingly, glucagon-mediated glycogenolysis appears to be routed through cAMP in the liver.

Glycogen metabolism is regulated by a balance between **glycogenesis** and **glycogenolysis**, which however do not take place at the same time, since a rise in concentration of cyclic AMP not only activates phosphorylase necessary for glycogenolysis but also converts the active into the inactive form of glycogen synthetase.

Abnormalities of glycogen metabolism are seen in the rare, inherited glycogen-storage diseases of various types according to the specific enzyme deficiency. In Von Gierke's type of glycogenosis due to the deficiency of glucose-6-phosphatase, the hepatic cells as well as the cells of the renal convoluted tubules are loaded with metabolically inert glycogen stores. Ketosis and hyperlipemia are also encountered.

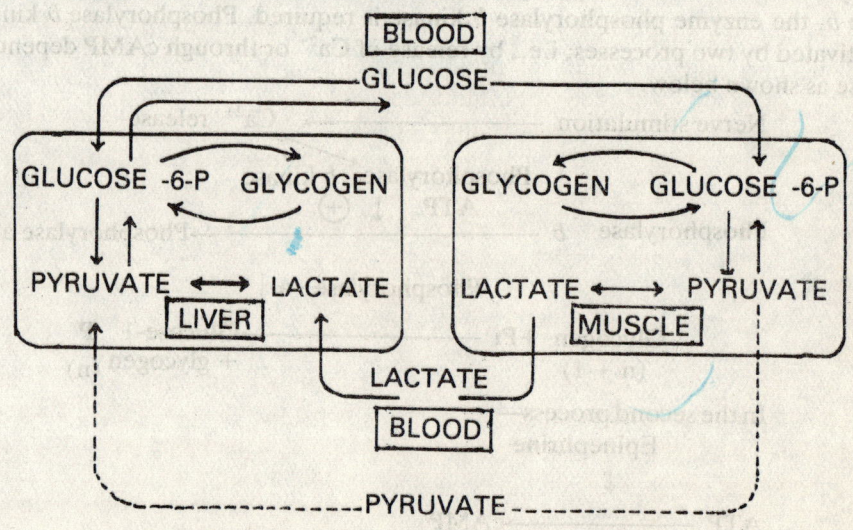

Cori's lactate – glucose interconversion

Glycolysis

The term 'glycolysis' refers to the oxidation of glucose or glycogen to pyruvate and lactate by the Embden-Meyerhof pathway. However, the fate of pyruvate depends on the redox state of the tissues. Glycolysis occurs virtually in all tissues. During muscular contraction in an anaerobic environment, glycogen undergoes degradation to pyruvate and lactate while adequate oxygen supply prevents the accumulation of lactic acid. Pyruvic acid is oxidised finally into carbon dioxide and water. Though glycolysis can occur under anaerobic conditions, it ordinarily occurs in the presence of oxygen and is by far more economical than what it would be under anaerobic surroundings when more glucose must undergo glycolysis to provide a given amount of energy (Pasteur effect).

The enzymes of the Embden-Meyerhof pathway are located in the extra-mitochondrial soluble fraction of the cell. The various steps involved are as follows: Glucose is first phosphorylated to glucose-6-phosphate catalysed by hexokinase and glucokinase in non-ruminants in the presence of ATP and magnesium ions. Glucokinase is different from hexokinase in that it has a higher Km for glucose, inducible by insulin, and not subject to feed-back inhibition by glucose-6-phosphate. Glucose-6-phosphate is the meeting point of diverse pathways of carbohydrate metabolism. Glycogen gives rise to the same derivative via glucose-1-phosphate. Glucose-6-phosphate is transformed into fructose-6-phosphate by the specific phosphohexose isomerase. An additional phosphorylation with ATP yields fructose-1, 6-diphosphate catalysed by phospho-fructokinase which is an inducible enzyme influencing the rate of glycolysis.

Fructose 1,6-diphosphate is acted upon by the specific aldolase into glyceraldehyde-3-phosphate isomeric with dihydroxy-acetone phosphate. This stage provides a point of linkage between the EM pathway and glycerol from the lipids. The next step in glycolysis involves the oxidation of glyceraldehyde-3-Phosphate and additional phosphorylation to 1, 3-diphospho glycerate. The oxidation is effected by NAD^+ dependent glyceraldehyde 3-phosphate dehydrogenase. Inorganic phosphate is added to form 1,3-diphosphoglycerate which retains a high energy phosphate in position 1. This is subsequently captured as ATP following the dephosphorylation of 1,3-diphosphoglycerate by phosphoglycerate kinase forming 3-phosphoglycerate. This stage is an example of phosphorylation at the substrate level whereby two molecules of ATP are generated per molecule of glucose.

3-phosphoglycerate is then converted into 2-phosphoglycerate by phospho-glyceromutase. Enolase is the next enzyme that catalyses the dehydration of 2-phosphoglycerate to phosphoenol pyruvate elevating the PO_4 in position 2 to a high energy state. Fluoride prevents glycolysis by inhibiting enolase and is, therefore, used for the collection of blood samples prior to the estimation of blood sugar. Again, another phosphorylation at the substrate level is encountered by way of capture of the high energy as ATP, due to dephosphorylation of phosphoenol pyruvate by pyruvate kinase, thereby generating two molecules of ATP per molecule of glucose undergoing glycolysis. Pyruvate kinase requires K^+ for maximum activity. The enol pyruvate formed as a result is spontaneously isomerised into the keto form.

The fate of pyruvate is decided by the redox state of the tissues. Under anaerobic conditions, reoxidation of NADH by hydrogen transfer through the respiratory chain cannot take place. However, glycolysis can still proceed by virtue of re-oxidation of NADH, by being coupled with the reduction of pyruvate to lactate, catalysed by lactate dehydrogenase, with NADH as the hydrogen donor. Under hypoxic or anoxic state, lactate accumulates in tissues. But the skeletal muscle has got the unique capacity to incur oxygen debt, if need be, for example, during strenuous muscular exercise, at the end of which there is hyperpnoea stimulated by the CO_2 excess, with the result that the oxygen debt is promptly repaid and the onset of muscular fatigue prevented.

The mammalian erythrocyte is also unique in that glycolysis meets most of its energy needs even under conditions of low ATP requirement, since there is a bypassing of phosphoglycerate kinase. It dose not also possess mitochondria the major site of oxidation. (See 'Rapoport cycle'.)

Role of fructose 2, 6 bisphosphate: Fructose 2, 6 bisphosphate is the most potent allosteric effector of phospho-fructokinase 1 and inhibitor of fructose 1, 6 bisphosphatase in the liver. It is formed by the phosphorylation of fructose-6 phosphate by phosphofructokinase-2. The same enzyme has fructose 2, 6 bisphosphatase activity and breaks down fructose 2, 6 bisphosphate.

When glucose is in abundance, say after a meal, the fructose 6 phosphate concentration will be high. This substance allosterically activates phospho-fructokinase 2. Hence the greater the glucose, greater the fructose 6 phosphate, and greater the fructose 2, 6 bisphosphate. The 2, 6 bisphosphate activates phospho-fructokinase 1 and augments glycolysis. Hence, the utilisation of glucose is promoted. Under conditions of glucose shortage, gluconeogenesis is stimulated by a decrease of concentration of fructose 2, 6 bisphosphate. This is due to glucagon acting through cAMP which promotes fructose 2, 6 bisphosphatase activity and hydrolyses 2, 6 bisphosphate. Decrease of fructose 2, 6 bisphosphate is not conducive to the utilisation of glucose.

The bioenergetics of the EM pathway is fascinating in that, for instance, the energy for muscular contraction does not originate in an oxidation. At the instant of muscular contraction there is no consumption of oxygen or elimination of carbon dioxide. The skeletal muscle

Embden-Meyerhof pathway of glycolysis

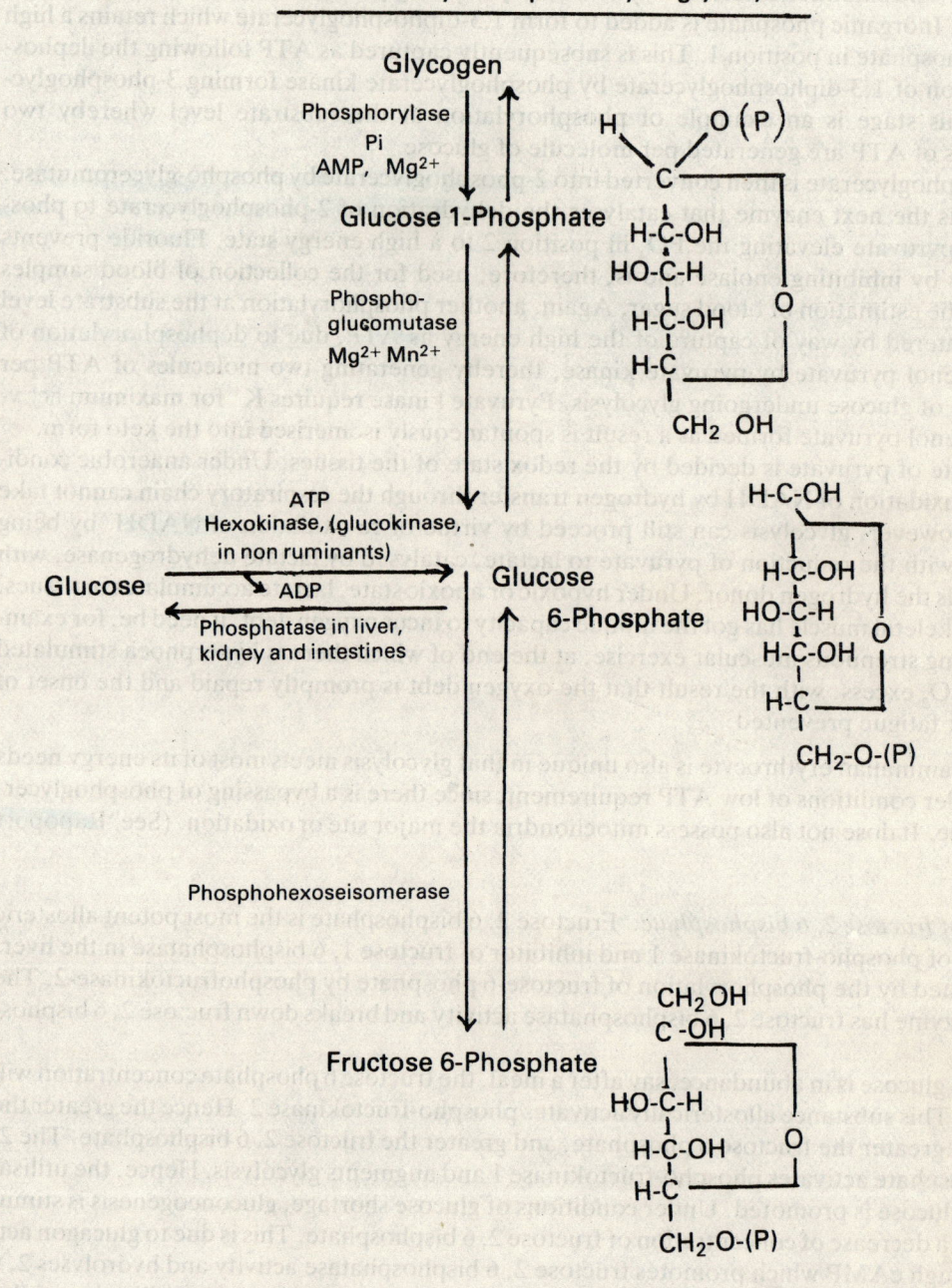

Glycogen

↓ Phosphorylase
   Pi
   AMP, Mg^{2+}

Glucose 1-Phosphate

↓ Phospho-
   glucomutase
   Mg^{2+} Mn^{2+}

```
         H      O (P)
          \    //
           C
           |
         H-C-OH
           |
        HO-C-H
           |         O
         H-C-OH
           |
         H-C
           |
          CH₂ OH
```

↓ ATP
  Hexokinase, (glucokinase, in non ruminants)

Glucose ⇌ Glucose 6-Phosphate
         ADP
Phosphatase in liver, kidney and intestines

```
         H-C-OH
           |
         H-C-OH
           |
        HO-C-H
           |         O
         H-C-OH
           |
         H-C
           |
         CH₂-O-(P)
```

↓ Phosphohexoseisomerase

Fructose 6-Phosphate

```
         CH₂OH
           |
         C -OH
           |
        HO-C-H
           |         O
         H-C-OH
           |
         H-C
           |
         CH₂-O-(P)
```

CARBOHYDRATE METABOLISM

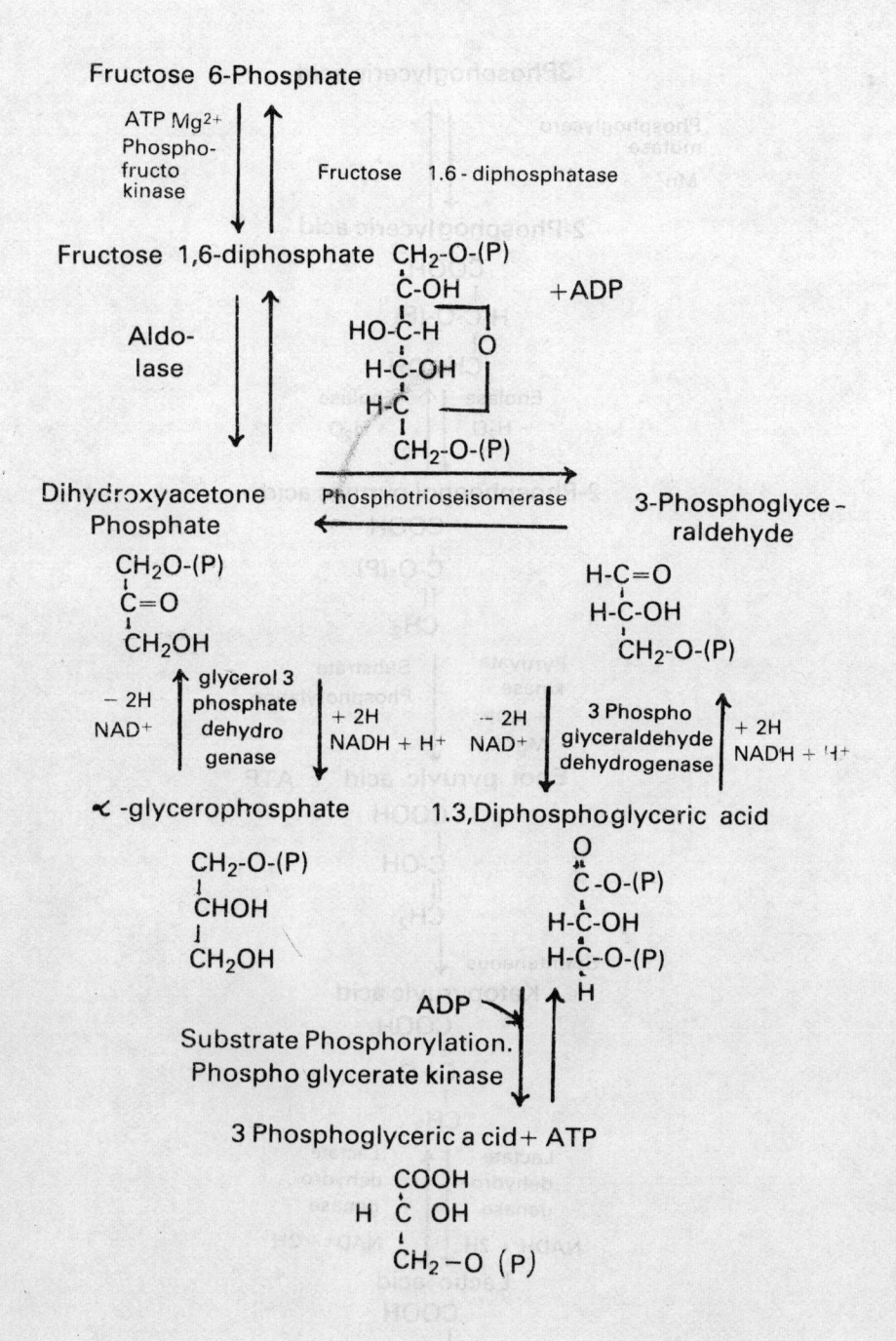

Fructose 6-Phosphate

ATP Mg^{2+}
Phospho-
fructo
kinase ↕   Fructose 1,6-diphosphatase

Fructose 1,6-diphosphate   CH$_2$-O-(P)
                               |
                               C-OH
                               |
                               HO-C-H   +ADP
                               | ⌐
                               H-C-OH   O
                               | ⌐
                               H-C
                               |
                               CH$_2$-O-(P)

Aldo-
lase ↕

Dihydroxyacetone Phosphate ⇌ Phosphotrioseisomerase ⇌ **3-Phosphoglyce-raldehyde**

CH$_2$O-(P)   H-C=O
| |
C=O   H-C-OH
| |
CH$_2$OH   CH$_2$-O-(P)

−2H ↑ glycerol 3   ↓ +2H   −2H ↑ 3 Phospho   ↓ +2H
NAD$^+$ phosphate   NADH + H$^+$ NAD$^+$ glyceraldehyde NADH + H$^+$
         dehydro   dehydrogenase
         genase

α-glycerophosphate   **1,3-Diphosphoglyceric acid**

CH$_2$-O-(P)   O
| ‖
CHOH   C-O-(P)
| |
CH$_2$OH   H-C-OH
                                               |
                                               H-C-O-(P)
                                               |
                                               H

                ADP ↘ ↑
Substrate Phosphorylation.
Phospho glycerate kinase ↓

3 Phosphoglyceric acid + ATP

COOH
|
H-C-OH
|
CH$_2$-O-(P)

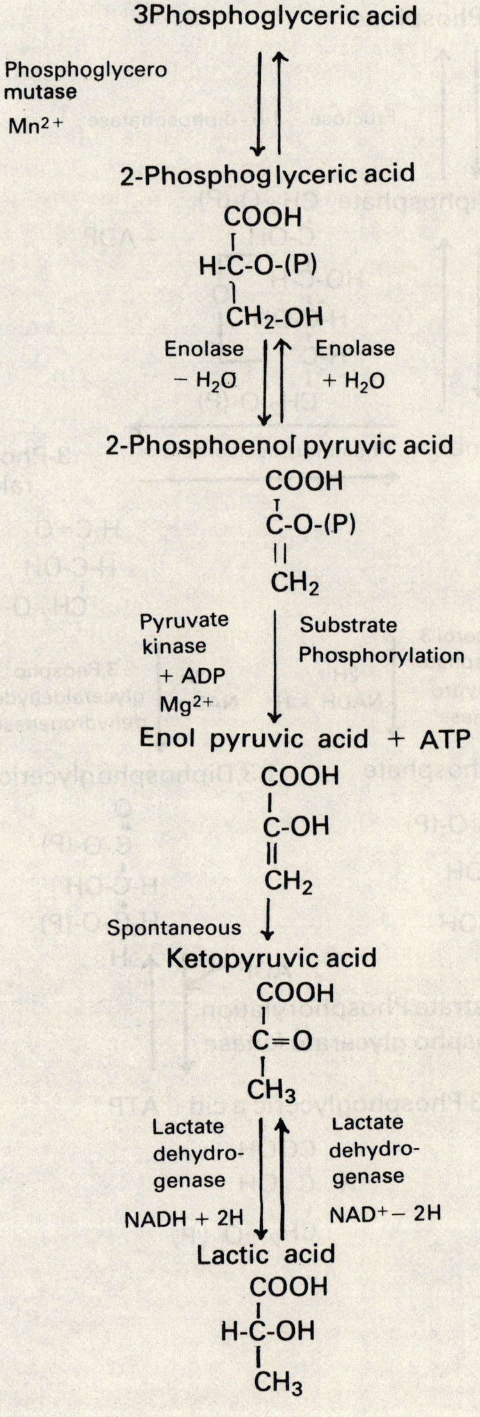

3Phosphoglyceric acid

Phosphoglycero mutase
Mn^{2+}

⇅

2-Phosphoglyceric acid

COOH
|
H-C-O-(P)
|
CH_2-OH

Enolase −H_2O ⇅ Enolase +H_2O

2-Phosphoenol pyruvic acid

COOH
|
C-O-(P)
||
CH_2

Pyruvate kinase + ADP Mg^{2+} | Substrate Phosphorylation ↓

Enol pyruvic acid + ATP

COOH
|
C-OH
||
CH_2

Spontaneous ↓

Ketopyruvic acid

COOH
|
C=O
|
CH_3

Lactate dehydrogenase NADH + 2H ↓ ↑ Lactate dehydrogenase NAD^+ − 2H

Lactic acid

COOH
|
H-C-OH
|
CH_3

unlike the cardiac muscle can contract for an appreciable length of time, though not indefinitely, in the absence of oxygen if necessitated by circumstances. But oxygen has to be supplied very soon so that fatigue may not supervene. There is a net generation of eight ATP molecules per molecule of glucose undergoing glycolysis and forming pyruvate. This is due to the fact that during the oxidation of glyceraldehyde-3-phosphate, the respiratory chain re-oxidation of 2 NADH* accounts for six ATP molecules in the muscle, while the two phosphorylations at the substrate level catalysed by phosphoglycerate kinase and pyruvate kinase account for four ATP molecules. From the total of ten, two have to be deducted on account of consumption of ATP in the endergonic reactions catalysed by hexokinase and phosphofructokinase. However, under anaerobic conditions, only two ATP molecules are generated because of the re-oxidation of two NADH being coupled with the reduction of pyruvate to lactate. If glycogen is converted to lactic acid, three ATP molecules are generated.

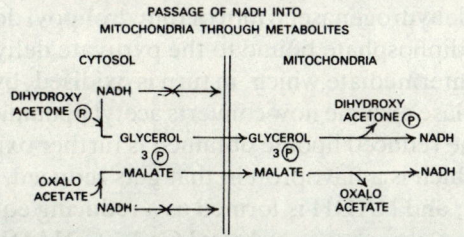

```
           PASSAGE OF NADH INTO
      MITOCHONDRIA THROUGH METABOLITES

     CYTOSOL          |        MITOCHONDRIA
                      |
   DIHYDROXY  NADH -->|
   ACETONE (P)        |         DIHYDROXY
         ↓            |         ACETONE (P)
       GLYCEROL    -->|-> GLYCEROL  --> NADH
         3 (P)        |      3 (P)
         ↑            |
       MALATE      -->|-> MALATE    --> NADH
   OXALO ↑            |      ↓
   ACETATE            |   OXALO
         NADH ->      |   ACETATE
```

The Rapoport-Luebering cycle: This cycle operates in the red blood cells of man and certain other mammals and can be considered as supplementary to glycolysis starting at the 1, 3 diphosphoglycerate level. The red blood cell utilises more glucose than it requires to maintain its cellular integrity. The excess energy has to be dissipated, as otherwise ATP and 1, 3 diphosphoglycerate would accumulate with a decrease of ADP and Pi and a consequent slowing of glycolysis. ATP inhibits Phosphofructokinase.

The Rapoport-Luebering cycle serves to 'waste' the energy not needed by the red cell. 1, 3 diphosphoglycerate is converted to 2, 3 bisphosphoglycerate with the enzyme phosphoglycerate mutase. The 2, 3 bisphosphoglycerate is converted to 3 phosphoglycerate by 2, 3 bisphosphoglycerate phosphatase. The 3 phosphoglycerate is added to the pool. There is a total energy loss of 14 Kcals: ten in the first reaction and four in the next.

Organic phosphates like 2, 3 bisphosphoglycerate, ATP and most of hexose-phosphates bind to Hb and decrease the affinity of Hb for oxygen. Such an effect is helpful in the delivery of oxygen by Hb to tissues in circumstances of low oxygen tension.

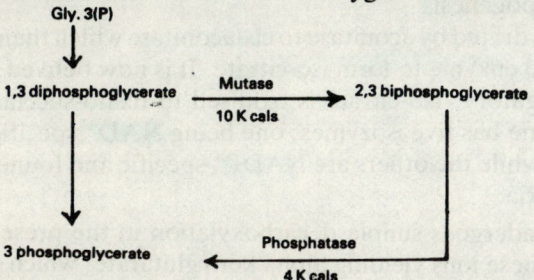

```
                    Gly. 3(P)
                       ↓
                              Mutase
   1,3 diphosphoglycerate  ------------->  2,3 biphosphoglycerate
                              10 K cals
         ↓                                        ↓
                              Phosphatase
   3 phosphoglycerate  <-------------------
                              4 K cals
```

*The NADH produced in the cytosol cannot pass through the mitochondrial membrane for re-oxidation and donation of ATP. It is transported through ∝ glycerophosphate, i.e., in the cytosol the NADH formed reduces dihydroxy acetone phosphate to ∝ glycerophosphate (glycerol 3 phosphate) which moves into the mitochondria. There it is re-oxidised with the help of NAD⁺ As a result NADH is formed inside the mitochondria. This is oxidised in the Electron Transport Chain to give ATP.

Instead of ∝ glycerophosphate, NADH can be transported towards inside of the mitochondria from outside as malate.

This cycle is also known as the citric acid cycle, Kreb's cycle, or the common metabolic pool as it serves to integrate carbohydrate, protein and lipid metabolisms. Unlike the enzymes of the EM pathway, those operating in the citric acid cycle are essentially located in the matrix of the mitochondria close to the enzymes of the respiratory chain; this is necessitated by the fact that this cycle generates many reducing equivalents in the form of reduced coenzymes which are immediately oxidised with the production of ATP.

Pyruvate is oxidatively decarboxylated to acetyl CoA catalysed by the pyruvate dehydrogenase complex present in the mitochondria. For this, the pyruvate has to be transported into the mitochondria by the pyruvate transporter (a symport mechanism whereby one H^+ is also co-transported). Pyruvate dehydrogenase is present as a multi-enzyme complex consisting of 29 mol of pyruvate dehydrogenase, 8 mol of dihydrolipoyl dehydrogenase and 1 mol of transacetylase. Thiamine diphosphate bound to the pyruvate dehydrogenase enzyme decarboxylates pyruvate to an intermediate which, in turn is oxidised, by oxidised lipoate to acetyl lipoamide. The transacetylase enzyme now converts acetyl lipoamide in the presence of coenzyme A to acetyl CoA. The reduced lipoate obtained is further oxidised to lipoate by lipoate dehydrogenase enzyme which is a flavoprotein that gets reduced. The reduced flavoprotein is then oxidised by NAD^+, and NADH is formed as a reducing equivalent in this system.

$$\text{Pyruvate} + NAD^+ + \text{CoA} \longrightarrow \text{Acetyl CoA} + NADH + H^+ + CO_2$$

The FAD, functioning together with lipothiamide pyrophosphate, precedes the NAD^+ system and is different from the Flavin system of the respiratory chain which usually succeeds NAD^+ system. NADH in turn transfers the reducing equivalents to the respiratory chain. Acetyl coenzyme A contains a high energy thiol ester bond. For all practical purposes, the conversion of pyruvate into acetyl coenzyme A is unidirectional, which accounts for the non-synthesis of carbohydrates from fatty acids having an even number of carbon atoms in their molecules.

Acetyl coenzyme A, obtained chiefly by the oxidation of carbohydrates or fatty acids, subsequently condenses with oxalo-acetate to form citrate. This is the sparking reaction which continually ignites the cycle thereby perpetuating it. This is made possible by the regeneration of oxalo-acetate from malate. The condensing enzyme catalyses the formation of citrate in the presence of ATP. This reaction is now believed to be reversible due to the extra-mitochondrial citrate cleavage enzyme or lyase whereby acetyl coenzyme A is regenerated and is available for lipogenesis.

Citrate is next dehydrated by aconitase to cis-aconitate which then undergoes a rehydration catalysed by the same enzyme to form iso-citrate. It is now belived that the formation of cis-aconitate is not obligatory. Iso-citrate is oxidised to oxalo-succinate by iso-citrate dehydrogenase. This enzyme has five isozymes, one being NAD^+ specific and present exclusively in the mitochondria while the others are $NADP^+$ specific and found in the cytosol and possibly in the mitochondria.

Oxalo-succinate undergoes simple decarboxylation in the presence of iso-citrate dehydrogenase and manganese ions yielding alpha keto glutarate, which is then oxidatively decarboxylated into succinyl coenzyme A, a reaction exactly analogous to the conversion of pyruvate into acetyl coenzyme A. There is thus the ∞ keto glutarate dehydrogenase complex. Succinyl coenzyme A is next converted to succinate by succinate thiokinase for which reaction GDP or IDP is required. They are converted in the presence of inorganic phosphate to either GTP or ITP respectively providing an example of generation of high energy phosphate at the substrate level, the only one of its kind in the citric acid cycle. GTP or ITP can directly be the source of high energy phosphate bonds or can enable the re-synthesis of ATP from ADP. Alternate enzyme systems for the formation of succinate from succinyl coenzyme A have been envisaged, namely, thiophorase in the extra-hepatic tissues and deacylase in the liver.

Succinate is dehydrogenated to fumarate by succinate dehydrogenase in the presence of FAD as the hydrogen acceptor, bypassing the NAD^+ system reflected in a P/O (or P/2e) ratio of 2:1. Succinate dehydrogenase is activated by calcium ions. Malonic acid, being a structural analogue of succinic acid, is a competitive inhibitor of the succinoxidase system. Fumarate is hydrated by fumarase to malate. The last step in the cycle involves the oxidation of malate by NAD^+ dependent malate dehydrogenase into oxalo-acetate which is thereby regenerated continually for the perpetuation of the citric acid cycle.

Some of the enzymes participating in the metabolism of glucose are inhibited by various reagents. They are given below:

Enzyme	Inhibitor
Glyceraldehyde-3 phosphate dehydrogenase	Iodo-acetate
Enolase	Fluoride
Phospho-glycerate kinase	Arsenate
Aconitase	Fluoro-acetate
Succinate dehydrogenase	Malonate

Bioenergetics and the physiological significance of the TCA cycle: The energy of oxidation that occurs in the tissues is mostly captured as ATP molecules preventing undue dissipation of energy which otherwise takes place in conditions like thyrotoxicosis due to the uncoupling of oxidative phosphorylation by thyroxine. Most of the ATP formed is due to oxidative phosphorylation resulting from re-oxidation of reduced coenzymes by the respiratory chain. The rest of ATP generated is by virtue of phosphorylation at the substrate level. As many as 30 high energy phosphate bonds are generated per molecule of glucose oxidised via the citric acid cycle. Six each are generated in the reaction catalysed by pyruvate dehydrogenase, iso-citrate dehydrogenase and alpha keto glutarate dehydrogenase followed by the respiratory chain re-oxidation of 2 NADH. Two high energy phosphate bonds are generated during the oxidation at the substrate level brought about by succinate thiokinase. The succinoxidase system involves the respiratory chain re-oxidation of 2 $FADH_2$ accounting for the generation of four high energy phosphate bonds. The malate dehydrogenase reaction accounts for the formation of six high energy phosphate bonds by way of re-oxidation of 2 NADH by the respiratory chain. Hence, the total number of ATP molecules generated per molecule of glucose oxidised under aerobic conditions is 38 (while it is 39 for glycogen): 30 ATPs from TCA and 8 from EM pathway.

As each molecule of ATP can give about 7400 calories, the total energy of oxidation of glucose is $38 \times 7400 = 2,81,200$ cals. But thermo-chemically, when a molecule of glucose is oxidised in vitro, it gives about 6,73,000 calories. Hence the available energy from a molecule of glucose is $\frac{2,81,200}{6,73,000} \times 100$ = about 42 per cent. The rest of the energy of oxidation is dissipated as heat.

Acetyl coenzyme A is considered the focal point of integration of the carbohydrate, protein and lipid metabolisms and it is the form in which all two carbon fragments irrespective of their

Tricarboxylic Acid Cycle

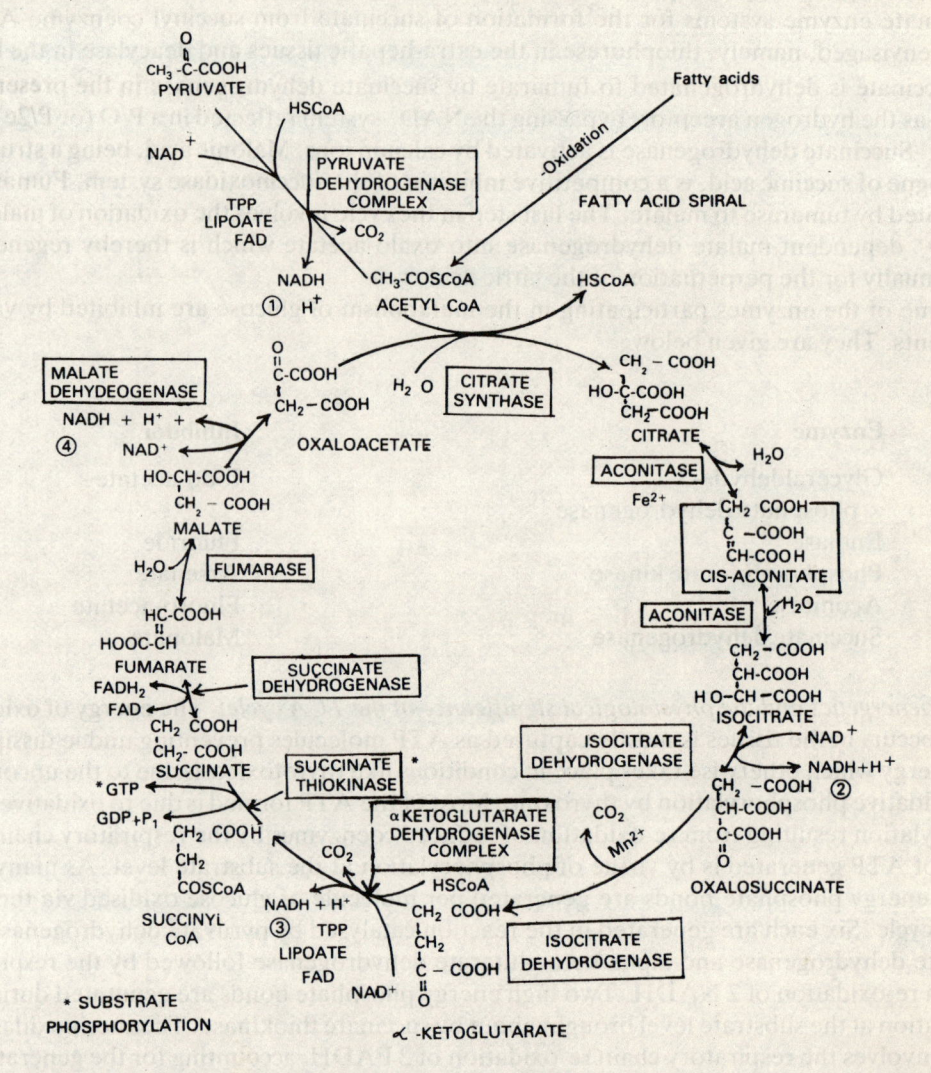

(Diagram of the TCA cycle showing the following components and reactions:)

- **Pyruvate** (CH_3-C(O)-COOH) + HSCoA + NAD^+ → via **Pyruvate Dehydrogenase Complex** (TPP, LIPOATE, FAD) → CO_2 + NADH + H^+ + CH_3-COSCoA (**Acetyl CoA**) ①
- **Fatty acids** → Oxidation → **Fatty Acid Spiral** → Acetyl CoA + HSCoA
- Acetyl CoA + Oxaloacetate (C(O)-COOH, CH_2-COOH) + H_2O → via **Citrate Synthase** → **Citrate** (CH_2-COOH, HO-C-COOH, CH_2-COOH)
- Citrate → via **Aconitase** (Fe^{2+}) → **Cis-Aconitate** (CH_2 COOH, C-COOH, CH-COOH) + H_2O
- Cis-Aconitate + H_2O → via **Aconitase** → **Isocitrate** (CH_2-COOH, CH-COOH, HO-CH-COOH)
- Isocitrate + NAD^+ → via **Isocitrate Dehydrogenase** → NADH + H^+ + **Oxalosuccinate** (CH_2-COOH, CH-COOH, C(O)-COOH) ②
- Oxalosuccinate → via **Isocitrate Dehydrogenase** (Mn^{2+}) → CO_2 + **α-Ketoglutarate** (CH_2 COOH, CH_2, C(O)-COOH)
- α-Ketoglutarate + HSCoA + NAD^+ → via **α-Ketoglutarate Dehydrogenase Complex** (TPP, LIPOATE, FAD) → CO_2 + NADH + H^+ + **Succinyl CoA** (CH_2 COOH, CH_2, COSCoA) ③
- Succinyl CoA + GDP + P_i → via **Succinate Thiokinase** → *GTP + **Succinate** (CH_2 COOH, CH_2-COOH)
- Succinate + FAD → via **Succinate Dehydrogenase** → $FADH_2$ + **Fumarate** (HC-COOH, HOOC-CH)
- Fumarate + H_2O → via **Fumarase** → **Malate** (HO-CH-COOH, CH_2-COOH)
- Malate + NAD^+ → via **Malate Dehydrogenase** → NADH + H^+ + **Oxaloacetate** ④

*Substrate Phosphorylation

origin get incorporated into a common metabolic pathway. Once they get into the pathway, they are no longer distinguishable regarding their origin. The citric acid cycle also provides additional linking points like alpha keto glutarate, which with aspartate can give rise to glutamate and oxalo-acetate by transamination, while the reverse reaction provides aspartate for the urea cycle. Fumaric acid is another point of linkage of the citric acid cycle with the urea cycle. Oxalo-acetate regenerated by the oxidation of malate can undergo a transamination with glutamate to yield alpha keto glutarate and aspartate. Malate can be oxidatively decarboxylated into pyruvate by the malic enzyme in the presence of $NADP^+$ providing thereby NADPH for extra-mitochondrial biosynthesis of steroids and lipids. The citric-acid cycle, by its reversibility as well as the transaminase reactions, contributes towards gluconeogenesis in conjunction with other gluconeogenic enzymes like pyruvate carboxylase and phosphoenol pyruvate carboxykinase, (phospho pyruvate carboxylase) which circumvent the concerned energy barriers.

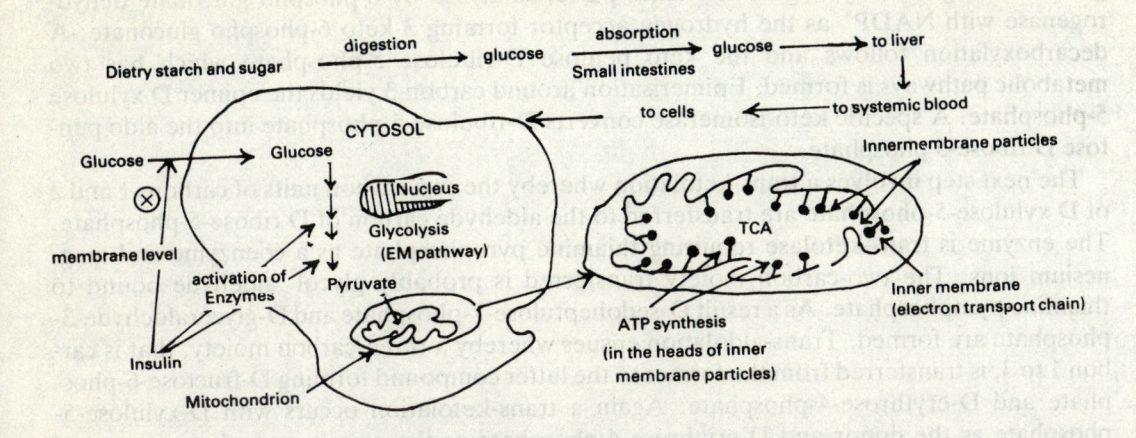

Thus, the starch and sugars in our food give us energy in the form of ATP through the metabolites of glucose oxidised in the EM-TCA pathway. It is for this purpose, inter alia, we take food. EM-TCA is a catabolic pathway for glucose to give energy.

Anaplerotic reactions: The TCA cycle operates continuously in all the cells. Some metabolic pathways end in a constituent of the cycle while other pathways originate from the cycle. These pathways concern the processes of gluconeogenesis, transamination, deamination and fatty acid synthesis.

Anaplerotic reactions lead to net transfer into the cycle as a result of several different reactions. Among the most significant is the formation of oxalo-acetate by the carboxylation of pyruvate catalysed by pyruvate carboxylase. This reaction is considered important in maintaining adequate concentrations of oxalo-acetate for the condensation reaction with acetyl CoA. If acetyl CoA accumulates, it acts as an allosteric activator of pyruvate carboxylase, thereby ensuring a continuous supply of oxalo-acetate.

The hexose monophosphate shunt (Pentose phosphate pathway)

This pathway is a direct oxidative pathway unlike the EM pathway of glycolysis. The enzymes of this pathway are located in the extra-mitochondrial soluble portion of the cell. The tissues concerned are the liver, lactating mammary glands, adipose tissue, erythrocytes, the lens, thyroid, testes and adrenal cortex. As in the EM pathway and citric acid cycle, here too oxidation is effected by dehydrogenation, but in the presence of $NADP^+$ and not NAD^+ The physiological and metabolic significance of the shunt pathway is that it provides adequate NADPH for the extra-mitochondrial biosynthesis of fatty acids as well as steroids. Another important purpose served by this pathway is the provision of pentoses for nucleotide and nucleic acid synthesis. Inter-conversion of hexoses and pentoses takes place in this pathway.

The pentose pathway is broadly divided into two phases. At first glucose-6-phosphate is dehydrogenated by glucose-6-phosphate dehydrogenase in the presence of $NADP^+$ and subsequently decarboxylated to form ribulose-5-phosphate. The second phase consists of the conversion of ribulose-5-phosphate back to glucose-6-phosphate by a series of reactions including trans-ketolation involving the role of thiamine and also trans-aldolation.

Glucose-6-phosphate is dehydrogenated by glucose-6-phosphate dehydrogenase which is $NADP^+$ dependent. 6-phospho glucono lactone that is formed is hydrolysed into 6-phospho gluconate. Again dehydrogenation takes place, catalysed by 6-phospho gluconate dehydrogenase with $NADP^+$ as the hydrogen acceptor forming 3-keto 6-phospho gluconate. A decarboxylation follows and the keto pentose D-ribulose 5-phosphate which has two metabolic pathways is formed. Epimerisation around carbon 3 yields the epimer D xylulose 5-phosphate. A specific keto-isomerase converts D-ribulose 5-phosphate into the aldo pentose D ribose-5 phosphate.

The next step involves a trans-ketolation whereby the two carbon units of carbons 1 and 2 of D xylulose-5-phosphate are transferred to the aldehyde carbon of D ribose-5-phosphate. The enzyme is trans-ketolase requiring thiamine pyrophosphate as a coenzyme and magnesium ions. The two-carbon moiety transferred is probably glycol aldehyde bound to thiamine pyrophosphate. As a result D-sedoheptulose-7-phosphate and D-glyceraldehyde-3-phosphate are formed. Trans-aldolation ensues whereby a three-carbon moiety, that is carbon 1 to 3, is transferred from the former to the latter compound forming D-fructose-6-phosphate and D-erythrose-4-phosphate. Again a trans-ketolation occurs with D-xylulose-5-phosphate as the donor and D-erythrose-4-phosphate as the acceptor, in the presence of thiamine pyrophosphate and magnesium ions, yielding D-fructose-6-phosphate and D-glyceraldehyde 3-phosphate.

Unlike the EM pathway, the shunt pathway is characterised by the production of CO_2. Most of the tissues in which the shunt operates are actively engaged in the utilisation of NADPH for the extra-mitochondrial biosynthesis of fatty acids and steroids. The skeletal muscle is capable of synthesising ribose despite the fact that the tissue contains negligible amounts of glucose-6-phosphate dehydrogenase and 6-phospho gluconate dehydrogenase. A reversal of the shunt pathway, involving fructose-6-phosphate, glyceraldehyde-3-phosphate and the enzymes trans-aldolase and trans-ketolase, is partly responsible for making ribose available to the skeletal muscle. The shunt pathway in the metabolism of the eye lens is known to exist and galactose is believed to inhibit the activity of glucose-6-phosphate dehydrogenase. The development of cataract is often associated with congenital galactosemia and galactosuria. Increased fragility of the erythrocyte and susceptibility to hemolysis is seen when there is hereditary deficiency of the oxidative enzymes of the shunt pathway, like glucose-6-phosphate dehydrogenase. The condition is met with in Favism when the person subsists on Fava beans. It is also responsible for **Primaquine** sensitivity.

Energetics of HMP shunt: One CO_2 molecule is eliminated during one cycle when 6-phospho gluconic acid is converted to ribulose-5-phosphate. Thus one cycle oxidises only carbon 1 of glucose so that six passages around are necessary for complete oxidation of a molecule of glucose. During one passage 2 $NADP^+$ are used and if oxidised they will give six ATP molecules. Hence for six passages, a total of 36 ATP molecules, i.e., 36×7400 cals, could be produced. But it has to be remembered that in many cases, the NADPH formed is used for lipogenesis.

William points out that there are two types of HMP shunt, the F-type (fatty tissues) and the L-type (liver and other tissues). In the L-type, due to aldolase participation, arabinose-5-phosphate, D-glycero-D-ido-octulose 1, 8 di-phosphate, D glycero - D ido-octulose 8-phosphate and sedoheptulose 1, 7 di-phosphate are additional intermediates. This suggestion has however been contradicted by some workers.

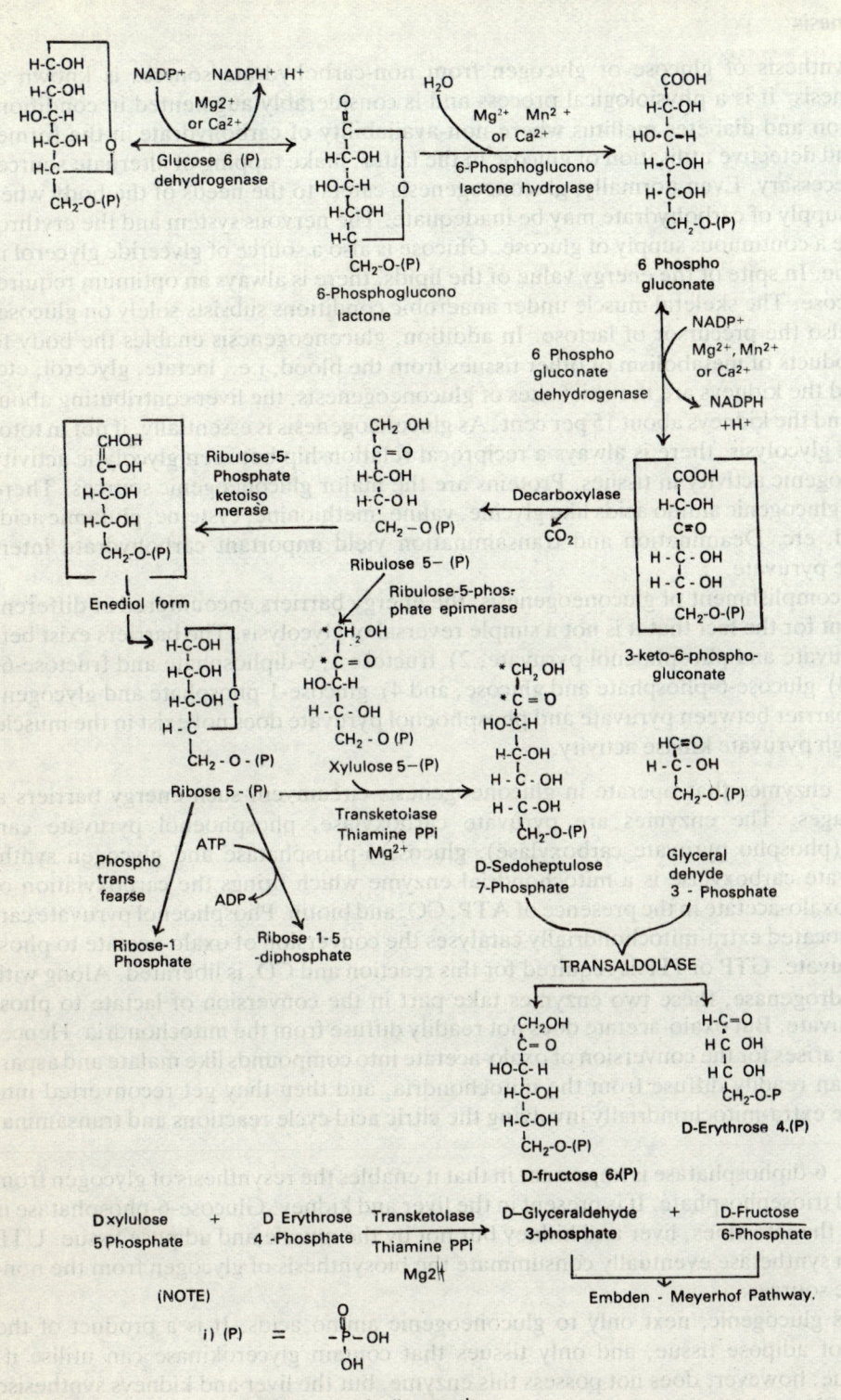

CARBOHYDRATE METABOLISM

(Pentose Phosphate Pathway / HMP Shunt diagram)

Glucose 6 (P) → [Glucose 6 (P) dehydrogenase, NADP⁺ → NADPH + H⁺, Mg²⁺ or Ca²⁺] → 6-Phosphogluconolactone → [6-Phosphoglucono lactone hydrolase, H₂O, Mg²⁺, Mn²⁺ or Ca²⁺] → 6 Phosphogluconate → [6 Phosphogluconate dehydrogenase, NADP⁺ → NADPH + H⁺, Mg²⁺, Mn²⁺ or Ca²⁺] → 3-keto-6-phosphogluconate → [Decarboxylase, −CO₂] → Ribulose 5−(P)

Ribulose 5−(P) ⇌ [Ribulose-5-Phosphate ketoisomerase, Enediol form] ⇌ Ribose 5−(P)

Ribulose 5−(P) → [Ribulose-5-phosphate epimerase] → Xylulose 5−(P)

Ribose 5−(P) + Xylulose 5−(P) → [Transketolase, Thiamine PPi, Mg²⁺] → D-Sedoheptulose 7-Phosphate + Glyceraldehyde 3-Phosphate

Ribose 5−(P) → [ATP → ADP, Phosphotransferase] → Ribose-1 Phosphate / Ribose 1-5 diphosphate

D-Sedoheptulose 7-Phosphate + Glyceraldehyde 3-Phosphate → [TRANSALDOLASE] → D-Erythrose 4-(P) + D-fructose 6-(P)

D-Xylulose 5-Phosphate + D-Erythrose 4-Phosphate → [Transketolase, Thiamine PPi, Mg²⁺] → D-Glyceraldehyde 3-phosphate + D-Fructose 6-Phosphate → Embden-Meyerhof Pathway

(NOTE)

i) (P) = $-\overset{O}{\underset{OH}{P}}-OH$

ii) NADPH is used for lipogenesis

iii) Ribose 5 (P) is utilized for synthesis of nucleotides and nucleic acids.

Gluconeogenesis

The biosynthesis of glucose or glycogen from non-carbohydrate sources is known as gluconeogenesis. It is a physiological process and is considerably augmented in conditions like starvation and diabetes mellitus where non-availability of carbohydrate in the former condition and defective utilisation of glucose in the latter, make tapping of alternate sources of energy necessary. Even normally, gluconeogenesis caters to the needs of the body when the dietary supply of carbohydrate may be inadequate. The nervous system and the erythrocytes require a continuous supply of glucose. Glucose is also a source of glyceride glycerol in adipose tissue. In spite of the energy value of the lipids, there is always an optimum requirement of glucose. The skeletal muscle under anaerobic conditions subsists solely on glucose. Glucose is also the precursor of lactose. In addition, gluconeogenesis enables the body to clear the products of metabolism of other tissues from the blood, i.e., lactate, glycerol, etc. The liver and the kidneys are the chief sites of gluconeogenesis, the liver contributing about 85 per cent and the kidneys about 15 per cent. As gluconeogenesis is essentially, if not in toto, a reversal of glycolysis, there is always a reciprocal relationship between glycolytic activity and gluconeogenic activity in tissues. Proteins are the major gluconeogenic sources. There are a host of glucogenic amino acids like glycine, valine, methionine, cysteine, glutamic acid, aspartic acid, etc. Deamination and transamination yield important carbohydrate intermediates like pyruvate.

For the accomplishment of gluconeogenesis, the energy barriers encountered at different stages account for the fact that it is not a simple reversal of glycolysis. The barriers exist between 1) pyruvate and phosphoenol pyruvate, 2) fructose 1, 6-diphosphate and fructose-6-phosphate, 3) glucose-6-phosphate and glucose, and 4) glucose-1-phosphate and glycogen. The energy barrier between pyruvate and phosphoenol pyruvate does not exist in the muscle which has high pyruvate kinase activity.

The chief enzymes that operate in gluconeogenesis circumvent such energy barriers at different stages. The enzymes are pyruvate carboxylase, phosphoenol pyruvate carboxykinase (phospho pyruvate carboxylase), glucose-6-phosphatase and glycogen synthetase. Pyruvate carboxylase is a mitochondrial enzyme which brings the carboxylation of pyruvate to oxalo-acetate in the presence of ATP, CO_2 and biotin. Phosphoenol pyruvate carboxykinase located extra-mitochondrially catalyses the conversion of oxalo-acetate to phosphoenol pyruvate. GTP or ITP is required for this reaction and CO_2 is liberated. Along with lactate dehydrogenase, these two enzymes take part in the conversion of lactate to phosphoenol pyruvate. But oxalo-acetate does not readily diffuse from the mitochondria. Hence, the necessity arises for the conversion of oxalo-acetate into compounds like malate and aspartate which can readily diffuse from the mitochondria, and then they get reconverted into oxalo-acetate extra-mitochondrially involving the citric acid cycle reactions and transaminations.

Fructose 1, 6-diphosphatase is important in that it enables the resynthesis of glycogen from pyruvate and triosephosphate. It is present in the liver and kidney. Glucose-6-phosphatase is produced by the intestines, liver and kidney but not by the muscle and adipose tissue. UTP and glycogen synthetase eventually consummate the biosynthesis of glycogen from the non-carbohydrate sources.

Glycerol is glucogenic, next only to gluconeogenic amino acids. It is a product of the metabolism of adipose tissue, and only tissues that contain glycerokinase can utilise it. Adipose tissue, however, does not possess this enzyme, but the liver and kidneys synthesise the enzyme. Glycerokinase catalyses the conversion of glycerol into alpha glycero phosphate

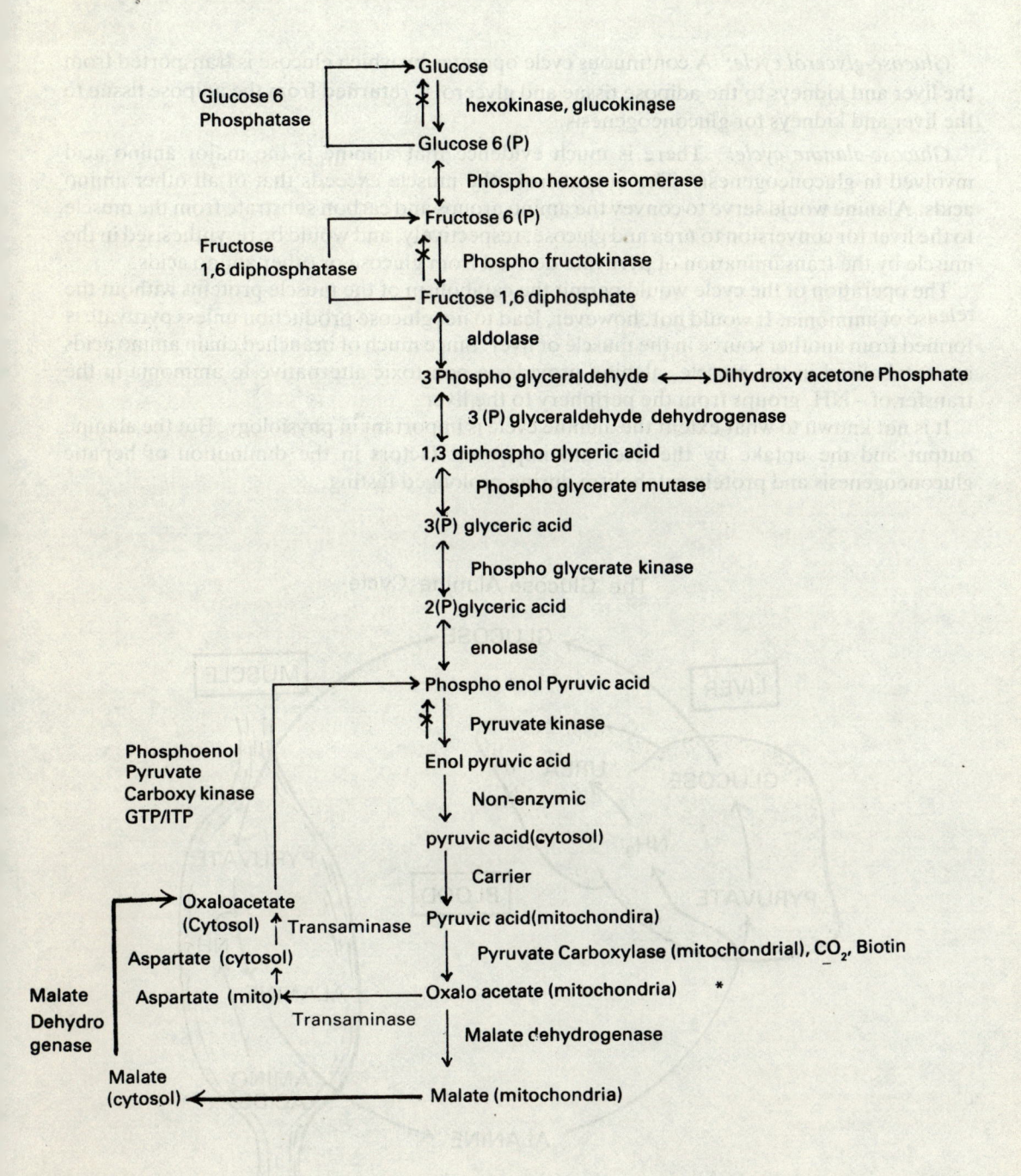

```
                              Glucose
Glucose 6            ⇅  hexokinase, glucokinase
Phosphatase
                              Glucose 6 (P)
                              ↓  Phospho hexose isomerase
                              Fructose 6 (P)
Fructose
1,6 diphosphatase    ⇅  Phospho fructokinase
                              Fructose 1,6 diphosphate
                              ↓  aldolase
                              3 Phospho glyceraldehyde  ←→  Dihydroxy acetone Phosphate
                              ↑↓  3 (P) glyceraldehyde  dehydrogenase
                              1,3 diphospho glyceric acid
                              ↓  Phospho glycerate mutase
                              3(P) glyceric acid
                              ↓  Phospho glycerate kinase
                              2(P)glyceric acid
                              ↓  enolase
                              Phospho enol Pyruvic acid
                              ⇅  Pyruvate kinase
Phosphoenol                   Enol pyruvic acid
Pyruvate                      ↓  Non-enzymic
Carboxy kinase
GTP/ITP                       pyruvic acid(cytosol)
                              ↓  Carrier
            Oxaloacetate      Pyruvic acid(mitochondira)
            (Cytosol) ↑ Transaminase
            Aspartate (cytosol)     Pyruvate Carboxylase (mitochondrial), CO₂, Biotin
Malate      Aspartate (mito) ←   Oxalo acetate (mitochondria)         *
Dehydro              Transaminase
genase                        ↓  Malate dehydrogenase
        Malate
        (cytosol) ←  Malate (mitochondria)
```

(Diagram as rendered above; equations/chemistry preserved.)

CO_2

connecting it with the triosephosphate stage of the EM pathway. Thus, the liver and kidney are able to synthesise glucose and glycogen from glycerol.

Glucose-glycerol cycle: A continuous cycle operates by which glucose is transported from the liver and kidneys to the adipose tissue and glycerol is returned from the adipose tissue to the liver and kidneys for gluconeogenesis.

Glucose-alanine cycle: There is much evidence that alanine is the major amino acid involved in gluconeogenesis. The output from the muscle exceeds that of all other amino acids. Alanine would serve to convey the amino groups and carbon substrate from the muscle to the liver for conversion to urea and glucose, respectively, and would be resynthesised in the muscle by the transamination of pyruvate derived from glucose or other amino acids.

The operation of the cycle would permit the catabolism of the muscle proteins without the release of ammonia. It would not, however, lead to net glucose production unless pyruvate is formed from another source in the muscle or liver. Since much of branched chain amino acids are catabolised in the muscle, alanine provides a non-toxic alternative to ammonia in the transfer of $-NH_2$ groups from the periphery to the liver.

It is not known to what extent the alanine cycle is important in physiology. But the alanine output and the uptake by the liver are important factors in the diminution of hepatic gluconeogenesis and protein catabolism during prolonged fasting.

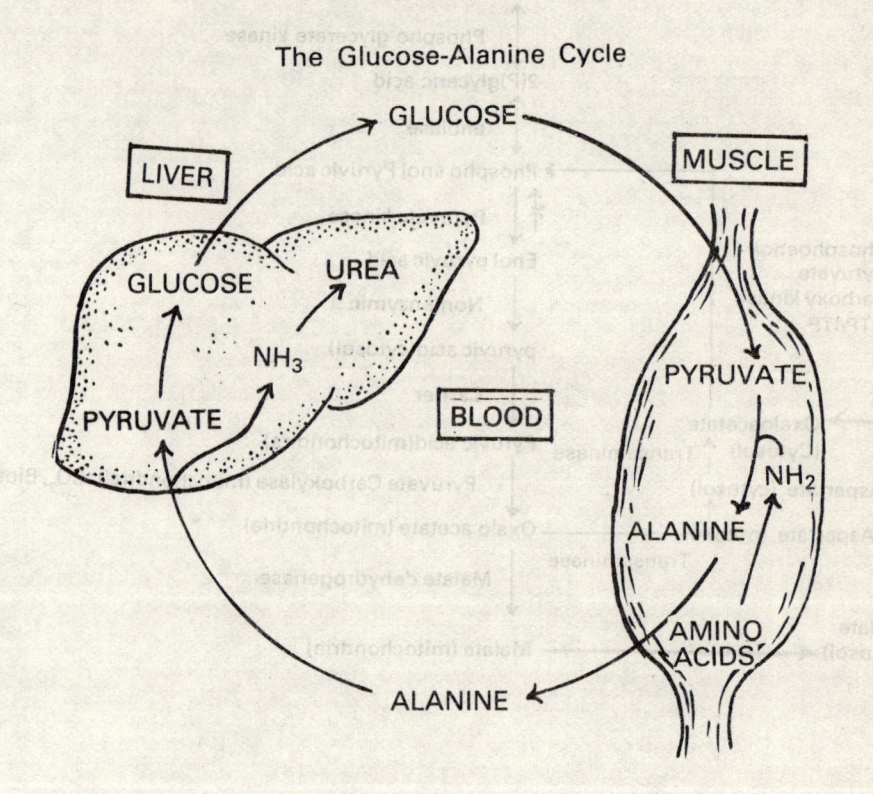

The Glucose-Alanine Cycle

As in other processes in carbohydrate metabolism, hormones play a vital role in regulating gluconeogenesis. Insulin is known to bring about a repression of important gluconeogenic enzymes while hormones like thyroxine, gluco-corticoids as well as ACTH of the adenohypophysis elevate the blood sugar level by the stimulation of gluconeogenesis. The availability of amino acids for gluconeogenesis is increased by gluco-corticoids which are protein-catabolic hormones.

The uronic acid pathway

This pathway provides an alternate direct oxidative pathway for glucose whereby it is convertible into pentoses via glucuronic acid. Ascorbic acid is also formed in this pathway in many species where the concerned enzyme system operates. But it is not found in humans, monkeys and guinea pigs. Glucose-6-phosphate is converted into glucose-1-phosphate by phospho glucomutase, and glucose-1-phosphate reacts with UTP to give UDP glucose, the reaction being catalysed by UDPG pyrophosphorylase. UDP glucose is next oxidised at carbon 6 to UDP glucuronic acid by NAD^+ dependent UDPG dehydrogenase. UDP glucuronic acid is the active form of glucuronic acid which gets incorporated into various acid mucopolysaccharides like chondroitin sulphate, hyaluronic acid and heparin. It also enables conjugation of bilirubin, steroid hormones and drugs by the liver as protective synthesis.

D-glucuronic acid derived from the UDP glucuronic acid is next reduced to L-gulonic acid in a NADPH-dependent reduction. L-gulonic acid is oxidised by an NAD^+ dependent dehydrogenase to 3 keto L-gulonic acid which is then decarboxylated to L-xylulose. A connection of the uronic acid pathway with the HMP shunt is made possible by the reduction of L-xylulose to D-xylitol by a NADPH dependent reductase reaction. If the enzyme which is responsible for this reduction is lacking, it results in an inborn error of metabolism characterised by an essential pentosuria where L-xylulose is excreted in urine. D-xylitol is then oxidised by an NAD^+ dependent reaction to D-xylulose which can next be phosphorylated in the presence of ATP to D-xylulose-5-phosphate, and this compound can be metabolised by the HMP shunt pathway. Drugs like barbital increase significantly the conversion of glucose into glucuronic acid.

L-gulonic acid is the precursor of ascorbic acid in species that possess the specific dehydrogenase for the oxidation of L-gulonolactone to 2 keto L-gulonolactone and the isomerase for converting the 2 keto gulonolactone to vitamin C. However, in human beings, monkeys and guinea pigs such an enzymatic machinery is lacking and hence the guinea pig, for example, is the animal of choice for quick experimental induction of scurvy.

Metabolism of amino sugars (Hexosamines)

Amino sugars are vital constituents of the widely distributed structural elements of the tissues, i.e., the glycosamino glycans. Many glycoproteins also contain amino sugars. Membrane receptors are believed to be mostly glycoproteins.

The biosynthesis of glucosamine is catalysed by the specific L-glutamine-D-fructose 6 phosphate transamidase, a reaction which is unique in that the required energy is derived not from ATP but by the cleavage of the amide bond of L-glutamine.

The liver is concerned with the biosynthesis of N-acetyl galactosamine by the epimerisation of UDP-N-acetyl glucosamine. N-acetyl neuraminic acid, another important amino sugar present in glyco-proteins and glycosaminoglycans is obtained by the condensation of N-acetyl mannosamine 6-phosphate with phosphoenol pyruvate. N-acetyl mannosamine 6-phosphate in turn is derived by the epimerisation of N-acetyl glucosamine 6-phosphate. The interrelationship in the metabolism of amino sugars is important since they are concerned with physiologically important compounds like, hyaluronic acid, heparin, chondroitin sulphates, sialic acid, glycoproteins, etc.

Metabolism of fructose and galactose

Fructose and galactose are phosphorylated by fructokinase and galactokinase, respectively, in the liver. However, while the non-specific hexokinases convert them into the corresponding hexose-6-phosphates, the specific kinases on the other hand, bring about the formation of the respective hexose-1-phosphates. Fructokinase is present in the kidney and intestines as well. Unlike glucokinase, fructokinase, activity is not affected by fasting or insulin and hence fructose seems to be better tolerated by the diabetic person. Fructose-1-phosphate is split by the specific aldolase into D-glyceraldehyde phosphate isomeric with dihydroxyacetone phosphate. When fructose-1-phosphate aldolase is deficient it causes hereditary fructose intolerance. The alternate metabolic pathway is the phosphorylation of fructose-1-phosphate to fructose-1-6-diphosphate catalysed by phospho-fructokinase in the liver and muscle. The conversion of fructose into glucose is prevented in the experimental animals from which the liver and the intestines are removed. Fructose bypasses the steps in glucose metabolism catalysed by glucokinase and phospho-fructokinase, since fructokinase activity is independent of insulin. This is responsible for the quick disappearance of fructose from the blood of the diabetic patient. Fructose is present in seminal plasma. Conversion of glucose to fructose can occur through sorbitol. Glucose is at first reduced to sorbitol by the NADPH dependent specific reductase, and sorbitol is oxidised to fructose in the presence of NAD^+ dependent sorbitol dehydrogenase. While fructose is converted to fructose-1-phosphate in the liver, it is directly converted to fructose-6-phosphate in the brain.

Galactose is readily converted in the liver to glucose. Any defect in this conversion results in intolerance to galactose. Galactose is phosphorylated by galactokinase in the presence of ATP to galactose-1-phosphate which next combines with UDP-glucose to form UDP-galactose and glucose-1-phosphate. This reaction is catalysed by galactose-1-phosphate-uridyl transferase and a deficiency of this enzyme results in congenital galactosemia and galactosuria. UDP galactose is epimerised in a reversible reaction into UDP glucose around carbon 4. Finally, glucose is obtained from UDPG as glucose-1-phosphate. In the biosynthesis of lactose by the mammary gland, UDP galactose condenses with glucose to form lactose, catalysed by lactose synthetase.

There is a large accumulation of galactose-1-phosphate in the erythrocytes in the galactosemic patients showing that galactokinase deficiency is not responisble for the condition. Epimerisation is also adequate and hence the exact biochemical lesion has been located in the deficiency of galactose-1-phosphate uridyl transferase.

Blood sugar homeostasis

The regulation of blood sugar is governed chiefly by the endocrine glands at the cellular and enzymatic levels. The liver plays a gluco-static role while the kidneys operate by way of filtration and re-absorption of glucose. The post-absorptive blood sugar level ranges from 80 to

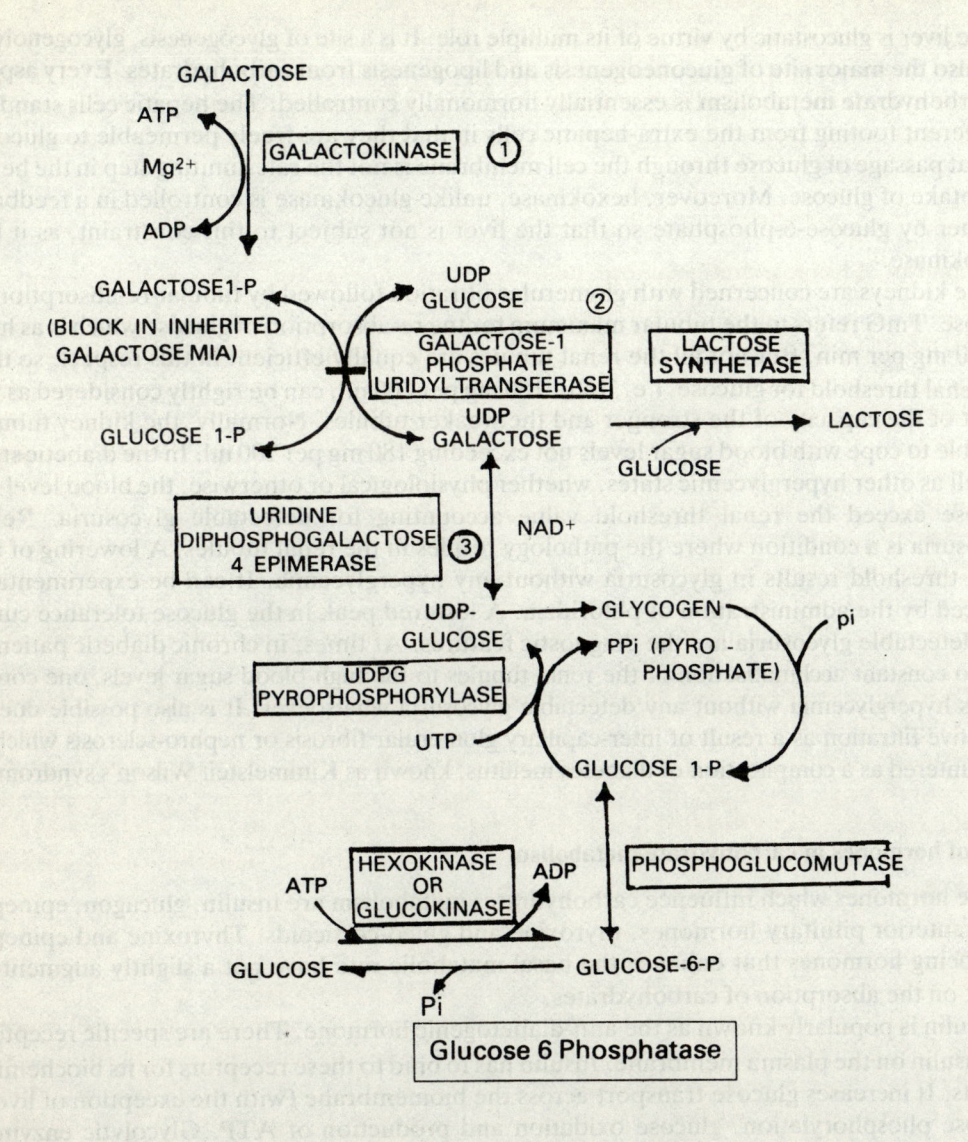

Interconversion of glucose and galactose and biosynthesis of lactose.

120 mg/100 ml (Folin-Wu) of whole blood. The fasting level varies between 70 and 90 mg per 100 ml. The liver, extra-hepatic tissues and several hormones contribute towards the homeostasis. In fact, there is a harmonious interplay between the chief anti-diabetogenic hormone, insulin, and the various diabetogenic hormones which act against insulin. However, in the diabetic state, the hyperglycemic hormones are not adequately countered due to the subnormal insulin status of the individual.

Dietary polysaccharides are slowly assimilated so that normally there is no tendency for hyperglycemia or detectable glycosuria. The alimentary absorption of carbohydrates is not believed to be under the influence of insulin though thyroxine and epinephrine, being hormones that enhance the basal metabolic rate, have got a slight augmenting effect on the absorption of carbohydrate.

The liver is glucostatic by virtue of its multiple role. It is a site of glycogenesis, glycogenolysis and also the major site of gluconeogenesis and lipogenesis from carbohydrates. Every aspect of carbohydrate metabolism is essentially hormonally controlled. The hepatic cells stand on a different footing from the extra-hepatic cells in that they are freely permeable to glucose, so that passage of glucose through the cell membrane is not the rate-limiting step in the hepatic uptake of glucose. Moreover, hexokinase, unlike glucokinase is controlled in a feedback manner by glucose-6-phosphate so that the liver is not subject to this constraint, as it has glucokinase.

The kidneys are concerned with glomerular filtration followed by tubular re-absorption of glucose. TmG refers to the tubular maximum for the re-absorption of glucose which is as high as 350 mg per min. But not all the renal tubules are equally efficient in this respect, so that the renal threshold for glucose, i.e., 160–180 mg per 100 ml, can be rightly considered as the result of the capacity of the stronger and the weaker tubules. Normally, the kidney tubules are able to cope with blood sugar levels not exceeding 180 mg per 100 ml. In the diabetic state as well as other hyperglycemic states, whether physiological or otherwise, the blood levels of glucose exceed the renal threshold value accounting for detectable glycosuria. Renal glycosuria is a condition where the pathology resides in the renal tubules. A lowering of the renal threshold results in glycosuria without any hyperglycemia. It can be experimentally induced by the administration of phloridzin. A lowered peak in the glucose tolerance curve and detectable glycosuria are the diagnostic features. At times, in chronic diabetic patients, due to constant acclimatisation of the renal tubules to the high blood sugar levels, one comes across hyperglycemia without any detectable glycosuria whatsoever. It is also possible due to defective filtration as a result of inter-capillary glomerular fibrosis or nephro-sclerosis which is encountered as a complication of diabetes mellitus, known as Kimmelsteil Wilson's syndrome.

Role of hormones in carbohydrate metabolism

The hormones which influence carbohydrates metabolism are insulin, glucagon, epinephrine, anterior pituitary hormones, thyroxine and gluco-corticoids. Thyroxine and epinephrine being hormones that enhance the basal matabolic rate have got a slightly augmenting effect on the absorption of carbohydrates.

Insulin is popularly known as the anti-diabetogenic hormone. There are specific receptors for insulin on the plasma membrane. Insulin has to bind to these receptors for its biochemical effects. It increases glucose transport across the biomembrane (with the exception of liver), glucose phosphorylation, glucose oxidation and production of ATP. Glycolytic enzymes-glucokinase, phosphofructo-kinase and pyruvate kinase-are induced by insulin, and pyruvate dehydrogenase is activated. Insulin increases glucose-6-phosphate dehydrogenase activity, promotes the HMP shunt pathway and the production of NADPH in this pathway. By this action, insulin augments lipid synthesis. Insulin represses gluconeogenic enzymes-pyruvate carboxylase, phospho pyruvate carboxylase and fructose 1, 6 disphosphatase-and inhibits gluconeogenesis. By converting glycogen synthetase into its active 'a' form, insulin promotes the synthesis of glycogen. By effectively oxidising acetyl CoA in the TCA cycle, insulin prevents the formation of ketone bodies, and it inhibits glycogenolysis by its antagonism to cAMP. In adipose tissue, insulin enhances lipid synthesis by providing adequate acetyl coenzyme A, ATP, NADPH as well as glycerophosphate for triacylglycerol synthesis (For details see Hormones.) Insulin, by decreasing the levels of cyclic AMP, counters the action of epinephrine and glucagon and thereby retards lipolysis from the triacyl glycerol stores of adipose tissue resulting in the fall of plasma-free fatty acid levels.

By increasing glucose utilisation and decresing glucose production, in the various ways listed above insulin brings down blood glucose.

Epinephrine raises blood sugar level by activating adenylate cyclase and raising the level of cyclic AMP which in turn activates the specific kinase that transforms the inactive form into the active form of phosphorylase. Epinephrine acts on the liver as well as muscle, but in the latter, lactic acid and not glucose is the end product. Glucagon, unlike epinephrine, acts on the liver but not on skeletal muscle though it acts on cardiac muscle. The hyperglycemic factor of the alpha cells of the islets, viz, glucagon is also credited with stimulation of gluconeogenesis to some extent. Thyroxine, like epinephrine has a dual role to play. Due to increased BMR, there is a greater utilisation of glucose by the tissues but this is masked by the promotion of gluconeogenesis brought out by the hormone. In advanced cases of hyperthyroidism there is hyperglycemia and glycosuria and the glucose tolerance of the person is reduced. Glycogenolysis is also to some extent increased by thyroxine.

The growth hormone decreases glucose uptake in the muscle and other tissues. Its effect is due to its augmentation of lipolysis and release of free fatty acids in blood. Such a release of FFA has a retarding effect on glycolysis.

The 11 Oxo-steroids of the adrenal cortex stimulate gluconeogenesis by making amino acids available through enhanced degradation of proteins. These amino acids are utilised for gluconeogenesis. ACTH has a similar effect as it stimulates the production of adrenal corticoids (the gluco-corticoids).

It can be seen that insulin is the only hormone that helps in the active utilisation of glucose and tends to bring down the blood sugar. Other hormones like epinephrine, glucagon, glucocorticoids and anterior pituitary hormones tend to elevate the blood sugar. By the interplay of various hormones, the homeostasis of blood sugar is maintained.

Fructose does not require insulin for its transport across the plasma membrane.

Diabetes mellitus

It is well known that diabetes mellitus is a disease due to impaired carbohydrate metabolism. The disease is caused by a quantitative or functional deficiency of insulin. If the patient has quantitative deficiency of insulin, the condition is referred to as IDDM (Insulin-dependent diabetes mellitus). If however, there is enough insulin in the body whicn is not utilised due to its functional lapse, the condition is known as NIDDM (non-insulin-dependent diabetes mellitus). Juvenile diabetes, the disease in young age, is usually IDDM, while maturity onset diabetes is usually NIDDM.

When there is insulin deficiency, glucose is not properly utilised and oxidised, and glucose transport across the biomembrane is not facilitated. Insulin is not available to induce glycolytic enzymes and activate pyruvate dehydrogenase. There is no repression of gluconeogenic enzymes. In insulin deficiency, glycogen synthetase 'a' is converted to 'b' which is inactive. Hence, glucose is not converted to glycogen. Because of the under-utilisation of glucose by the cells, blood sugar level is increased. Hence, in diabetes mellitus, the diagnostic test is the evaluation of blood sugar. Normal post-absorptive levels of the blood sugar are 80-120 mg/dl if the test is done by the Folin and Wu method. This method estimates not only blood glucose levels but also those of the other reducing substances in the blood like glutathione and vitamin C. Glucose oxidase, and Nelson and Somogyi methods estimate only blood glucose which is referred to as 'true sugar'. The value of glucose is only 60-95 mg/dl. Increase in blood sugar is called hyperglycemia. It is a pathological condition as excess of glucose in the blood causes osmotic disturbance and affects the life of the red blood cells. The body gets rid of the excess glucose in the blood by the excretion of glucose in urine. This condition is known as glycosuria. It is known that normal urine does not contain glucose. As glucose is to be eliminated with water, the diabetic patient has to pass urine frequently, a condition which is called 'polyuria'. Glycosuria and polyuria happen only if the blood sugar level exceeds about 180 mg per cent, which is the renal threshold for glucose. Upto this value in blood, all the glucose filtered at the glomerulus is re-absorbed in the kidney tubules, so much so there is no glucose in urine. Only when the blood sugar exceeds the renel threshold value will there be glycosuria.

Though diabetes mellitus arises out of a primary derangement of the glucose metabolism, metabolisms of proteins and lipids are also adversely affected soon enough so that there is a pan-metabolic involvement.

If not controlled, proteins will be degraded in the body and amino acids utilised for gluconeogenesis. Hence blood urea and non-protein nitrogen levels of the blood increase. Since glycolysis is decreased due to lack of insulin, less of ATP is generated, and consequently it has an adverse impact on the biosynthesis of proteins in all tissues. Fatty acid and lipid synthesis is decreased as well, due to decreased production of ATP, acetyl coenzyme A, NADPH and glycerophosphate. However, triacyl glycerol stores in adipose tissue in the diabetic condition are converted into more of free fatty acids due to augmented lipolysis caused by the elevation of cyclic AMP levels. A rise in the plasma non-esterified fatty acids in turn has a profound impact at several metabolic steps. Fatty acids reaching the liver in high concentrations block further fatty acid synthesis by a feedback control at the acetyl coenzyme A carboxylase level.

Increased acetyl coenzyme A coming from fatty acid degradation activates the gluconeogenic enzyme, pyruvate carboxylase. Fatty acids also promote gluconeogenesis by increasing the formation of oxaloacetate through the above mechanism. They finally retard the TCA cycle at the level of citrate synthetase. The effect of the various events occurring sequentially is that, acetyl coenzyme A, instead of being largely metabolised via the TCA cycle or being used for the biosynthesis of fatty acids, is shunted instead in the diabetic organism into production of excess of keto acids like aceto-acetic acid and beta hydroxy butyric acid and acetone. These are known as ketone bodies which accumulate in blood (ketonemia) and are therefore excreted in urine (ketouria). Ketone bodies are tested by Rothera's test. With sodium nitroprusside and sodium hydroxide, a pink colour will develop if ketone bodies are present in urine. As keto acids accumulate in uncontrolled diabetes mellitus, the condition is called 'diabetic keto acidosis' which has to be treated immediately. Otherwise the patient with high blood sugar and ketone bodies may go into a coma. Prompt therapy with insulin should be instituted to overcome the crisis.

In diabetic ketoacidosis, beta hydroxy butyric acid and aceto-acetic acid react with sodium bicarbonate of the bicarbonate buffer pair and deplete it. This means diminished defence towards acidosis. Again, when the acids are excreted, they take away the sodium of the body. The hyponatremia so produced will affect the electrolyte and water balance of the body. A low plasma pH stimulates the respiratory centre producing deep respiration referred to as "air hunger" or 'Kussmaul breathing' Dehydration may also set in due to glycosuria.

Again, the liver has to get rid of the free fatty acids coming into it, and converts them to Very Low Density Lipoproteins rich in triacyl glycerol. Hence, triacyl glycerol will also increase in the blood. In short, important lipid substances like triacyl glycerol, free fatty acids and cholesterol increase in the blood.

Cholesterol increase in the blood is due to diminished catabolism and vascular changes. Degradation of the blood capillary wall may also contribute to the increase of blood cholesterol.

Uncontrolled diabetes mellitus is characterised by the following chemical pathology:
1. Increase of blood sugar i.e., hyperglycemia
2. Excretion of sugar in urine i.e., glycosuria
3. Polyuria (frequent urination)
4. Increased urea and non-protein nitrogen in blood
5. Increased triacyl glycerol and cholesterol in blood
6. Increased ketone bodies, i.e., aceto-acetic acid, beta hydroxy butyric acid and acetone in blood in severe conditions, i.e., diabetic ketoacidosis
7. Excretion of ketone bodies in urine
8. Dehydration
9. Lowered pH, hyperventilation and Kussmaul breathing
10. Lowered HCO^-_3 and alkali reserve, disturbance in acid base equilibrium
11. Lowered sodium in blood (hyponatremia), and disturbance in fluid and electrolyte balance.

Prolonged disease affects the kidney function (diabetic nephropathy), eyes (diabetic retinopathy) and nervous system (diabetic neuropathy). Hence, diabetes mellitus as a syndrome rather than a disease and should be continuously monitored, once diagnosed.

If there is insulin deficiency as in IDDM, insulin should be given by injection. But in NIDDM, the patient has enough insulin in the body which does not bind to the receptor sites. Hence, these patients are treated by diet control, exercise, oral anti-diabetic tablets, traces of chromium, etc., which improve the functioning of insulin. Only if absolutely necessary, insulin should be given to NIDDM patients. If the patient is obese, the body weight should be reduced. The increased peripheral resistance to insulin in obesity is at least partially the result of decreased insulin receptors on the target cell membranes and partly also of impaired intracellular metabolic events. This is not a genetic characteristic since receptors quickly increase after weight loss. The present trend of treatment of NIDDM is *caloric restriction,* i.e., taking food with restriction of the total quantity and having enough fibre in the diet, as fibre prevents rapid absorption of glucose in the intestine.

A genetic basis for the causation of the condition has been postulated. In the juvenile diabetic, there is almost an atrophy of the islet tissue. Hence the oral anti-diabetic drug like sulphonylureas fail to have any effect since there is no dormant islet tissue to get stimulated.

Diminished glucose tolerance indicates the diabetic process, and a similar finding is seen also hyperpituitarism, hyperthyroidism and hyperadrenalism of the cortex or the medulla. In the opposite state like myxoedema, Addison's disease, and Simmond's disease, a high insulin sensitivity is encountered. (For the screening of diabetes mellitus as well as the pre-diabetic state, see Pancreas Function Tests.)

Insulin antibodies: High levels of antibodies to insulin are occasionally found in plasma in insulin-resistant patients. Antibodies formed in the system in response to exogenously administered insulin can also bind in vivo and inactivate the endogenous insulin which is thereby rendered metabolically unavailable thus aggravating the diabetic state. At times, an unexpected, spontaneous release of insulin from its bondage may precipitate repeated and alarming bouts of hypoglycemia particularly when the person is already being treated with large doses of insulin, as in diabetic coma.

PHOTOSYNTHESIS

Prithivya Oshadhayaha	..	From Earth the plants
Oshadhibhyo Annam	..	From plants the food
Annat purushaha	..	From food the Man

(Vedas)

Photosynthesis is the vital food-giving process as it is only this reaction which gives us nutrients like carbohydrates, fats and proteins. Usually photosynthesis is related with the production of starch in plants. This is not strictly correct as fats and proteins are also formed in this process. The first products are ATP and NADPH which are used for the synthesis of the first sugar derivative, fructose-6-phosphate.

Photosynthesis takes place in what are called the 'Biochemists' Green Mansions' — the chloroplasts which house the green pigments, the chlorophylls. The chloroplasts of plant cells are the counterparts of the mitochondria of animal cells. Not all chloroplasts are involved in this process. Again, there are two chlorophylls, chlorophyll 'a' and chlorophyll 'b'. The difference between the two is only in a substituent, 'a' contains $-CH_3$ while 'b' has $-CHO$.

Moreover, even the carotenoids can participate in photosynthesis, but their efficiency is only 40-50 per cent. In this process, the pigments absorb the photons of sunlight and use easily available raw materials like water and carbon-dioxide, and give us the valuable macromolecules of carbohydrates, proteins and lipids of bewildering complexity. Incidently, this explains why all plants and trees have green leaves.

Though photosynthesis uses light quanta and converts light energy into chemical energy, it is actually a combination of two distinct reactions: 1) Light reaction or Hill reaction which gives ATP and NADPH; and 2) dark of thermochemical reaction which uses ATP and NADPH and gives fructose-6-phosphate, and thereby other nutrients.

It has been shown that even in the first light reaction, there are actually two photo acts — system I and system II linked by a thermo-chemical event.

Formation of ATP in light reaction (photo-phosphorylation)

As ATP is formed from ADP and Pi using light energy, the process is called photo-phosphorylation. Just as in oxidative phosphorylation in the inner membrane of the mitochondria and the inner membrane particles, here too the electron moves from the redox (reduction-oxidation) system of cytochrome 'b' with E_o of $-.03$ V to cytochrome 'f' system of $E_o + 0.375$ V. Cytochrome b_6 is associated with plastoquinone while cytochrome 'f' is associated with plastocyanin. As the voltage difference between the two redox systems is greater than 0.15V, ΔF is greater than 7400 cals. Hernce, ATP is formed thus:

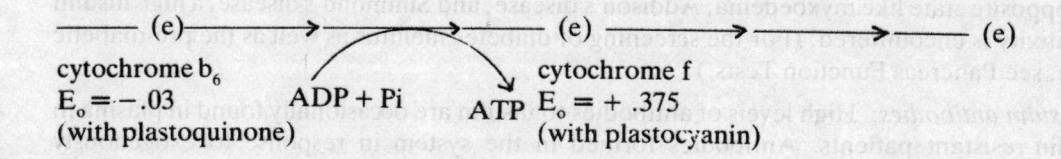

$$\xrightarrow{\quad(e)\quad} \xrightarrow{\quad(e)\quad} \xrightarrow{\quad(e)\quad}$$

cytochrome b_6 cytochrome f
$E_o = -.03$ ADP + Pi $\searrow$ ATP $E_o = +.375$
(with plastoquinone)   (with plastocyanin)

Arnon suggests that in cyclic photophosphorylation only chlorphyll 'a' takes part with the light of a single wave-length. According to him, when light falls on chlorophyll 'a' the latter absorbs photons and ejects electrons. The electrons move through the cytochrome systems and come back to the excited chlorophyll and neutralise the free redical a$^{(+)}$

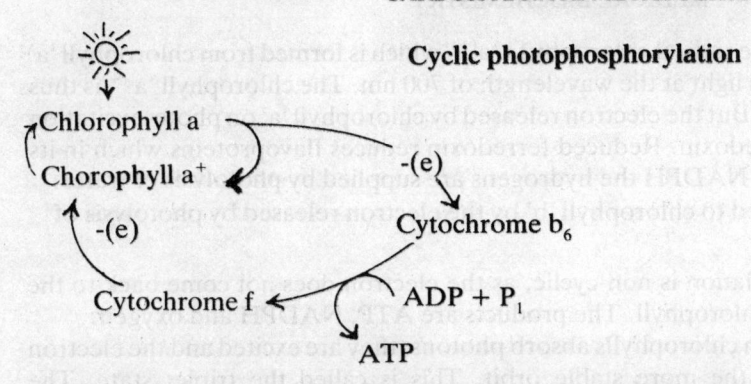

Cyclic photophosphorylation

```
Chlorophyll a ──────────┐
Chlorophyll a⁺ ◄──  -(e) │
      ▲                  ▼
     -(e)           Cytochrome b₆
      │                  │
Cytochrome f ◄───────────┤
                    ADP + Pᵢ
                         ▼
                        ATP
```

This type of cyclic photophosphorylation is, however, uneconomical to the plants and did not explain the Emerson Enhancement Effect. Emerson and Lewis made the crucial observation that efficient photosynthesis required the simultaneous excitation of more than one photosynthetically active pigment namely chlorophyll 'a' and 'b'.

Non-cyclic photophosphorylation

According to Hill and Bendall, 'light reaction' is made of two oxidation and reduction reactions, a photo-act driven by the light of shorter wavelength with chlorophyll 'b' and another photo-act driven by the light of longer wavelength, the far red, with chlorophyll 'a', and these two photo-acts are linked by a thermo-chemical (dark) event. In photo-act I, a weak oxidant cytochrome 'f' and plastocyanin and a strong reductant ferredoxin are involved, while in photo-act II a strong oxidant, a weak reductant cytochrome 'b_6' and plastoquinone take part yielding molecular oxygen. There is thus a collaborative endeavour of both chlorophyll 'a' and 'b' and utilisation of the lights of two different wavelengths – the far red and the light of shorter wave length. This explains the Emerson Enhancement Effect.

As oxidised pyridine nucleotide $NADP^+$ and ferredoxin are available, the electron from chlorophyll 'a' is used to reduce $NADP^+$ to give NADPH in the non-cyclic photo-phosphorylation.

The sequence of events is as follows:

When the light of relatively shorter wavelength, i.e., 680 nm, falls on chlorophyll 'b' it is raised to an excited state of free radical chlorophyll 'b⁽⁺⁾' releasing electron. This electron moves through the cytochrome 'b_6' and cytochrome 'f' effecting photo-phosphorylation. It is

```
         680 nm
           │
           ▼
      Chlorophyll                    Cytochrome    -(e)    Cyto f    -(e)
         (b) ◄──────────────────►       b₆      ─────►    (C₅₅₂)  ─────►
         II                            (b₅₅₉)      │        │
      Chlorophyll                                  │        │
         (b⁺)        O₂                          ADP+Pi    ATP
  -(e)                           -(e)   Chlorophyll (a)⁺  -(e) Ferredoxin -(e)  FP  -(e)
                                          │                ─────►         ────►   ────►
                                       Chlorophyll (a)
        OH⁻ + H⁺                                       -(e)  NADP⁺  H⁺
                                                      ─────►       ────► NADPH
           ▲
           │                         700 nm (far red)
      Photolysis                           │
         H₂O                               ▼
```

then accepted by activated free radical chlorophyll 'a$^{(+)}$' which is formed from chlorophyll 'a' which simultaneously absorbs light at the wavelength of 700 nm. The chlorophyll 'a $^{(+)}$is thus neutralised to chlorophyll 'a'. But the electron released by chlorophyll 'a' on photo-excitation is in its turn taken up by ferredoxin. Reduced ferredoxin reduces flavoproteins which in its turn reduces NADP$^+$ to form NADPH the hydrogens are supplied by photolysis of water. Chlorophyll 'b$^{(+)}$' is neutralised to chlorophyll 'b' by the electron released by photolysis of water.

Thus the photo-phosphorylation is non-cyclic, as the electron does not come back to the same excited free redical of chlorophyll. The products are ATP, NADPH and oxygen.

It has been shown that when chlorophylls absorb photons, they are excited and the electron in their molecules shifts to the more stable orbit. This is called the triplet state. The chlorophyll in the triplet state accepts the light quanta to yield the second-triplet excited state.

Very recently, it has been shown that light at wavelengths larger than 700 nm is ineffective in driving photosynthesis, while a mixture of light at 700 nm and 680 nm, applied together or alternately, was more effective than the sum of the two applied alone. It is also shown that it is not chlorophyll 'a' proper but the pigment designated P_{700} which is one in about 30 molecules of chlorophyll 'a' that forms System I. This P_{700} is in the immediate proximity of the specific molecules that serve as its electron donor and acceptor when photo-activated. For photo-act II, it is a complex of 'a', 'b' and protein. Chlorophyll is not necessarily the immediate oxidant of water but no other reactant has been identified so far. In System II, light less than 680 nm is absorbed. The immediate reductant of P_{680} is cytochrome b_{559}. A substance designated Q is in close proximity. It is also known as C_{552}. This, with b_{599}, is associated with a plastoquinone. The electron from cytochrome b_{559}, is received by cytochrome C_{552} (cytochrome 'f') in the viscinity of plastocyanin and FeS. Membrane-bound iron-sulphur proteins (Fe S)$_4$ are also the ferredoxin-reducing substances. The E_o between cytochrome b_{559} and C_{552} is about 0.4 V.

Differences between cyclic and non-cyclic photo-phosphorylation:

Cyclic	Non-cyclic
1. Chlorophyll 'a'	Both chlorophylls 'a' and 'b'
2. Light of far red, 700 nm	Light of far red 700 nm and shorter wavelengths 680 nm
3. No NADPH	Formation of NADPH
4. No photolysis of water	Photolysis of water and liberation of electrons, O_2 and H^+
5. Efficiency limited – so called red-drop	Efficiency greater – Emerson Enhancement Effect

Differences between mitochondrial oxidative phosphorylation and photophosphorylation :

Mitochondrial Oxidative Phosphorylation	Non-cyclic photophosphorylation
1. P:O 3 or 4	P:O 1
2. O_2 is consumed.	O_2 is liberated.
3. NADH is oxidised.	NADP$^+$ is reduced.

Dark reaction

With the products of the light reactions, viz, ATP and NADPH, CO_2 is reduced to hexose and this takes place in the ribulose 1,5 diphosphate cycle. It was Calvin who using $^{14}C-CO_2$, demonstrated unambiguously that carbon dioxide was used only in the dark reaction. Thus the oxygen that it liberated during photosynthesis is not from CO_2, but only from H_2O in the light reaction.

CARBOHYDRATE METABOLISM

Ribulose 1, 5 diphosphate takes up CO_2 to form 3-phosphoglycerate. This with ATP gives 1, 3 diphosphoglycerate. This is reduced with NADPH to 3 phosphoglyceraldehyde. Part of the 3 phosphoglyceraldehyde isomerises to dihydroxy acetone phosphate. Both of them condense to form fructose-6-phosphate, which is the gain and the first sugar product of photosynthesis. Fructose-6-phosphate and 3-phosphoglyceraldehyde with trans-ketolase form xylulose 5P and erythrose 4P. Again fructose-6-phosphate and erythrose 4P with trans-aldolase form Sedoheptulose 7P and glyceraldehyde 3P which in their turn give xylulose 5P on trans-ketolation. The xylulose 5P formed by two steps is epimerised to ribulose 5P. This with ATP regenerates the ribulose 1, 5 diphosphate and the cycle goes on till CO_2, ATP and NADPH are available.

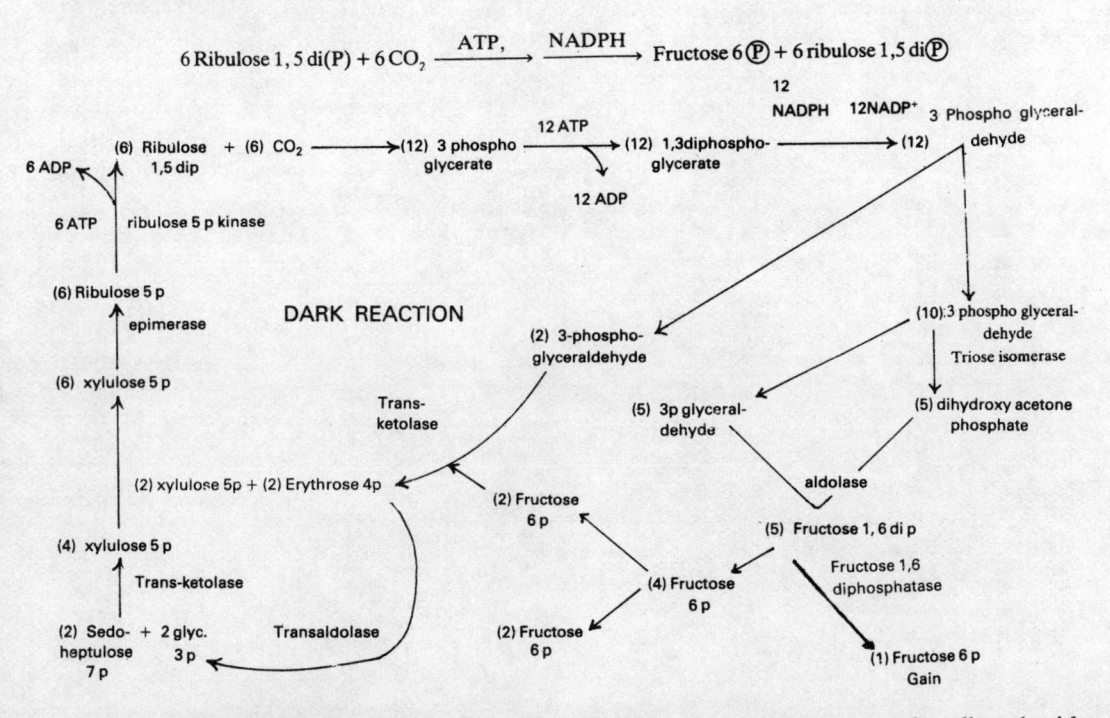

$$6 \text{ Ribulose } 1,5 \text{ di}(P) + 6 CO_2 \xrightarrow{ATP,\ NADPH} \text{Fructose } 6\text{\textcircled{P}} + 6 \text{ ribulose } 1,5 \text{ di}\text{\textcircled{P}}$$

```
                                                        12
                                               12 ATP        NADPH   12NADP⁺   3 Phospho glyceral-
        (6) Ribulose  + (6) CO₂ ──→(12) 3 phospho ──→(12) 1,3diphospho- ─────────→ (12)  dehyde
6 ADP ←     1,5 dip              glycerate            glycerate
6 ATP    ribulose 5 p kinase                       12 ADP
(6) Ribulose 5 p
     epimerase          DARK REACTION           (2) 3-phospho-           (10) 3 phospho glyceral-
(6) xylulose 5 p                                glyceraldehyde                dehyde
                                                                              Triose isomerase
                                   Trans-         (5) 3p glyceral-       (5) dihydroxy acetone
                                   ketolase       dehyde                     phosphate
 (2) xylulose 5p + (2) Erythrose 4p ← (2) Fructose         aldolase
                                      6 p                 (5) Fructose 1,6 di p
(4) xylulose 5 p                                                    Fructose 1,6
     Trans-ketolase                    (4) Fructose                 diphosphatase
                                           6 p
(2) Sedo- + 2 glyc.   Transaldolase   (2) Fructose
 heptulose    3 p  ←                      6 p                  (1) Fructose 6 p
 7 p                                                               Gain
```

The fructose which is the first sugar formed in the dark reaction is converted to disaccharides and polysaccharides by the following reactions:

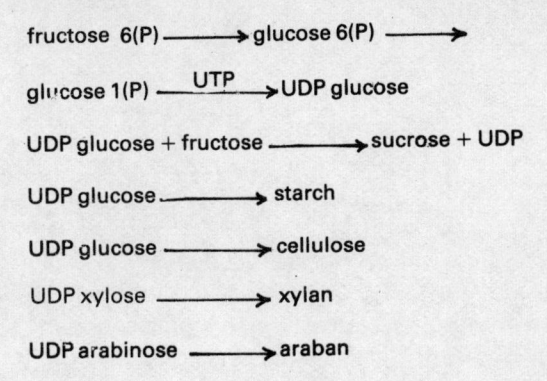

fructose 6(P) ──→ glucose 6(P) ──→

glucose 1(P) $\xrightarrow{UTP}$ UDP glucose

UDP glucose + fructose ──→ sucrose + UDP

UDP glucose ──→ starch

UDP glucose ──→ cellulose

UDP xylose ──→ xylan

UDP arabinose ──→ araban

The 3-phospho-glyceric acid can be converted to amino acids, proteins and fats as follows:

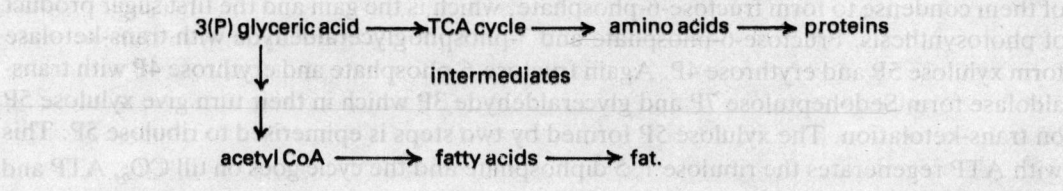

3(P) glyceric acid ⟶ TCA cycle ⟶ amino acids ⟶ proteins
                         ↓
                    intermediates
                         ↓
                    acetyl CoA ⟶ fatty acids ⟶ fat.

13 Lipid Metabolism

Lipids comprise 18–20 per cent of the total body weight in the human adult. Neutral fats (triglycerides, triacyl glycerols) are the forms in which fat is stored in the body in adipose tissue, the subcutaneous fat pads, peri-renal fat depots, inter-muscular connective tissue and the fat depots of mesentery and omentum. About 70–80 per cent of the lipids in adipose tissue is due to triacyl glycerols.

Lipids are also present in other areas of the body. In the decreasing order of lipid content in per cent, the tissues rank as follows: skeleton 25, heart 17, pancreas 13, brain 12, kidney 7, muscle 6, liver 3 and blood 0.57.

Phospholipids, cholesterol and cerebrosides are not stored in the adipose tissue. They are present, however, in all other tissues and have certain structural functions and physiological roles in the cells. The lipids of the brain are mostly cholesterol, phospholipids, and cerebrosides. But they are low in neutral fat. The adrenal cortex, testes, ovary and corpus luteum have a high content of cholesterol. The latter acts as the precursor for the biosynthesis of steroid hormones in these glandular tissues.

The major part of the lipids of our diet consists of neutral fats with smaller amounts of phospholipids, steroids and glycolipids.

Physiological value of fats

1) Fats are poor conductors of heat and their presence in the subcutaneous tissues prevents not only loss of heat from the body but also protects the body from the cold. Thus persons having thick layers of fat in their bodies appear more comfortable in winter and less comfortable in summer than thinner persons.

2) Fats are the best heat producers in the body. While one gram of carbohydrate or protein, on complete oxidation in the body, gives four calories of heat, one gram of fat yields more than twice that value, i.e. 9.2 calories. This is because a molecule of fat contains more carbon and hydrogen in relation to oxygen than a molecule of carbohydrate or protein.

3) Fats in the diet act as sources for fat-soluble vitamins A, E, D and K and also carotenoids which are converted to vitamin A in the animal body.

4) Fats are needed in the diet as sources of certain essential fatty acids which the body is unable to synthesise, and which have at the same time an important physiological role in the tissue. Essential fatty acids like arachidonic acid are the precursors of prostaglandins in our tissues.

5) Fats in the diet spare the requirements of carbohydrate, protein, thiamine, etc.

6) Increase in the intake of fat results in the retention of calories by preventing the loss of heat from the body.

7) Fats have a high satiety value and the inclusion of fats in the diet reduces the consumption of both proteins and carbohydrates.

8) An adequate intake of fat is necessary for sex maturation, pregnancy and lactation.

Digestion and absorption of fat: (See Chapter on 'Biochemistry of Digestion and Absorption'.)

Transport: A fat-rich meal is accompanied by a characteristic 'alimentary hyperlipemia' which reaches its peak around 2 – 2½ hrs following the meal. The turbid or milky appearance (lactescence) of plasma noticed after the intake of a fat-rich meal is due to microscopic particles of fat about 1 μ in diameter called 'chylomicrons'. The plasma may get cleared off in about 5 – 6 hrs when the chylomicrons get converted to other lipoproteins. The triacyl glycerols being insoluble in the aqueous phase are stabilised by combining with different proportions of phospholipids, cholesterol and proteins to form chylomicrons which are lipoprotein complexes. The formation of these chylomicrons takes place in the mucosal cells during absorption from intestine and thus they find their way first into the lymph and finally to the blood plasma.

The turbid plasma gets cleared of the milkiness during its passage through various organs and tissues by virtue of hydrolysis of the triacyl glycerol portion by the enzyme lipoprotein lipase or the 'clearing factor'. This clearing factor requires heparin as a cofactor. Insulin also activates lipoprotein lipase. The lipoprotein lipase action releases some free fatty acids which are then stored in fat depots as triacyl glycerols or are used for metabolic purposes.

The free fatty acids (FFA) or non-esterified fatty acids (NEFA) are present in plasma in combination with albumin and are the form in which they are transported. The adipose tissue is rich in the enzyme hormone-sensitive triacyl glycerol lipase and is a site from which free fatty acids are released into plasma and transported to various sites for metabolic purposes. The adipose tissue also contains another lipoprotein lipase which helps in storing neutral fat.

Blood lipids

As the lipids are transported through blood plasma, from their sites of origin to their sites of utilisation or storage, the level of plasma lipids at any time can be considered to represent the net balance between production, utilisation and storage. Lipids, though hydrophobic compounds, are present in plasma as stable hydrophilic lipoprotein complexes. These complexes are combinations of triacyl glycerols and phospholipids, with cholesterol and plasma proteins, in the following forms:

1) Chylomicrons: The fat derived from intestinal absorption is transported to storage depots as chylomicrons which consist of one per cent protein and 99 per cent lipids. The lipid fraction contains triacyl glycerols (88 per cent), which are kept as stable emulsions by combining with phospholipids (8 per cent) and cholesterol (4 per cent).

2) Very low density lipoproteins (VLDL): The triacyl glycerols derived from the liver are transporated as very low density lipoproteins consisting of proteins (7 – 10 per cent) and total lipids (90 – 93 per cent). This density is 0.96 – 1.006. The total lipids are made up of 56 per cent triacyl glycerols, 20 per cent phospholipids, 15 per cent cholesterol esters, 8 per cent free cholesterol and one per cent free fatty acids. The Svedberg flotation units (Sf units) of VLDL are 20 – 400.

3) Low density lipoproteins (LDL): These are formed from VLDL during circulation in blood. Their Sf values are 2 – 20 and a density of 1.019 – 1.063. LDL has 21 per cent proteins and 79 per cent lipids. The total lipids are made up of 13 per cent triacyl glycerols, 28 per cent phospholipids, 48 per cent cholesterol esters, 10 per cent free cholesterol and 1 per cent free fatty acids. LDL is rich in cholesterol (58 per cent) and the carriers of cholesterol from the liver to the peripheral cells (direct cholesterol transport)

4) High density lipoproteins: There are three fractions of High Density Lipoproteins, HDL 1, HDL 2 and HDL 3 of density 1.063 – 1.210. HDL 2 contains 33 per cent protein and 67 per cent total lipids in which phospholipids and cholesterol fractions are greater than triacyl glycerols. It has 16 per cent triacyl glycerols 43 per cent phospholipids and 41 per cent cholesterol. HDL 1 is very low. Changes in HDL run parallel to those of HDL 2. Cholesterol from the peripheral cells is transported to the liver by HDL (reverse cholesterol transport).

5) Free fatty acids: The free fatty acids derived from adipose tissue and released by lipolytic hormones are transported as FFA-albumin complex (contains 99 per cent albumin and one per cent fatty acid). Some 25 – 30 molecules of free fatty acids are present in combination with a molecule of albumin. The FFA of plasma are metabolically the most active and have a half-life of only 2 – 3 minutes.

The lipoproteins are characterised also by the presence of protein or polypeptide (apolipoprotein or apoprotein) linked to it. Thus LDL has Apo-B as the main apoprotein which is also present in VLDL and chylomicrons. The Apo-B of chylomicrons is smaller in size (B-48) as compared to that of LDL or VLDL (B-100). HDL has two apoproteins A-I and A-II; chylomicrons, HDL and VLDL also contain apoproteins C-I, C-II and C-III. In addition to Apo-A, Apo-B and Apo-C, another apoprotein, Apo-E, has been isolated from VLDL.

The apoproteins differ from each other in their amino acid content and immuno-chemical properties and hence can be easily identified. The apoproteins can be separated from the original lipoproteins by de-lipidation methods and later purified by gel filtration, ion exchange chromatography and other modern techniques.

The distribution of lipids in plasma or serum has been studied by extraction with suitable lipid solvents (alcohol-ether mixture, alcohol-acetone mixture, chloroform-methanol mixture, petroleum ether, etc.) and by estimating the triacyl glycerol, cholesterol, phospholipids, etc. The different fatty acids and phospholipids have been estimated by thin-layer chromatography and gas-liquid chromatography. The composition of lipids is given in Table 1.

Table 1

LIPIDS OF HUMAN BLOOD PLASMA

Lipid fraction	Mean level mg/100 ml
Total lipids	570
Triacyl glycerols	142
Phospholipids	216
Total cholesterol	200
Free cholesterol	55
Ester cholesterol	145
Free fatty acids	12

Lipoproteins of plasma: The lipoproteins can be separated by subjecting the plasma to ultracentrifugation. For this, the plasma is mixed with a solution of NaCl (specific gravity 1.063) and centrifuged for 13 – 16 hrs at 80,000 g. The top lipoprotein fraction (that rises to the surface) is removed, mixed with NaCl solution and spun in an analytical centrifuge at 215,000 g.

Pure fat is less dense than water. If the ratio of lipid to protein in lipoprotein increases, the density decreases. The rate at which each of the lipoprotein floats up through a solution of NaCl (specific gravity 1.063) may be expressed in Svedberg units (Sf). One Sf unit = 10^{-13} cm/second/dyne/g at 26°C. As the protein, cholesterol and phospholipid content decreases, the density also decreases, but the triacyl glycerol content increases. Thus the levels of low density lipoproteins (LDL) reflect the extent to which the cholesterol are carried from the liver to the various tissues while the VLDL reflects the levels of the triacyl glycerols.

Five groups of lipoproteins have been identified in blood plasma:
1) High density or alpha-lipoproteins (Sf 0-2), density 1.063 – 1.21.
2) Low density or ß-lipoproteins (Sf 2-20), density, 1.006 – 1.063.
3) Very low density or pre-ß (alpha$_2$ - lipoproteins (Sf 20-400), density 0.96 – 1.006.
4) Chylomicrons (Sf > 400), density 0.96
5) Albumin-FFA

Recently, the plasma lipoproteins have also been separated by paper electrophoresis as well as by agarose gel electrophoresis.

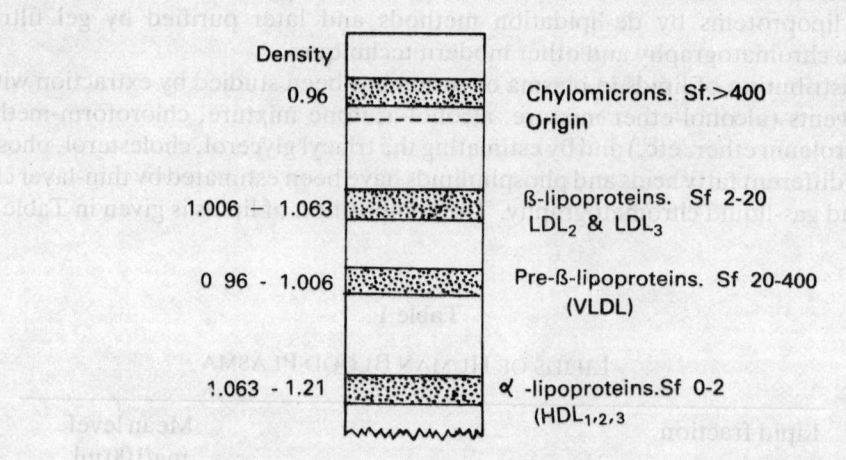

Density		
0.96		Chylomicrons. Sf.>400
------		Origin
1.006 – 1.063		ß-lipoproteins. Sf 2-20 LDL$_2$ & LDL$_3$
0 96 – 1.006		Pre-ß-lipoproteins. Sf 20-400 (VLDL)
1.063 – 1.21		α -lipoproteins.Sf 0-2 (HDL$_{1,2,3}$)

Body lipids

Dynamic state of body lipids: The lipids in the body are not inert materials merely stored but are being constantly mobilised for the body's use and are replaced continuously. This has been confirmed in recent years by experiments with animals, using ^{14}C labelled fatty acids. The body lipid content at any time represents a balance between the processes of breakdown or utilisation and the processes of synthesis or deposition.

The lipids in the body can be broadly divided into two types: 1. protoplasmic lipid (elemente constant), and 2. depot lipid (elemente variable).

If an animal is starved, there is depletion in the content of stored lipids. This diminution occurs mostly in the adipose tissue and is due to the increased use of lipid by the body as a source of energy during starvation. The lipid that is diminished during starvation and increased during feeding is called the depot lipid (depot fat) or the 'elemente variable'. It consists mainly of triacyl glycerols and is mostly of fuel value. It is also useful as supporting fat pads to keep internal organs in their position and protect them from shock and damage. On the other hand, the protoplasmic lipid which is present in all tissues is almost unaffected immediately following starvation or increased food intake. Even after prolonged starvation these lipids are not depleted from tissues. These are therefore called the 'elemente constant'. The protoplasmic lipid consists mostly of phospholipids and cholesterol and is essential to the life of the cells, having a structural and functional role in the cells.

Fate of lipids in the body: The triacyl glycerols and fatty acids in the body are:
1) oxidised to provide heat and energy;
2) stored for future utilisation;
3) secreted in milk in lactating women;
4) utilised for the synthesis of phospholipids and glycolipids;
5) broken down, and intermediates used for synthesis of cholesterol and new fatty acid molecules; and
6) excreted in feces.

The phospholipids have important physiological functions in the cells and are not easily depleted from the body like the triacyl glycerols. However, they also do have flux, though at different rates in different tissues, ultimately being broken down as triacyl glycerols.

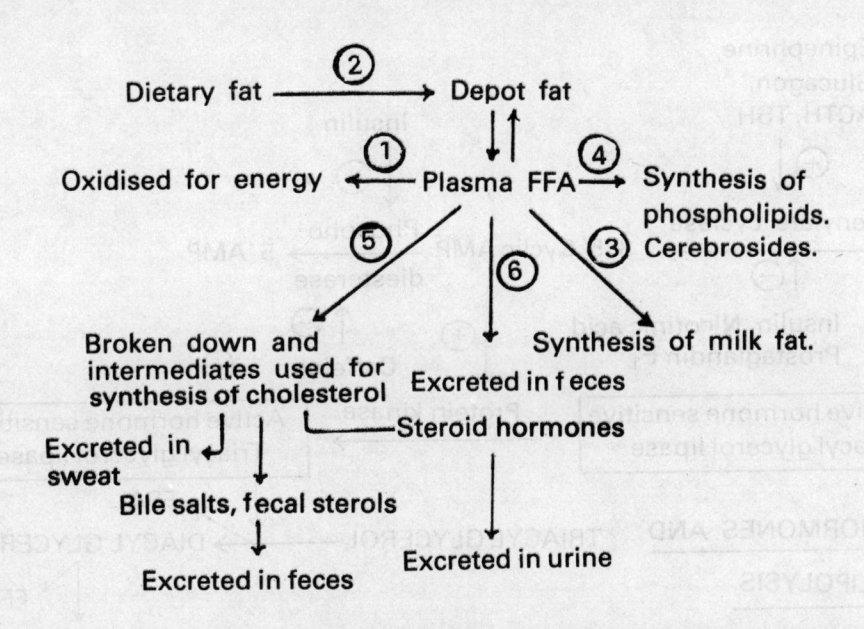

Fig. Fate of fat in the body

The chief precursor of cholesterol synthesis is acetyl CoA. The cholesterol from the liver is mainly converted into bile acids and is excreted in bile. The fecal lipids consist of unabsorbed sterols, bile salts and small quantities of fatty acids. The cholesterol of skin is converted to vitamin D_3. Steroid hormones are biosynthesised from cholesterol in certain endocrine glands like adrenal cortex, testes, etc.

Adipose tissue fat

Each species of animal stores in its adipose tissue a characteristic mixture of fats. The characteristics of these fats like melting point, unsaturation, chain length of the fatty acids, etc. depend to a great extent on the environment in which they live, as well as on other factors. Thus pork fat and mutton fat have a higher melting point than cod liver oil. Feeding of linseed oil to dogs results in decreasing the melting point of their body fat from 20°C to 0°C. Any species of animal modifies the dietary fatty acids it receives by lengthening or shortening the chain-length as well as by saturation or desaturation of the fatty acids so as to produce its own specific fat for storage and use. Unwanted fatty acids are oxidised and degraded quickly while the required fatty acids are retained as storage fat.

Influence of hormones on adipose tissue fat: The stored fat from adipose tissue is mobilised to free fatty acids by the help of various lipolytic hormones, most important of which are epinephrine and nor-epinephrine. Certain anterior pituitary hormones like TSH, ACTH, the growth hormone and the pancreatic hormone glucagon have also a similar lipolytic action on the adipose tissue. In recent years, it has been reported that the lipolytic effect of the growth hormone is enhanced by gluco-corticoids and that the former is not effective in the absence of the latter. It is now believed that the lipolytic action of some hormones is mediated via 3' 5'-cyclic AMP whose level is enhanced by them. The 3'5'-cyclic AMP stimulates the cAMP dependent protein kinase that activates the inactive hormone sensitive or mobilizing lipase to the active form and helps lipolysis. Experiments have shown that 3'5'-cyclic AMP itself can stimulate

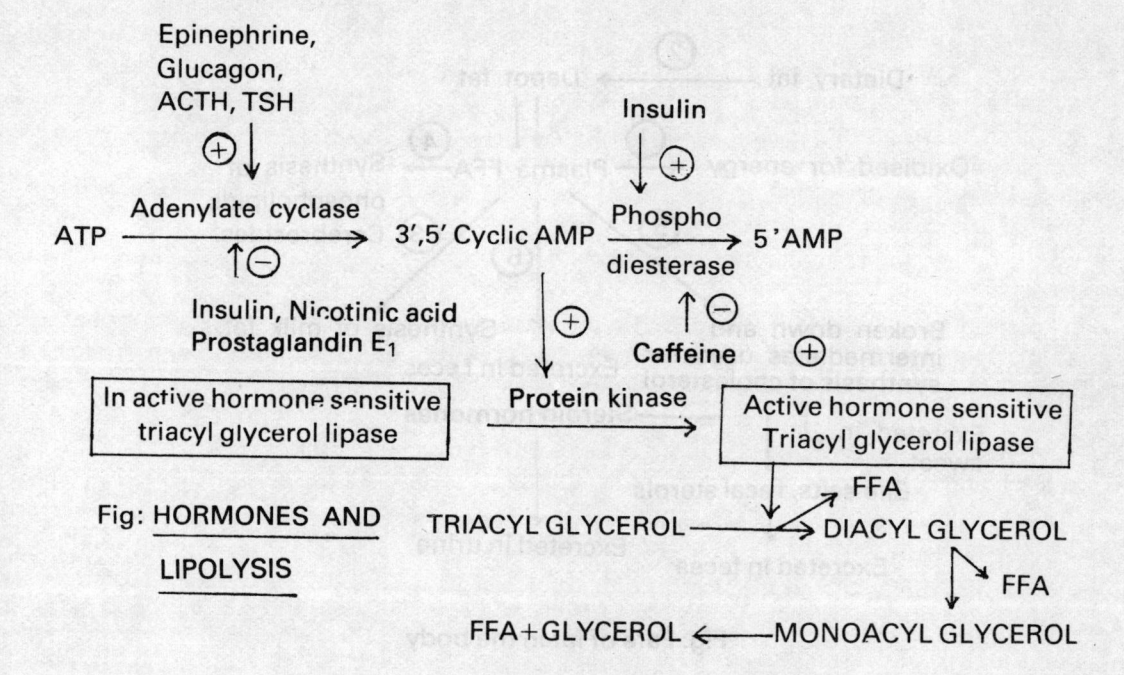

Fig: HORMONES AND LIPOLYSIS

lipolysis in adipose tissue. The hormone which prevents lipolysis is insulin which decreases the levels of 3'5'-cyclic AMP by inhibiting adenylate cylase and enhancing the activity of phospho-diesterase. The latter enzyme is concerned with the destruction of 3'5'-cyclic AMP. A decrease of cAMP has been demonstrated with prostaglandin E and nicotinic acid which inhibit adenylate cyclase. Caffeine, theophylline and other methyl xanthines enhance lipolysis by inhibiting the enzyme phospho-diesterase and thus elevating the cyclic AMP levels.

In the normal animal, lipolysis is mostly regulated and controlled by epinephrine and insulin which maintain the homeostasis.

Fats from carbohydrates

Fat in our body is not only made from the assimilated dietary fat but also from several other non-fat substances like carbohydrates. Chaikoff fed glucose labelled with ^{14}C to mice and found that ^{14}C was found in the palmitic acid of tissues. When a high carbohydrate diet is fed to animals, a greater percentage of it is diverted for fatty acid synthesis. For conversion of carbohydrate to fat thiamine is needed.

$$\text{Carbohydrate} \longrightarrow \text{Glucose} \longrightarrow \text{Pyruvic acid} \xrightarrow{\text{Thiamine required}} \text{Acetyl CoA} \downarrow \text{Fatty acid}$$

Role of liver in lipid metabolism

The liver is the most important organ dealing with the metabolism of lipids. It is not a storage organ for fat, but has a great role to play in the disposal of fatty acids entering it, by bringing about some modification in the molecule by metabolic degradation, oxidation, synthesis to other fatty acids or phospholipid and even synthesis of fat. The role of liver in lipid metabolism can be summarised as follows. Liver is the main site:

1) for the synthesis of phospholipids,
2) for the synthesis of lipoproteins and thus the disposal of the triacyl glycerols that reach the liver,
3) for saturation and desaturation of fatty acids and also the lengthening and shortening of fatty acid chains,
4) for the synthesis, esterification and degradation of cholesterol. Cholesterol is oxidised to bile acids, secreted in bile and excreted into intestines,
5) for ketogenesis, i.e., the production of ketone bodies. The ketone bodies are, however, not oxidised further by liver but by extra-hepatic tissues,
6) for the biosynthesis of fatty acids by both mitochondrial and cytoplasmic (extra-mitochondrial) systems, and
7) for synthesis of triacyl glycerols from fatty acids either synthesised in this tissue or brought in by plasma as FFA from adipose tissue.

METABOLISM OF FATS

As far carbohydrates there is both catabolism and anabolism of lipids.

Catabolism of fat

Fat, whether ingested or obtained from fat storage organs like the adipose tissue, is lipolysed to fatty acids and glycerol. Fatty acids are usually long-chain molecules with an even number of carbon atoms. All fatty acids, whatever be their chain length, are broken down to the same final product having just two carbon atoms, i.e., acetyl coenzyme A.

(1) OXIDATION OF FATTY ACIDS

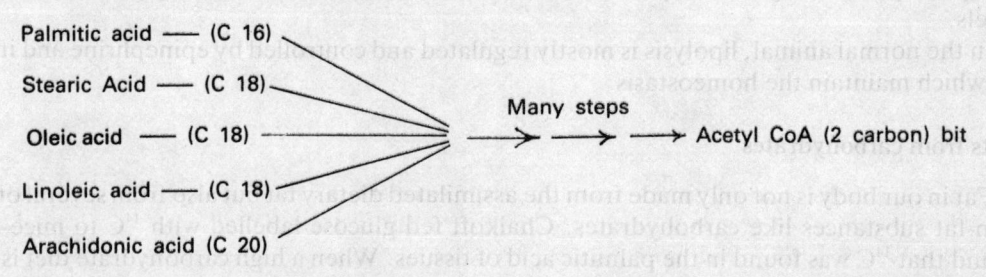

Palmitic acid —— (C 16)
Stearic Acid —— (C 18)
Oleic acid —— (C 18) → Many steps → → → Acetyl CoA (2 carbon) bit
Linoleic acid —— (C 18)
Arachidonic acid (C 20)

This breakdown occurs in the process called beta oxidation, as successive oxidations take place in the beta carbon atoms, step by step. For each oxidation, there are a number of steps, but the first reaction is the same for all the fatty acids, i.e., the conversion of fatty acids into an active fatty acyl CoA with coenzyme A and ATP. This sparking reaction is catalysed by the enzyme acyl CoA synthetase (thiokinase).

Many enzymes participate in the beta oxidation. The 16 carbon acid gives one 2 carbon acetyl CoA and becomes 14 carbon acid, which, in turn, gives one 2 carbon acetyl CoA and becomes 12 carbon acid. This goes on till the 16 carbon acid is broken down to a pool of 8 acetyl CoA molecules.

$$\beta - \alpha$$
C–C–C–C–C–C–C–C–C–C–C–C–C–C–C–COOH
↑
↓ Beta carbon
  ATP, CoA SH

C–C–C–C–C–C–C–C–C–C–C–C–C–C–C–COSCoA
↓ Beta oxidations

C–C|^{7}C–C|^{6}C–C|^{5}C–C|^{4}C–C|
^{3}C–C|^{2}C–C|^{1}C–COSCoA

———————→ 8 Acetyl Coenzyme A (CH_3—CO—SCoA) molecules

The acetyl CoA thus formed enters the TCA cycle and is oxidised to CO_2 and water to give energy in the form of ATP.

FATTY ACID SPIRAL

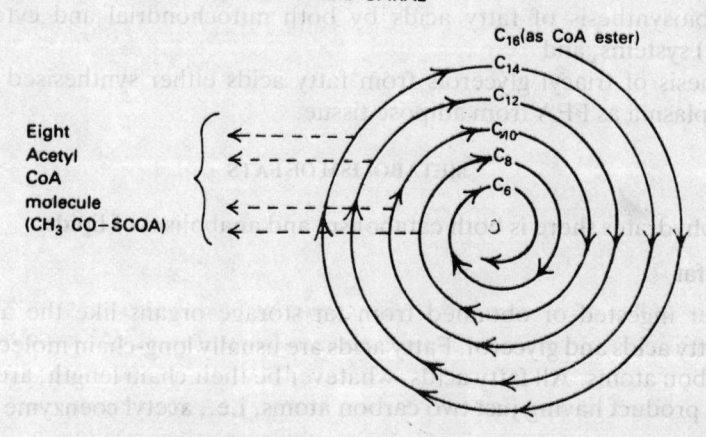

C_{16} (as CoA ester)
C_{14}
C_{12}
C_{10}
C_8
C_6

Eight Acetyl CoA molecule (CH_3 CO–SCOA)

It is clear that acetyl CoA, the fuel for TCA, is provided by glucose, and by fatty acids also. Whether it is from glucose or fatty acid, it is meant to give ATP mostly through TCA cycle. Hence the saying, fat is cooked in the fire of carbohydrate.

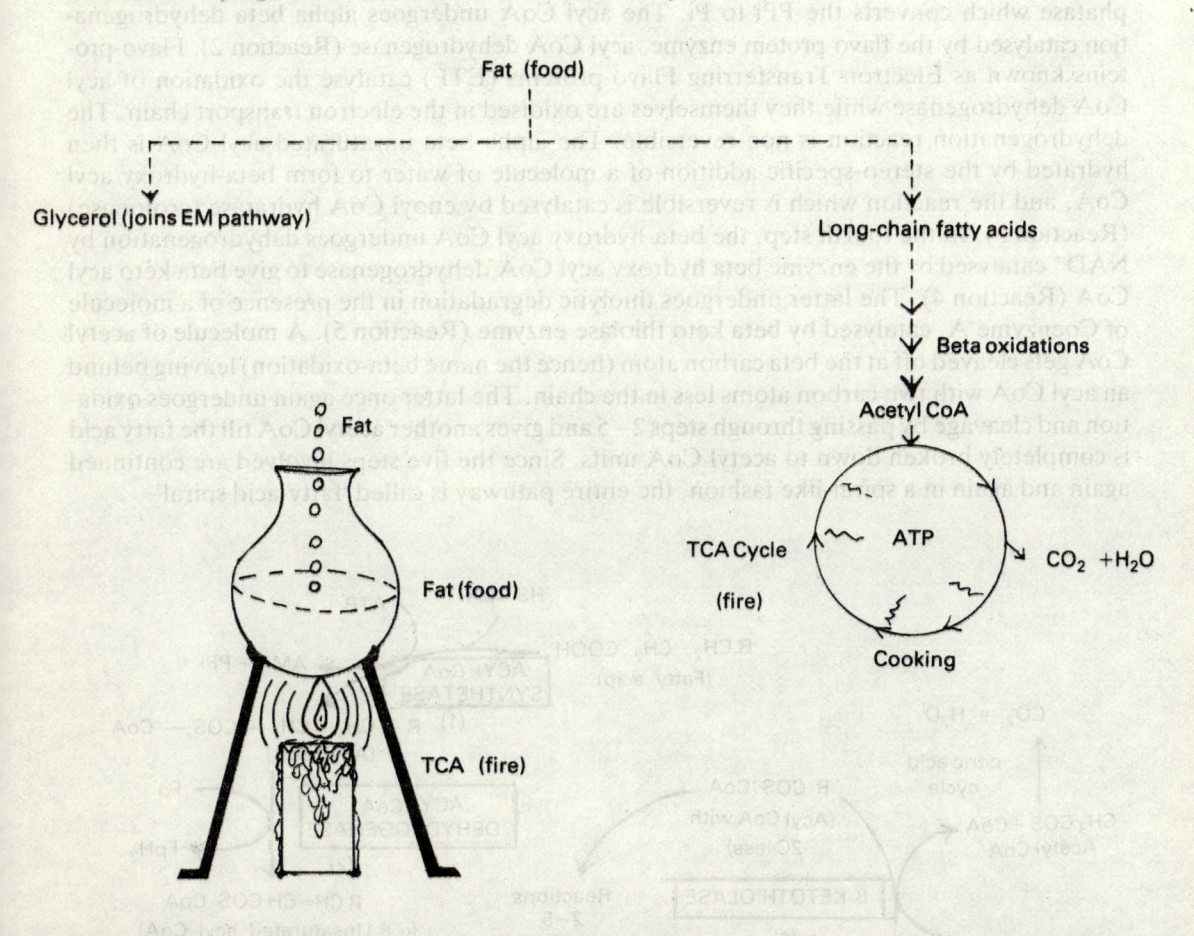

```
                        Fat (food)
          ┌─────────────────┴─────────────────┐
          ▼                                   ▼
Glycerol (joins EM pathway)         Long-chain fatty acids
                                              │
                                              ▼
                                      Beta oxidations
                                              │
                                              ▼
                                        Acetyl CoA
                                              │
                                              ▼
                                   ┌──── ATP ────┐
                              TCA Cycle          → CO₂ + H₂O
                                (fire)
                                   └── Cooking ──┘
```

(Fat being heated over TCA fire diagram)

Oxidation of fatty acids

The most common fatty acids that are utilised in our body have an even number of carbon atoms in their molecules. It has been recognised that fatty acids undergo beta oxidation with the help of enzymes and get degraded as acetyl CoA units, stage by stage. This oxidation takes place mainly in the liver, and to a lesser extent in the heart, kidneys, muscles, lungs, testes, brain and adipose tissue. The enzymes concerned with fatty acid oxidation are found to be present in the mitochondria of the liver associated with the enzymes of the respiratory chain. While fatty acid is broken down in the liver to acetyl CoA units, there is simultaneous production of energy as ATP due to oxidative phosphorylation.

For the initial activation of fatty acid or the 'sparking reaction', a molecule of ATP with Mg^{2+} is required along with a molecule of coenzyme A. The 'activated fatty acid' or acyl CoA is a labile thioester, the esterification having taken place at the − COOH group of the fatty acid. The enzyme concerned with this activation reaction (see reaction 1 in figure) is acyl CoA synthetase which a thiokinase. Once activated, further reactions in the oxidative pathway go on with great regularity without need of energy. Since ATP which is required for this reaction is split into AMP and PPi, it is necessary to remove the PPi immediately in order to ensure that the activation reaches completion. This takes place efficiently by the enzyme pyrophosphatase which converts the PPi to Pi. The acyl CoA undergoes alpha beta dehydrogenation catalysed by the flavo-protein enzyme, acyl CoA dehydrogenase (Reaction 2). Flavo-proteins known as Electron Transferring Flavo-proteins (ETF) catalyse the oxidation of acyl CoA dehydrogenase while they themselves are oxidised in the electron transport chain. The dehydrogenation reaction is non-reversible. The alpha-beta unsaturated acyl CoA is then hydrated by the stereo-specific addition of a molecule of water to form beta-hydroxy acyl CoA, and the reaction which is reversible is catalysed by enoyl CoA hydratase (crotonase) (Reaction 3). In the fourth step, the beta-hydroxy acyl CoA undergoes dehydrogenation by NAD^+ catalysed by the enzyme beta hydroxy acyl CoA dehydrogenase to give beta keto acyl CoA (Reaction 4). The latter undergoes thiolytic degradation in the presence of a molecule of Coenzyme A, catalysed by beta keto thiolase enzyme (Reaction 5). A molecule of acetyl CoA gets cleaved off at the beta carbon atom (hence the name beta-oxidation) leaving behind an acyl CoA with two carbon atoms less in the chain. The latter once again undergoes oxidation and cleavage by passing through steps 2–5 and gives another acetyl CoA till the fatty acid is completely broken down to acetyl CoA units. Since the five steps involved are continued again and again in a spiral-like fashion, the entire pathway is called 'fatty-acid spiral'.

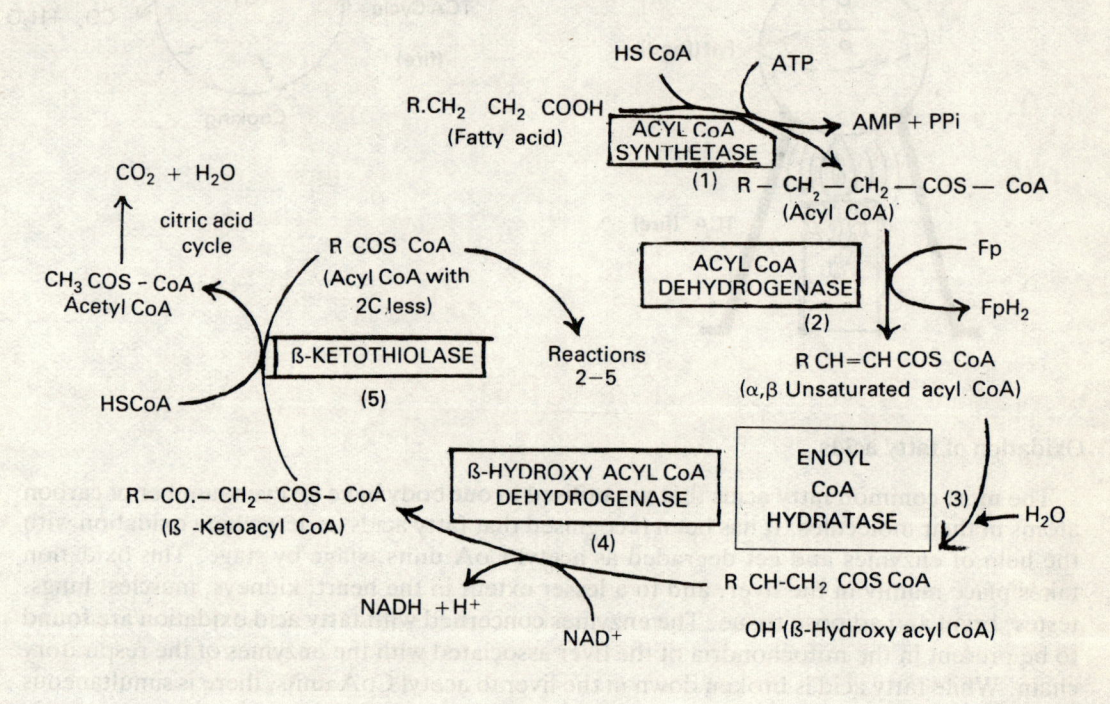

Fatty acid spiral

The acetyl CoA goes to the general pool and is finally oxidised to CO_2 and water by the extra-hepatic tissues, by the operation of the citric acid cycle. Hence, for proper utilisation of fatty acids, the citric acid cycle should function normally.

LIPID METABOLISM

Role of carnitine

Carnitine is gamma-trimethyl ammonium-beta-hydroxybutyrate, and its derivatives are acetyl-carnitine and acyl-carnitine.

$$H_3C-\overset{\overset{CH_3}{|}}{\underset{\underset{CH_2}{|}}{N^+}}-CH_3$$

$$CH_3-COO-\underset{\underset{COOH}{|}}{\overset{\overset{|}{CH_2}}{CH}}$$

Acetyl group

Acetyl carnitine

It is known that the mitochondrial membrane is impermeable to acetyl CoA and acyl CoA. Fatty acids in the cytosol are transported across the inner membrane of the mitochondria to the matrix for oxidation with the help of carnitine as shown below:

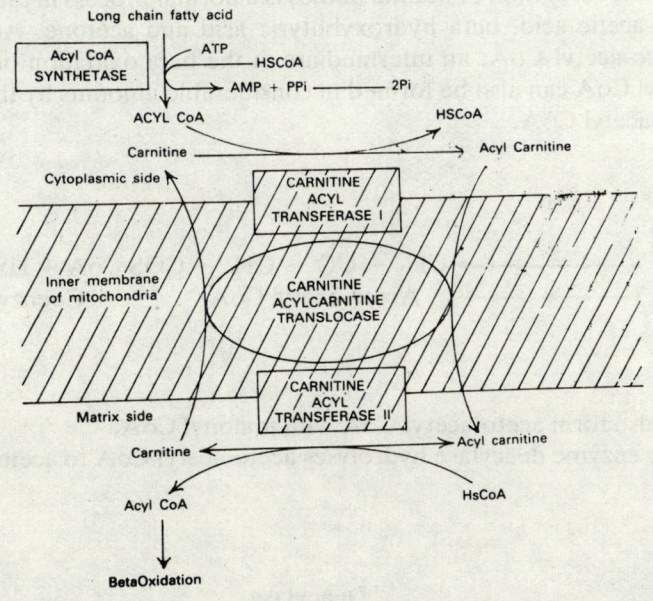

The acyl CoA synthetase enzyme (thiokinase) is found both inside and outside the mitochondria. While the activation of short-chain fatty acids can occur inside the mitochondria, that of long-chain fatty acids occurs outside (mostly in the endoplasmic reticulum and also on the outer membrane of the mitochondria) and this needs as carnitine. Long-chain fatty acyl CoA as such cannot penetrate the inner membrane of the mitochondria. In the presence of carnitine, and the enzymes carnitine-acyl transferase I and II (present outside and inside the mitochondrial membrane, respectively) and carnitine-acyl carnitine translocase (present on the inner membrane) the transport of the acyl groups into the matrix is effected. Carnitine acts as a transporter for the transport of the acetyl groups also and thus facilitates fatty acid oxidation.

Energetics of fatty acid oxidation

The transport of electrons in the respiratory chain from reduced flavo protein and NAD$^+$ will lead to five ATP molecules, two from the re-oxidation of FADH$_2$ and three from the re-oxidation of NADH for each beta oxidation. For a typical fatty acid, like palmitic acid C$_{15}$H$_{31}$ – COOH, seven such beta oxidations are required for complete oxidation and the formation of eight acetyl CoA units, thus giving a total of $7 \times 5 = 35$ ATP molecules.

One molecule of acetyl CoA when completely oxidised in the citric acid cycle gives 12 ATP molecules. Thus eight molecules of acetyl CoA yield $12 \times 8 = 96$ ATP molecules, making a total of 131 ATPs.

For the initial activation of palmitic acid, one molecule of ATP is split up to give AMP and PPi. Conversion of ATP to AMP may be considered equivalent to the utilisation of two ATPs. Thus the net ATP molecules obtained is $131-2 = 129$ per molecule of palmitic acid. Since one ATP molecule gives 7400 calories, the total energy due to 129 ATPs is equal to 9,54,600 calories. Thermo-chemically, palmitic acid on complete combustion, gives 23,40,000 calories. Hence the energy captured by the process of oxidation of palmitic acid in the body is 40.8 percent of the total energy of combustion.

Ketogenesis

Ketogenesis or the formation of ketone bodies is a normal process in the liver. The ketone bodies are aceto-acetic acid, beta hydroxybutyric acid and acetone. Aceto-acetic acid is formed from aceto-acetyl CoA, an intermediate in the beta oxidation of fatty acids in the liver. Aceto-acetyl CoA can also be formed in considerable amounts by the condensation of two molecules of acetyl CoA.

$$2CH_3 \; COS \; CoA \; \rightleftarrows \; CH_3 - CO - CH_2 - COS \; CoA + HS \; CoA$$

Acetyl CoA → Aceto-acetyl CoA → Coenzyme A

Acetyl CoA can also form aceto-acetyl CoA with malonyl CoA.

In the liver, the enzyme deacylase hydrolyses aceto-acetyl CoA to aceto-acetic acid.

$$CH_3 - CO - CH_2 - COS \; CoA + H_2O \xrightarrow{\text{Deacylase}} CH_3 - CO - CH_2 - COOH$$

Aceto-acetyl CoA → Aceto-acetic Acid + HS CoA coenzyme A

Recent evidences indicate that a more important pathway for the formation of ketone bodies involves beta-hydroxy-beta-methyl glutaryl CoA (HMG CoA) as an obligatory intermediate.

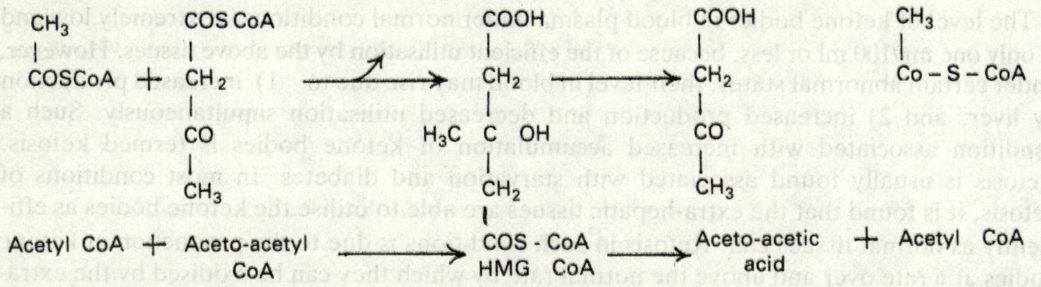

LIPID METABOLISM

$$\begin{array}{c}CH_3 \\ | \\ COSCoA\end{array} + \begin{array}{c}COSCoA \\ | \\ CH_2 \\ | \\ CO \\ | \\ CH_3\end{array} \rightleftharpoons \begin{array}{c}COOH \\ | \\ CH_2 \\ | \\ H_3C\;C\;OH \\ | \\ CH_2 \\ | \\ COS\text{-}CoA \\ HMG\;CoA\end{array} \rightarrow \begin{array}{c}COOH \\ | \\ CH_2 \\ | \\ CO \\ | \\ CH_3\end{array} + \begin{array}{c}CH_3 \\ | \\ Co-S-CoA\end{array}$$

Acetyl CoA + Aceto-acetyl CoA → → Aceto-acetic acid + Acetyl CoA

Aceto-acetic acid formed in the liver may undergo reduction by NADH to give beta-hydroxy butyric acid, or decarboxylation to give acetone.

$$CH_3 - CO - CH_2\;COOH \xrightarrow{NADH + H^+} CH_3 - \underset{OH}{CH} - CH_2 - COOH$$

$$CH_3 - CO - CH_2\;\boxed{COO}\;H \xrightarrow{-CO_2} CH_3 - CO - CH_3$$

In the mitochondria of the liver, deacylase activity is very high while the activities of thiokinase and thiophorase required for the utilisation of aceto-acetic acid are low. Hence much of the free aceto-acetic acid passes into circulation and is transported to the muscles and other peripheral tissues which reconvert it to aceto-acetyl CoA with succinyl CoA catalysed by aceto-acetyl succinic thiophorase.

Aceto-acetic acid + succinyl CoA ⇌ aceto-acetyl CoA + succinic acid

The enzyme is also known as succinyl CoA-aceto-acetate CoA transferase.

Another reaction involves the activation of aceto-acetate with ATP in the presence of CoA catalysed by aceto-acetate thiokinase. This reaction also gives aceto-acetyl CoA. Aceto-acetyl CoA is then split into acetyl CoA by cleavage enzyme thiolase.

Acetoacetate $\xrightarrow{\text{Thiokinase, ATP}}$ Acetoacetyl CoA

Aceto-acetyl CoA + HS CoA $\xrightarrow{\text{THIOLASE}}$ 2 Acetyl CoA

The oxidation of the ketone bodies after being converted to acetyl CoA is carried out very efficiently in the extra-hepatic tissues like the muscle, heart, kidneys, brain and testes, which contain appreciable amounts of the enzymes aceto-acetyl succinic thiophorase and aceto-acetate thiokinase which are very low in the liver.

Ketosis

The level of ketone bodies in blood plasma under normal conditions is extremely low and is only one mg/100 ml or less, because of the efficient utilisation by the above tissues. However, under certain abnormal states, their level in blood may rise due to 1) increased production by liver, and 2) increased production and decreased utilisation simultaneously. Such a condition associated with increased accumulation of ketone bodies is termed ketosis. Ketosis is usually found associated with starvation and diabetes. In most conditions of ketosis, it is found that the extra-hepatic tissues are able to utilise the ketone bodies as efficiently as normal tissues can. Ketosis in such conditions is due to the production of ketone bodies at a rate over and above the normal rate by which they can be oxidised by the extra-hepatic tissues. In diabetes and starvation, an increased breakdown of fatty acids occurs resulting in the over-production of ketone bodies which may give rise to ketonemia and ketonuria. Hence, in uncontrolled diabetes mellitus, there could be ketoacidosis which may lead to coma. Prompt insulin therapy would correct the ketoacidosis.

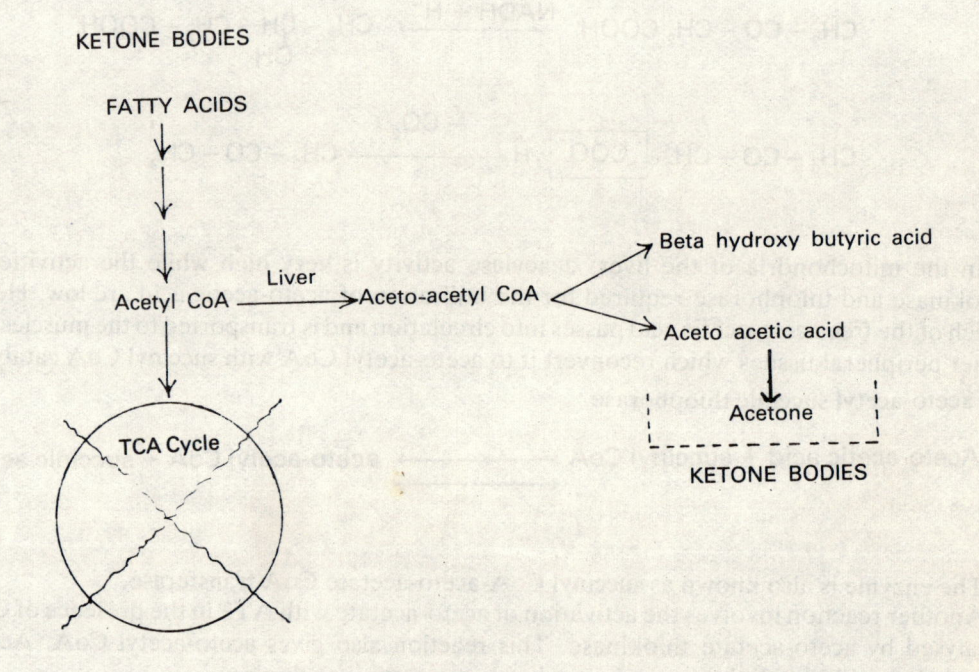

KETONE BODIES

FATTY ACIDS
↓
↓
↓
Acetyl CoA —Liver→ Aceto-acetyl CoA → Beta hydroxy butyric acid
↓   → Aceto acetic acid
TCA Cycle   ↓
                                              Acetone
                                            KETONE BODIES

For some time, it was belived that the oxidation of aceto-acetic acid was catalysed by the presence of glucose. Shaffer grouped substances into ketogenic and anti-ketogenic groups and believed that the ratio between these two was a deciding factor in ketosis. Fat is ketogenic and is broken down to ketone bodies at increased rates during starvation and diabetes mellitus. As the ketogenesis in these conditions exceeds the rate at which ketone bodies can be oxidised by extra-hepatic tissues, ketosis results. It is known now that the beneficial effect of glucose in preventing ketosis is not because it increases the utilisation of ketone bodies, but because it prevents their overproduction by decreasing the breakdown of fatty acids. Glucose and glycogen act as the main energy source, and fatty acids are conserved. In diabetes mellitus, the increased breakdown of fatty acids gives rise to ketosis. Administration of insulin to the diabetic alleviates ketosis by improving carbohydrate metabolism and decreasing the breakdown of fatty acids and proteins. Apart from the breakdown of fatty acids, three of the amino acids, i.e., leucine, phenyl alanine and tyrosine are ketogenic in character. The anti-ketogenic substances are glucose, and glucose-producers like the glycerol portion of fat and the glucogenic amino acids like glycine, glutamic acid, alanine, serine, etc. During ketosis of starvation, any of these anti-ketogenic substances can control the over-production of ketone bodies.

Since beta-hydroxybutyric acid and aceto-acetic acid are relatively strongly dissociated organic acids, their accumulation in tissues and body fluids can cause metabolic acidosis. The bicarbonate of the blood decreases considerably. Due to increased excretion of ketone bodies, water and Na^+ are lost from the body. This leads to a total electrolyte and Na^+ deficiency and dehydration. These result in symptoms like depression, thirst, fatigue, hyperpnoea and coma often seen in diabetic acidosis.

Ketone bodies in urine are identified by Rothera's test. With sodium nitroprusside and alkali, a red colour is obtained.

Biosynthesis of fatty acids

It was earlier supposed that the biosynthesis of fatty acids was merely a reversal of the process of beta-oxidation in liver. The mitochondrial system involving the enzymes of the fatty acid spiral is now known to be only of limited use for the elongation of the chain rather than an overall synthesis of fatty acid. The de novo synthesis of a fatty acid from acetyl CoA is now proved to be taking place by the extra-mitochondrial system. Both are considered here.

Extra-mitochondrial system for *de novo* synthesis of fatty acids (Lipogenesis)

While the dietary fat, on digestion by pancreatic lipase and on hydrolysis by the lipoprotein lipase of blood, can give us fatty acids, the latter could also be synthesised in the body. The liver, kidneys, brain, lungs, adipose tissue and mammary glands take part in significant synthesis of fat.

The main site of synthesis of fatty acids inside the cells is outside the mitochondria, (extra-mitochondrial), in the cytosol. All the enzymes required for the de novo synthesis of fatty acids are present in the cytosol. In the interests of the economy of the cell, all the enzymes are clubbed together as a 'kit'. If the enzymes were separate, and if even one enzyme was deficient, the other enzymes could not synthesise the fatty acids, and their presence in such a situation would be a waste for the cell.

In the lower organisms the six enzymes are associated as acyl carrier protein (ACP) whose functional units are 4 phosphopantetheine (-SH) and cysteine (SH).

In higher animals, it is not precisely the ACP but a 'multi-enzyme' complex, that is present, but this incorporates ACP. This multi-enzyme complex is a dimer, i.e., made up of two identical units aligned 'head to foot' on either side. One end has the 4 phospho pantetheinyl group of ACP while the other end has cysteinyl SH. Each monomer has seven enzymes required for the total synthesis of fatty acids from acetyl CoA.

The enzymes are :
1) acetyl transacylase,
2) malonyl transacylase,
3) beta ketoacyl synthetase,
4) beta ketoacyl reductase,
5) hydratase,

6) enoyl reductase, and
7) thioesterase (deacylase).

This system has been found in the soluble (cytosolic) fraction of many tissues including the liver, kidney, brain, lungs, mammary gland and adipose tissue. The co-factor requirements are NADPH, ATP, Mn^{2+}, and HCO_3^- (as a source of CO_2). Acetyl CoA is the starting substance and free palmitate is the end product. The first reaction is the conversion of acetyl CoA to malonyl CoA catalysed by the enzyme acetyl CoA carboxylase. This is the rate-limiting or key enzyme of lipogenesis.

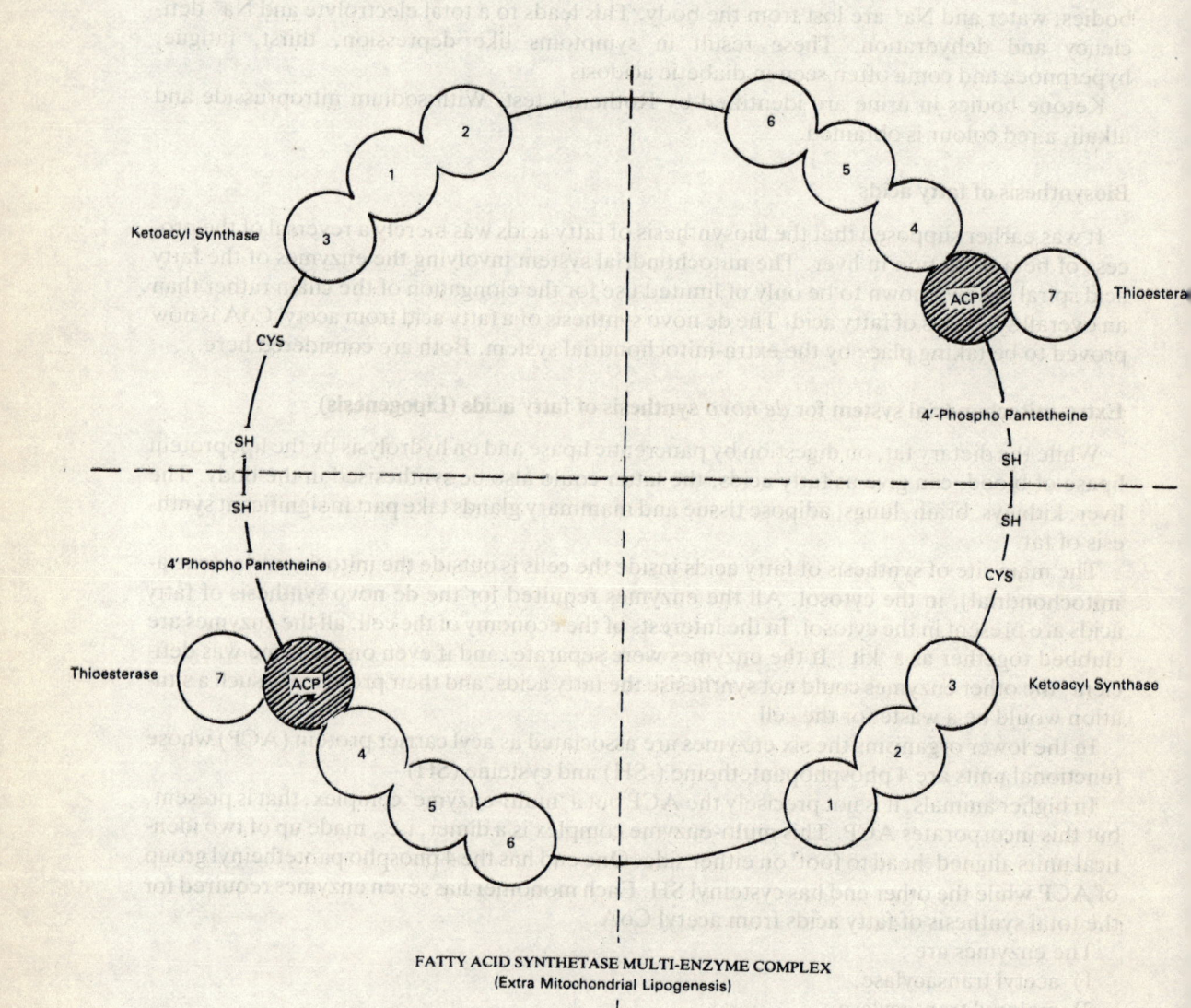

FATTY ACID SYNTHETASE MULTI-ENZYME COMPLEX
(Extra Mitochondrial Lipogenesis)

LIPID METABOLISM

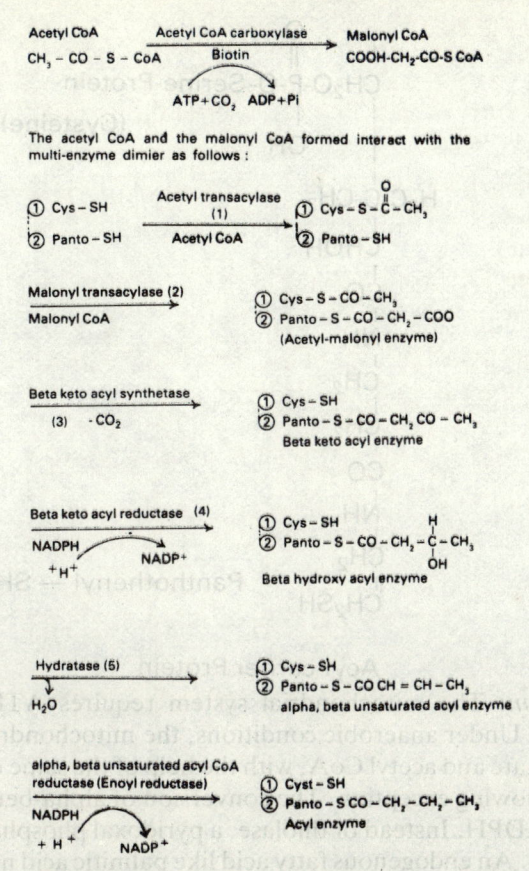

$$\text{Acetyl CoA} \xrightarrow[\text{Biotin}]{\text{Acetyl CoA carboxylase}} \text{Malonyl CoA}$$

$$CH_3-CO-S-CoA \quad\quad ATP+CO_2 \quad ADP+Pi \quad\quad COOH-CH_2-CO-S\,CoA$$

The acetyl CoA and the malonyl CoA formed interact with the multi-enzyme dimier as follows:

$$\begin{array}{l}\text{(1) Cys-SH}\\ \text{(2) Panto-SH}\end{array} \xrightarrow[\text{Acetyl CoA}]{\text{Acetyl transacylase (1)}} \begin{array}{l}\text{(1) Cys-S-}\overset{\overset{O}{\|}}{C}\text{-CH}_3\\ \text{(2) Panto-SH}\end{array}$$

$$\xrightarrow[\text{Malonyl CoA}]{\text{Malonyl transacylase (2)}} \begin{array}{l}\text{(1) Cys-S-CO-CH}_3\\ \text{(2) Panto-S-CO-CH}_2\text{-COO}\end{array}$$
(Acetyl-malonyl enzyme)

$$\xrightarrow[\text{(3)}\quad -CO_2]{\text{Beta keto acyl synthetase}} \begin{array}{l}\text{(1) Cys-SH}\\ \text{(2) Panto-S-CO-CH}_2\text{-CO-CH}_3\end{array}$$
Beta keto acyl enzyme

$$\xrightarrow[\substack{NADPH \\ +H^+}]{\text{Beta keto acyl reductase (4)}} \xrightarrow{NADP^+} \begin{array}{l}\text{(1) Cys-SH}\\ \text{(2) Panto-S-CO-CH}_2\text{-}\underset{OH}{\overset{H}{C}}\text{-CH}_3\end{array}$$
Beta hydroxy acyl enzyme

$$\xrightarrow[H_2O]{\text{Hydratase (5)}} \begin{array}{l}\text{(1) Cys-SH}\\ \text{(2) Panto-S-CO CH=CH-CH}_3\end{array}$$
alpha, beta unsaturated acyl enzyme

$$\xrightarrow[\substack{NADPH \\ +H^+}]{\text{alpha, beta unsaturated acyl CoA reductase (Enoyl reductase)}} \xrightarrow{NADP^+} \begin{array}{l}\text{(1) Cyst-SH}\\ \text{(2) Panto-S CO-CH}_2\text{-CH}_2\text{-CH}_3\end{array}$$
Acyl enzyme

A new malonyl CoA can now enter the panto –SH and the various reactions follow. Each time a malonyl CoA enters the panto –SH, chain elongation takes place by two carbon atoms. Seven malonyl CoA molecules are therefore required for a complete cycle of operations as palmitic acid is always the end product. This is hydrolysed by thioesterase (deacylase) (7) to give palmitic acid.

It may be noted that NADPH is needed at two steps, in fatty acid synthesis. This is supplied by the HMP shunt, isocitrate dehydrogenase and Malic enzyme. It is for this reason that in tissues which indulge in considerable fat synthesis, like the liver, mammary gland, etc., the HMP shunt pathway is pronounced.

The CO_2 taken up in the acetyl CoA carboxylase reaction is discarded in the beta keto acyl synthase step. It is not incorporated in the fully grown palmitic acid.

As there is fixation of CO_2, biotin is needed. Biotin is antagonised by the protein, avidin of the raw egg. Hence, the avidin of the raw egg will deplete biotin and inhibit extra-mitochondrial lipogenesis.

Acyl-carrier protein (ACP)-SH: This was first discovered in the fatty acids synthesis system of E.coli. It replaces coenzyme A in the synthetic reactions. It has a molecular weight of 8,847 and possesses one –SH group and has the prosthetic group 4'-phosphopantetheine. During the reactions of fatty acid biosynthesis, the acetyl as well as the acyl groups are covalently linked to the sulphur of the –SH group and are carried as acetyl–ACP and acyl–ACP (see figure). The 4-phosphopantetheine of ACP is attached to the serine residue of the protein on its hydroxyl group by a phospho-diester linkage.

```
              O
              ‖
    4  CH₂O-P-O-Serine-Protein
              |
              OH                    (Cysteine) — SH
       H₃C-C-CH₃
            |
           CHOH
            |
           CO
            |
           NH
            |
           CH₂
            |
           CH₂
            |
           CO
            |
           NH
            |
           CH₂              Panthothenyl — SH
            |
           CH₂SH
```

Acyl-carrier Protein

Mitochondrial system: The mitochondrial system requires ATP, pyridoxal phosphate NADH and NADPH. Under anaerobic conditions, the mitochondria catalyse the synthesis of stearate from palmitate and acetyl CoA, with the help of the same enzymes of the fatty acid spiral, but with the following exception. The conversion of alpha-beta unsaturated acyl CoA reductase requires NADPH. Instead of thiolase, a pyridoxal phosphate-dependent condensing enzyme is required. An endogenous fatty acid like palmitic acid needs initial activation by CoA SH and ATP to acyl CoA. It is found that the mitochondrial system is only for elongation of fatty acids and as such has limited significance.

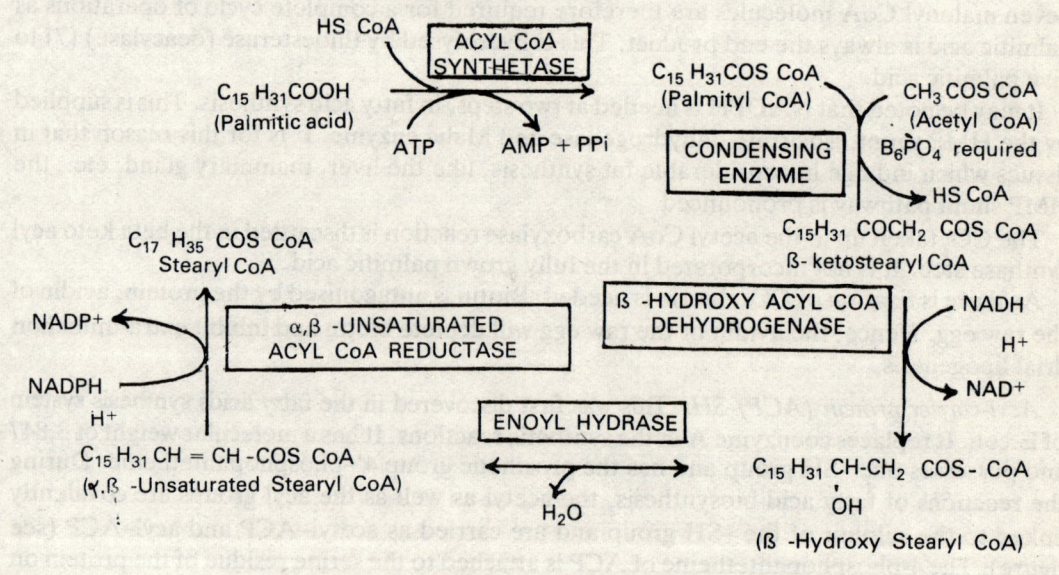

Mitochondrial system for fatty acid synthesis.

Microsomal synthesis: When microsomes are added to soluble synthetase preparations, there is a stimulation of fatty acid synthesis. Microsomes also catalyse the conversion of malonyl CoA to fatty acids.

The differences between mitochondrial and extra-mitochondrial synthesis are given below:

Mitochondrial	Extra-mitochondrial
1) Elongation of existing acids (formation of stearic acid, etc.)	1) Synthesis of fatty acid from acetyl CoA (formation of palmitic acid)
2) No formation of malonyl CoA	2) Formation of malonyl CoA
3) Bicarbonate not required	3) Bicarbonate required for CO_2
4) Biotin not required	4) Biotin required
5) Avidin insensitive	5) Avidin sensitive

As stated earlier, the extra-mitochondrial lipogenesis is of great significance.

Control of fatty acid synthesis

Fatty acid synthesis is depressed in starvation, diabetes mellitus and thiamine-deficiency.

Decreased carbohydrate metabolism in such conditions leads to decreased production of acetyl CoA, NADPH and NADH, and is responsible for decreased lipogenesis. The formation of acetyl CoA from pyruvate is depressed in thiamine deficiency. In starvation and diabetes, the activity of acetyl CoA carboxylase, the rate limiting enzyme, is diminished.

The addition of citrate stimulates fatty acid synthesis through the formation of acetyl CoA by the cleavage of citrate with the help of the cleavage enzyme (ATP citrate lyase).

$$\text{Citrate} + \text{ATP} + \text{HS CoA} \xrightarrow{\text{Citrate Cleavage Enzyme}} \text{Acetyl CoA} + \text{Oxalo-acetate} + \text{ADP} + \text{Pi}$$

Citrate also activates acetyl CoA carboxylase in an allosteric manner.

The long chain fatty acyl CoA inhibits fatty acid synthesis by a feedback control. Deficiencies of vitamins, thiamine and biotin also result in depressing the synthesis of fatty acids.

Biosynthesis of diacyl glycerols and triacyl glycerols (triglycerides)

Di and triacyl glycerols are formed in the liver, adipose tissue and intestines.

The first step is the formation of L-alpha glycerophosphate which may be formed by the action of glycero kinase on glycerol and ATP.

$$\text{ATP} + \text{glycerol} \xrightarrow{\text{glycero kinase}} \begin{array}{c} CH_2OH \\ | \\ HOCH \\ | \\ CH_2-O-PO_3H_2 \end{array} + \text{ADP}$$

L alpha glycerophosphate

Glycerokinase is active in the liver, kidneys and intestines, but not in many other tissues including adipose tissue. In such tissues, glycerophosphate is supplied by glycolysis through dihydroxy acetone phosphate. Glycerophosphate combines with acyl CoA to form phosphatidic acid. This is hydrolysed by a phosphatase to form diacyl glycerols. Reaction with another molecule of acyl CoA will give tracyl glycerols. The enzyme required is diacyl glycerol acyl transferase.

$$\begin{array}{c} CH_2OH \\ | \\ HOCH \\ | \\ CH_2\text{-}O\text{-}PO_3H_2 \end{array} + R\,COS\,CoA \xrightarrow{enzyme}$$

$$\begin{array}{c} CH_2\text{-}O\text{-}COR \\ | \\ R\text{-}COO\text{-}CH \\ | \\ CH_2\text{-}O\text{-}PO_3H_2 \end{array} + 2HS\text{-}CoA$$

$$\begin{array}{c} CH_2\text{-}O\,COR \\ | \\ RCOO\text{—}C\text{—}H \\ | \\ CH_2\text{-}O\text{-}PO_3H_2 \end{array} \xrightarrow{H_2O} \begin{array}{c} CH_2\text{-}O\,COR \\ | \\ RCOO\text{-}C\text{-}H \\ | \\ CH_2OH \end{array}$$
diacyl glycerol

$$\xrightarrow[\text{diacyl glycerol acyl transferase}]{RCOSCoA} \text{Triacyl glycerols}$$

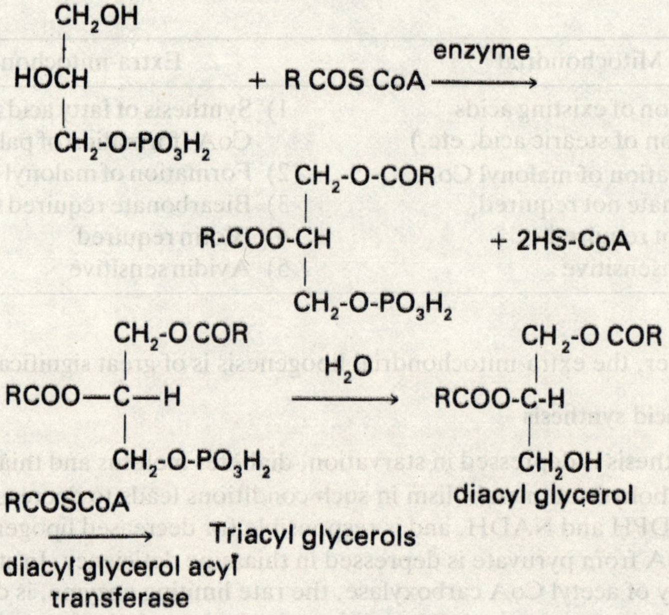

Metabolism of unsaturated and essential fatty acids'

The physiologically important essential fatty acids (EFA) are the following:

Linoleic acid $CH_3(CH_2)_4 CH = CH\,CH_2\,CH = CH\,(CH_2)_7\,COOH\,(C_{17}H_{31}COOH)$

Linolenic acid $CH_3\,CH_2\text{-}CH = CH\,CH_2\,CH = CH\text{--}CH_2\text{---}CH = CH\text{--}(CH_2)_7\,COOH$
$(C_{17}H_{29}COOH)$

Arachidonic acid $CH_3(CH_2)_4(CH = CH\,CH_2)_4(CH_2)_2\,COOH\,(C_{19}H_{31}\,COOH)$

There are a few other polyunsaturated fatty acids, mostly derived from linolenic acid by chain elongation.

The liver is able to synthesise palmitoleic and oleic acids among the unsaturated fatty acids but not the essential fatty acids. The desaturation of a saturated fatty acid to a mono-unsaturated fatty acid takes place in the liver microsomes. Oxygen, NADPH or NADH are necessary for this.

Rats maintained on a diet deficient in essential fatty acids develop dermatitis, scaly skin and necrosis of the tail. Lesions in the reproductive system also occur. The symptoms resemble pyridoxine deficiency. It is now known that pyridoxal phosphate is required as a co-factor for the condensing enzyme in the mitochondrial system of fatty acid synthesis.

Although linoleic acid or any of the other EFA cannot be completely synthesised in our body, it is possible to covert linoleic acid to linolenic acid and arachidonic acid by the enzyme systems in the liver mitochondria as follows:

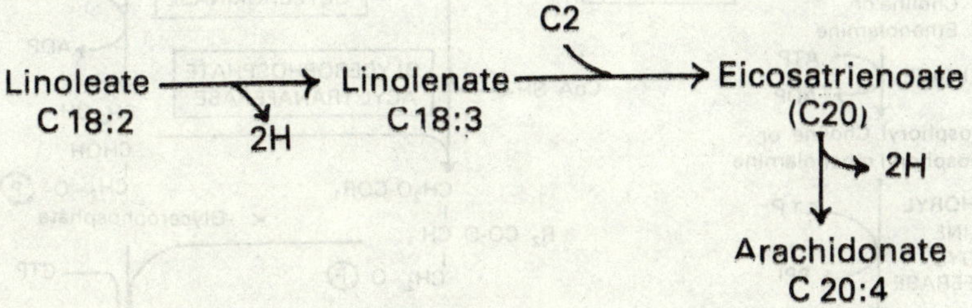

$$\text{Linoleate} \xrightarrow{-2H} \text{Linolenate} \xrightarrow{C2} \text{Eicosatrienoate} \xrightarrow{-2H} \text{Arachidonate}$$
$$C\,18{:}2 \qquad\qquad C\,18{:}3 \qquad\qquad (C20) \qquad\qquad C\,20{:}4$$

The desaturation and chain elongation require insulin and are depressed in fasting and diabetic conditions.

In recent years the functions of essential fatty acids (EFA) have become clearer. Their occurrence in most of the phospholipids indicates their great role concerned with the structural integrity of the mitochondrial membrane. EFA are required for the synthesis of phospholipids and thereby in exerting a lipotropic effect in the prevention of fatty liver. EFA occur in high concentration in the reproductive organs and may be concerned with reproduction. Arachidonic acid is established now as the precursor for the synthesis of prostaglandins which have diverse functions in the body in almost all tissues.

Apart from EFA, certain unsaturated fatty acids are needed in our body, i.e., palmitoleic and oleic acids. These are efficiently synthesised in the liver from stearic acid.

The unsaturated fatty acids are responsible for decreasing the melting point of body fat and for the formation of soft adipose tissue fat. The unsaturated fatty acids are better oxidised and more easily utilised in the body than the saturated fatty acids. They also tend to regulate blood cholesterol levels.

METABOLISM OF PHOSPHOLIPIDS

Phospholipids are present in all tissues and cells. They are biosynthesised efficiently by the intestines, liver, brain and other tissues and are not needed in the diet. However, the dietary phospholipids are absorbed and assimilated in the body as efficiently as those produced endogenously.

Biosynthesis of lecithins and cephalins

The phospholipids are biosynthesised from fatty acids, diacyl glycerols and triacyl glycerols and require certain factors like choline, ethanolamine and serine or their precursors.

Fatty acids are activated by Acyl CoA synthetase to acyl CoA. Two molecules of acyl CoA combine with $\propto$-glycerophosphate to give alpha beta diacyl glycerol phosphate. The latter, also called phosphatidic acid, is the simplest of phospholipids. Choline, ethanolamine or serine get activated as CDP choline and other CDP derivatives which then combine with the phosphatidic acid to give lecithins and cephalins. From phosphatidyl ethanolamine, lecithin can also be formed by progressive methylation of the N-atom. Inositol can also be incorporated to alpha beta diacyl glycerol as shown in the diagram. The phosphatidic acid on conversion to diacyl glycerol can take up a molecule of acyl CoA to form triacyl glycerol.

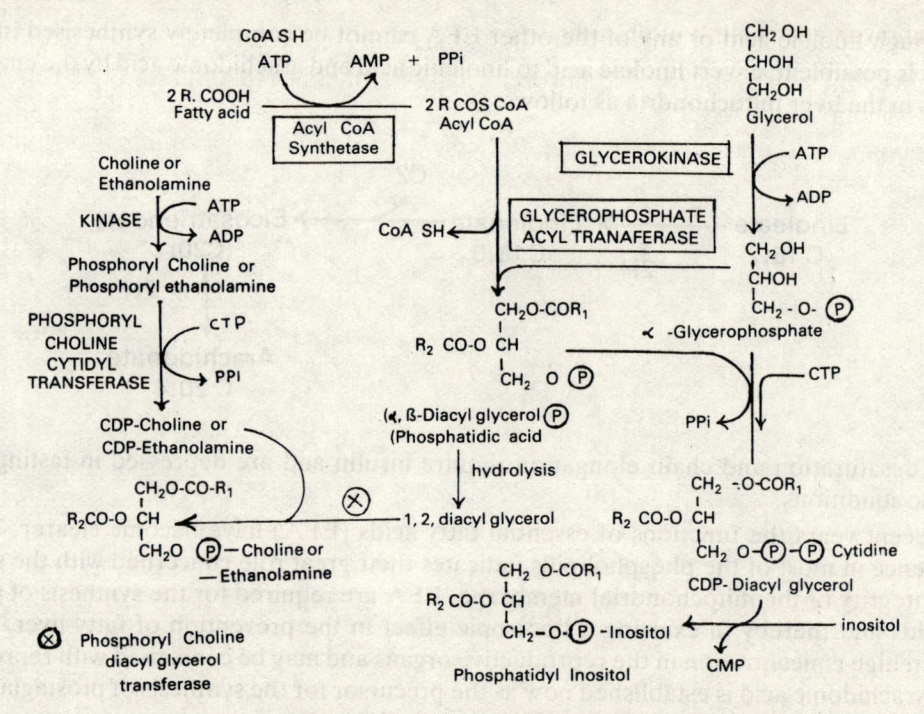

Biosynthesis of phospholipids.

Biosynthesis of sphingomyelin

Sphingomyelins do not contain glycerol. They contain fatty acid, phosphoric acid, choline and sphingosine. The sphingosine, a complex amino alcohol is synthesised from palmityl CoA and serine. The scheme for biosynthesis of sphingomyelin is as follows:

Degradation and catabolism of phospholipids

Enzymes that split and degrade phospholipids are called phospholipases. There are four such phospholipases A_1, A_2 C and D.

Phospholipase A_1 present in rice hulls, *Penicillium notatum* and *Aspergilus oryzae*, attacks the ester linkage at position 1 of phospholipids while phospholipase A_2 present in pancreatic juice (and also in snake venom) attacks and cleaves the ester linkage at position 2 of phospholipids. The lysophospholipid formed by the action of the latter is attacked by lysophospholipase (originally called phospholipase B) to form glycerophosphoryl base. Phospholipase C is present in *Clostridium welchii* and it splits phosphoryl choline from phospholipids giving rise to 1,2 diacyl glycerol. Phospholipase D is present in various plant sources like cotton seed, cabbage, etc., and splits choline from phospholipid giving rise to phosphatidic acid.

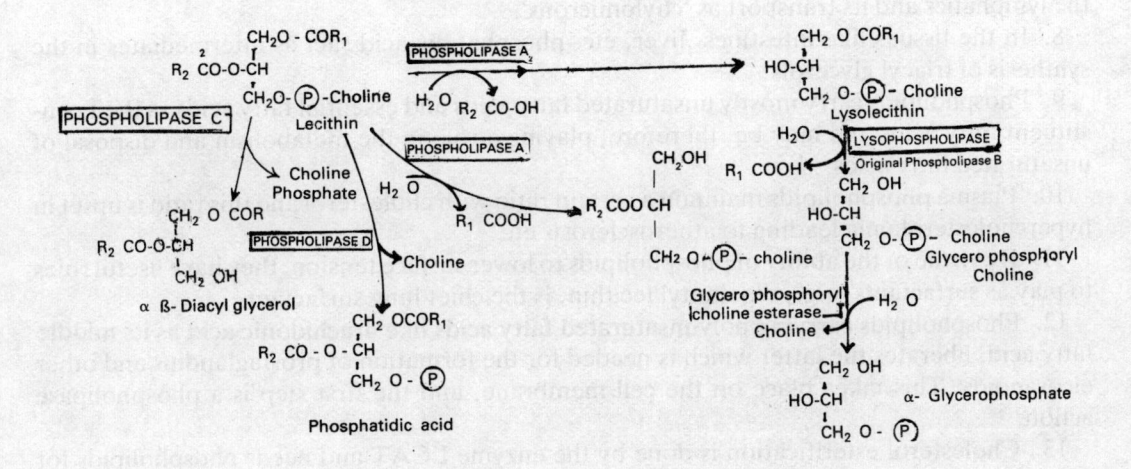

The products of the degradation of phospholipids enter the metabolic pool of the intermediates of carbohydrate and lipid metabolisms, and may be used over and over again.

Functions of phospholipids

1. Phospholipids are present in all cells and tissues. They are not depleted during starvation and are thus considered the 'elemente constant' or protoplasmic fat. By virtue of their ability to combine with polar and non-polar compounds equally efficiently, they play an important role in the physico-chemical organisation of protoplasm.

2. As components constituting the lipoprotein matrix of the cell wall (animal cells), they have an important function in regulating membrane permeability. Phosphatidyl choline and sphingomyelin are concentrated to a great extent on the surfaces that face ECF while phosphatidyl ethanolamine and phosphatidyl inositol are on the cytoplasmic side of the cell membrane.

3. They are present in the myelin sheath of the nerve cells along with cholesterol where they have a stuctural role to perform.

4. They are constituents of mitochondria and are required for the integrity of the mitochondrial membrane. The mitochondria contains a high proportion of lipids — 90 per cent

as phospholipids, consisting of 35 per cent phosphatidyl choline, 35 per cent phosphatidyl ethanolamine and 16 per cent cardiolipin. The presence of this amount of cardiolipin is specific for mitochondria. The mitochondrial phospholipids help to position and maintain various proteins participating in the respiratory chain in conformations active for electron transport. They also maximise their productive interactions with other essential components like CoQ (lipid soluble) and the coupling factors required to link oxidation with phosphorylation.

5. That phospholipids are extremely important metabolically as well is proved by the fact that they have a high rate of 'turnover' in most of the tissues like the liver, kidney, etc. In the brain and nervous tissue, however, their turnover is slow, where they have a structural role.

6. Phospholipids are necessary for the formation of lipoproteins and this is important in the disposal of triacyl glycerols in the liver. When there is a shortage of phospholipids or factors necessary for their formation, it leads to fatty infiltration in the liver. Phospholipids act as efficient lipotropic agents in preventing fatty liver.

7. Phospholipids are involved in the absorption of fat from the intestine, its disposal into the lymphatics and its transport as 'chylomicrons'.

8. In the tissues like intestines, liver, etc. phosphatidic acids act as intermediates in the synthesis of triacyl glycerols.

9. Phospholipids carry mostly unsaturated fatty acids and essential fatty acids as their constituent fatty acids and may be, therefore, playing a role in the metabolism and disposal of unsaturated fatty acids.

10. Plasma phospholipids maintain a certain ratio with cholesterol and this ratio is upset in hypercholesterolemia leading to atherosclerosis etc.

11. By virtue of the ability of phospholipids to lower surface tension, they have useful roles to play as surfactants, e.g., dipalmityl lecithin, is the chief lung surfactant.

12. Phospholipids carrying polyunsaturated fatty acids like arachidonic acid as its middle fatty acid, liberates the latter which is needed for the formation of prostaglandins and other eicosanoids. This takes place on the cell membrane, and the first step is a phospholipase action.

13. Cholesterol esterification is done by the enzyme LCAT and needs phospholipids for the donation of the central fatty acid (unsaturated). Hence, phospholipids play an important role in the metabolic disposal of cholesterol and thus decrease hypercholesterolemia.

14. Cephalins are involved in blood coagulation.

METABOLISM OF CHOLESTEROL

Cholesterol is essential to life and is present in all animal cells. Brain, egg yolk, liver and meat are good sources of cholesterol in the diet. A vegetarian diet, however, does not contain any cholesterol. Plant sterols present in such a diet are not absorbed from the intestine. The vegetarian gets all his cholesterol in the body by biosynthesis from acetyl CoA, from fatty acids, glucose and, to a limited extent, amino acids through their metabolites. An average non-vegetarian diet provides only 0.3 g of cholesterol per day although it is known that the turnover of cholesterol in the body is about 1.3 g per day. It is therefore concluded that the remaining amount of one g per day is produced in the body from various precursors. Most tissues are capable of synthesising cholesterol. Liver, adrenal cortex, skin, intestinal wall, testis, ovary and aorta are all involved in the synthesis of cholesterol, although the rates of synthesis are different in different tissues.

Biosynthesis of cholesterol

Experiments by Schoenheimer, Rittenberg, Block and others have shown that cholesterol is synthesised from simple precursors like acetate and water. It can be synthesised from many of the intermediates, from carbohydrate, protein and fatty acid metabolisms. Pyruvic acid aceto acetic acid, alcohol, leucine, alanine and octanoic acid labelled with ^{14}C or ^3H are found to act as precursors for cholesterol synthesis in liver *in vitro*. It is also found that all these compounds at some stage pass through acetyl CoA as well as HMG CoA.

Tissues known to sythesise cholesterol efficiently in the body include the liver, intestines, adrenal cortex, skin and aorta. The enzymes concerned with the synthesis of cholestrol are present in the microsomal fraction.

That acetyl CoA is the source of all the 27 C atoms of cholesterol has been confirmed by the use of labelled acetyl CoA in which the methyl C atom and the acyl carbon atom were labelled with ^{13}C and ^{14}C, respectively. In the final cholesterol molecule synthesised, the ^{13}C and ^{14}C were found to be incorporated alternatively on the carbon skeleton. The methyl groups on C_{10}, C_{13} and C_{25} were found to be the same as the methyl group of acetyl coenzyme A.

● Methyl C of acetyl CoA
○ Acyl C of acetyl CoA

Incorporation of methyl C and acyl C of acetyl CoA on cholesterol skeleton.

The synthesis of the cholesterol molecule takes place in several stages as follows:
1) Biosynthesis of mevalonate from acetyl CoA and other precursors.
2) Formation of iso-prenoid units from mevalonate by loss of CO_2.
3) Formation of squalene by condensation of 6-iso-prenoid units.
4) Formation of lanosterol from squalene by ring closure.
5) Conversion of lanosterol to cholesterol by oxidation of 3-methyl groups to CO_2

Stage 1: Biosynthesis of mevalonate: Acetyl CoA condenses with aceto-acetyl CoA to form beta hydroxy beta methyl glutaryl CoA, the reaction being catalysed by HMG CoA synthetase. The HMG CoA is converted to mevalonate by reduction by NADPH in two stages catalysed by HMG CoA reductase. (See figure.)

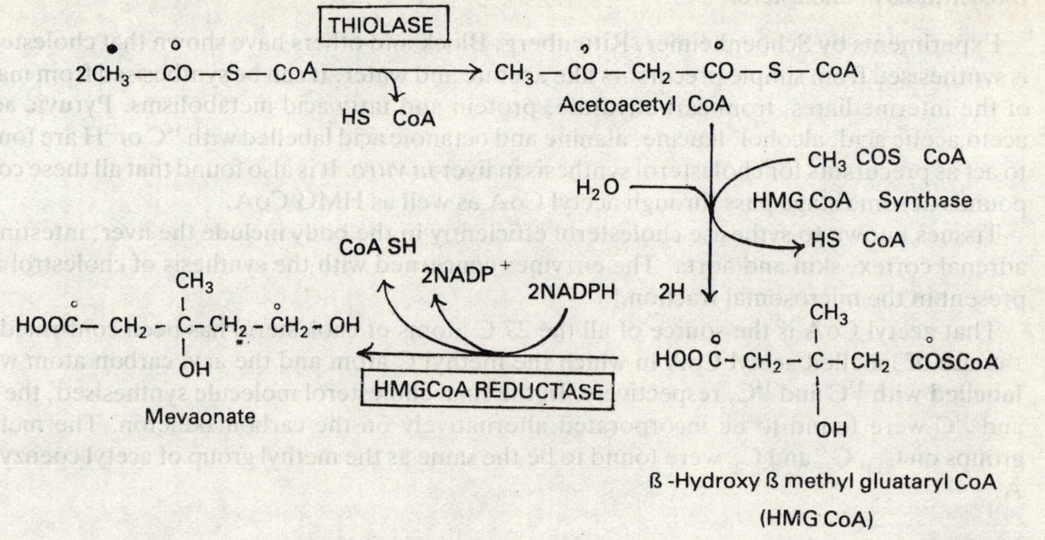

Biosynthesis of mevalonate from acetyl CoA.

$$2 CH_3-CO-S-CoA \xrightarrow[\text{HS CoA}]{\text{THIOLASE}} CH_3-CO-CH_2-CO-S-CoA$$

Acetoacetyl CoA

$$\xrightarrow[H_2O]{} \xrightarrow[\text{HS CoA}]{CH_3 \text{ COS CoA, HMG CoA Synthase}}$$

$$HOOC-CH_2-\underset{\underset{OH}{|}}{\overset{\overset{CH_3}{|}}{C}}-CH_2-CH_2-OH \xleftarrow[\text{HMGCoA REDUCTASE}]{2NADPH \to 2NADP^+ + 2H, CoASH} HOOC-CH_2-\underset{\underset{OH}{|}}{\overset{\overset{CH_3}{|}}{C}}-CH_2-COSCoA$$

Mevaonate

β-Hydroxy β methyl gluataryl CoA
(HMG CoA)

Biosynthesis of mevalonate from acetyl CoA.

Stage 2: Formation of isoprenoid units from mevalonate: The mevalonate gets phosphorylated in three stages to give 3-phospho mevalonate-5-pyrophosphate, catalysed by kinases (see figure)

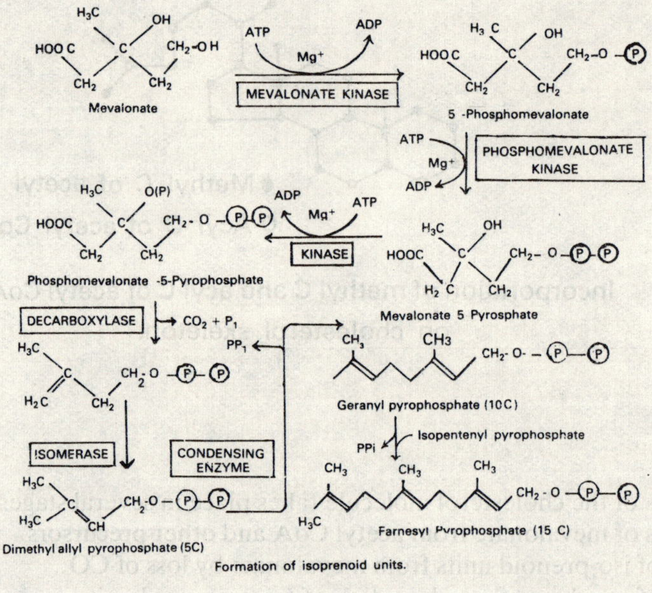

Formation of isoprenoid units.

This compound undergoes decarboxylation followed by splitting of one Pi radical to give isopentenyl pyrophosphate. Isopentenyl pyrophosphate (5C) isomerises to give dimethyl allyl pyrophosphate (5C) and these two molecules condense to form geranyl pyrophosphate (10C). One molecule of geranyl pyrophosphate (10C) then condenses with one molecule of isopentenyl-pyrophosphate to form farnesyl pyrophosphate (15C).

Stage 3: Formation of squalene from isoprenoid units: Two molecules of farnesyl pyrophosphate condense at the pyrophosphate end with the elimination of two PPi radicals to give squalene, a straight chain unsaturated hydrocarbon. This reaction is catalysed by squalene synthetase which also needs NADPH for reduction and Mn^{2+} or Mg^{2+} for activation. (see diagram).

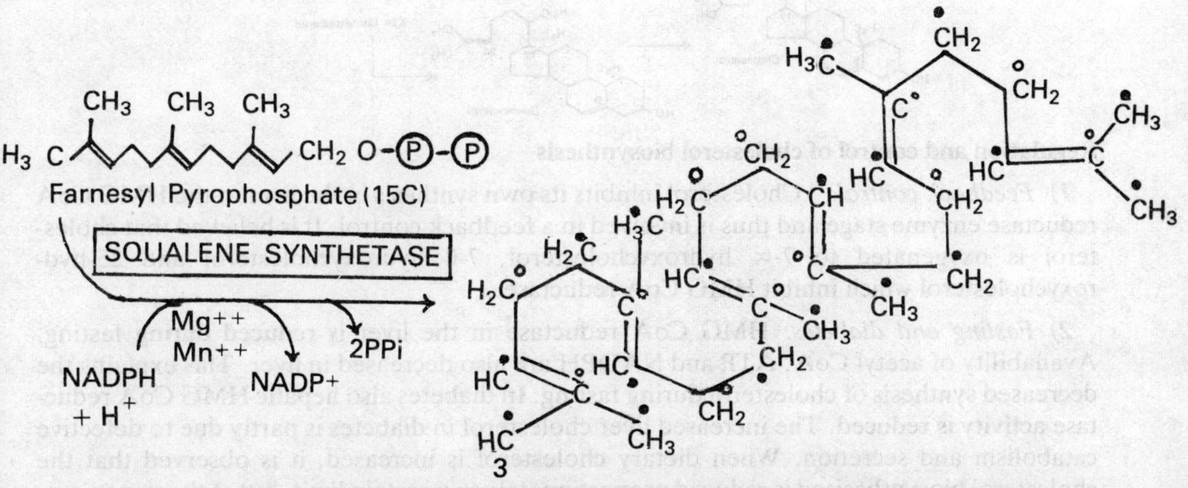

Formation of squalane from farnesyl pyrophosphate

Stage 4: Biosynthesis of lanosterol from squalene: The squalene gets attached to a specific protein called squalene-binding protein before it undergoes further transformation. This is necessary, as squalene is a lipid soluble hydrocarbon, and it has to be carried in a bound form with hydrophylic protein. Squalene undergoes ring closure following oxidation in presence of a cyclase and epoxidase. The methyl group on C_{14} moves on to C_{13} and that from C_8 gets away and occupies the C_{14} position C_3 gets hydroxylated with the help of a hydroxylase enzyme requiring molecular oxygen. Squalene and lanosterol have been detected in liver fat in small quantities.

Stage 5: Formation of cholesterol from lanosterol: The methyl groups on C14 and the gem-methyl groups on C4 get oxidised to CO_2 to form zymosterol, a sterol which is also present in yeast. Zymosterol gives rise to $\triangle^{7,24}$ cholestadienol by shifting of a double bond from C8 to C7 and to desmosterol by the shifting of the double bond again from C7 to a position between C5 and C6. The C24 double bond of desmosterol gets reduced to give cholesterol. In the three steps, i.e., oxidation of lanosterol to zymosterol, zymosterol to desmosterol and desmosterol to cholesterol, NADPH is required as co-factor. The two oxidative steps earlier also require molecular O_2.

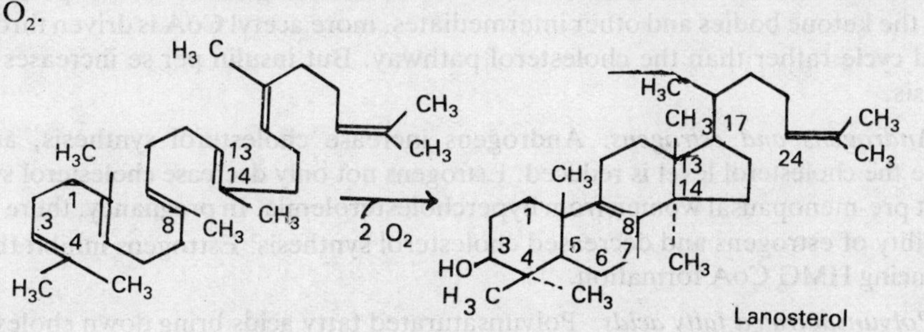

Formation of lanosterol from squalene

Regulation and control of cholesterol biosynthesis

1) Feedback control: Cholesterol inhibits its own synthesis in the liver at the HMG CoA reductase enzyme stage and thus is involved in a feedback control. It is believed that cholesterol is oxygenated to 7-α hydroxycholesterol, 7-ß hydroxycholesterol and 25-hydroxycholesterol which inhibit HMG CoA reductase.

2) Fasting and diabetes: HMG CoA reductase in the liver is reduced during fasting. Availability of acetyl CoA, ATP and NADPH are also decreased in liver. This explains the decreased synthesis of cholesterol during fasting. In diabetes also hepatic HMG CoA reductase activity is reduced. The increased liver cholesterol in diabetes is partly due to defective catabolism and secretion. When dietary cholesterol is increased, it is observed that the cholesterol biosynthesised is reduced proportionately to a certain limit, but at the same time, it is not completely suppressed.

3) Species variation: It is now believed that extra-hepatic synthesis of cholesterol, mainly in the intestines, is also proceeding in certain species especially in man. This may be regulated by factors other than cholesterol in the diet alone. There is a species variation regarding the assimilation of exogenous cholesterol. Cholesterol feeding causes hypercholestrolemia and the onset of atherosclerotic lesions in rabbit, pig, monkey and man but not in dog, rat and cat, which are resistant to it. In dogs and cats, thyroidectomy or thiouracil treatment is necessary to produce hypercholesterolemia and increased deposition of cholesterol on the aortic walls.

4) Hypophysectomy: Cholesterol biosynthesis is depressed in hypophysectomised animals.

5) Hyper-and hypothyroidism: In hypothyroidism, BMR is decreased, the oxidative processes in the body get depressed and the intermediary metabolites are diverted towards cholesterogenesis resulting in hypercholesterolemia. In hyperthyroidism, on the other hand, although the rate of cholesterol synthesis is increased, its rate of turnover is much higher, resulting in the lowering of the cholesterol to normel levels. Thyroid hormones facilitate the synthesis as well as the degradation of cholesterol, the latter effect being more predominant.

6) Insulin: As insulin increases the overall oxidation of glucose and the peripheral oxidation of the ketone bodies and other intermediates, more acetyl CoA is driven through the citric acid cycle rather than the cholesterol pathway. But insulin per se increases cholesterol synthesis.

7) Androgens and estrogens: Androgens increase cholesterol synthesis, and in their absence the cholesterol level is reduced. Estrogens not only decrease cholesterol synthesis but protect pre-menopausal women from hypercholesterolemia. In pregnancy, there is increased availibility of estrogens and decreased cholesterol synthesis. Estrogens inhibit the synthesis by reducing HMG CoA formation.

8) Polyunsaturated fatty acids: Polyunsaturated fatty acids bring down cholesterol in the blood by esterifying, solubilising and mobilising the cholesterol to the liver. LCAT is required for this. The polyunsaturated fatty acids are routed through phospholipids. The oxidation of cholesterol and conversion to bile acids in the liver is thus enhanced. Vegetable oils like corn oil and cotton seed oil are very good sources for polyunsaturated fatty acids, while animal fats like butter fat and beef fat are very poor sources.

9) *Drugs and other agents:* Choloxin, atromids S^R and neomycin have a beneficial effect in decreasing cholesterol levels in the blood by increasing the fecal excretion of coprosterol and bile acids which arise from the body's cholesterol reserves. Triparanol blocks the biosynthetic pathway of cholesterol at the desmosterol stage. Nicotinic acid, vanadium salts and ethyl-p-chlorophenoxy-isobutyrate are also cholesterol lowering substances. Cholestyramine (questran) is effective in lowering both total cholesterol and LDL, while probucol (Lorelco), a new drug, lowers both LDL and HDL and keeps the HDL/LDL ratio lower. Other new drugs, Compactin and Mevinolin, control cholesterol synthesis by blocking HMG CoA reductase, and at the same time stimulating the LDL receptor sites in cells and activating the feedback mechanism. A combination of Compactin and Cholestyramine is found to be most useful in the treatment of hyperlipidemia and hypercholesterolemia.

Catabolism of cholesterol and fecal excretion

The liver is the main site of cholesterol catabolism. It is found that 60 per cent of the total cholesterol of liver is converted to bile acids and excreted in bile. The rest 40 per cent is excreted as a solution in bile into the intestine from where a part is reabsorbed and another part is reduced by intestinal bacteria to coprosterol. The coprosterol is excreted in feces along with the bile salts as neutral sterols. Some of the bile salts may be reabsorbed, but this will depress the further catabolism of cholesterol in the liver and the formation of more bile acids by a feedback control.

In the gonads, adrenal cortex, corpus luteum and placenta, cholesterol is the precursor of the biosynthesis of various steroid hormones (see chapter on 'Hormones'). These hormones are released into the circulating blood and are modified or destroyed in the liver and excreted mostly th ough urine with the steroid ring intact. Compared to the amount of cholesterol catabolised by the liver, this amount converted to steroid hormones is quantitatively negligible but physiologically very important.

A small quantity of cholesterol is oxidised to 7-dehydrocholesterol and finally converted into cholecalciferol (vitamin D_3) in the skin by irradiation. Vitamin D_3 is finally excreted mostly in feces, or through the skin in small amounts along with a little cholesterol and fat.

METABOLISM OF THE PLASMA LIPOPROTEINS

It has already been mentioned that plasma contains the following lipoproteins:
1) Chylomicrons derived from intestinal absorption of triacyl glycerol.
2) Very low density lipoproteins (VLDL) derived from the liver.
3) Low density lipoproteins (LDL) derived from VLDL during its metabolism while in circulation.
4) High density lipoproteins (HDL) produced in the liver and intestines and involved in the metabolism of VLDL and cholesterol.
5) Free fatty acids present in combination with plasma albumin.

Chylomicrons are formed in the intestine during the digestion of dietary fat. They contain mostly triacyl glycerol. In the intestinal mucosal cells they are modified at the endoplasmic reticulum and secreted into the lymphatic vessels. Chylomicrons have a half life of only five minutes. The metabolism of chylomicrons involves a few steps during which the nascent chylomicrons with Apo-B-48 take up Apo-A, followed by combination with Apo-C and Apo-E The lipoprotein lipase then acts on the mature chylomicrons and degrades to fatty acids and chylomicron remnants (after releasing both Apo-A and Apo-C from it). The fatty acids released are taken up by the tissues, while the chylomicron remnants are taken up by the liver. The enzyme lipoprotein lipase is present on the walls of blood capillaries in many tissues like the lungs, adipose tissue, heart, lactating mammary gland, spleen, etc., and anchored to the capillary membrane with the proteoglycan chains of heparan sulphate. If heparin is injected, there will be a rapid release of lipoprotein lipase into the circulation. Both Apo-CII and some phospholipids act as co-factors for lipoprotein lipase. Insulin also activates lipoprotein lipase.

The liver secretes VLDL which is formed from endogenous sources, the hepatic triacyl glycerol being the main source. This is contributed by the triacyl glycerol reaching the liver as chylomicron remnants, the fatty acids reaching the liver via the plasma from adipose tissue during lipolysis and the triacyl glycerol synthesised in the liver (extra-mitochondrial de novo synthesis). The source differs in the fed state from that in the fasting state. Conditions associated with increased triacyl glycerol synthesis and secretion of VLDL are feeding of high carbohydrate diet in the presence of high insulin status (especially high source feeding), and ethanol ingestion in higher amounts (in moderate and severe alcoholics).

Nascent VLDL that contains triacyl glycerol and cholesterol along with Apo-B-100 takes up Apo-C and Apo-E to form VLDL. By the action of the enzyme lipoprotein lipase in extra-hepatic tissues, Apo-C and some fatty acids cleave off, along with glycerol forming LDL passing through an intermediate stage of VLDL remnants or IDL and fatty acids. The LDL is finally destroyed in the liver or in the extra-hepatic tissues, fibroblasts, lymphocytes, etc.

The liver and intestines also secrete a high density lipoprotein HDL. The LCAT (lecithin-cholesterol acyl transferase) enzyme transfers the fatty acids from phospholipids lecithin to free cholesterol thus forming cholesterol esters. Discoid nascent HDL becomes spherical HDL, which transfers the cholesteryl ester thus formed to the other low density lipoproteins, by virtue of the presence of a cholesteryl ester transfer protein, and to the liver (reverse cholesterol transport). HDL helps to dispose of the cholesterol by esterification which then enters the liver and gets oxidised to bile acids. Thus HDL scavenges the tissue cholesterol including that of the smooth muscle cells of the arteries.

METABOLISM OF GLYCOLIPIDS (CEREBROSIDES), SULFATIDES AND GANGLIOSIDES

Cerebrosides or glycolipids contain a carbohydrate or its derivative in combination with ceramide (a combination of sphingosine with fatty acid). They differ from phospholipids in having a carbohydrate (usually galactose) instead of phosphoryl choline residue. The cerebrosides contain C24 fatty acids like lignoceric acid, nervonic acid, cerebronic acid, or oxynervonic acid in various types. The cerebrosides are mostly present in the cerebral lipids although small quantities also occur in the liver, spleen, etc. The sulfatides and gangliosides are related to cerebrosides. All of them are sphingolipids. The enzymes necessary for the biosynthesis of cerebrosides, sulfatides and gangliosides occur in brain tissue. The sphingosine needed for biosynthesis is derived from palmityl CoA as in phospholipids, and the carbohydrate part is derived from glucose metabolism.

The cerebrosides of the brain do not have a rapid turnover and are believed to have some structural and functional role to play in brain cells, since these compounds, though lipids, have a higher affinity towards water like the phospholipids and may be concerned with permeability processes in brain cells.

LIPID METABOLISM

```
Glucose ──────→ glucose-6-(P) ──────→ glucose-1-(P)
                                              │ ┌── UTP
                                              ├─┤
Palmityl CoA + "Active" Serine                │ └─→ PPi
                   │                          ↓
              Dihydrosphingosine         UDP-Glucose
                   │                          │ UDPGlucose
              Sphingosine                     │ epimerase
                   │                          ↓
Galactosyl ←──────┘                     UDP-Galactose
sphingosine
(Psychosin)
    │
    │    ┌── R-COS  CoA          '3' PAPS
    │    │   AcylCoA
    ↓    ↓
Cerebroside ──────────────────────────→ Sulfatides
                    ┌─────────────────┐
                    │ BRAIN SULFATASE │
                    └─────────────────┘
    ┌── CMP NANA
    │   (a combination of CMP with
    │   N-acetyl neuraminic acid
    │   or sialic acid)
    │ ─→ CMP
    ↓
  Gangliosides
```

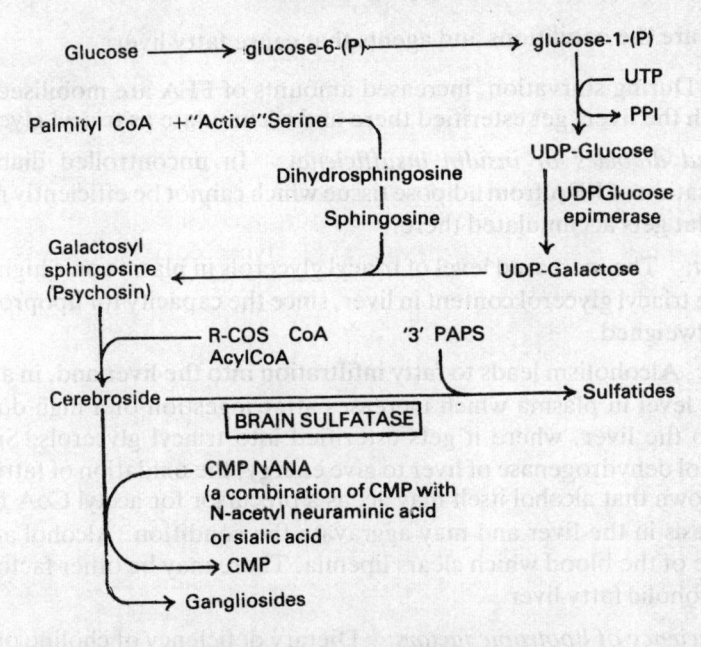

DISORDERS OF LIPID METABOLISM

Fatty liver, obesity, atherosclerosis, hyperlipoproteinemias, Niemann Pick's disease and Gaucher's disease are some of the disorders of lipid metabolism.

Fatty liver

Neutral fat is stored in the body mostly in the adipose tissue of skin pad, perirenal fat pad, omentum, etc. The liver, though not a storage organ for fat, may accumulate lipids – mainly as triacyl glycerols in exceptional circumstances and it is considered pathological. It is usually referred to as 'fatty liver' or 'fatty infiltration' in the liver. As against a normal content of about five per cent of lipid, the lipid content can go up to 25 or 30 per cent in fatty livers. When this conditions becomes chronic, fibrotic changes and extensive necrosis in the parenchyma cells occur, leading to cirrhosis and impairment of liver function. In fatty liver itself, there is impairment of liver function and a slowing down of metabolism in general. Fatty livers may be considered from two angles:

1) *Due to the FFA being brought to the liver in increased amounts,* to an extent more than what can be metabolised by the organ. Here, the production of plasma lipoproteins by the liver does not keep pace with the influx of FFA and the triacyl glycerols accumulate. This may occur during starvation and uncontrolled diabetes and also on continued intake of high-fat diets (physiological fatty liver).

2) *Due to some defect in the formation of lipoproteins:* The synthesis of lipoproteins requires phospholipids and also the apoprotein. A defect in the synthesis of phospholipids, apoprotein, or the lipoprotein itself may thus result in the accumulation of triacyl glycerols. Deficiencies of essential fatty acids or of certain lipotropic factors, like choline, result in decreased formation of phospholipids. A deficiency of methionine may result in deficiency of the labile methyl groups needed for the formation of choline and betaine. Fatty liver may also result from a defect in the secretion of lipoproteins by the liver for want of energy as ATP. In addition, other factors like increased synthesis of fatty acids, decreased oxidation of fatty acids and decreased ability of the liver to form VLDL from triacyl glycerol and to secrete it into plasma, can also contribute to the production of fatty livers in some instances.

The following are the conditions and agents that cause fatty liver:

1) Starvation: During starvation, increased amounts of FFA are mobilised from adipose tissue which reach the liver, get esterified there and accumulate as triacyl glycerols.

2) Uncontrolled diabetes or insulin insufficiency: In uncontrolled diabetes, there is increased mobilisation of FFA from adipose tissue which cannot be efficiently metabolised by liver, and hence fat gets accumulated there.

3) High fat diet: The increased level of triacyl glycerols in plasma on a high fat diet brings about a rise in the triacyl glycerol content in liver, since the capacity for lipoprotein formation by the liver is outweighed.

4) Alcoholism: Alcoholism leads to fatty infiltration into the liver and, in a few cases, cirrhosis. The FFA level in plasma which increases after ingestion of a high dose of ethanol, finds its way into the liver, where it gets esterified into triacyl glycerols. Since alcohol is oxidised by alcohol dehydrogenase of liver to give energy, the oxidation of fatty acid is depressed. It is also known that alcohol itself may act as a precursor for acetyl CoA for endogenous fatty acids synthesis in the liver and may aggravate the condition. Alcohol also inhibits the lipoprotein lipase of the blood which clears lipemia. There may be other factors also operating in causing alcoholic fatty liver.

5) Dietary deficiency of lipotropic factors: Dietary deficiency of choline or precursors of choline, i.e., the amino acids, glycine, serine and methionine, may cause fatty liver. Choline can be biosynthesised if methionine and the other amino acids are present, e.g., on a moderate protein intake. But in severe protein deficiency and malnutrition, choline deficiency occurs, leading to fatty infiltration of liver.

6) Injection of certain hormones which increase lipolysis: Growth hormone, ACTH, TSH and other lipolytic hormones, if injected into fasting animals, produce fatty liver by mobilising fat from adipose tissue and depositing the same in the liver.

7) Deficiency of essential fatty acids: A deficiency of essential fatty acids results in decreased formation of phospholipids and thereby decreased ability for disposal of triacyl glycerols from the liver as lipoprotein.

8) High cholesterol diet: Excessive amounts of cholesterol in the diet compete for essential fatty acids for esterification and may cause fatty liver apart from its accumulation in this organ along with triacyl glycerols.

9) Deficiencies of vitamin E, pyridoxine, pantothentic acid and proteins: Vitamin E deficiency causes fatty liver and necrosis of the parenchyma cells. An organic compound of selenium (factor 3) has a protective action in such a condition. Deficiencies of pyridoxine and pantothenic acid interfere with metabolism and decrease the availability of ATP needed for protein synthesis. Protein deficiency in general (e.g., kwashiorkor) may cause fatty liver.

10) Experimental fatty liver could be induced in animals by using chemicals like carbon tetrachloride, ethionine, orotic acid, phosphorus, etc. Carbon tetrachloride interferes with the conjugation of apoprotein with lipid to form the secretory lipoprotein, while ethionine traps adenine as S-adenosyl ethionine and prevents the formation of ATP required for the synthesis of lipoprotein. Orotic acid interferes with the glycosylation of the apoprotein part of the lipoprotein. Carbon tetrachloride and other compounds also are believed to give rise to the formation of free radicals that bring about lipid peroxidation of the membranes of the endoplasmic reticulum, and thus their disruption. Anti-oxidants however can bring about some amount of protection in these cases.

Obesity

Obesity is the state in which the accumulation of reserve fat becomes so extreme that the normal functions of the individual are interfered with. The cause is mostly over eating which means ingestion of excess calories than is normally needed. Every seven calories of excess consumption would mean 1 g of fat. Most of the accumulated fat is deposited in the adipose tissue, in the subcutaneous fat pads, peri-renal fat depots and in the fat depots of mesentery and omentum. Men are considered obese when more than 20 per cent of their weight is due to fat (adipose tissue) and women are considered obese when more than 25 percent of their weight is due to fat.

In some cases obesity might occur due to diminished utilisation of the ingested calories because of lowered basal metabolic rate. This could be pathological or due to lack of physical exercise.

As a person grows older, his basal metabolic rate gradually diminishes and he needs less calories. If the food consumption is not reduced, the excess calories consumed may find their way to storage as fat and may lead to obesity. In puerperal and post-menopausal women, obesity may set in due to deficiency or lack of estrogens. Some other endocrinological disorders could also cause obesity. Genetic and environmental factors may also be a contributory factor to obesity. It has been suggested recently that people become obese because of a decreased ability to convert metabolic fuels to heat and this is due to having significantly less of brown adipose tissue in them. Brown adipose tissue is the site for diet-induced thermogenesis. In obese individuals, carbohydrate, and fat are stored as glycogen or fat rather than being oxidised. Lack of exercise and physical activity may be one of the causes of obesity. Resorting to a new routine with regular food habits and exercise, intermittent fasting and avoiding 'excessive' eating have been reported to be helpful in maintaining body weight and preventing certain types of obesity, with the exception of the hereditary type. In the dietary method of treatment for obesity, emphasis is given to reducing caloric intake from fat and carbohydrate but maintaining the protein, vitamin and mineral intake at normal levels. Correction of obesity is mostly by restriction of food intake and adequate physical exercise.

Atherosclerosis

This is a condition associated with hardening of the walls of the aorta and arterial walls (sclerosis) due to a 'variable combination of changes in the intima of arteries'. This consists of local accumulation of lipids, complex carbohydrates, blood and blood products, fibrous tissues and calcium deposits, and is associated with medial changes. Deposition of cholesterol esters needs special mention.

Atherosclerosis can occur in diseases with elevated levels of cholesterol and LDL, lowered levels of HDL and an increse of VLDL. This is because HDL and VLDL have a reciprocal relationship, and a lowered level of HDL may cause an increase of VLDL. There is also a decrease of glycosamino glycans of the aortic tissue. The various diseases which could cause atherosclerosis are diabetes mellitus, nephrotic syndrome, hypothyroidism and familial hyperlipoproteinemias.

The process of ageing brings about significant changes in the blood vessel wall as the metabolism of cholesterol slows down with advancing age. There is increased utilisation of glucose in the HMP pathway with the production of NADPH which is conducive to increased lipogenesis. The enzymes elastoproteinase and elatomucase, required for solubilising and lipolysis of fat are decreased with age. Again, as age advances, the elasticity of the vessel wall decreases thereby decreasing the lateral pressure while there is a turbulent flow of blood due to greater axial pressure. For these reasons, ageing invites atherosclerosis.

Obesity, lack of exercise, soft drinking water with deficiency of minerals, high-fat diet containing saturated fat, emotional stress, excessive smoking, alcoholism and high blood pressure are all conducive to atherosclerosis. In advanced conditions it may not be possible to correct the defect. In the earlier stages, the probable cause should be investigated and suitable therapy instituted. In most cases, the levels of blood cholesterol and LDL cholesterol should be brought down to normal and those of HDL cholesterol should be increased. If smoking is found responsible, it should be stopped.

A strong correlation exists between hypertension, hypercholesterolemia, serum VLDL and LDL and atherosclerosis and coronary heart diseases.

Hypercholesterolemia, increase of LDL, decrease of HDL and hypertension are all direct risk factors for atherosclerosis.

As a decrease of HDL causes an increase of VLDL, the latter is an indirect risk factor.

LDL carries cholesterol in plasma and deposits it in tissues, and thus may be a contributing factor in atherogenesis. While LDL represents bad cholesterol in atherogenesis, HDL represents good cholesterol to safeguard against cholesterol deposition and prevent its accumulation. HDL efficiently removes cholesterol from tissues, esterifies it and transports it back to the liver where it gets excreted as bile salts.

Prostaglandins and prostacyclins have a beneficial effect as they cause disaggregation of platelets. Many agents and drugs that reduce serum cholesterol and hyperlipemia are beneficial by preventing the incidence of atherosclerosis. Among dietary methods, a diet low in fat and sugar is found helpful. Substitution of animal fat with vegetable oils like groundnut oil in the diet, which is increasingly being adopted in recent years to prevent atherosclerosis, is believed to depend upon the beneficial value of the polyunsaturated fatty acids. These polyunsaturated fatty acids held by phospholipids esterify cholesterol with the help of LCAT. Ester cholesterol is brought to the liver by HDL and efficiently catabolised. The fats containing unsaturated acids have relatively lower melting points than saturated fats which have, therefore, to be avoided.

HDL has the capacity of even stripping the deposited cholesterol from the smooth muscle cells of the arteries. Hence cardiologists and lipid biochemists working on atherosclerosis place a lot of importance on this class of lipoproteins because HDL scavenges the body cholesterol.

Investigation of total cholesterol, triacyl glycerol, lipoprotein electrophoretic profile, HDL-cholesterol, and HDL/LDL ratio are very helpful in the diagnostic biochemistry of atherosclerosis.

The incidence of atherosclerosis is several times higher in affluent countries like America when compared to the economically backward countries of Africa and South East Asia.

Hyperlipoproteinemias

Hyperlipoproteinemias are of two major types, hereditary or familial, and acquired.

There are as many as five types of familial hyperlipoproteinemias with two sub-types of Type II. In these, there is an elevation of one or more of lipoproteins in the blood, like chylomicrons, LDL, VLDL, VLDL remnants (IDL) due to hereditary deficiency of enzymes. Some types could cause atherosclerosis.

The acquired hyperlipoproteinemia is a sequel of some other disease or habit in which there could be an increase of blood cholesterol and LDL and lowering of HDL, e.g., uncontrolled diabetes mellitus, nephrotic syndrome, hypothyroidism (myxoedemia), alcoholism, etc.

Niemann Pick's disease

This is due to the accumulation of sphingomyelins in the cells of the liver and spleen due to enzyme deficiency. This is an inborn error of metabolism.

Gaucher's disease

In this hereditary disease, the cerebroside content of the spleen, liver and lymph nodes is increased, and consequently they are enlarged. This is also an inherited error of metabolism.

In addition to the above-mentioned defects of lipid metabolism, cholelithiasis or biliary calculi is another abnormality associated with the deposition of cholesterol in the gall bladder or in the bile ducts. Such calculi could cause obstructive jaundice and have to be surgically removed.

ALCOHOLISM AND ALDEHYDISM

According to Walker, alcohol can be a food, a drug or a poison depending on the dosage, the frequency of intake, the nature of the beverage and the condition of the stomach. In small quantities it is completely oxidised to acetyl CoA, and finally CO_2 and water like glucose, and gives ATP. At permissiable dosages under medical supervision, alcohol is most effective to relieve anxiety, tension, stress and irritability. Alcohol (seven per cent) stimulates gastric secretion.

According to Goldberg, Chairman of the Alcohol Research Institute, Sweden, the safe limit of alcohol intake is 7g/hour/70kg body weight. This has been endorsed by other workers in this field. If a person takes just 7g or less of alcohol per hour, his liver will metabolise it completely so that there is no alcohol going into systemic blood circulation. On this basis and taking into account the concentration of alcohol in beverages, the safe limits of intake of beer (about two per cent) are 4.5 litres, toddy and wine (10 per cent) 1.5 litres, and distilled spirits like arrack, rum, brandy and whisky (30 – 50 per cent) just 500 ml, and the quantities approved should be spaced for 24 hours.

Once the safe limits are exceeded, the liver cannot cope with the oxidation and removal of alcohol. Therefore, alcohol and aldehyde both get mixed with the systemic blood. Constant high levels of blood alcohol and aldehyde lead to alcoholism and aldehydism. Both exhibit psycho-dietetic effects, affect the brain and the health of the body. This is so because the alcohol reaching the brain through the blood decomposes the metalloporphyrin complexes which are semi-conductors and pulls back the synapses a little. Aldehyde, on the other hand, reacts with biogenic amines like nor-epinephrine, serotonine, etc., in view of its aldehyde group, and depletes these neuro-transmitters. Incidentally, alkaloids known as TIQ (tetre hydro isoquinoline) alkaloids are formed in the brain. Hence, the alcoholic experiences psychic effects. Alcohol is a depressant, but because it seems to remove inhibitions it appears to be a stimulant. In fact when the blood alcohol level rises to 700 mg/100 ml, the person can die.

Alcohol is first oxidised to acetaldehyde with alcohol dehydrogenase, an NAD^+ enzyme. The aldehyde in its turn is oxidised by aldehyde dehydrogenase, also a NAD^+ enzyme to acetyl CoA. Depending on the extent of oxidation, the formation of NADH and acetyl CoA alcohol and aldehyde exhibit dietetic effects. Continued consumption of alcohol leads to fatty liver. This may be due to increased synthesis is acetyl CoA and thereby fat, diminished oxidation of fatty acids, increased mobilisation of fat from fat depots and augmented esterification of fatty acids to triacyl glycerol. There is also fat accumulation in the kidneys, heart and even in the brain. There is also accumulation of cholesterol in the heart, aorta and brain. The accumulation of cholesterol in the arota leads to atherosclerosis and that in the brain causes a disturbance in the sodium channels and thereby in the brain activity.

Alcoholic fatty liver may become alcoholic hepatitis and ultimately alcoholic cirrhosis. Aldehyde also brings about these changes. The pulmonary surfactant of the lungs, thyroid function and testicular function are affected, and protein and carbohydrate metabolisms are impaired. Excess intake of alcohol leads to excessive production of NADH. Once the ratio of NADH/NAD increases there is accumulation of lactic acid. Because of a competition for renal clearance for lactic and uric acids, there will be hyperuricemia, which might lead to gout. Due to deranged fat metabolism there will be hypercholesterolemia and hypertriacylglycerolemia in alcoholism and aldehydism.

When excess alcohol is ingested, a significant amount is oxidised by an alternate pathway — the Microsomal Ethanol Oxidising System (MEOS) in the liver which is NADPH dependent. There is no production of energy. Hence, this is a wasteful process.

1) $CH_3CH_2OH \xrightarrow[(NAD^+)]{\text{alcohol dehydrogenase}} CH_3-CHO \xrightarrow[NAD^+]{\text{aldehyde dehydrogenase}} CH_3COS\,CoA$

Ethyl alcohol acetaldehyde acetyl CoA

2) MEOS Pathway

$CH_3CH_2OH + NADPH + H^+ + O_2 \rightarrow CH_3CHO + NADP^+ + 2H_2O$

In view of their deleterious manifestations, alcoholism and aldehydism should be prevented rather than cured. On no account should a person exceed the safe limits and allow his blood alcohol level to reach 700 mg/100 ml, when he would collapse.

Even though the safe limits of alcohol are fixed as 7g/hr/70kg, alcoholic beverages contain toxic substances like methanol, isopentanol, acetaldehyde and congeners, like low mol. wt alcohols, aldehydes, esters, and phenols. Hence, even if these beverages are considered safe with respect to their alcohol content if taken in prescribed quantities, they cannot be considered absolutely harmless as they contain these toxic substances.

14 Protein and Amino Acid Metabolism

Protein, the primary constituents of the body, may be structural or functional. All enzymes some hormones are proteins. Hence, the metabolism of proteins has a unique place in biochemistry as it is concerned with growth, repair and reconstruction of tissues, regulation of biochemical processes, transport of oxygen to the cells and functioning as antibodies for the defence of the body. While carbohydrate and lipid metabolisms are chiefly concerned with the supply of energy to the body, protein metabolism has a relatively minor role in this respect.

Though all proteins have 20 – 22 common amino acids, they might differ in the following:
1) The presence of a few amino acids,
2) the quantity of amino acids and thereby the molecular weight of proteins, and
3) the sequence of amino acids in the molecules.

Because of such differences, the proteins of the blood differ from each other; so also do the proteins of tissues and the proteins of different species. In fact, the proteins of one species injected into an animal of another species cause the production of antibodies in the animal receiving the foreign protein. Subsequent injection of the foreign protein may result in a serious toxic state, i.e., anaphylaxis, and prove fatal to the receiver. However, in spite of their diversity, proteins are noted for one common property – they all contain amino acids linked to one another by peptide bonds. Hence, the metabolism of proteins is essentially the metabolism of amino acids.

The proteins of food are converted by digestion in the alimentary canal to amino acids which are then absorbed. To this, amino acids from tissue destruction and those synthesised in the liver are added to a lesser extent. Thus the body has what is commonly known as the 'amino acid pool'. From this pool, the necessary amino acids are drawn for the synthesis of the required type of proteins, enzymes and hormones. Valuable non-protein nitrogenous substances like creatine, choline and heme are synthesised from the amino acids. Again if the availability of amino acids is more than that required for the functions referred to above, they are converted to keto acids in the liver and used for energy. In the growing child, protein is required for the synthesis of fresh tissue, while in the adult it is required for the compensation of the wear and tear of the body tissues. An animal is said to be in nitrogenous equilibrium or nitrogen balance if the nitrogen ingested, usually as proteins, is equal to the nitrogen excreted. It is in negative nitrogen balance, if more nitrogen is lost than is consumed, and in positive nitrogen balance if the nitrogen lost is less than the amount of intake.

Urinary nitrogen is the chief end product of protein metabolism. Urea represents about 90 per cent of the nitrogen excreted in man; animals excreting urea as the end product of protein metabolism are referred to as 'ureotelic'. Man and mammals are ureotelic. Other excretory products are creatinine, uric acid, ammonia and very little of amino acids. Birds and reptiles excrete uric acid as the end product of protein metabolism and are called 'uricotelic', while the teleostei fishes excreting ammonia as the end product of protein metabolism are called 'ammonotelic'. The sulphur of urine arises from the metabolism of the sulphur-containing amino acids. The phosphorus of urine has its origin in the metabolism of phospho-proteins and nucleo-proteins.

Amino acid pools

The body has a number of amino acid pools, e.g., the blood amino acid pool and the tissue amino acid pool. The composition of the pools will be different and is decided by the requirement of amino acids by the concerned tissue for the synthesis of the necessary type of protein or non-protein nitrogenous compounds. There is inter-communication between the different pools. However, the transport of amino acids into the cells takes place by an active transport mechanism which requires energy.

The source of amino acids, to a large extent, is the digestion of dietary proteins. The amino acids absorbed from the intestines reach the liver through the portal circulation and subsequently enter the systemic blood. Amino acids are synthesised chiefly by the liver while the breakdown of tissue proteins by cathepsins adds more amino acids to the pool. All the amino acids are L-amino acids and alpha amino acids (the human body cannot utilise D-amino acids while the bacterial cell wall contains some D-amino acids). However, in the course of metabolism, ß-amino acids like ß-alanine and ß-amino isobutyric acid, and gamma amino acids like gamma amino butyric acid are produced. The amino acids obtained from the three sources are added to the blood and reach the various tissues. The blood level of amino acid nitrogen in the fasting state is 4–4.5 mg per cent which may be elevated after a meal to about 7.0 mg per cent.

The amino acids from the pools are utilised for 1) the synthesis of body proteins like plasma proteins, tissue proteins, enzymes, hormones and milk proteins. They may also be used for the growth of tissues and formation of blood cells; 2) The synthesis of non-protein, nitrogenous substances of biological importance like choline, creatine, purine and pyrimidine, nucleotides, heme, melanins, non-protein hormones like epinephrine and small peptides like glutathione; and 3) energy production. Utilisation of amino acids for energy production takes place if energy cannot be derived from carbohydrates and fat. In such cases, the amino acids are deaminated in the liver to give keto acids and ammonia. The toxic ammonia is immediately converted to urea in the liver in 'ureotelic' animals and excreted in the urine, while the keto acids are converted into glucose and used for energy purposes.

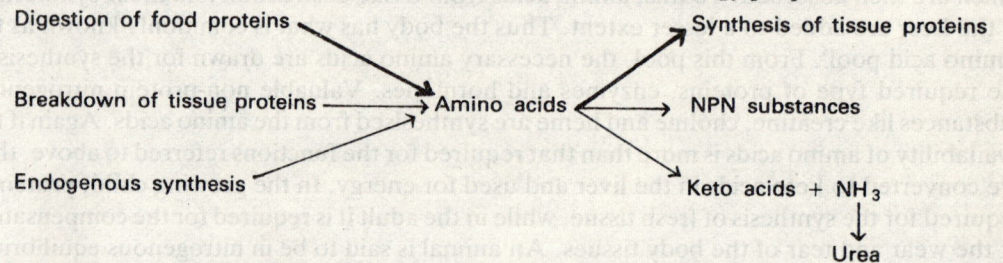

```
Digestion of food proteins ──────┐              ┌──→ Synthesis of tissue proteins
                                 │              │
Breakdown of tissue proteins ────┼─→ Amino acids ←──→ NPN substances
                                 │              │
Endogenous synthesis ────────────┘              └──→ Keto acids + NH₃
                                                         │
                                                         ↓
                                                        Urea
```

The question arises as to why there should be day to day synthesis of proteins in the body when the body itself has a stock of proteins. This is explained by the flux of the body proteins.

Dynamic flux of body proteins

There is at present overwhelming evidence to show that body proteins, like many other body constituents, are in a state of dynamic flux, i.e., constant synthesis and breakdown.

The classical view about protein metabolism proposed by Folin has now been discarded. According to Folin, the amounts of neutral sulphur and creatinine excreted in urine, which are rather constant, are an index of endogenous protein metabolism while the urea in urine, which is variable is an index of exogenous protein metabolism. Endogenous protein metabolism meant the metabolism of tissue proteins (the wear and tear quota), while exogenous protein metabolism was concerned with that of ingested protein. As the amount of creatinine in urine was a very small percentage of total nitrogen, Folin concluded that tissue proteins were relatively inert.

But in 1935, Borsook and Keighley indicated that endogenous protein metabolism was very active and there was a constant flux of body proteins. Schoenheimer in 1939 fed ^{15}N leucine and ^{15}N glycine to animals. Only a fraction of ^{15}N was found in the urine and the remainder was incorporated into tissue proteins. Subsequently, studies with tritium (^3H) and ^{35}S have also proved unequivocally that there is a constant synthesis and breakdown of body proteins. The retention of the isotopes in the tissue proteins isolated could be explained by a constant replacement of tissue proteins in normal subjects. Again the amino nitrogen also was found to change from one amino acid to another except lysine.

Dynamic equilibrium between tissue and plasma proteins (Whipple)

In addition to dynamic flux of body proteins, there is also a dynamic equilibrium between plasma proteins and tissue proteins. One can be converted into the other. This has been proved by using ^{15}N labelled plasma protein. ^{15}N labelled plasma protein injected intraperitoneally in dogs soon appears in blood. Again, labelled plasma protein injected into the circulation can be detected in lymph in about 20 minutes. Thus, in addition to the dynamic flux of body proteins, there is also a dynamic equilibrium between the plasma and tissue proteins.

Nitrogen equilibrium

As there is a constant synthesis and breakdown of proteins in the body, the anabolism and catabolism of proteins are equally important and should go hand in hand with proper equilibrium to the full benefit of the individual. As growth means addition of cells, which in turn means addition of proteins (nitrogen containing substances) in the body, retention of nitrogen in the body indicates growth. Between ingested nitrogen in the diet and nitrogen excreted in the urine and feces, less of nitrogen is excreted in growth, which shows that nitrogen is retained in the body. In this case, the nitrogen balance is said to be positive, e.g., in growth and pregnancy. After growth has ceased, the individual should have nitrogen equilibrium, i.e. the amount of nitrogen ingested and that excreted should be equal. If in any case the amount of nitrogen excreted is greater than that consumed, it means that tissue proteins are catabolised to a greater extent than they are formed. As a result, the body will lose weight consequent to the loss of tissue proteins. Such a condition is called *negative nitrogen balance* which is encountered in starvation, wasting diseases like tuberculosis, in post-operative conditions and in burns.

Storage of proteins and amino acids in the body

The body cannot store proteins in the sense in which it stores glucose as glycogen, or lipid as neutral fat. However, during periods of high protein intake, a considerable amount of proteins are stored in the liver and smaller quantities in other tissues. This type of stored protein is called *reserve protein or labile protein*. During starvation, after glycogen is depleted, about 85 per cent of body fat and 15 per cent of labile proteins are tapped for energy purposes. The utilisation of labile proteins for energy does not affect the functioning of the organ.

The body cannot store amino acids as well. Hence, even if one amino acid is lacking, protein synthesis is affected.

Nutritionally dispensable and nutritionally indispensable amino acids

From the nutritional point of view, amino acids are classified into essential and non-essential amino acids. The terms indispensable for essential, and dispensable for non-essential are in use now. An essential or indispensable amino acid cannot be synthesised by the body and has to be supplied in the diet. Deficiency of an indispensable amino acid would cause impairment of growth and lead to negative nitrogen balance or in some cases specific diseases. Nutritionally essential amino acids for humans are lysine, tryptophan, phenylalanine, methionine, threonine, leucine, isoleucine and valine (eight amino acids). Arginine and histidine are semi-dispensable for humans because, though arginine is synthesised in the body, it is not produced in adequate amounts while histidine is required only during the growth period.

The other amino acids like alanine, tyrosine and glutamic acid are non-essential. Non-essential does not mean not essential for the body but non-essential only in the diet since they can be synthesised in the body. Perhaps they are more essential for the body than even some of the *so-called* essential ones so that the body has learnt the trick of synthesising them. For example, tyrosine is termed a non-essential amino acid in the diet but it is known that it is very essential for the body (which can synthesise it from phenylalanine) for the production of hormones like thyroxine, epinephrine and nor-epinephrine. A similar argument holds good for glycine which is non-essential in the diet but is highly essential for the body for the synthesis of hemoglobin, purine nucleotides, creatine, glutathione, etc. (Glycine which is dispensable for man is an essential amino acid for the chick.)

Proteins and amino acids in urine

There are practically no proteins excreted in urine as they are not filtered by the glomerulus. However, in diseases of the kidneys, as in nephrotic syndrome, there will be proteinuria and chiefly albuminuria. Very small amounts of amino acids which have escaped metabolism are excreted in urine. In adults, about 150 – 200 mg of amino acid nitrogen are excreted in the urine in 24 hours. Hier and associates have reported an excretion of about 1,600 mg of amino acids per 24 hours in humans.

METABOLISM OF AMINO ACIDS (GENERAL)

Protein biosynthesis

As has been described earlier under protein and amino acid metabolism, the foremost use of amino acids is in the construction of body proteins. The amino acids are carried by the tRNAs to the site of protein synthesis, namely, polysomes (also called ergosomes), and the biosynthesis of proteins takes place with the assistance of mRNA and rRNA. This role of amino acids could be referred to as the general metabolism of all the amino acids or protein anabolism or biosynthesis of protein. (For more details see chapter on 'Biochemical genetics and biosynthesis of proteins'.)

Role of amino acids in the synthesis of non-protein nitrogenous compounds

Certain amino acids are required for the formation of non-protein nitrogenous compounds which are vital for biochemical functions.

i) Creatine: This is produced from the amino acids — glycine, arginine and methionine. This is required for functioning as energy reservoir of the muscle in the form of creatine phosphate. (For details refer 'Transmethylation of amino acids'.)

ii) Choline: This is another non-protein nitrogenous compound and can be synthesised from serine after being decarboxylated to ethanolamine. Choline is a valuable substance for the formation of phosphatidyl choline (lecithin) and is a lipotropic agent. It is also required for the production of acetyl choline which is needed for the transmission of nerve impulses. (For further details refer 'Transmethylation and vitamins of B group.)

iii) Heme: Glycine is required for the synthesis of the heme of hemoglobin (further details under 'Hemoglobin').

iv) Purine and pyrimidine nucleotides: Glycine, aspartic acid, and glutamine are required for the de novo synthesis of purine nucleotides while aspartic acid and glutamic acid are required for the synthesis of pyrimidine nucleotides (further details under 'Metabolism of Purine and Pyrimidine Derivatives').

v) Synthesis of melanins: Tyrosine is used for the synthesis of melanin pigments which give the characteristic colour to skin and hair.

vi) The above-mentioned substances are strictly non-protein nitrogenous substances. However, compounds like glutathione and small peptide hormones, for which amino acids are used, are also grouped under NPN compounds; though, however, these polypeptides are derived proteins.

Glutathione contains glutamic acid, glycine and cysteine. The synthesis of hormones like glucagon, oxytocin, vasopressin, adrenocorticotropic hormone, melanocyte stimulating hormones ($\propto$ MSH and ß MSH), para-thyroid hormone, calcitonin, ß lipotropin, endorphins, etc., requires amino acids.

Energy from amino acids (Glucogenic and ketogenic amino acids)

When an amino acid is not required for incorporation into protein and is catabolised, the carbon framework of the amino acid may be converted to glucose by gluconeogenesis and may meet the glucose needs to a limited extent. In prolonged starvation, glucose is synthesised from the carbon skeleton of amino acids to a significant extent. The glucogenic amino acids are glycine, alanine, serine, threonine, cysteine, methionine, valine, glutamic acid, aspartic acid, arginine, histidine, proline and hydroxy-proline.

If, instead of glucose, aceto-acetic acid is produced, the amino acid is referred to as a ketogenic amino acid, e.g., leucine and lysine.

Some are both glucogenic and ketogenic. They are isoleucine, phenylalanine and tyrosine. From glucose, produced or spared, and keto acids, the body can derive energy. Tryptophan is neither ketogenic nor glucogenic. As it gives alanine in the course of its metabolism, it can be considered glucogenic. Lysine is a relatively metabolically inert amino acid and is unclassified by some authors.

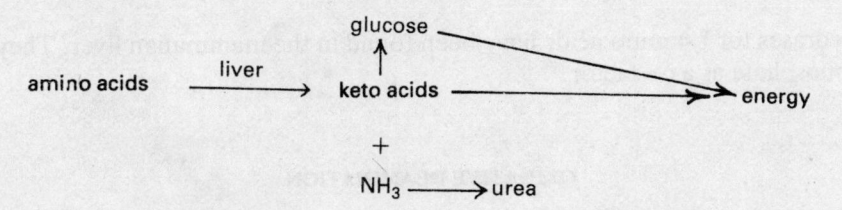

```
                                    glucose
                                       ↑
             liver
amino acids  ────→  keto acids ──────────────→ energy
                        +
                       NH₃ ──→ urea
```

The amount of glucose that could be obtained from 100 g of protein under conditions like starvation, treatment with phlorhizin, etc., has been worked out. The GN ratio, formerly called the DN ratio, is that between glucose (dextrose) and nitrogen. It was found to be 3.65.

G (glucose from the protein)
─────────────────────────── = 3.65
N (nitrogen from the protein)

As the nitrogen of the protein is about 16 per cent
G/16 = 3.65 So, G = 58.4
So, 100 g of protein (containing 16 g of nitrogen) could give about 58 g of glucose.

The liver and, to a lesser degree, the kidneys are the organs engaged in gluconeogenesis.

Biochemical transformations of amino acids

In the course of metabolism, the amino acids have to undergo different types of chemical transformations. These are deamination (non-oxidative and oxidative), desulphuration, transamination, transamidination, transpeptidation, decarboxylation and transmethylation.

Most of these reactions take place in the liver. In the course of deamination, desulphuration and transamination with glutamine or asparagine, ammonia is produced as a necessary evil. As ammonia is highly toxic, it is immediately converted to urea by the liver in many animals and excreted in urine.

DEAMINATION

Removal of the amino group of amino acids is referred to as deamination.

1) Non-oxidative deamination: This is met with in the deamination of hydroxy amino acids like serine and threonine. The enzyme involved is an amino acid dehydrase.

$$HO-CH_2-CH(NH_2)-COOH \xrightarrow{-H_2O} CH_2=C(NH_2)-COOH \longrightarrow$$
$$\text{L Serine}$$

$$CH_3-C(NH)-COOH \xrightarrow{+H_2O} CH_3-C(O)-COOH + NH_3$$
$$\text{pyruvic acid}$$

The dehydrases for L-amino acids have been found in the mammalian liver. They require pyridoxal phosphate as a co-factor.

OXIDATIVE DEAMINATION

The liver and the kidneys of many animals contain enzymes that liberate ammonia from amino acids and simultaneously oxidise the amino acid to a keto acid (hence the name oxidative deamination). The enzymes required for this are the L-amino acid dehydrogenases which are flavo-proteins (Fp) containing FMN as a co-factor.

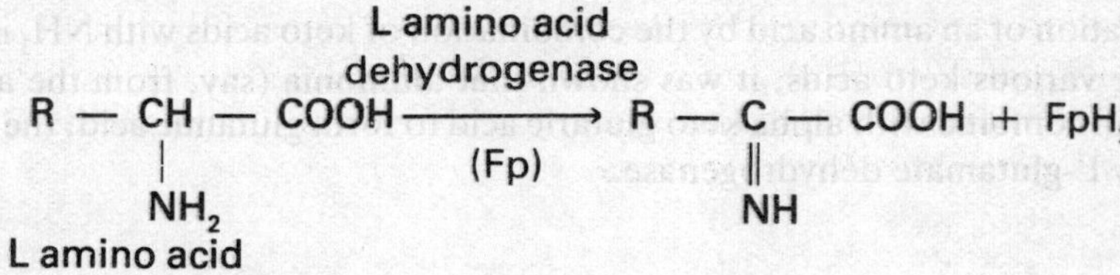

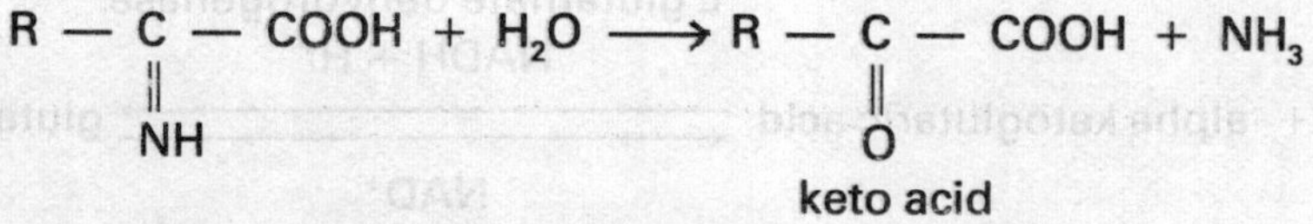

PROTEIN AND AMINO ACID METABOLISM

$$\text{R}-\underset{\underset{\text{NH}_2}{|}}{\text{CH}}-\text{COOH} \xrightarrow[\text{(Fp)}]{\text{L amino acid dehydrogenase}} \text{R}-\underset{\underset{\text{NH}}{\|}}{\text{C}}-\text{COOH} + \text{FpH}_2$$

L amino acid

$$\text{R}-\underset{\underset{\text{NH}}{\|}}{\text{C}}-\text{COOH} + \text{H}_2\text{O} \longrightarrow \text{R}-\underset{\underset{\text{O}}{\|}}{\text{C}}-\text{COOH} + \text{NH}_3$$

keto acid

The hydrolytic reaction takes place spontaneously.

$$\text{FpH}_2 + \text{O}_2 \longrightarrow \text{Fp} + \text{H}_2\text{O}_2$$

$$2\text{H}_2\text{O}_2 \xrightarrow{\text{catalase}} 2\text{H}_2\text{O} + \text{O}_2.$$

It is now believed that the oxidation of FpH_2 takes place through the formation of super oxide ion.

L-glutamic acid is subjected to oxidative deamination significantly not by L-amino acid dehydrogenase but by L-glutamate dehydrogenase which is widely distributed. Glutamate dehydrogenase uses either NAD^+ or NADP^+ as co-substrate and the reaction is reversible. However, the equilibrium constant favours glutamate formation. The reaction leading to ammonia release and formation of ($\propto$ KG) $\propto$ ketoglutarate is favoured in the liver for urea formation, as the product NH_4^+ is immediately used up for the formation of carbamoyl phosphate. Glutamate dehydrogenase reaction links the transamination of other amino acids with channelling of their nitrogen for urea formation. The C-skeleton of $\propto$KG gets oxidised in the TCA cycle.

Though tissues do not contain D-amino acids, they contain higher amounts of D-amino acid dehydrogenase than L-amino acid dehydrogenase. D-amino acid dehydrogenase also is a flavoprotein but contains FAD instead of FMN. Its function is probably to catabolise any D-amino acids of the diet to keto acids which may be converted subsequently to L-amino acids.

In both oxidative and non-oxidative deaminations, ammonia is the immediate toxic by-product which requires elimination.

DEAMIDATION

Amino acid amides like glutamine and asparagine undergo deamidation by the enzymes present in the liver called deamidases.

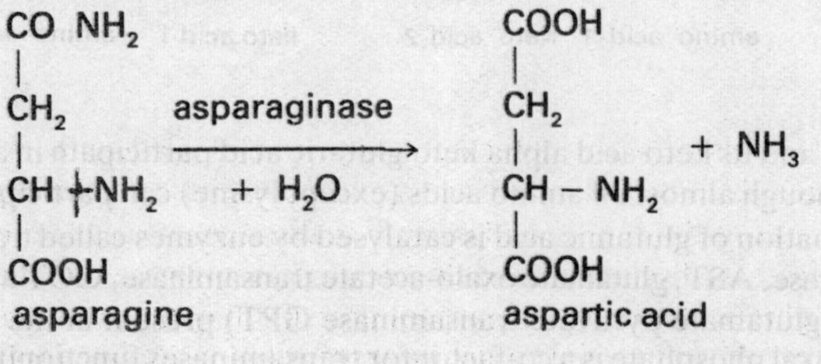

$$\underset{\text{asparagine}}{\overset{\text{CO NH}_2}{\underset{|}{\underset{\text{CH}_2}{\underset{|}{\underset{\text{CH}-\text{NH}_2}{\underset{|}{\text{COOH}}}}}}}} \xrightarrow[+ \text{H}_2\text{O}]{\text{asparaginase}} \underset{\text{aspartic acid}}{\overset{\text{COOH}}{\underset{|}{\underset{\text{CH}_2}{\underset{|}{\underset{\text{CH}-\text{NH}_2}{\underset{|}{\text{COOH}}}}}}}} + \text{NH}_3$$

AMINATION

The formation of an amino acid by the combination of keto acids with NH_3 is called amination. Of the various keto acids, it was shown that ammonia (say, from the ammonium salt ingested) will combine with alpha keto glutaric acid to form glutamic acid, the reaction being catalysed by L-glutamate dehydrogenase.

$$NH_3 + \text{alpha ketoglutaric acid} \underset{NAD^+}{\overset{\substack{\text{L glutamate dehydrogenase}\\ NADH + H^+}}{\rightleftharpoons}} \text{glutamic acid}$$

Thus, ingested ammonia is only converted to the amino group of glutamic acid to begin with. But since glutamic acid can easily transaminate with keto acids to form other amino acids (except lysine), all amino acids except lysine can be formed in this manner.

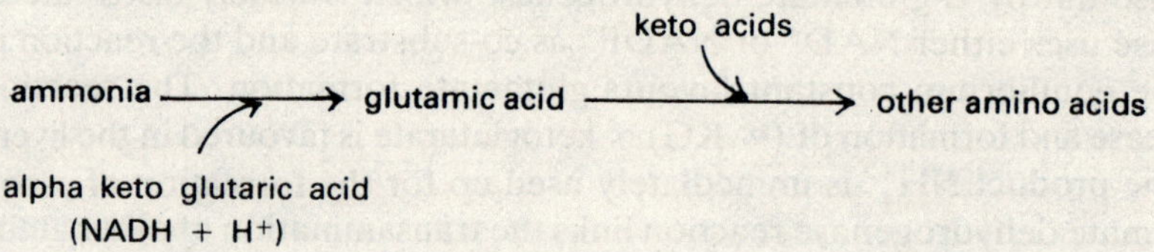

$$\text{ammonia} \longrightarrow \text{glutamic acid} \xrightarrow{\text{keto acids}} \text{other amino acids}$$
$$\underset{(NADH + H^+)}{\text{alpha keto glutaric acid}}$$

TRANSAMINATION

An amino acid and a keto acid participate in this type of reaction. The amino group of the amino acid is reversibly exchanged with the keto group of the keto acid. The general reaction may be represented as follows:

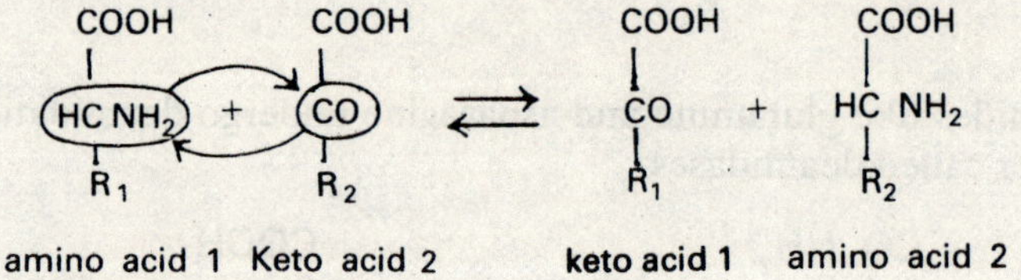

$$\underset{\text{amino acid 1}}{\overset{COOH}{\underset{R_1}{|}}\!\!\!\overset{|}{HCNH_2}} + \underset{\text{Keto acid 2}}{\overset{COOH}{\underset{R_2}{|}}\!\!\!\overset{|}{CO}} \rightleftharpoons \underset{\text{keto acid 1}}{\overset{COOH}{\underset{R_1}{|}}\!\!\!\overset{|}{CO}} + \underset{\text{amino acid 2}}{\overset{COOH}{\underset{R_2}{|}}\!\!\!\overset{|}{HCNH_2}}$$

Glutamic acid and its keto acid alpha keto glutaric acid participate in a large number of transaminations, though almost all amino acids (except lysine) can participate in transaminations. The transamination of glutamic acid is catalysed by enzymes called transaminases, Aspartate amino transferase, AST, glutamate oxalo-acetate transaminase, GOT and Alanine amino transferase, ALT (glutamate pyruvate transaminase GPT) present in the liver, muscle and other tissues. Pyridoxal phosphate is a co-factor for transaminases functioning as an amino carrier.

TRANSDEAMINATION

The glutamate dehydrogenase reaction is considered even more important than L-amino acid dehydrogenase reactions in the conversion of amino acids to keto acids. While L-amino acid dehydrogenase is present in the liver and kidneys, its activity is very low. L-glutamate dehydrogenase is widely distributed in the tissues and its activity is very high. Glutamic acid will be acted upon by glutamate dehydrogenase to form NH_3 and alpha keto glutarate. The other amino acids could transaminate with alpha keto glutarate to give glutamate which will further be acted upon by the glutamate dehydrogenase. Thus, the elimination of NH_2 of amino acids could be considered to be routed through glutamic acid.

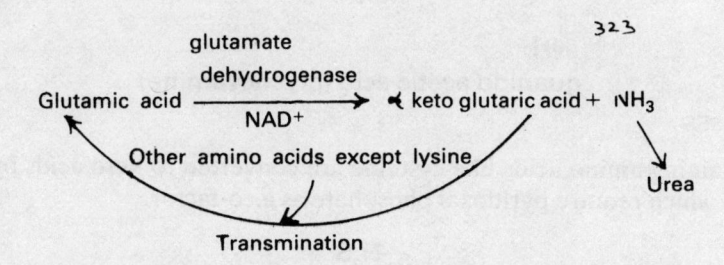

As this scheme combines the transamination of amino acids and the deamination of glutamic acid, this is refered to as transdeamination.

DECARBOXYLATION

Enzymes which bring about decarboxylation of amino acids have been found in the liver, kidneys and brain. These are called decarboxylases. Pyridoxal phosphate acts as the co-factor for this enzyme.

$$\text{Histidine} \xrightarrow[\text{liver}]{\text{decarboxylation}} \text{Histamine} + CO_2$$

$$\text{Glutamic acid} \xrightarrow[\substack{\text{brain} \\ \text{(glutamate decarboxylase)}}]{\text{decarboxylation}} \text{gamma amino butyric acid} + CO_2$$
$$\text{GABA}$$

TRANSPEPTIDATION

The synthesis of peptide bonds by the enzymes by inserting an amino acid into the peptide chain or by adding one amino acid to another is called transpeptidation. In mammalian liver, kidney and brain, there is a gamma glutamyl transpeptidase that transfers this group from glutathione to various amino acids.

$$\text{Glutathione} + \text{L Leucine} \longrightarrow \text{Glutamylleucine} + \text{Cysteinyl glycine}$$

This means the transfer of $-\underset{\underset{NH}{\|}}{C} - NH_2$ group to a molecule. The trans-amidinase system has been found to occur in the kidney and pancreas.

Arginine and glycine react under the influence of the enzyme in the kidneys to form glycocyamine or guanido acetic acid. The amidino group of arginine is transferred to glycine.

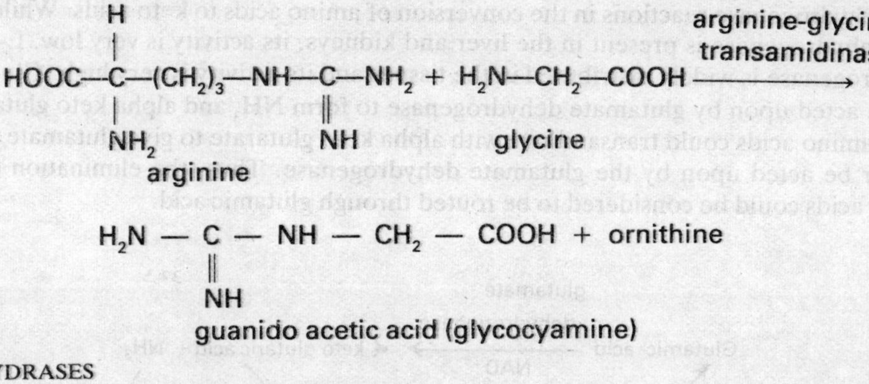

$$\underset{\text{arginine}}{HOOC-\underset{NH_2}{\overset{H}{C}}-(CH_2)_3-NH-\underset{NH}{\overset{\|}{C}}-NH_2} + \underset{\text{glycine}}{H_2N-CH_2-COOH} \xrightarrow{\text{arginine-glycine transamidinase}}$$

$$\underset{\text{guanido acetic acid (glycocyamine)}}{H_2N - \underset{NH}{\overset{\|}{C}} - NH - CH_2 - COOH} + \text{ornithine}$$

DESULPHYDRASES

Sulphur-containing amino acids like cysteine are converted to keto acids by the action of desulphydrases which require pyridoxal phosphate as a co-factor.

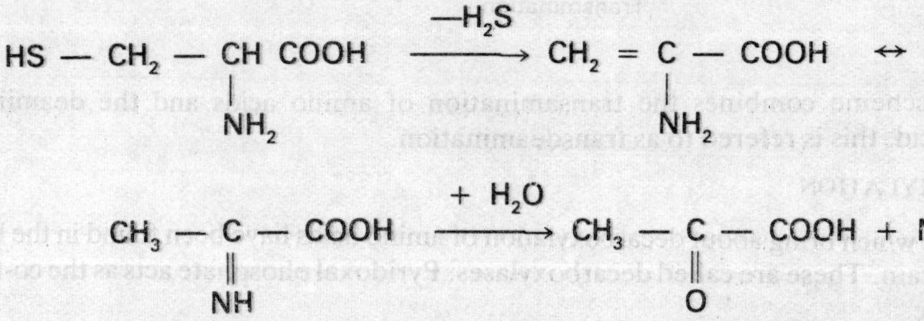

$$HS-CH_2-\underset{NH_2}{\overset{}{C}H}\,COOH \xrightarrow{-H_2S} CH_2=\underset{NH_2}{\overset{}{C}}-COOH \leftrightarrow$$

$$CH_3-\underset{NH}{\overset{\|}{C}}-COOH \xrightarrow{+H_2O} CH_3-\underset{O}{\overset{\|}{C}}-COOH + NH_3$$

PROTEIN AND AMINO ACID METABOLISM

Disposal of ammonia (urea cycle): It can be noted from the foregoing biochemical transformations of amino acids, that ammonia is produced in deamination, deamidation, desulphuration, transamination involving amino acid amides and transdeamination. These reactions take place chiefly in the liver. If the ammonia produced in these reactions mixes with the systemic blood, it can bring about highly toxic reactions. Even levels as low as 5 mg per cent ammonia in the blood are found to be fatal in rabbits. Hence, it is imperative that the ammonia produced during the metabolism of amino acids be removed immediately. This is the most important detoxication mechanism of the body and is effected by the conversion of toxic ammonia into harmless water-soluble urea in the liver by the operation of the Krebs-Henseleit urea cycle.

While the citric acid cycle is 'revenue' to the body, the Krebs-Henseleit urea cycle is 'expenditure'. ATP molecules are produced in the former while they are spent in the latter. For each molecule of ammonia to be converted to urea, three ATP molecules-four high energy bonds are used up. As many as five amino acids participate in this cycle. They are arginine, ornithine, citrulline, glutamic acid and aspartic acid.

UREA CYCLE

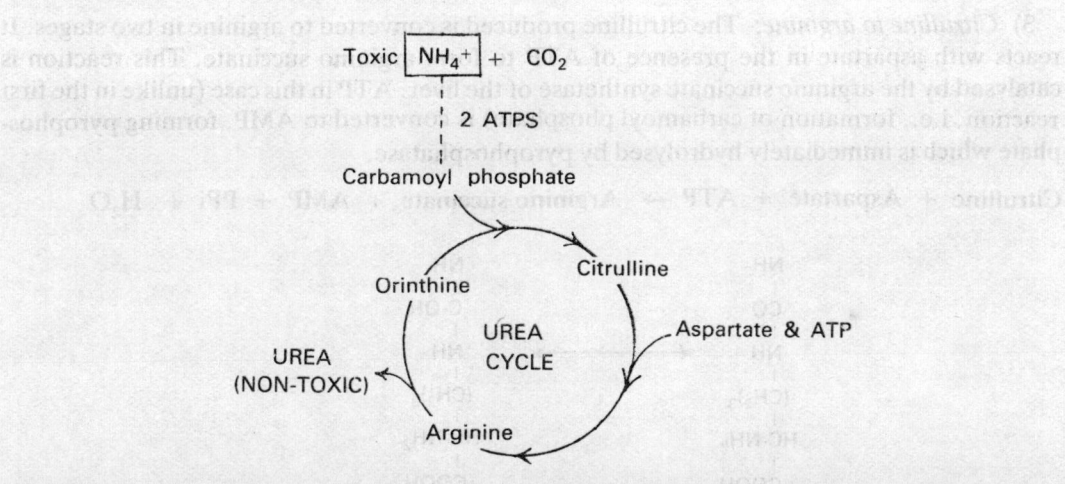

Toxic NH_4^+ + CO_2

2 ATPS

Carbamoyl phosphate

UREA CYCLE: Orinthine → Citrulline → Aspartate & ATP → Arginine → UREA (NON-TOXIC)

Formation of carbamoyl phosphate: 1) The initial step in the detoxication of ammonia is its conversion to carbamoyl phosphate by its union with carbon dioxide. Carbon dioxide condenses with one mol each of ammonium ion and phosphate derived from ATP to form carbamoyl phosphate. This reaction is catalysed by the enzyme carbamoyl phosphate synthetase present in the mitochondria. Actually two mol of ATP are needed to provide the driving force for this reaction. Mg^{2+} and N-acetyl glutamate (AGA) are also needed.

AGA is the allosteric activator for the enzyme. In its absence the second molecule of ATP will not bind to carbamoyl phosphate synthetase.

Just as acetyl CoA is the fuel for the TCA cycle, carbamoyl phosphate is the fuel for the urea cycle.

$$CO_2 + NH_4^+ \xrightarrow[\text{* 2 Mg}^{++}\text{ ADP}]{2 \text{ Mg}^{++}, \text{ATP, NGA}} \text{Carbamoyl phosphate}$$

* enzyme carbamoyl phosphate synthetase

2) Carbamoyl phosphate + L-Ornithine ⟶ Citrulline + Pi

The carbamoyl phosphate so formed enters the urea cycle and reacts with ornithine (produced in the cycle) to form citrulline, in the presence of ornithine carbamoyl transferase (also called ornithine transcarbamoylase). Ornithine transcarbamoylase is present in the mitochondria and the subsequent steps take place in cytosol. So, the citrulline formed has to be transported to cytosol by a transporter system.

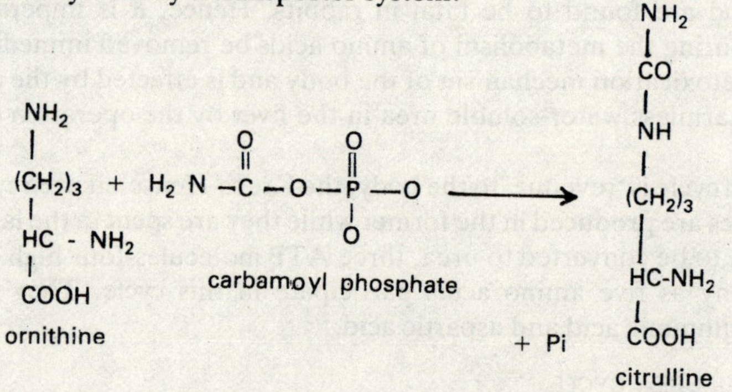

$$\text{ornithine} + \text{carbamoyl phosphate} \longrightarrow \text{citrulline} + P_i$$

3) Citrulline to arginine: The citrulline produced is converted to arginine in two stages. It reacts with aspartate in the presence of ATP to form arginino succinate. This reaction is catalysed by the arginino succinate synthetase of the liver. ATP in this case (unlike in the first reaction, i.e., formation of carbamoyl phosphate) is converted to AMP, forming pyrophosphate which is immediately hydrolysed by pyrophosphatase.

Citrulline + Aspartate + ATP → Arginino succinate + AMP + PPi + H_2O

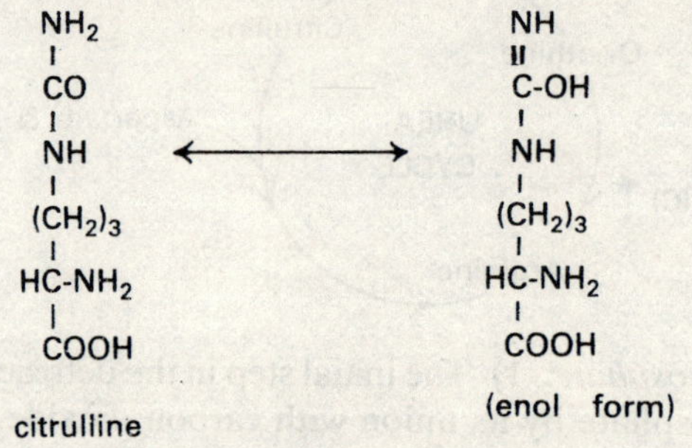

citrulline ⇌ (enol form)

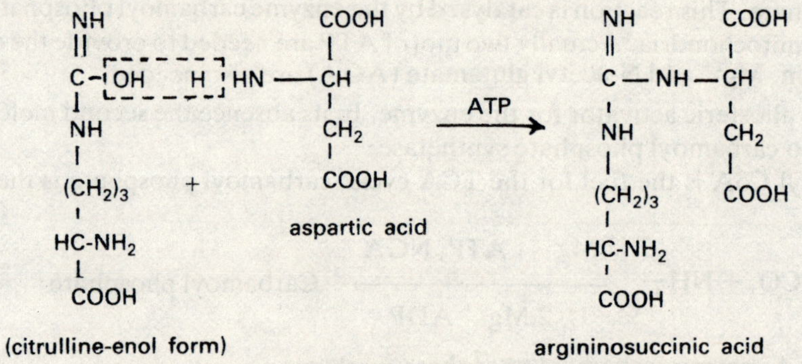

(citrulline-enol form) + aspartic acid —ATP→ argininosuccinic acid

The enzyme argininosuccinate synthetase is a cytosolic enzyme, and this, and subsequent steps take place in the cytosol part of the liver cells.

PROTEIN AND AMINO ACID METABOLISM

Arginino succinate is hydrolysed to arginine and fumarate by the enzyme arginino succinase.

```
   NH      COOH              NH
   ||       |                ||
   C—NH—CH                   C—NH₂           HOOC—C—H
   |       |                 |                    ||
   NH      H-CH      →       NH              H-C—COOH
   |       |                 |
   (CH₂)₃  COOH              (CH₂)₃     +
   |                         |
   HC-NH₂                    HC-NH₂
   |                         |                fumaric acid
   COOH                      COOH
Argininosuccinic acid        arginine
```

(The fumarate enters the citric acid cycle and is converted to oxalo-acetate which on transamination gives aspartate. This is required for the reaction of citrulline in the formation of arginino succinate.)

INTEGRATION OF TCA & UREA CYCLES

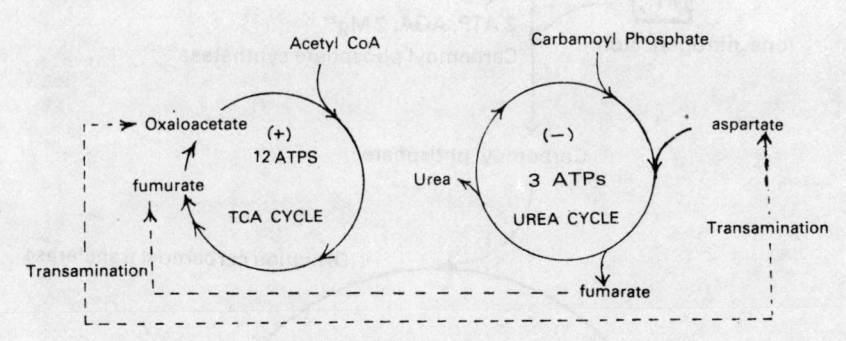

```
              Acetyl CoA              Carbamoyl Phosphate
         ┌─→ Oxaloacetate                          aspartate
         │         (+)                   (-)          ↑
         │       12 ATPS                            
         │   fumarate         Urea ←  3 ATPs
         │      ↑                     UREA CYCLE   Transamination
         │   TCA CYCLE
   Transamination
         │                                       fumarate
         └─────────────────────────────────────────┘
```

Conversion of arginine to ornithine and urea: Arginine is hydrolysed to ornithine and urea by the enzyme arginase present in the livers of the ureotelic animals.

$$\text{arginine} \xrightarrow{\text{arginase}} \text{ornithine} + \text{urea}$$

The formation ornithine completes and perpetuates the cycle. In a second turn of the cycle ornithine again taps carbamoyl phosphate into the cycle.

```
        NH₂                          NH₂               NH₂
         |                            |                 |
   HO    C = NH     arginase          C = NH    +      (CH₂)₃
         |           ──────→          |                 |
   H     NH           H₂O             OH               HC — NH₂
         |                            ↑                 |
         (CH₂)₃                      NH₂               COOH
         |                            |               ornithine
         CH — NH₂                     C = O
         |                            |
         COOH                         NH₂
         arginine                     urea
```

It may be noted that of the two nitrogen atoms of urea, one nitrogen atom comes from ammonia and the other from aspartic acid. Three molecules of ATP are utilised for the formation of one molecule of urea. As one of them is converted to AMP, it is equivalent to utilising four high energy bonds.

The integrated picture of the Krebs-Henseleit urea cycle is given in the figure.

The overall reaction is represented by this equation:

$$NH_4^+ + CO_2 + \text{aspartic acid} + H_2O + 3\,ATP \xrightarrow{2\,Mg^{2+}} \text{urea} + \text{fumaric acid} + 2\,ADP + AMP + PPi + 2\,Pi + H^+$$

Energetics of urea cycle: The direct reaction of CO_2 and NH_3 to form urea cannot take place as it is an endergonic reaction with increase of free energy. Hence, it is coupled with the 'group transfer' of ATP to make it exergonic with decrease of free energy.

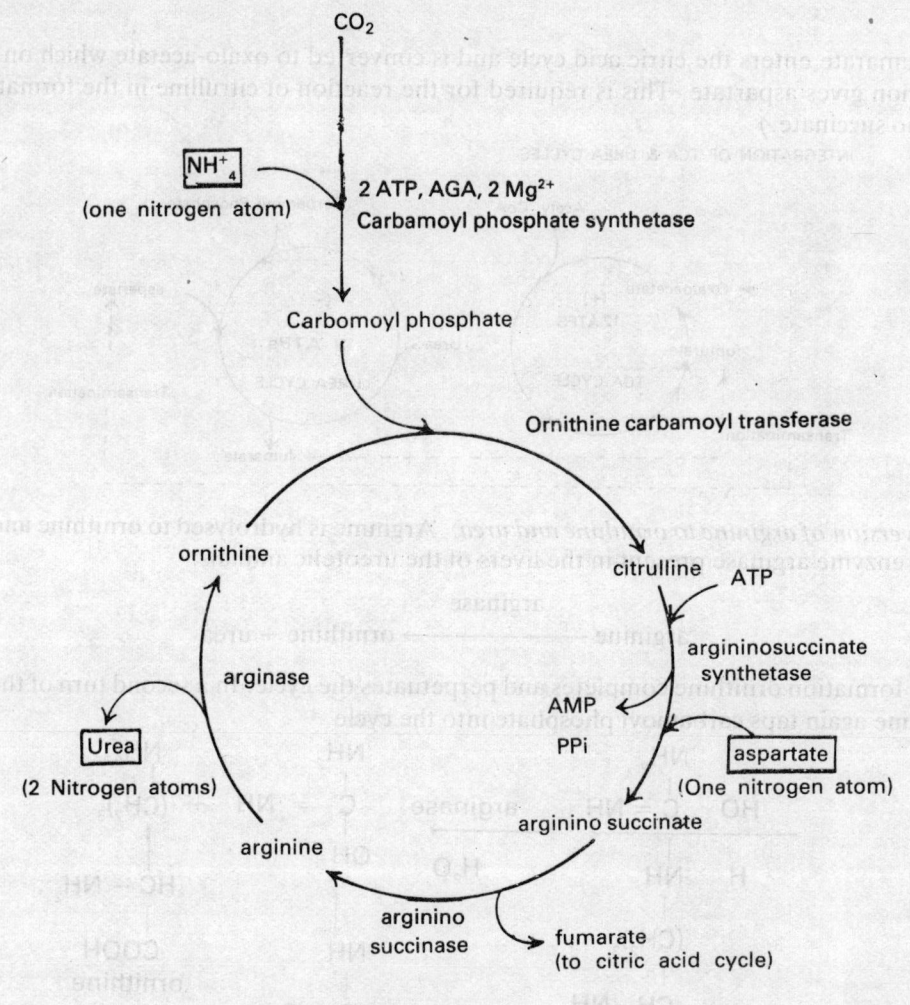

The urea produced from ammonia during the metabolism of amino acids is excreted in the urine. The daily excretion of urea is 20-30 g. Blood urea is an index of kidney function and is about 15-40 mg per ml of blood. When the kidney function is impaired as in acute nephritis, the blood urea level rises due to retention.

While the liver is undoubtedly the major organ in eliminating ammonia as urea, it has been shown that the brain and mammary glands can also convert small amounts of NH_3 to urea. However, the brain converts any ammonia reaching it into glutamine rather than urea.

$$\text{Glutamic acid} + NH_3 \xrightarrow{ATP} \text{glutamine}$$

The detoxication of NH_3 in the brain is thus effected through the formation of glutamine.

Disorders of urea cycle: Deficiency of one or the other of the five enzymes of the urea cycle results in rare disorders associated with ammonia intoxication, vomiting, ataxia, irritability and mental retardation. Most of them are fatal during infancy. They are of the following types:

1) Hyperammonemia Type I: This is a familial disorder associated with the deficiency of carbamoyl phosphate synthase. Only one case has been reported so far.

2) Hyperammonemia Type II: This is an X-chromosome-linked hereditary condition associated with the deficiency of ornithine transcarbamoylase. The glutamine levels get elevated in the blood, since the NH_4^+ has to be detoxicated. The ammonia level in serum rises and symptoms of ammonia toxicity develop. Infants develop an aversion for protein foods.

3) Citrullinemia: This is also an inherited disease and is associated with citrulline excretion in urine. The enzyme argininosuccinate synthase is absent.

4) Arginino-succinic aciduria: This is due to absence of arginino succinase in the liver.

TRANSMETHYLATION

The transfer of the methyl group from methionine to the acceptors is called transmethylation. Creatine and choline are the two very important non-protein nitrogenous substances synthesised by transmethylation by methionine. Before donating its methyl group, methionine has to be converted to S-adenosyl methionine (Cantoni), also called 'active methionine', in the presence of ATP.

$$CH_3-S-CH_2-CH_2-\underset{\underset{NH_2}{|}}{CH}-COOH \text{ (This cannot give its } -CH_3)$$

$$\text{Adenine} - \text{Ribose} - CH_2 - \underset{\underset{CH_3}{|}}{S^+} - CH_2 - CH_2 - \underset{\underset{NH_2}{|}}{CH} - COOH$$

(active methionine)

$$\text{Methionine} + ATP \xrightarrow[\text{Methionine adenosyl transferase}]{H_2O \quad\quad Pi + PPi} \text{S-adenosyl methionine}$$

(active methionine)

Glycocyamine is methylated by 'active methionine' to form creatine phosphate in the liver, catalysed by guanido-acetate methyl transferase (a specific transmethylase). The reaction requires ATP and GSH.

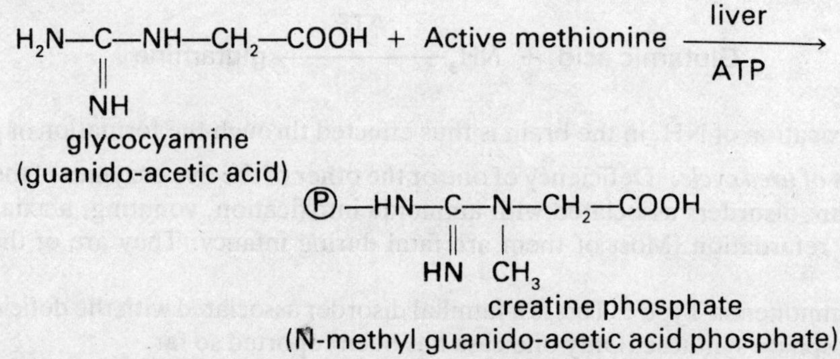

$$H_2N-\underset{\underset{NH}{\|}}{C}-NH-CH_2-COOH + \text{Active methionine} \xrightarrow[ATP]{liver}$$

glycocyamine
(guanido-acetic acid)

$$\text{\textcircled{P}}-HN-\underset{\underset{HN}{\|}}{C}-\underset{CH_3}{N}-CH_2-COOH$$

creatine phosphate
(N-methyl guanido-acetic acid phosphate)

It is obvious that creatine synthesis in the body takes place in two steps, first by the guanido-acetic acid in the kidneys with the participation of arginine and glycine and the subsequent methylation of guanido-acetic acid to N-methyl guanido-acetic acid (creatine) in the liver. When methionine labelled with deuterium was fed to animals, the deuterium could be traced in creatine. Recently it has been shown that the pancreas can do the functions of both the kidneys and the liver and can perform the total synthesis of creatine. Its role, however, is not significant.

Creatine phosphate easily undergoes non-enzymatic conversion to creatinine in the muscle.

Under the utilisation of amino acids it has been indicated that they are required for the synthesis of non-protein nitrogenous compounds like choline.

Choline (Trimethyl amino ethanol)

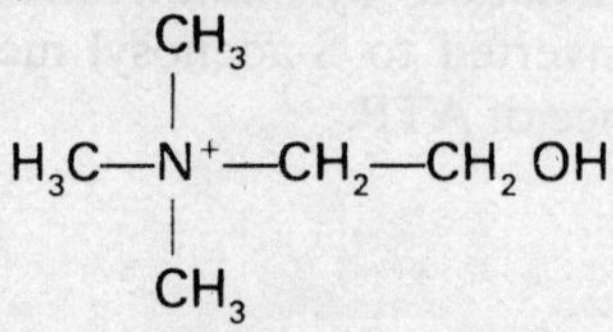

$$H_3C-\underset{\underset{CH_3}{|}}{\overset{\overset{CH_3}{|}}{N^+}}-CH_2-CH_2OH$$

Choline is synthesised from the amino acid serine which is decarboxylated in a pyridoxal phosphate – dependent reaction to ethanolamine. The ethanolamine is progressively methylated to choline with S-adenosyl methionine by the process of transmethylation. When methionine labled with deuterium is fed to animals, the label is traced in choline in addition to creatine.

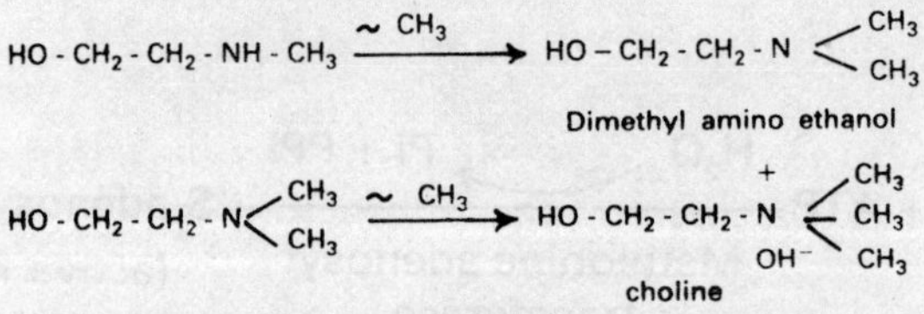

$$HO-CH_2-\underset{NH_2}{CH}-COOH \xrightarrow{-CO_2} HO\ CH_2-CH_2-NH_2$$
Serine   ethanolamine

$$HO-CH_2-CH_2-NH_2 \xrightarrow[\text{(active methionine)}]{\sim CH_3} HO-CH_2-CH_2-NH-CH_3$$
methyl amino ethanol

$$HO-CH_2-CH_2-NH-CH_3 \xrightarrow{\sim CH_3} HO-CH_2-CH_2-N\overset{CH_3}{\underset{CH_3}{<}}$$
Dimethyl amino ethanol

$$HO-CH_2-CH_2-N\overset{CH_3}{\underset{CH_3}{<}} \xrightarrow{\sim CH_3} HO-CH_2-CH_2-\overset{+}{N}\overset{CH_3}{\underset{CH_3}{\underset{OH^-}{<}}}CH_3$$
choline

Again, epinephrine (adrenaline) is obtained by transmethylation of nor-epinephrine (noradrenaline).

The transmethylations involving methionine can be summed up in the following diagram:

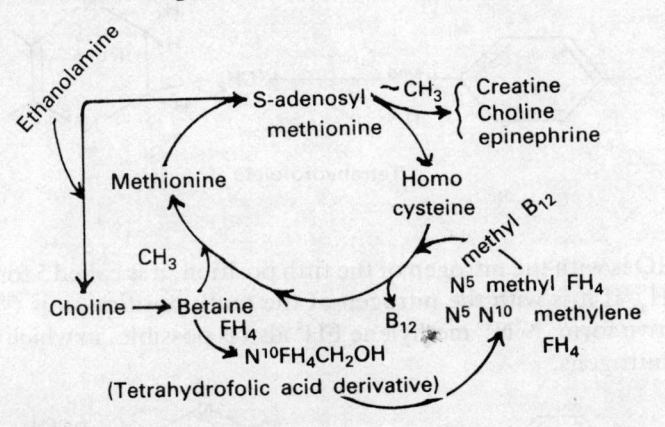

[Diagram: Methionine cycle showing Ethanolamine → Methionine → S-adenosyl methionine → (−CH₃ to Creatine, Choline, epinephrine) → Homocysteine → (methyl B₁₂, N⁵ methyl FH₄, N⁵ N¹⁰ methylene FH₄, B₁₂) → N¹⁰FH₄CH₂OH (Tetrahydrofolic acid derivative) → Choline → Betaine (FH₄) → back to Methionine with CH₃]

TRANS-SULPHURATION

The transfer of the —SH group from homo-cysteine to serine to form cysteine, with pyridoxal phosphate as the co-factor, is called trans-sulphuration.

Amino acids in detoxication

Most detoxications (removal of toxic substances) take place in the liver. The amino acid glycine detoxicates benzoic acid. Phenyl acetic acid is detoxicated by glutamine in man, glycine in dog and ornithine in birds. Cysteine detoxicates bromo benzene.

One carbon metabolism

The one carbon metabolism includes changes in which units with one carbon are removed or added in the course of metabolism. This could be discussed either under folic acid or amino acid metabolism, as both folic acid and amino acids are involved in one carbon metabolism.

The untis of one-carbon (1-C) are of two types:

a) $- CH_2OH; - CHO, - CH_2 -, - CH =, - CH = NH$ and $> CH - SH$

b) $- CH_3$

All the groups referred to under (a) are transferred to acceptors with the help of tetra-hydrofolic acid (FH_4).

After the introduction of $- CH_2OH$ in FH_2, the $- CH_2OH$ group can be converted to $- CH_3$ which can be donated by the FH_4 to other compounds. In other cases, the $- CH_3$ is donated by S-adenosyl methionine and betaine obtained by the oxidation of choline. Only the first group of betaine is removed as $- CH_3$ while the second and the third $- CH_3$ of betaine are removed as $—CH_2OH$ and have to be carried on by FH_4.

The 'one carbon' groups can be carried either by nitrogen in the fifth position or the tenth position or jointly by the nitrogens of the fifth and tenth positions of FH_4.

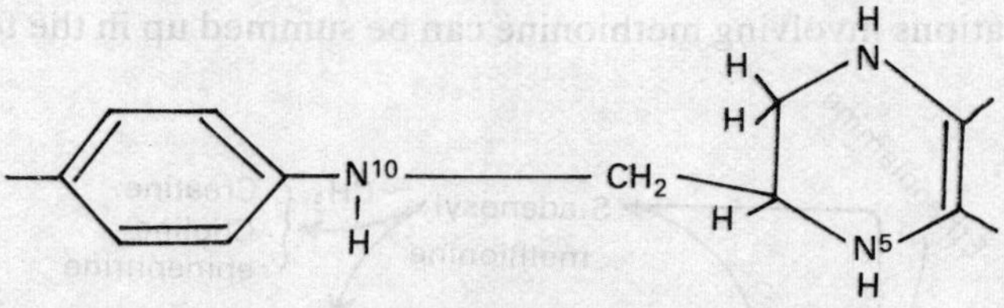

Tetrahydrofolate

For e.g., if —CHO is with the nitrogen of the fifth position, it is called 5 formyl (f^5) tetra-hydrofolic acid, f^5FH_4. If it is with the nitrogen of the tenth position, it is $f^{10}FH_4$. $f^{10}FH_4$ is the metabolically active form. N^5N^{10} methylene FH^4 also is possible, in which —CH_2— is jointly held by the two nitrogens.

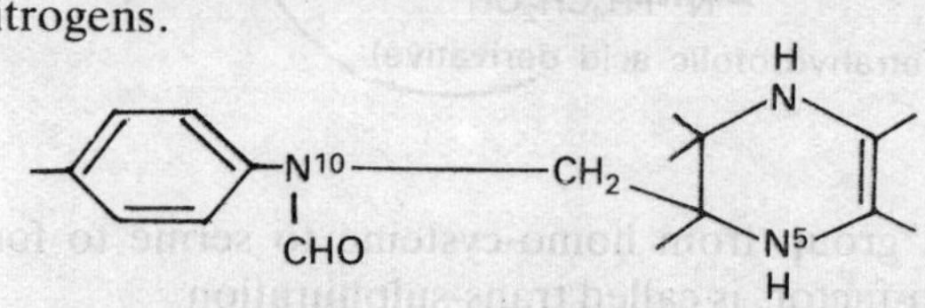

$f^{10} FH_4$ (active)

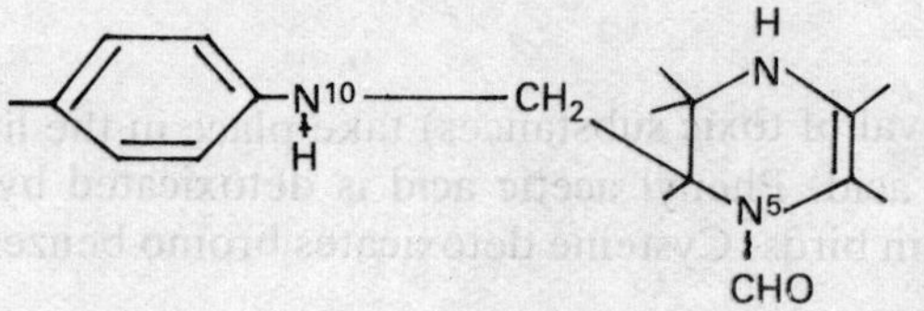

$f^5 FH_4$

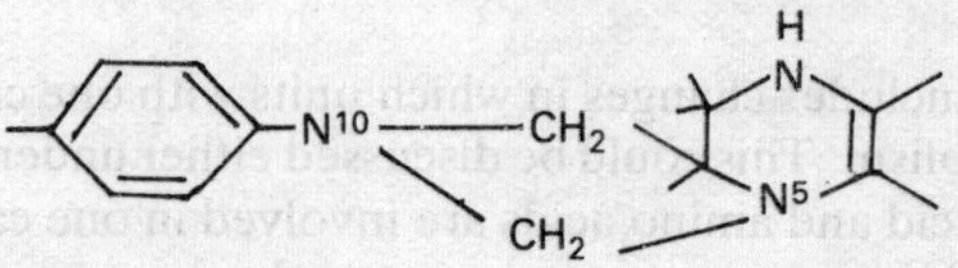

$N^5 N^{10}$ methylene FH_4

f^5FH_4 can however be converted to $f^{10}FH_4$. Again the groups — CH_2OH, — CHO and — COOH borne on FH_4 are inter-convertible. — CH_2OH on FH_4 can also be reduced to — CH_3. The various substances which donate 1-carbon to FH_4 are listed below:

Formate
Formaldehyde normally does not occur in the body.
Methanol
Acetone
Glycine
Serine
Histidine (f^5)
Tryptophan
Choline → Betaine (2nd and 3rd steps)
Thymine
Hydroxy methyl Sarcosine
Hydroxy methyl glycine
δ amino levulinic acid

PROTEIN AND AMINO ACID METABOLISM

(AMP, GMP, Inosinic acid)   (from δ ALA)

```
acetone ──┐        FH₄      ┌── glycine
          │                 ├── ß carbon of sarcosine
          ├──→ (C) ────────→ serine
CH₃OH ────┘         ↑       ├── ß C of serine
                    │       └── Thymine    ← Uracil
                    │
                    │           homocysteine
                    │               ↓ B₁₂
                    │           → methionine
glycine ──┐         │
histidine ├──→ (CHO)
tryptophan┘     ↕ [o]           → N formyl methionine
             (CH₂OH)
                ↕ [o]                    ethanolamine ← serine
glycine ←   (CH₃)  ~CH₃ →  methyl amino ethanol
        FH₄                         ↓ ~CH₃
        CH₂OH
hydroxy methyl glycine          dimethyl amino ethanol
    [o] ↑                                ↓ ~CH₃
    sarcosine
        ↑    FH₄ CH₂ OH                Choline
    FH₄ ↑                                ↓ − [2H]
hydroxy methyl sarcosine
    [o] ↑
    dimethyl glycine
        ↑  ~CH₃    Betaine  ←  Betaine aldehyde
                              [o]
```

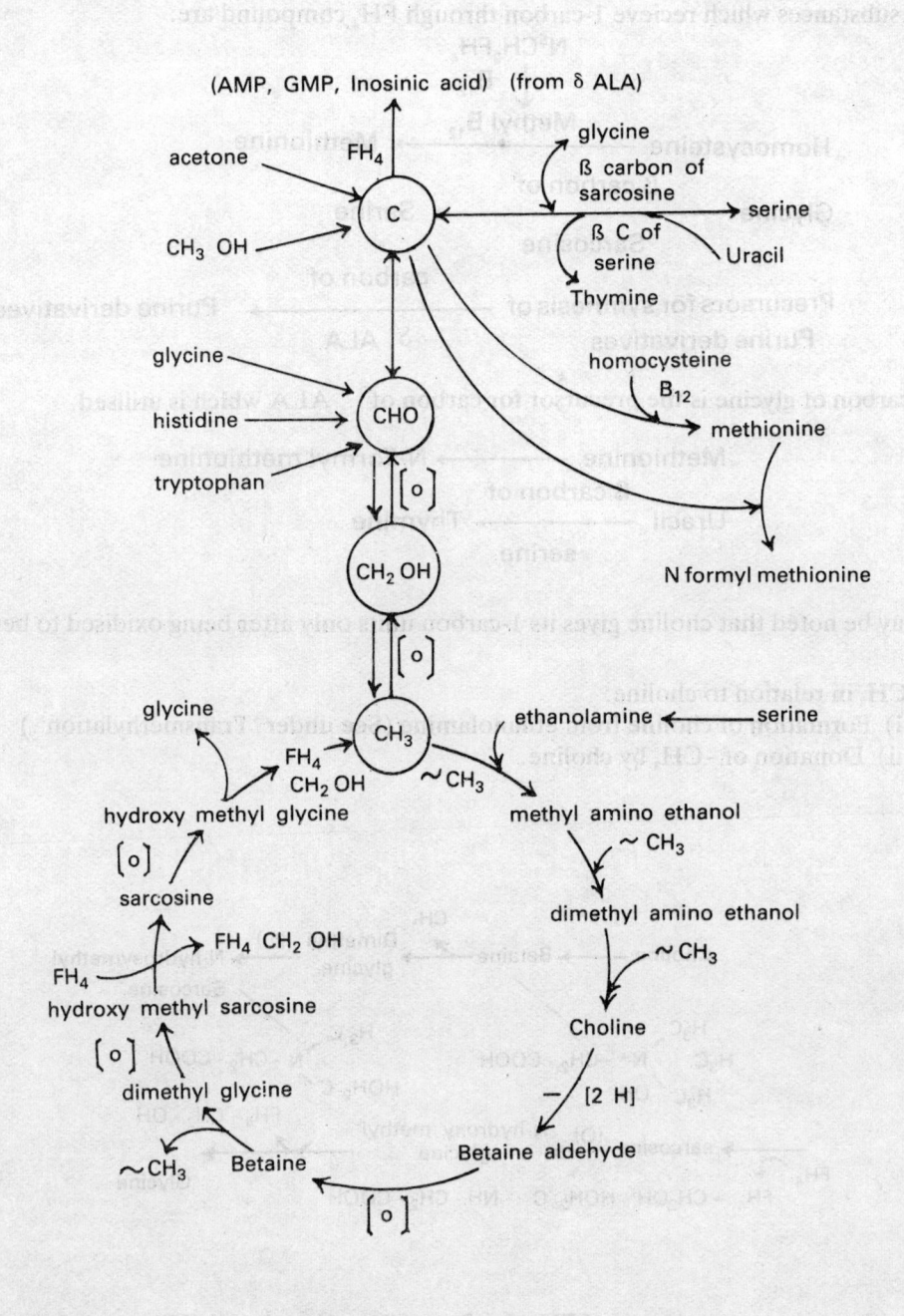

(In most cases the compound f^{10} or N^{10} hydroxymethyl is formed. In the case of histidine, it is N^5.)

The substances which recieve 1-carbon through FH_4 compound are:

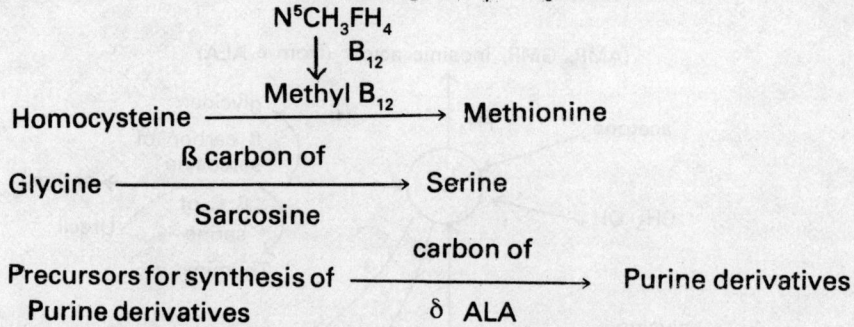

$$N^5CH_3FH_4 \downarrow B_{12}$$

Homocysteine $\xrightarrow{\text{Methyl } B_{12}}$ Methionine

Glycine $\xrightarrow{\text{ß carbon of Sarcosine}}$ Serine

Precursors for synthesis of Purine derivatives $\xrightarrow{\text{carbon of } \delta \text{ ALA}}$ Purine derivatives

a) carbon of glycine is the precursor for carbon of ALA which is utilised.

Methionine $\longrightarrow$ N-Formyl methionine

Uracil $\xrightarrow{\text{ß carbon of serine}}$ Thymine

It may be noted that choline gives its 1-carbon units only after being oxidised to betaine.

b) CH_3 in relation to choline:
  i) Formation of choline from ethanolamine (See under 'Transmethylation'.)
  ii) Donation of $-CH_3$ by choline:

Choline $\longrightarrow$ Betaine $\xrightarrow{CH_1}$ Dimethyl glycine $\xrightarrow{(O)}$ N-hydrosymethyl Sarcosine

H_3C
H_3C — N^+ —CH_2 - COOH
H_3C OH^-

H_3C
  N - CH_2 - COOH
$HOH_2 C$

FH_4 - CH_2 - OH

$\xrightarrow{}$ sarcosine $\xrightarrow{(O)}$ N-hydroxy methyl glycine

$FH_4 \downarrow$
$FH_4 - CH_2OH$ $HOH_2 C$ - NH - CH_2 - $COOH$

Glycine

Thus without the requirements of FH_4, the first $-CH_3$ of betaine could be transferred.

c) $-CH_3$ in relation to methionine:
  i) Formation of S-adenosyl methionine:
    In the case of methionine, as has been already indicated, it has to be converted to active methionine.

PROTEIN AND AMINO ACID METABOLISM

$$\text{Methionine} \xrightarrow{\text{ATP, methionine adenosyl transferase}} \text{Active methionine}$$

ii) Donation of $\sim CH_3$ by active methionine.

$$\text{Glycocyamine} \xrightarrow{\sim CH_3} \text{Creatine}$$

$$\text{Ethanolamine} \xrightarrow{\sim CH_3} \text{Methyl amino ethanol}$$

$$\text{Methylamino ethanol} \xrightarrow{\sim CH_3} \text{Dimethyl amino ethanol}$$

$$\text{Dimethyl amino ethanol} \xrightarrow{\sim CH_3} \text{Choline}$$

$$\text{Nor-epinephrine} \xrightarrow{\sim CH_3} \text{Epinephrine}$$

$$\text{Nicotinamide} \xrightarrow{\sim CH_3} \text{N methyl nicotinamide}$$

$-CH_3$ can also be oxidised to other 1 carbon fragments on FH_4.

The relationship of choline, betaine, methionine and active methionine is given in the following figure:

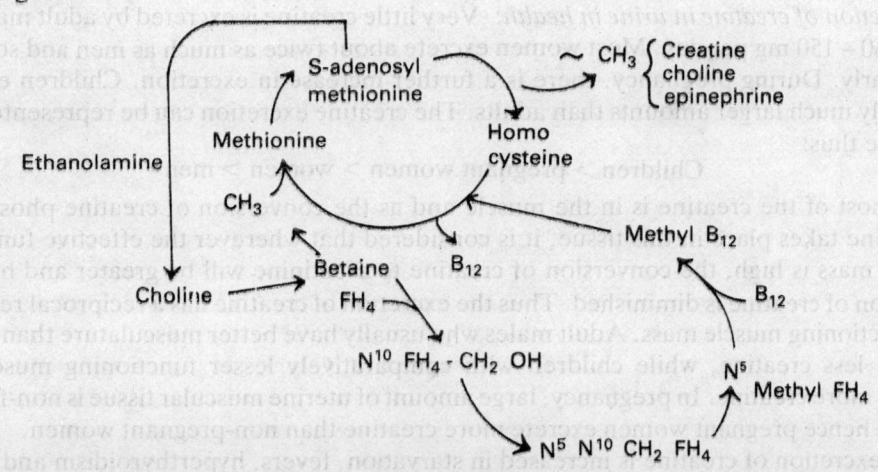

Vitamin B_{12} and 1-C metabolism: 1) The neogenesis of the methyl group:

N^5, N^{10} methylene tetra-hydrofolate is reduced to N^5 methyl tetrahydrofolate by a NAD^+-dependent reductase. This production of the methyl group is called the neogenesis of the same. N^5 methyl tetrahydrofolate is the prevalent form of folate transported in the blood.

2) The conversion of homocysteine to methionine:

N^5methyl FH_4 and B_{12} are required,

$$N^5 \text{ methyl } FH_4 + \text{deoxy adenosyl } B_{12} \longrightarrow \text{methyl } B_{12}$$

$$\text{Methyl } B_{12} + \text{homocysteine} \longrightarrow \text{methionine} + B_{12}$$

These reactions are important in relieving the 'folate-trap' as N^5 methyl FH_4.

In the absence of the above reaction, folate is trapped as N^5 methyl FH_4 and is not available for 1-carbon transfers.

Creatine and creatinine

Creatine is widely distributed in animal tissues. About 120 grams of creatine and creatine phosphate are present in the human body. About 98 per cent of the total creatine in the body, is in the muscles and occurs as creatine phosphate. In the body creatine phosphate is efficiently converted to creatinine non-enzymically in the muscle.

Creatinine cannot be converted back to creatine and hence it is a waste product and is excreted in urine. Creatinine is present in the muscle in very small amounts. While 100 g of skeletal muscle may contain 300 – 500 mg of creatine phosphate, it may have only about 10 mg of creatinine.

$$HN=C\begin{array}{c}NH(P)\\N-CH_2-C=O\\|\\CH_3\end{array}\begin{array}{c}OH\\|\\\end{array} \xrightarrow{(P)} HN=C\begin{array}{c}NH\\\diagdown\\N-CH_2\\|\\CH_3\end{array}CO$$

Creatine phosphate   Creatinine

Excretion of creatine in urine in health: Very little creatine is excreted by adult males, i.e., about 60 – 150 mg per day. Most women excrete about twice as much as men and somewhat irregularly. During pregnancy, there is a further increase in excretion. Children eliminate regularly much larger amounts than adults. The creatine excretion can be represented quantity-wise thus:

Children > pregnant women > women > men

As most of the creatine is in the muscle and as the conversion of creatine phosphate to creatinine takes place in this tissue, it is considered that wherever the effective functioning muscle mass is high, the conversion of creatine to creatinine will be greater and hence the excretion of creatine is diminished. Thus the excretion of creatine has a reciprocal relation to the functioning muscle mass. Adult males who usually have better musculature than women, excrete less creatine, while children with comparatively lesser functioning muscle mass excrete more creatine. In pregnancy, large amount of uterine muscular tissue is non-functioning and hence pregnant women excrete more creatine than non-pregnant women.

The excretion of creatine is increased in starvation, fevers, hyperthyroidism and diseases affecting musculature, e.g., amyotonia congenita. Creatinine excretion will correspondingly decrease. The total amount of creatine and creatinine is almost constant.

While creatine is easily biosynthesised from glycine, arginine and methionine creatinine in the diet can supplement the creatine of the tissues. Non-vegetarians get creatine in the diet in the form of meat, meat gravy and soups. On a creatine-free diet, tissue creatine is the sole source of urinary creatinine.

Excretion of creatinine in urine: Creatinine is one of the normal non-protein nitrogenous compounds in urine excreted to the extent of 1 – 1.5 g per day. It is the end product of creatine catabolism. The amount of creatinine in urine is almost constant for an individual, provided the diet does not contain creatine or its derivatives. Hence, creatinine estimation is done to ensure that complete 24 hour samples of the patient's urine is being collected in the ward. The creatinine-coefficient is the amount of creatinine excreted expressed as mg per 24 hours per kg of body weight. Normal values for men are between 18 – 32 with an average of about 25, and for women the values are 9 – 26 with an average of 18. Thus creatinine excretion has a reciprocal relation to creatine excretion. The better the musculature, the greater is the creatine coefficient. Men, having better musculature, therefore excrete more of creatinine than women.

$$\text{Children} < \text{women} < \text{men}$$

The formation of creatinine from creatine phosphate depends predominantly on the functioning muscle mass and not on muscular activity. An athlete with well-developed muscles would excrete more creatinine than an ordinary man of the same weight. Again, the creatinine output of a person is about the same on a day of intense muscular exercise as on a day of rest, proving that muscular exertion does not significantly affect the daily excretion of creatinine.

Creatinine cannot be converted back to creatine in the body.

Sulphur metabolism

Sulphur is an important element in protein in addition to nitrogen. It is present in the sulphur-containing amino acids cysteine, cystine and methionine.

Sulphur is ingested in the form of the sulphur-containing amino acids present in food proteins, which contain on an average one per cent of sulphur. Small amounts are obtained as mucoproteins and sulphatides while the amount of inorganic sulphate ingested is very small.

Sulphur is required for the synthesis of tissue proteins and proteins like the keratins of hair and nail, the protein hormone insulin, vitamins like thiamine, lipoic acid and biotin, the important peptide, glutathione, coenzyme A and other substances like taurine, tauro cholic acid (bile acid), ergothionine and thiocyanates (of saliva). Sulphur, as active sulphate, is used for the synthesis of muco-polysaccharides like chondroitin sulphates and heparin. It is present in the cornea, connective tissues, tendons and vitreous humour. The excess sulphur is oxidised to sulphuric acid in the liver and excreted as sulphates in the urine. Sulphur is lost also through loss of hair, nails, bile and saliva.

Sulphur is excreted in three forms in urine to the extent of about 1 g/day.

$$\left. \begin{array}{l} \text{Inorganic sulphate} \\ \text{Ethereal sulphate} \\ \text{Neutral sulphur} \end{array} \right\} \text{Total sulphates}$$

The inorganic sulphate constitutes about 80 per cent (0.8 g/day) of total urinary sulphur and represents the sulphates of K^+, Ca^{2+} and NH_4^+

(Inorganic sulphate gives a white precipitate of $BaSO_4$ with $BaCl_2$ and HCl.)

Ethereal sulphates constitute about 8 per cent urinary sulphur (at .08 g/day) and consist of the sulphuric esters of certain phenols. The salts of sodium and potassium phenol sulphuric acid and indoxyl sulphuric acid are important ethereal sulphates. The potassium salt of indoxyl sulphuric acid is *indican*.

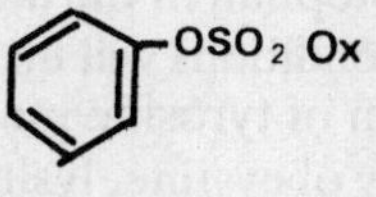

Salt of phenol sulphate
(x = Na or K)

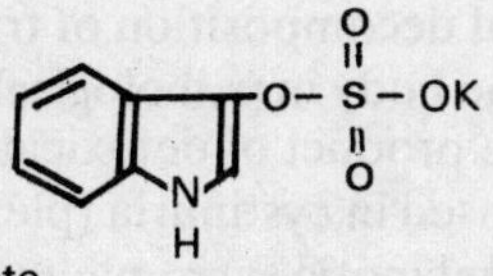

Indican

Ethereal sulphates do not give a precipitate with BaCl$_2$ and HCl. But on boiling with HCl, hydrolysis takes place and inorganic sulphate is released. Hence, after boiling with HCl ethereal sulphates give a precipitate with BaCl$_2$.

Neutral sulphur represents about 12 per cent of urinary sulphur (0.12 g/day). This is sometimes referred to as unoxidised sulphur. This fraction comprises sulphur-containing amino acids which have escaped metabolism, urochromes, thiocyanates, taurocholic acid, etc.

Quantities of inorganic and ethereal sulphates in urine run parallel with nitrogen excretion.

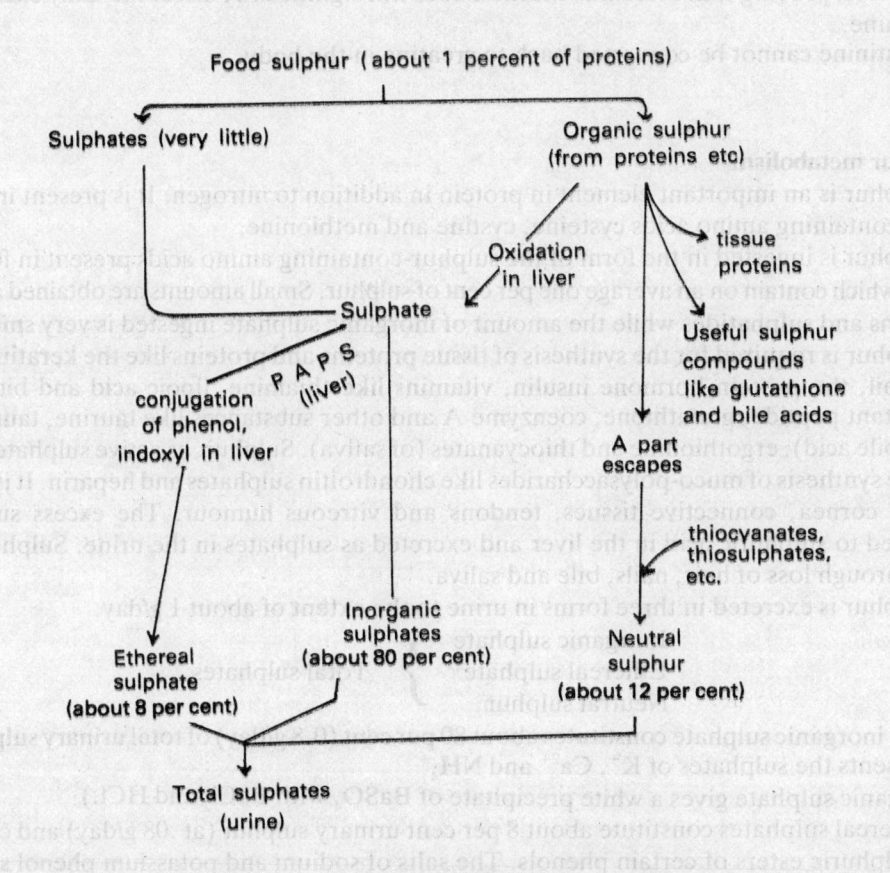

```
                Food sulphur (about 1 percent of proteins)
               /                                    \
    Sulphates (very little)              Organic sulphur
                                         (from proteins etc)
                          Oxidation              → tissue
                          in liver                 proteins
                        Sulphate
                    P                        → Useful sulphur
                    A                          compounds
                    P                          like glutathione
                    S                          and bile acids
           conjugation (liver)
           of phenol,                         A part
           indoxyl in liver                   escapes
                                              → thiocyanates,
                                                thiosulphates,
                                                etc.
        Ethereal         Inorganic         Neutral
        sulphate         sulphates         sulphur
        (about 8         (about 80         (about 12
         per cent)        per cent)         per cent)
                    \       |       /
                     Total sulphates
                        (urine)
```

Clinically, the ethereal sulphates and neutral sulphur are related to diseases. In acute intestinal obstruction due to the production and absorption of putrefactive products, carcinoma of liver, cholera, typhus, etc. the excretion of the etheral sulphates is increased. In cholera and typhus, sufficient indican is excreted to cause the urine to assume a bluish tinge on the surface, on standing. Bacterial decomposition of tryptophan in the large intestines or decomposition of pus anywhere in the body in pathological conditions will increase the excretion of indican. Phenol sulphate is the product of detoxication of tyrosine metabolites.

Neutral sulphur is increased in cystinuria (plenty of cystine, lysine, arginine), Fanconi syndrome (amino-aciduria) melanotic sarcoma, cyanide poisoning (cyanide being converted to thiocyanate) and even in chloroform anesthesia.

In general, sulphate excretion is increased when the catabolism of tissue protein is increased.

METABOLISM OF INDIVIDUAL AMINO ACIDS

Glycine

Glycine does not exhibit optical activity as it does not possess an asymmetric carbon atom unlike other L-amino acids. Glycine can be incorporated into protein. It is also readily deaminated by a specific enzyme glycine-oxidase which is flavoprotein to glyoxylic acid (CHO – COOH). Glyoxylic acid may be aminated back to glycine by transamination. It can be oxidised rapidly to formic and oxalic acid and CO_2 in the liver.

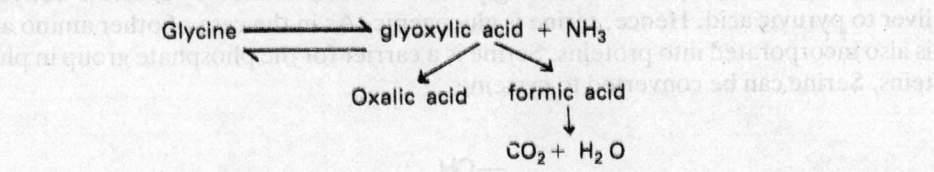

$$\text{Glycine} \rightleftharpoons \text{glyoxylic acid} + NH_3$$
$$\text{Oxalic acid} \quad \text{formic acid} \rightarrow CO_2 + H_2O$$

Glycine can be converted to serine and vice versa. (Refer 'One-Carbon Metabolism'.)

$$FH_4 + \text{Serine} \rightleftharpoons N^5N^{10} \text{methylene } FH_4 + \text{glycine}$$

Glycine is used for the synthesis of various substances of biological importance.

Purine derivatives: The carbon atoms 4,5 and nitrogen 7 of purine derivatives are supplied by glycine in the de novo synthesis of purine nucleotides.

Glycine is required for the synthesis of creatine, heme, glutathione and glycocholic acid. It is also used for the detoxication of benzoic acid as benzoyl glycine or hippuric acid.

Glycine is one of the glucogenic amino acids. The fact that glycine is converted to serine which can in turn be converted to pyruvic acid explains the incorporation of glycine carbon atoms into glucose and glycogen. Glycine stimulates gluconeogenesis and this role is called the 'protein effect' on gluconeogenesis.

Therapeutic value of glycine: As glycine is required for the synthesis of creatine, the feeding of glycine was tested for therapeutic purposes in certain conditions involving the musculature. In myasthenia gravis in which disturbances of creatine metabolism appear to be of secondary importance, the feeding of glycine has sometimes favourable results but not in all cases.

Summary of glycine metabolism

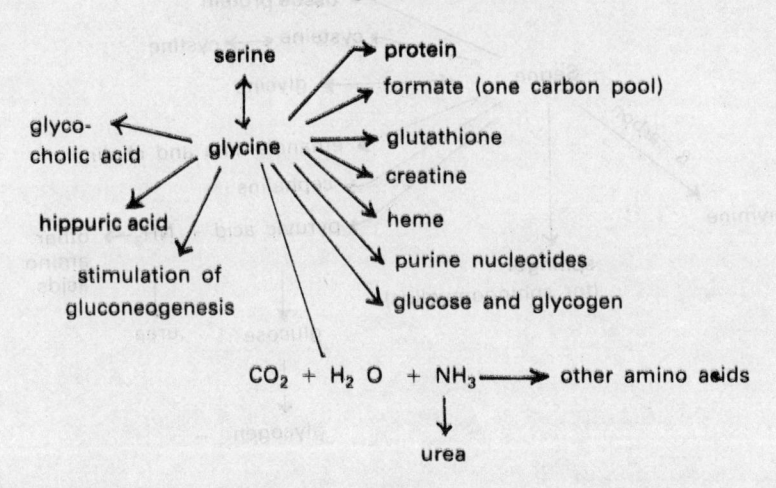

glycine →
- serine → protein
- formate (one carbon pool)
- glutathione
- creatine
- heme
- purine nucleotides
- glucose and glycogen
- glycocholic acid
- hippuric acid
- stimulation of gluconeogenesis
- $CO_2 + H_2O + NH_3 \rightarrow$ other amino acids → urea

The diseases associated with glycine metabolism are glycinuria and primary hyperoxaluria. In glycinuria there is excessive urinary excretion of glycine with a tendency to form oxalate renal stones. This disease is inborn and very rare. Primary hyperoxaluria is a metabolic disease characterised biochemically by a continuous and high excretion of oxalate in urine, the excess oxalate probably coming from glycine. There is recurrent infection of the urinary tract. In vitamin B_6 deficiency also, increased quantities of oxalate are excreted.

Serine

Serine can be easily formed from glycine. It is deaminated by L-serine dehydrase of the liver to pyruvic acid. Hence, serine is glucogenic. As in the case of other amino acids, serine is also incorporated into proteins. Serine is a carrier for the phosphate group in phospho-proteins. Serine can be converted to cysteine.

$$\text{Methionine} \xrightarrow{-CH_3} \text{Homocysteine}$$
$$\text{Homocysteine} + \text{Serine} \longrightarrow \text{Cystathionine}$$
$$\longrightarrow \text{Homoserine} + \text{Cysteine}$$

While the sulphur of cysteine comes from methionine, the rest of the molecule comes from serine.

Serine undergoes decarboxylation to form ethanolamine.

$$HO - CH_2 - \underset{\underset{NH_2}{|}}{CH} - COOH \xrightarrow{\text{decarboxylase}} HO - CH_2 - CH_2 - NH_2 + CO_2$$
$$\text{ethanolamine}$$

This is a very important conversion as the ethanolamine can be used for the formation of phosphatidyl ethanolamine. The ethanolamine can be progressively methylated to choline with S-adenosyl methionine. Thus, serine is useful for the synthesis of choline.

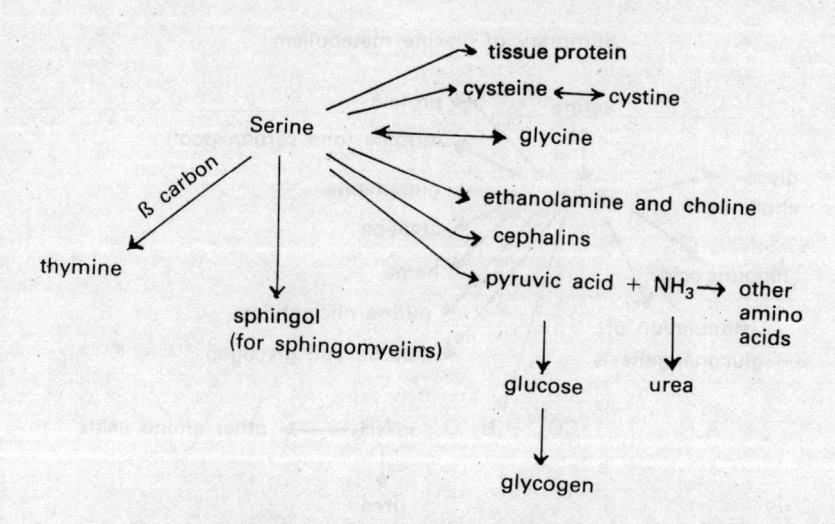

```
                            → tissue protein
                          → cysteine ⇌ cystine
              Serine    ←
                          → glycine
        ß carbon
                          → ethanolamine and choline
                          → cephalins
   thymine                 → pyruvic acid + NH₃ → other
                  ↓                    ↓         amino
              sphingol             glucose    urea  acids
         (for sphingomyelins)          ↓
                                   glycogen
```

The ß carbon atom of serine comes from sarcosine through the one carbon pool.

Alanine

Like other amino acids alanine occurs as L-amino acid in proteins. It is readily synthesised by the body. Alanine is glucogenic as it can be converted to pyruvic acid.

D-isomer of alanine is present as a polypeptide in the cell wall of *Lactobacillus casei*.

Alanine is deaminated by L-amino acid dehydrogenase and also by transamination with alpha keto glutaric acid.

L alanine + alpha ketoglutaric acid ⟶ pyruvic acid + L glutamic acid

The enzyme transaminase required for the above reaction is alanine amino transferase (glutamic pyruvic transaminase GPT). The serum ALT (GPT) level is increased in certain liver and heart diseases.

Together with glycine, it makes up a considerable fraction of the amino nitrogen in human plasma (see alanine cycle).

Beta alanine is a constituent of pantothenic acid and an end product of pyrimidine metabolism.

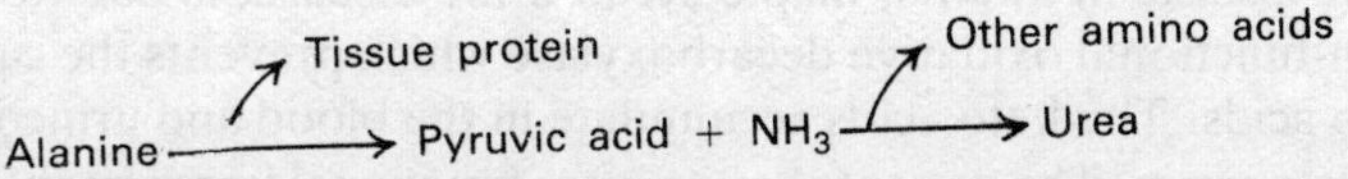

```
                    ↗ Tissue protein          ↗ Other amino acids
   Alanine ─────→ Pyruvic acid + NH₃ ─────→ Urea
```

Threonine

This amino acid has two asymmetric carbon atoms and exists in four isomeric forms. Threonine is an indispensable amino acid to humans though it is synthesised in yeast, *E. coli*, etc. from aspartic acid.

Threonine is deaminated by threonine dehydrase. It does not undergo transamination. Threonine is split into glycine and acetaldehyde. Dehydrogenation of threonine and subsequent spontaneous decarboxylation yields amino acetone (amino acetone can be converted to pyruvic acid in a number of steps). Threonine is therefore a glucogenic amino acid.

Like other amino acids threonine is required for incorporation into the proteins. Like serine, threonine is a carrier for the phosphate group in phospho proteins.

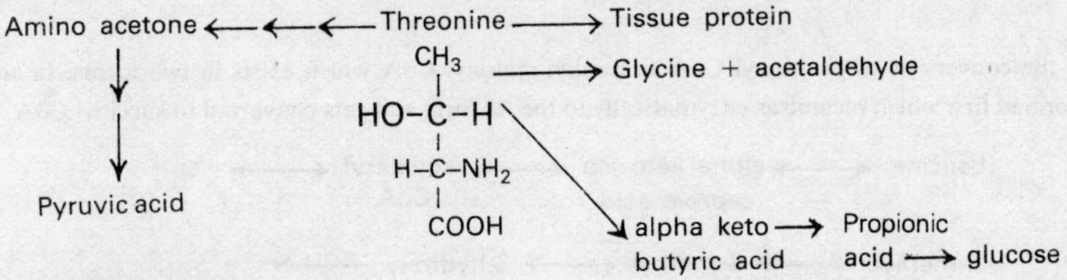

```
Amino acetone ←←← Threonine ──→ Tissue protein
      ↓            CH₃
      ↓             |       ──→ Glycine + acetaldehyde
      ↓          HO-C-H
      ↓             |
  Pyruvic acid   H-C-NH₂
                    |
                  COOH    ↘ alpha keto ──→ Propionic
                            butyric acid    acid ──→ glucose
```

Valine, leucine and isoleucine

These three are indispensable amino acids. They all contain branched-chain aliphatic groups with common metabolic features. All the three form the corresponding keto acids by transamination. These are biosynthesised in micro-organisms from pyruvic acid and active acetaldehyde.

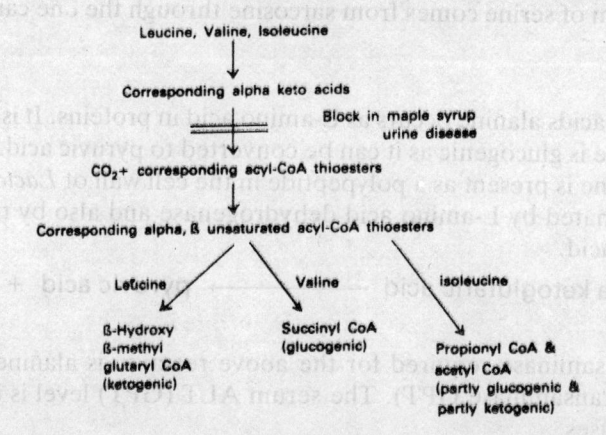

```
           Leucine, Valine, Isoleucine
                      │
                      ▼
           Corresponding alpha keto acids
                      ╪ ══ Block in maple syrup
                      │        urine disease
                      ▼
           CO₂ + corresponding acyl-CoA thioesters
                      │
                      ▼
        Corresponding alpha, β unsaturated acyl-CoA thioesters
            │             │              │
         Leucine        Valine        isoleucine
            ▼             ▼              ▼
       β-Hydroxy     Succinyl CoA    Propionyl CoA &
       β-methyl      (glucogenic)    acetyl CoA
       glutaryl CoA                  (partly glucogenic &
       (ketogenic)                   partly ketogenic)
```

The three amino acids are required for incorporation into proteins. Valine is glucogenic while leucine is ketogenic and isoleucine partly glucogenic and partly ketogenic. Leucine is the most ketogenic of all amino acids. Upto a certain stage, the catabolism of the three amino acids is the same in view of their structural similarity.

A rare genetic disease in infants, maple syrup urine disease, is due to a metabolic block because of a non-functional oxidative decarboxylase which prevents the catabolism of all the three alpha keto acids. The keto acids accumulate in the blood and urine imparting to urine the odour of maple syrup. The patients have severe functional impairment of the central nervous system.

The individual steps in the metabolism of the three amino acids are given in the flow chart below:

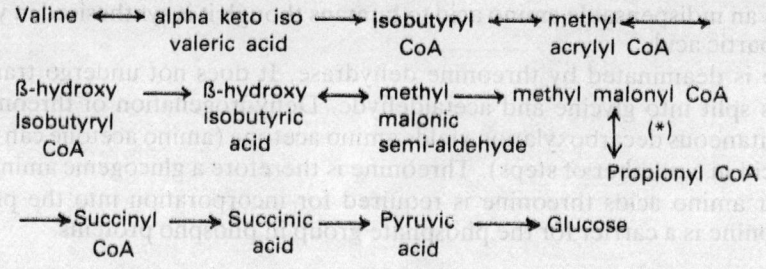

```
Valine ⟷ alpha keto iso ⟶ isobutyryl ⟷ methyl
         valeric acid        CoA         acrylyl CoA

β-hydroxy   ⟶  β-hydroxy  ⟷  methyl    ⟶ methyl malonyl CoA
Isobutyryl     isobutyric     malonic              ↑↓ (*)
CoA            acid           semi-aldehyde
                                         Propionyl CoA

    ⟶ Succinyl ⟶ Succinic ⟶ Pyruvic ⟶ Glucose
      CoA          acid        acid
```

(*) In the conversion of propionyl CoA to methyl malonyl CoA which exists in two forms, (a and b) the 'a' form is formed first which racemises enzymatically to the 'b' form and gets converted to succinyl CoA.

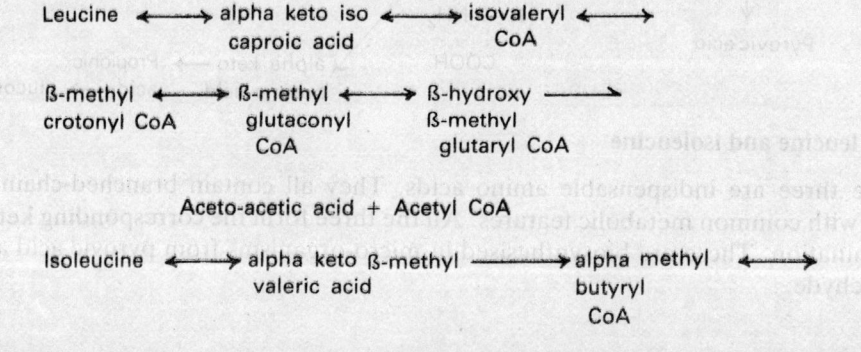

```
Leucine ⟷ alpha keto iso ⟷ isovaleryl ⟷
          caproic acid       CoA

β-methyl    ⟷  β-methyl   ⟷  β-hydroxy    ⟶
crotonyl CoA   glutaconyl     β-methyl
               CoA            glutaryl CoA

         Aceto-acetic acid + Acetyl CoA

Isoleucine ⟷ alpha keto β-methyl ⟶ alpha methyl ⟷
             valeric acid            butyryl
                                     CoA
```

alpha methyl crotonyl ⟷ alpha methyl ⟷ alpha methyl
CoA   ß-hydroxy   aceto acetyl
(tiglyl CoA)   butyryl   CoA
                        CoA

⟶ Propionyl CoA + Acetyl CoA
         ↓   ↓
      glucose   Aceto-acetic acid

Lysine

L-lysine is an indispensable amino acid and supports growth. Though indispensable to human beings and higher organisms, micro-organisms can synthesise it from acetate or aspartate.

Lysine is required for the synthesis of proteins. It is the only amino acid which, once present in the tissues, does not exchange its nitrogen with other amino acids of the pool. However, lysine can give up its nitrogen to other amino acids but the lost amino nitrogen cannot be regained. The keto acid corresponding to lysine does not undergo transamination. The final product of catabolism of lysine is aceto-acetyl CoA which may give rise to aceto-acetic acid. Hence, lysine is ketogenic.

The catabolism of lysine is given below:

$$-NH_3$$
L lysine → alpha keto ⟶ pipecolic ⟶ piperidine 2
          ε amino   acid   carboxylic acid
          caproic acid

⟶ alpha amino ⟶ L alpha ⟶ alpha keto ⟶ glutaryl
  adipic   amino   adipic   CoA
  ε semi aldehyde adipic   acid
                  acid

⟶ glutaconyl ⟶ crotonyl ⟶ aceto-acetyl CoA
  CoA   CoA

Many cereal proteins are deficient in lysine.

Hydroxy lysine

5-hydroxy lysine is present in collagen and derivatives of collagen like gelatin. It is probably absent in other mammalian proteins.

$$H_2N-H_2{}^6C-{}^5CH-{}^4CH_2-{}^3CH_2-{}^2CH-{}^1COOH$$
$$\qquad\qquad\quad |\qquad\qquad\qquad\qquad\quad\ |$$
$$\qquad\qquad\ OH\qquad\qquad\qquad\qquad\ NH_2$$

(5 hydroxy lysine; alpha, γ-diamino δ-hydroxy caproate)

Hydroxy lysine is not incorporated as such in the synthesis of collagen. On the other hand, only lysine is incorporated and hydroxylation takes place subsequently.

Aspartic and glutamic acids

These amino acids have much in common and hence are grouped under the same heading as in the case of valine, leucine and isoleucine. Aspartic and glutamic acids are synthesised in the body and are dispensable. Of the various amino acids, these are metabolically the most active.

They are present in proteins and contribute much to the electrical charge on the proteins. Differences in the electrophoretic mobilities of the isozymes of lactate dehydrogenase are due to different amounts of these twin acidic amino acids.

Glutamic acid is most important in the fixation of dietary ammonia if any and the amination of keto acids. Ammonia can combine only with alpha keto-glutaric acid (keto acid corresponding to glutamic acid) and form glutamic acid, the reaction being catalysed by L-glutamate dehydrogenase. This amino group of glutamic acid can be transferred to most other keto acids which are thus aminated with the help of glutamic acid. Again, glutamic acid itself reacts with ammonia in the presence of glutamine synthetase and ATP to form glutamine. Glutamine is an important source of urinary ammonia as glutamine can be converted back to glutamic acid and NH_3, *by glutaminase* action. This glutamic acid, glutamine interconversion is important in the acid-base balance of the blood. Glutamine is the storehouse of ammonia.

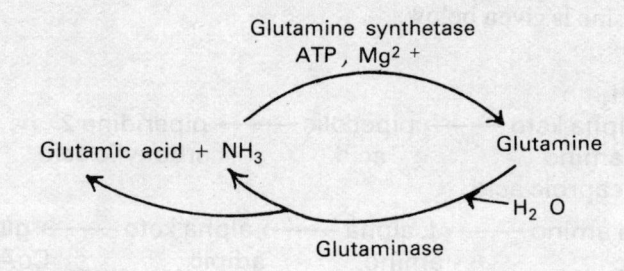

$$\text{Glutamic acid} + NH_3 \underset{\underset{H_2O}{\text{Glutaminase}}}{\overset{\overset{\text{Glutamine synthetase}}{ATP, Mg^{2+}}}{\rightleftarrows}} \text{Glutamine}$$

Glutamine formation is a defence mechanism in the brain and is important in the detoxication of ammonia in the brain.

Glutamic acid is decarboxylated in the brain to GABA by glutamate decarboxylase. GABA is required for proper neuronal function. It is interesting to note that glutamic acid does not normally cross the blood-brain barrier while glutamine does. Glutamic acid in the form of N-acetyl glutamate, is required in the Krebs-Henseleit urea cycle. The oxidative deamination of glutamic acid leads to the formation of alpha keto glutaric acid, an intermediate of the citric acid cycle, and can be converted to glucose. So glutamic acid is anti-ketogenic. Glutamic acid is required for the synthesis of glutathione.

Aspartic acid, on oxidative deamination, gives oxalo-acetic acid, an intermediate in the citric acid cycle, and hence aspartic acid also is anti-ketogenic. Aspartic acid is required in the Krebs-Henseleit urea cycle for giving the second nitrogen for the formation of urea. The brain contains large amounts of N-acetyl aspartic acid.

Both glutamic acid and aspartic acid take part in transamination reactions which are catalysed by AST and ALT (for details refer 'Transamination').

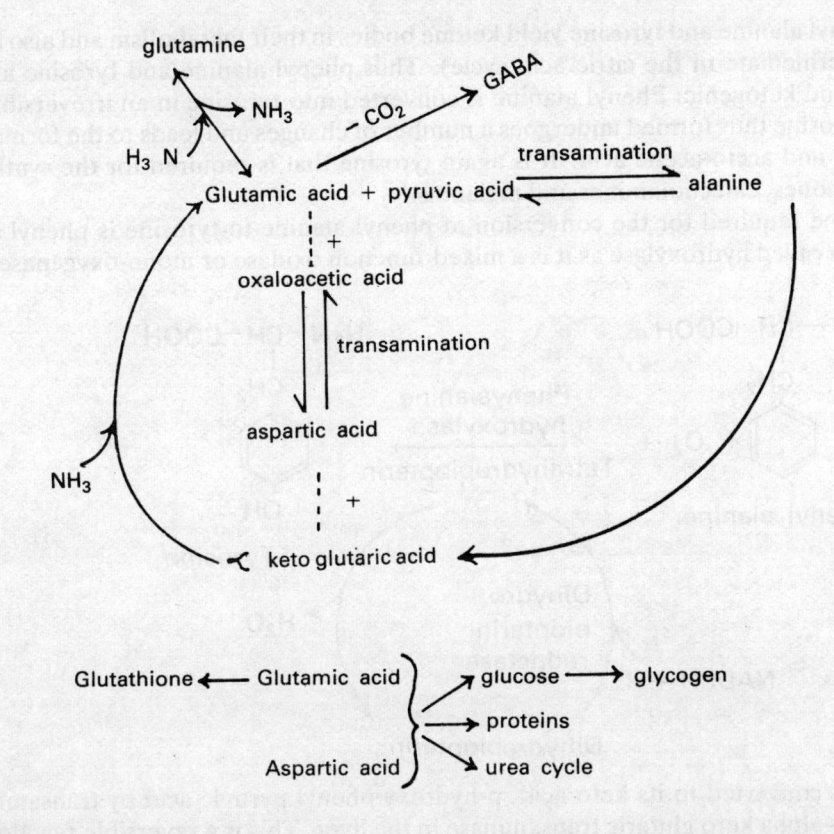

```
                    glutamine
                      ↑
                      |              GABA
                      ← NH₃       ↗
                         ↖  -CO₂ ↗
              H₃N  ↘        ↘  ↗        transamination
                  Glutamic acid + pyruvic acid ⇌ alanine
                           +
                    oxaloacetic acid
                           ↑
                      transamination
                           ↓
                    aspartic acid
     NH₃                   +
                           ↓
                    ←  keto glutaric acid  ←

    Glutathione ← Glutamic acid  ⎫ → glucose → glycogen
                                 ⎬ → proteins
                  Aspartic acid  ⎭ → urea cycle
```

Glutamine

This is synthesised in tissues from glutamic acid and ammonia with glutamine synthetase, ATP and Mg^{++}. It is converted back to glutamic acid by the action of glutaminase. Thus, glutamine and glutamic acid are readily interconvertible.

This amide of glutamic acid constitutes a large proportion of the available free amino nitrogen of tissues and blood and of the metabolic nitrogen pool. It is an important reservoir of ammonia nitrogen in tissues. Amide nitrogen is made use of in the synthesis of hexosamines, purine derivatives, histidine and NAD^+. Glutamine formation in the kidneys and its degradation constitute one of the factors in maintaining the acid-base balance of the blood and the levels of blood and tissue ammonia. Glutamine also takes part in transamination. It is of special importance in brain metabolism.

Phenyl alanine and tyrosine

Phenyl alanine and tyrosine are aromatic amino acids and structurally closely related. Phenyl alanine is an indispensable amino acid while tyrosine is not, as the latter can be formed from phenyl alanine. Tyrosine in diet has a sparing action on phenyl alanine and decreases the dietary need of it.

Phenyl alanine and tyrosine are very important amino acids involved in diverse metabolic events. In addition to their being incorporated into proteins, they are required for the synthesis of thyroid hormones, catecholamines epinephrine and nor-epinephrine and the melanin pigments. A study of these amino acids is fascinating as it gives an explanation for some inborn errors of metabolism.

Both phenyl alanine and tyrosine yield ketone bodies in their metabolism and also fumaric acid (an intermediate in the citric acid cycle). Thus phenyl alanine and tyrosine are both glucogenic and ketogenic. Phenyl alanine is converted into tyrosine in an irreversible reaction. The tyrosine thus formed undergoes a number of changes and leads to the formation of fumaric acid and aceto-acetic acid. It is again tyrosine that is required for the synthesis of thyroid hormones, catecholamines and melanins.

The enzyme required for the conversion of phenyl alanine to tyrosine is phenyl alanine oxidase (also called hydroxylase as it is a mixed-function oxidase or mono-oxygenase).

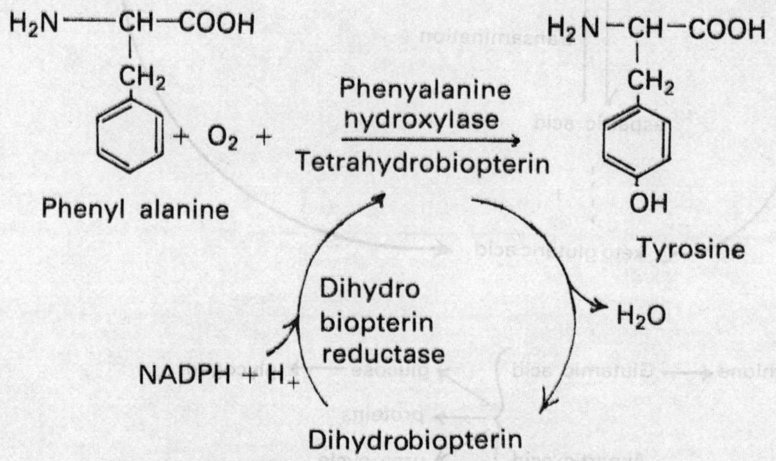

Tyrosine is converted to its keto acid, p-hydroxy phenyl pyruvic acid by transamination with tyrosine–alpha keto glutaric transaminase in the liver. This is a reversible reaction.

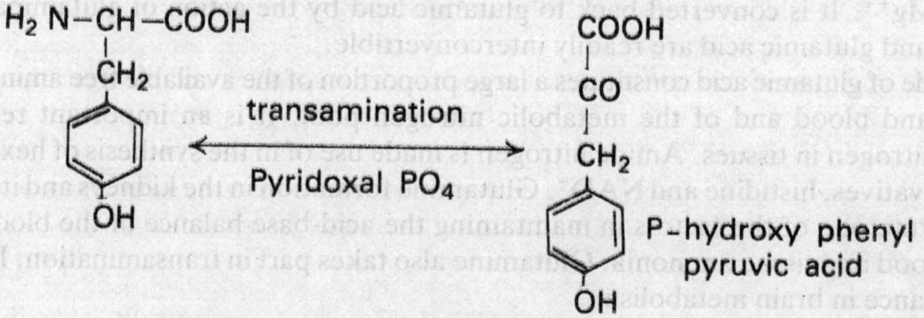

The keto acid is oxidised to 2, 5, dihydroxy phenyl acetic acid by the enzyme para hydroxy phenyl pyruvic oxidase in the presence of vitamin C.

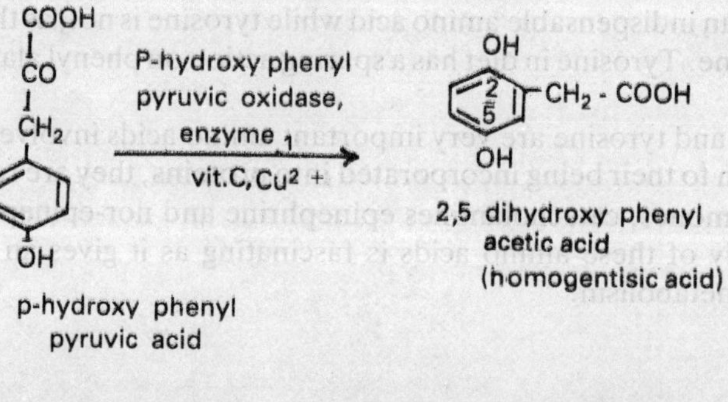

(In the formation of the product, the original side chain $-CH_2-CO-COOH$ shifts to the next carbon atom to the right.)

Homogentisic acid is acted upon by homogentisic oxidase (dioxygenase) in the presence of Fe^{++} and oxygen to form maleyl aceto-acetic acid.

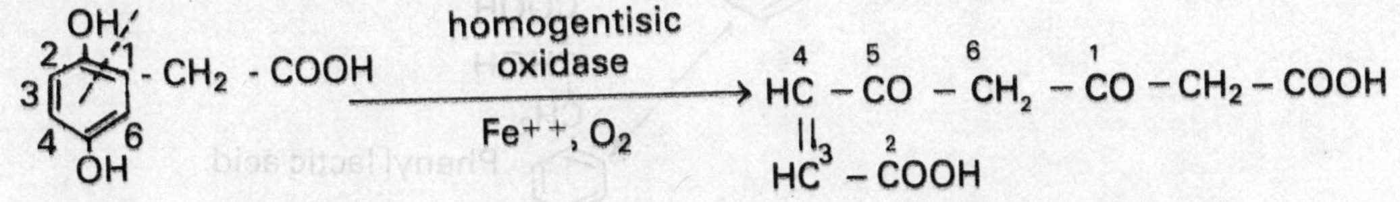

$$\underset{\text{}}{\overset{OH}{\underset{OH}{\bigcirc}}}-CH_2-COOH \xrightarrow[Fe^{++}, O_2]{\text{homogentisic oxidase}} \underset{\text{maleyl aceto-acetate}}{\overset{4}{HC}-\overset{5}{CO}-\overset{6}{CH_2}-\overset{1}{CO}-CH_2-COOH \atop \overset{3}{HC}-\overset{2}{COOH}}$$

Maleyl aceto-acetate isomerises to fumaryl aceto-acetate in the presence of maleyl aceto-acetate cis-trans isomerase and glutathione.

$$\underset{\text{maleyl aceto-acetate}}{\overset{H-C-CO-CH_2-CO-CH_2-COOH}{\underset{H-C-COOH}{\|}}} \xrightarrow[\text{glutathione}]{\text{isomerase,}} \underset{\text{fumaryl aceto acetate}}{\overset{H-C-CO-CH_2-CO-CH_2-COOH}{\underset{HOOC-C-H}{\|}}}$$

$$\underset{HOOC-CH}{\overset{H-C-CO-CH_2-CO-CH_2-COOH}{\|}} \xrightarrow{\text{fumarate aceto acetate hydrolase}} \underset{\text{fumaric acid}}{\overset{H-C-COOH}{\underset{HOOC-C-H}{\|}}} + \underset{\text{aceto-acetic acid}}{CH_3-CO-CH_2-COOH}$$

The fumaryl aceto-acetate is hydrolysed by fumaryl aceto-acetic hydrolase to fumaric acid and aceto-acetic acid.

$$\text{Aceto acetate} \xrightarrow{\text{CoASH, ß keto thiolase}} \text{Acetyl CoA}$$

The various enzymes involved in the metabolism of phenyl alanine are found in the liver.

Incidentally, the transformation of phenyl alanine to phenyl pyruvic acid, by the transamination and phenyl acetic and phenyl lactic acids could take place. Normally, phenyl pyruvic acid and phenyl lactic acid are converted back to phenyl alanine which is metabolised to fumaric acid. Phenyl acetic acid is conjugated with glutamine in man and excreted in urine.

Main Pathway of phenyl alanine and tyrosine metabolism

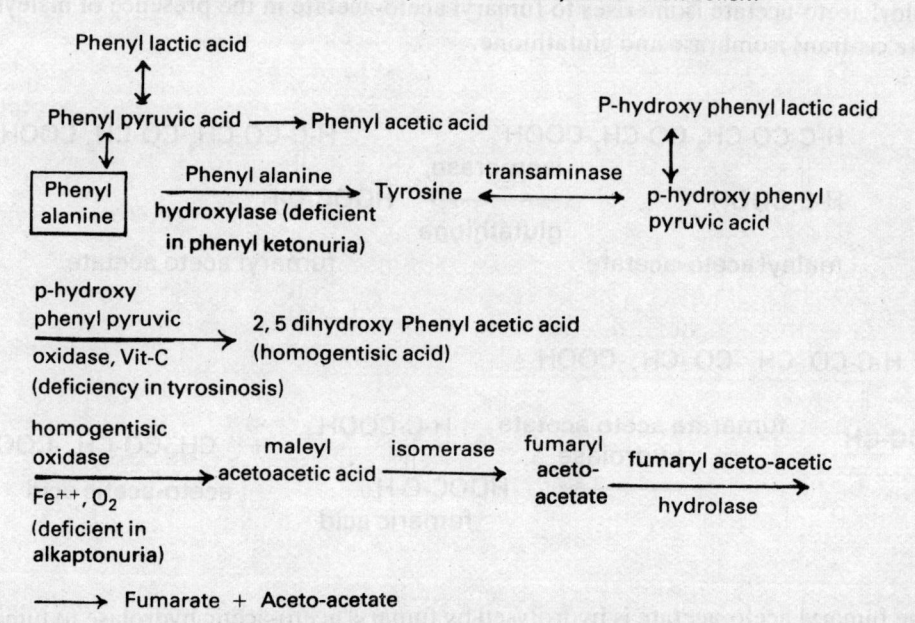

```
                    NH₂-CH-COOH           COOH        COOH
                         |                 |           |
                        CH₂                CO          CH₂       Phenyl acetic acid
                         |   transaminase  |           |         (conjugated with
                        ⬡   ⟷             CH₂         ⬡            glutamine)
                                           |         
                                           ⬡          COOH
                                                       |
                                                      CHOH
                                                       |
                                                       CH₂
                                                       |
                                                       ⬡       Phenyl lactic acid
```

p-hydroxy phenyl pyruvic acid is reversibly reduced to p-hydroxy phenyl lactic acid.

```
                    Phenyl lactic acid
                          ↕
             Phenyl pyruvic acid ──→ Phenyl acetic acid           P-hydroxy phenyl lactic acid
                          ↕                                                   ↕
         ┌─────────┐    Phenyl alanine              transaminase
         │ Phenyl  │ ─────────────────→  Tyrosine  ⟷              p-hydroxy phenyl
         │ alanine │  hydroxylase (deficient                        pyruvic acid
         └─────────┘  in phenyl ketonuria)

         p-hydroxy
         phenyl pyruvic  ─────────→  2, 5 dihydroxy Phenyl acetic acid
         oxidase, Vit-C              (homogentisic acid)
         (deficiency in tyrosinosis)

         homogentisic
         oxidase            maleyl           isomerase    fumaryl         fumaryl aceto-acetic
         ──────────→  acetoacetic acid  ─────────────→   aceto-    ─────────────────────────→
         Fe⁺⁺, O₂                                        acetate          hydrolase
         (deficient in
         alkaptonuria)

                          ───→  Fumarate + Aceto-acetate
```

In the above sequence of reactions, three hereditary defects due to the deficiency of enzymes have been observed. They are phenyl alaninemias, tyrosinosis and alkaptonuria.

The major metabolic disorder of phenyl alanine metabolism is hyper-phenylalaninemia of different types due to defective conversion of phenyl alanine to tyrosine.

Hyper-phenyl alaninemia Type I is commonly known as phenyl ketonuria (PKU, phenyl pyruvic oligophrenia). PKU is associated with mental retardation occuring in late childhood. There could be seizures, psychosis and eczema.

In this inborn error of metabolism, the enzyme phenyl alanine hydroxylase is deficient. As phenyl alanine cannot be efficiently converted to tyrosine, it is metabolised to phenyl pyruvic acid, phenyl lactic acid and phenyl acetic acid. Phenyl alanine accumulates in blood. The urine of such patients will contain phenyl pyruvic acid and excessive amounts of phenyl alanine with smaller quantities of phenyl lactic acid. With ferric chloride, the urine gives a green to blue colour. It can have a mousy odour due to conjugated phenyl acetic acid. The disease is characterised by mental retardation. The incidence of this disease is 1 in 10,000 births.

Hyper-phenyl-alaninemia Types II and III are due to the deficiency of dihydro biopterine reductase. Type IV and V are due to deficient dihydro biopterin synthesis.

Alkaptonuria: This inborn error is due to the complete lack of homogentisate oxidase. Hence, homogentisic acid is not metabolised to maleyl aceto-acetic acid. Excessive amounts of homogentisic acid and other intermediates of tyrosine metabolism like p-hydroxy phenyl lactic acid are excreted. The urine, on standing, gradually becomes darker in colour and may finally turn black. The substance responsible for the black colour is oxidised homogentisic acid. It also has a strong reducing action and gives violet colour with ferric chloride. The cartilage of the body may also darken in alkaptonuria probably due to the deposition of oxidised homogentisic acid. The condition is called ochronosis.

Tyrosinosis: Tyrosinosis is an inherited metabolic disorder. There are three types.

Tyrosinemia Type I, the tyrosinosis proper, is due to the deficiency of fumaryl aceto-acetyl hydrolase. Both acute and chronic forms of tyrosinosis are known. In acute tyrosinosis, infants exhibit diarrhoea, vomiting, a 'cabbage-like' odour and failure to thrive. Death within 6 – 8 months from liver failure is certain in untreated acute tyrosinosis. In chronic tyrosinemia, similar but milder symptoms lead to death by the age of 10 years. Treatment consists of a diet low in tyrosine and phenyl alanine.

Tyrosinosis Type II (Richnar-Hanhart syndrome) is due to the paucity of the enzyme tyrosine transaminase. There may be raised levels of plasma tyrosine. Characteristic skin and eye lesions and moderate mental retardation, self-mutilation and disturbance of fine co-ordinations have been reported.

Neo-natal tyrosinosis is due to a hereditary deficiency of para-hydroxy phenyl pyruvate oxidase.

Tyrosine in the synthesis of melanins – albinism : In addition to the above three inherited disorders, there is also albinism, a disorder associated with tyrosine metabolism. In one of its alternate pathways of metabolism, tyrosine is converted in the melanocyte cells to DOPA (3,4 dihydroxy phenyl alanine), DOPA quinone and melanin which is the pigment of the skin, hair and choroid.

Conversion of tyrosine to DOPA is brought about by the enzyme tyrosinase, a copper-containing enzyme. If this enzyme is absent in the melanocyte cells, pigments of the skin, hair and choroid are not formed and are hence absent, totally or partly.

Tyrosine for the synthesis of epinephrine and nor-epinephrine (catecholamines): In the adrenal medulla, tyrosine is oxidised to 3, 4 dihydroxy phenyl alanine (DOPA) by tyrosine hydroxylase requiring tetra-hydro biopterin (THB) in the presence of oxygen. DOPA is decarboxylated by DOPA decarboxylase to hydroxy tyramine (Dopamine). Pyridoxal phosphate is needed for this decarboxylation. This is hydroxylated to give nor-epinephrine (nor-adrenaline). Nor-epinephrine is converted to epinephrine (adrenaline) by transmethyltion.

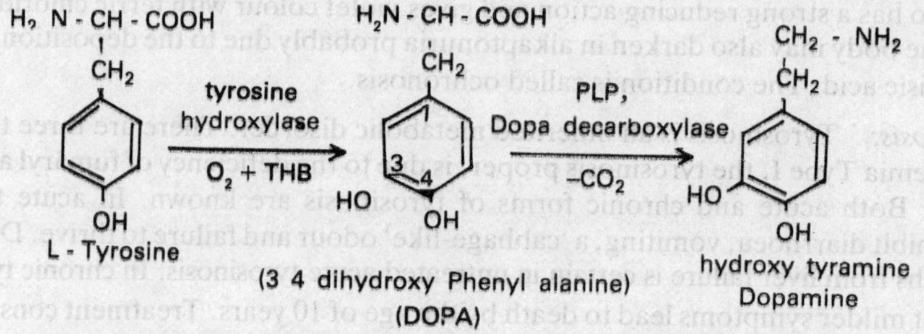

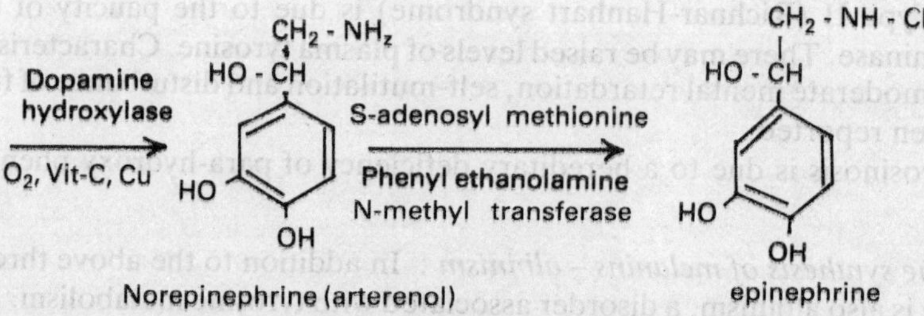

(The chief pathway of metabolism of epinephrine and nor-epinephrine is through the formation of 3-0-methyl derivatives formed by the action of S-adenosyl methionine. The 0-methyl derivative is conjugated with glucuronic acid and excreted in urine.)

Tyrosine for the synthesis of thyroid hormones: Thyroxine, the active hormone of the thyroid gland is synthesised from tyrosine of the protein thyroglobulin. Tyrosine (of thyroglobulin) is iodinated to mono-iodo-tyrosine and di-iodo-tyrosine. Di-iodo-tyrosine undergoes oxidative coupling with the elimination of alanine and forms thyroxine (T_4).

Another hormone tri-iodo-thyronine (T_3) having the same physiological function as thyroxine, and more active, may be formed by the removal of one iodine atom from thyroxine. T_3 could also be formed by the coupling of mono-iodo-tyrosine with di-iodo-tyrosine.

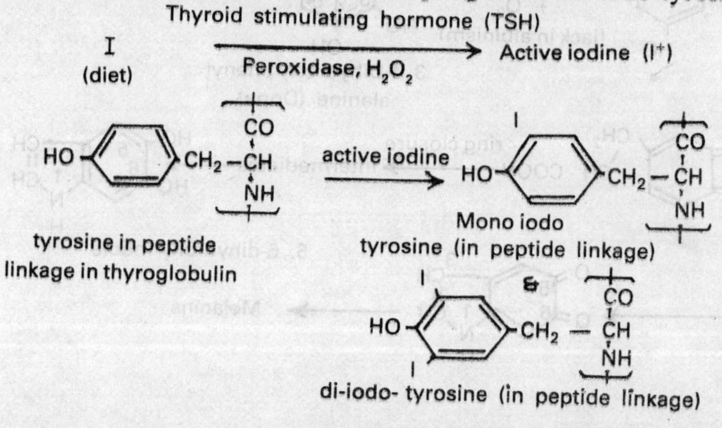

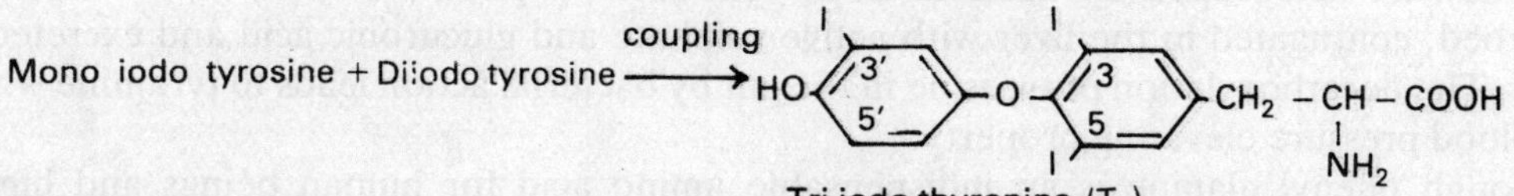

Mono iodo tyrosine + Diiodo tyrosine $\xrightarrow{\text{coupling}}$ Tri-iodo thyronine (T_3)

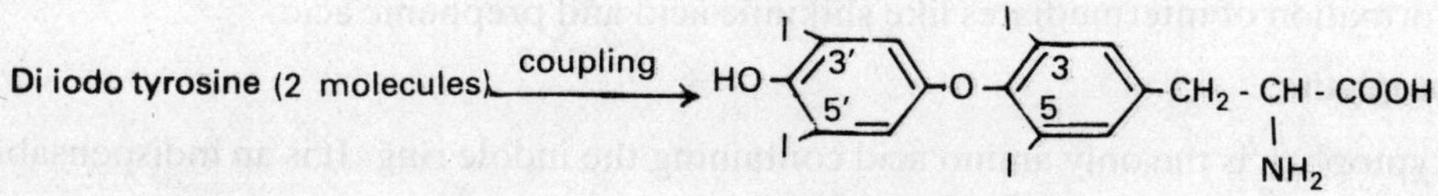

Di iodo tyrosine (2 molecules) $\xrightarrow{\text{coupling}}$ Thyroxine (T_4)

The important metabolic role of phenyl alanine and tyrosine is summarised below:

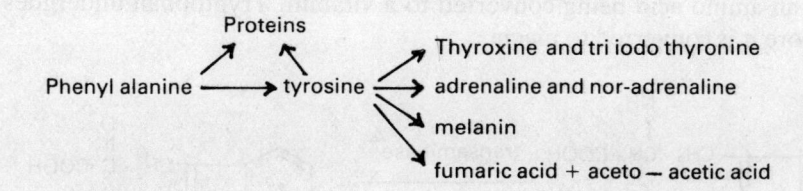

Phenyl alanine → tyrosine →
- Proteins
- Thyroxine and tri iodo thyronine
- adrenaline and nor-adrenaline
- melanin
- fumaric acid + aceto – acetic acid

Again, a summary of the inborn errors associated with phenyl alanine and tyrosine metabolism is given below:

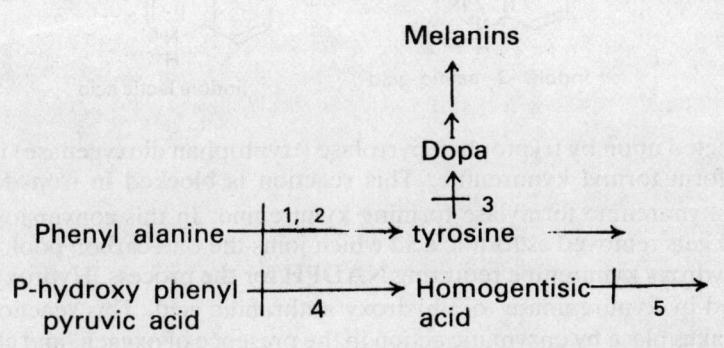

Phenyl alanine $\xrightarrow{1, 2}$ tyrosine $\xrightarrow{3}$ Dopa → Melanins

P-hydroxy phenyl pyruvic acid $\xrightarrow{4}$ Homogentisic acid $\xrightarrow{5}$

Inborn error	Enzyme deficiency
1. Hyper phenyl alaninemia Type I (PKU)	Phenyl alanine hydroxylase
2. Hyper phenyl alaninemia Type II and III	Dihydrobiopterin reductase
3. Albinism	Tyrosinase
4. Tyrosinosis	p-hydroxy phenyl pyruvic oxidase
5. Alkaptonuria	homogentisic oxidase

Phenols and tyramine from tyrosine in the gut: Tyrosine and phenyl alanine (through tyrosine) are acted upon by bacteria in the gut to form phenol and p-cresol. These are absorbed, conjugated in the liver with active sulphate and glucuronic acid and excreted in urine. The decarboxylation of tyrosine in the gut by bacterial action leads to tyramine which has blood pressure elevating property.

Though phenyl alanine is an indispensable amino acid for human beings and higher organisms, it can be synthesised by micro-organisms from D-erythrose 4-phosphate through the formation of intermediates like shikimic acid and prephenic acid.

Tryptophan

Tryptophan is the only amino acid containing the indole ring. It is an indispensable amino acid.

Tryptophan is required for incorporation into proteins. It can be subjected to oxidative deamination or transamination to indole pyruvic acid. Indole pyruvic acid can be reversibly converted to indole lactic acid. It can also be converted to indole acetic acid.

One important metabolic pathway of tryptophan is its conversion to niacin giving a good example of an amino acid being converted to a vitamin. Tryptophan undergoes a series of changes before it is converted to niacin.

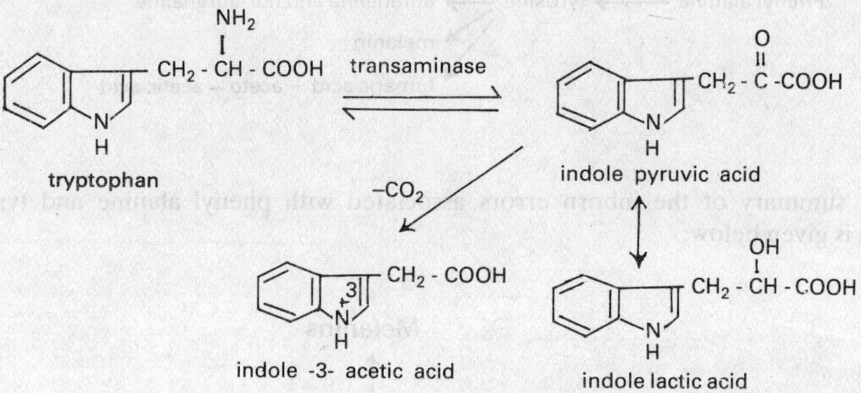

Firstly, it is acted upon by tryptophan pyrrolase (tryptophan dioxygenase) in the presence of oxygen to form formyl kynurenine. This reaction is blocked in iron deficiency. It is acted upon by kynurenine formylase forming kynurenine. In this conversion carbon atom No. 2 of indole gets removed as formic acid which joins the one-carbon pool. Kynurenine is oxidised to 3-hydroxy kynurenine requiring NADPH for the process. Hydroxykynurenine is then hydrolysed by kynureninase to 3-hydroxy anthranilic acid. This reaction requries B_6. Ring opening takes place by enzymatic action in the presence of oxygen, and closure, in a different way, forming quinolinic acid. This undergoes decarboxylation giving niacin, or nicotinic acid ribonucleotide (NMN). When ATP is available, NAD^+ will be formed.

3-hydroxy anthranilic acid can also be converted through a number of steps to picolinic acid, alpha keto adipic acid, glutaryl CoA and finally acetyl CoA and CO_2.

Tryptophan is neither glucogenic nor ketogenic. However, the three-carbon atoms of tryptophan are removed as alanine which is glucogenic.

In the above scheme kynurenine has to be converted to 3-hydroxy kynurenine and 3 hydroxy anthranilic acid. In pyridoxin (B_6) deficiency, instead of this conversion, kynurenine is converted to a yellow compound xanthurenic acid (3-hydroxy kynurenic acid).

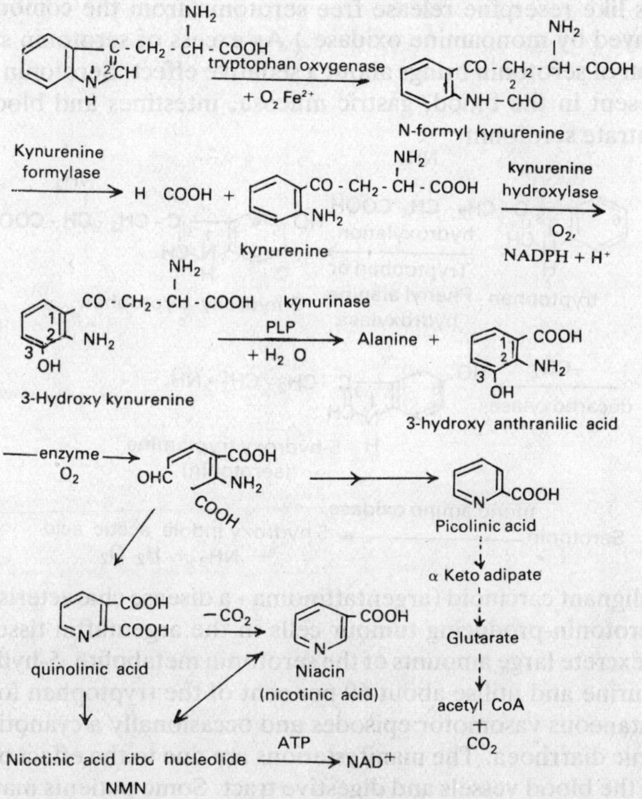

(Xanthurenic acid is excreted as a greenish yellow substance in the urine of man, monkey and rat in B_6 deficiency.) Relatd compounds of kynurenine like kynurenic acid and quinaldic acid also may be excreted in small amounts.

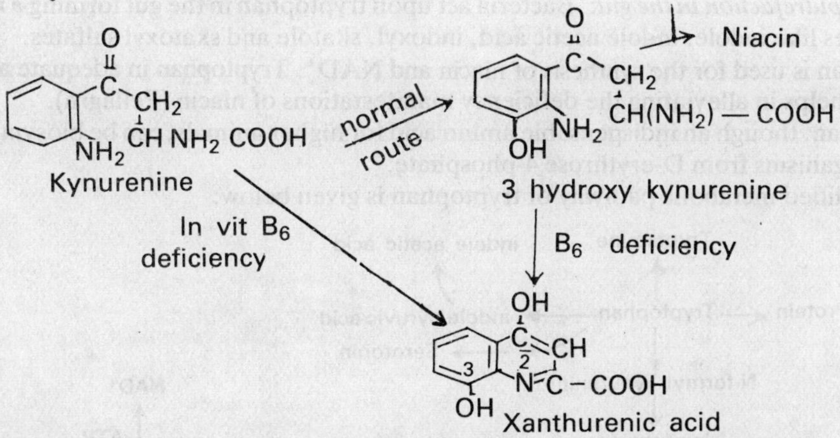

Serotonin (enteramin, thrombocytin): An important metabolite of tryptophan is serotonin which is 5-hydroxy tryptamine (5HT). Normally one per cent of tryptophan is converted to serotonin. Serotonin has a potent effect on the metabolism in the brain. Serotonin, when first produced in the brain, is in a bound form which is not acted upon by monoamine oxidase.

(Depressant drugs like reserpine release free serotonin from the combination. This freed serotonin is destroyed by monoamine oxidase.) As excess of serotonin stimulates cerebral activity, destruction of serotonin brings about a sedative effect. Serotonin is also a vaso-constrictor and is present in the blood, gastric mucosa, intestines and blood platelets which adsorb and concentrate serotonin.

$$\text{tryptophan} \xrightarrow[\text{Tryptophan or Phenyl alanine hydroxylase}]{\text{hydroxylation}} \text{5-hydroxy tryptophan}$$

$$\xrightarrow[\text{decarboxylase}]{-CO_2} \text{5-hydroxy tryptamine (serotonin)}$$

$$\text{Serotonin} \xrightarrow{\text{mono amino oxidase}} \text{5 hydroxy indole acetic acid} + NH_3 + H_2O_2$$

Patients with malignant carcinoid (argentaffinoma - a disease characterised by a widespred development of serotonin-producing tumour cells in the argentaffin tissue throughout the abdominal cavity) excrete large amounts of the serotonin metabolite, 5-hydroxy indole acetic acid (5-HIAA) in urine and utilise about 60 per cent of the tryptophan for this conversion. Patients exhibit cutaneous vasomotor episodes and occasionally a cyanotic appearance and there may be chronic diarrhoea. The manifestations are due to the effect of serotonin on the smooth muscles of the blood vessels and digestive tract. Some patients may have respiratory distress and others right-sided heart failure. The serotonin reaching the right side of the heart is destroyed by the monoamine oxidase of the lungs so that the blood returning to the left side of the heart is free from serotonin. This explains the selective right-sided heart failure.

Bacterial putrefaction in the gut: Bacteria act upon tryptophan in the gut forming a number of substances like indole, indole acetic acid, indoxyl, skatole and skatoxyl sulfates.

Tryptophan is used for the synthesis of niacin and NAD^+. Tryptophan in adequate amounts in the food helps in alleviating the deficiency manifestations of niacin (Pellagra).

Tryptophan, though an indispensable amino acid for higher animals, can be biosynthesised in micro-organisms from D-erythrose 4-phosphate.

The simplified metabolic pathway of tryptophan is given below:

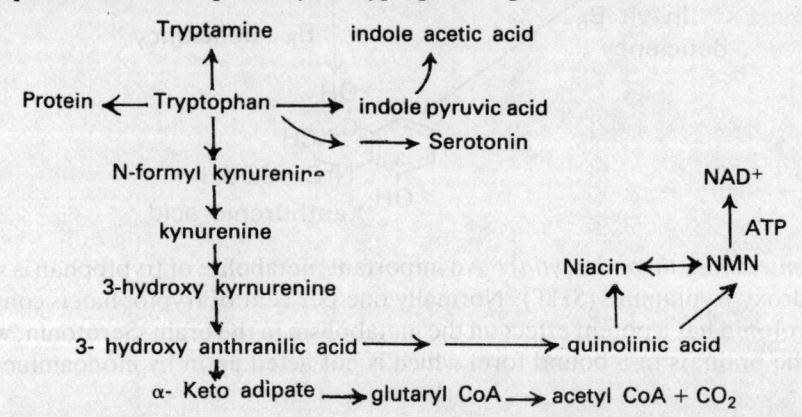

$$\text{Protein} \leftarrow \text{Tryptophan} \rightarrow \begin{cases} \text{Tryptamine} \rightarrow \text{indole acetic acid} \\ \text{indole pyruvic acid} \\ \text{Serotonin} \end{cases}$$

Tryptophan → N-formyl kynurenine → kynurenine → 3-hydroxy kynurenine → 3-hydroxy anthranilic acid → α-Keto adipate → glutaryl CoA → acetyl CoA + CO_2

3-hydroxy anthranilic acid → quinolinic acid → Niacin ← NMN $\xrightarrow{ATP}$ NAD^+

Melatonin : N-acetyl 5-methoxy serotonin is a hormone elaborated by the pineal body and peripheral nerves of man and some other higher mammals. This hormone lightens the colour of melanocytes in the skin of frog and blocks the action of MSH (melanocyte stimulating hormone) and ACTH.

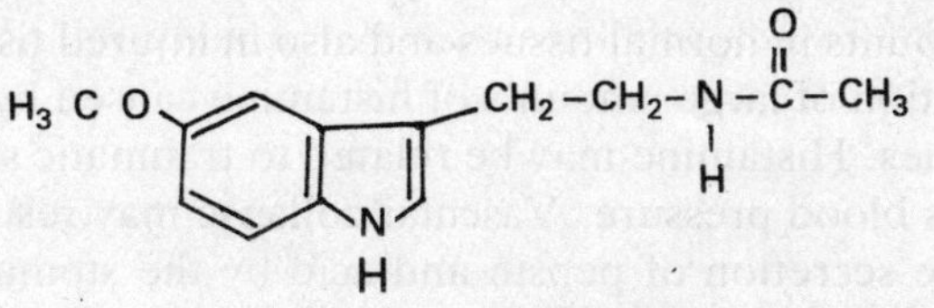

$$H_3CO-\text{(indole)}-CH_2-CH_2-NH-\overset{O}{\underset{}{C}}-CH_3$$

Melatonin

Hartnup's disease: This is an inborn error associated with the metabolism of tryptophan. The clinical symptoms are skin rash, intermittent cerebellar ataxia and mental retardation. The urine of patients with Hartnup's disease contains increased amounts of indole acetic acid and tryptophan.

Hartnup's disease is named after the family in which it was discovered and is due to a deficiency of tryptophan dioxygenase in the liver. It is also suggested that the disease could be due to a defective indole transport and a consequent impairment of tryptophan metabolism.

Histidine

Histidine is recognised from among the amino acids by the presence of the imidazole ring. Histidine is an indispensable amino acid as it is required for growth. However, nitrogen balance can be maintained in adult humans for short periods even in the absence of dietary histidine.

Histidine is required for incorporation into proteins. In mammalian liver, it is converted to glutamic acid, through a sequence of reactions in which urocanic acid, imidazolone propionic acid and N-formimino L-glutamic acid (figlu) are produced. Many enzymes participate in this reaction.

$$\text{Histidine} \xrightarrow{\text{histidase}} \text{Urocanic acid} \xrightarrow{\text{Urocanase}} $$

$$\xrightarrow{} \text{imidazolone propionic acid} \xrightarrow{\text{Urocanase}} \underset{\text{(figlu)}}{\text{N Formimino glutamic acid}} \xrightarrow{FH_4} \text{glutamic acid} + (HN=CH)^5 FH_4$$

$$(HN=CH)^5 FH_4 \xrightarrow{H_2O} f^5 FH_4 + NH_3$$

$$f^5 FH_4 \longrightarrow f^{10} FH_4$$

$$\searrow N^5 \text{ methyl } FH_4$$

Deamination by histidase is an example of non-oxidative deamination. As folic acid is required for the conversion of figlu into glutamic acid, figlu is detected in large amounts in the urine of patients with folic acid deficiency. A chemical test for lack of folic acid is to give the patient a loading dose of histidine and study the excretion of figlu in urine.

As f^5FH_4 is formed during the catabolism of histidine, the latter contributes to the one-carbon pool, after conversion to $f^{10}FH_4$.

As histidine gives rise to glutamate, it is a glucogenic amino acid.

Histidine undergoes transamination with pyruvate to form imidazole pyruvate and alanine catalysed by a specific transaminase.

Histamine is formed by the bacterial decarboxylation of histidine in the gut. It is continually formed in small amounts in normal tissues and also in injured tissues. Allergic reactions are related to the liberation of large amounts of histamine caused by the entry of sensitising substances into the tissues. Histamine may be related to traumatic shock. It is a vasodilator and markedly depresses blood pressure. Vascular collapse may result from large doses. As histamine stimulates the secretion of pepsin and acid by the stomach, it is used in gastric analysis. Histamine is catabolised and detoxified by histaminase in mammalian tissues and also through the formation of 3-methyl histamine.

Histamine is biosynthesised in lower organisms from phosphoribosyl pyrophosphate and ATP.

Histidinuria in pregnancy : When compared with other amino acids, histidine is excreted in relatively larger amounts in urine. In normal pregnancy, there is significant increase of histidine in urine. However, in toxemic conditions associated with pregnancy such an increased excretion does not happen. It was suggested that histidine metabolism might be altered in pregnancy with a possible deficiency of the histidase of liver. But Page and others explain that the increased excretion of histidine is due to changes in renal function which occur during pregnancy and due to pregnancy toxemias.

Again, in addition to histidine, the excretion of other amino acids is also affected in pregnancy.

The important histidine compounds found in the body are ergothionine in the red blood cells and liver, carnosine, a dipeptide of histidine and ß-alanine and anserine (methyl carnosine).

Inherited errors of histidine metabolism are imidazole amino aciduria, in which there is celebro-vascular degeneration and histidinemia in which speech development may be affected. In histidinemia, there is increased excretion of imidazole pyruvic acid in urine and this gives the same colour test with $FeCl_3$ as phenyl pyruvic acid in phenyl ketonuria. Hence, phenyl ketonuria can be mistaken for histidinuria and vice versa.

In imidazole amino aciduria, large amounts of histidine, carnosine, anserine and methyl histidine are excreted.

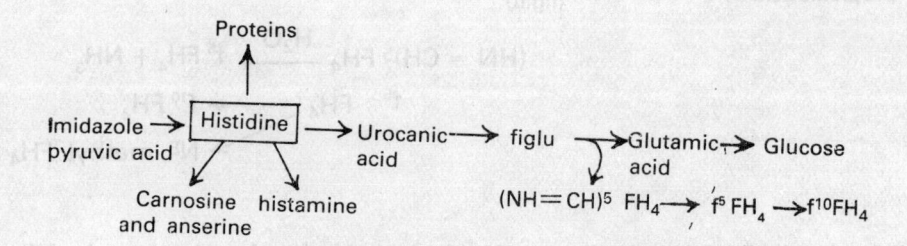

Imidazole pyruvic acid → Histidine → Urocanic acid → figlu → Glutamic acid → Glucose
Histidine → Proteins
Histidine → Carnosine and anserine
Histidine → histamine
$(NH=CH)^5 FH_4 → f^5FH_4 → f^{10}FH_4$

Summary of histidine metabolism

Arginine, ornithine, proline and hydroxy proline

Arginine and ornithine are basic amino acids. While arginine is a constituent of proteins, ornithine is not. Arginine and ornithine take part in the Krebs-Henseleit urea cycle. The biosynthesis of creatine requires arginine. Though arginine is formed in the body in the urea cycle, it is considered semi-dispensable, and the rate of its synthesis is too slow to provide for normal growth and maintenance.

Proline is also a constituent of proteins. Ornithine and proline are inter-convertible through the formation of glutamic γ semi-aldehyde which can also be obtained from glutamic acid.

```
Arginine                                    CH₂—CH₂
   │                                        │     │
   │ Urea cycle                             CH₂   CH – COOH
   ↓       transamination                    \   /
H₂C - NH₂ ─────────────────→  CHO             NH      Proline
   │                           │              ↕↑
   CH₂                         CH₂
   │                           │            H₂C───CH₂
   CH₂          ←──────────→   CH₂          │     │
   │                           │            HC    CH–COOH
   CH - NH₂                    CH NH₂         \\ /
   │                           │               N
   COOH                        COOH
                               Glutamate     Δ¹ Pyrroline
   Ornithine                   γ semi-aldehyde  5 carboxylic acid
                                    ↓
                               glutamic acid
```

Hydroxy proline is not found in appreciable amounts in body proteins. It is, however, present in collagen. It has been shown that proline is hydroxylated after incorporation into tissue proteins. Vitamin C and oxygen are required for this. Hydroxy proline is broken down through γ hydroxy glutamic acid semi-aldehyde and γ hydroxy glutamic acid to form alanine.

An outline of the metabolic reactions of arginine, ornithine, proline and hydroxy proline is given below:

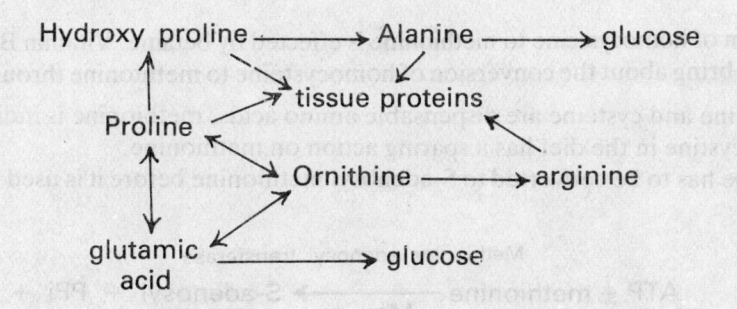

```
Hydroxy proline ──────→ Alanine ──────→ glucose
       │             ↗
       │         ↗ tissue proteins
     Proline ↙                    ↖
       │  ↘                         ↖
       │     ↘  Ornithine ──────→ arginine
       ↓      ↗       ↖
    glutamic ↙          ↘
      acid  ──────────→ glucose
```

The sulphur-containing amino acids

Methionine and cystine represent most of the sulphur of proteins while cysteine accounts for a small amount of sulphur in proteins. The presence of free -SH of cysteine in enzyme protein is required for efficient enzyme action.

Cystine and cysteine are readily inter-convertible by non-enzymatic oxidation and reduction.

$$\text{Cystine} \underset{-2H}{\overset{+2H}{\rightleftarrows}} 2\text{ Cysteine}$$

Cysteine can also be oxidised to cystine by glutathione present in the cells.

$$2\text{ Cysteine} + G-S-S-G \rightleftarrows 2GSH + \text{Cystine}$$

Likewise, homocysteine and homocystine are inter-convertible. Homocysteine can be obtained by demethylation of methionine.

$$\text{Methionine} \xrightarrow{-CH_3} \text{homocysteine}$$

Homocysteine with serine can be converted to cystathionine which cleaves to homoserine and cysteine. In this reaction the sulphur of cysteine comes from methionine while the rest of the molecule comes from serine.

$$\text{Homocysteine + serine} \xrightarrow{\text{cystathionine reductase}} \text{cystathionine}$$

$$\text{Cystathionine} \longrightarrow \underset{\substack{\text{(rest of} \\ \text{methionine)}}}{\text{Homoserine}} + \underset{\substack{\text{(S of} \\ \text{methionine)}}}{\text{cysteine}}$$

Thus cysteine, cystine, homocysteine, homocystine and methionine metabolisms are inter-linked.

$$\text{Methionine} \longleftrightarrow \text{Homocysteine} \longleftrightarrow \text{Homocystine}$$
$$\downarrow$$
$$\text{cysteine} \longleftrightarrow \text{cystine}$$

Conversion of homocysteine to methionine is effected by betaine. Vitamin B_{12} and N^5 methyl FH_4 can also bring about the conversion of homocysteine to methionine through methyl B_{12}.

While cystine and cysteine are dispensable amino acids, methionine is indispensable. The presence of cystine in the diet has a sparing action on methionine.

Methionine has to be converted to S-adenosyl methionine before it is used for transmethylation.

$$\text{ATP + methionine} \xrightarrow[Mg^{++}]{\text{Methionine adenosyl transferase}} \text{S-adenosyl methionine} + PPi + Pi$$

S-adenosyl methionine donates the methyl group for the formation of compounds like creatine, choline, epinephrine and N-methyl nicotinamide. Most of the active methionine in the body is formed from the methionine in the diet; however, it can also be formed from the one-carbon pool under unusual circumstances.

It has been shown that some amount of the methyl group of active methionine is destroyed to CO_2.

Methionine which is an indispensable amino acid for higher animals is, however, synthesised in micro-organisms from aspartic acid through the formation of homoserine. Homoserine and cysteine, which cannot combine in higher animals, react to form cystathionine which gives homocysteine and then methionine.

Methionine can undergo transamination with alpha ketoglutaric acid to form alpha keto methyl thiobutyric acid.

The fate of methionine could be summarised as follows:

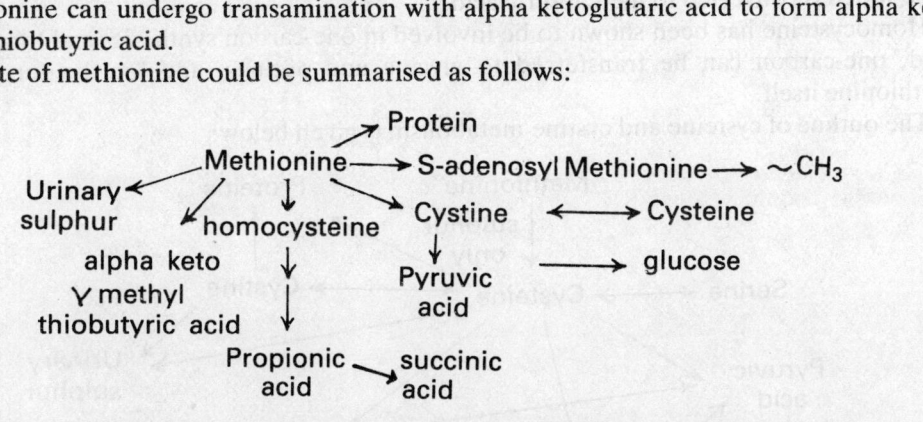

```
                          → Protein
              Methionine → S-adenosyl Methionine → ~CH₃
Urinary ←          ↓
sulphur   ↙ homocysteine → Cystine   ←→   Cysteine
   alpha keto      ↓         ↓
   γ methyl                Pyruvic   ———→  glucose
   thiobutyric acid ↓       acid
              Propionic  ↘  succinic
              acid         acid
```

As glucose can be formed from methionine, the latter is a glucogenic amino acid.

Cysteine and cystine: (The inter-conversion of cysteine and cystine has already been described.)

Cysteine can be formed from serine:

$$\text{Serine} + H_2S \xrightarrow{\text{serine sulphhydrase}} \text{Cysteine} + H_2O$$

Cysteine can be converted to pyruvic acid by the cysteine desulphhydrase of the liver.

$$\text{Cysteine} \longrightarrow \text{pyruvic acid} + NH_3$$
$$\downarrow$$
$$\text{glucose}$$

Cysteine is thus glucogenic.

Cysteine and cysteine constitute some urinary sulphur.

Cysteine undergoes transamination to ß mercaptopyruvic acid.

$$HS-CH_2-CO-COOH$$

This compound has valuable functions. If detoxifies cyanide to thiocyanate (CN^- to SCN^-), sulphite to thiosulphate and is decomposed to pyruvic acid. (The thiosulphate formed also detoxifies CN^- by converting it to SCN^-.)

Cysteine is also converted enzymatically to cysteine sulphinic acid. This compound

$$\underset{\displaystyle\|}{O} \quad \underset{\displaystyle|}{NH_2}$$
$$HO-S-CH_2-CH-COOH$$ can be converted to pyruvic acid or cysteic acid or hypotaurine.

The cysteic acid $HO-\underset{\underset{O}{\|}}{\overset{\overset{O}{\|}}{S}}-CH_2-\underset{NH_2}{CH}-COOH$ and hypotaurine can be converted into taurine

$HO-\underset{\underset{O}{\|}}{\overset{\overset{O}{\|}}{S}}-CH_2-CH_2-NH_2$ which is required for the formation of tauro cholic acid (bile acid).

Taurine has also an important role in neuronal function. It is an inhibitory neuro-transmitter.

Cysteine is also used for the detoxication of aromatic halogen compounds, like bromobenzene which is converted to mercapturic acid.

Homocysteine has been shown to be involved in one-carbon synthesis, and through folic acid, one-carbon can be transferred to glycine and purines, and for the formation of methionine itself.

The outline of cysteine and cystine metabolism is given below:

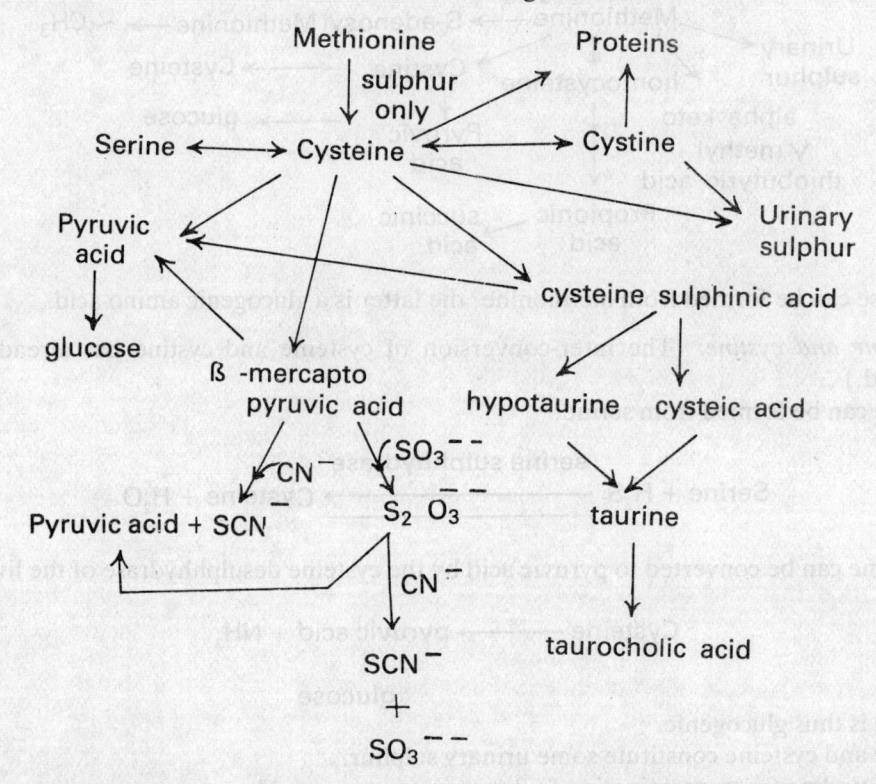

```
                    Methionine              Proteins
                      │sulphur              ↗
                      ↓ only               ↗
   Serine ←——→ Cysteine ←————→ Cystine
              ↙    ↓    ↘           ↘
   Pyruvic  ↙      ↓      ↘          → Urinary
    acid           ↓        ↘             sulphur
     ↑             ↓         → cysteine sulphinic acid
     │             ↓              ↓        ↓
   glucose      ß-mercapto    hypotaurine  cysteic acid
                pyruvic acid       ↓        ↓
                  ↓  CN⁻  ↙SO₃⁻⁻       ↘   ↙
              Pyruvic acid + SCN⁻   S₂O₃⁻⁻   taurine
                                     ↓         ↓
                                    CN⁻        ↓
                                     ↓    taurocholic acid
                                   SCN⁻
                                    +
                                   SO₃⁻⁻
```

It may be observed from the above that cystine metabolism is almost entirely through cysteine which is referred to as 'half cystine' by some authors. Cystine is a prominent amino acid in the proteins of hair, horn and skin.

There are three important hereditary diseases associated with the metabolism of the sulphur-containing amino acids.

1) Cystinuria is of historical importance in that it was one of the four inborn errors of metabolism originally described by Garrod. In this condition there is an excretion of large amounts of the amino acids, cystine, lysine, arginine, ornithine and isoleucine. Though the disease is called cystinuria, there is a greater excretion of lysine and arginine than cystine

itself. Cystine being relatively less soluble in water, there may be cystine calculi in urine and also formation of cystine calculi in the kidneys and the rest of the urinary tract. It has been shown recently that, in addition to cystine, there is the mixed sulphide compound of L-cysteine and L-homocysteine in the urine of cystinuric patients. The disease was ascribed to the defective transport mechanism for the reabsorption of these amino acids in the kidneys. But strangely enough, the transport mechanism of cystine itself was normal while only that of lysine and arginine was defective. This was shown in experiments with tissues from cystinuric patients and this observation suggests some other etiological factors also for cystinuria. It has been suggested that in addition to defective renal transport there could also be a defect in the intestinal transport of the amino acids concerned.

2) Cystinosis (or cystine storage disease) is another inborn error characterised by the distribution of cystine crystals in many tissues and organs throughout the body, with generalised amino aciduria. This is a more serious disease than cystinuria which is compatible with a normal existence.

3) A disorder of methionine metabolism is homocystinuria in which there is a high level of homocystine and methionine in the blood and increased excretion of homocystine in urine. The disease is due to the deficiency of cystathionine synthetase. The clinical findings especially in children are mental retardation, bilateral posterior dislocation of the lens, fair complexion with blue eyes and fine hair.

TERMINAL PATHWAY OF CARBOHYDRATE, LIPID AND PROTEIN METABOLISM

The integration of the diverse metabolic pathways is a meaningful physiological process since homeostasis is a composite of various biochemical transformations and inter-conversions in the body. Such a linkage is made possible by various enzymes and hormones as well as mineral elements including trace elements.

There are conditions which warrant the diversion of metabolites out of a physiological necessity to cope with the altered situation in the organism. In a fasting person, for example, there is a definite sequence of events necessitated by the condition. Carbohydrates are depleted, to start with, by way of depletion of liver glycogen. But the glycogen of the muscle is not depleted. Lipids with a little of proteins are tapped next, resulting in ketosis which has a profound impact on acid-base homeostasis and fluid and electrolyte balance. Major proteins are finally mobilised for gluconeogenesis mostly from glucogenic amino acids so as to meet the energy requirements for which normally proteins are not intended.

Though there is primarily a derangement of carbohydrate metabolism in the diabetic organism, there is, ultimately, a universal metabolic involvement brought about by a logical sequence of biochemical events in untreated cases.

The citric acid cycle or the common metabolic pathway provides a fascinating channel of integration into a metabolic pool. The focal point of such a get-together of diverse metabolites is provided by active acetate or acetyl coenzyme A, which is the form in which all the two carbon fragments get incorporated into a common pool. Once they get into it, they are no longer distinguishable as to their origin. Acetyl coenzyme A that is formed from pyruvate as well as fatty acids again and again condenses with oxalo-acetate thereby perpetuating the citric acid cycle.

The citric acid cycle also provides additional points of linkage. One such point is the formation of alpha keto glutaric acid by the decarboxylation of oxalo succinate. Alpha keto-glutarate can transaminate with aspartate to yield glutamic acid and oxalo-acetate. The reverse reaction provides aspartic acid for urea formation in the liver. Succinate formed is a point of lin-

MEDICAL BIOCHEMISTRY

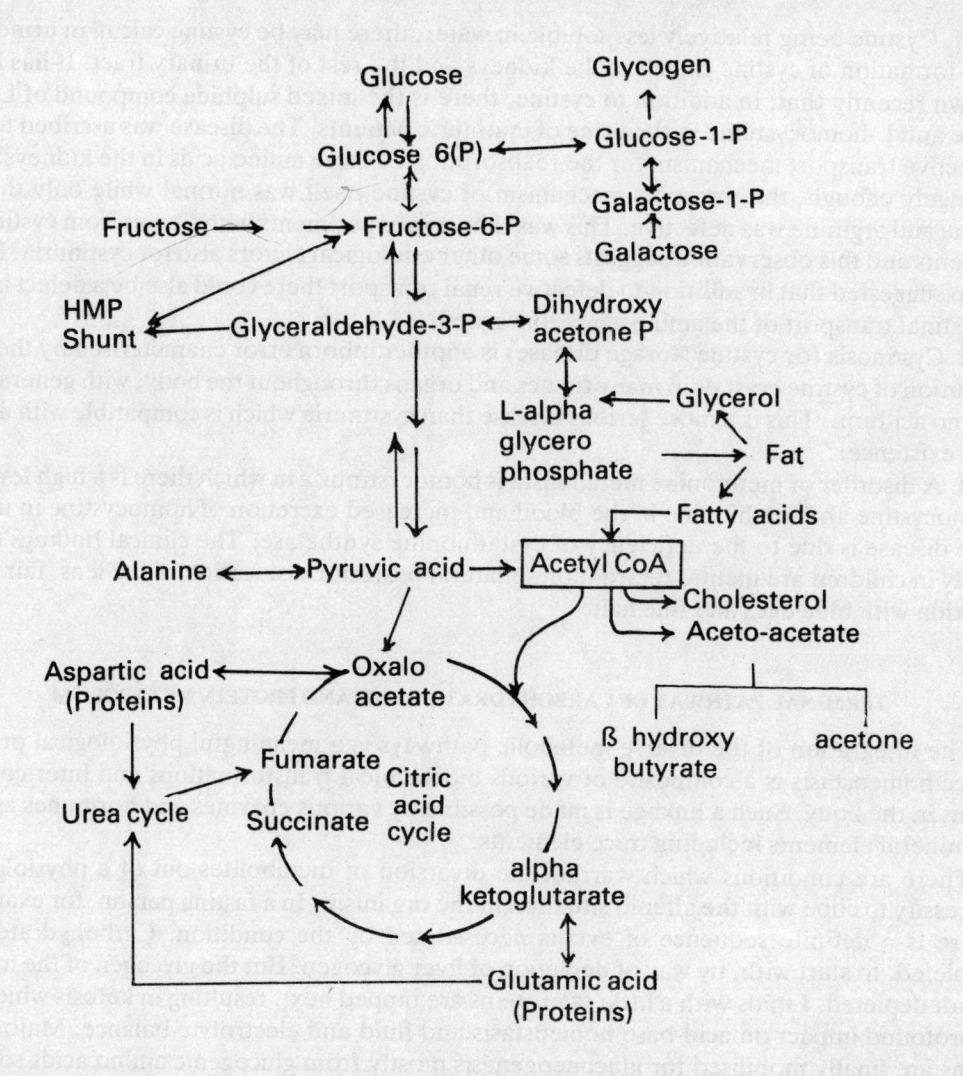

```
                              Glucose        Glycogen
                                ↑              ↑
                           Glucose 6(P) ←→ Glucose-1-P
                              ↑ ↓             ↑
                                            Galactose-1-P
         Fructose →      Fructose-6-P         ↑
                           ↑ ↓             Galactose
                           ↑ ↓
                           ↑ ↓           Dihydroxy
         HMP     Glyceraldehyde-3-P ←→  acetone P
         Shunt                            ↑ ↓
                           ↓                       Glycerol
                           ↓              L-alpha
                           ↓              glycero    →    Fat
                           ↓             phosphate
                           ↓
                                         Fatty acids
         Alanine ←→  Pyruvic acid →    Acetyl CoA
                                         → Cholesterol
                                         → Aceto-acetate
         Aspartic acid ←→ Oxalo
         (Proteins)       acetate
                            ↓                ß hydroxy    acetone
                         Fumarate             butyrate
         Urea cycle        |         Citric
                         Succinate    acid
                                     cycle
                                               alpha
                                            ketoglutarate
                                                ↑ ↓
                                          Glutamic acid
                                           (Proteins)
```

kage with the urea cycle through fumarate. Oxalo-acetate obtained by the oxidation of malic acid can give rise to asparate and alpha keto glutaric acid on transamination with glutamic acid. Alanine, on oxidative deamination, gives pyruvic acid and is thus connected with the citric acid cycle.

Glyceraldehyde 3-phosphate, isomeric with dihydroxy acetone phosphate, connects glycerol from triacyl glycerols with the glycolytic pathway, which accounts for the glucogenic value of glycerol, though it is next only to proteins. Such a linkage is necessary on account of the non-secretion of glycerokinase by adipose tissue. The EM pathway and the HMP shunt are connected through 3-phospho glyceraldehyde and also through fructose-6-phosphate.

These instances are a few of the host of examples of integrated metabolic pathways which have been unravelled by tracer studies. The diabetic organism provides an obvious example of the response of the system to aberrant circumstances by way of shunting, for example, of acetyl CoA into more of ketone body formation. Several hormones, enzymes, coenzymes and inorganic co-factors facilitate the integration by a harmonious interplay. A lack of such a harmony is seen in many pathological states.

METABOLISM IN STARVATION

Starvation is the deprivation of food and thereby exogenous supply of calories to meet the energy demands of the body for basal metabolism and other activities. Fasting could be for a day or two, many days or even for one or two months. Depending on the availability of the endogenous stores of triacyl glycerols, proteins, carbohydrates, hormonal influences and the body's adaptation, there is a definite sequence in the metabolic events that take place in a total fast. To appreciate the biochemical changes during starvation, it is worthwhile to know the extent of fuel stores in the body. In a normal 70 kg individual the weight of fuel in grams/kg is as follows:

Triacyl glycerols	15
Proteins	6
Glycogen, liver and muscle	0.225
Circulating fuels (glucose, fatty acids and TAG)	0.023

It may be seen that the amount of TAGs and proteins is much larger than that of carbohydrates. TAG alone can, theoretically, provide the basal requirement for over 70 days in a normal adult.

In the first one or two days, the liver glycogen is tapped for the supply of glucose. It falls to about 10 per cent of its level and remains approximately constant at this level subsequently. Muscle glycogen and the free glucose in the tissues are also used up, but the blood glucose is relatively constant at about 80 mg per cent for about four weeks or more. Muscle glycogen is slightly decreased but not considerably. The urinary nitrogen level falls during the first two days.

After the second day and during the first week, the utilisation of triacyl glycerols from the fat depots in the abdominal and subcutaneous areas starts. Low levels of insulin and increased levels of glucagon and the growth hormone promote lipolysis. cAMP plays the role of a 'second messenger'. Ketone bodies are also formed in the liver which esterifies some free fatty acids also. The RQ falls from the normal 0.82 to 0.7 to 0.8. Ketosis starts after two days and gets aggravated as the condition continues. However, the ketosis of starvation is of a mild type compared to that of diabetes mellitus since the mechanisms for oxidative utilisation are not impaired here. The functional decalcification that occurs serves to combat, to some extent, the keto-acidosis of starvation. The ketone bodies are partly excreted as ammonium salts, or salts with Na^+ and later with K^+. It is interesting to note that in spite of the availability of free fatty acids and ketone bodies, proteins also are drained, and in the very first week, as high as about 100 grams of proteins a day are used. This is reflected in the increase of urinary nitrogen after two days. The question arises as to why proteins are used, when fat is available for energy purposes. The answer is that there are some tissues and cells which can use only glucose for energy. The brain, though it constitutes only about two per cent of the body weight utilises nearly 140 grams of glucose a day. There are the RBCs, again, which use only glucose. Of course, the lactic acid produced by the RBCs during glycolysis can be converted to glucose in the liver. The glycerol from lipolysis meets a substantial demand for glucose through gluconeogenesis, but its supply of glucose is inadequate. Hence, glucogenic protein amino acids are used for gluconeogenesis to supply glucose to the brain. There is a definite chronology of the nature of proteins used for this purpose. The first proteins lost are the digestive enzymes secreted by the stomach, pancreas and the small intestines, as these enzymes are no longer needed. The enzymes of the liver which are concerned with the metabolism of incoming nutrients are also lost early. These are the so-called 'labile proteins' the loss of which does not affect the integrity of the tissues. If need be, even the plasma albumin may fall resulting in edema in the form of distended abdomen, swelling of the extremities, puffy face, etc. At this stage, muscle proteins are not used as the amino acids from them, chiefly alanine, are not released. Corticosteroids, which are increased, augment gluconeogenesis. Lowered insulin level also makes the metabolism swing from glycolysis to gluconeogenesis. Increased fat oxidation results in increased acetyl CoA which inhibits the pyruvate dehydrogenase step while it stimulates allosterically pyruvate carboxylase. Increased acetyl CoA causes an increase of citrate which suppresses phosphofructokinase. There is also inhibition of hexokinase by glucose 6-phosphate which also tends to accumulate. The decrease of insulin causes a de-repression of the gluconeogenic enzymes like phospho-pyruvate carboxylase and fructose 1, 6 diphosphatase. Thus glycolysis is inhibited and gluconeogenesis is enhanced to maintain blood sugar and supply glucose to the brain. Urinary nitrogen and blood urea levels are increased showing the liberal utilisation of proteins.

But, in the course of the second to the fourth week, the brain starts adapting itself to the changed environment, uses ß hydroxy butyrate and decreases its demand for glucose without affecting the mental faculties. This, with the added constraint of the stock of 'labile' proteins, makes a favourable turn in the tapping of proteins. As against the initial 100 grams of proteins a day, only about 12 – 15 g of proteins are used in 4 – 6 weeks of fasting, the energy being derived from FFA, glycerol and ketone bodies so long as fat is available. Muscle proteins are not used at this stage. The urinary excretion of nitrogen falls during this period.

When total fasting goes on to the most severe phase, the TAG stores are completely exhausted and muscle proteins, the largest single source of proteins are heavily drained. Not only those of contractile proteins but also the glycolytic enzymes of sarcoplasm are lost. The body loses weight though the weight loss is not evenly distributed. Subcutaneous tissue and extra-cellular water are lost first, followed by intra cellular water. While most of the weight loss is due to that of muscle, spleen, liver and G.I.T., there is only about 20 per cent loss of weight by the kidneys, 2 – 8 per cent by gonads and thyroid, and five per cent by CNS. Ammonia is increasingly eliminated in the urine. Creatine in the blood increases while creatinine falls. The urinary excretion of creatine is also increased but the total amount of creatine and creatinine excreted is almost constant. Just 2 – 3 days before death, there could be a primordial rise, in some cases, of nitrogen, due to the accelerated breakdown of tissue proteins. P and S show an initial rise but undergo a gradual decline. Sulphur-containing amino acids are conserved. The urinary excretion of Cl^-, Na^+, K^+ and Mg^{2+} are less due to poor intake. Antibody production is also affected in prolonged fasting with a reduction of the γ globulins.

The BMR, body temperature, pulse rate and BP may fall progressively. All these should caution the individual for prompt action to set right the metabolism.

15 Biochemical Genetics

Biochemical Genetics is the study of biochemistry of the genetic material, i.e., the substance constituting the genes, and the properties of such material. Genes are located at specific loci on the chromosomes. The transmission of hereditary characters takes place through the genes which pass over from the parents to the offspring through the chromosomes. Human beings have 23 pairs of chromosomes of which one pair constitutes the sex chromosomes (XX for female and XY for male), the remaining 22 pairs being the autosomal pairs. The child gets one chromosome each from its father and mother for each pair and thus gets the genes of both its parents.

Genetic material

The genetic material has been shown to be DNA in most cases for the following reasons:

Location: DNA is located in the chromosomes as nucleo-protein which is also the site of genes.

Hereditary information: It has been shown that the transforming substances of bacteria which carry hereditary information consist of DNA. It is DNA which determines the morphological, chemical and metabolic characteristics of an individual cell or organism.

Mutation: Genes can undergo mutation by exposure to ultra-violet light and some chemical reagents like nitrous acid. DNA also undergoes mutations.

There are other properties of DNA which help to consider DNA as the genetic material. DNA undergoes replication. The replication of viruses has been shown to be due to the nucleic acid of such viruses. The DNA content of a set of chromosomes is an absolute constant for each species. DNA has a poor metabolic turnover. It directs the formation of enzymes and proteins. All these go to identify DNA with the gene.

The human genome consists of 46 chromosomes. Each chromosome contains one very large DNA molecule. A double stranded DNA molecule is estimated to contain 150 million nucleotide pairs (150×10^6).

The gene or cistron of DNA can have 75 – 5,500 nucleotide pairs. One DNA molecule can code for about 27,000 – 60,000 genes and a genome of 46 chromosomes can code for 1.25 – 3 million gene pairs.

The science of Genetics developed from the study of Mendel, while Biochemical Genetics from that of Meselson and Stahl, Kornberg-Nirenberg. Ochoa, Matthaei, Khorana, Sanger and others. Beadle suggested that one gene could influence one step in metabolism by controlling the synthesis of a single enzyme (one gene-one enzyme hypothesis). However, it has now been shown that one gene may be involved in the synthesis of one polypeptide chain rather than a whole protein. In hemoglobin, the synthesis of the alpha chains of the globins is under the control of one gene while that of the beta-chain is under the control of another gene. Very recently, one gene has been shown to control the synthesis of more than one polypeptide in Ø × 174 bacteriophage.

While DNA is the genetic material in most cases, RNA can serve the purpose when DNA is absent, as for example in certain plant viruses like the tobacco mosaic virus (TMV).

Types of genes

With regard to the role of genes (DNA) in the synthesis of proteins, they can be divided into three types in prokaryotes: 1) *Structural genes* which determine the molecular organisation of enzymes and other proteins, 2) *operator genes* which initiate protein synthesis, and 3) *regulator genes* which control the rate of protein synthesis. These three types of genes work in close liason in what is called Operon for proper synthesis of proteins of required quality and quantity. In eukaryotes, there are the sensor, integrator and receptor genes.

Properties of DNA

Three important properties of DNA are to be considered in biochemical genetics. They are 1) duplication or replication of DNA, 2) transcription, and 3) translation through RNA.

Replication of DNA: The DNA molecule duplicates itself. This is designed primarily to allow for hereditary transmission by the formation of exact copies of the original DNA. If there is any change in the chemical nature of DNA when it duplicates, the hereditary characteristics will change, which should not happen.

Genetic information should be scrupulously transmitted to the progeny and hence the replication of DNA should be carried out with high fidelity to maintain genetic stability within the organism and the species. This is done by the method of semi-conservative replication. The two strands unwind partly for this purpose and each strand acts as a template for the synthesis of a new strand. When the new strand is formed, its nitrogen bases will be complementary to those in the parent template strand. The work of Meselson and Stahl with *E.coli* has provided proof for the semi-conservative replication.

Both the strands can function as templates.

The new DNA molecule formed will have one DNA strand of the parent and will be identical with the parent DNA.

The substrates are ATP, GTP, CTP, and TTP, and Mg^{2+} as the co-factor. Enzyme in the prokaryotes is DNA polymerase III* (Pol III*) with Co-polymerase III* (Copol III*), while in the eukaryotes it is DNA polymerase alpha (maxi). The enzyme can bring about the synthesis of new DNA in the 5' → 3' direction only. When synthesis takes place, both strands are used as independent templates and two new molecules are formed. As the enzyme can effect the synthesis of the new DNA in 5' → 3' direction, the leading strand will be continuous while the lagging strand is discontinuous as the enzyme has to 'turn its back', as it were, to do the replication of the other parent.

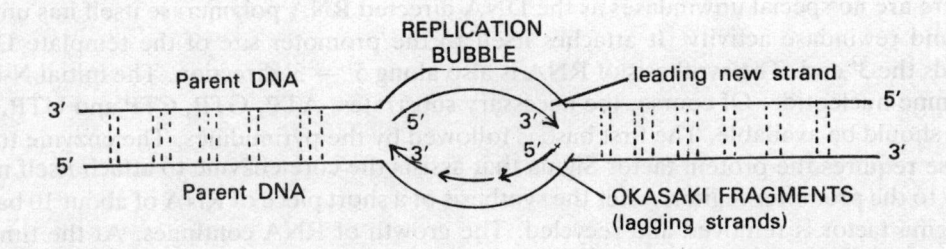

It does not require the entire DNA helix to separate. Certain portions unwind with the help of enzymes 'swivelases', and unwinding proteins, which produce nicks which are later sealed. Stabilising proteins attach themselves to the active region. As unwinding takes place, more stablishing proteins attach themselves to it and there is positive co-operativity.

DNA synthesis requires the formation of short RNA strands. The new DNA synthesised will be a continuation of these short RNA strands. In the eukaryotes the short RNA strand is about 10 nucleotides in length. The RNA and short DNA (about 150 bases) combination is known as Okasaki fragment.

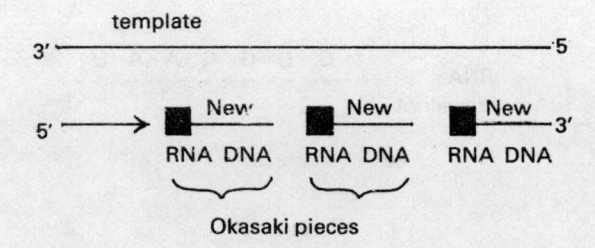

Subsequently, the head RNA pieces are removed by endo-nucleases. The small DNA pieces are ligated with DNA ligase. In the prokaryotes, the head RNA pieces will have 50 – 100 pieces and the short pieces of DNA of Okasaki fragments have 1000 – 2000 bases each.

In the mitochondrial DNA, the head piece of RNA is retained and the DNA molecule is circular and double-stranded.

Transcription: RNA synthesis: It is interesting to note that for DNA synthesis, the RNA head piece formation is a 'must', while for RNA synthesis, DNA template is essential. The formation of m RNA from DNA is called *transcription*.

The enzyme that brings about transcription is transcriptase or DNA directed RNA polymerase II. Other forms of RNA, ribosomal RNA and t RNA, are also formed from the DNA template with RNA polymerase I and III, respectively.

Unlike in DNA synthesis, during which both the strands of the parent DNA are active, in RNA synthesis, only one strand acts as a template at a time. This active strand is called the template (or sense) strand. But the so-called anti-template strand can function as a template strand on a different occasion.

There are no special unwindases as the DNA directed RNA polymerase itself has unwindase and rewindase activity. It attaches itself to the promoter site of the template DNA towards the 3' end. The synthesis of RNA is also along 5' → 3' direction. The initial N-base is guanine nucleotide. Of course, the necessary substrates, ATP, GTP, CTP and UTP, and Mg^{2+}, should be available. The first base is followed by the pyrimidines. The enzyme transcriptase requires the protein factor Sigma that assists the core enzyme to attach itself more tightly to the promoter region. After the synthesis of a short piece of RNA of about 10 bases, the Sigma factor is removed and recycled. The growth of RNA continues. At the time of release, another protein factor 'Rho' (ρ) has to attach itself to the enzyme. It is released along with the processed RNA molecule. The 'Rho' factor is also recycled. Post-transcriptional modifications, like the removal of transcripts of 'introns' and splicing of 'exons-transkripts', attachment of poly-A-tail, formation of pseudo-uridine, dihydro-uridine and thymine, take place in the RNA molecules produced wherever applicable.

```
                    5' anti template
DNA         - - - - - - - - - - - - - 3
             G T G G A A T

             3' C A C C T T A 5'
DNA         - - - - - - - - - - - - -
                        template

             G U G G A A U
RNA         - - - - - - - - - - - - -
transcript  5'                    3'
```

Though transcription is confined to the formation of mRNA from DNA, other RNAs like tRNA and rRNA also are produced from DNA.

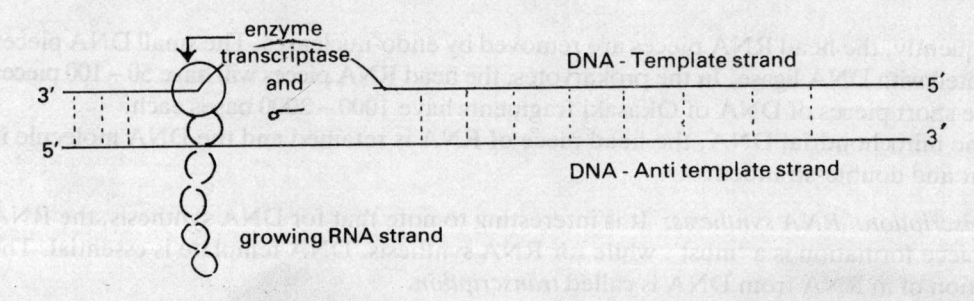

```
              enzyme
              transcriptase         DNA - Template strand
         3' ─────────○──────────────────────────── 5'
                     │ and
                     │ σ
         5' ─────────○──────────────────────────── 3'
                     ()              DNA - Anti template strand
                     ()
                     () growing RNA strand
```

The transcription of DNA has been proved by the work of Cohen, Volkin and Spiegleman on *E.coli* infecting it with T_2 bacteriophage. The base ratio of mRNA was similar to that of DNA (of T_2 phage) with uracil in the place of thymine. Again, hybrid helices were formed between the DNA of T_2 bacteriophage and mRNA produced in *E.coli* infected with T_2 phage. When the same mRNA was gently heated with genetically unrelated DNA, no hybrid was formed. Again, RNA synthesised from the single stranded DNA of ∅ X 174 bacteriophage had the same base composition for complementary bases as shown by Hurwitz.

DNA (single-stranded)	RNA
A = 1	U = 1
T = 1.33	A = 1.33
G = 0.98	C = 0.98
C = 0.75	G = 0.75

Thus the transcription of DNA has been demonstrated unequivocally. Only about three per cent of DNA is used for transcription of mRNA while the rest could be used for producing tRNA and rRNA. As mRNA is short-lived, every time protein synthesis has to take place, fresh transcription should occur.

The question now arises as to how the required information is passed on to mRNA by DNA.

Codons of mRNA (Genetic code): The message for the incorporation of suitable amino acids in a particular sequence is contained in the portions of DNA called 'cistrons'. The entire length of DNA may not be involved and there may be commas and pauses in the molecule. As deoxy sugar phosphate and the quality of the N-bases are the same for the DNA molecules of different species, it should be only the sequence of N-bases that should be concerned with the actual transmission of information regarding the sequence of amino acids in a protein. The sequence of amino acids in a protein synthesised in a cell therefore depends on the sequence of N-bases in the cistrons of DNA (the genes). This has ushered in the concept of codons for mRNA which are decided by the sequence of the N-bases in mRNA. Nearly 20 different amino acids have to be incorporated into protein. If each one is to have a codon, there should be a minimum of 20 codons. Of the four bases in mRNA, if the sequence of two bases is considered a codon, there could be only 4^2 i.e., 16 codons which fall short of the minimum of 20 codons required. Hence, triplet codons were suggested, i.e., sequence of three N-bases, meaning one codon. According to this, there could be $4^3 = 64$ codons. There is experimental support for the concept of triplet codons. In addition to an availability of 20 codons for 20 amino acids there are actually 44 extra codons. It has been shown that the same amino acid can have even six codons (degenerate codes). There are two chain initiating codons and three chain terminating codons, which start and stop protein synthesis, respectively. The codes are universal and non-overlapping. A recent finding of one gene controlling the synthesis of more than one polypeptide in ∅ × 174 bacteriophage suggests rare overlapping of codons.

Precise information about the particular codon for each amino acid came from the work of Nirenberg and Matthaei. Using the homo-polynucleotide (i.e., something like an RNA having only uracil and no other bases), Poly U, it was shown that the polypeptide synthesised was only polyphenyl alanine. -U-U-U-U-U-U-U- coded only Phen-ala-Phen-ala-Phen-ala. So, the codon for phenyl alanine could be only UUU of mRNA. This is a breakthrough in biochemical genetics. Extending the work with the poly UG (i.e., something like an RNA containing only uracil and guanine, a co-polynucleotide), codons for many other amino acids were deciphered. For this, the ratio of UUU to UUG, UGG and GGG was matched with the ratio of phenyl alanine to other amino acids in the protein synthesis. As an example, the ratio of UUU to UUG in the Poly UG was 5:1. The ratio of phenyl alanine (coded by UUU) and valine was 5:1. Hence, valine should have UUG as its codon.

In addition, it was found that AUG is the chain initiating code, and AUA in the mitochondria. The chain terminating codes are UAA, UAG and UGA.

As UUU of mRNA codes phenyl alanine and the complementary base for U is adenine, this UUU should have been the result of the transcription of that region of DNA in cistron, containing a sequence of AAA.

$$\begin{array}{ccc} \text{AAA} & = & \text{UUU} & = & \text{Phenyl alanine} \\ \text{(in DNA)} & & \text{(in mRNA)} & & \text{(in protein)} \end{array}$$

Thus the overall control in fixing the amino acid sequence in a protein is by DNA. Not only is mRNA itself synthesised from the DNA template, the codons of mRNA also are decided by the sequence of N-bases in the cistrons of DNA. The Cistron of DNA is the gene. A human DNA molecule may have 27,000 – 60,000 genes or even more.

Translation (patternisation): The actual role of mRNA in protein synthesis is referred to as translation (patternisation). As DNA decides everything, it can be considered to have the third function, i.e., translation which is an indirect function.

Anticodons of tRNA: tRNAs also are produced from DNA. The sequence of the N-bases in tRNAs is also decided by the N-base sequence of DNA. As in the formation of RNA, only complementary bases are involved. AGCTA in DNA will lead to a sequence of UCGAU in tRNA. The sequence of three N-bases in tRNA in a particular location is called 'anti-codon'. In the actual protein synthesis, the codons of mRNA and the anti-codons of tRNA amino acid complex must match with each other. For a codon of UUU in mRNA, the anti-codon of tRNA would be AAA.

Mutation

Genetic variants are known in man and other species in which either certain proteins are lacking or some abnormal proteins are present (for details refer chapter on 'Inborn Errors of Metabolism'). This is explained on the basis of gene (DNA) mutations. Mutation of regulator or operator gene results in changes in the quantity of protein synthesised, and that of structural genes results in the production of proteins different in their amino acid sequence from natural proteins. Mutations occurring in the base sequence of DNA usually affect only one base and are called point mutations. They can occur by the replacement of one base by another, deletion of one or more bases (frame-shift mutation) or insertion of new bases (also frame-shift mutation). If a pyrimidine base is changed to another pyrimidine, or a given purine is changed to another purine it is called transition, while a change from a given purine is either of the pyrimidines or a change from a given pyrimidine to either of the purines is called transversion. Point mutations can occur due to the influence of ionising radiations, high energy radiations or chemical mutagens like mustard gas, nitrous acid, etc. Mutations have also been found to be due to the absence of the whole or parts of chromosomes. The change of position of a gene in a chromosome may also result in an abnormality.

Thus the mutation of DNA (gene) causes inherited errors of metabolism like phenyl ketonuria, galactosemia, hemoglobinopathies and diabetes mellitus.

Regulation of gene expression—the Operon

Diseases may be caused when the quality and the quantity of protein synthesised are affected. The quantity can be affected by the rate of its synthesis or degradation. For example, in diabetes mellitus, the disease is due to decreased amounts of insulin (protein-hormone) synthesised. Thus it is necessary that the quantity of protein synthesised is properly controlled. Other examples are thalassemia and Heavy Chain disease. A regulation of the rate of protein synthesis (which includes enzyme proteins) is also under genetic control. The smallest unit of gene expression is a cistron. Two types of mechanisms control the regulation of gene expression. They are 1) induction, and 2) repression. With these mechanisms, much of the additional genetic information is obtained in mammalian cells which possess only about 1000 times more genetic information than that in *E.coli* or the bacterial cell.

Genes may be of two major types, the *inducible* genes and *constitutive* genes. The expression of the inducible gene increases in response to an inducer. Inducers are small molecules. The expression of some genes is constitutive, i.e., they are expressed at a reasonably high rate in the absence of any specific regulatory signal.

The enzymes also are referred as inducible if their activity increases in response to an inducer. Those independent of inducer are called constitutive enzymes.

Induction: Induction is a process by which the synthesis of an enzyme is increased in an organism. Subsequently, the organism shows a changed pattern of metabolism.

E.coli grown on glucose will not ferment lactose as there is absence of ß-galactosidase enzyme required for fermentation, a specific permease for lactose and an acetylase, though the organism has the genes to synthesise the same. If now lactose or certain other ß-galactosides are added to the medium, the three enzymes are induced after glucose is exhausted. The organism can now ferment lactose. Inducers serve as the substrate for the inducible enzymes, in this case, ß-galactosidase, the permease and the acetylase. Compounds similar to the substrate may also be inducers and are termed 'gratuitous inducers'. They are, however, not substrates.

A few additional examples of inducible enzymes are tryptophan pyrrolase, threonine dehydrase, sucrase, the enzymes of the Krebs-Henseleit urea cycle and tyrosine keto-glutarate transaminase. The enzymes independent of inducers are constitutive enzymes.

Induction permits a micro-organism to respond to the presence of a given nutrient in the surrounding medium by producing enzymes for the catabolism of that nutrient. In the absence of the inducer, little or no enzyme is produced. The ability to avoid the synthesis of the unnecessary enzymes in the absence of the nutrient offers biochemical economy to the organism.

The mode of action of inducers is discussed under 'Lac Operon'.

Repression and de-repression: If the synthesis of a substance, which is normally possible, is curtailed by its presence, the process is called repression, e.g., *E.coli* is able to synthesise an enzyme required for the metabolism of tryptophan only in the *absence* of tryptophan. Tryptophan is said to repress the synthesis of the enzyme required for its own metabolism, so also histidine and leucine, which are examples for product feed-back repression. Repression may affect many biosynthetic pathways.

Catabolic repression is the ability of a product or intermediate to repress the synthesis of some or all of the catabolic enzymes concerned, e.g., if glucose is added to a culture of *E.coli*, it represses the synthesis of enzymes concerned with the catabolism of any other carbon source. When exposed to both lactose and glucose as sources of 'carbon', the organism first metabolises glucose, then temporarily ceases to grow before starting to metabolise lactose.

The catabolites of glucose have caused catabolic repression mediating through a catabolite gene activator protein (CAP) and cAMP. The recent view is that it is the deficiency of cAMP that is mostly responsible for catabolic repression. The addition of cAMP relieves the repression.

The repressor substance is a macromolecule such as protein, nucleic acid or nucleoprotein. It may require some small molecules as co-repressors before effecting repression (modified repression), e.g., tryptophan, histidine or leucine are co-repressors in the above example. In the absence of co-repressor, some repressors may not bring about repression.

Some other small molecules can combine with repressor macromolecules and change their nature. In that case, the repressor loses its ability to repress gene expression. This phenomenon is called 'de-repression'. Inducers bring about de-repression.

The lac operon

Lactose metabolism *E.coli* is effected by the 'Lac Operon'. The combination of regulator, operator and structural genes is called the Operon (Jacob and Monod). These different genes are present in the same molecule of DNA and are situated in such a way that they are physically associated. The Lac Operon consists of an 'i' gene (regulator gene), an operator gene, and three structural genes, 'Z' gene to form ß-galactosidase, 'Y' gene for permease and 'A' gene for acetylase.

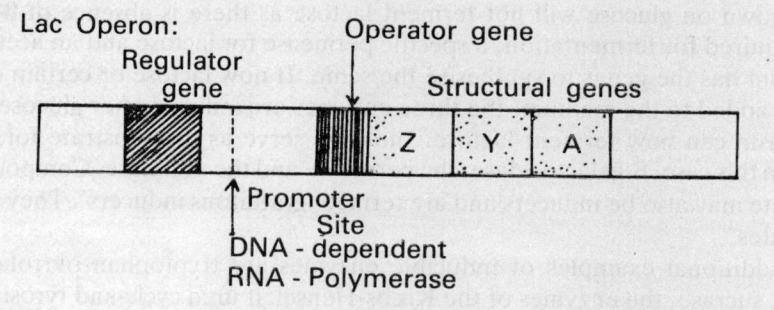

Lac Operon:

Regulator gene | Operator gene | Structural genes Z Y A

↑ Promoter Site
DNA - dependent RNA - Polymerase

The expression of the normal 'i' gene is constitutive and independent of the inducer. It produces sub-units of the lac repressor. These assemble into a lac repressor molecule (protein), with a molecular weight of about 40,000. This repressor protein has a high affinity for operator locus. The operator locus is between the promotor site in which the DNA-dependent RNA polymerase acts and the beginning of the Z-gene.

The repressor protein attaches itself to the operator locus and prevents the transcription of the structural genes Z, Y and A which are inducible. The enzymes for the catabolism of lactose are not translated when there is repression. There are normally 20 – 40 repressor molecules and one or two operator loci per cell.

The addition of lactose or a gratuitous inducer results in prompt de-repression and induction. The inducer molecules react with the repressor molecules attached to the operator loci as well as in cytosol. The repressor is detached from the operator locus. The DNA-dependent RNA polymerase already attached to the Operon at the promotor site causes transcription and generates polycistronic mRNA from the Z, Y and A genes. Thus, the inducer de-represses the Lac Operon.

It has been shown recently that in order that DNA-dependent RNA polymerase attaches to promotor site, the CAP-cAMP complex must be present. So long as glucose is available, cAMP is not produced. Hence the polymerase cannot cause transcription.

BIOCHEMICAL GENETICS

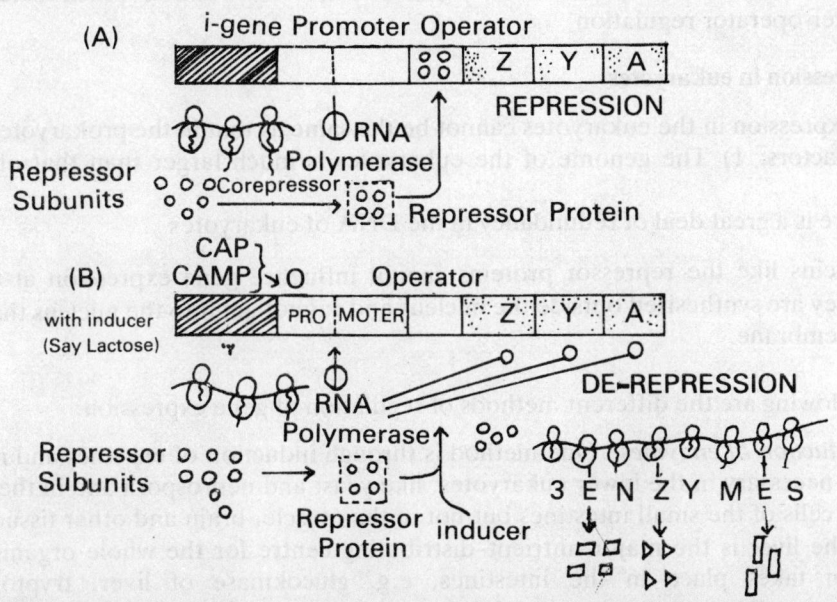

(A)

i-gene Promoter Operator

Z Y A

REPRESSION

Repressor Subunits — Corepressor — RNA Polymerase — Repressor Protein

(B) with inducer (Say Lactose)

CAP cAMP — O — PROMOTER — Operator — Z Y A

DE-REPRESSION

Repressor Subunits — RNA Polymerase — Repressor Protein — inducer — 3 ENZYMES

In summary, for effecting de-repression and induction, the CAP-cAMP complex and the inducer molecules are required. The inducer molecules remove the repressor protein which is a negative regulator while the CAP-cAMP complex is a positive regulator. Most of the knowledge about Operon is limited to the prokaryotes. In the eukaryotes, different mechanisms operate.

Regulation of gene expression in Lambda bacteriophage

The lambda bacteriophage has a double-stranded DNA. It has the right operator OR, and Repressor and Cro genes on the left and the right of the right operator. The expression is such that when the Repressor gene is on, the 'Cro' gene is off, and vice versa. At the lysogenic phase of the bacteriophage, repressor dimers bind to OR-1 and OR-2. These block the transcription of the 'Cro' promoter. When the E.Coli infected by Lambda is exposed to UV, the repressor dimer dissociates due to proteolysis. Hence, there is no binding to OR-1. Now, the enzyme DNA-directed RNA polymerase casues the transcription of the 'Cro' gene (control of repressor and other genes). Thus only one of the two, i.e., repressor or 'Cro' gene, will be expressed.

Attenuation

This type of regulation of gene expression is encountered in *E.coli*. In this, the position occupied by the ribosome brings about a change in the secondary structure of RNA which may or may not allow transcription. The RNA polymerase transcribes TrpL and pauses at nucleotide 90. Now, a ribosome attaches itself at initiator codon AUG at 27 – 29 and commences translation of 14 amino acid leader peptide. When tryptophan is deficient, ribosome stalls at nucleotide 54, where there are two tryptophan codons. Because of this position of the ribosome, the secondary structure of RNA forms a particular hair-pin shape and transcription proceeds beyond TrpL to other genes. On the other hand, if the ribosome positions itself in such a way that a different hair-pin secondary structure is formed, the transcription stops and an immature leader transcript is released. This happens when tryptophan is available in the medium. Such a transcriptional termination is termed 'attenuation' which is independent of promoter-operator regulation.

Gene expression in eukaryotes

Gene expression in the eukaryotes cannot be the same as that in the prokaryotes because of three factors: 1) The genome of the eukaryotes is much larger than that of the prokaryotes.

2) There is a great deal of redundancy in the DNA of eukaryotes.

3) Proteins like the repressor proteins cannot influence gene expression at the DNA level as they are synthesised outside the nucleus and cannot get into the nucleus through the nuclear membrane.

The following are the different methods of regulation of gene expression:

1. Production of enzymes: One method is through induction of enzymes and repression wherever necessary in the lower eukaryotes, like yeast and neurospora and in the liver and epithelial cells of the small intestines but not in the muscle, brain and other tissues. This is because the liver is the major nutrient–distributing centre for the whole organism, while absorption takes place in the intestines, e.g. glucokinase of liver, tryptophan 2,3 dioxygenase, and tyrosine amino transferase.

Hormones like estrogens, androgens, gluco-corticoids, thyroxine, etc. bind to specific receptor proteins in chromatin and induce gene expression, e.g., gluco-corticoids promote gluconeogenesis by increasing the transcription of DNA and translation of required enzymes. Ovalbumin and lysozyme synthesis are augmented by this method.

2. Regulation of gene expression during the cell cycle: (a) During the cell cycle, DNA synthesis is increased only at one phase, i.e., the S-phase. But during mitosis, the synthesis of DNA and RNA is turned off. But RNA and protein synthesis are increased in phases G_1, S and G_2, i.e., there is restriction of the replication of the DNA of genome but segments of the genome are available for transcription.

It is suggested that the cellular concentration of cAMP and cGMP might be responsible for this kind of gene expression. The cAMP levels are low in the rapidly dividing cells but are increased when the proliferation of cells stops. The cGMP and cAMP concentrations have a reciprocal relationship.

(b) *Regulation by histones and non-histone proteins:* It is suggested that the histones of nucleosomes can also regulate gene expression. Though the molecular species of histones is too small in number to regulate 1.25 to 3 million genes in an eukaryote genome, it has been found that during the S-phase of cell cycle there are many post-translational modifications of the different histones. Such modified histones can regulate gene expression.

3. Britten and Davidson's model: These scientists suggest specific sites known as sensor sites adjacent to the integrator gene. The sensor sites are recognised by various molecular signalling agents which bind to them. The signal is picked up by the integrator gene which produces an activator RNA. The activator RNA moves to the receptor sites located elsewhere in the genome on the same or other chromosomes. The structural gene situated adjacent to the receptor site is transcribed. Presumably, many different activator RNAs are produced, and it is in this context that the small nuclear RNA from hnRNA may have a role in the regulation of gene expression.

There could be multiplicity of receptor and integrator genes to influence different combinations of structural genes.

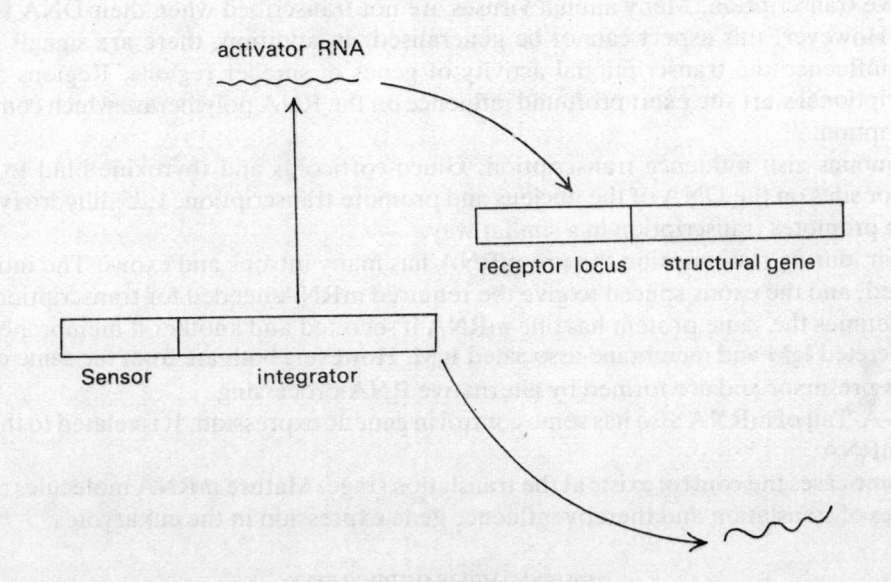

```
                    activator RNA
                       ~~~~~~~
                          |         \
                          |          \
                          ↑           ↓
                    ┌──────────┬──────────────┐
                    │          │              │
                    └──────────┴──────────────┘
                    receptor locus  structural gene

    ┌──────────┬──────────────┐
    │          │              │
    └──────────┴──────────────┘
     Sensor      integrator                    ~~~~~~
```

4) *Gene amplification:* There could be a sudden increase in the number of genes required for the transcription of specific molecules, e.g., in the fruit fly (Drosophila) there occurs during oogenesis an amplification of a few pre-existing genes such as those for the chorion (egg shell) proteins S36 and S38.

5) *Rearrangement of immunoglobulin genes:* Each light-chain is encoded by three distinct segments; the variable (VL), the joining (JL) and the constant (CL) elements (L for light chain). The mammalian haploid genome contains over 500 VL segments, five or six JL segments and perhaps 10 or 20 CL segments. During the differentiation of a lymphoid B-cell, a VL segment is brought from a distant site on the same chromosome to a position in the same chromosome close to the region of the genome containing the JL and CL segments. This DNA rearrangement then allows the VL, JL and CL segments to be transcribed as a single mRNA precursor and later the mRNA proper to produce a specific antibody light-chain. This DNA rearrangement is referred to as the VJ joining of the light-chain.

The heavy chain is encoded by four gene segments, the VH, the D (diversity), the JH and the CH, DNA segments. (H for heavy chain). There is VDJ joining for the heavy-chain as there is VJ joining for light-chain.

Class-switching, i.e., changing from one class to another is seen in immunoglobulins. IgM is the first synthesised but this can switch on to IgA to IgG.

Once the VH-D-JH rearrangement has taken place, no further DNA rearrangement is necessary for IgM as the JH segment of the gene is next to the $C\mu$ gene. The other classes of immunoglobulins as IgA, IgG are obtained by switching from IgM, i.e., by rearranging VDJ from the $C\mu$ to $C\alpha$ gene or $C\gamma$, etc.

Others: The regulation of gene expression may be effected through the control of transcription, i.e., by controlling where and when the transcription has to commence. There are large regions of chromatin that are transcriptionally inactive either constitutively or facultatively, while other areas of chromatin are potentially active. There is evidence that the methylation of the deoxy cytidine residues in DNA may effect gross changes in chromatin so as to preclude its active transcription. Many animal viruses are not transcribed when their DNA is methylated. However, this aspect cannot be generalised. In addition, there are signals in DNA which influence the transcriptional activity of genes in smaller regions. Regions 5' to the transcriptional start site exert profound influence on the RNA polymerase which commences transcription.

Hormones also influence transcription. Gluco-corticoids and thyroxine bind to specific receptor sites on the DNA of the nucleus and promote transcription. 1,25 dihydroxy vitamin D_3 also promotes transcription in a similar way.

Again, during transcription the pre-mRNA has many introns and exons. The introns are removed, and the exons spliced to give the required mRNA needed for transcription.

Sometimes the same protein has one mRNA if secreted and another if membrane bound, e.g., secreted IgM and membrane-associated IgM. However, both are from the same common mRNA presursor and are formed by alternative RNA processing.

Poly-A-Tail of mRNA also has some control in genetic expression. It is related to the stability of mRNA.

In some cases the control exists at the translation stage. Mature mRNA molecules can alter the rates of translation and thereby influence gene expression in the eukaryotes.

BIOSYNTHESIS OF PROTEINS

The biosynthesis of proteins has to take place constantly in the body in view of the dynamic flux of body proteins. Like the anabolism and catabolism of carbohydrates and lipids, the biosynthesis of proteins is nothing but protein anabolism and should come under protein metabolism. However, the biosynthesis of proteins requires separate treatment and that too after a discussion of nucleic acids and the genetic code, because a knowledge of the latter is required for the understanding of protein biosynthesis. It is for this reason that biosynthesis of proteins is discussed after biochemical genetics in this book. It is the polypeptide which is first synthesised, and this transforms itself into the actual protein.

Active transport of amino acids

The building stones required for the construction of proteins are amino acids. As the site of protein synthesis is the cytoplasm (chromatin protein is an exception as it is synthesised in the nucleus), the amino acids from the pools have to reach the actual site, i.e., the ribosomes on the endoplasmic reticulum. The required amino acids get into the cells by active transport as shown by Christensen. Pyridoxal phosphate promotes this active transport. Insulin and growth hormone enhance this process. Water and sodium move along with amino acids while potassium moves out. Insulin stimulates protein synthesis by increasing the amino acid uptake and subsequent incorporation into protein. It also increases the protein synthesis at the level of mRNA by enhancing translation. The growth hormone also is known to stimulate protein synthesis by promoting the transport of amino acids into the cells and also by stimulating the synthesis of DNA as well as the various types of RNAs involved in this process.

Endocrine influence

The growth hormone and insulin increase the rate of protein synthesis. Thyroxine also increases protein synthesis by speeding up metabolic reactions in general. Androgens stimulate protein anabolism. Cortisone inhibits protein synthesis.

The site of protein synthesis inside the cell is the polysomes of the rough endoplasmic reticulum as shown below:

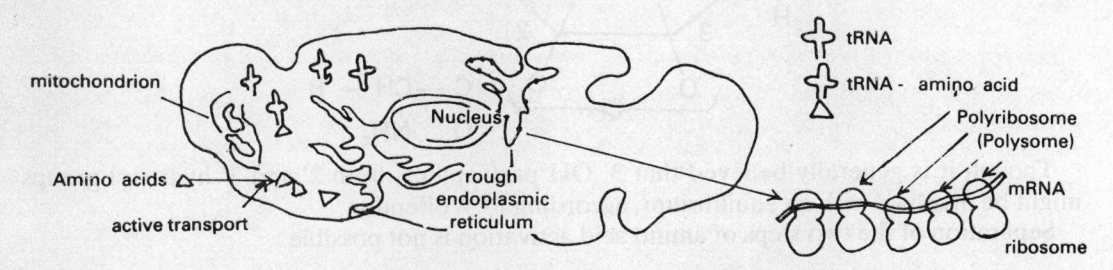

The requirements for protein synthesis are:
1) Amino acids
2) DNA and RNAs
3) Polyribosomes (Polysomes)
4) ATP
5) GTP
6) Mg^{2+}
7) Amino acyl tRNA synthetase
8) Initiation factors IF_1, IF_2 and IF_3
9) Elongation factors EF_1, EF_2
10) Folic acid as FH_4 for lower organisms
11) Peptide synthetase (Peptide transferase)
12) Release factor

Novikoff and Potter showed that the nucleic acids of developing chick embryos had the highest concentration at the time of most rapid protein synthesis. While the DNA of the nucleus controls the synthesis of chromatin proteins, the RNAs synthesise cytoplasmic proteins, under the control of DNA. All the types of RNA are involved (these RNAs are synthesised with the help of DNA in the nucleus, but move out of the cytoplasm to engage themselves in protein synthesis).

tRNA amino acids

Before any molecule of amino acid could be used for polypeptide synthesis, it has to be activated and converted to tRNA amino acid. Each of the 20 amino acids should have a definite tRNA for the formation of tRNA-AA, the formation of which is catalysed by the enzyme tRNA amino acyl synthetase for both the steps.

i. amino acid + enzyme + ATP $\xrightleftharpoons{Mg^{2+}}$ amino acyl AMP enzyme + PPi

ii. amino acyl AMP enzyme + tRNA $\rightleftharpoons$ + amino acyl tRNA + AMP + enzyme
(or tRNA-AA)

In tRNA amino acid, the —COOH of amino acid is linked to the hydroxyl groups of the 3' terminal AMP residue of tRNA

tRNA-Cp-Cp-Adenosine - Amino acid
(Common 3' terminus for all tRNAs)

Though it is generally believed that 3' OH participates, both 2' and 3' hydroxyl groups might be involved with an equilibrium, according to Wolfendon.

Separation of the two steps of amino acid activation is not possible.

Recognition of tRNA by amino acyl tRNA synthetase: As all the amino acids have to be taken by the common 3' terminus —CMP—CMP—AMP of tRNAs, the acceptor specificity (i.e., which amino acid should be tagged on) does not lie in the terminus. It is found that the acceptor specificity lies in the double-stranded regions of different tRNAs which are different in the nucleotide sequences. Recognition of the tRNA by the amino acyl tRNA synthetase rests in the dihydrouracil (DHU) loop of tRNA while the thymine-pseudouridine-cytosine (TψC) loop of tRNA binds tRNA-AA to ribosomal surface.

Transfer specificity of tRNA-AA: Once the amino acid has been converted into tRNA-AA complex, even if the amino acid is subsequently changed (say, cysteine to cysteic acid) it could be transferred to a growing peptide. The 'transfer specificity' of tRNA carrying AA is located in the single-stranded loop in the centre of the molecule. This is the coding area called the 'anti-codon' of tRNA which is read from the 3' end that carries the amino acid. So, the actual transfer of amino acid is decided by the anti-codon. Thus, in tRNA alanine, the anti-codon which is located in the single-stranded loop is CGI.

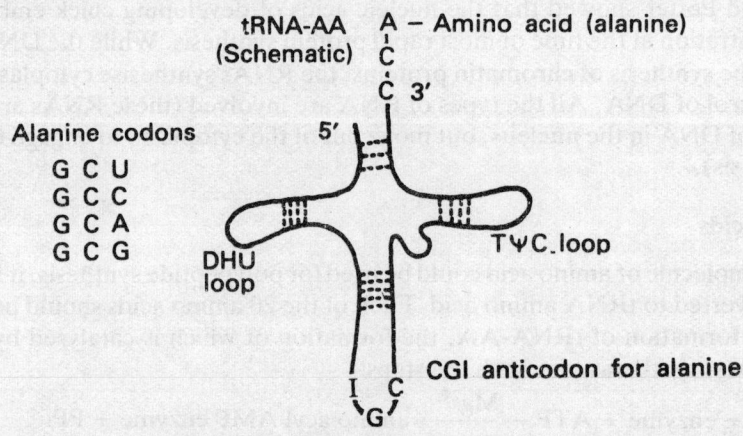

Alanine codons

G C U
G C C
G C A
G C G

Heterogeneous nuclear RNA (h$_n$ RNA), pre-mRNA and messenger RNA

Very recently it has been shown that the mRNA proper required for protein synthesis in the cytosol is obtained by the cleavage of a precursor, i.e., h$_n$ RNA synthesised in the nucleus, having a half-life of about 23 minutes. This is 10 – 100 times bigger than mRNA, has 400 – 4000 nucleotides and is bound to macromolecular proteins called 'informers'. The h_n RNA contains a unique sequence and repetitive sequences. A few h_n RNA molecules may function as pre-messenger RNA which gives rise to mRNA. The pre-mRNA has regions known as introns (intervening sequences not required for translation) and exons (active regions). The introns are excised and exons are spliced together to give the mRNA proper. When ready for further protein synthesis, mRNA has, at its 5' end, a cap of guanosine triphosphate with methyl group in position 7(G^mTP) (this cap might have come from pre-mRNA) and possibly a 2-oxy-methyl group in the second nucleotide. mRNA also has a poly-A-tail either from its precursor or added in the cytosol. The capping of G^mTP is required for the translocation of the messenger and for binding the mRNA to the 40s of the ribosome. Poly-A-tail may maintain the intra-cellular stability of specific mRNA. Some mRNAs including those for histones do not contain poly-A.

Ribosomes and rRNA

It is in a combination of ribosomes (polysomes) that the biosynthesis of protein takes place. Poly-ribosomes attached to the rough endoplasmic reticulum synthesise 'transport' proteins while the free polysomes in cytosol synthesise cytoplasmic proteins. A single mammalian ribosome can translate about 200 codons in two minutes and produce a protein with a mol. weight of 20,000

In the eukaryotes, each ribosome is 80s made of two pieces, one 40s and the other 60s. There is a 18s portion of 40s in which binding is initiated. The ribosomes have the ribosomal RNA and 'interferance factors' which offer specificity to them. They also take up 'initiation factors' (proteins, IF_3 and IF_2).

In *E.coli* the pieces are 30s and 50s to make one of 70s pieces. The 60s piece of eukaryote ribosome has got two sites, one called the amino acyl site (A) and the other the peptidyl site (P). The two pieces of ribosomes are separate at the beginning of peptide synthesis.

How DNA actually prescribes the sequence of amino acids in a peptide

The cistrons of DNA give the required information to the mRNA synthesised by DNA during transcription. mRNA gets the instruction imprinted in the codons. The codon of mRNA would fit in only with a particular anti-codon of tRNA bearing the AA. When the codon of mRNA and the anti-codon of tRNA-AA fit in with each other, the amino acid from tRNA-AA complex will be transferred for the formation of the peptide chain and its growth. As the actual formation and growth of peptide has to take place in the ribosomes, all the three main participants, i.e., ribosomes, mRNA and tRNA-AAs, should come together. Only then, the peptide synthesis will take place. In addition, there should be trigger for the initiation, elongation, translocation, termination, energy for the real process, enzymes and a control of the rate of production of protein.

It has already been shown that each amino acid should be having a definite codon in mRNA which also has chain initiating and chain terminating codons. The codons are all a sequence of three nucleotides. For example, UUU is the codon for phenyl alanine. The mRNA message is read from left to right, from the 5' end, i.e., the codon on the left of mRNA codes the N-terminal amino acid, and one on the far right, the C-terminal.

DNA which spells out the genetic message is in the nucleus, and cannot cross the nuclear membrane. Hence, it uses messenger RNA(mRNA) for translating the message of protein synthesis, i.e., the sequence of amino acids (which amino acid is first, which is second and so on) for a particular protein. We know that each protein has a definite sequence which decides its structure and function

The DNA acts as a template to synthesise a mRNA. In so doing, the mRNA formed will have the complementary bases to those in the DNA. In other words, if there is adenine in DNA it will be transcribed as uracil in the mRNA. If there is guanine it will be transcribed as cytosine. We know that the genetic information is contained in the cistrons or the genes of DNA and that the sequence of three nucleotides constitutes a codon of mRNA. If, for example, the cistron of DNA has the following sequence,

— A A T G C T A T T G C G T A —

the mRNA synthesised from this DNA (template) would be,

— U U A C G A U A A C G C A U —

(That is, wherever there is A in DNA there would be U in mRNA; wherever there is T in DNA, there would be A in mRNA, wherever there is C in DNA, there would be G in the mRNA and wherever there is G in DNA, there would be C in mRNA). This is how the genetic message of DNA in its cistrons or genes is transcribed into the codons of mRNA. Hence, the formation of mRNA from DNA is called transcription. Each triplet codon of mRNA codes for a particular amino acid in the protein synthesis, e.g., UUU codes for phenyl alanine, GCU codes for alanine, AUG for methionine, (chain initiator codon), AAA for lysine and UAG (chain terminating codon). mRNA, with the help of the codons obtained by transcription of DNA, translates the coded message into the specific sequence of the amino acids of a protein or enzyme.

T A C A A A C G A T T T A T C (in DNA-cistron)
transcription ↓
A U G U U U G C U A A A U A G (mRNA) codons
translation ↓
     AUG   UUU   GCU   AAA   UAG
Start-methionine - phenyl alanine - alanine - lysine - Stop.

mRNA, though formed from the DNA template in the nucleus, is formed as a bigger molecule of RNA known as pre mRNA molecule which may undergo cleavage and splicing, giving mRNA. The mRNA so formed comes out of the nucleus. It has a 'cap' of methyl guanosine triphosphate (G^mTP and a 'tail' of poly adenylate (A – A – A – A –). Four or five ribosomes of 80s associate with the processed mRNA like beads on a string. Each ribosome of 80s is made up of two units: one of 40s and the other of 60s. The ribosomal RNA is involved in the drama of protein synthesis as a component of ribosomes. Protein synthesis takes places in all the five ribosomes, at the same time.

Wobble theory

Eck and Jukes showed that in a codon of mRNA, the sequence of the first two bases was more important than the third. This is called the 'Shared Doublet Concept'. AGA and AGG are codons of arginine. It may be noted tht the first two nucleotides are the same. The anti-codon is UCU. It can fit in with AGA fully as all the three bases are complementary. It can also fit in with AGG with 'wobbling'. Likewise, GGU, GGC and GGA are codons for glycine while CCI is the anti-codon which can work by the 'wobbling mechanism' wherever necessary. The Wobble theory was put foward by Crick.

Rich's scheme

Since a single ribosome is only about 250 Å and the length of a strand of mRNA is one of the order of 1500 Å, Rich and co-workers suggested that peptide synthesis will be taking place at the same time in more than one ribosome, i.e., polysome (ergosome). The individual ribosomes are held by mRNA, like beads in a string. Normally, it is a pentamer and occasionally a tetramer or hexamer.

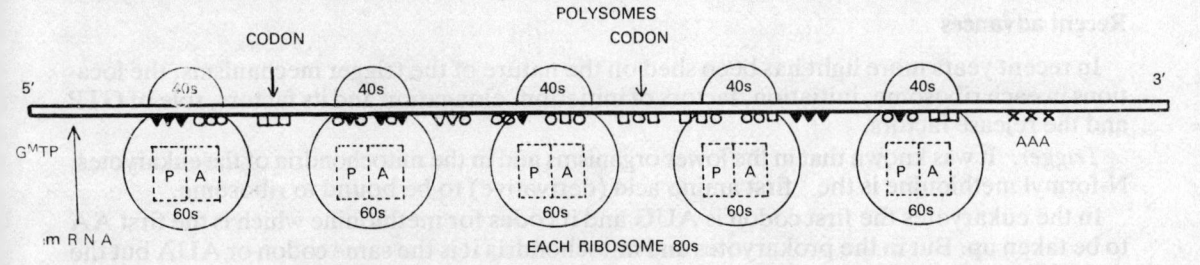

(This has been verified by electron microscopy and treatment of polysome with RNAse when the mRNA alone gets hydrolysed and individual ribosomes come off separately.)

According to Rich, peptides are synthesised by a sequential addition of amino acids beginning from the N-terminal end. Individual ribosomes start at the 5' end of the mRNA and gradually move along the mRNA strand. As the ribosome moves over codon after codon in mRNA, the peptide chain increases in length by the sequential addition of specific amino acids. The amino acid to be selected is decided by the codon. The codon of mRNA and the anti-codon of the required tRNA-AA will base-pair and the AA will be incorporated in the growing polypeptide. At the end of the mRNA molecule, there is the chain terminating codon. There are release factors. Hence, protein synthesis stops, and ribosome bales out of the mRNA. The same ribosome with its ribosomal RNA (rRNA) can be used over and over again. mRNA is read from the 5' end.

As the gap between each ribosome is 50 – 150 Å (equivalent to a distance of about 80 nucleotides), a peptide growing on one ribosome cannot be transferred to the neighbour.

A typical mRNA is hydrolysed within a day.

rRNA is believed to bind mRNA and the particular enzymes necessary to catalyse peptide bond synthesis.

Recent advances

In recent years more light has been shed on the nature of the trigger mechanisms, the locations in each ribosome, initiation, factors of initiation, elongation and its factors, role of GTP and the release factors.

Trigger: It was known that in the lower organisms and in the mitochondria of the eukaryotes, N-formyl methionine is the first amino acid (derivative) to be bound to ribosome.

In the eukaryotes the first codon is AUG and it codes for methionine which is the first AA to be taken up. But in the prokaryotes and mitochondria it is the same codon or AUA but the AA is N-formyl methionine. It is for formylation, folate as FH_4 is needed in the prokaryotes. However, in the proteins formed, methionine is not a common N-terminal amino acid, as an enzyme removes the same in the prokaryotes even before peptide synthesis is complete. Two types of methionyl tRNA (F and M) are known. In the eukaryotes, it is methionyl tRNA that is accomodated first in the synthesis of proteins in the endoplasmic reticulum.

GTP plays many roles, its function as the G^mTP cap being one of them.

Initiation: Firstly, mRNA, the 40s ribosomal sub-unit and the initiation factor 3 (IF-3) join together. GTP, 1F2 and tRNA methionine join separately to form a complex. The mRNA-40s-IF3 complex and the tRNA - meth - GTP IF_2 complex interact in the presence of IF_1 to form a bigger complex. This with the 60s ribosomal unit brings abou the fusion of 40s with 60s and gives the mRNA-80s-tRNA-meth, the last one occupying the peptidyl site.

Energy for fusion is given by GTP hydrolysis.

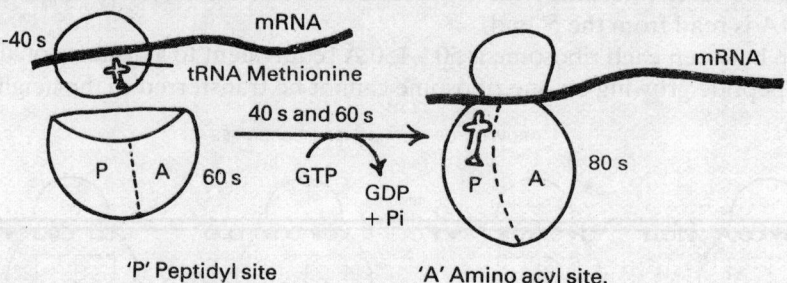

'P' Peptidyl site   'A' Amino acyl site.

Accommodation: The A side is vacant, and tRNA valine, if the second codon is for valine, will join with GTP and EF_1 (EF-T in prokaryotes) and enter the amino acyl site, energy being given by GTP. Now the tRNA meth, and tRNA valine are occupying the adjacent sites 'P' and 'A' in the 60s part.

Elongation: The two amino acids now interact and form a peptide bond. This catalysed by peptidyl transferase. Now, in the 'A' site we have the tRNA peptide but in the 'P' site there is only tRNA without AA as the latter has crossed the floor and joined the second amino acid. This free tRNA vacates the 'P' site.

Translocation: GTP and EF_2 (EG-G in prokaryotes) react with the above 'kit' GTP undergoes hydrolysis, and the energy of hydrolysis makes the ribosome move along the mRNA from the 5' end towards the 3' side. In such a movement (mechano-chemical process), the tRNA peptide combination which was in the 'A' site occupies the 'P' site, Again, the 'A' site falls vacant and the third codon can now direct the acceptance of specific tRNA-AA by base-pairing with the anti-codon of tRNA. As before, GTP and EF_1 are needed for this accommodation.

Elongation again: Peptide elongation takes place. Every time the ribosome moves and a new codon comes against the 'A' site, an amino acid is added. The prescribed sequence of the amino acids dictated by DNA is established by the codon-anti-codon interaction. The

313

BIOCHEMICAL GENETICS

INITIATION

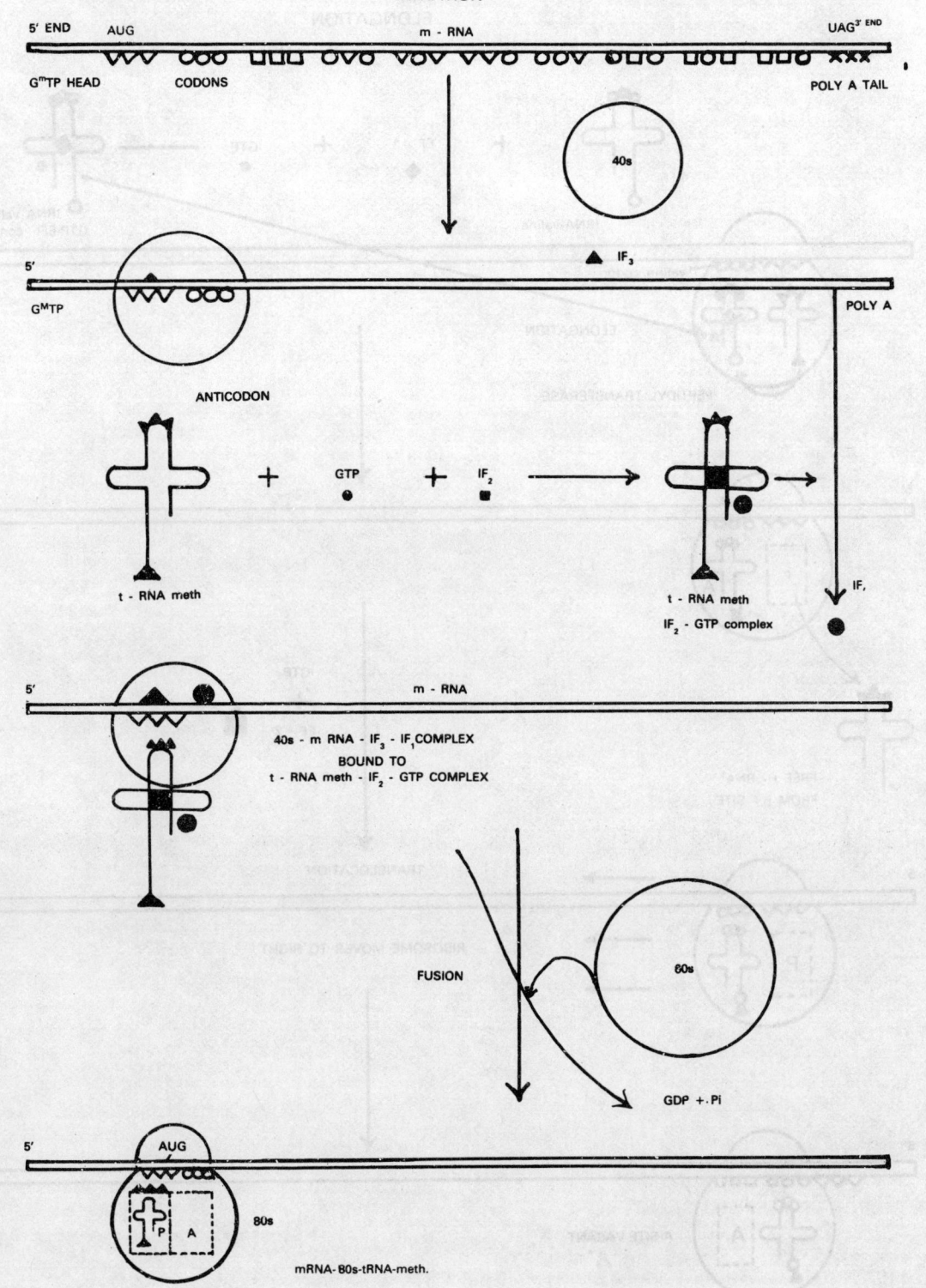

5' END AUG   m - RNA   UAG 3' END

G^mTP HEAD   CODONS   POLY A TAIL

▲ IF_3

G^MTP   POLY A

ANTICODON

t - RNA meth   +   GTP   +   IF_2   →   t - RNA meth   →   IF_1
                                         IF_2 - GTP complex

m - RNA

40s - m RNA - IF_3 - IF_1 COMPLEX
BOUND TO
t - RNA meth - IF_2 - GTP COMPLEX

60s

FUSION

GDP + Pi

AUG

P A   80s

mRNA - 80s - tRNA - meth.

314   MEDICAL BIOCHEMISTRY

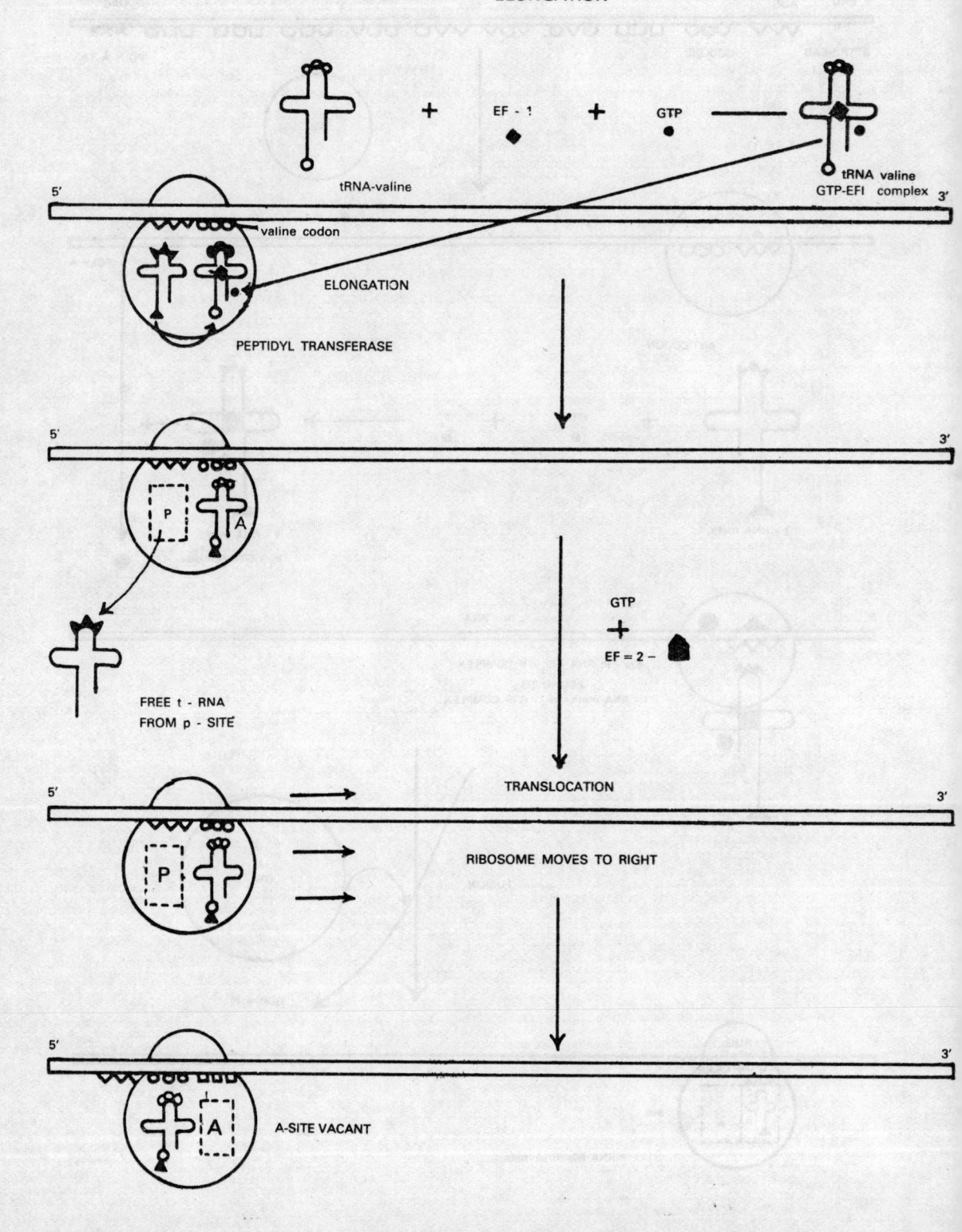

ELONGATION

tRNA-valine + EF-1 + GTP → tRNA valine GTP-EFI complex

valine codon

ELONGATION

PEPTIDYL TRANSFERASE

FREE t-RNA FROM p-SITE

GTP + EF = 2 –

TRANSLOCATION

RIBOSOME MOVES TO RIGHT

A-SITE VACANT

unwanted amino acds will not be accepted as their tRNA-AA complexes will have different anti-codons.

Termination: After the polypeptide has grown to the stipulated length, protein synthesis has to stop. This takes place by the appearance of the chain terminating codons like UAG, UGA, UAA which are also known as non-sense codons appearing in the 'A' site. At this stage, the release factor in conjunction with GTP and the peptidyl transferase promotes the hydrolysis of the bond between the peptide and the tRNA occupying the 'P' site.

Thus, hydrolysis releases the protein and tRNA from the 'P' site. The 80s ribosome, upon hydrolysis and release also dissociates into its 40s and 60s sub-units which are then recycled. The release factors are proteins which hydrolyse the peptidyl tRNA bond when a non-sense codon comes against the 'A' site.

As indicated earlier, a single mammalian ribosome is capable of translating in two minutes about 200 codons into a protein with a molecular weight of approximately 20,000.

The peptide formation can be explained in the form of an equation

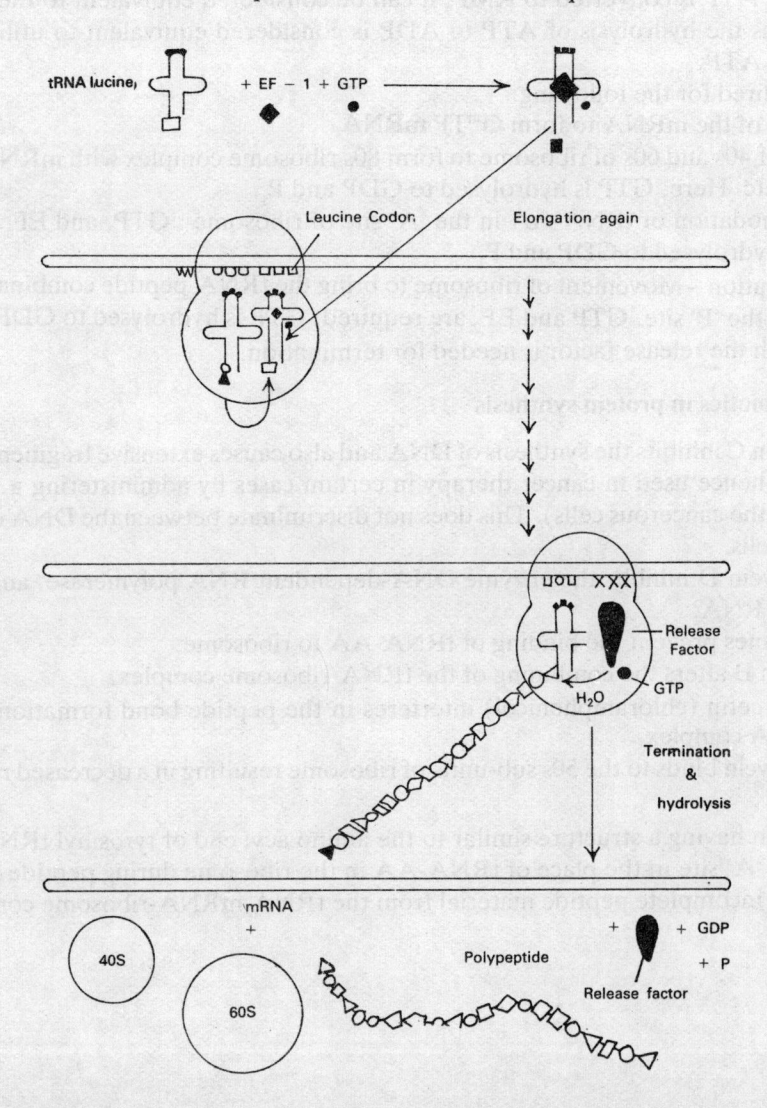

It is not necessary that all the ribosomes of the cell should be engaged in protein biosynthesis in as much as the entire DNA is not required for transcription. In hemoglobin synthesis, only about five per cent are active in protein biosynthesis.

Secondary, tertiary and quaternary structures

Once the polypeptides are synthesised, these are changed to full-fledged proteins by various types of forces as covalent bonds between two sulphur atoms (disulphide), hydrogen bonds, hydrophobic and van der Waal's forces, and polar (salt) linkages. Except the primary structure, the other levels of protein structure are not under genetic control. This is clearly shown by the denaturation of proteins like hemoglobin and bringing them back to the native state.

Energetics of protein synthesis

ATP is required for the activation of amino acid and formation of tRNA-AA. As one molecule of ATP is converted to AMP, it can be considered equivalent to the utilisation of two ATPs as the hydrolysis of ATP to ADP is considered equivalent to utilisation of one molecule of ATP.

GTP is required for the following:
1) Capping of the mRNA to form G^mTP mRNA
2) Fusion of 40s and 60s of ribosome to form 80s ribosome complex with mRNA, methionyl tRNA, etc. Here, GTP is hydrolysed to GDP and P_1.
3) Accommodation of tRNA-AA in the 'A' site of ribosome : GTP, and EF_1 are required. GTP is hydrolysed to GDP and P_1
4) Translocation – Movement of ribosome to bring the tRNA-peptide combination from the A site to the 'P' site. GTP and EF_2 are required. GTP is hydrolysed to GDP and P_1.
5) GTP with the release factor is needed for termination.

Role of antibiotics in protein synthesis

Mitomycin C inhibits the synthesis of DNA and also causes extensive fragmentation of nuclear DNA (hence used in cancer therapy in certain cases by administering a dosage which affects only the cancerous cells). This does not discriminate between the DNA of normal and cancerous cells.

Actinomycin D inhibits the enzyme DNA-dependent RNA polymerase, and thereby the synthesis of RNA.

Tetracyclines prevent the binding of tRNA-AA to ribosome.

Neomycin B alters the combining of the tRNA-ribosome complex.

Chloromycetin (chloramphenicol) interferes in the peptide bond formation on the ribosome-mRNA complex.

Streptomycin binds to the 50s sub-units of ribosome resulting in a decreased rate of protein synthesis.

Puromycin having a structure similar to the amino acyl end of tyrosinyl tRNA is incorporated in the 'A' site in the place of tRNA-AA in the ribosome during peptide synthesis and releases the incomplete peptide material from the tRNA-mRNA-ribosome complex.

BIOCHEMICAL GENETICS

Diagrammatic representation of effect of antibiotics at different stages of protein synthesis and on DNA.

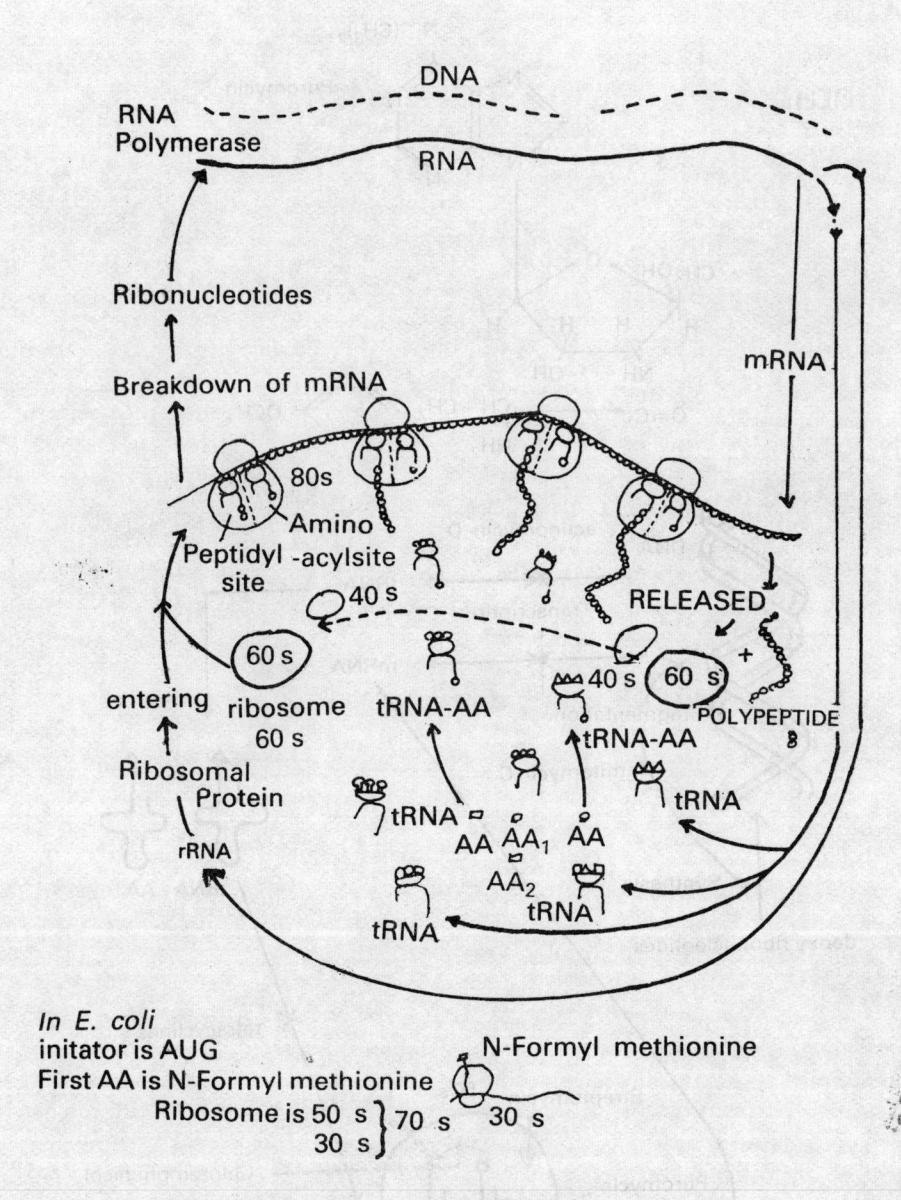

BIRD'S EYE VIEW OF PROTEIN BIOSYNTHESIS

16 Metabolism of Nucleoproteins, Purines and Pyrimidines

CATABOLISM

Catabolism of nucleoproteins

Nucleoproteins in the diet are broken down step by step during digestion in the alimentary tract. This is brought about by gastric, pancreatic and intestinal enzymes. It is the pancreatic nucleases, chiefly ribonuclease and deoxyribonuclease and phospho-diesterases, which hydrolytically depolymerise nucleic acids into their constituent nucleotides. Hence, these enzymes are designated as polynucleotidases. The next step invloves the action of intestinal nucleotidases which are phosphatases concerned with the hydrolytic dephosphorylation of nucleotides, removing phosphoric acid in the process and forming nucleosides. Digestion is completed by the intestinal nucloesidases which are phosphorylases catalysing the phosphorylation of the sugar moiety of nucleosides, setting free the purine and the pyrimidine bases. Adenosine is largely absorbed as such.

Tracer studies indicate that guanine, uracil, cytosine and thymine are largely catabolised after absorption, while labelled adenine is found to be incorporated into the tissue nucleoprotein and some of it is transformed into guanine. It is inferred, therefore, from such experimental evidences that, except for adenine, none of the free purines or pyrimidines of the diet are direct precursors of tissue nucleic acids. However, similar findings were not observed when the purines or pyrimidines were administered as nucleotides. Labelled cytidine and uridine are incorporated into RNA and to a smaller extent in DNA as cytidine and thymidine, and even into the purines of DNA but not of RNA. Injected thymidine is however incorporated into DNA. This has proved to be a valuable technique for labelling newly produced DNA in biological materials. Tritiated thymidine is used for this purpose.

Catabolism of purines

Since adenase is present in negligible amounts in the human liver, Christman has envisaged an alternate pathway for the catabolism of adenine according to which adenosine deaminase secreted by the liver catalyses the deamination of adenosine forming inosine which is the nucleoside of hypoxanthine. Inosine is acted upon by the specific nucleoside phosphorylase yielding the ribose phosphate and hypoxanthine. On the other hand, guanine is straightaway deaminated by the hepatic guanase to xanthine. Hence, xanthine provides the meeting point between the metabolism of adenine and that of guanine.

Xanthine as such, and also xanthine derived from hypoxanthine by xanthine dehydrogenase is further oxidised by the very same enzyme present in the liver. Milk is another source of this enzyme. It is known as 'Schardinger's enzyme'. It is an enzyme complex and an aerobic dehydrogenase having riboflavin in the form of FAD, molybdenum and non-heme iron as the non-protein prosthetic factors. Th oxidation of xanthine yields uric acid.

Extremely low xanthine dehydrogenase activity can result in inhertited metabolic defects characterised by a quick glomerular filtration of xanthine. But there is less reabsorption by the renal tubules so that very little of xanthine is oxidised to uric acid. Non-radio opaque xanthine calculi may be formed by this.

In sub-primate mammals like the rat and the dog, uric acid is further oxidised by uricase (urico oxidase) to allantoin. However, in the Dalmatian coach dog, more uric acid is excreted than in man. This is ascribed to a failure in the renal reabsorption of uric acid and not because of absence of uricase. Thus the Dalmatian dog is believed to have a 'leaky kidney' for uric acid!

Tracer studies with ^{15}N labelled uric acid have revealed the miscible pool of uric acid by dilution of the injected labelled compound. In normal persons the miscible pool is on an average slightly higher than 1 g, the plasma content being higher than in other fluids. But in patients suffering from gout, the pool of uric acid may be four times as high or even higher and the serum levels of uric acid are correspondingly high, the normal level being 3 – 6 mg per 100 ml.

Tracer studies have also enabled the elucidation of the turnover of uric acid in the body which denotes the rate of biosynthesis and degradation in the body. It is found that as much as 20 per cent of uric acid lost from the body cannot be accounted for by the amount excreted in urine. The labelled nitrogen is detectable in urea and ammonia and to some extent in bile. However, uricolysis does take place in our body, possibly in the liver itself, and intestinal and bacterial degradation is not believed to be responsible for the uricolytic process.

The 24-hour urinary excretion of uric acid ranges from 0.5 – 1 g and even on an exclusively purine-free diet like milk, there is a constant output of uric acid due to endogenous origin.

Uric acid is actively reabsorbed by the renal tubules, and in this glycine competes with uric acid. It is also significant that there is an active secretion of uric acid by the renal tubules. Lactic acid also competes with uric acid in renal excretion. Uricosuric drugs like salicylates, cinchophen, 11-oxosteroids and ACTH increase the excretion of uric acid by blocking the tubular reabsorption of uric acid.

Uric acid is sparingly soluble in water and soluble in alkali. In acid urine, it tends to precipitate on standing and form renal calculi.

A familiar, though rare, aberration in the metabolism of purines is seen in gout where there is a pronounced hyperuricemia and also elevated excretion of uric acid in urine. Insoluble mono-sodium urates get deposited in this condition as 'tophi' in the ear lobes and joints accounting for arthritic symptoms like stiffness and pain. In this condition, the miscible pool of uric acid is considerably enhanced. Tracer studies have revealed that the rate of synthesis of uric acid from glycine is considerably increased in gout, particularly when the protein content of the diet is increased. Other conditions causing hyperuricemia include polycythemia, chronic leukemia and other blood dyscrasias resulting in a secondary metabolic gout, brought about by increased degradation of nucleic acids and not inherited, unlike the primary metabolic condition.

Catabolism of pyrimidines

The liver is the principal site for the catabolism of pyrimidines. A major metabolic pathway of uracil is the release of carbon-2 of the pyrimidine nucleus as respiratory carbon dioxide. Experiments on rats fed with diet rich in thymine or DNA have shown excretion of large amounts of beta amino iso-butyric acid in the urine while a similar result is not obtained on the administration of RNA. Beta alanine is the end product of cytosine as well as uracil, while beta amino iso-butyric acid is the end product of the degradation of thymine. The excretion of beta amino iso-butyric acid is found to be increase in leukemias and after X-ray irradiation, showing the high destruction of cells. Familial high beta amino iso-butyric acid excretion is reported to be due to a recessive gene. Beta amino iso-butyric acid is found to be commonly excreted in the urine of people of Chinese and Japanese origin, but not in Caucasians. In experimental animals a reversible transamination is shown to be responsible for the conversion of beta amino iso-butyric acid to methyl malonic semi-aldehyde which is also an intermediate in the catabolism of valine. Propionate and, in turn, succinate can be formed from methyl malonic semi-aldehyde. However, in metabolism in man such a degradation is not established, hence the metabolism stops at the stage of beta amino iso-butyric acid. In the absence of other factors, an increased excretion of beta amino iso-butyric acid in urine can take place due to hepatic dysfunction (see chart).

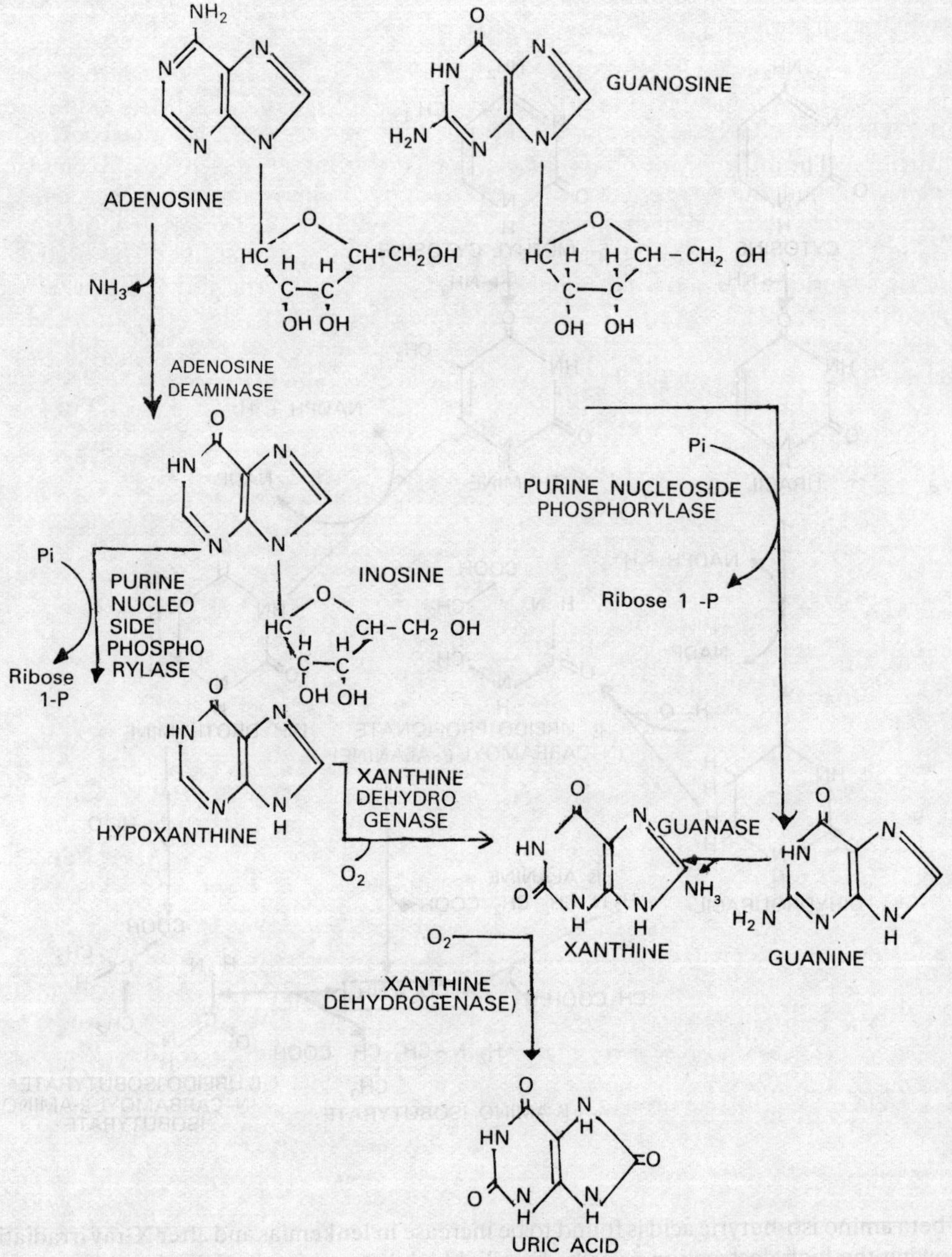

BIOSYNTHESIS

Biosynthesis of purines

It has been conclusively established by tracer studies that the nucleic acids in tissues are mostly synthesised from endogenous sources. While RNA is being constantly synthesised in all the cells and the labelled precursors appear in RNA quickly, DNA is formed only prior to cell division. The incorporation of the labelled precursors is very slow in the cells that are not dividing. The building stones for the biosynthesis of purine derivatives have been identified. Glycine provides the carbons in positions 4 and 5 of the purine nucleus while the nitrogen in position 7 is accounted for by the nitrogen of glycine. Thus, glycine is fully incorporated. The nitrogen at position 1 is derived from the amino nitrogen of aspartic acid while the nitrogens at positions 3 and 9 are derived from the amido nitrogen of glutamine. The carbon atom in position 6 is obtained from respiratory carbon dioxide while the carbons in positions 2 and 8 originate from the one-carbon compound through FH_4. The delta carbon of delta amino levulinic acid derived from the alpha carbon of glycine may be utilised as a carrier molecule for the transfer of the alpha carbon of glycine to the purine ring.

BUILDING STONES FOR BIOSYNTHESIS OF PURINE DERIVATIVES

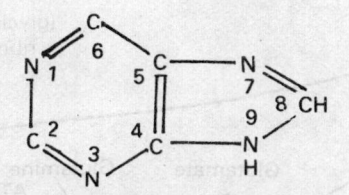

N_1 : From amino nitrogen of aspartate
C_2 and C_8 : from $\propto$ carbon of glycine and FH_4
N_3 and N_9 : from amino nitrogen of glutamine
C_4, C_5 and N_7: from glycine
C_6 : from respiratory carbon dioxide.

The biosynthesis of purine derivatives involves many important metabolic transformations. A nucleotide is formed at first with glycine, 5-phosphoribosyl-1-pyrophosphate being the phosphate donor. The same compound takes part in the synthesis of many nucleotides like NAD^+. The first reaction takes place in the liver as follows: Ribose-5-phosphate, in the presence of ATP and magnesium ions, forms adenosine-5-phosphate and 5-phosphoribosyl-1-pyrophosphate. The ribonucleotides are reduced into the deoxyribonucleotides catalysed by the ribonucleotide reductase dependent on dimethyl benzimidazolyl cobamide

Ribose 5 -P   (5, Phosphoribosyl 1 Pyrophosphate)

MEDICAL BIOCHEMISTRY

P-RIBOSYL-PP + Glutamine → (Glutamate, PPi) → 5-phosphoribosylamine
(GLUTAMYL AMIDOTRANSFERASE)

Glycine + ATP, Mg^{2+} → ADP + Pi → glycinamide ribosyl-5-P

Methenyl FH_4 → FH_4 (FORMYL TRANSFERASE) → Formyl glycinamide ribosyl-5-P

Formyl glycinamide ribosyl-5-P + Glutamine, ATP, Mg^{2+} → Glutamate → Formyl glycinamidine ribosyl 5-P

Ring closure, ATP, Mg^{2+}, H_2O → Aminoimidazole ribosyl-5-P

Aminoimidazole ribosyl-5-P + CO_2, ATP, Mg^{2+} → Aminoimidazole Carboxylate ribosyl 5-P

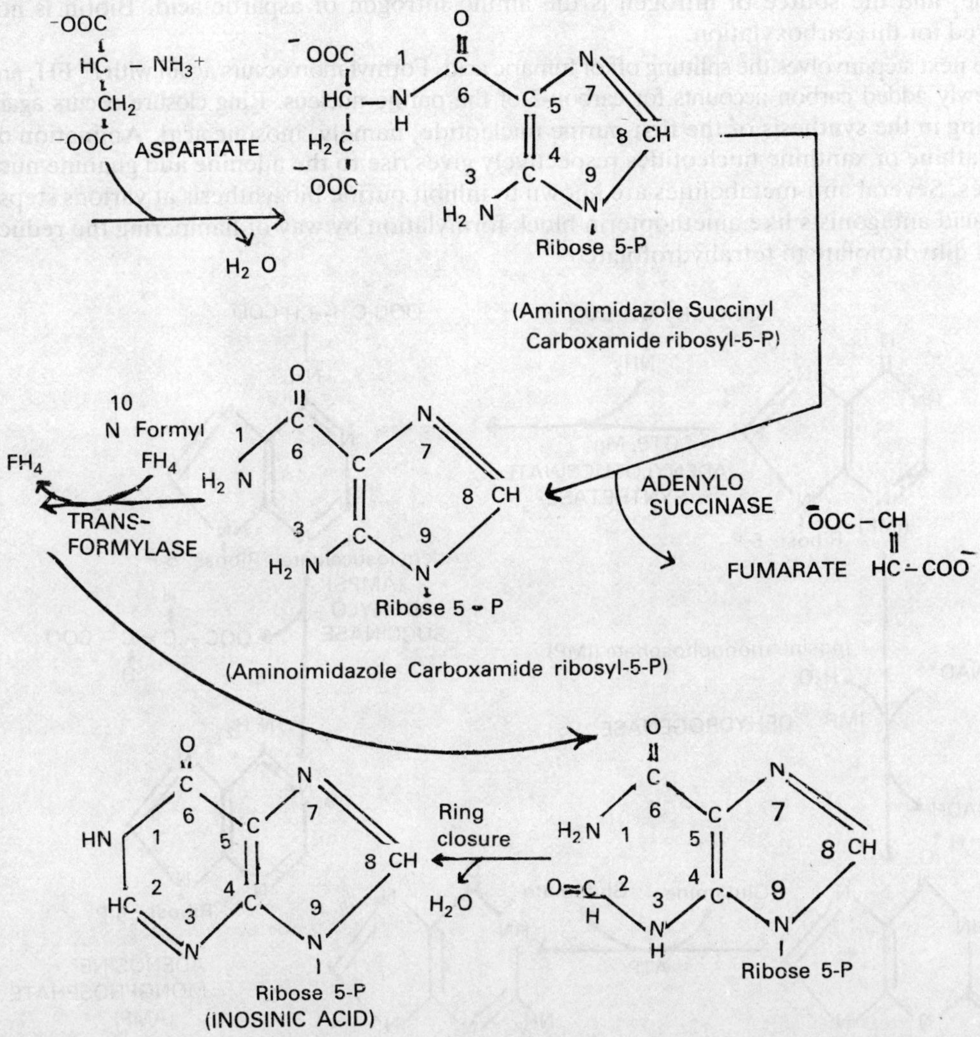

5'deoxyadenosine as a co-factor in the hydrogen transfer in prokaryotic organisms. However, in mammals, the reduction of ribonucleoside diphosphates to deoxyribonucleoside diphosphates is a complex reaction requiring ribonucleotide reductase, thioredoxin (a protein cofactor), thioredoxin reductase (a flavoprotein), with NADPH as a co-factor.

The next step consists of the phosphoribosyl pyrophosphate reacting with glutamine, catalysed by phosphoribosyl pyrophosphate glutamine amido transferase to form 5-phosphoribosyl amine, which then reacts with glycine to form glycinamide ribosyl-5-phosphate which is the source of positions 4,5,7 and 9 of the purine nucleus. A feed-back inhibition is seen in the inhibition of phosphoribosyl pyrophosphate glutamine amido transferase activity, particularly by AMP, GMP and IMP.

Glycinamide ribosyl phosphate is next formylated in which reaction tetrahydrofolic acid (N^5N^{10} Methenyl FH_4) and a transformylase take part, by which the one-carbon moiety is transferred to become position 8 of the purine nucleus. Amination occurs at carbon-4 of the formylated glycinamide, glutamine being the amino donor, the added nitrogen forms position 3 in the purine nucleus. Subsequently, ring closure takes place forming an amino imidazole followed by carbamoylation, the source of the carbon being respiratory carbondioxide, and the source of nitrogen is the amino nitrogen of aspartic acid. Biotin is not required for this carboxylation.

The next step involves the splitting off of fumaric acid. Formylation occurs again with $f^{10} FH_4$ and the newly added carbon accounts for carbon-2 of the purine nucleus. Ring closure occurs again resulting in the synthesis of the first purine nucleotide, namely, inosinic acid. Amination of hypoxathine or xanthine nucleotides respectively gives rise to the adenine and guanine nucleotides. Several anti-metabolities are known to inhibit purine biosynthesis at various steps. Folic acid antagonists like amethopterin block formylation by way of hampering the reduction of dihydrofolate to tetrahydrofolate.

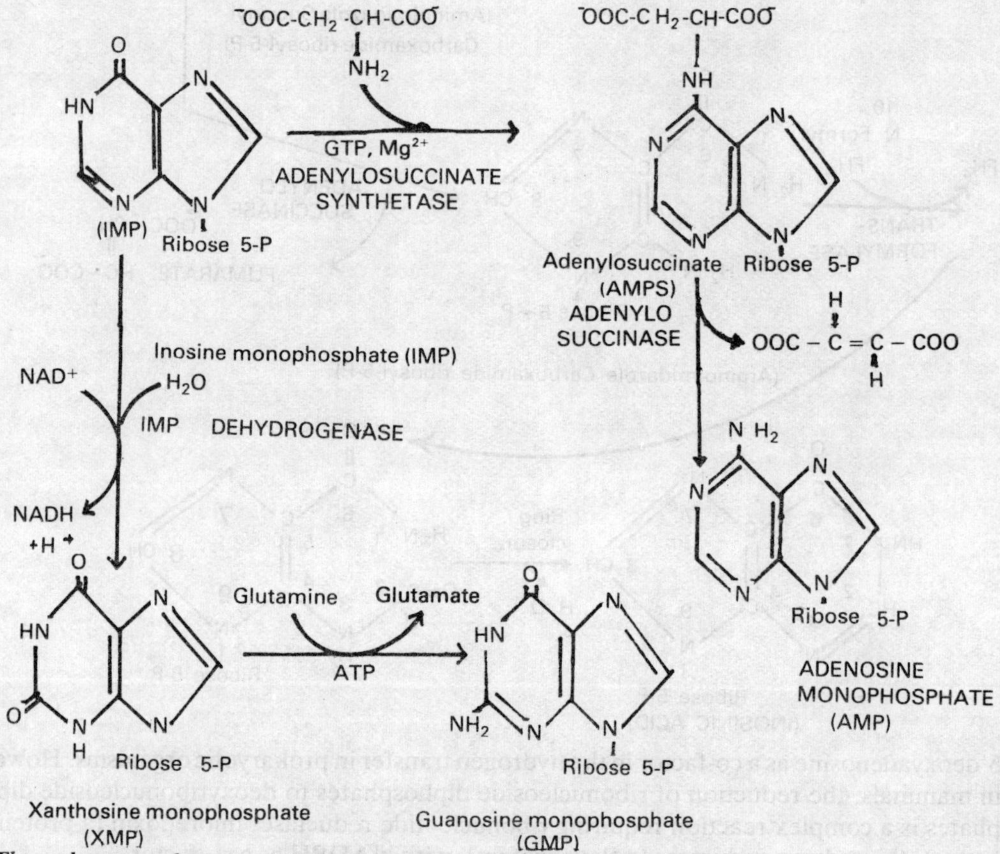

The salvage pathways: All tissues are not capable of *de novo* synthesis of purine nucleotides. Erythrocytes, neutrophils and the brain are examples. The liver is the major site for providing purines to be salvaged by such tissues.

The salvage of pre-formed purines or purine nucleosides occurs mostly by way of phosphoribosylation of the free purine bases by specific enzymes, or alternately by the phosphorylation of purine nucleosides at their 5'-hydroxyl groups. The major salvage pathway is catalysed by enzymes like adenine phosphoribosyl transferase (APRTase), while the second pathway is catalysed by hypoxanthine guanine phosphoribosyl transferase (HGPRTase). The enzyme adenosine kinase can catalyse the salvage of purine nucleosides to purine nucleotides.

In addition, there is a cycle in which IMP and GMP as well as their deoxyribonucleotides are converted into their respective nucleosides by a purine 5'nucleotidase, the level of which in serum is assayed as a liver function test. The nucleosides so formed can be hydrolytically phosphorylated producing the corresponding sugar phosphates and setting free the N-bases. The hyproxanthine and guanine can then be phosphoribosylated again to complete the cycle.

In the human organism, the consumption of phosphoribosyl pyrophosphate by the salvage cycle is greater than that seen in the synthesis of purine nucleotides *de novo*. Mammalian cells have also active salvage pathways for converting pyrimidine nucleosides into their respective nucleotides.

At least three enzymes compete for the same substrate PRPP.
They are
1. PRPP glutamine amidotransferase,
2. adenine phosphoribosyl transferase, and
3. hypoxanthine guanine phosphoribosyl transferase.

APRTase is the best competitor as its Km is the lowest. HGPRTase is the next and PRPP amido-transferase is the last. But the availability of adenine is low. Hence the HGPRTase reaction is more significant than the APRTase reaction. The relative affinity of various systems for the substrate PRPP establishes a single priority system within the cells that could very well explain the preferential utilisation of pre-formed purines for nucleotide formation. Exogenous purines will deplete PRPP and inhibit the *de novo* synthesis of purines as PRPP is utilised by APRTase and HGPRTase in preference to PRPP amido-tranferase.

The formation of purine nucleotides from pre-formed purines has, until recently, been regarded merely as a salvage mechanism for the thrifty re-utilisation of pre-formed purines of endogenous or dietary origin. Since the discovery of APRTase and HGPRTase enzymes, pre-formed purines also are considered to make contributions to urinary uric acid and to the serum urate concentration. More than that, the AMP, GMP and IMP formed exhibit a significant feed-back inhibition on PRPP glutamine amido-transferase and decrease the de novo synthesis of uric acid.

Biosynthesis of pyrimidines

The initial step consists of the formation of carbamoyl aspartate, otherwise known as ureido succinate. Carbamoyl phosphate is the source of the carbamoyl group for the carbamoylation of aspartate. Dihydro orotic acid is next formed as a result of ring closure, and orotic acid is obtained by oxidation of dihydro orotic acid by the NAD^+ dependent specific dehydrogenase. Orotic acid then reacts with 5-phosphoribosyl 1-pyrophosphate forming orotidine-5-phosphate. Decarboxylation gives rise to the primary pyrimidine nucleotide, namely, uridine-5-phosphate or uridylic acid. Uridine monophosphate may be aminated through uridine triphosphate with glutamine and CTP synthetase to yield CTP or, after conversion into the deoxy uridine monophosphate, may be methylated to yield thymidylic acid. The carbon required for methylation is derived from the beta carbon of serine or from the alpha carbon of glycine, after conversion to serine through tetahydrofolic acid as N^5N^{10} methylene FH_4. The enzyme required is thymidylate synthetase.

Feed-back inhibition of the biosynthesis of pyrimidines is illustrated by uridylic acid inhibiting aspartate transcarbamoylase activity. Uridylic acid is also a competitive inhibitor of orotodylic acid decarboxylase. Studies with *Escherichia coli* show the control exercised by cytidine triphosphate on the biosynthetic pathway of pyrimidines. It is found that cytidine triphosphate also acts at an allosteric site of aspartate transcarbamoylase bringing about a conformational change at the active site of the enzyme. It is an example of allosteric feedback inhibition. The halogenated derivatives of pyrimidine compounds are powerful antimetabolites, e.g., 5-fluoro-uracil which is a structural analogue of thymine. Such compounds are known to have potent anti-viral activity.

THE BIOSYNTHETIC PATHWAY FOR PYRIMIDINES

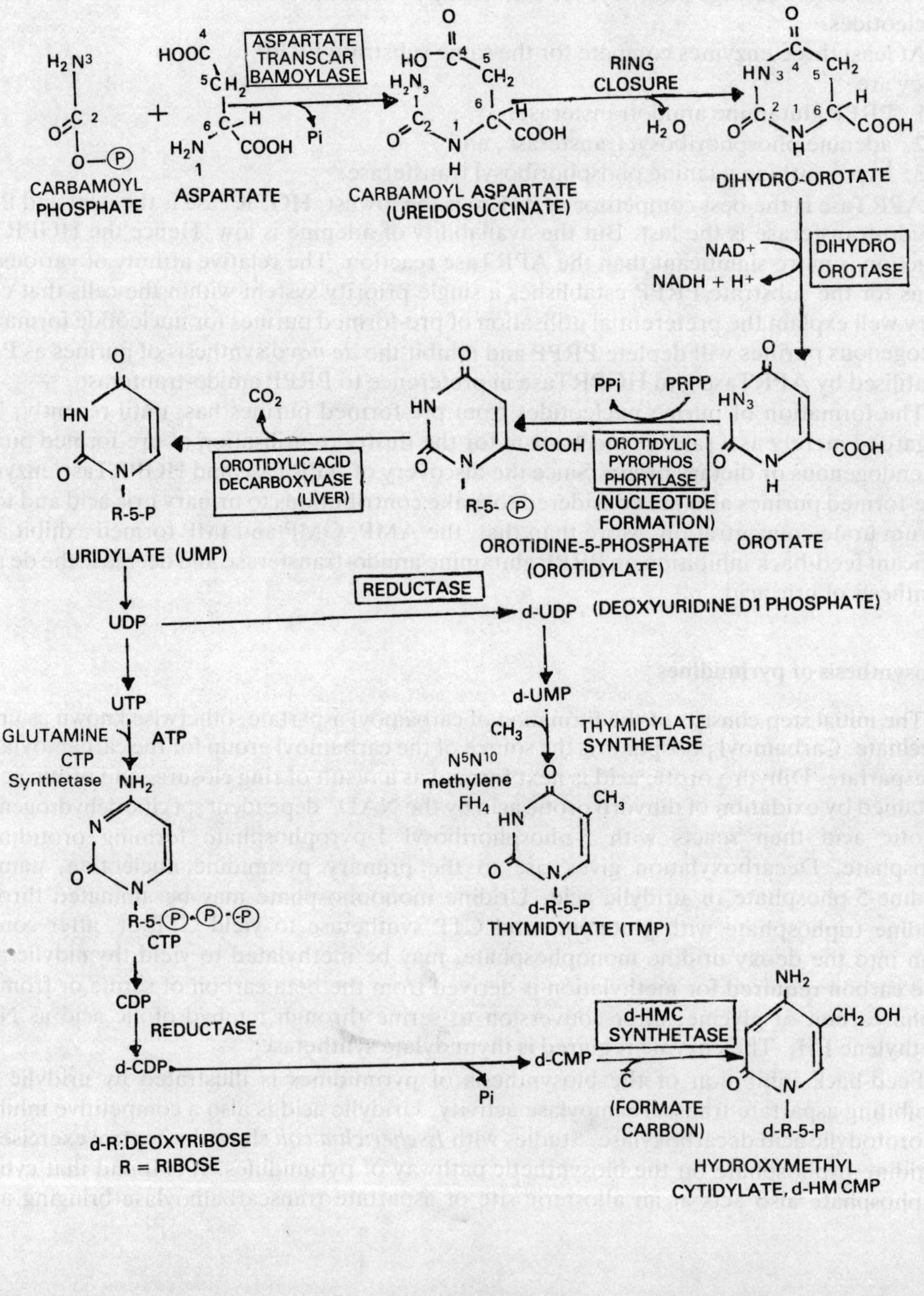

Biosynthesis of nucleic Acids

It has been shown that DNA and mRNA are synthesised in the body by the processes of replication and transcription, respectively. tRNA and rRNA also are synthesised from DNA.

While RNAs are being synthesied constantly in all the cells and have an appreciable flux, DNA has a poor flux in cells which are not dividing. However, DNA is formed prior to cell division. The deoxyribonucleotides required for DNA are obtained by the reduction of ribonucleotides without rupture of N–base ribose bond.

In vitro synthesis of DNA: DNA directs its own synthesis. The enzyme required is DNA polymerase III* with co-polymerase III* in prokaryotes. It is DNA polymerase alpha in enkaryotes. In addition, DNA primer, DNA template and trinucleotides are required.

$$\text{DNA Primer: and template} + \begin{matrix} n\,TTP \\ n\,dGTP \\ n\,dATP \\ n\,dCTP \end{matrix} \xrightarrow{Mg^{2+}} \text{DNA}-\begin{bmatrix} pT \\ pdG \\ pdA \\ pdC \end{bmatrix}_n + 4n\,PPi$$

d = stands for deoxy
p = phospho
PP = pyrophosphate

The DNA produced by this method by Kornberg has the base composition and the nearest neighbour frequencies as the DNA primer. Hence, by this method, DNA with a stipulated sequence cannot be synthesised though the DNA-resembling primer could be synthesised.

On the other hand, Khorana has been successful in synthesising DNA of his choice with a prescribed sequence. He is able to synthesise polynucleotides with the desired sequence of N-bases from simple nucleotides, using the synthetic polynucelotides, the enzymes DNA polymerases and DNA ligase and the ingenious method of 'sticky end technique'. Khorana was able to synthesise DNA corresponding to alanine tRNA. Kornberg's and Khorana's contributions have made a tremendous impact in the field of genetic engineering. Their contributions were 'firsts' and got them the Nobel Prize.

In vivo synthesis of DNA: It has already been described that DNA synthesis in the body takes place by the process of semi-conservative replication. RNA piece formation is a prerequisite for DNA synthesis (Refer 'Biochemical genetics'). Short DNA strands with RNA head piece called 'Okasaki fragments' are first formed and they are then fused with the help of DNA ligase after excision of the RNA heads.

Synthesis of RNA in vitro: Interestingly enough, the synthesis of RNA also requires the DNA primer in one method. Nucleotide triphosphates (UTP instead of TTP in this case), the enzyme DNA-dependent RNA polymerase (transcriptase), and Mg ions are required. The RNA formed will have its base sequence decided by the DNA primer.

$$\begin{matrix} n\,ATP \\ n\,GTP \\ n\,UTP \\ n\,CTP \end{matrix} \xrightarrow[\text{DNA template}]{\substack{\text{DNA - directed} \\ \text{RNA polymerase} \\ Mg^{++}}} RNA + 4n\,PP$$

(In this, the sugar in the nucleotides is ribose. PP stands for pyrophosphate.)

Even single-stranded DNA can be used as the template. When DNA is used for the transcription of mRNA, it is reported that both strands can function and give information in *in vitro*-synthesis and in *in vivo* transcription. Only certain regions, of the DNA strands are used. The strand engaged in transcription is termed the template strand. The other one is the anti-template strand. Both strands can function as template strands in *in vitro* synthesis but in *in vivo* synthesis, only at different times.

In the method using DNA-dependent RNA polymerase, the sequence and composition of the bases in the RNA formed are decided by the DNA template and not influenced by the composition of the reaction mixture.

On the other hand, RNA could be synthesised *in vitro* by using another enzyme, polynucleotide phosphorylase (Ochoa). This enzyme acts on diphosphates (and not triphosphates) and the RNA formed has a base composition decided by the composition of the nucleoside diphosphates in the reaction mixture. The enzyme can be used to produce homo-polynucleotides as poly-U and co-polymers like poly-UG using only required nucleoside diphosphates.

There is another enzyme, RNA-dependant RNA polymerase (replicase), and in this, RNA is the template for the synthesis of another RNA.

In some viruses, RNA is the template for the synthesis of DNA. The enzyme in such cases is RNA-dependant DNA polymerase or Reverse Transcriptase.

In vivo synthesis of RNA: It has already been shown that mRNA synthesis takes place by way of transcription of DNA. The 'template' strand of DNA is used for this. The enzyme catalysing this *in vivo* synthesis is the DNA-dependent RNA polymerase (transcriptase). (For details refer chapter on 'Biochemical' genetics.)

17 Vitamins

Till the beginning of the twentieth century, it was believed that our diet need contain only some of the proximate principles, i.e., carbohydrates, fats, proteins, minerals and water. But when experiments were carried out with animals maintained on a chemically defined diet containing purified protein, carbohydrate, fat and mineral salts, it was found that the sustenance of life was rendered difficult. The presence of some 'accessory food factors' in natural food was recognised and many of these factors were characterised as 'organic compounds'. The first such substance isolated from rice husk which was a potent-growth factor, was found to be an amine and was therefore called 'vitamine'. This name had to be changed to 'vitamin' – dropping the last letter 'e' as many were not amines. Vitamins differ from each other chemically, but they share some common general functions as growth factors as well as agents promoting metabolic reactions, The deficiency or lack of a vitamin produces symptoms of hypovitaminosis which may lead to pathological conditions. The pathological, anatomical and biochemical lesions in many vitamin-deficient states have been clearly studied, and the functions of each of these vitamins have also been now established. Most of the vitamins needed by man have to be obtained through his diet, with the exception of vitamin D and choline. The bacterial flora living in the intestinal tract are able to synthesise a few of the B-vitamins in small amounts.

Vitamins may be defined as organic molecules required in small amounts by mammals, including man, in their diet, for metabolic purposes. Most of the B-vitamins function as coenzymes or co-factors in many enzymatic reactions.

Vitamins are classified as water soluble and fat soluble, depending upon their solubility in water or fats and oils, as well as in fat solvents.

FAT SOLUBLE VITAMINS

The fat soluble vitamins are vitamins A, D, E and K.

Vitamin A (Retinol) (Anti-xerophthalmic vitamin)

Chemistry: Vitamin A is a complex primary alcohol, with the formula $C_{20}H_{29}OH$. It contains a beta ionone ring. Carotenes are the precursors (or pro-vitamins) of vitamin A. Beta-carotene is converted into vitamin A in the tissues, i.e., in the intestine and liver.

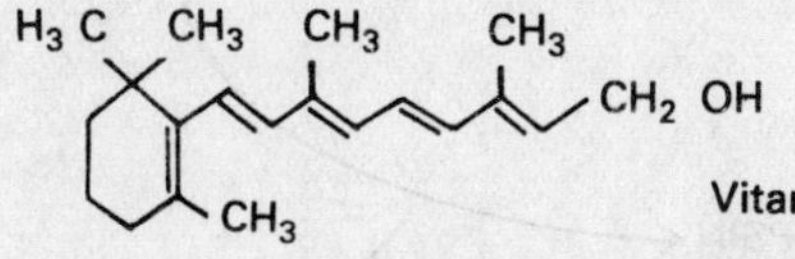

Vitamin A

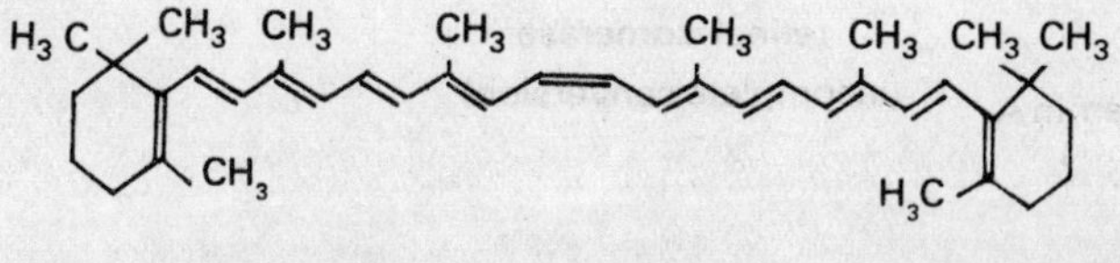

ß -Carotene

Vitamin A_1 and vitamin A_2 alcohols are known to exist. Vitamin A_2 has one double bond more than vitamin A_1.

Vitamin A is soluble in fats and oils and in fat solvents like chloroform. Heat does not destroy it, but heating in the presence of oxygen may oxidise it slowly. Vitamin E by its antioxidant action spares vitamin A. Ultra violet light destroys this vitamin. Vitamin A has a characteristic absorption spectrum at 324 mμ.

Sources: The best sources of this vitamin are the fish liver oils, shark, halibut and cod liver oils. Animal tissues, egg yolk, cheese and butter contain moderate amounts of this vitamin. Plants do not contain any pre-formed vitamin A. However, carrots, yellow vegetables spinach, etc. contain provitamins A, viz., alpha, beta and gamma carotenes and cryptoxanthine. Beta-carotene is the most potent of all and is readily converted to vitamin A in the intestinal wall. Its absorption requires the presence of bile and fat in the intestine.

Biochemical action: Vitamin A is an alcohol and is also called 'retinol.' Its oxidation products are vitamin A aldehyde, i.e., retinal and vitamin A acid (retinoic acid). These three together are called 'retionoids'. Retinol probably serves as a hormone and its action is similar to that of the steroid hormone. It is bound to cellular retinol-binding protein and taken to the nucleus where it gets bound to the nuclear proteins. Retinol supports the normal function of the reproductive system in males and females.

Retinal has an important role in the visual process. The photo-sensitive pigment 'rhodopsin' (visual purple) of the rods of the retina contains 11-cis vitamin A aldehyde and the protein opsin. On exposure to light (chiefly blue light) this gets bleached and is converted to all-trans retinal (visual yellow) and opsin. All-trans retinal has to be converted to 11-cis retinal to be used again. This conversion is brought by retinal isomerase.

11-cis retinal then combines with opsin in the dark to reform rhodopsin. As the rods are meant for vision in dim light (night vision), rhodopsin should be available adequately for night vision. It is to be noted that all-trans retinal is incompletely converted to 11-cis retinal. Hence, a constant supply of vitamin A is needed in the diet. In vitamin A deficiency, the retinal pigment of the rods is not regenerated efficiently and hence night-blindness or nyctalopia results.

WALDS VISUAL CYCLE

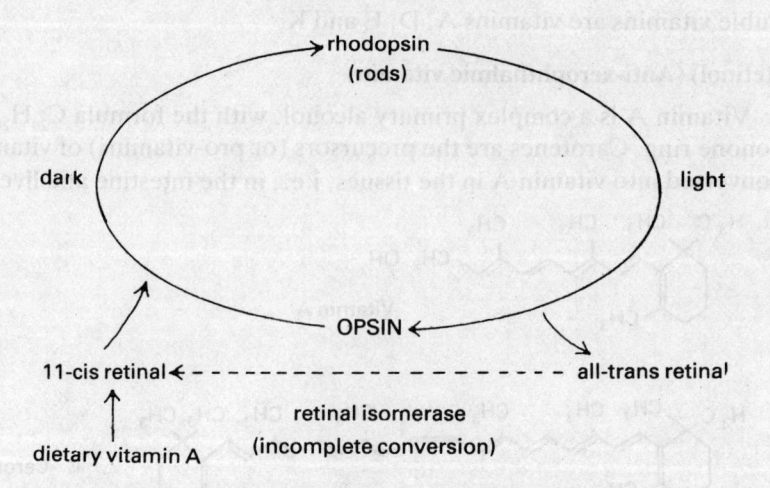

WALD'S VISUAL CYCLE

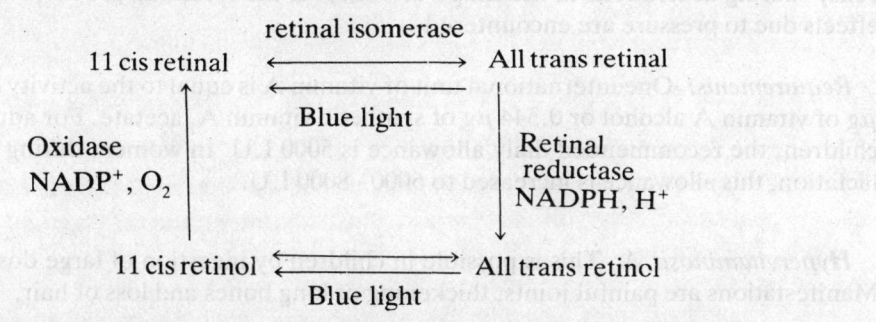

```
                    retinal isomerase
  11 cis retinal  ←─────────────────→  All trans retinal
        ↑              Blue light              │
   Oxidase                                 Retinal
   NADP⁺, O₂                               reductase
        │                                  NADPH, H⁺
                                                ↓
  11 cis retinol ←─────────────────→  All trans retinol
                     Blue light
```

Retinoic acid is required for supporting growth and differentiation of epithelia. Its distinct biochemical function is in the synthesis of glycoproteins, as carriers of oligosaccharides. There is cellular retinoic acid-binding protein also but there is no corresponding nuclear protein. Retinoic acid has many other biological and biochemical *in vitro* responses from cells which are 1) increase of epidermal growth, 2) differentiation of embryonal carcinoma cells, and 3) inhibition of collagenase. Retinoic acid keeps the mucous membrane healthy and moist because it is necessary for the synthesis of glycoproteins. In vitamin A deficiency, keratinisation of the mucous membrane occurs and the skin becomes dry, scaly and rough.

Gluconeogenesis is also promoted by vitamin A. This is considered to be through adrenal corticoids.

In blood vitamin A is carried by the retinol – binding protein cemented to it by pre-albumin.

Colour vision: Colour vision is related to three retinal-containing pigments of the cone cells, porphyropsin, iodopsin and cyanopsin. Depending on the colour of the light falling on the retina, one or more of the above pigments undergo bleaching and get converted to ali-trans retinal and the protein moiety (opsin) gets released. The nerve impulses given out during this reaction are read out as colour by the brain.

Porphyropsin is related to red colour while iodopsin and cyanopsin are linked up with green and blue colours, respectively. The congenital deficiency of any of these pigments may be related to colour blindness.

Effects of deficiency: In general, vitamin A deficiency results in keratinising metaplasia, i.e., replacement of the normal secretory epithelia by dry non-secretory keratinised epithelia. The various deficiency manifestations are as follows:
1) Failure of growth in the young. The collagenous tissue is affected.
2) Night-blindness (nyctalopia) due to defective resynthesis of rhodopsin.
3) Dryness of the eye (xerophthalmia) due to decreased lacrymal secretion (tears)
4) Destruction of the cornea (keratomalacia)
5) Drying of skin and atrophy of sebaceous glands, appearance of pustules around hair follicles.
6) Drying of the mucous membrane, keratinisation and loss of cilia on the epithelium of the mucous membrane of the respiratory tract resulting in infections.
7) Degeneration of germinal epithelium, thus affecting reproduction. This causes sterility in men and cornification of the vaginal epithelium in women. Though ovulation and implantation may occur, the placentas are defective and abnormalities in fetus occur, mostly resulting in fetal death.
8) Defective formation of the enamel of teeth.
9) Imbalance between osteoblasts (bone-forming cells) and osteoclasts (bone-destroying

cells) causing aberrations in the shape of bones. If the foramina are invloved, neurological effects due to pressure are encountered.

Requirements: One international unit of vitamin A is equal to the activity contained in 0.3 μg of vitamin A alcohol or 0.344 μg of synthetic vitamin A_1 acetate. For adults and growing children, the recommended daily allowance is 5000 I.U. In women, during pregnancy, and lactation, this allowance is increased to 6000 - 8000 I.U.

Hypervitaminosis A: This is possible in children by ingestion of large dose of vitamin A. Manifestations are painful joints, thickening of long bones and loss of hair.

Estimation of vitamin A: Vitamin A present in foods, such as fish liver oils, is determined colorimetrically using the Carr-Price reaction. When a chloroformic solution of $SbCl_3$ is added to a dilute solution of vitamin A extract from the oils, a blue colour appears which is immediately measured at 620 mμ. The colour may fade away quickly and so readings have to be taken immediately after the addition of the reagent.

Spectro-photometric methods are also available for the estimation of vitamin A at 436 mμ.

Vitamin D (Anti-rachitic vitamin)

Vitamin D is the fat soluble factor present in animal fats, and is related to calcium and phosphate metabolism. Its absence leads to rickets in infants, in which there is failure of calcification of bones. This vitamin is present in the non-saponifiable fraction of fish liver oils and is related to the sterols from which they are derived by ultra-violet irradiation. These sterols are called pro-vitamins D.

Chemistry: Two most important D vitamins are known. The natural vitamin D present in fish liver oil and animal fats is called cholecalciferol. Vitamin D_2 is not present in animal fats but may be obtained artificially by irradiation of ergosterol. The D vitamins are not, strictly speaking, sterols since the B ring is broken up between C_9 and C_{10} and the methyl group on C_{10} is converted to a methylene group, during irradiation of the pro-vitamin sterol.

Ergosterol (Provitamin D_2) → Ergocalciferol (Vitamin D_2)

7-dehydrocholesterol (Provitamin D_3) → Cholecalciferol (Vitamin D_3)

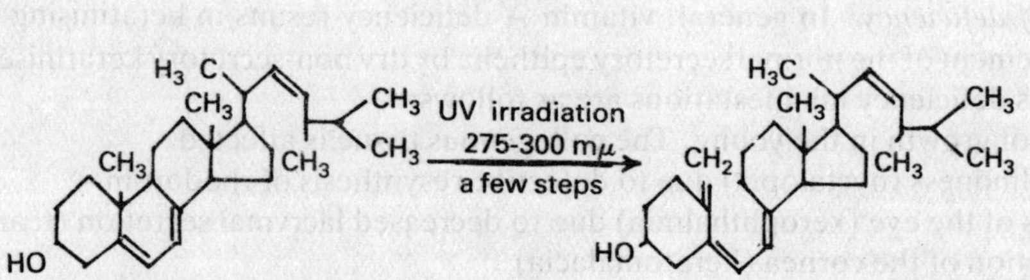

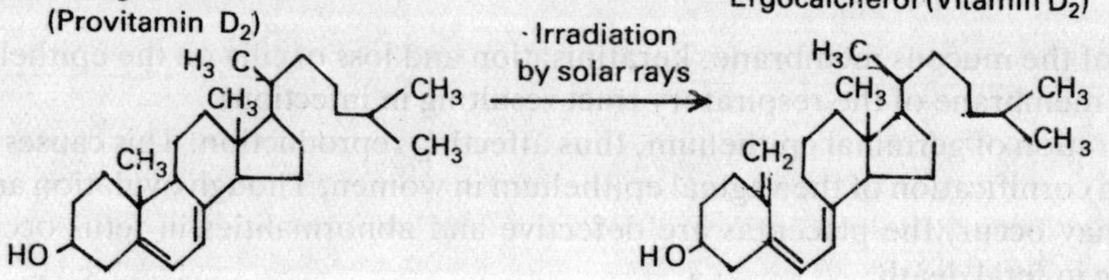

Vitamin D_2 and D_3 are quite stable compounds. They are not destroyed by heat or oxidation. They are soluble in fat solvents. They have been purified as white crystalline substances. In many of their properties they resemble sterols.

Sources: Vitamin D_3 is readily synthesised in the body from cholesterol with 7-dehydrocholesterol as an intermediate. Irradiation of 7-dehydrocholesterol of skin lipids with ultra-violet exposure of the body to sunlight results in the synthesis of this vitamin. All animal fats contain vitamin D_3, the best source being the fish liver oils. Egg yolk fat contains moderate amounts; milk contains only small quantities of vitamin D_3 but irradiated milk contains more of the anti-rachitic vitamins. Commercially, vitamin D_2 is produced by the irradiation of ergosterol obtained from ergot. Many other sterols also exhibit anti-rachitic properties after ultra-violet irradiation.

Biochemical effects: Vitamin D_3, as such, has no biochemical effect. It is converted to 25-hydroxy vitamin D_3 in the liver and further hydroxylated to 1,25 dihydroxy vitamin D_3 in the kidneys. This is the most potent active form of vitamin D_3 which regulates the metabolism of calcium and phosphorus.

Adequate phosphate in serum along with Ca should be available for proper bone mineralisation and bone formation. When the serum phosphate level is abnormally low, the formation of 1,25 dihydroxy vitamin D_3 in the renal tubule is stimulated. This active form of vitamin D_3 is bound to the cytoplasmic receptor molecule and then translocated to the nucleus. It brings about an increase in the synthesis of the calcium-binding protein in the intestines. Thus, the intestinal absorption of both Ca and P is increased. The increased absorption of P is both indirect through increased Ca absorption and direct by an unknown mechanism. In the kidneys, 1,25 dihydroxy D_3 enhances the re-absorption of filtered tubular phosphate but this is usually masked by the inhibition of phosphate re-absorption by the para-thyroid hormone.

1,25 dihydroxy vitamin D_3 affects the cross linking of bone collagen and increases the synthesis of the vitamin K-dependent calcium-binding protein, osteocalcin of the bone, thereby influencing the mineralisation of bone tissues. This would mask the reported permissive role of 1,25 dihydroxy D_3 in the parathyroid hormone – mediated mobilisation of calcium and phosphate from the bone.

The active metabolite of vitamin D_3 seems to prevent myopathy also. 25 hydroxy D_3 also gives, in a minor pathway, 24,25 dihydroxy vitamin D_3. This also increases intestinal Ca absorption but decreases serum Ca and P though, strangely, it promotes normal bone mineralisation. PTH causes a decrease in the formation of 24, 25 dihydroxy vitamin D_3. The biochemical function of the two hydroxy compounds are reciprocally related.

Effects of deficiency: The deficiency of vitamin D leads to rickets in children and osteomalacia in adults. Rickets is characterised by defective ossification leading to soft and pliable bones, bow-legs, knock-knees, bead- like swellings at the rib junctions (rachitic rosary), enlargement of the epiphyses and contracted pelvis. X-ray of the bone reveals abnormal ossification. The rachitic bone has different composition from the normal bone. The content of calcium phosphate (bone mineral) decreases while the organic matter and water content increase in the bone during vitamin D deficiency. A lowering of inorganic phosphate level in the blood is observed and a marked rise in the alkaline phosphatase, although the serum Ca level may remain more or less normal. The product of Ca x P of normal blood which is usually around 50 – 60 in growing children and 30 – 40 in adults, decreases to a lower value in the case of rickets. In this condition, dentition is also delayed.

Osteomalacia occurs in adults suffering from a deficiency of vitamin D. In osteomalacia or adult rickets, the bones become very soft and the Ca x P ratio falls to below 30. Serum calcium is lowered, unlike in rickets. In the bone, more Ca is lost than P, and sometimes there is a relative increase in the Mg content. The ionic serum calcium is reduced to low levels resulting in neuro-muscular irritability and tetany. Osteomalacia is seen in a condition known as celiac disease, which is associated with malabsorption of vitamin D from the intestine.

In both rickets and osteomalacia, a decreased serum phosphate increases the formation of 1, 25 dihydroxy vitamin D_3. A decreased serum Ca^{2+} stimulates the secretion of the parathyroid hormone which also increases the formation of 1,25 dihydroxy vitamin D_3. This increases the intestinal absorption of calcium and phosphorus and enhances calcium reabsorption from the kidneys. Thus, there is replenishment of serum Ca^{2+} and PO_4^{3-} 1,25 dihydroxy vitamin D_3 increases the synthesis of osteocalcin and increases bone mineralisation Thus, rickets and osteomalacia are corrected by vitamin D_3. If the hydroxylation of vitamin D_3 does not take place, rickets cannot be cured by intake of vitamin D, e.g. vitamin D resistant rickets (renal rickets).

Requirements: For infants and children, a requirement of 400 I.U. per day has been proposed. Pregnancy and lactation demand 600 – 800 I.U. per day.

Excess of vitamin D causes toxicity, characterised by nausea, anorexia, digestive disturbances, metastatic calcification of soft tissues, calculi formation, etc. and therefore should be avoided.

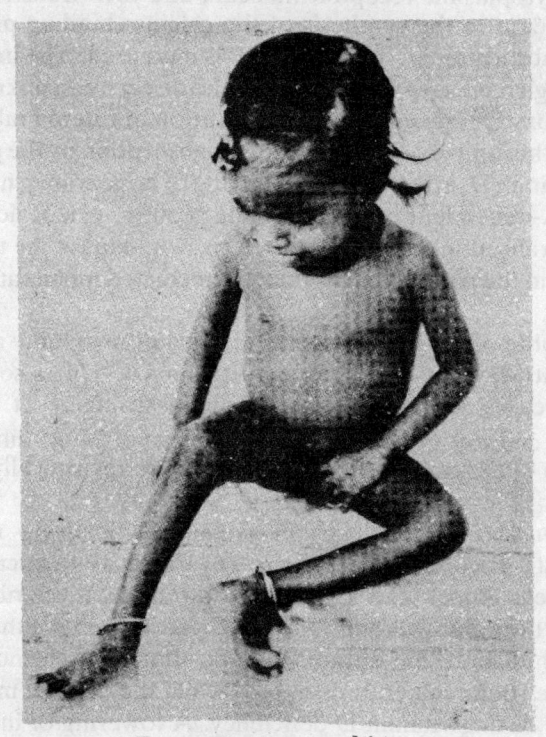

Rickets

Courtesy: Department of Paediatrics, Jawaharlal Institute, Pondicherry, 605 006.

Vitamin D as hormone: Recent research on the mechanism of action of vitamin D has revealed an interesting fact that vitamin D functions more as a hormone than as a vitamin for the following reasons:

1) Vitamin D can be synthesised by the human body.
2) Vitamin D proper is inactive and is only a storage form. It has to be converted to 25 hydroxy and 1,25 dihydroxy compounds for its functions.
3) It has definite target organs bone, kidney, and small intestines (property of hormone).
4) The formation of active forms of vitamin D_3 is subject to feed — back control (property of hormone).
5) The active forms of vitamin D_3 maintain calcium homeostasis along with two other hormones, the parathyroid hormone and calcitonin. The parathyroid hormone is even considered a tropic hormone for 1,25 dihydroxy vitamin D_3.
6) 1,25 dihydroxy D_3 resembles steroid hormones in its mode of action. In the intestine, it enters the cell and is bound to a cytoplasmic receptor molecule. This complex is translocated to the nucleus. By some mechanism it effects an increase in the synthesis of the intestinal calcium-binding protein necessary for the intestinal absorption of calcium.

Vitamin E (Tocopherols) (Anti-sterility vitamin)

Vitamin E activity is exhibited by a class of fat soluble compounds – tocopherols. Alpha, beta, gamma and delta tocopherols have been obtained from natural sources, and their relationship with fertility, and prevention of muscular dystrophy has been recognised.

Chemistry: The alpha tocopherol is the most active form. It has a chromane ring system containing four methyl groups, one phenolic group and a phytyl side chain. The beta and gamma tocopherols have one methyl group less in the ring than the alpha tocopherol.

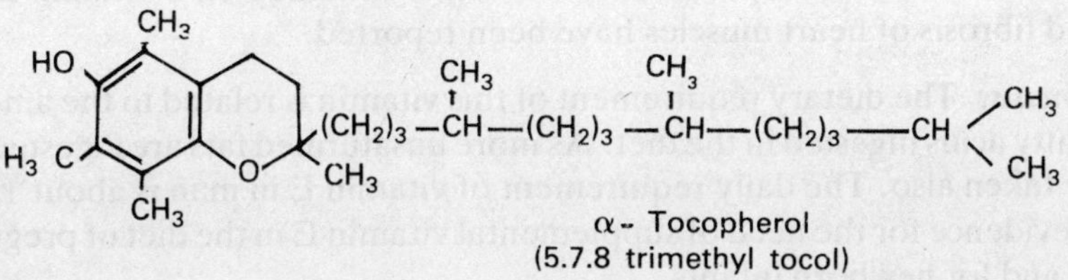

α – Tocopherol
(5.7.8 trimethyl tocol)

The tocopherols are very stable to heat, acids, alkalis and mild oxidising agents. However, they are destroyed on exposure to ultra-violet light for a prolonged time. The tocopherols are oily liquids and are insoluble in water.

Sources: Among the vegetable sources of tocopherol, wheat germ oil and cotton seed oil have been found to be the richest. However, all vegetable oils and animal fat contain at least small quantities of this vitamin. Good sources of animal foods include eggs, meat, fish and liver. Milk is a poor source of vitamin E.

Biochemical action: Vitamin E deficiency manifestations occur in certain species like rabbit, rat, guinea pig but not significant in man. These manifestations are related to 1) gonadal and reproductive functions, and 2) muscle metabolism and structure.

The protective effect of vitamin E on reproduction and prevention of sterility is due to its physiological function in the gonadal structures like the germinal epithelium, and the embryo. All the three layers of the embryo — ectoderm, endoderm and mesoderm — are preserved by vitamin E.

Vitamin E is required in higher animals such as poultry and cattle for fertility. There is no reliable evidence that vitamin E is necessary for fertility in humans. However, in severely impaired intestinal fat absorption, vitamin E deficiency can occur in man. This is because vitamin E has to be absorbed with fat. The signs of vitamin E deficiency in humans are muscular weakness, creatinuria and fragile erythrocytes. Vitamin E is required for the preservation or storage of creatine in the muscles and is a most potent fat soluble anti-oxidant. It offers the first line of defence against the peroxidation of cellular and sub-cellular phospholipids. This explains the fragility of erythrocytes in vitamin E deficiency. Selenium also offers protection against the peroxidation of membrane lipids. Selenium and vitamin E spare each other by co-operative functioning. The role of selenium as an anti-oxidant is through its presence in glutathione peroxidase. Cystine, selenium-containing organic compound, and vitamin E act synergistically in the prevention of hepatic necrosis. Vitamin E being the most potent anti-oxidant also protects vitamin A from oxidative destruction.

Vitamin E protects the unsaturated fatty acids in the erythrocyte cell membrane from oxidation and thus prevents hemolysis. In infants, some types of macrocytic anemia respond to vitamin E therapy. Vitamin E is also one of the inducers of ALA synthetase and ALA dehydratase enzymes involved in heme synthesis.

Effects of deficiency: The deficiency of this vitamin causes resorption of the fetus in female rats and atrophy of spermatogenic structures in male rats leading to permanent sterility. The anti-sterility value is not established in human beings.

The lack of this vitamin in the diet can also cause degenerative changes in the muscles. The muscle fibres atrophy, and are replaced by connective tissue and fat. The glycogen as well as creatine content of the muscle decreases in vitamin E deficiency, while cholesterol and phospholipids are increased. In experimental animals maintained on a vitamin E deficient diet, necrosis and fibrosis of heart muscles have been reported.

Requirements: The dietary requirement of this vitamin is related to the amount of polyunsaturated fatty acids ingested in the diet. As more unsaturated fats are ingested, more vitamin E has to be taken also. The daily requirement of vitamin E in man is about 15 – 30 mg.

There is evidence for the need of supplemental vitamin E in the diet of pregnant and lactating women and for newborn infants.

VITAMIN K

Chemistry: A large number of compounds are found to exhibit vitamin K activity. Two important ones among them are vitamin K_1 or 2-methyl, 3-phytyl,-1 4-naphthoquinone, and K_2 or 2-methyl, 3-difarnesyl 1, 4-napthoquinone. The synthetic vitamin K is K_3 (Menadione). It is 2-methyl 1, 4-naphthoquinone. This is also fat soluble. They all have a common ring system, the difference being only in the side chain.

The K vitamins are fat soluble and are stable in heat although destroyed by strong acids and oxidising agents. They are also destroyed by sunlight.

Sources: Green leafy vegetables, spinach and alfalfa are the best sources for vitamin K_1. Vitamin K_2 is synthesised by the intestinal flora. Commercially, vitamin K_2 is produced from putrefied fish meal.

Biochemical action: Vitamin K has been known to be required for the maintenance of normal levels of the blood clotting factors II (prothrombin), VII, IX and X, all of which are synthesised in the liver. These are produced as inactive precursors and require gamma carboxy glutamate for their modification into the active form. The main function of vitamin K is to serve as an essential co-factor for the carboxylase enzyme that forms gamma carboxy glutamate (GLA) from glutamate residues in specific proteins. As vitamin K is involved in the formation of the active forms of clotting factors including prothrombin, it is required for normal blood clotting. Dicoumarol is antagonistic to vitamin K and prolongs clotting time. It inhibits an enzyme required for vitamin K in its function of gamma carboxylation.

An important function of vitamin K, recognised recently, is its role as a co-factor in oxidative phosphorylation being associated with mitochondrial lipids. Dicoumarol which is an antagonist to vitamin K is known to uncouple oxidative phosphorylation. Ubiquinones resembling vitamin K in structure are known to take part in electron transport.

Effects of deficiency: Deficiency of vitamin K, e.g. in fat malabsorption syndrome or in the sterilization of large intestines with inadequate dieatry intake results in the prolongation of clotting time and a tendency to bleed profusely. Vitamin K deficiency symptoms are observed in man in obstructive jaundice because this vitamin cannot be absorbed from the intestines in the absence of bile salts.

In the first few days of life, hypo-prothrombinemia may occur in infants, as in the immediate post-natal period vitamin K is not synthesised by the intestinal flora. This can be prevented by giving vitamin K to the mother before parturition or to the infant in small amounts and this prevents melena neonatorum. (Large doses of vitamin K should be avoided as it may lead to hyperbilirubinemia.)

Requirements: The minimum requirement of vitamin K in human adults is about 2 mg. Deficiency usually does not occur on a mixed diet. Severe deficiency does not usually occur because the intestinal flora are able to produce this vitamin in small quantities.

WATER SOLUBLE VITAMINS

The water-soluble vitamins consist of the members of the B-group (or B-complex), vitamin C and certain other compounds possessing vitamin activity. These are characterised by their high solubility in water and their insoluble nature in fats and oils.

B-GROUP OF VITAMINS

The B-group of vitamins or the vitamin B complex group consists of 12 important members. They are:
Thiamine or aneurin (anti-beriberi vitamin), B_1
Riboflavin B_2
Pantothenic acid B_3
Niacin and Niacinamide (anti-pellagra vitamin) B_8, Pellagra preventive factor

Pyridoxine B_6
Biotin (anti-egg white injury factor) B_7
Folic acid (pteroyl glutamic acid) and folinic acid, B_9
Lipoic (thioctic acid) acid
Para-aminobenzoic acid (PABA)
Inositol
Choline
Cyanocobalamine (anti-pernicious anemia factor) B_{12}

Some of the members of the B vitamins are present as natural constituents in yeast, cereals and/or liver. All the B-vitamins are growth promoting factors for micro-organisms and they function as coenzymes in cellular reactions.

Thiamine or vitamin B_1
(anti-beriberi vitamin, anti-neuritic vitamin)

Chemistry: The vitamin is present in crystalline form as thiamine hydrochloride, $C_{12}H_{17}ClN_4OS \cdot HCl$. The molecule consists of two important parts, i.e., pyrimidine part and a thiazole part, joined together by a methylene bridge. There is a quaternary nitrogen atom on the thiazole moiety. The alcoholic group on the thiazole moeity gets esterified with phosphoric acid to form the thiamine coenzyme, TPP (thiamine pyrophosphate).

Thiamine is a white crystalline compound highly soluble in water. It has the odour of yeast. It is stable on heating upto 100°C in the dry state and in the presence of acids but heating in moist conditions and alkalies destroys it.

Thiamine chloride hydrochloride

Sources: Thiamine is widely distributed in many plant and animal foods. Whole grains, rice bran, wheat bran, nuts and yeast are some of the best sources while liver, eggs and fish are reasonably good sources. Polishing of the rice removes the vitamin B_1 content by 80 per cent, since most of the vitamin is present in the bran. During cooking of rice, the water in which it is cooked carries almost all the thiamine in the form of a solution.

Biochemical functions: Thiamine in the form of thiamine pyrophosphate (TPP) and in conjunction with lipoic acid is the coenzyme for the decarboxylation of alpha ketoacids like pyruvic and alpha ketoglutaric acid. Hydroxy ethyl thiamine pyrophosphate is an integral part of the pyruvate dehydrogenase complex which functions in conjunction with lipoamide. Thus thiamine is involved in carbohydrate metabolism in all the cells of the body. It acts as co-carboxylase in the conversion of pyruvic acid to acetyl CoA and alpha keto glutaric acid to succinyl CoA, thus helping the oxidation of glucose to CO_2 and also converting glucose to fatty acid. TPP also acts as a coenzyme transketolase in the HMP shunt pathway of glucose

VITAMINS

metabolism. Thus, in thiamine deficiency, there is an accumulation of pyruvic acid and other alpha keto acids in the blood. The neuritic symptoms are due to an excess of pyruvic acid.

The addition of thiamine to the diet or administration of thiamine intra-peritoneally restores the condition to normal.

Effects of deficiency: One of the first symptoms of thiamine deficiency is loss of appetite (anorexia) which leads to decreased food intake and loss of weight as well as arrested growth. The arrested growth may also be independent of decreased food intake.

Normal albino rat

Vitamin B_1 – deficient albino rat of same age
Courtesy: Department of Biochemistry, Jawaharlal Institute, Pondicherry 605 006.

In man, the symptoms of thiamine deficiency were first observed in the rice-eating population in Java and the condition was called beriberi. Beriberi also occurs in many parts of South Asia and Japan. This disease is characterised by polyneuritis with muscular atrophy, edema and cardio-vascular changes. Weakness, fatigue, anorexia, headache, insomnia, gastro-intestinal disorders and tachycardia are usually seen in beriberi, in its early stages. Beriberi is of four types:

a) dry beriberi, in which nervous symptoms or polyneuritis predominate.
b) wet beriberi, in which the symptons are also associated with edema and serous effusions,
c) acute pernicious beriberi in which the heart is involved, and
d) mixed beriberi, where all these symptoms are seen together.

Requirements: The requirement of thiamine is dependent upon the carbohydrate intake. The more the carbohydrate content of the diet, the more is the thiamine requirement. The addition of fat in diet decreases the thiamine requirement. The recommended allowance for thiamine is 1.5 – 2 mg daily for men and 1 – 1.2 mg for women. Although thiamine is ordinarily non-toxic, it has ocassionally produced anaphylactic shock after repeated intravenous injections.

Some of the anti-vitamins for thiamine are pyrithiamine and oxythiamine.

Estimation: With alkaline potassium ferricyanide, it is oxidised to a blue compound, thiochrome, which is estimated fluorimetrically.

Riboflavin (Vitamin B_2, Lactoflavin)

Chemistry: Riboflavin is a yellow orange compound which is soluble in water to a limited extent. Its aqeuous solutions exhibit greenish-yellow fluorescence. Riboflavin is heat-stable but alkali-labile and photo-labile. Exposure to sunlight destroys this vitamin. It is stable towards acids and oxidising agents. Chemically, it contains the 6,7-dimethyl isoalloxazine ring which is linked by the central nitrogen atom to a D-ribitol moiety.

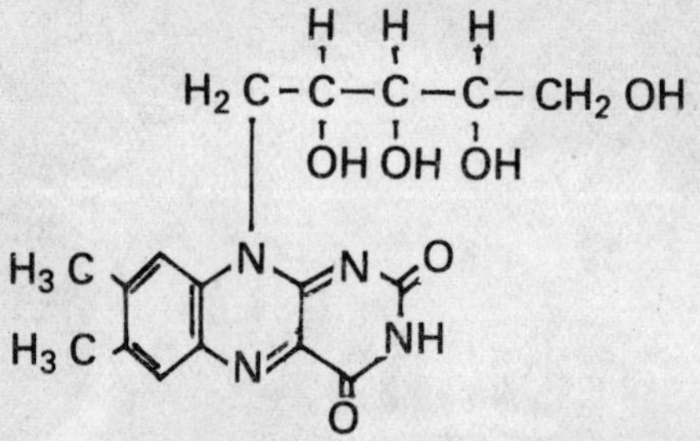

Sources: Riboflavin is widely distributed in plant and animal sources. Milk, liver, kidney and eggs are the best sources of this vitamin. Vegetables, fruits and roots also contain moderate quantities of this vitamin. Cereals, grain and seeds do not contain much of riboflavin.

Biochemical functions: In combination with proteins, riboflavin coenzymes, like flavin mononucleotide (FMN) and flavin adenine dinucleotide (FAD), function in various dehydrogenation reactions. Many of these flavoproteins or yellow enzymes are distributed widely and take part in tissue respiration, where FAD and FMN act as components of the electron transport system. FAD is the coenzyme for diaphorase, D-amino acid dehydrogenase, glycine oxidase, xanthine dehydrogenase and fatty acyl CoA dehydrogenase, while FMN is a constituent of cytochrome C-reductase and L-amino acid dehydrogenase. The flavoproteins are therefore called flavo-enzymes which are oxidation-reduction enzymes. FMN and FAD are tightly but not covalently bound to proteins and hence function as prosthetic groups. Many flavoproteins contain metal atoms also and are actually metallo-flavoproteins.

Effects of deficiency: Pellagra is associated with a lack of niacin and sometimes also riboflavin. Many of the symptoms like dermatitis could be cured by combined therapy with niacin and riboflavin.

In man, cheilosis (fissures at the angles of the mouth or angular stomatitis), glossitis, inflammation of cornea, blood-shot eyes, dimness of vision, photophobia, itching, burning and dryness of eyes and redness of the conjunctiva are reported. The increase in the blood supply to the eye (the increased vascularisation of cornea) is intended to furnish more oxygen to avascular tissues due to diminution of oxidations in such deficient state.

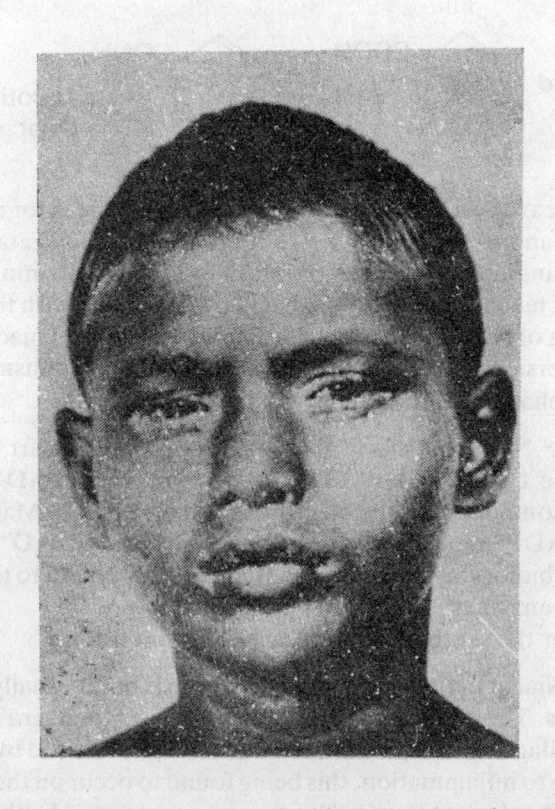

Angular Stomatitis and Angular Conjunctivitis in Vitamin B_2 Deficiency

Courtesy: Padmashri Prof. G. Venkataswamy, Professor of Ophthalmology and Vice Dean, Madurai Medical College.
(Director, Aravind Eye Hospital, Madurai)

Requirements: For children and adults, the recommended daily allowance of riboflavin is 1.6-2.0 mg.

Some of the anti-vitamins are galactoflavin and diethyl derivative of riboflavin.

Estimation is done 1) either by using Fluorimeter, taking advantage of the fluorescence of riboflavin solution, or 2) by microbiological assay using *Lactobacillus casei*.

Niacin and Niacinamide

Niacin is the pellagra-preventing factor, which is now recognised as the constituent of two of the important coenzymes, i.e., NAD^+ and $NADP^+$ concerned with a cellular oxidation reactions.

Chemistry: Niacin is nicotinic acid and niacinamide is nicotinamide. Both these compounds contain a pyridine ring system. Originally, niacin or nicotinic acid was obtained by the oxidation of nicotine, the alkaloid present in tobacco leaves. Niacin and Niacinamide are white crystalline, water-soluble compounds.

Sources: Liver, fish, yeast, beans and peanuts are the best sources for niacin, in which it is present mostly as niacinamide. Wheat and rice bran as well as vegetables are moderate sources but milling of grains and loss of bran results in loss of this vitamin. Milk and fish contain only a low amount of niacin. However, supplementing the diet with these foods, helps to provide an ample amount of tryptophan which is readily converted to niacin in the liver. (For a full discussion on conversion of tryptophan to niacin, refer 'Metabolism of Proteins'.)

Corn is poor in tryptophan and hence in the available niacin.

Biochemical functions: Niacin in the form of niacinamide is a part of the coenzymes, NAD^+ (DPN^+, coenzyme 1) and $NADP^+$ (TPN^+, coenzymes II). NAD^+ is concerned with cellular respiration as a component of the electron transport system. Many dehydrogenases use NAD^+, some use $NADP^+$, and a few enzymes can use either. NAD^+ and $NADP^+$ function as 'electron-sinks' in biological oxidation. They are loosely bound to the enzymes, dialysable and hence typical coenzymes.

NADPH is required for the synthesis of fatty acids and cholesterol.

Effects of deficiency: Niacin deficiency symptoms are exhibited usually only by man, and dogs among the mammals. The deficiency of this vitamin causes 'pellagra' in man and 'canine black tongue' in dogs. Pellagra (= rough skin) is a disease characterised by bronzing and thickening of the skin leading to inflammation, this being found to occur on the back of the hands, forearms and the neck. The dermatitis of pellagra can be prevented by the administration of niacin, the response being spectacular. Other manifestations of pellagra are diarrhoea, angular stomatitis glossitis and dementia. The three 'd's associated with symptoms of pellagra are dermatitis, diarrhoea and dementia.

Requirements: The recommended allowance for children is 6 – 14 mg per day and for adults 17 – 20 mg. The requirement is increased in women during pregnancy and lactation. The niacin requirement is influenced by the protein content of the diet, since niacin can be synthesised in the liver from the amino acid tryptophan. It has been calculated that 1 g of good quality protein contains about 60 mg of tryptophan from which 1 mg of niacin can be synthesised.

Hypervitaminosis leads to flushing due to vaso-dilatation and G.I. distress. Blood pressure is reduced. So, heart patients with already low blood pressure should not be given niacin though niacin offers beneficial effects in atherosclerosis as it lowers serum cholesterol, while the amide does not.

Some of the anti-vitamins are pyridine 3-sulphonic acid and 3-acetyl pyridine.

Estimation is done by: 1) treating with cyanogen bromide and coupling with aniline, and 2) by microbiological assay using *Lactobacillus arabinosus*.

Detoxication: Excess of niacin is detoxified by methylation of its pyridine nitrogen (protective synthesis) and excretion.

Pyridoxine (Vitamin B_6)

Chemistry: Pyridoxine, pyridoxal and pyridoxamine, are all inter-convertible in the body and can act as vitamin B_6. Their structures are given below:

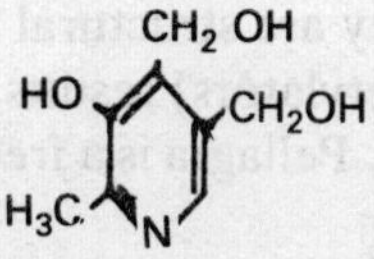

Pyridoxine, pyridoxal and pyridoxamine are all water soluble. They are also soluble in alcohol and slightly soluble in certain fat solvents. Pyridoxine is resistant to heat, but sensitive to light, ultra-violet rays and alkali. Pyridoxal and pyridoxamine are less stable in heat and are liable to be destroyed if heated to high temperatures.

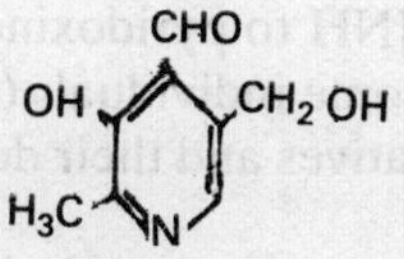

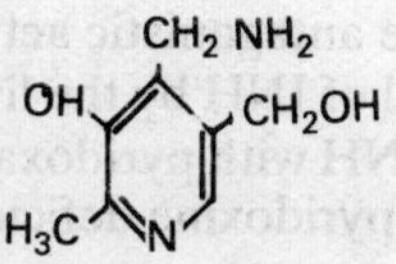

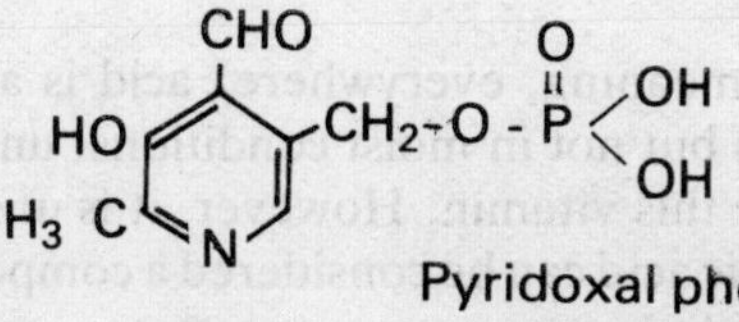

Pyridoxal phosphate
(Coenzyme)

Sources: Egg yolk, fish, milk, and meat are the richest sources. Vegetables like cabbage and legumes and foods such as whole grains, crude molasses, etc., contain moderate amounts of this vitamin.

Biochemical functions: Pyridoxine, pyridoxal and pyridoxamine are inter-convertible in the body. They all function as their phosphorylated derivatives as coenzymes for certain enzyme systems, the most important of them being the transaminases. The latter enzymes are important in the conversion of amino acids to ketoacids and vice versa and in integrating protein and carbohydrate metabolisms. Biosynthesis of several physiologically important compounds like porphyrins from succinyl CoA and glycine, nicotinic acid from tryptophan and gamma aminobutyric acid (GABA) from glutamic acid requires pyridoxal phosphate. In the case of metabolism of tryptophan, its conversion to nicotinic acid does not take place in pyridoxine deficiency. Xanthurenic acid, however, appears in increased amounts in urine in such cases. Pyridoxal phosphate also acts as co-decarboxylase for decarboxylation of certain amino acids and also as a coenzyme for desulfhydration and transulphuration of reactions in cysteine metabolism. Dehydrases (serine and threonine) also require this vitamin. Pyridoxal phosphate is a component of phosphorylases.

There is a strong relationship between unsaturated fatty acids and pyridoxine. The symptoms seen in essential fatty acid deficiency resemble those seen in pyridoxine deficiency, and one has a sparing action on the other. It is thought that pyridoxine can help in the utilisation of the unsaturated fatty acids with great economy.

Effects of deficiency: In many species, pyridoxine deficiency causes a hypochromic microcytic anemia and retardation of growth.

In man, the deficiency of pyridoxine gives rise to certain symptoms gradually. The first effect is a fall in hemoglobin level and anemia followed by lethargy, depression and mental confusion. Unlike adults, the deficiency symptoms appear quickly in infants, when pyridoxine is lacking in their milk. General irritability, vomiting, diarrhoea and convulsions are observed.

The symptoms of pyridoxine deficiency have been observed in patients receiving isonicotinic acid hydrazide (INH) in the treatment of tuberculosis. This is now explained to be due to the antagonistic action of INH to pyridoxine, as they are structural analogues. A slow removal of INH by the liver in some individuals (slow acetylators) results in a chemical reaction of INH with pyridoxal derivatives and their depletion. Pellagra is a frequent accompaniment of pyridoxine deficiency.

Requirements: The adult requirement of this vitamin is between 2 – 3 mg per day and in the case of pregnant women 6 – 7 mg per day.

Some anti-vitamins of pyridoxine are deoxy pyridoxine and isoniazid.

Estimation is done by 1) bioassay using rat, and 2) microbiological assay, using *Saccharomyces carlsbergensis (L.casei* for pyridoxal only).

Pantothenic acid

Chemistry: Pantothenic (*Greek* meaning, everywhere) acid is a yellow viscous oil; it is destroyed by heat in dry conditions but not in moist conditions, unlike thiamine. Oxidising and reducing agents do not destroy this vitamin. However, it is unstable in the presence of strong alkalies and acids. Pantothenic acid can be considered a compound of pantoic acid with beta alanine connected by a peptide linkage.

$$\text{HOCH}_2-\underset{\underset{\text{CH}_3\text{OH}}{|}}{\overset{\overset{\text{CH}_3}{|}}{\text{C}}}-\text{CH-CO-NH-CH}_2\text{-CH}_2\text{-COOH}$$

Pantothenic acid
Beta-alanine
Pantoic acid part

It is soluble in water, and when food is cooked and the water discarded, some vitamin is lost.

Sources: Pantothenic acid is distributed in almost all foods but its amount differs with various foodstuffs. The best and richest sources of this vitamin are liver, kidney, eggs, beef, milk, etc. among animal foods, and cabbage, cauliflower, peanuts and peas among the plant foods. Yeast is another good source of pantothenic acid.

Biochemical functions: Pantothemic acid functions in a number of biochemical reactions in the form of its coenzyme, i.e., Coenzyme A (Lipman). The thiol group (-SH) of CoA -SH is a carrier of the acyl group. Acetyl CoA and acyl CoA have high energy bond equivalent to that of ATP. Reactions involving Coenzyme A are important in various metabolic processes involving carbohydrates, lipids, proteins and various physiologically important molecules. In carbohydrate metabolism, it is involved in the conversion of pyruvic acid to acetyl CoA and alpha keto glutaric acid to succinyl CoA and thereby in the operation of the TCA cycle. Coenzyme A is also involved in the production of acetyl CoA from acetic acid and also by the oxidative breakdown of fatty acids. The oxidation of fatty acids requires coenzyme A at the initial step of activation to fatty acyl CoA, as well as at the thiolytic cleavage step in the fatty acid spiral. In the form of acetyl CoA, it transfers the acetyl group to choline to synthesise acetyl choline. Succinyl CoA is required for the synthesis of porphyrins. Recently, it has also been established that a protein called acyl carrier protein (ACP) has an important role in the biosynthesis of fatty acids and contains pantothenic acid moiety, 4-phophopantetheine. Cholesterol and steroid hormones are all synthesised in the body from acetyl CoA.

Effects of deficiency: No definite lesions due to the deficiency of pantothenic acid had been observed in man. As pantothenic acid is universally present in almost all types of foodstuffs, deficiency diseases associated with the lack of this vitamin are not manifested. However, it has been reported that the burning feet syndrome (Gopalan's syndrome) observed in certain malnourished cases in India, responded only when pantothenic acid was given along with thiamine, riboflavin and niacin. Pantothenic acid does not appear to play any role in the prevention of grey hair in humans.

Requirements: The average content of pantothenic acid in an ordinary mixed diet is about five mg which is more than sufficient for the adult requirement. No recommended daily allowance is known about this vitamin.

Some of the anti-vitamins are pantoyl taurine and ω - methyl pantothenic acid.

Estimation is by microbiological assay using *L. arabinosus*.

Lipoic acid (Thioctic acid)

Chemistry: Lipoic acid is 6,8-dithio-octanoic acid which is chemically a sulphur containing fatty acid. It is fat-soluble as such, but forms water-soluble complexes in combination with other substances. It readily becomes water-soluble when it combines with thiamine pyrophosphate to form LTPP (i.e., lipothiamide pyrophosphate). LTPP takes part in many decarboxylation reactions.

$$H_2C-CH_2-CH-(CH_2)_4COOH$$
$$||$$
$$S\text{---------}S$$

Alpha - Lipoic acid
(Oxidized form)

$$H_2C-CH_2-CH-(CH_2)_4COOH$$
$$||$$
$$SHSH$$

Alpha – Lipoic acid
(reduced form)

Biochemical function: Attempts to produce lipoic acid deficiency in animals have not been successful. However, many of the physiological functions of lipoic acid are known, especially those concerned with the oxidative decarboxylation of pyruvate and alpha keto glutarate, producing acetyl CoA and succinyl CoA, respectively. The reactions take place in conjunction with TPP.

Biotin (anti-egg white injury factor)

Biotin was first recognised as an essential growth factor for micro-organisms. Later on, it was observed that experimental biotin deficiency could be produced by feeding raw egg white, which contains avidin as an antogonist or anti-vitamin to biotin.

Chemistry: Biotin is a heat stable, crystalline substance soluble in water and in ethyl alcohol. It contains a sulphur atom in its molecule. However, if the sulphur is replaced by oxygen, as in oxybiotin, it is active in curing biotin deficiency. Biotin occurs both in a free and in a combined state in natural foods. In combination with lysine it may occur as biocytin.

$$\begin{array}{c} O \\ \parallel \\ C \\ HN \quad NH \\ | \quad | \\ HC - CH \\ | \quad | \\ H_2C \quad CH-(CH_2)_4 COOH \\ \diagdown S \diagup \end{array}$$ Biotin

Sources: Biotin is widely distributed in both animal and plant sources. Animal foods like liver, kidney and milk are very good sources of this vitamin. Vegetables contain moderate amounts, tomatoes and yeast being good sources of this vitamin A. Large amount of biotin is supplied by intestinal bacteria.

Biochemical functions: The multi sub-unit enzymes which are involved in carboxylation reactions are pyruvate carboxylase; acetyl CoA carboxylase etc. Biotin is linked to the epsilon – NH_2 of lysine of the apoenzyme. CO_2 is then added to the biotin nitrogen of the enzyme as a whole. The functions of biotin in the form of its coenzyme, N-carboxy biotin, in various carboxylation reactions have been understood only recently. Carbon dioxide fixation results in the biosynthesis of compounds incorporating one more carbon atom and is one of the first steps in the synthesis of fatty acids by the extra mitochondrial system. The N-carboxybiotin enzyme complex transfers its 'active' CO_2 to compounds like acetyl CoA to give rise to malonyl CoA. The biotin enzyme plays an active role in the formation of oxaloacetate from pyruvate and CO_2.

Effects of deficiency: When rats are fed with a diet containing raw egg white in large quantities, characteristic symptoms like dermatitis and nervous manifestations start appearing. From regions around the eyes, the hair falls in a fashion resulting in the 'spectacled – eye syndrome'. All these symptoms can be avoided by supplementing the diet with biotin or by feeding cooked egg white. The egg-injury factor is now identified as a heat labile protein known as avidin and is believed to exert its effect by combining with biotin and inducing a deficiency of the latter by hampering absorption.

In human volunteers, biotin deficiency resulted in symptoms like scaly desquamation of skin, muscular pain, loss of appetite, anemia and lethargy.

Children develop dermatitis, alopecia, and loss of muscular control, and have retarded growth in biotin deficiency.

Requirements: The biotin requirement of man is uncertain, although an intake of 150–300 μg per day has been suggested. The intestinal biosynthesis of biotin by microflora contributes to some extent to the requirement of this vitamin. Even in deficient states, the excretion of biotin occurs in urine, more than what can be accounted for by the dietary content.

Folic acid (Pteroyl glutamic acid)

Chemistry: Folic acid can be considered as consisting of three important portions: the pteridine nucleus, p-amino-benzoic acid (PABA) and glutamic acid. The number of glutamic acid residues may be one, three or seven, in different natural forms of folic acid. In plants, it is one of seven glutamates while liver was shown to contain penta glutamate.

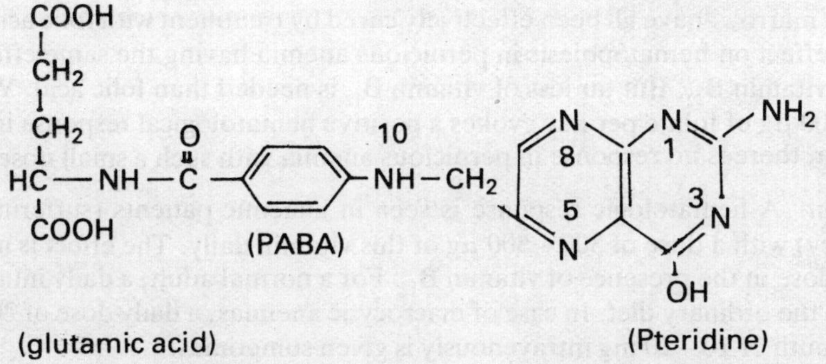

The simplest one is the pteroyl monoglutamic acid (PGA) shown to be necessary for the growth of *Lactobacillus casei*. It is the monoglutamate that can be absorbed. Intestinal enzymes hydrolyse polyglutamate to mono glutamate. A major portion is reduced to tetra hydro folate and methylated as N^5-methyl FH_4 within the intestinal cell as an integral part of the absorption process. Folate absorption is impaired in idiopathic steatorrhoea, tropical sprue and in various other diseases of the small intestines.

In blood, approximately two-thirds of folate is bound to proteins.

Sources: Folic acid is widely distributed in nature and is abundantly present in the green foliage of plants (hence the name 'folic'). It is a yellow substance, slightly soluble in water. Liver, kidney, yeast, cauliflower and cabbage are the best sources.

Biochemical functions: Folic acid functions in the body as folate coenzymes. Folic acid is converted into tetra hydrofolic acid with the aid of NADPH catalysed by folate reductase. Folic acid antagonists like amethopterin block the reduction of FH_2 to FH_4 at the dihydrofolate reductase step. (Hence they are used for producing remissions in leukemias.)

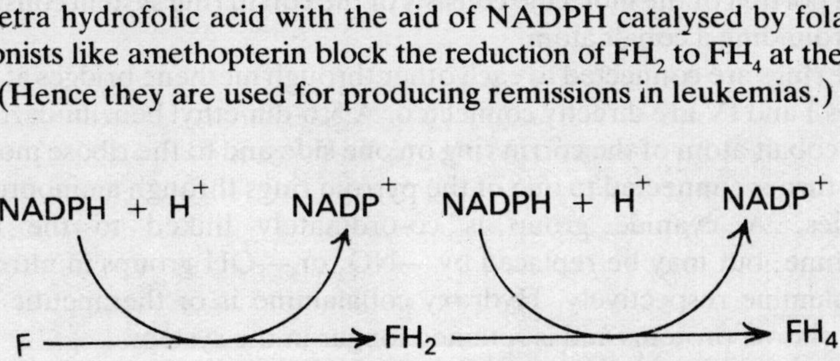

$$F \xrightarrow{NADPH + H^+ \quad NADP^+} FH_2 \xrightarrow{NADPH + H^+ \quad NADP^+} FH_4$$

The tetra hydrofolic acid then acts as the one-carbon acceptor and forms derivatives like formyl tetra hydrofolic acid. The one-carbon moiety may be —CHO (formyl), —CH$_2$OH (hydroxy methyl), formate (H—COO—), —HC = NH (formimino) or methyl (—CH$_3$). The complexes are inter convertible due to a NADP$^+$ dependent hydroxy methyl dehydrogenase system and thus any of these groups can be added to or removed from compounds, in the various reactions involved in one-carbon metabolism. N^5 formyl FH$_4$ is called folinic acid. N^5 methyl FH$_4$ is the major form of folate derivative in blood. Folic acid has its biochemical role in the metabolism of glycine, serine, glutamic acid, histidine, betaine and choline. It is also involved in several biosynthetic processes, in the incorporation of formyl carbon into the purine skeleton and for the synthesis of thymine. Folic acid is required for the formation of N – formyl methionine required to initiate biosynthesis of proteins in lower organisms. In fact, the macrocytic anemia in folic acid deficiency may be attributed to the decreased formation of new red blood cells due to interference with purine and thymine synthesis.

Effect of deficiency: The deficiency of this vitamin results in a macrocytic anemia. Many cases of macrocytic anemia in infants, nutritional macrocytic anemia in adults with a megaloblastic marrow, have all been effectively cured by treatment with folic acid. It also has a favourable effect on hematopoiesis in pernicious anemia having the same effect on blood formation as vitamin B$_{12}$. But far less of vitamin B$_{12}$ is needed than folic acid. While a small dose of 300-500 µg of folate per day evokes a positive hematological response in folate deficiency anemia, there is no response in pernicious anemia with such a small dose.

Requirement: A hematologic response is seen in anaemic patients (suffering from folic acid deficiency) with a dose of 300 – 500 µg of this vitamin daily. The effect is more marked with a lower dose in the presence of vitamin B$_{12}$. For a normal adult, a daily intake of 500 µg is available in the ordinary diet. In case of macrocytic anemias, a daily dose of 200 mg of this vitamin by mouth or 10 – 20 mg intravenously is given sometimes.

Some of the anti-vitamins are aminopterin and amethopterin.

Estimation is done by 1) microbiological assay using *Streptococcus lactus* and 2) estimation of 'figlu' (formimino glutamic acid) in urine. In deficiency of folic acid there will be increased excretion of 'figlu' in urine on a loading dose of histidine.

Cobalamines, Vitamin B$_{12}$ (Cyanocobalamine)

Chemistry: Vitamin B$_{12}$ or the anti pernicious anemia factor (extrinsic factor of Castle) is a red crystalline compound containing cobalt and phosphorus. Its structure is shown in the figure.

The central portion of the molecule consists of the corrin ring system consisting of four pyrrole rings surrounding a cobalt atom.

The pyrrole rings are connected to each other through methene bridges at three places only while the rings I and IV are directly connected. A 5,6-dimethyl benzimidazole moiety is connected to the cobalt atom of the corrin ring on one side and to the ribose moiety at the other. The ribose in turn is connected to one of the pyrrole rings through aminopropanol and phosphate moieties. A cyanide group is co-ordinately linked to the cobalt atom in cyanocobalamine, but may be replaced by —NO$_2$ or —OH groups in nitrocobalamine and hydroxy cobalamine respectively. Hydroxy cobalamine is of therapeutic use, for it binds strongly with plasma proteins and is retained longer in the system.

VITAMINS

[Structure of Vitamin B₁₂ (cyanocobalamin) showing the corrin ring with Co center coordinated to CN, four pyrrole nitrogens (rings I, II, III, IV) with various CH₃, CH₂CH₂CONH₂, and NH₂COCH₂ substituents; lower axial ligand through aminopropanol-phosphate-ribose linked to 5,6-dimethyl benzimidazole part]

Vitamin B_{12} is cobalamine containing cyanide, but in its coenzyme forms the cyanide is substituted by adenosyl moiety or CH_3 group. So far, 4 cobamide coenzymes are known. The coenzymes are 1) 5,6-dimethyl benzimidazole cobamide (DBC), 2) Benzimidazole cobamide (BC), 3) adenyl cobamide (AC), and 4) a coenzyme in which methyl group is attached to cobalt instead of the adenosyl moiety. These coenzymes do not contain the cyanide group and are called corrinoid coenzymes.

[Simplified structure showing Co with Benzimidazole (BC) or 5,6 dimethyl Benzimidazole (DBC) or adenyl (AC) on top axial position and adenosyl moiety or –CH₃ on bottom axial position]

Cobalamines are water soluble, heat stable and stable in the presence of dilute acids at pH 4.0.

Sources: Vitamin B_{12} is almost absent in plant materials. It is produced by the *Streptomyces* group of organisms and is a by-product in the manufacture of Streptomycin. Animal tissues are good sources of this vitamin. Egg yolk and milk contain adequate amounts.

Biochemical function: Cobalamin binding proteins, known collectively as 'R-proteins', are secreted by the salivary glands and the stomach and bind the cobalamins at the acid pH. The 'R-proteins' are normally degraded by pancreatic proteases. Subsequently, the cobalamins bind to the intrinsic factor of Castle. Therefore, the absorption of vitamin B_{12} by the ileum requires previous release of the bound vitamin B_{12} in the stomach in the presence of HCl. A low gastric acidity interferes with the avilability of B_{12} for absorption. In pancreatic insufficiency, cobalamine molecules are not released from the 'R-proteins'.

The intrinsic factor of Castle is a low molecular weight specific glycoprotein and is not protease-sensitive. It is secreted by the parietal cells of the gastric mucosa of the Cardia and the fundus of the stomach. After release from the R-proteins, the cobalamins bind to the intrinsic factor. The complex crosses the ileal mucosa. The intrinsic factor is released and the vitamin is transferred to a plasma transport protein, transcobalamin II. In addition there is transcobalamin I which exists in the liver and plasma. This protein functions as the storage form of cobalamin, a unique situation for the storage of water soluble vitamins.

The major circulating vitamin is methyl cobalamin. Cobalamin binds to the receptors on the plasma membrane and gets internalised. Once inside, it is converted to hydroxy cobalamin and then methyl cobalamin.

In the absence of the intrinsic factor, there is no absorption of vitamin B_{12}. Hence, very often, B_{12} is parenterally administered for treatment.

B_{12} functions as cobamide coenzyme and is involved in many biochemical processes. One of the reactions catalysed in the presence of B_{12} coenzyme is the enzymatic conversion of methyl malonyl coA to succinyl CoA in animal tissues. The enzyme is L-methyl malonyl CoA mutase and the coenzyme, 5-deoxy adenosyl cobalamin. The reaction takes place in the mitochondria. Methyl malonic acid appears in excess in the urine of patients with pernicious anemia but disappears on treatment with B_{12}. The neurological symptoms of pernicious anemia due to B_{12} deficiency are attributed to the accumulation of methyl malonic acid.

In addition, the neurological manifestations are also due to a deficiency of methionine required for transmethylation for the metabolism in the myelin sheath and deranged fatty acid metabolism, chiefly propionic acid metabolism as B_{12} is required for the conversion of propionic acid to succinic acid. As homocysteine is not effectively converted to methionine for want of Methyl B_{12}, there is deficiency of methionine and excretion of taurine. Taurine is a metabolite of homocysteine.

The other important reaction of vitamin B_{12} is to release the folate trap and the methylation of homocysteine to methionine. This takes place in the cytosol. Only vitamin B_{12} can convert N^5-methyl FH_4 to FH_4 and in this process, methyl B_{12} is formed.

$$N^5 \text{ methyl } FH_4 \xrightarrow{B_{12} \otimes} \text{methyl } B_{12} + FH_4$$
$$\downarrow$$
$$\text{Homocysteine} \longrightarrow \text{methionine}$$

$\otimes$ Methyl transferase apoenzyme binds cobalamin and N^5-methyl FH_4, and effects the transfer of the methyl group.

In B_{12} deficiency, the above reactions do not take place, and folate is permanently trapped as N^5-methyl FH_4 (Folate-trap) and is therefore not available for one-carbon transfers. This would result in diminished synthesis of thymidylate and DNA. Hence, the macrocytic megaloblastic anemia in B_{12} deficiency. It is not the direct manifestation of B_{12} deficiency, but the accompanying deficiency of folate. Hence, the hematological manifestations of pernicious anemia are not the primary effect of B_{12} but of secondary deficiency of FH_4 and therefore cured with folate. The neurological manifestations however can be cured only with B_{12}. Vitamin B_{12} is required to convert D-ribonucleotides to deoxy D-ribonucleotides functioning in a specific DBC dependent ribonucleotide reductase in the prokaryotes. Thus it helps in the formation of DNA, and, thereby, proteins.

The mechanism involved in erythropoiesis and increase in the number of erythrocytes after vitamin B_{12} therapy is believed to be due to its effect on the DNA synthesis by releasing the folate from folate trap. The megaloblastic and macrocytic anemia seen in B_{12} deficiency disappears on treatment with this vitamin.

Effects of deficiency: Vitamin B_{12} is the anti-pernicious anemia factor and is found to dramatically cure the hematological defects (megaloblastic and macrocytic anemia) and the neurological manifestations of pernicious anemia. It is recognised as the extrinsic factor of Castle required for hemopoiesis. Manifestations of B_{12} deficiency include loss of appetite and failure of growth, macrocytic and magaloblastic anemia, pernicious anemia (involving subacute combined degeneration of the cord) and deranged metabolism in nervous tissue. In most cases, the deficiency of B_{12} is due to deficiency of the intrinsic factor rather than the vitamin itself.

Requirements: The average requirements for adults is $1-2\,\mu$ of vitamin B_{12} per day. Vegetarians do not get the required amount through their diet, unless they include milk in it. Some of the fermented foods contain vitamin B_{12} in adequate amounts. Natural drinking water from the rivers and wells may contain the required amounts of this vitamin.

Estimation is by microbiological assay using *L. lactis*.

Para-aminobenzoic acid

It was observed that a diet containing all the known vitamins like thiamine, riboflavin, niacin, pyridoxine and pantothenic acid was still lacking in a factor concerned with lactation in rats. The hair of black rats on such a diet turned grey and this factor was called 'anti-grey hair' factor.

This is now identified as para-aminobenzoic acid. This compound is already present as a constituent part of folic acid and is believed to be necessary for the biosynthesis of folic acid by micro-organisms in the intestines. PABA by itself has been recognised as a growth factor in rats. PABA is a structural analogue of sulfanilamide and the bacteriostatic effect of the sulfa drug is explained to be due to its inhibition (competitive inhibition) of the phenolase system which is required for the life of certain micro-organisms.

Some authors do not consider PABA as a vitamin although it is a substance which cannot be biosynthesised in our body. Perhaps there is no need for this compound in our body except for the intestinal flora.

Inositol

Chemistry: This compound exists in the natural form as mesoinositol (myo-inositol) which is biologically effective as a growth factor. There are nine stereo-isomers of which the biologically active form is the optically inactive form.

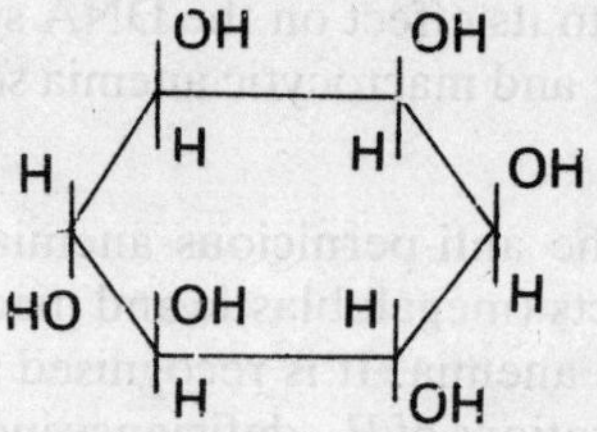

myo - inositol

Biochemical functions: It acts as a lipotropic agent along with choline in experimental animals. It converts neutral triacyl glycerols and phosphatidic acids to inositol phosphatides (lipositols). Some hormones use the latter as their second messengers to release Ca^{2+}. Deficiency of inositol results in alopecia and failure of growth. Inositol is required for the formation of inositol phospholipids in the brain. Inositol probably forms a complex with tocopherols needed for the proper storage of creatine in the muscle. It is widely distributed in most natural foodstuffs of plants and animal origin. Yeast, milk, nuts and fruits are the best sources.

Phytic acid is inositol hexaphosphoric acid. The calcium and magnesium phytins are present in corn.

Choline

Chemistry: Choline is trimethyl-hydroxyethyl-ammonium hydroxide, It is synthesised in the body from glycine, and is related to one-carbon metabolism and folic acid. Some authors do not include choline under vitamins as it could be synthesised by the human system from serine through ethanolamine in required amounts.

Sources: Meat, egg yolk, bread, cereals, beans and peanuts are good sources.

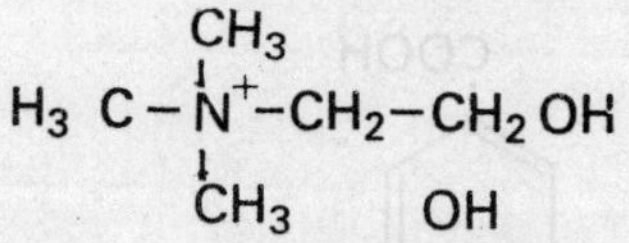

Choline
(Trimethyl N - ethanolamine hydroxide)

Biochemical functions: Choline as a lipotropic substance: Choline is a constituent of various phospholipids like the lecithins and sphingomyelins which have important physiological functions of the body. Choline helps in the formation of phospholipids in the liver, and thereby the disposal of triacyl glycerols as lipoprotein complexes, and preventing of fatty infiltration of liver. It is also reported to enhance oxidation of fatty acids in the liver. For these reasons choline is a potent lipotropic substance.

Choline after being oxidised to betaine is methylating agent (refer 'Protein Metabolism'). Choline is also required for the formation of acetyl choline, one of the chemical mediators of nerve activity.

Effects of deficiency: Choline deficiency results in fatty infiltration into the liver. On a low choline diet, puppies develop lack of appetite, failure of growth and fatty liver. Rats maintained on low choline diet develop fatty livers, cirrhosis of the liver, hemorrhages in the kidneys and eyes. In chicks and turkeys, choline deficiency causes perosis or slipped tendon diesease.

Requirement: The human requirement of choline has not been established. The presence of adequate amounts of methionine in the diet has a sparing action on the exogenous requirement of choline.

Vitamin P

The colouring pigments in plants like rutin and hesperidin are referred to as vitamin P. They are bioflavonoids. They influence capillary permeability and potentiate vitamin C. They are present in the peels of oranges, lemons, etc.

Anti-vitamins (Vitamin antagonists)

Anti-vitamins are synthetic compounds which are either structural analogues of the vitamins (in many cases having opposite effects physiologically) or compounds which increase the requirement of a vitamin thereby inducing an artificial deficiency. Pyrithiamine and oxythiamine are antagonists to thiamine while iso-riboflavin is the anti-vitamin for riboflavin. Niacin is antagonised by pyridine sulphonic acid and pantothenic acid by pantoyltaurine. Isonicotinic hydrazide (Isoniazid), a potent drug used in the treatment of tuberculosis, is antagonistic to pyridoxine and over-dosage of this drug produces convulsive episodes characteristic of pyridoxine deficiency. Aminopterin and amethopterin are anti-vitamins for folic acid and are used in the treatment of cancer. The sulpha drugs antagonise para-aminobenzoic acid and folic acid. Vitamin K is antagonised by dicoumarol.

Most of these anti-vitamins function by inhibiting the enzyme systems requiring the vitamin as a coenzyme, this being a type of competitive inhibition. Such inhibition can be overcome by using higher doses of the vitamin.

Vitamin C (L-ascorbic acid)

The water soluble vitamins include vitamin C also, the anti-scorbutic vitamin.

Chemistry: L-ascorbic acid is related chemically to glucose and glucuronic acid. It is highly soluble in water, has a sour taste and is easily oxidised in aqueous solution by atmospheric oxygen and oxidising agents. The oxidation is enhanced in the presence of minute traces of copper. It is the least stable of all the water soluble vitamins. Alkalies completely decompose it, while in acidic solution it is quite stable. Metaphosphoric acid and trichloro-acetic acid act as negative catalysts and depress the oxidative decomposition of ascorbic acid in vitro.

On oxidation under mild conditions, ascorbic acid is converted to dehydro-ascorbic acid.

```
O=C ─┐              O=C ─┐
 |                   |
HO-C                O=C
 ‖                   |
HO-C   O            O=C    O
 |                   |
 HC ─┘              H-C ─┘
 |                   |
HO-CH               HO-C-H
 |                   |
 CH₂ OH              CH₂ OH
```
L - ascorbic acid (Reduced form)   Dehydro L - ascorbic acid (Oxidized form)

Sources: Ascorbic acid is distributed widely in many plant and animal foods. Fruits and vegetables are excellent sources of this vitamin. Citrus fruits, guava, gooseberry and green peppers are the richest sources, while cabbage and spinach are reasonably good sources. Among animal tissues, the highest amounts are present in the adrenal cortex, gonads and glandular organs. Cow's milk does not contain appreciable amounts and has no value as an anti-scorbutic food.

Biochemical functions: Ascorbic acid has been recognised as an important substance in the body involved in the redox mechanisms. In the presence of glutathione, the ascorbic acid can be converted from its oxidised form to the reduced form easily and is believed to play a role in many of the respiratory processes. Cytochrome C, pyridine and flavin necleotides may need vitamin C.

The presence of a high concentration of ascorbic acid in the gonads and adrenal cortex shows that it plays a role in the production of the steroid hormones from cholesterol, by activating certain enzyme systems involved in the process. A biochemical reaction to stress is depletion of ascorbic acid from the adrenal cortex.

The formation of bile pigments from the breakdown of hemoglobin also requires vitamin C.

The metabolism of phenyl alanine and tyrosine requires vitamin C. The excretion of certain phenyl pyruvic acid derivatives in infants is abolished by vitamin C therapy.

The absorption of iron from the intestines is facilitated by vitamin C which converts the ferric form to the ferrous form. Inter-conversion of ferrous to ferric forms later on in the tissues may also be helped by the redox systems in our body and vitamin C may play a role in that. The mobilisation of iron from the bones is also augmented.

Vitamin C is involved in the formation of the inter-cellular cementing substances and collagen. In its deficiency, although the organic matrix and inorganic calcium phosphate are available, the cementing substance is lacking and this leads to bone disorders. This also affects the formation of dentine and is needed for the maintenance teeth. Collagen contains hydroxy proline to the extent of 12 per cent which is formed from the amino acid proline. It also contains hydroxy lysine. The conversion of proline to hydroxy proline and lysine to hydroxy lysine requires vitamin C. In the absence of this vitamin, an abnormal collagen is produced and causes lesions like spongy gums, hemorrhages, fracture of bones, faulty cartilage formation leading to pain in joints, etc.

The major excretory products of vitamin C are ascorbic acid itself and dehydro-ascorbic acid with a small amount of oxalic acid.

Effects of deficiency: Ascorbic acid is biosynthesised in many animal species except in man, monkey and guinea-pigs. When guinea pigs are maintained on a scorbutic diet, growth ceases in two weeks and symptoms start appearing. The joints become tender, get swollen and the animal lies on its back with its legs kept sprawled (scurvy position). It winces when pressed because of pain in the joints due to subcutaneous and sub-periosteal hemorrahages. The gums become tender and show hemorrhages. The teeth and nails become loose. In man, ascorbic acid deficiency may occur due to continued sub-normal intake of this vitamin.

Decreased resistance to infection, slow healing of wounds and union of fractures, hemorrhages inside the muscles and under the skin (petechial hemorrhages) occur as the deficiency progresses. Subcutaneous bleeding results in the formation of red patches under the skin; many endocrine functions become sluggish on a low vitamin C intake especially in the gonads and adrenal cortex.

Deficiency of vitamin C causes a reduction in the amount of inter-cellular substance and weakens the endothelial wall of the capillaries. In scurvy due to the deficiency of vitamin C, there is anemia, pains at the joints and hemorrhages from the mucous membranes of the mouth and gastro-intestinal tract. There is swelling and bleeding of gums with ulceration and even gangrene.

There are no known toxic effects of vitamin C. Potential complications of chronic massive overdose include calcium oxalate stones and detrimental effects of vitamin C on the absorption of other vitamins like vitamin B_{12} and some drugs.

Clinical importance: In febrile conditions, infections and stress, vitamin C is lost heavily from the body. Hence, adequate quantities of this vitamin should be taken in such conditions.

The usefulness of this vitamin in many pathological conditions like atherosclerosis, deep vein thrombosis, diabetes mellitus, cancer, skin diseases, frost-bite and in common cold and maintenance of youthfulness invites one to call it the 'versatile' vitamin.

Requirements: On an average, about 60 mg of ascorbic acid per day is required in the diet by normal adults. The requirement is increased during infection, fevers and toxic conditions. Presently some workers in this field consider that 60 mg per day is inadequate for many beneficial effects of this vitamin, and that about 1 – 2 grams should be taken daily.

Ascorbic acid saturation test: The determination of merely the plasma levels or urinary levels of ascorbic acid may not always be indicative of the ascorbic acid status of an individual. A satisfactory method for assessing it, is the ascorbic acid saturation test. Orally or intravenously administered ascorbic acid will be taken up by the tissues which get saturated with the vitamin. The excretion of this vitamin in urine is then determined after four hrs and six hrs, and again the next day. On administering 700 mg of ascorbic acid in the test, at least 200 mg will be excreted in the urine within four hrs in normal subjects. On the subsequent day, not less than 50 mg should be present in the urine. In ascorbic acid deficient states, lesser amounts are excreted. This test is useful in screening sub-clinical scorbutic states.

Estimation: Estimation is done by a) titration with 2,6 dichlorophenol indophenol which is reduced by this vitamin, and b) colorimetrically by treatment of the dehydro form with 2,4-dinitrophenyl hydrazine.

18 Hormones, their Chemistry and Functions

Hormones belong to the class of substances which, although present in minute quantities in circulating blood, influence the rate of cellular biochemical reactions. They play a role in regulating and controlling a reaction already catalysed by cellular enzymes. All the hormones are recognised to be organic compounds which are synthesised in a set of specialised organs called the endocrine glands. These ductless glands pour their secretions directly into the bloodstream or into the immediate fluid environment. The circulating blood carries these secretions to sites away from the glands where they exert their influence thereby acting as 'chemical messengers'.

Hormones are defined as substances secreted by ductless glands and these substances act on specific target organs. Insulin is secreted by the beta cells of the islets of Langerhans of the pancreas and acts chiefly on the liver, muscle and adipose tissue. The para-thyroid hormone is elaborated by the para-thyroid glands and acts on the bone, kidneys and the intestines. This is the classical picture of hormones.

But recent research suggests that this definition needs modification. Many findings indicate that some organs other than the ductless glands synthesise and secrete substances with hormonal functions. For example, the hypothalamus of the brain elaborates 'regulatory hormones' or 'release factors' like the thyrotropin releasing hormone (TRH), somatostatin, etc. The gut produces hormones like vaso-active intestinal polypeptide (VIP). Prostaglandins which are considered as 'local hormones' are produced almost all over the body. Hence, the idea that hormones are produced *only by glands* is not correct.

Many hormones, especially small peptide hormones, have a neuro-transmitter function. The synthesis, secretion and action of typical neuro-transmitters involve processes similar to those of hormones. Hence, it is becoming increasingly difficult to distinguish between some hormones and neuro-transmitter substances. It can therefore be safely concluded that hormones are substances synthesised in the body in small quantities but have a profound biochemical effect in the control and regulation of metabolic events, and contribute, in some cases, to intercellular and intracellular communication. For this latter function, they may use second messengers like the cyclic AMP, phospho inositides and third messengers like calcium, the first messenger being the hormone itself.

Because of their role in regulating metabolic processes, the hormones are related to one another, some of them being synergistic while others are antagonistic to each other. In the normal steady state, endocrine equilibrium is maintained. Derangement of this endocrine balance, either due to hypofunction or hyperfunction of these glands, results in a variety of metabolic aberrations or syndromes. This is seen in many clinical conditions like diabetes mellitus, myxoedema, Addison's disease, etc. Some of these syndromes can be induced experimentally by surgical removal of the gland or by treatment with chemical compounds having a specific effect on the gland and its secretion.

HORMONES, THEIR CHEMISTRY AND FUNCTIONS

Hormones are present in circulating blood in extremely minute amounts and they act at very low concentrations. The effect of a hormone on certain biochemical processes may differ from tissue to tissue. Some of the hormones like those of the anterior pituitary have specific effect on their target glands which are stimulated by them to produce their own hormones. However, the hormones produced by these target glands have the ability to depress the production of the anterior pituitary hormones by a negative feed-back mechanism. Although at one time, the pituitary was considered as the head of the endocrine system, recent knowledge about neuro-hormones and their mechanism of action has replaced the pituitary with the hypothalamus in endocrine regulation. The stimulation of the hypothalamus in stress produces several neuro-hormones called regulatory factors which are carried by hypothalamo-hypophyseal system to the anterior pituitary. These regulatory factors stimulate the pituitary to produce the trophic hormones which in turn trigger the production of hormones by target glands like the adrenal cortex, thyroid and gonads.

Classification of hormones

Depending on their chemical nature, hormones can be classified as follows:
1. Proteins of molecular weight less than 30,000, e.g., insulin.
2. Small polypeptides, e.g., anti-diuretic hormone
3. Single amino acids, e.g., thyroxine
4. Steroid hormones, e.g., adrenal cortical hormones.

Hormones can also be classified depending on their mode of action:
1. Hormones which act by binding to their receptors on the plasma membrane, e.g., insulin.
2. Hormones which act though the second messenger, cAMP, e.g., glucagon, epinephrine.
3. Hormones which bind to high affinity receptor proteins in the cytosol, move to the nucleus as a complex, interact with chromatin there and increase the production of mRNA and thereby proteins, e.g., steroids.
4. Hormones which straightaway move to the nucleus and interact with specific receptor proteins in the nucleus and increase transription and translation, e.g., tri-iodo-thyronine (T_3).
5. Hormones which increase the extent of translation without increasing transcription, e.g., insulin, ACTH.

General characteristics of hormones

In Greek, 'hormo' means to 'excite' or 'arouse'. Hormones are chemically defined molecules secreted by the endocrine cells directly into the bloodstream, and which excite or arouse the activity of other specific target cells located away from the site of origin. Hormones exhibit chemical diversity. While the steroid hormones like testosterone, cortisol, estradiol and progesterone are derived from cholesterol, there are also compounds like prostaglandins which are derived from polyunsaturated fatty acids like arachidonic acid. Catecholamine hormones like epinephrine, nor-epinephrine and dopamine are small molecules derived from the amino acid, tyrosine. Tyrosine is also the precursor for the thyroid hormones T_3 and T_4. Some hormones, like Thyroid hormone regulatory factor (TRF) and oxytocin, are simple peptides. Insulin, glucagon and ACTH, are polypeptides while some others like prolactin, growth hormone and parathyroid hormone are proteins containing a large number of amino acids. There are also some glycoprotein hormones more complex than the earlier ones. They are the thyroid stimulating hormone (TSH), follicle stimulating hormone (FSH), luteinising hormone (LH) and human chorionic gonadotrophin (hCG).

The steroid hormones of the adrenal cortex and gonads are all derived from cholesterol by a common initial pathway which gets ramified or diversified after a certain point in its biosynthetic pathway, under the influence of the pituitary trophic hormones acting in their corresponding target glands. Apart from their chemical similarities, their functions are also comparable and support the fact that very often physiological action is related to the chemical groups and structural orientation of molecules. Certain protein and peptide hormones are produced initially as inactive or less active precursor molecules from which the active hormone is formed as and when needed. Thus pro-insulin gives rise to insulin in the beta islet cells by peptide cleavage brought about by a trypsin-like protease. The precursor molecule pro-opio melano cortin gives rise to ACTH, α-MSH, and ß-lipotropin by specific enzymatic cleavage of the molecule. ß-lipotropin also gives rise to ß-MSH, and the endorphins. It is also interesting to note that insulin produced by the ß cells of the islets of Langerhans, and the insulin-like growth factors (IGF_1 and IGF_2) produced elsewhere have certain common chemical similarities both in their chemical structures and also in some of the anabolic effects.

Certain glycoprotein hormones produced by the pituitary and placenta are dimers consisting of both α and ß sub-units, in which the α or ß-sub-units are identical, the difference between them being located between the α and ß-sub-units.

Thyroxine (T_4) can be converted to a physiologically more active form (T_3) by partial de-iodination in target tissues like the liver. Testosterone undergoes conversion to the physiologically more active dihydrotestosterone (DHT) in the secondary sex tissues. So also dehydro-epiandrosterone produced in the adrenal cortex is converted to androstene-dione in the liver and finally to testosterone or estrone. Vitamin D_3 of skin fat undergoes modification to give 25-hydroxy vitamin D_3 in the liver, which is further hydroxylated to 1,25-dihydroxy vitamin D_3 (calcitriol) in the renal tissue.

The hormones that are peptides or proteins, being hydrophilic molecules, are carried in plasma as such while the hydrophobic steroid hormones are bound to specific transport globulins such as the cortico-steroid binding globulin (CBG) and the sex hormone binding globulin (SHBG) and transported in blood. These bound hormones cannot be metabolised or excreted and hence act as reservoirs, the free hormone which is present in equilibrium with this being responsible for all the physiological activity. The plasma half-life of steroid hormones is about a few hours.

Thyroxine and T_3 are also bound to certain plasma proteins like the thyroxine-binding globulin (TBG) and pre-albumin, and thus these hormones have a half-life of a few days. However, catecholamines like epinephrine and nor-epinephrine are not bound to any plasma protein and are destroyed and excreted within minutes after they exert their physiological action.

Hormones and receptors

Most of the hormones exert their biological effects in the target cells after getting bound to specific receptors there. The receptors are certain molecules present in the cell which have specific ability to bind selectively to the hormones and they may be present in the plasma membrane or inside the cell in organelles like the nucleus. For most of the protein hormones, the receptors are located on the cell membrane, while for thyroid hormones as well as cortico-steroids and sex hormones, they are located in the endoplasmic reticulum as well as in the nucleus. The binding of hormones to receptors is against a concentration gradient and is saturable at physiological levels. The number of receptors for hormones in a certain cell is altered during pathological conditions. Antibodies (IgG) acting against a specific hormone receptor can also block hormone binding and can lead to a pathological state. For example, insulin antibodies can bring about resistance to insulin and aggravate the diabetic state. In the obese diabetic (Type II diabetes), the number of receptors in proportion to body size is less than that of normal individuals and hence even with a normal insulin production, there is an apparent insulin insufficiency and manifestation of diabetes.

Receptors are mostly proteins or glycoproteins and sometimes also lipoproteins and the binding sites appear to have either disulphide bonds or phospholipid and other moieties.

Homeostatic mechanisms and feed-back control

A variety of homeostatic mechanisms operte to keep the plasma hormone levels at physiological limits, both at the site of secretion and at the site of target cells. This is called negative feed-back mechanism. Such feed-back controls operate between the hypothalamus and pituitary and also between the pituitary and the target glands like thyroid, adrenal cortex, gonads, etc. The plasma level of Ca^{++} has a feed-back effect on the secretion of the parathyroid hormone. The plasma glucose level regulates the secretion of insulin by B-cells and or secretion of glucagon by $\propto$-cells of the islets of Langerhans. Whenever this feed-back control is disturbed, pathological conditions set in.

Positive feed-back control is also observed in certain cases. For example, progesterone and estrogen levels stimulate LH secretion but inhibit FSH secretion and thus bring about the ovulation-follicular luteinisation cycles (menstrual cycle).

MODE OF ACTION OF HORMONES

The mode of action of hormones is fascinating. The different modes of action are listed below:

1. Induction of enzyme synthesis at the nuclear level: e.g. thyroxine, steroid hormones. These hormones bind to specific receptor proteins in the nucleus of the cell and interact with chromatin, i.e. nuclear DNA. By this, they augment the production of mRNA, i.e. (transcription), and thereby the synthesis of specific proteins (translation) needed for regulating metabolism.

While T_3 binds directly to the specific receptor proteins in the nucleus, the steroid hormones first bind to the high affinity receptor proteins in the cytosol. This complex moves to the nucleus, interacts with chromatin and nuclear DNA, and augments transcription perhaps by conformational changes.

2. Some hormones do not influence transcription, i.e. mRNA production, but augment directly the protein or enzyme synthesis at the ribosomal level, i.e., translation, e.g., insulin, ACTH.

3. Some hormones act as the level of the bio-membranes first. They may have no activity in membrane-free preparation, e.g., catecholamines, insulin.

There are specific receptors usually glycoproteins on the membranes. The hormone binds to the receptors and affects the future course of action by different methods, like phosphorylation of membrane proteins, kinases, production of small peptides, etc.

4. Many hormones will be functionless if cAMP does not serve them, hence cAMP is called the second messenger of those hormones which are themselves first messengers. The hormone binds to specific receptor sites of the membrane and activates the inner membrane-bound enzyme, adenylate cyclase. A GTP activated protein may be required for mediation.

cAMP is produced and this activates the inactive protein kinase, which with the catalytic portion executes the further chain of events.

5. **Action through calcium:** Now it is being realised that many a hormone discharges its function through calcium. This is because, the action of most protein hormones is inhibited in the absense of calcium, though their ability to increase or decrease cAMP is comparatively unimpaired. Thus calcium appears to be a more direct signal than cAMP for hormonal action. However cAMP can help in mobilising tissue bound calcium. In addition, protein hormones may increase the uptake of extra cellular calcium by the cells.

The calcium, so increased inside the cell, can combine with an ubiquitous protein, calmodulin. The calcium binding of calmodulin can result in conformational changes leading to rapid variations in enzymatic and membrane activity. This is because calmodulin is distributed throughout the cells and, in particular, is found associated with cellular membranes and many enzymes. Thus calcium can be considered the 'third messenger' of some hormones or even the 'second messenger', if it is used by the hormones independent of cAMP.

Communication between the hormone receptor on the cell membrane and the intra-cellular Ca^{++} stored in the mitochondria is achieved by phospho-inositide. When a hormone like epinephrine binds to the α-receptor, it enhances the breakdown of the phospholipid, i.e., phosphatidylinositol to diacyl glycerol and myoinositol 1,4,5-triphosphate. Myoinositol 1,4,5-triphosphate produced liberates the stored Ca^{++} in the mitochondria and endoplasmic reticulum (sarcoplasmic reticulum in the case of the muscle). Several hormones like ACTH, LH, and other pituitary trophic hormones help to increase the turnover of phospholipids like phosphatidyl inositol in their target glands like the adrenal cortex, ovary and testes, and help in the binding of cholesterol to cytochrome P_{450} there which is a major step in the production of the steroid hormones.

Certain hormones like insulin-like growth factors (IGF_1 and IGF_2), growth hormone and prolactin appear to have no known intra-cellular messenger for their action.

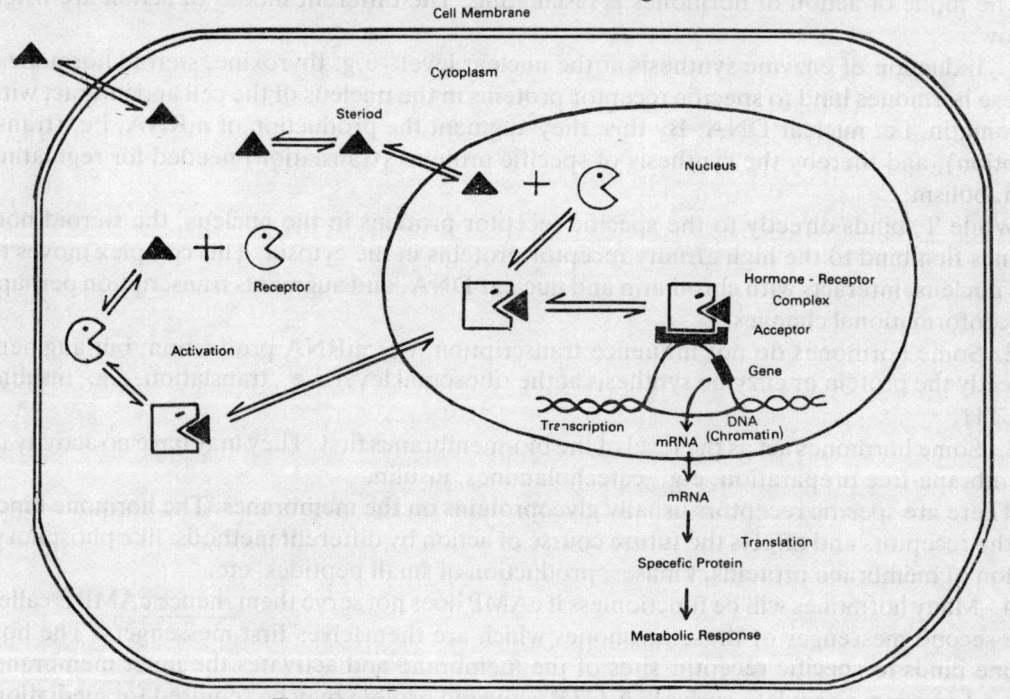

Fig-1 Mechanism of action of steroid Hormones

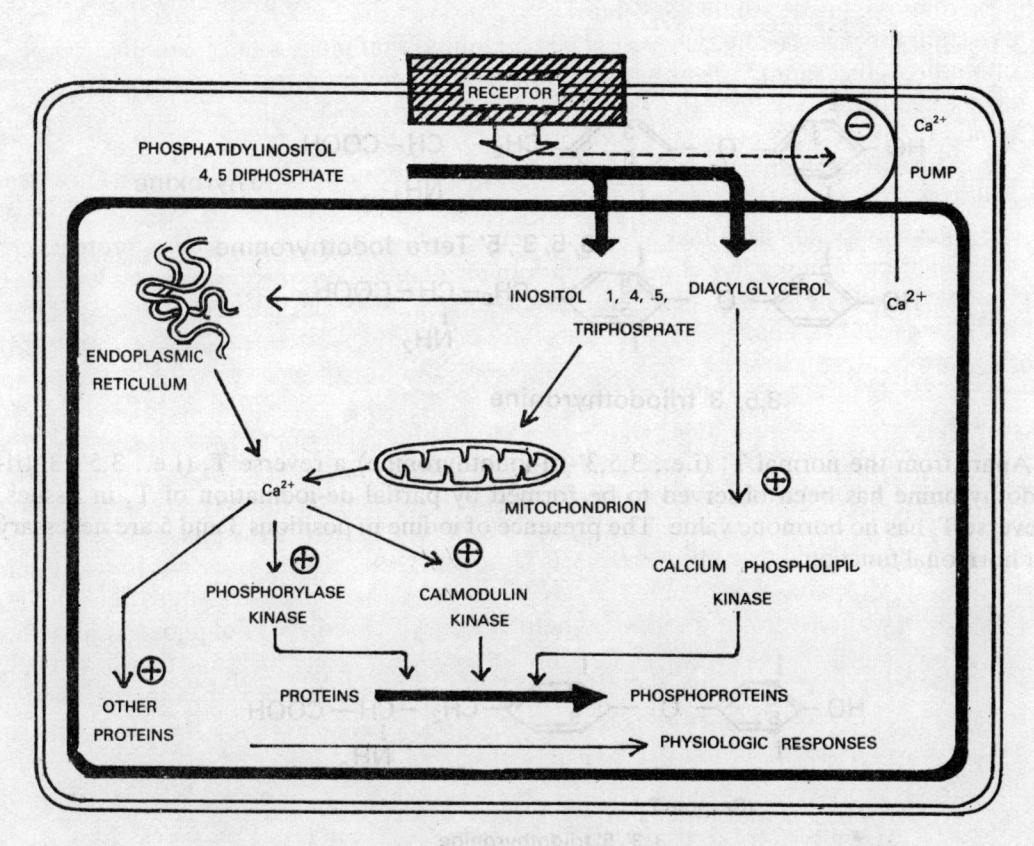

Fig 2 : HORMONE ACTION REGULATION BY Ca^{++} ions

The various individual hormones, their synthesis and action are described below.

THE THYROID AND ITS HORMONES

The thyroid is a small gland weighing about 25 g in the adult. It is made up of closed vesicles lined with a single layer of epithelial cells and filled with a colloid material which contains a protein known as thyroglobulin. Thyroglobulin is an iodine-containing glycoprotein (molecular weight of 650,000) which undergoes hydrolysis by intra-cellular proteases, releasing the thyroid hormones into the blood of capillaries surrounding the cells.

The main hormone of the thyroid gland, thyroxine was first isolated by Kendall. In addition to thyroxine, other compounds such as mono-iodotyrosine, 3,5-di-iodotyrosine and 3,5,3'-tri-iodothyronine are present in the thyroid. Physiologically, only thyroxine and 3,5,3'-tri-iodothyronine are active. The activity of the latter is about 5 – 10 times more than that of thyroxine, although the former is present in greater quantity in circulating blood. The chemical structure of these hormones is given in the next page.

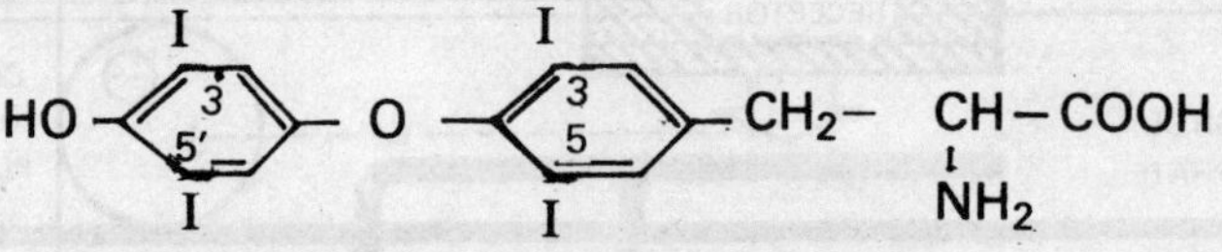

Thyroxine
3, 5, 3', 5' Tetra iodothyronine

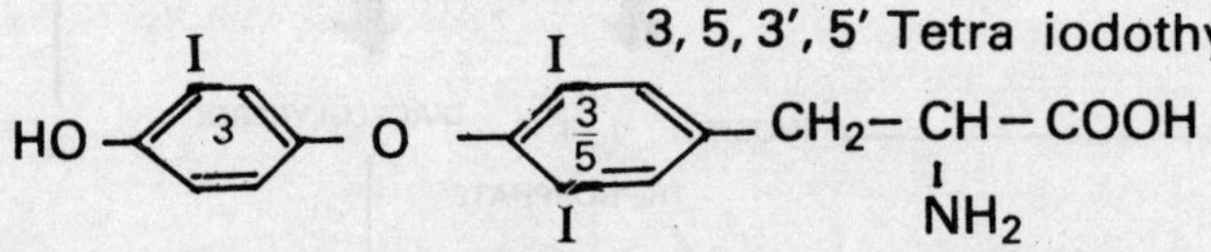

3,5, 3' triiodothyronine

Apart from the normal T_3 (i.e., 3,5,3'-tri-iodothyronine) a reverse T_3 (i.e., 3,5', 3'-tri-iodothyronine has been observed to be formed by partial de-iodination of T_4 in tissues. Reverse T_a has no hormone value. The presence of iodine in positions 3 and 5 are necessary for hormonal function.

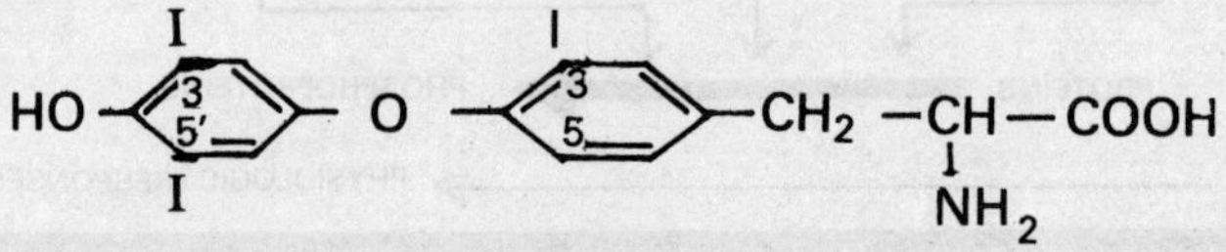

(Reverse T_3)
3, 3', 5' triidothyronine

Biosynthesis and secretion

The thyroid gland contains more than half of the total iodine content in the body. It has a remarkable capacity to concentrate iodide, brought to it by the circulating blood. The entry of iodide into the colloid occurs against a concentration gradient of 20:1 and is energy dependent, being followed by a K^+ influx and Na^+ efflux. It is enhanced by the thyrotropic hormone but inhibited by anti-thyroid agents like thiocyanate and perchlorate.

The iodide entering the thyroid cells is rapidly converted to the iodinated derivatives of tyrosine of the thyro-globulin molecule which is a glycoprotein having 115 tyrosine residues. This requires the conversion of iodide to 'active iodine' by a peroxidase catalysed reaction requiring H_2O_2 which is formed by the auto-oxidation of flavoproteins.

$$FADH_2 + O_2 \longrightarrow FAD + H_2O_2$$
$$2 H_2O_2 + I^- \longrightarrow 2 H_2O + I^+ + O_2$$

Thyroid is the only tissue which can oxidise I^- to a higher valency state (I^+). The organification of iodide is assisted by thyroperoxidase (a heme peroxidase) at the luminal surface of the follicular cell. This reaction requires H_2O_2. Thio-urea prevents organification. The thyrotropic hormone is also active in stimulating this reaction. The 'I⁺ first iodinates at position 3 of the tyrosine nucleus and then at position 5 to give mono-and di-iodotyrosine. Coupling of two molecules of di-iodotyrosine results in the formation of thyroxine.

Coupling of mono-iodotyrosine and di-iodotyrosine gives tri-iodothyronine which can also be formed from de-iodination of thyroxine. Coupling within the thyroglobulin molecule is effected by the same thyroperoxidase.

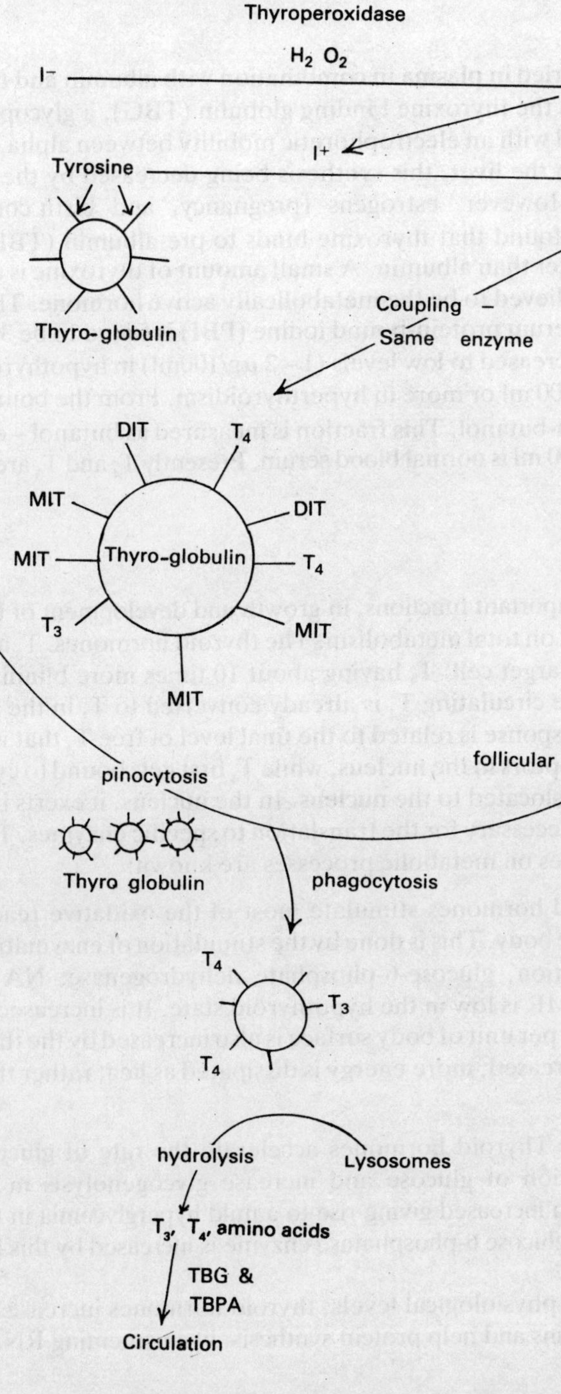

$$\text{I}^- \xrightarrow[\text{H}_2\text{O}_2]{\text{Thyroperoxidase}} \text{I}^+$$

Tyrosine $\xrightarrow{\text{I}^+}$ MIT, DIT → Thyro-globulin (MIT, MIT, DIT, DIT)

Coupling — Same enzyme →

Thyro-globulin (DIT, MIT, MIT, T_3, MIT, DIT, T_4, MIT)

follicular cell

pinocytosis

Thyro globulin phagocytosis → T_4, T_3, T_4

hydrolysis ← Lysosomes

T_3, T_4, amino acids

↓

TBG & TBPA

↓

Circulation

Thyroglobulin is broken down by proteolytic enzymes believed to be present in the vesicular membranes. Thyroxine and tri-iodothyronine are released into the blood, the proteolysis being stimulated by the thyrotropic hormone and by stress conditions like exposure to cold. The T_4, T_3 ratio is 7:1. About 50 μg of thyroid hormone iodide is synthesised per day. Nearly 70 per cent of the iodide in thyroglobulin is mono and diodotyrosines and only about 30 per cent is T_4 and T_3.

Transport

The thyroid hormones are carried in plasma in combination with albumin and two specific plasma proteins. One of them is the thyroxine binding globulin (TBG), a glycoprotein with a molecular weight of 50,000 and with an electrophoretic mobility between $alpha_1$ and $alpha_2$ globulins. TBG is synthesised in the liver, this synthesis being decreased by the androgens and gluco-corticoid therapy. However, estrogens (pregnancy, and birth control pills) increase its synthesis. It is also found that thyroxine binds to pre-albumin (TBPA) having an electrophoretic mobility greater than albumin. A small amount of thyroxine is always present in the free state, which is believed to be the metabolically active hormone. The total circulating hormone measured as serum protein-bound iodine (PBI) is found to be 3 – 7 μg/100 ml in normal individuals. It is decreased to low levels (1 – 2 μg/100ml) in hypothyroidism and increased to high levels of 9 μg/100 ml or more in hyperthyroidism. From the bound protein, the hormone is extractable with n-butanol. This fraction is measured as butanol – extractable This value is equal to 2 – 6.5 μg/100 ml is normal blood serum. Presently T_4 and T_3 are estimated by radio immunoassay.

Mechanism of action

Thyroid hormones have two important functions, in growth and development of the body as well as having a stimulating effect on total metabolism. The thyroid hormones. T_3 and T_4 bind to high affinity receptors in the target cell, T_3 having about 10 times more blinding affinity than T_4. About 80 per cent of the circulating T_4 is already converted to T_3 in the peripheral tissues, and thus the biological response is related to the final level of free T_3 that is bound to the receptors. T_3 binds to the receptors in the nucleus, while T_4 first gets bound to cytoplasmic core receptors and then gets translocated to the nucleus. In the nucleus, it exerts its effect in the transcription to the mRNAs necessary for the translation to specific enzymes. The following effects of the thyroid hormones on metabolic processes are known:

1) Calorigenic action: Thyroid hormones stimulate most of the oxidative reactions and regulate the metabolic rates in the body. This is done by the stimulation of enzymatic systems, e.g., enzymes of glucose oxidation, glucose-6-phosphate dehydrogenase, $NADP^+$-cytochrome C-reductase, etc. The BMR is low in the hypothyroid state. It is increased in hyperthyroidism. Oxygen consumption per unit of body surface is also increased by the thyroid hormones. Although oxidation is increased, more energy is dissipated as heat rather than stored as high energy compounds.

2). Carbohydrate metabolism: Thyroid hormones accelerate the rate of glucose oxidation, promote intestinal absorption of glucose and increase glycogenolysis in the liver. Gluconeogenesis in the liver is also increased giving rise to a mild hyperglycemia in the hyperthyroid state. The activity of liver-glucose 6-phosphatase enzyme is increased by this hormone.

3) Protein metabolism: At the physiological levels, thyroid hormones increase the incorporation of amino acids into proteins and help protein synthesis, by augmenting RNA production at the nuclear level and also by the translation of the message contained in the messenger RNA at the ribosome. They also stimulate the production of the growth hormone. Above the physiological levels, thyroxine has the opposite effect on protein metabolism, i.e., increasing the breakdown of amino acids, depressing protein synthesis and bringing about a negative N-balance.

4) *Lipid metabolism:* Thyroxine increases the turnover of lipids. Both the deposition of fat in adipose tissue as well as its release and oxidation are increased by this hormone. The cholesterol level in blood increases in hypothyroidism.

5) *Effect on mitochondria:* Thyroxine uncouples oxidative phosphorylation by causing swelling of mitochondria. Hence, although oxidation proceeds, no ATP is produced and the energy of oxidation is given off as heat. This action of thyroxine is observed only at a high concentration of the hormone and not at physiological levels.

6) *Effect on Na^+/K^+ ATPase pump:* Thyroid hormones enhance the function of Na^+/K^+ ATPase pump, thus increasing the ATP utilisation. Along with the increased oxygen consumption and ATP utilisation, the metabolic processes are also made to proceed at a faster rate.

Anti-thyroid agents

Substances that inhibit the normal functioning of thyroid are termed anti-thyroid agents. They are:

1) Agents which retard the synthesis of thyroid hormone, e.g., thiocynate, thiocarbamide, sulpha drugs and perchlorate. Thiocyanate and perchlorate inhibit the iodide uptake while thiocarbamide and sulpha drugs interfere with the iodination process.

2) Synthetic analogues of the thyroxine like 2',6'-di-iodothyronine bring about competitive inhibition.

3) Deep X-ray therapy destroys the thyroid tissue and this depresses the thyroid activity. ^{131}I in high doses also has a similar effect.

4) Cobaltous chloride administration interferes with thyroid hormone synthesis, and clinical myxoedema and goitre may result.

5) Thyroxine itself can inhibit the production of TSH of the anterior pituitary and thus check the release of thyroxine by a feed-back mechanism.

6) Certain organic compounds are present in vegetables like cabbage and turnip which act as natural goitrogens and depress thyroid activity. These are present as goitrins in the raw foods and are however destroyed on cooking.

Metabolic fate and excretion

Thyroxine undergoes the following alterations in the liver:

1) Transamination and deamination: Transamination with the alpha keto acids gives the corresponding pyruvic acid analogues of the hormones.

2. Conjugation with glucuronic acid and sulphate: The phenolic group of the iodothyronines gets conjugated with glucuronic acid or sulphate and the conjugates are excreted through bile into the intestines.

3. De-iodination: De-iodination also may take place in the liver in the presence of the enzyme, de-iodinase. The de-iodinated thyronines are metabolised by transamination followed by conjugation with glucuronic acid and excretion by the kidney.

Most of the thyroid iodine finds its way into the urine in which it is present as inorganic iodide. One per cent of organic iodide, mostly as thyropyruvic acid is also present in urine. A small amount of thyroxine is destroyed in saliva by de-iodination and this is another channel for the excretion of iodine.

Hypothyroidism

Insufficient development of thyroid in embryonic life results in the child becoming a cretin. Cretinism is associated with slow growth, dwarfism and mental retardation. Dry skin, scanty hair, saddled nose, puffy lips and vacant expression are some of the features of cretinism.

Hypothyroidism in the adult results in myxoedema. Here, the BMR and body temperature are lowered. Sensitivity to cold, puffiness of face, anemia, slowing of physical and mental functions, hypercholesterolemia and low serum PBI. T_3 and T_4 levels are the features of this condition.

Hyperthyroidism

Increased acitivity accompanied by excessive secretion of the thyroid hormone occurs in Grave's disease (exophthalmic goitre) and toxic adenoma. Hyperthyroidism may occur with or without goitre. Nervousness, irritability, loss of weight, increased body temperature, increased heart rate, fatigue, increased appetite, protrusion of the eyeballs (exophthalmos) are found associated with the hyperthyroid state. The BMR is raised and serum PBI, T_3 and T_4 levels are increased in this condition.

Goitre

Enlargement of the thyroid gland is called goitre. Simple endemic goitre (colloid goitre) is caused by deficiency of iodine in the diet. This occurs in regions far away from the sea coast, where the soil and water are low in iodine. Iodised salt (containing 0.002 per cent of iodide) is used to check the incidence of goitre. Naturally occurring goitrogenic factors are present in raw cabbage, cauliflower and turnip. These factors are activated by intestinal digestion but destroyed on cooking these foods.

In toxic goitre, enlargement of the gland is accompanied by increased hormone secretion. This condition is treated by removing a part of the gland surgically. Treatment with ^{131}I is also effective.

CALCITONIN

This new hormone was discovered in 1962 by Copp and others who believed it to be a calcium lowering hormone produced by the para-thyroids. Recently, however, it has become known to be of ultimo-branchial origin and elaborated by the C-cells of the thyroid. The name 'thyrocalcitonin', given to this hormone by earlier workers is not in current use, instead the term 'calcitonin' approved by WHO and Index Medicus is used. The name 'calcitonin' implies that the hormone helps to maintain the 'tone' of calcium in the body.

Calcitonin release is stimulated by high Ca^{++} levels in serum. This hormone has a direct effect on the bone, but its effect is opposite to that of the para-thyroid hormone. It lowers the serum Ca levels and prevents the release of bone Ca. The action of calcitonin is found to be through its action on pyrophosphatase and iso-citrate dehydrogenase enzyme systems influencing them through the concentration of intracellular calcium ions (second messenger for calcitonin). By **antagonism** to para-thyroid hormone and possibly by a calcium pump mechanism, calcitonin keeps down intracellular calcium which, in its turn, affects the two enzyme systems referred to above. The inhibition of pyrophosphatase by para-thyroid hormone in the presence of calcium is corrected by calcitonin, while at a particular concentration of calcium, there is a direct stimulation of this enzyme by calcitonin. The inhibition of isocitrate dehydrogenase brought about by increase of intracellular calcium (action of para-thyroid hormone) is also corrected by calcitonin. In addition, metaphyseal osteoclastic acid phosphatase activity stimulated by the para-thyroid hormone is inhibited by calcitonin.

Calcitonin leads to phosphaturia and lowered serum phosphorus. Calcitonin can act in the absence of the para-thyroid hormone as well as 25-hydroxy cholecalciferol or 1,25 dihydroxy cholecalciferol. This hormone is chemically characterised as a polypeptide of molecular weight 3,600 consisting of a straight chain containing 32 amino acids.

The secretion of calcitonin and the para-thyroid hormone are inversely related and both are controlled by the plasma Ca^{++} levels. Also, glucagon and pentagastrin are potent secretagogues of calcitonin.

Calcitriol

Calcitriol is the derivative obtained by hydroxylation of the 25-hydroxy cholecalciferol in the mitochondria of the renal proximal convoluted tubule. The enzyme 1-hydroxylase is a mono-oxygenase that requires molecular O_2, NADPH, cytochrone P_{450} and ferredoxin (iron-sulphur protein). Calcitriol is now defined clearly as a hormone which stimulates intestinal absorption of calcium and phosphate and also the renal reabsorption of calcium. Most of the actions of vitamin D_3 in our body appear to be routed through its conversion to calcitriol. In vitamin D resistant rickets, calcitriol production is impaired. It is an inherited autosomal recessive trait characterised by a defect in the conversion of 25-hydroxy vitamin D_3 to calcitriol. There is also another type of vitamin D-resistant rickets in which the receptors for calcitriol are absent.

In man, vitamin D_3 is available both from dietary source (fish oils, egg yolk) as well as by endogenous synthesis from 7-dehydrocholesterol in skin, on exposure to ultra-violet light. The enzymatic photolysis reaction takes place in the epidermis and is available to the body. In plasma, vitamin D_3 is carried from the skin as well as from the intestine to the liver by a D_3 binding protein. 25-hydroxylation takes place in the liver and 1-hydroxylation in the renal tubules by mono-oxygenase reactions requiring molecular O_2, Cytochrome P_{450} and NADPH. The calcitriol formed reaches various cells of the intestinal lumen and bone from the renal tubules and exerts its influence on calcium absorption and calcium translocation. Calcitriol production is under feed-back control. Its own level can control its production, by inhibiting 1-hydroxylase but diverting the 25-hydroxy vitamin D_3 towards formation of 24,25-dihydroxy vitamin D_3.

The mechanism of action of calcitriol resembles that of most of the steroid hormones. In the intestinal mucosal cell, there are receptors for calcitriol located in the chromatin of the nucleus with which the hormone gets bound. This stimulates gene-transcription and production of specific mRNAs that are then translated into calcium binding protein (CBP). The CBP then binds to Ca^{++} and helps to transport these ions along with an equivalent amount of phosphate.

PARA-THYROID HORMONE

Closely associated with the thyroid gland are two pairs of small glands called the para-thyroids. In man, the total mass weighs only 50–300 mg. These glands produce a polypeptide hormone having a profound influence on calcium and phosphate metabolisms. The maintenance of serum Ca^{++} levels within normal physiological limits is now believed to be the primary function of the parathyroid hormone. It also controls the renal excretion of calcium and phosphate.

Biochemical effects

Administratoin of the para-thyroid hormone gives rise to the following effects:
1. The serum calcium is raised but serum phosphorus is lowered.
2. The excretion of both calcium and phosphorus is increased.
3. Calcium from the bone gets mobilised and added on to the serum.
4. The serum alkaline phosphatase activity is increased.

The para-thyroid hormone (PTH) is a single-chain peptide containing 84 amino acids and having a molecular weight of 9,500. It is synthesised as a precursor molecule or pro-hormone containing 115 amino acids which is believed to come from a pre-prohormone. The pre-prohormone gets converted in two steps to the active para-thyroid hormone by peptide cleavage in the Golgi apparatus of the para-thyroid cell and then gets immediately secreted. The pro-parathyroid hormone is produced in spite of any changes in Ca^{++} concentration, but its conversion to para-thyroid hormone and secretion of the latter is increased when the blood Ca^{++} levels are low, and decreased when the Ca^{++} levels are higher. It is also known that ionized Mg^{++} level also may affect the secretion of PTH. The PTH level in plasma increases in hyper-para-thyroidism when the Ca^{++} levels are also high. PTH secretion is stimulated by a wide variety of compounds like cyclic AMP, prostaglandins, E_1 and E_2, dopamine, etc.

The mechanism by which the para-thyroid hormone acts on the bone and brings about resorption of calcium is due to various factors. It increases the production of lactic and citric acid in bone tissue which can solubilise the bone. The para-thyroid hormone specifically inhibits osteoblasts but stimulates the osteoclasts. It activates membrane-bound adenylate cyclase which increases the levels of cyclic AMP. This leads to increased permeability for the calcium ions across the membrane. The levels of intra-cellular calcium influence the enzymes, pyrophosphatase and isocitrate dehydrogenase possibly through phosphorylation of specific intracellular kinases. The para-thyroid hormone inhibits pyrophosphatase in the presence of calcium (a condition unfavourable for bone formation). Again, iso-citrate dehydrogenase is also inhibited due to the accumulation of calcium. This leads to the accumulation of citric acid which has a solubilising effect on bone. Metaphyseal osteoclastic acid phosphatase activity is also stimulated by the para-thyroid hormone. Thus, this hormone acts through the 'second messengers', cyclic AMP and calcium ions. It is known that cyclic AMP itself causes resorption of calcium. The para-thyroid hormone is also required for the conversion of 25-hydroxy vitamin D_3 to 1, 25-dihydroxy vitamin D_3 in the kidneys. The dihydroxy vitamin D_3 promotes the synthesis of the calcium binding protein and increases the influx of calcium. Increased intracellular Ca^{2+} brings about de-mineralisation as described above.

A direct effect of the parathyroid hormone on the kidney is on the tubular reabsorpton of calcium and phosphate. It facilitates tubular reabsorption of Ca^{++} but inhibits the reabsorption of phosphate. This results in increased serum Ca^{++} but decreased serum phosphorus levels. There is thus phosphaturia.

Control of secretion

The secretion of the para-thyroid hormone is not under the control of the pituitary. It is controlled by serum Ca^{++} levels by a negative feed-back mechanism which is also regulated by calcitonin. The removal of the para-thyroid accidentally during the removal of a part of the thyroid during surgery results in symptoms of hypo-para-thyroidism. In the latter case, hypocalcemia leading to muscular spasms and tetany may ensure. The treatment of this condition is usually done by administration of calcium salts along with vitamin D.

HORMONES OF THE ADRENAL MEDULLA

The medullary portion of the adrenals elaborates two hormones, epinephrine and nor-epinephrine. The common secretion from the adrenal medulla is a mixture, 1/10 to 1/3 of which is due to nor epinephrine and the rest due to epinephrine. Both these hormones belong to the group of catecholamines as they possess the catechol residue (ortho-dihydroxy benzene) in their structure. Epinephrine differs from nor-epinephrine by having a methyl group attached to the N-atom.

These hormones are related to the aromatic amino acid tyrosine and are derived from it. The naturally occurring hormones are L-isomers. L-epinephrine produced synthetically is found to be as active as the natural hormone.

Biosynthesis and secretion

Epinephrine is elaborated by the adrenal medulla as a result of sympathetic stimulation. It is produced readily during 'fight, fright and flight' and in emergencies like cold, fatigue, shock, etc. Experiments have shown that isotopically labelled tyrosine and phenyl alanine are converted to epinephrine by the adrenal medulla. (For biosynthetic steps refer 'Metabolism of Phenyl Alanine'.) The chromaffin granules of the adrenal medulla are concerned with the biosynthesis, storage and secretion of catecholamines. The hormone is released from these granules by the exocytosis of cells which is a calcium-dependent process. Neural stimulation, fright, emotional conditions like anger etc. are responsible for a quick release of the catecholamine hormones.

Physiological and biochemical functions

1. Sympathomimetic function: Epinephrine causes a rise in blood pressure due to arteriolar vaso-constriction particularly in the skin, mucous membranes and splanchnic viscera. However, the arterioles of the skeletal muscles undergo vaso-dilatation. The overall effect is a rise in blood pressure, increase in pulse rate, heart rate and cardiac output.

Nor-epinephrine causes rise in blood pressure by increasing peripheral resistance. It is an overall vaso-constrictor and has no effect on cardiac output. Nor-epinephrine is therefore used in the treatment of hypotensive shock, particularly since it does not produce tachycardia.

2. *Action on smooth muscle:* Epinephrine dilates bronchial musculature relaxes the musculature of the gastro-intestinal tract and contracts the pyloric sphincter. These effects are exhibited very weakly by nor-epinephrine.

3. *Effects on carbohydrate metabolism:* Epinephrine promotes glycogenolysis in the muscle and liver and produces an increase in blood lactic acid level as well as blood glucose levels, respectively. The primary effect of epinephrine is to activate the enzyme adenylate cyclase and raise the cyclic AMP levels. Cyclic AMP in turn activates the inactive phosphorylase b-kinase to active phosphorylase b kinase which converts the inactive phosphorylase 'b' to the active phosphorylase 'a'. The active phosphorylase 'a' catalyses the breakdown of glycogen to the phosphorylated derivatives of glucose and to lactic acid. In the liver, dephosphophosphorylase kinase is activated to convert dephosphophosphorylase to phosphophosphorylase which catalyses the breakdown of liver glycogen to glucose. In the muscle, epinephrine depresses glycogen synthesis by activating the cyclic AMP dependent proteinkinase which then inactivates the glycogen synthase by phosphorylating it (see 'Carbohydrate metabolism').

Nor-epinephrine has only ⅛th activity of epinephrine in glycogen breakdown.

Epinephrine decreases the glucose uptake in the heart, but whatever glucose enters the cardiac muscle is shunted towards glycogen production, thus increasing the cardiac glycogen. It prevents glucose uptake in the peripheral tissues, making it available for the central nervous system.

4. *Effects on lipid metabolism:* Epinephrine and nor-epinephrine act as lipolytic hormones They stimulate lipolysis in adipose tissue and release free fatty acids into the blood, thereby raising the blood FFA level. The blood cholesterol and phospholipids are also increased. The circulating plasma brings to the heart free fatty acids which are used as very good fuel by this organ.

5. *Pheochromocytoma:* Tumours of the chromaffin tissue result in this condition observed in man. Pheochromocytoma is charactersied by hyperglycemia, glycousuria, hypertension and hyperlipemia. The plasma catecholamine levels rise several times above the normal value, and urinary excretion of epinephrine metabolites increases remarkably.

6. *Metabolic fate of epinephrine and nor-epinephrine:* The adrenal medulla contains 1 – 3 mg/g tissue of epinephrine and only 0.2 – 0.6 mg of nor-epinephrine. However, in plasma, the concentration of nor-epinephrine is much higher than that of epinephrine. These hormones occur in urine both in the free state and in conjugated forms with sulphate or glucuronic acid.

Catecholamine, once elaborated and circulated in blood, are very quickly destroyed and excerted in urine. The enzymes that act on catecholamines are catechol-O-methyl transferase (COMT) and mono-amine oxidase (MAO). These enzymes are found in many tissues. The metabolic products, metanephrine and vanillyl mandelic acid (VMA) are inctive and are useful indices for the adrenal medullary function. These metabolites are excreted in large amounts in patients with pheochromocytoma (tumours of the adrenal medulla).

HORMONES OF THE PANCREATIC ISLET CELLS

The endocrine part of the pancreas comprises the islets of Langerhans. The islet cells are of four types, A-cells, B-cells, D-cells, and F-cells. The A cells or $\propto$-cells produce glucagon, the B-cells or ß-cells produce insulin and the D-cells produce somatostatin. The F-cells are concerned with the production of pancreatic polypeptide whose function is not clearly known at present. Somatostatin is also produced by the hypothalamus but in the pancreas, the locally produced somatostatin is involved in the regulation of secretion of both glucagon and insulin.

INSULIN

Insulin is an important hormone concerned primarily with carbohydrate metabolism. Its deficiency results in deranged carbohydrate metabolism and diabetes mellitus. Some other metabolisms are also impaired.

Chemistry: Insulin is a protein hormone which has been otbained in crystalline form. The crystalline insulin contains traces of Zn, which is believed to be present in pancreatic tissues in greater quantities than in many other tisses. Insulin obtained from the pig, whale and dog is identical, while that from the sheep, horse and cow differs in three amino acids at 8, 9 and 10 positions of the A-chain. Porcine insulin almost resembles human insulin, except for the difference in one amino acid, alanine, at the C-terminus of the B-chain. On treatment with carboxy peptidase, this amino acid is removed, but the hormone will retain its activity. This hormone is commercially available as de-alanated insulin and is used in the treatment of human diabetes because of its low antigenicity. Protamine Zn insulin (PZI), a combination with protamine, is absorbed more slowly than crystalline (regular) insulin and may exert its effect only for 24 hrs.

Biosynthesis of insulin: Insulin is synthesised as a pre-prohormone, i.e., pre-proinsulin (mol wt 11,500). The initial portion of 23 amino acids is called the leader sequence and directs the synthesis of pro-insulin. The leader sequence is cleaved off. The pro-insulin (mol wt 9,000) which is synthesised has the conformation necessary for forming the proper disulphide bridges between the A-and B-chains. It undergoes a series of site-specific cleavages at the Golgi with trypsin like enzymes forming the mature insulin and an equi-molecular amount of connecting peptide (C-peptide) of 31 amino acids.

Secretion: Human pancreas secrets 40 – 50 units of insulin daily which is 15 – 20 per cent of the hormone in the gland stored. The insulin formed is packed into secretary granules at the Golgi and released into circulation by mediators like glucose. Both the absolute plasma concentration of glucose and the rate of change of plasma glucose trigger the release of insulin from the pancreas. The release is biphasic. There is an immediate rise of plasma insulin within a minute of plasma glucose increase which lasts for 5 – 10 minutes. This is followed by a gradual but prolonged second phase reaching a peak in about 60 minutes. This terminates soon after the glucose stimulus is removed. The amino acids, arginine, lysine and particularly leucine, the gastro-intestinal hormones secretin, cholecystokinin, and gastrin, glucagon, epinephrine and xylitol also cause insulin release. Drugs like tolbutamide (sulphonyl ureas) stimulate insulin secretion and have achieved widespread use in the treatment of Type II diabetes mellitus. The biguanides are also available now for the purpose.

It is suggested that glucose binds to receptors possibly located on the B-cell membrane of the pancreas; alternately, intracellular increase of Ca^{2+} brought about by K^+-induced depolarisation of the membrane or an increased $NADPH/NADP^+$ ratio through the HMP shunt of glucose metabolism could cause insulin release. Increased intracellular cAMP due to the influence of hormones for instance can also independently cause release of insulin from the pancreas. cAMP may phosphorylate a component of microfilament-microtubule through a protein kinase and play its role. It may also increase intracellular Ca^+. In recent times, phosphatidyl inositol is also implicated. The hydrolytic product, myo-inositol triphosphate, may release calcium from the intracellular stores. The released ionic Ca^{2+} can bring about the effect.

Effects of insulin on metabolism: Insulin has an important role in carbohydrate metabolism in insulin-sensitive tissues. It also affects protein, lipid, nucleoprotein and mineral metabolisms. Insulin is very active in the skeletal muscle, diaphragm, heart muscle, adipose tissue, leukocytes and lens of the eye. It is believed that it is almost inactive in the erythrocytes and intestinal tissue.

1. Carbohydrate metabolism: Administration of insulin decreases the level of glucose in the extra-cellular fluids, while it increases the formation of hexose phosphates and other metabolic products in the intra-cellular fluid. It increases the glycogen content of the liver and muscle. It is now established beyond doubt that in the muscle and adipose tissue, insulin facilitates the transport of glucose across the cell membrane by increasing the permeability which leads to an increase in all the pathways of glucose metabolism, i.e., oxidation to CO_2, indicated by increased O_2 consumption and CO_2 production, glycogen deposition and increased conversion of glucose to fatty acids.

Insulin brings about its effect through its action at the membrane level directly, and inside the cells through messengers.

At the membrane level, there are receptors for insulin in the plasma membrane of the muscle and adipose tissue cells. The receptors are specific glycoproteins made of two $\propto$ and two ß sub-units and hence are heterodimers. The $\propto$ and ß units are linked by disulphide bonds. The insulin molecule binds to the $\propto$-sub-units. Insulin receptor binding activates a cAMP - independent protein kinase in the ß-unit which phosphorylates tyrosine of the receptor protein. This leads to a relay of intracellular events like the activation of intracellular threonine and serine kinases, peptide messengers, etc. The result is phosphorylation in many cases and dephosphorylation of endogenous proteins and enzymes to bring about the desired metabolic effect. While enzymes like glycogen synthetase and pyruvate dehydrogenase are dephosphorylated and activated, phosphodiesterase and acetyl CoA carboxylase are phosphorylated. Apart from the tyrosine, threonine and serine kinases, some other intracellular messengers have also been suggested. They are calcium, cAMP, cGMP, membrane derived peptides, membrane phospholipids, monovalent cations, etc.

Insulin increases the permeability of glucose across the biomembrane by the following mechanism: There are glucose (hexose) carriers in the plasma membrane and in the intracellular organelles like Golgi. When insulin binds to its receptors on the plasma membrane, the glucose carriers of the intra-cellular organelles are translocated to the plasma membrane

Translocation of glucose transporters by insulin

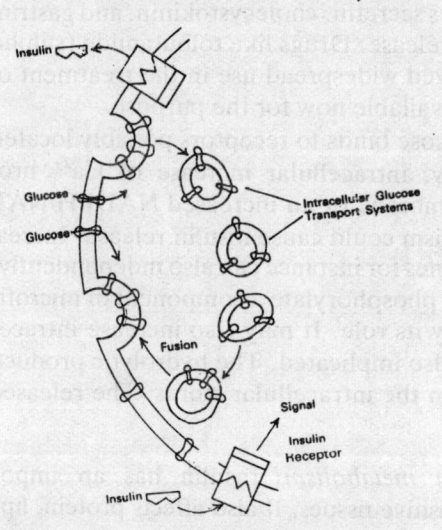

through some signal. This increases the total number of glucose carriers in the plasma membrane.

Consequently, there is increased permeability and transport of glucose across the plasma membrane. However, this does not operate in hepatic cells. The transport of glucose across the membrane is referred to as 'carrier mediated', facilitated transport. Sugars like galactose, D xylose and L-arabinose, having a similar configuration as glucose in respect of three carbon atoms around the functional one, are transported by the same mechanism.

Other effects of insulin on enzymes are achieved through messengers like protein kinases or other substances, and phosphorylation or dephosphorylation as the case may be.

Insulin receptors are synthesised and degraded in 7 – 12 hours. When the plasma insulin levels are high, e.g., as in obesity, the number of insulin receptors is decreased and target tissues become less sensitive to insulin. This is referred to as 'down regulation' which explains part of the insulin resistance in obesity and Type II diabetes mellitus. Reducing body weight and physical exercise increase the number of receptors, as also supplementing insulin therapy with oral tolbutamide or traces of trivalent chromium salts.

Insulin is not required for the entry of glucose into the liver and kidney cells which are easily permeable to glucose. It appears that the entry of fructose into the cells is not dependent on insulin.

Apart from the major effect on the transport of glucose for its metabolism, insulin induces the enzymes glucokinase, phosphofructokinase and pyruvate kinase. By its effect on glucokinase, it promotes the phosphorylation of glucose. The oxidation of glucose is increased in view of the action of insulin on the different enzymes of glycolysis. The respiratory rate of diabetic mammalian muscle is depressed due to deficiency of insulin. Administration of insulin restores it to normal levels.

The effect of insulin on hexokinase activity is only indirect. Bessman (1966) has suggested that the hormone may be concerned with the translocation of hexokinase. The inhibition of hexokinase activity which is observed with the growth hormone and cortico-steroids is reversed by insulin.

Insulin activates pyruvate dehydrogenase.

Insulin increases glycogenesis as it promotes the formation of glycogen synthetase 'a' by stimulating synthetase phosphatase. In addition, the conversion of glycogen synthetase 'a' to 'b' is inhibited by insulin indirectly by antagonism to cAMP.

The formation of glucose through gluconeogenesis is suppressed as the gluconeogenic enzymes like glucose-6-phosphatase, pyruvate carboxylase, phospho pyruvate carboxylase and fructose 1.6 diphosphatase are repressed by insulin.

Insulin also increases the HMP shunt pathway as it promotes the oxidation of C-1 carbon atom of glucose, incidentally making available NADPH for lipogenesis.

2. Protein metabolism: Insulin helps protein synthesis from amino acids and acts as an anabolic hormone, this being dependent on glucose utilisation. Experimental studies indicate that insulin stimulates both the entry of amino acids into the cell and the subsequent reactions which bring about protein synthesis. Insulin augments the ribosomal activity of the liver where it increases its capacity of translation of information from mRNA to protein synthesising machinery. The polyribosomes are disaggregated in the diabetic cell, but administration of insulin restores them to the normal aggregated form. Insulin acts to bring about the synthesis of a specific protein called the 'translation factor' which acts on the ribosomes to form polysomes.

3. Lipid metabolism: Insulin by promoting the HMP shunt pathway in the adipose tissue as well as liver, provides NADPH required for the reductive synthesis of fatty acids and thus promotes lipogenesis. This is blocked in diabetes but administration of insulin restores it to normalcy.

In the adipose tissue, insulin increases the synthesis of fatty acids, resulting in an increase in the fat content. Another effect of insulin on lipid metabolism is its inhibitory action on lipolysis. Lipolysis is depressed by insulin by maintaining diminished levels of cyclic AMP, by its action on phospho diesterase. It thereby decreases the plasma FFA levels and acts antagonistically towards many other hormones like the growth hormone, epinephrine, etc.

4. Nucleic acid metabolism: The incorporation of ^{32}P labelled phosphate, adenine, etc. into the nucleic acids is increased by insulin. Probably the synthesis of all types of RNA is increased and is concerned with many anabolic effects of this hormone.

5. Mineral metabolism: Insulin increases the transport of K^+ and phosphate into the cells. Injection of insulin causes a lowering of serum potassium and inorganic phosphorus.

Most of the physiological effects of insulin can be summarised thus:

1) Increase of glucose phosphorylation, glucose oxidation, and production of energy (ATP) required for various endergonic processes.
2) Increase of glucose-6-phosphate dehydrogease activity, promotion of the HMP pathway favouring formation of NADPH, pentoses, tc. required for fatty acid and nucleic acid synthesis respectively.
3) Inhibition of gluconeogenesis, by repressing glucose-6-phosphatase, phosphopyruvate carboxylase, fructose 1,6-diphosphatase and pyruvate carboxylase. This also helps in diverting glucose-6-phosphate towards glycogen synthesis.
4) Stimulation of glycogen synthetase and thus promotion of glycogen deposition.
5) Increase of glycerol-3-phosphate dehydrogenase activity, favouring formation of alpha glycerol phosphate from triosephosphates and helping triacyl glycerols synthesis.
6) Stimulation of acetyl CoA synthesis from products of glucose metabolism, thereby promoting synthesis of fatty acids and triacyl glycerols.
7) Inhibition of lipolysis and decreasing the plasma FFA levels, by maintaining low levels of cyclic AMP.
8) Promotion of RNA synthesis and thereby stimulation of protein synthesis from amino acids.
9) Increasing the oxidative processes, and preventing accumulation of acetyl CoA, inhibiting the ketogenesis and ketosis.

Insulin deficiency: Insulin deficiency is mainfested in the clinical condition, diabetes mellitus. This is due to 1) inadequate insulin production, 2) insulin not binding to the receptors, 3) accelerated insulin destruction, or 4) insulin inhibitors and antagonists.

Inadequate insulin production may be due to the degeneration of the pancreatic islet cells by exhaustion. The hyperglycemia due to the excessive secretion of the growth hormone or ACTH or adrenal steroids may stimulate or augment insulin secretion continuously, resulting in the exhaustion of the beta islet cells in the long run.

The liver contains the enzyme, insulinase, which destroys insulin both *in vivo* and *in vitro*, by bringing about a reductive cleavage of the -S-S- bonds of insulin in the presence of reduced glutathione. This enzyme is also present in the kidney and muscle.

The condition arising from partial or total deficiency of insulin or its non-availability for physiological needs is called diabetes mellitus. Diabetes mellitus is of two types, one due to real deficiency or shortage of insulin in the circulating plasma, and the other because of resistance to the hormone due to lack of receptors or their decreased number in proportion to body size. About 90 per cent of the diabetics have the adolescent or Type II (non-insulin dependent diabetes, NIDDM) while the remaining 10 per cent have juvenile or Type I diabetes (insulin dependent diabetes, IDDM). The Type-II diabetic patients have often normal or even higher levels of plasma insulin. They need not take insulin injections, but can manage their condition by dietary regulation (decreasing the intake of carbohydrates and free sugar) or by taking oral hypoglycemic drugs. In obese diabetics of this category, dieting and consequent weight reduction are useful to restore normalcy, as the number of insulin receptors in proportion to body size may be increased by this method. Type II diabetes is associated with a decreased number of receptors which cannot handle the insulin efficiently. Oral hypoglycemic agents may enhance the action of these receptors. Resistance to one type of insulin (body's insulin), in some cases, is due to insulin antibodies, and this can be overcome only by substituting insulin from another species. In Type I diabetes, regular administration of insulin at intervals is required to help, prevent or avoid all the metabolic derangements.

Insulin-like Growth Factors, IGF_1 and IGF_2: There are two insulin-like growth factors known — IGF_1 and IGF_2, which resemble insulin in structure and function, but are not of pancreatic origin. It is believed that IGF_1 and IGF_2, which are single-chain polypeptides containing 70 and 67 amino acids, respectively, have about 50 per cent of the amino residues in common with insulin, and their physiological effects are very similar and overlapping. While insulin is more potent as a hormone which controls metabolism, the IGFs are more concerned with growth stimulation and cell proliferation. Other hormones resembling insulin are the nerve growth factor (NGP), relaxin (produced by the corpus luteum), etc.

Oral hypoglycemic agents: Many drugs when taken orally are effective as hypoglycemic agents. Some of these drugs are sulphonyl ureas like tolbutamide (orinase) and chloropropamide (diabenese). These compounds are effective only if at least a part of the pancreas is functional. These drugs are ineffective in alloxan-diabetic animals and pancreatectomised animals or in juvenile diabetic patients. The primary action of these drugs appears to be in stimulating insulin secretion. These drugs are widely used in treating maturity onset diabetes.

Apart from sulphonyl ureas, another group of drugs of the phenethyl biguanidine (phenformin) group is now being used. The biguanides are effective substances which mimic many of the actions of insulin, like increasing glucose uptake by the muscle, increasing glycolysis and increasing the oxidation of glucose by the peripheral tissues.

Glucagon

Glucagon is the hyperglycemic hormone produced by the alpha cells of the islets of Langerhans. It has been purified and crystallised from pancreatic extracts. It is characterised as a linear polypeptide with a molecular weight of 3485 and containing 29 amino acid residues. The sequence of amino acids has been worked out and the molecule has been completely synthesised in the laboratory. It contains histidine as the N-terminal amino acid and threonine as the C-terminal amino acid. It does not contain cysteine, proline or iso-leucine and has no disulphide bridge.

Glucagon has an action exactly opposite to that of insulin. In the liver, it enhances the breakdown of glycogen, releasing it into blood as glucose. In the adipose tissue, glucagon increases the breakdown of lipids to fatty acids and glycerol. The glycogenolysis in liver and lipolysis in adipose tissue are both mediated through the activation of adenylate cyclase, and the increase in the levels of cyclic AMP by this hormone. Glucagon is ineffective in enhancing glycogenolysis in the skeletal muscle. It is suggested that it cannot enter the muscle cells. In the liver, the action of glucagon on glycogenolysis is much greater than that of epinephrine.

Glucagon is destroyed by the liver. This destruction is, however, prevented by insulin. Glucagon is often used to treat cases of hyper-insulinism (endogenous hypoglycemia) arising out of beta islet cell tumour.

Glucagon secretion is stimulated by low glucose levels (hypoglycemia) and by certain gastro-intestinal hormones. High glucose levels and the hormone somatostatin are inhibitors of glucagon secretion.

Glucagon acts in the cells of the liver by its ability to elevate cyclic AMP through its action on adenylate cyclase. It enhances various processes like gluconeogenesis and glycogenolysis in the liver, and lipolysis in the adipose tissue. One of the gluconeogenic enzymes, PEP carboxykinase, is an inducible enzyme that responds to glucagon through gene transcription and helps to elevate its levels. In the adipose tissue, it activates the hormone-sensitive triacyl glycerol lipase through cyclic AMP and a protein kinase. This action of glucagon is much more pronounced in the absence of insulin (as in diabetes mellitus) as it can act unopposed in that condition.

Somatostatin

This hormone is also produced by the hypothalamus, but the hormone produced by the D-cells of the pancreatic islet cells and the hypothalamic hormone are chemically identical and have 14 amino acids in the form of a cyclic peptide with a molecular weight of 1640. It is an inhibitor of both insulin and glucagon secretion in the islet cells, and is believed to have side-effects in other sites.

Pancreatic polypeptide

The F-cells of the pancreatic islets which occur randomly distributed between the alpha and beta cells produce pancreatic polypeptide whose function as a hormone has not yet been established. It is a peptide containing 36 amino acids and is released when the cells are stimulated by hypoglycemia, high protein meal, exhaustion, etc. It is possible that in a small way it may function like glucagon in maintaining homeostasis by its effect on glycogenolysis in the liver and by contributing glucose to blood. Its secretion is also inhibited by somatostatin.

HORMONES OF THE ADRENAL CORTEX

The adrenal cortex is essential to life, and the surgical removal of the adrenal cortex (adrenalectomy) produces the same symptoms as Addison's disease which finally is fatal. The symptoms are:

1) Low blood sugar, especially on fasting.
2) Increased sensitivity to insulin.
3) Loss of appetite, gastro-intestinal disturbances, vomitting and diarrhoea.
4) Rapid loss of weight.
5) Weakness and prostration.
6) Low resistance to infection.
7) Low blood urea.
8) Increased pigmentation (bronzing) of the skin, being more prominent in regions already having some pigmentation: e.g., lips, nipples, ankles, around the eyes, etc.
9) Loss of NaCl and water in urine resulting in hyperkalemia, dehydration and hemoconcentration.

In the case of adrenalectomised animals, the survival time can be prolonged by giving them NaCl and water and by injection of adrenal cortical extracts.

In 1937, Kendal, Pfiffner and Wintersteiner in the U.S.A., and Reichstein and co-workers in Switzerland isolated from the adrenal cortical extracts active lipid soluble substances some of which could be crystallised. About 30 different steroids were isolated consisting of cholesterol (C_{27} steroids), cortico-steroids (C_{21} steroids) and androgenic steroids (C_{19} steroids) with small quantities of estrogenic compounds (C_{18} steroids). Six of the cortico-steroids have been found to possess marked physiological activity. They are shown below:

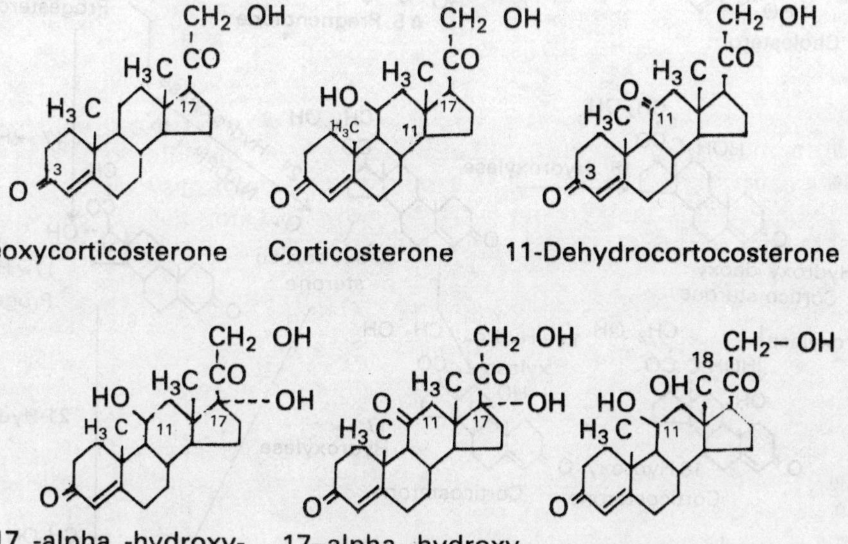

Deoxycorticosterone   Corticosterone   11-Dehydrocortocosterone

17-alpha-hydroxy-corticosterone (Costisol)   17–alpha hydroxy 11-dehydro corticosterone (Cortisone)   Aldosterone

Biosynthesis of adrenal cortico-steroids

Although all the six steroid hormones are biosynthesised from cholesterol by the adrenal cortical tissue of man, only two of them, cortico-sterone and cortisol, are released into the bloodstream with small amounts of aldosterone. It has been calculated that on an average 10–30 mg of cortisol, 2–4 mg of cortico-sterone and 0.3–0.4 mg of aldosterone are produced per day in the human adrenal cortex. The adrenal cortex contains a maximum concentration of cholesterol as compared to other tissues. On stimulation by ACTH, this concentration of cholesterol diminishes with simultaneous release of cortico-steroid hormones. The enzyme which brings about the oxidative cleavage of cholesterol is the cytochrome P_{450}-side chain cleavage enzyme. This is facilitated by cyclic AMP which is liberated from ATP by the action of ACTH on adenylate cyclase. ACTH may also help to bind cholesterol with a specific protein and then facilitate the cytochrome P_{450} to take part in the cleavage process. This is inhibited by amino-glutathemide which can control steroid biosynthesis. Spironolactone is also an inhibitor of steroidogenesis.

Many of these enzyme systems require NAD^+ or $NADP^+$ as the coenzyme for dehydrogenases, hydroxylases, etc. The hydroxylases require molecular oxygen. 11 ß-hydroxylase is inhibited by a compound known as metyrapone and thus cortisol synthesis can be controlled its administration.

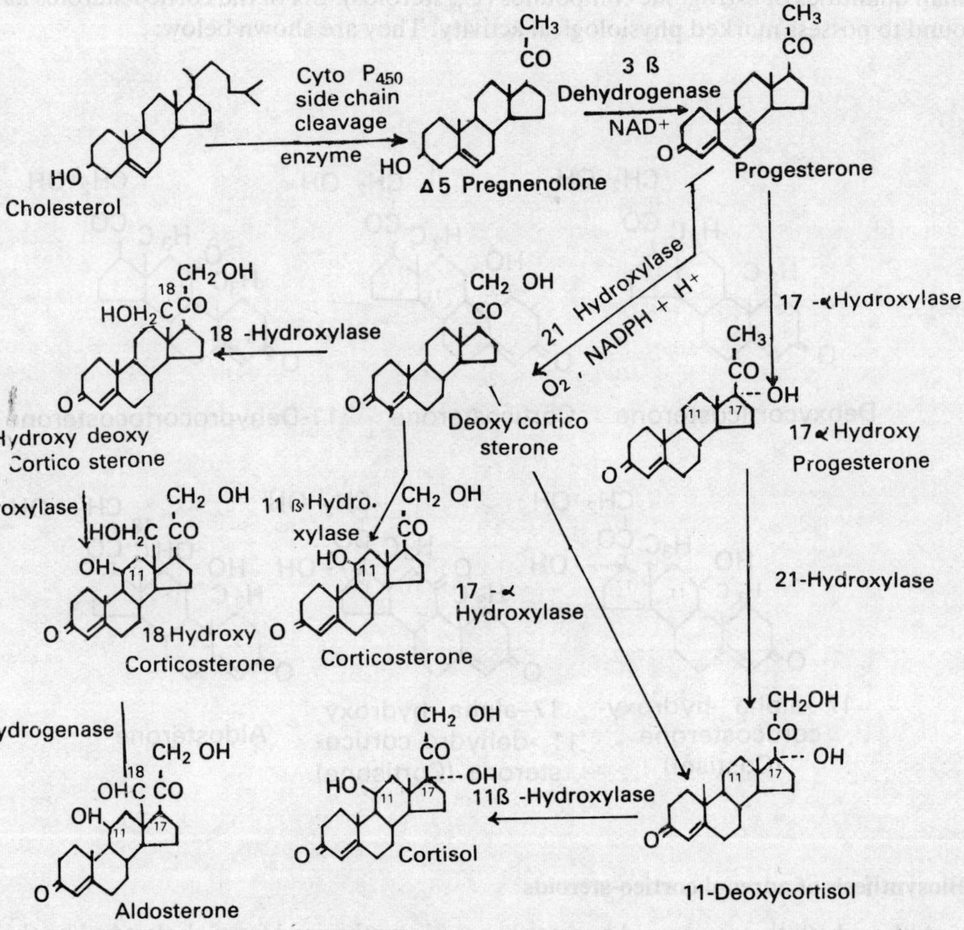

Biochemical effects of cortico-steroids

1. Carbohydrate metabolism: Some of the cortico-steroids exert a profound action on carbohydrate, lipid and protein metabolism. The hormones with oxygen at position 11 of the steroid skeleton, cortisone and cortisol (aldosterone is an exception) are called as gluco-corticoids and are concerned with increased release of glucose from the liver and accentuated gluconeogenesis. In adrenalectomised animals, liver glycogen and blood glucose fall to low levels during fasting, but increase after administration of cortisone. Adrenalectomy ameliorates experimental diabetes mellitus and increases insulin sensitivity. Gluco-corticoids restore the condition to normal but deoxycortico-sterone and aldosterone (mineralo-corticoids) have a negligible effect. Gluco-corticoids enhance gluconeogenesis by increasing the activity of some of the key enzymes like PEP carboxykinase and glucose-6-phosphatase of the gluconeogenic pathway and induce the active enzymes of amino-acid metabolism like alanine amino-transferase, tyrosine amino-transferase, etc. All these are effected by gene transcription and translation effected at the nuclear level. These effects are mostly due to cortisol and not the other cortico-steroids. The glucocorticoids make amino acids available for gluconeogenesis as they are protein catabolic hormones.

2. *Protein and nucleic acid metabolism*: The gluco-corticoids promote gluco-neogenesis from amino acids by enhancing the synthesis of gluconeogenic enzymes and suppressing the glycolytic enzymes. Protein breakdown is increased resulting in increased urea excretion. While cortisol and other gluco-corticoids have protein catabolic effects in the muslces and other tissues, they have a protein anabolic role in the liver as far as the liver enzymes are concerned. They promote gene transcription etc. in the liver and also enhance nucleic acid synthesis.

3. *Lipid metabolism*: Lipolysis in the adipose tissue is enhanced by corticosteroids. This results in hyperlipemia, mild hypercholesterolemia, ketonemia and ketonuria. Gluco corticoids increase the level of free fatty acids in plasma and prevent their re-esterification to triacyl glycerol, by decreasing the production of glycerol.

4. *Action on digestive secretions*: HCl production and pepsinogen secretion by the cells of the gastric mucosa are enhanced by cortisone. In the exocrine pancreas, it stimulates increased secretion of trypsinogen.

5. *Hematological changes*: Cortico-steroids bring about the destruction of lymphocytes and produce lymphocytopenia. This happens usually during the hypertrophy of the adrenal cortex. Hypofunctioning of adrenal cortex, on the other hand, results in lymphocytosis.

6. *Electrolyte and water metabolism*: Deoxycortico-sterone and aldosterone regulate the concentration of Na^+ and K^+ in the extra-cellular fluids by causing increased reabsorption of Na^+ and Cl^- from the renal tubules, gastro-intestinal mucosa, salivary glands and sweat glands. The Na^+, Cl^- and water pass from the cells into the extracellular fluid, increase BP, and increase the excretion of urine. In Addison's disease, Na^+, Cl^+ and water are excreted in increased amounts. These may lead to hemo-concentration and increased serum K^+ levels. The viscosity of the blood increases, the blood pressure falls and the cardiac output decreases as a result of this effect.

7. *Bone and Ca-metabolism:* Bone formation is affected by the cortico-steroids. By retardation of protein synthesis (osteoid), the organic matrix on which $Ca_3(PO_4)_2$ is to be deposited is not available. Hence, there is increased excretion of Ca^{++} and phosphate in urine.

8. *Immune response and anti-inflammatory response*: Cortisol in high doses (higher than the physiological levels, i.e., after exogenous administration) suppresses the host immune response.The lymphoid tissue gets involuted, it also kills the lymphocytes, and thus the antibody production by these cells.

Another effect of cortisol is its anti-inflammatory response. It brings about the destruction and lysis of lymphocytes, eosinophils and monocytes.

Metabolic fate and excretion

Cortico-sterone and cortisol are carried in the blood by the ß-lipoprotein fraction of plasma called cortico-steriod binding globulin(CBG) or transcortin. With the help of 3-keto dehydrogenases, the 3-keto group is converted to 3-OH and the 4–5 double bond gets saturated to form uro-cortisone and uro-cortisol, which are conjugated with glucuronic acid in the liver before they get excreted through the kidneys. About 70 per cent of the cortico-steroids is excreted in urine, while 20 per cent is eliminated through feces and five per cent is converted to 17-ketosteroids by cleavage of the 2C-side chain.

HORMONES OF THE GONADS

The sex hormones or the gonadal hormones are elaborated by the testes, ovary and corpus luteum mostly, and also in small quantities by the placenta and adrenal cortex. They are all steroid compounds related to cholesterol and are actually biosynthesised from that precursor. Sex hormones are also related to the adrenal cortical hormones both in chemical nature and in the common biosynthetic pathway and inter-conversions. The sex hormones are of three types:
1) androgens or male hormones,
2) estrogens or female hormones, and
3) gestogens or progestational hormones.

Androgens

The male sex hormones or androgens are produced by the testes and have been isolated from testicular extracts as well as from urine.

Chemistry of androgens: The androgens belong to the group of C_{19} steroids and are related to the hypothetical parent substance, androstane. They have no side chain on the 17-carbon of the steroid skeleton.

Androstane (Hypothetical)

Testosterone (from testes)
(17 α - Hydroxyandrost - 4 - ene - 3 - one)

Androsterone (from urine)
(3α - Hydroxy - 5α - androstane - 17 - one)

Androstene - 3 17 - dione (from testes)

Dehydroepiandrosterone (from urine)
(3ß - Hydroxyandrost -5- ene -17- one)

Testosterone is the principal male hormone synthesised by the interstitial (Leydig) cells of the testes from cholesterol. It is ten times more potent than androsterone. The latter occurs in urine. The testes also produce another androgen known as androstene 3,17 dione. The metabolic products of testosterone are present in urine as epiandrosterone, androsterone, dehydro-epiandrosterone, androstene 3, 17-dione and etiocholane 3-ol-17-one. Epiandrosterone is five times more active than androsterone, while etiocholane-ol-one is physiologically inactive. It is believed that testosterone is converted to its more active form dihydro-testosterone in the target tissues. It is then converted to androsterone and other 17-ketosteroids by the liver, where they are further conjugated with sulphate and excreted through urine.

Dihydrotestosterone (DHT) is formed from testosterone by the reduction of the A-ring through the action of the enzyme $5 \propto$ reductase. Human testes secrete about $50-100$ μg of DHT per day but most DHT is derived from peripheral conversion.

Biosynthesis of androgens: The androgens are synthesised from cholesterol by the gonadal tissues, chiefly the testes, and to a small extent by the adrenal cortex. In the testes, cholesterol is converted to pregnenolone, progesterone and hydroxy progesterone. The latter is then converted to androstene-dione. The whole pathway is triggered into action by the stimulating effect of the ICSH of the anterior pituitary on the Leydig cells.

$$\text{Cholesterol} \longrightarrow \Delta^5 \text{ pregnenlone} \longrightarrow \text{Progesterone}$$
$$\text{Testosterone} \longleftarrow \text{Androstenedione} \longleftarrow \text{Hydroxyprogesterone}$$
$$\downarrow$$
$$\textbf{Androsterone}$$

Small amounts of androgens are also produced in the ovaries and adrenal cortex, this amount being increased under exceptional conditions. The adrenal cortex produces a male hormone called adrenosterone (androstene 3, 11, 17-trione), whose production is increased in adrenal tumours and is believed to be responsible for the virilism and masculinising effects in women with the adreno-genital syndrome. Ovaries transplanted to a site under the ears in a castrated mouse are known to produce male hormones rather than estrogens.

In normal males, $4-12$ mg testosterone are secreted per day. The testosterone level in the blood plasma of males is about 0.6 μg/100 ml while in females it is only 0.1 μg/100 ml. It is believed that this amount seen in female blood plasma has its origin in the ovaries.

Most mammals have a plasma ß globulin that binds testosterone with specificity and fairly high affinity. This is called the sex hormone binding globulin (SHBG) or testosterone-estrogen binding globulin (TEBG) produced in the liver.

Its production is increased by estrogens and women have twice the serum concentration of SHBG as men. It is also increased in certain types of liver diseases and hyperthyroidism; it is decreased by androgens, advancing age and hypothyroidism.

Biochemical effects of androgens: 1. Protein metabolism: Androgens exert a striking anabolic effect on protein conservation and retention of nitrogen, and thereby muscle growth and maintenance of muscle mass. It is also observed that androgens help in increasing the storage of creatine in the muscles. Amino acid incorporation into protein by prostatic ribosomes is enhanced by this hormone and is dependent on the production of messenger RNA. Excretion of urea decreases without any increase of NPN in blood.

2. Mineral metabolism: Androgens stimulate the growth of bones before the closure of the epiphyseal cartilage. Testosterone exhibits a mineralo-corticoid effect by promoting the reabsorption of Na^+, Cl^- and water by the kidney tubules.

3. Carbohydrate metabolism: Androgens increase the fructose production by seminal

vesicles and the utilisation of this sugar by the seminal plasma by enhancing the activity of both aldose reductase as well as keto-reductase.

$$D\text{-glucose} \underset{NADPH\ H^+}{\overset{\text{aldose reductase}}{\rightleftharpoons}} Sorbitol \underset{NAD^+}{\overset{\text{Sorbitol dehydrogenase}}{\rightleftharpoons}} D\text{-fructose} + NADH + H^+$$

4. Other effects: The citric acid cycle and fatty acid synthesis are stimulated by androgens. The activities of the glycolytic enzymes are also enhanced by this hormone, by an effect mediated through increased RNA production and protein synthesis.

Metabolic fate and excretion: The androgens excreted in urine are classed as 17-ketosteroids. In the case of females, it gives an idea of the condition of the adrenal cortex and its function. In males, 17-ketosteroids arise from the testes (1/3 of the total), while the major amount arises from the adrenal cortex (2/3 of total). In females, the 17-ketosteroids are almost entirely of adrenal cortical origin. In 24-hr excretion of urine, normal adult females excrete 5 – 17 mg while normal adult males excrete 9 – 24 mg of neutral 17-ketosteroids.

ESTROGENS

Estrogens or ovarian hormones are produced by the Graafian follicles of the ovary. They are responsible for the production of estrus (heat) in animals and the estrus cycle. In women, these hormones are responsible for the regulation of menstrual cycle as well as the reproductive cycle.

estrone   estradiol   estriol

Biosynthesis: Cholesterol, Progesterone and testosterone all can act as precursors for estrogen biosynthesis.

Cholesterol ⟶ Pregnenolone ⟶ Progesterone
Δ^4 Androstene 3, 17-dione ⟵ Hydroxy Progesterone
Δ^4 Androstene 3, 17-dione, 19-ol
    ↓ Oxidation
19 - Hydroxy Progesterone
    ↓ $-CH_2O$
Estradiol, 17 ß ⟶ Estrone
    NADPH
    $+ H^+$

Oxidation of C_{19} methyl group and its removal results in the ring A becoming aromatic and the OH at 3 becoming phenolic ring

Chemistry: Estrogens are produced from the cholesterol present in the follicles by a pathway similar to the production of androgens, but stimulated by FSH of the anterior pituitary.

Estrone is the hormone produced in the follicles but is released into the blood as estradiol. In the liver, it is converted to estriol. Estradiol is ten times more potent than estrone and three hundred times more potent than estriol.

Biochemical effects of estrogens: Estrogens stimulate the development, maturation and functions of the female sex organs and thereby the secondary sex characteristics. The biochemical effects of estrogens are as follows:

1) Proliferation of vaginal epithelium as well as the endometrium. This depends upon the increase in alkaline phosphatase and increase in glycogen concentration in the endometrium and vagina brought about by this hormone.

2) **Augmented secretion of mucus by the cervical glands.**

3) Anabolic effect especially seen in the growth of the uterine tissue and mammary gland. Increase in the RNA polymerase activity and synthesis in these tissues, leading to protein synthesis.

4) In humans, estrogens bring down the hyperlipemia including hypercholesterolemia and may be related to the prevention of atherosclerosis. Coronary heart disease is less common in women as compared to men. Experimentally, administration of estrogens is known to bring down the severity of atherosclerosis.

5) Estrogen administration causes elevation of calcium and phosphorus levels in serum followed by calcification and hyper-ossification of long bones. The bone marrow may decrease and may disappear thereby resulting in anemia. It is known that decalcification of bone may take place during post-menopausal period leading to osteoporosis and fractures.

6) Estrogens have a stimulatory effect on iso-citrate dehydrogenase. They also activate the enzyme transhydrogenase capable of transferring H^+ from NADPH to NAD^+ forming NADH. It is because of this reaction that after delivery and menopause some women become obese, as NADPH is not eliminated, but continues to be utilised for lipogenesis.

Metabolic fate and excretion: Inter-conversion of estrone, estradiol and estriol can take place in the body in tissues like the liver. The circulatory estrogen in the blood is mostly beta estradiol. Estrone gets conjugated with sulphate and is excreted as estrone sulphate in urine, while estradiol and estriol are mostly excreted as conjugates with glucuronic acid in urine. In liver disorders like cirrhosis of the liver, the conjugation process is hampered and estrogen levels in the blood rise, as their excretion is not possible.

Synthetic estrogens: A number of synthetic estrogens are available, of which ethenyl estradiol and diethyl stilbesterol are important. Ethenyl estradiol is thirty times more potent than estradiol benzoate and can be given orally. The chemical structure of synthetic estrogens is given below:

Ethenyl estradiol

Diethyl stilbesterol

Progestational hormones

Progesterone, the progestational hormone, is secreted by the corpus luteum during the latter half of the menstrual cycle. It acts on the endometrium previously prepared by the estrogenic hormones and helps to induce secretion of mucus. It also contributes to the development of the mammary tissue and maintenance of the uterus in a quiescent state during the gestation period. Progesterone secretion continues till full term. By suppressing estrus and ovulation, it regulates the menstrual cycle as well as the female sex cycles by an antagonism and synergism mechanism with the pituitary hormones FSH and ICSH, respectively.

Biosynthesis, fate and excretion: In the corpus luteum, placenta and adrenal cortex, progesterone is efficiently biosynthesised with cholesterol as the starting material. Progesterone is a C_{21} steroid compound and is related to the cortico-steroids. It is an intermediate common to the biosynthesis of all the steroid hormones from cholesterol.

The ultimate fate of progesterone in the liver is its conversion to the inactive product pregnanediol which gets conjugated with glucuronic acid before it is excreted in urine. About 1-10 mg of pregnanediol is excreted per day during the latter half of the menstrual cycle.

GASTRO-INTESTINAL HORMONES

The presense of certain gastro-intestinal hormones concerned with the production of digestive juices in the intestinal wall and pyloric mucosa has already been mentioned in an earlier chapter (see chapter on 'Biochemistry of Digestion and Absorption').

Gastrin

Gastrin is the hormone produced by the antral gastric mucosa. It stimulates gastric secretion, if injected intravenously. Gastrin has been chemically characterised as a hepta-peptide, the active portion of it being a tetra-peptide with a C-terminal sequence of Try-Met-Phe-Asp.NH_2.

Secretin

Secretin stimulates the pancreas to produce an increased volume of pancreatic juice. Secretin itself is produced by the presence of acid chyme in the duodenum brought from the stomach during digestion. An acidic extract of duodenum injected into dogs stimulates the flow of pancreatic juice.

Secretin is a polypeptide containing 27 amino acids and it resembles glucagon in many ways, 14 of the amino acid residues being common to both. In fact, glucagon is also produced in the gastric and duodenal cells apart from the pancreatic α cells and can also be considered a gastro-intestinal hormone.

Cholecystokinin

Cholecystokinin is secreted by the I-cells of the duodenal mucosa and the proximal part of the jejunum. Its secretion is stimulated by the entry of acid chyme and fatty foods into the duodenum. Cholecystokinin is chemically similar to gastrin in some respects, but has the function of stimulating the contraction of gall bladder and enhancing the flow of bile into the duodenum. It is a small polypeptide with the molecular weight of 5,000-10,000.

Gastric inhibitory peptide (GIP)

This hormone is released from the duodenal mucosa, after intake of food, when the glucose level rises above the physiological levels (post-absorptive state). It inhibits gastric motility and secretion. It enhances insulin release in ß-cells.

Vaso-active intestinal peptide (VIP)

This is another hormone produced by the intestinal wall and is a potent stimulator for pancreatic secretion. It also acts as a smooth muscle relaxant and is involved in gut motility and blood flow in the intestinal wall.

Motilin

This is also identified as an intestinal hormone and is believed to play a role in regulating the motility of the intestine.

HYPOTHALAMIC AND PITUITARY HORMONES

Hypothalamic hormones

The hypothalamus releases at least six hormones which are also called 'releasing factors'. Some of them stimulate the release of the pituitary hormones while a few of them can also inhibit their release, in which case they are called 'release-inhibiting hormones'. The releasing factors are produced and secreted at the nerve-fibre endings of the hypothalamic-hypophyseal portal system which is present between the pituitary and hypothalamus in the pituitary stalk. These hormones have their function in the cells of the adeno-hypophysis (anterior lobe of the pituitary), where they exert a tonic control on the production and release of the pituitary hormones.

The following six hormones of the hypothalamus have been recognised:
1) Thyrotropin releasing hormone, TRH or TRF
2) Corticotropin releasing hormone, CRH or CRF
3) Growth hormone releasing hormone, GHRH or GHRF

4) Growth hormone release inhibiting hormone, GHRIH or GHRIF (also known as somatostatin)

5) Gonadotropin-releasing hormone, GnRH (includes LHRH and FSHRH which affects LH and FSH of the pituitary)

6) Prolactin-release-inhibiting hormone, PRIH or PRIF (This is believed to be the same as dopamine.)

All these hormones have been isolated, chemically characterised and their properties and function studied in recent years. The TRH is known to be a tripeptide containing pyroglutamic acid, histidine and proline. PRIH is found to be the same as dopamine, a hormone derived from tyrosine in the hypothalamus. GnRH and somatostatin are both polypeptides containing 10 and 13 amino acids, respectively. Somatostatin has a disulphide bridge between its two cysteine residues present in the single chain. Both CRH and GHRH are proteins containing 41 and 44 amino acids, respectively, having single straight-chains.

The hypothalamic hormones are released not continuously but in a pulsatile manner and are under feed-back control by the pituitary hormones. GnRH controls the secretion of FSH or LH in the pituitary, while the latter is also controlled by the levels of estrogens, progesterone and androgens. CRH controls the secretion of ACTH, which again is controlled by the levels of cortisol in the circulating blood reaching the pituitary gland. Similarly, TRH controls the secretion of TSH of the pituitary and the latter is regulated by the levels of T_3 and T_4. The secretion of the growth hormone by the pituitary is under the tonic control of both GHRH and GHRIH (balanced effect of the stimulating and inhibiting factors). For the release of prolactin in the pituitary, no releasing hormone or feed-back mechanism is known to operate. However, it is believed that an inhibiting factor operates for its release and this factor is identified to be dopamine.

Most of the hypothalamic hormones appear to exert their effects by a Ca^{2+}-phospholipid mediated mechanism and not through the mediation of cyclic AMP. It is interesting to note that some of the hypothalamic hormones are also found to be produced elsewhere outside the hypothalamus, e.g., somatostatin is produced in the D-cells of the pancreas. TRH and CRH are also known to be produced at other nerve structures, apart from the hypothalamus.

ADENO-HYPOPHYSIS AND ITS HORMONES

The anterior lobe of the pituitary (adeno-hypophysis) is of glandular origin and is supplied with a large network of blood capillaries. It elaborates mainly six hormones usually called trophic hormones. They are:

1. Growth hormone, GH or somatotropin.
2. Thyroid-stimulating hormone, TSH or thyrotropin.
3. Adreno-corticotropic hormone, ACTH or corticotropin.
4. Interstitial cell-stimulating hormone, ICSH or luteinising hormone (LH).
5. Follicle-stimulating hormone (FSH).
6. Lactogenic hormone, or prolactin or luteotropin (LTH).

Among these, the first three have an effect on metabolism and are called metabolic hormones. The latter three have their physiological effects mostly on gonads and hence are considered gonadotrophic hormones. These six trophic hormones are secreted continuously by the glandular cells of the adeno-hypophysis in small quantities.

However, the rate of secretion is increased by nervous stimuli from the hypothalamus through the hypothalamic releasing factors, which reach the glandular cells of the anterior lobe of the pituitary by the capillaries of the hypothalamic-pituitary portal system. Some of the anterior pituitary hormones like gonadotrophins and prolactin are also produced by the placenta and may supplement the action of these hormones during gestation. These hormones are called chorionic gonadotropin and somato-mammotropin, respectively and will be discussed along with the anterior pituitary hormones.

GROWTH HORMONE

The growth hormone is produced by the acidophilic cells of the anterior pituitary (somatotropes) which occur in great abundance when compared to other cells of the pituitary. The growth hormone is a single-chain polypeptide with molecular weight of 22,000, containing 191 amino acids and with two internal S-S-bridges.

Chah Ho Li (1970) has studied the full sequence of the amino acids in the human growth hormone and has achieved the total synthesis of this hormone, which possesses all the physiological effects of the natural hormone.

Regulation of growth hormone secretion: The production and release of the growth hormone are regulated by a number of factors. Two hypothalamic factors are directly involved in the production and release of the growth hormone and its control. They are the growth hormone releasing hormone GHRH which is stimulatory and the growth hormone release inhibiting hormone (GHRIH), also called somatostatin, which is inhibitory. This is a peptide of 14 amino acids. Apart from these two, there is an extra-hypothalamic factor, i.e., somatomedin-C (insulin-like growth factor IGF_1) which regulates it by inhibiting the release of GHRH and stimulating the release of somatostatin at the hypothalamus. In addition, there are other hormones like estrogens, dopamine, serotonin, glucagon and endorphins which regulate the release of the growth hormone by operating a feed-back control system.

Biological effects: The growth hormone is species-specific. Only the monkey and human growth hormones have been reported to be effective in man. The abnormalities or dysfunction of the hypophysis have been known to be responsible for symptoms like gigantism, dwarfism and acromegaly. The biological effects of the growth hormone depend upon its synergism and antagonism with the other hormones. Thus the growth hormone enhances the anabolic effect of androgens. It is also synergistic with insulin as anabolic hormone but while insulin causes hypoglycemia, GH has a net diabetogenic action and may cause hyperglycemia.

Metabolic effects: 1. Growth and protein anabolism: Administration of the growth-hormone produces gain in body weight and increases the total body substances, by the retention of N as protein (anabolic effect). GH also increases the entry of the amino acids into the cell through the cell membrane. It helps in the synthesis of DNA and RNA as well as proteins. The growth effect on the long bones is uniform and normal before the closure of the epiphyses. But after the closure of the epiphyseal ends, growth is not symmetrical and uniform. The chondrogenesis and osteogenesis at the endochondral portion results in the formation of abnormally shaped bones, especially the cheek bones, shoulder bones, etc., resulting in acromegaly. The growth hormone is now believed to enhance the growth of tissues including cartilage and bone and this action is mediated by somatomedin-C or IGF_1. The latter was previously called the 'sulfation factor' because the stimulatory effects of this factor on the synthesis of chondroitin sulphate and collagen had been well known for some time. Somatomedin-C is produced in the liver.

In pygmies and dwarfs, it has been observed that the growth hormone levels are usually normal or even higher than normal, but the levels of IGF_1 have been found to be low which indicates the role played by the latter in helping the overall growth along with the growth hormone. In actual growth hormone-deficient dwarfs also, the levels of IGF_1 have been found to be low.

Protein synthesis in bone-matrix, diaphragm, muscle, kidneys, liver, and connective tissues is enhanced by the growth hormone. The growth effect is very clearly demonstrated in young hypophysectomised animals when they are maintained on daily injections of the growth hormone. The effects of the growth hormone can be summarised thus: retention of nitrogen, lowering of blood amino acids, increase of protein and decrease of fat content in carcass, increase in alkaline phosphatase and inorganic P, enlargement of liver and other organs, decrease in urinary N and slight increase in liver RNA.

In many respects, the action of the growth hormone resembles that of insulin as far as growth and protein anabolism are concerned.

2. **Carbohydrate metabolism: Diabetogenic, pancreatotropic and glycostatic effects:** The surgical removal of the anterior lobe of the pituitary (hypophysectomy) in an animal results in hypoglycemia and decreased glycogen levels in the muscle and liver on fasting and increased sensitivity to insulin. In a normal animal, fasting does not bring about hypoglycemia or diminution in muscle glycogen. This is called the glycostatic effect and it is due to GH. This glycostatic effect of GH can be clearly demonstrated in the hypophysectomised animal.

Pancreatectomy results in diabetes mellitus. The diabetic condition is ameliorated by hypophysectomy. Administration of the growth hormone produces hyperglycemia, glycosuria and other symptoms of diabetes, this effect being known therefore as the diabetogenic effect.

Thus the growth hormone has an antagonistic effect towards insulin on these parameters.

The growth hormone enhances gluconeogenesis in the liver from amino acids and thereby increases glycogen and also brings about hyperglycemia by inhibiting or suppressing glycolysis and decreasing the transport and peripheral utilisation of glucose. The inhibition of glycolysis is thought to be due to the increased lipolysis in the adipose tissue and the consequent rise of FFA in plasma which depresses glycolysis.

The growth hormone has a direct stimulatory effect on the pancreatic islet cells. Both the alpha cells and beta cells are more or less equally stimulated to produce glucagon and insulin, respectively. The growth hormone appears to be necessary for the growth, development and functioning of the pancreas.

The production of both the growth hormone and insulin are inhibited by somatostatin in the pituitary and the beta islets, respectively.

3. **Lipid metabolism:** The effect of growth hormone as a lipolytic hormone has been demonstrated *in vitro* in the adipose tissue incubated in laboratory conditions. However, its effect *in vivo* is a much slower process. In the diabetic condition, its lipolytic effect is much more clearly exhibited, because there is no opposition from insulin in that condition. It is dependent on the formation of proteins involved in the synthesis of cyclic AMP. The lipid mobilisation from the adipose tissue results in the increase of plasma FFA.

4. **Renotropic effect:** The size of the kidney as well as its function is increased by the growth hormone. Renal clearance is also augmented.

5. **Lactopoietic effect:** The growth hormone has been demonstrated to have a marked effect in enhancing milk production in lactating animals. The proliferation of the mammary tissue is regulated by the growth hormone, prolactin and estrogens. The human growth hormone is believed to be closely related to prolactin since they are shown to be chemically and immunologically identical.

6. **Mineral metabolism:** The growth hormone promotes a positive calcium and phosphate balance along with magnesium, and thus helps bone mineralisation and bone growth. The growth of the long bones is enhanced by the growth hormone, this effect being conspicuous before the closure of the epiphyseal cartilage. In growing children, the formation of cartilage is also stimulated by the growth hormone. The growth hormone causes retention of Na^+, K^+ and chloride in the body as well.

Most of the effects of GH are due to its action in enhancing the synthesis of mRNA, and tRNA, and protein synthesis, i.e., at transcription as well as the translation levels.

ADRENO-CORTICOTROPIC HORMONE

Hypophysectomy results in the atrophy of the adrenal cortex but not the medulla. In normal animals, administration of pituitary extracts causes hyperplasia of the adrenal cortex. The hormone responsible for this is recognised as ACTH. ACTH has been isolated, purified and characterised as a simple polypeptide consisting of 39 amino acids and having a molecular weight of 4,500. It contains no cysteine. This hormone has noe been chemically synthesised. The synthetic hormone contains only 19 amino acids but possesses full activity.

ACTH or the adreno-corticotropic hormone (also called corticotropin) is first synthesised as a precursor molecule containing 285 amino acids known as pro-opiomelanocortin (POMC) in the pituitary. This precursor molecule can give rise to ACTH, ß-lipotropin (ß-LPH), ∝-MSH, ß-MSH, and endorphins, through cleavage by trypsin-like enzymes. Approximately, five per cent of the cells of the anterior pituitary produces pro-opiomelanocortin and its production and release is controlled by the cortico-tropin releasing hormone (CRH) of the hypothalamus.

Biological and metabolic effects: The most important action of ACTH is to increase the synthesis and release of cortico-steroids by enhancing the conversion of cholesterol to pregnenolone by cleavage of a 6-carbon fragment, followed by hydroxylation at various positions. ACTH has multiple biological effects, some of its effects being mediated through the adrenal cortex. Its effect mediated through the adrenal cortex is more or less the same as that of the cortico-steroids, viz., augmentation of gluconeogenesis, retardation of protein synthesis, increased lipid mobilisation from the adipose tissue and mild hypercholesterolemia.

ACTH has a direct effect on the melanocytes. In Addison's disease, the adrenal cortex does not function properly; however, the level of circulating ACTH is high and may be responsible for the increased pigmentation (bronzing of the skin).

ACTH and the melanocyte stimulating hormones (∝-and ß-MSH) are related to each other. ∝-MSH is a part of the ACTH molecule while ß-MSH is a part of ß-LPH, which along with ACTH is present in the precursor molecule of pro-opiomelanocortin. Perhaps the bronzing of the skin (hyperpigmentation) seen in Addison's disease may be due to the augmentation of secretion ACTH and ß-LPH secretion both of which have ∝-MSH and ß-MSH activities, respectively.

A tumour of the basophil cells of adeno-hypophysis results in Cushing's disease, a syndrome characterised by over-production of ACTH. Obesity of neck and trunk, face, buttocks, abdomen, edema, sexual dystrophy, hypertension, increased pigmentation and hirsutism are usually observed in the Cushingoid state. Hyperglycemia, decreased glucose tolerance, glycosuria, retention of Na^+, Cl^- and water, depletion of K^+ and alkalosis are the biochemical abnormalities noticeable in such a condition. The depletion of K^+ affects the mitochondria in the cells of the collecting tubule and may lead ultimately to severe renal damage (pyelonephritis).

ACTH is elaborated through the stimulation of the adeno-hypophysis by a hypothalamic regulatory factor known as corticotropin releasing hormone (CRH) consisting of three types, α_1 CRH, α_2 CRH, and ß CRH all being polypeptides. CRH release is regulated by norepinephrine, serotonin and acetylcholine and usually is increased during trauma, emotional stress, etc.

THYROID-STIMULATING HORMONE

The thyroid-stimulating hormone has been isolated but not completely purified. The purest preparation contains ICSH as a contaminant. TSH has a molecular weight of 26,000 – 30,000 and contains the usual amino acids with a high content of cysteine. The carbohydrate part has been identified as glucosamine, galactosamine, mannose and fucose.

Biological effects

1) The most important function of TSH is to increase the rate of uptake of iodide from the circulating blood by the thyroid.
2) TSH also enhances the conversion of iodide to thyroid hormones.
3) Proteolysis of thyroglobulin and the release of thyroid hormones into the blood also need TSH.

Thyroidectomy and goitrogenic drugs enhance the secretion of TSH. TSH secretion is also stimulated by the thyrotropin releasing hormone (TRH) from the hypothalamus. TRH is a tri-peptide amide, i.e., pyroglutamyl-histidyl-prolinamide and is effective even in minute quantities. Administered thyroxine inhibits TSH secretion by a negative feed-back mechanism while estrogens potentiate the action of TSH. Women respond better to the action of TRH when compared to men, and can tolerate cold-induced stress better.

In the absence of hypophysis or TSH, iodide cannot be properly incorporated into the thyroid hormones. In hypophysectomised animals, serum PBI values are decreased by 50 per cent.

Long acting thyroid stimulator (LATS)

In the serum of patients with Grave's disease, there is thyroid-stimulating protein factor immunologically different from TSH. It disappears from circulation more slowly than TSH. It has most of the effects of TSH but its maximal thyroid-stimulating effect occurs many hours after that of TSH, which explains its name. Recent work shows that LATS is an antibody developed as an auto-immune phenomenon against thyroid protein. Many of the phenomena of Grave's disease, particularly the exophthalmos, may be related to the auto-immune reaction. The site of formation is not the pituitary.

MELANOCYTE-STIMULATING HORMONE

The intermediate lobe of the hypophysis produces a hormone called the melanocyte-stimulating hormone (MSH). This hormone is concerned with the control of the dispersion of pigment granules and is present in a very active state in species able to change the skin colour. It is present in the mammalian and human pituitary.

The human pituitary contains ß-MSH as part of the ß- and γ-LPH from which it is derived and as α-MSH as a part of the ACTH molecule. Both α-MSH and ß-MSH are contained in the bigger precursor molecule pro-opiomelanocortin. α-MSH has 13 amino acids while ß-MSH has 18 amino acids present in their polypeptide chains.

ß-Lipotropin (ß-LPH)

ß-Lipotropin is derived from the precursor molecule of pro-opiomelanocortin and contains 93 amino acids in the form of a straight-chain polypeptide. γ-LPH containing 60 amino acids is a part of ß-LPH. ß-LPH is also the precursor for the three types of endorphins (α, ß and γ) and also for the ß-MSH. However, in the human pituitary, only ß-LPH and γ-LPH and ß-endorphin have been found. ß-MSH does not seem to be present in man.

The function of ß-LPH is mainly as a precursor for the production of ß-endorphins, the naturally occurring opiates in the pituitary. ß-LPH has some lipolytic effects and may bring about fatty acid mobilisation from the adipose tissue. However, this effect is very much less than the lipolytic effects of most of the other hormones known.

Endorphins and encephalins

Endorphins are a group of polypeptides which influence the transmission of nerve impulses. They are also known as opioids because they bind to these receptors which bind opiates like morphine. The opioids first discovered were two pentapeptides in the brain and were named encephalins. They are methionine encephalin and leucine encephalin. Endorphin is the generic name employed for all opioid peptides.

The precursor for endorphins is the Pro-opiomelano-cortin peptide (POMC peptide) which is the parent for beta lipotropin. The sequence of 31 amino acids at the C-terminus of beta lipotropin, i.e., AA 104 – 134, gives beta endorphin. α endorphin (104 – 117) contains 17 AAs less than the ß from the C-terminus and the γ (104 – 118) contains 16 AAs less than the ß from the C end.

The endorphins in the central nervous system are free and are neuro-transmitters or neuro-modulators. However, the peptides found in the pituitary are acetylated and hence not active.

As endorphins bind to the same CNS receptors as do the morphine opiates, they may play a role in the endogenous control of pain perception. They have higher analgesic potency than morphine.

The endorphins are by no means the exclusive possession of the brain. Several of the CNS peptides are found in the intestinal mucosa. Somatostatin is present in the pancreas and intestines. Gastrin, cholecystokinin and vaso-active intestinal peptide — the gut hormones — are present in the brain. Presumably, the molecular mechanisms by which these polypeptide hormones effect responses in their target cells closely resemble those operating at the synapses where the peptides function as neuro-transmitters or neuro-modulators.

PROLACTIN

Prolactin or mammotropin is produced by the lactotropes or acidophilic cells of the anterior pituitary and is believed not to be under the control of any releasing factor from the hypothalamus. However, its release is inhibited by the prolactin release inhibiting hormone (PRIH) from the hypothalamus which is the same as dopamine. Prolactin release is enhanced by estrogens and oxytocin. Nipple stimulation, sexual intercourse and emotional states in women all enhance prolactin release. The production of prolactin is increased during pregnancy.

Prolactin or the lactogenic hormone is concerned with the initiation of lactation in mammals after parturition. Both the growth hormone and prolactin have a similar effect on mammary growth. It is synergistic in this action with estrogens. Prolactin also functions as the luteotropic hormone and promotes the growth of corpus luteum during the luteal phase and is responsible for its functional activities like the production of progesterone.

Prolactin is characterised as a pure protein with a molecular weight of 23,000.

GONADOTROPHIC HORMONES (Gonadotropins)

The removal of the pituitary gland before the onset of maturity causes cessation of sexual development. Hypophysectomy results in the atrophy of the gonads. The testes become flabby and decrease in activity. Degenerative changes take place in the follicles of the ovary which ultimately get atrophied. All these are now recognised as due to the lack of the gonadotrophic hormones in the absence of the pituitary. The gonadotrophic hormones are the follicle-stimulating hormone (FSH), the interstitial-cell stimulating hormone (ICSH) or luteinsing hormone (LH) and prolactin or luteotrophin (LTH). FSH and ICSH are sereted by the pituitary. The release of FSH and LH are controlled by the gonadotropin-releasing hormone (GnRH) of the hypothalamus which is a stright-chain polypeptide containing 10 amino acids including pyroglutamic acid and histidine. Prolactin secretion is regulated by the prolactin-release-inhibiting-hormone (PRIH) of the hypothalamus (dopamine).

Biological effects: FSH exerts its effect in the female by enhancing the ripening of the Graafian follicles (follicular phase), increasing the weight of the ovaries and the production of estrogens. In the male, it is concerned with the growth of testes and stimulation of the epithelium of the seminiferous tubules, thereby inducing spermatogenesis. ICSH or LH in the female is concerned with the final ripening of the Graafian follicles, production of estrus, rupture of follicles and the formation of corpus luteum (luteal phase). In the male, ICSH stimulates the Leydig cells to produce testosterone. The action of both FSH and LH are effected through their stimulation of adenylate cyclase and the consequent production of increased amounts of cyclic AMP.

LH may bind to the Sertoli cells and induce the synthesis of a specific androgen binding protein (ABP) which is concerned with the transport of testosterone to the seminiferous tubules which is necessary for spermatogenesis.

Chemically, both FSH and ICSH are characterised as glycoproteins with a molecular weight of 26,000. They have a carbohydrate content of 7.4 per cent and contain sialic acid.

HORMONES FROM PLACENTA

The placenta produces certain hormones resembling those of the anterior pituitary and having the same functions as the latter. During the period of gestation, in fact, the placental hormones may supplement the action of the pituitary trophic hormones as a natural precaution so that the fetus and mother do not suffer from any block in the hormonal control.

Two types of placental hormones which resemble the pituitary trophic hormones have been identified. They are: 1) chorionic somatomammotropin (CS) or placental lactogen, and 2) human chorionic gonadotropin (hCG).

Both these hormones are produced by the syncytio-trophoblast cells of the placenta.

Chorionic somato-mammotropin has both growth promoting and lactogenic activities. It resembles the growth hormone and prolactin in some of their chemical structural features, amino acid content, presence of two disulphide linkages and immunological similarities. It is believed that the growth hormone, prolactin and chorionic mammotropin might have come from a common ancestral gene.

The human chorionic gonadotropin resembles LH in its chemical nature.

There are human chorionic gonadotrophin (hCG) from the placenta and pregnant mare serum gonadotrophin (PMSG). hCG is detected in urine early in pregnancy, i.e., one week after the first missed menstrual period, and is detectable by two common pregnancy tests, Ascheim-Zondek test and Friedmann's test. They give false positive tests in conditions associated with tumour of the placental tissue (chorionepithelioma), in cystic degenerative disease of the chorionic tissue (hydatidifom mole), and in testicular tumours. hCG is a glycoprotein with a molecular weight of 100,000. Administration of hCG has been shown to stimulate the production of androgens in male.

HORMONES OF THE POSTERIOR PITUITARY (Neuro hypophyseal hormones)

Two active substances have been isolated from the posterior pituitary (neuro-hypophysis) and these are now purified and characterised as peptides. They are oxytocin and vasopressin. Oxytocin is synthesised in the para-ventricular nucleus and vasopressin in the supra-optic nucleus of the hypothalamus. Both these hormones are then transported by axoplasmic flow to the nerve endings in the posterior pituitary where they get released. The axonal transport of these hormones is possible because of two carrier proteins, neuro-hypophysin I and II. Oxytocin is the prinicipal uterus-contracting hormone and is used in routine obstetrics to initiate labour. It has also the function of contracting the smooth muscles of the mammary gland and bringing about milk ejection during suckling. The release of oxytocin from the neuro-hypophysis is stimulated by impulses from the nipple (tactile stimulation).

Vasopressin, the other hormone of the neuro-hypophysis, is also called the anti-diuretic hormone (ADH). It is concerned with the regulation of water balance by reabsorption of water in the distal tubules of the kindney (facultative reabsorption). In the absence of this hormone, diabetes insipidus with polyuria results, in which large volumes of urine 15 – 20 litres per day are excreted. The specific gravity of urine decreases to that of pure water in such cases. It is believed that the specific manner in which vasopressin exerts its ADH effect is through stimulation of the synthesis of 3', 5' cyclic AMP which promotes the reabsorption of water. Vasopressin raises blood pressure by its vasopressor effect on the peripheral blood vessels. This property is made use of in employing vasopressin in surgical shock as an adjuvant in elevating blood pressure.

The chemical structure of these peptide hormones is similar except that two of the amino acids iso-leucine and leucine of oxytocin are replaced by phenyl alanine and arginine in vasopressin.

Both these hormones have been chemically synthesised, and the synthetic hormone is widely used in general medicine and obstetrics.

Oxytocin

```
    S—————————S
    |         |
Cys-Tyr-Ileu-Glu-Asp-Cys-Pro-Leu-Gly–NH2
             |   |
            NH2 NH2
```

Vasopressin

```
    S—————————S
    |         |
Cys-Tyr-Phen-Glu-Asp-Cys-Pro-Arg-Gly–  NH2
             |   |
            NH2 NH2
```

ENDOCRINE INTER-RELATIONSHIPS

The endocrine system as a whole functions in a well organised and co-ordinated way by the synergism and antagonism with one another being responsible for growth, metabolism, reproduction, maintenance of homeostasis and repetition of events in a regular fashion.

Adeno-hypophyseal ovarian inter-relationships

The sexual cycle: The menstrual cycle begins immediately after puberty in the human female and lasts for 28 days, this cycle being repeated thereafter till menopause at the age of 45 – 50 years. The menstrual cycle stops during pregnancy. It is regulated by three of the adeno-hypophyseal hormones, FSH, ICSH (LH) and prolactin, and two of the steroid hormones, estrogens and progesterone as indicated below.

The secretion of FSH, synergistically with a small amount of ICSH, stimulates the development and maturation of the ovarian follicles. The *theca interna* cells of the ovary then produce estrogens. The latter is responsible for the secretion of follicular fluid and sensitisation of the follicles. These estrogens soon begin to inhibit the FSH production and at the same time stimulate the production of ICSH and prolactin. ICSH or LH initiates the release of the ovum by rupture of the follicles. This process is called ovulation. The ruptured follicles are converted into the corpus luteum by the ICSH, which is then maintained by the beta-prolactin or luteotropic hormone. The corpus luteum now starts producing progesterone which later on inhibits the secretion of ICSH as well as prolactin. The hypophysis once again starts producing FSH and returns to the initial stage.

The endometrium of the uterus is prepared and made ready to receive a fertilised ovum and implant the same with the help of progesterone. Progesterone at the same time stimulates the growth and development of the mammary gland. In case fertilisation does not take place, the corpus luteum regresses and a sloughing of the endometrium results and menstruation takes place. However, if pregnancy occurs, the corpus luteum continues to function, being sustained by the human chorionic gonadotrophin (hCG) that appears soon. The corpus luteum goes on producing progesterone for some months after which the placenta takes over this function.

Thyroid-ovarian inter-relationships

The thyroid gland enlarges in size during pregnancy, especially towards the later months. The serum PBI level is increased in pregnancy without an increase in the metabolic rate. It is found that the thyroid hormone facilitates the formation of the ovum. Infertility is common in hypothyroidism. Estrogens diminish thyroid activity. The requirement of the thyroid hormone is increased during pregnancy.

HORMONES OF UBIQUITOUS ORIGIN

In recent times, a hormonal status is attributed to certain compounds of ubiquitous origin such as acetyl choline, serotonin and prostaglandins.

Acetylcholine

This is considered a neuro-hormone which is a chemical mediator of nerve impulse in parasympathetic and involuntary nerves to the skeletal muscles. This substance is easily broken down by acetyl choline esterase found in the nerve endings as well as the nerve fibres. It is synthesised by the enzyme choline acetylase from choline and acetyl CoA in the presence of ATP.

Serotonin

Serotonin or 5-hydroxy tryptamine is present in the blood platelets, brain and gastro-intestinal tract. It is a powerful pressor substance and raises the blood pressure by vaso-constriction of the peripheral blood vessels. It is an important transmitter of impulses in the sympathetic nervous system. Rage, fear, aggressiveness and catatonia ensue when serotonin is administered to animals, e.g. the cat.

Gamma amino butyric acid (GABA)

Gamma amino butyric acid is produced from glutamic acid by decarboxylation in the brain. It acts as a normal regulator of neuronal activity, as an inhibitor of neural transmisson.

25-hydroxy cholecalciferol and 1,25 dihydroxy cholecalciferol (Calcitriol)

These compounds formed by the oxidation of cholecalciferol (vitamin D_3) are present in circulating blood and intestinal mucosal cells and are concerned with a hormonal function in promoting mRNA transcription for the synthesis of protein for the Ca transport system.

Prostaglandins

The existence of prostaglandins was first hinted at by two U.S. Gynaecologists, Kurzrok and Lieb, in 1930 in the finding that the instillation of human semen causes the uterus either to contract or relax. Von Euler gave the name 'prostaglandins' to these substances coining the word from 'prostate' and sheep vesicular 'glands' from which PGs were originally extracted. It is present in high concentrations in human semen. There are as many as 17 individual members with slight variations in their structures. They differ from one another in the number and position of the double bonds and –OH groups. They belong to PGE, PGF, PGA, series, etc. The important members are PGE_1, PGE_2, $PGF_{2\alpha}$, PGA, etc. Considerable research work has been done by Bergstrom of Sweden. Prostaglandins belong to the group of lipid-soluble substances. They are C_{20} unsaturated fatty acids and have a five-membered ring, the parent fatty acid being called prostanoic acid. They are white powdery substances.

While PGs are ubiquitous in the body, fairly high concentrations are present in human seminal plasma, uterine tissue of the female, menstrual fluid, lung tissue and iris. The concentration in human seminal fluid is about 226 µg/ml. They are synthesised in the body from unsaturated acids like arachidonic acid.

Biosynthesis: For the biosynthesis of PGs, glutathione, FH_4, and 6, 7 dimethyl tetra-hydropteridine are the co-factors. Anti-oxidants like hydroquinone are necessary for the optimum activity. Eicosa tri-enoic acid, Eicosa tetra-enoic acid and Eicosa penta-enoic acid are the substrates. The mechanism of biosynthesis is the same in all tissues.

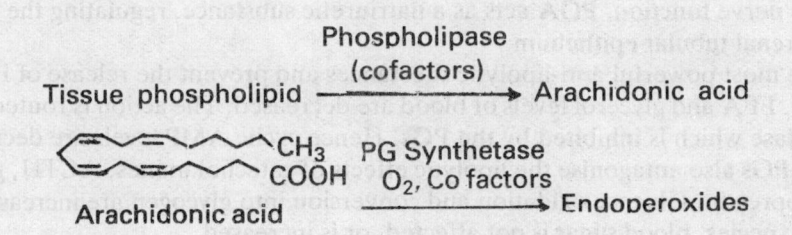

$$\text{Tissue phospholipid} \xrightarrow{\text{Phospholipase (cofactors)}} \text{Arachidonic acid}$$

$$\text{Arachidonic acid} \xrightarrow{\text{PG Synthetase, } O_2, \text{ Co factors}} \text{Endoperoxides}$$

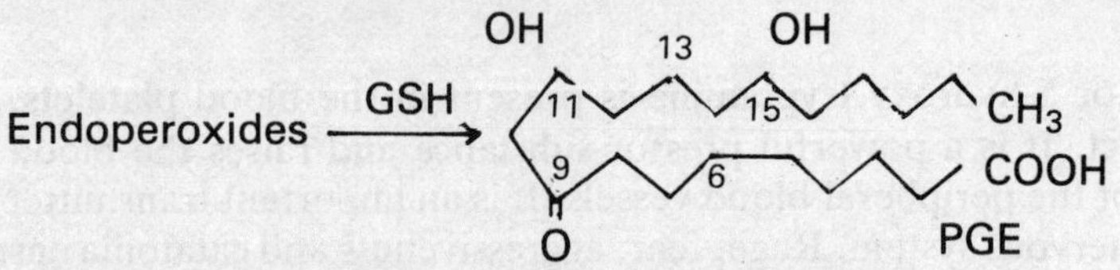

Endoperoxides $\xrightarrow{GSH}$ PGE

The structure of PGE$_1$ and PGF$_2\alpha$ are given below

PGE$_1$

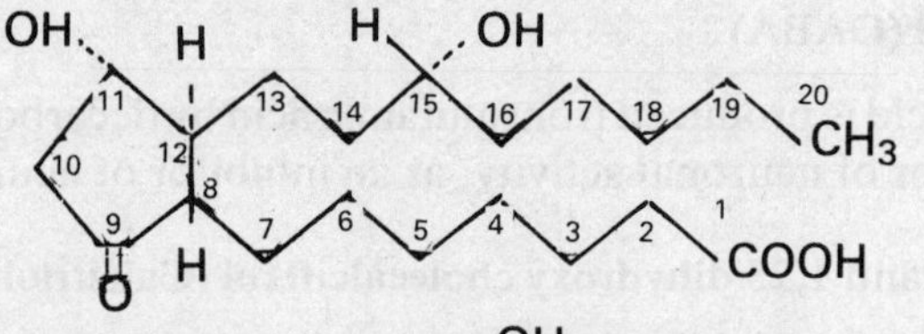

PGF$_2\alpha$

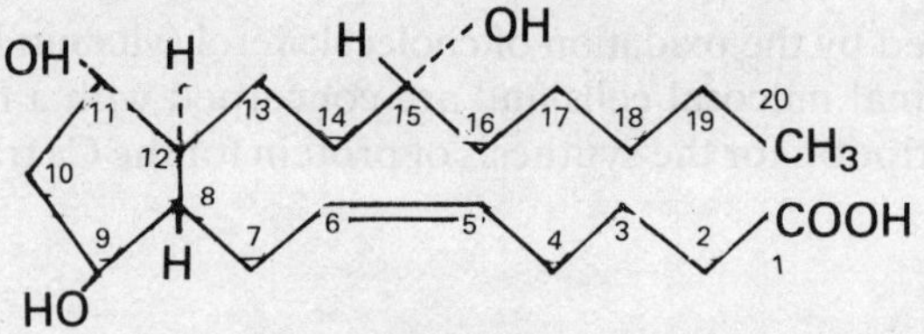

Urinary PGE level may reflect renal synthesis. PGE in 24 hour human urine = 0.2 to 1.2 micro gm.

Normal level of plasma (man and rat) PGs:

PGA	... ...	1062 ± 107 pg/ml.
PGE	... ...	385 ± 30 pg/ml.
PGF	... ...	141 ± 15 pg/ml.

PGs have several pharmacological effects. They cause inflammation, induce fever and produce pain (defensive role). Aspirin has been shown to inhibit the enzyme, prostaglandin synthetase (also known as fatty acid cyclo oxygenase), required for the synthesis of PGE$_2$ and PGF$_2\alpha$ from arachidonic acid.

The most important physiological effect of the PGs is their contraction of smooth muscles in general and the uterus in particular. This property is employed in using PGs to induce labour, abortions and as a contraceptive (Sultan Karim of Uganda). By controlled contraction of the uterus, PGs are of use in fertilisation as they increase the motility of the spermatozoa and direct them towards the uterus. PGs prevent the aggregation of the spermatozoa in semen. They also help in the implantation of the fertilised ovum in the uterus. PGE$_1$ and E$_2$ have beneficial effects in asthma as they decrease airway resistance while PGF$_2\alpha$ aggravates asthma by constricting the airway. The PGs decrease gastric secretion in the stomach (Andre Robert of Upjohn) and markedly increase intestinal motility. They also lower the blood pressure temporarily. The PGs may be of help in averting thrombosis as they inhibit the aggregation of platelets. As the PGs occur in all parts, they can affect membrane permeability. They also influence nerve function. PGA acts as a natriuretic substance, regulating the excretion of Na$^+$ at the renal tubular epithelium.

PGs are the most powerful anti-lipolytic substances and prevent the release of FFA from adipose tissue. FFA and glycerol levels of blood are decreased. The action is routed through adenylate cyclase which is inhibited by the PGs. Hence cyclic AMP levels are decreased by the PGs. The PGs also antagonise the lipolytic effects of catecholamines, ACTH, glucagon, TSH and vasopressin. Glucose oxidation and conversion into glycogen are increased by the PGs. In some species, blood sugar is not affected, or is increased.

PGs' usefulness in therapeutics is handicapped by their tendency to cause diarrhoea, lower the blood pressure and increase the heart rate.

Bergstrom considers the PGs as members of a new hormonal system. Though they are not typical hormones which are released from one organ and distributed through the blood to distant sites of action, they are considered as a 'local hormone'(or humoral mediator) which is released in the active form in many places and works near the spots. They fuction as mediators in many parts of the body including the male and female reproductive systems, digestive tract, air ways, spleen, blood vessels, skin, eye and central nervous system. Collier considers the PGs the coins of the body. As the same coins can be used for the purchase of different commodities, the same PGs discharge various functions in different locations.

Prostacyclins

Arachidonic acid derived from phospholipids is the precursor for prostagladin endo-peroxides. Prostaglandins or prostacyclins can be formed from these endo-peroxides. Prostacyclin synthetase, a microsomal enzyme, converts the endo-peroxides to prostacyclins.

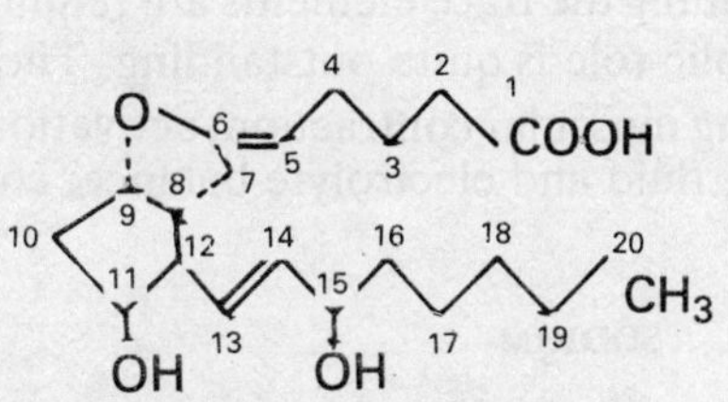

Prostacyclin

Prostacyclin has the important physiological function as a potent inhibitor of platelet aggregation and hence may be useful as an agent to prevent intra-vascular clotting of blood. It also acts as an effective vasodilator and may bring down blood pressure.

The heart tissue is believed to produce prostacyclins continuously and it may be playing a role in protecting coronary circulation against the formation of blood platelet aggregates. A direct synthesis of prostacyclins from arachidonic acid in the heart tissue is now known to be possible.

Prostacyclins are destroyed by the liver where they undergo mostly ß-oxidation and w-oxidation giving rise to short-chain compounds which are further degraded or excreted in urine.

19 Metabolism of Minerals

Sodium (Na), potassium (K), calcium (Ca), magnesium (Mg), sulphur (S), phosphorus (P), iron (Fe), copper (Cu), iodine (I), zinc (Zn), molybdenum (Mo), cobalt (Co), selenium (Se) manganese (Mn), fluorine (F), etc. are minerals. They exist as ions, free or bound. If they are bound, it may be by either covalent-binding or binding to proteins. Among the above-stated minerals, Na, K, Ca, Mg, S and P are present in greater quantities and are called *bulk elements*, while Fe, Cu, I, Zn, Mo, Co, F, Se, etc. are present in minute amounts and are called *trace elements*.

Only about eight per cent of the body weight of man is due to minerals. Unlike carbohydrates, proteins and lipids, minerals including the trace elements are required for the organism in very small amounts but their metabolic role is quite outstanding. They are indispensable for diverse metabolic processes including muscular contraction, activation of enzymes, transport of oxygen, acid-base homeostasis, fluid and electrolyte balance, coagulation of blood, etc.

SODIUM

Sodium is the predominant cation of the extra-cellular fluid. In the form of sodium chloride it takes part in fluid and electrolyte balance, and in the form of sodium bicarbonate (buffer salt) in acid-base equilibrium. Sodium chloride accounts for the maximum osmotic pressure as it gives rise to many ions. Through its contribution to the crystalloid osmotic pressure, sodium chloride influences the distribution of water in the body and is the back-bone of body water. Hence salt and water go together. Sodium chloride keeps the globulins in solution. The entry of the sodium ion into the cells causes action potential. These ions are also involved in the transmission of nerve impulses.

The daily requirement of sodium chloride for adults is 5–15 g. The excess NaCl over the body's requirements is excreted in urine and sweat.

The plasma level of Na is 130–150 m Eq/L. The cells, however, contain only about 37 m Eq/L. Mineralo-corticoids and aldosterone in particular of the adrenal cortex are chiefly concerned with the tubular reabsorption of sodium chloride taking place in the proximal convoluted tubules of the kidney. When sodium chloride is absorbed, water will also be absorbed as this process is obligatory.

In a deficiency of adrenal corticoids, there will be loss of sodium through urine. This happens in Addison's disease in which there is adrenal cortical hypofunction. As there is a reciprocal relationship between sodium and potassium, loss of body sodium means retention of potassium. Hence, in Addison's disease, there is hyponatremia (decrease of sodium) and hyperkalemia (increase of potassium in the blood). There may be dehydration incident to loss of NaCl.

Sodium depletion is common in chronic renal diseases also. Excess of perspiration is a well known cause of sodium deficiency. The manifestations of sodium deficiency are headache, nausea, muscular cramps, retarded growth, deficiency of osteoid tissue and muscular and testicular atrophy.

The biochemical findings in Cushing's syndrome are diametrically opposite of what is seen in Addison's disease, as in the former there is hyperactivity of the adrenal cortex. Hence, due to the availability of excessive amounts of corticoids, there is increased tubular reabsorption of sodium chloride and therefore retention of sodium. This causes hypernatremia and reciprocal hypokalemia. Retention of sodium is accompanied by retention of water and so in Cushings' syndrome, there will be edema (accumulation of intercellular water).

In the edematous states as well as after infusion with plenty of fluids, the serum sodium levels are misleading because of dilution hyponatremia. In fact, the total body sodium may be really increased due to retention of sodium chloride as in congestive cardiac failure or primary renal disease. Apart from the salt-retaining type of renal disease, a salt-losing type of nephrosis due to deficient tubular reabsorption is also encountered.

Hypernatremia may be due to overzealous saline infusion or Cushing's syndrome in which there is hyperactivity of the adrenal cortex. It can also be induced by administration of ACTH.

Progesterone exerts a slight sodium retaining effect so that even a normal pregnant woman without any toxemia may develop a transient occult edema which accounts for the amelioration of Addison's disease when the woman is pregnant.

There seems to be a relationship between sodium intake and diastolic blood pressure. In susceptible individuals excessive and wasteful intake of Na^+ as $NaCl$ may lead to or aggravate the pre-existing hypertension. On the other hand, when someone loses a lot of Na^+ in his sweat (e.g., miners), he should be given salt water for drinking, as otherwise sodium deficiency may cause muscular cramps.

Carrot, cauliflower, egg, milk, nuts and radish are rich in sodium. Animal foods are richer in sodium than plant foods. Due to the ease with which sodium is absorbed, fecal elimination is insignificant except in diarrhoea.

POTASSIUM

Potassium is the principal cation of the intra-cellular fluid. Inside the cell, only a part of K is in ionic form. The rest is bound to protein. As cell protein undergoes degradation, K is released into the extra-cellular fluid in a rather fixed proportion, i.e., about 110mg/g of nitrogen. Intra-cellular K, just like extra-cellular Na, plays the same role in acid-base homeostatis and possibly fluid and electrolyte balance. However, the extra-cellular potassium is equally important physiologically due to its promotion of relaxation of the skeletal as well as cardiac muscle. Its significant action on the depolarisation and contraction of the heart deserves special mention.

The intra-cellular potassium level ranges from 100–120 m Eq/L while the plasma K varies from 3–5 m Eq/L.

The resting membrane potential is mainly due to the difference in K^+ diffusion cell and is about -70 mv.

During transmission of nerve impulses, there is sodium influx and K^+ efflux with a reversal of polarisation. After transmission, the original state of affairs is restored.

K^+ is related to carbohydrate metabolism. Plasma K^+ rises and falls with lactic acid and glucose in blood. It enters the cells with glucose which is converted to glycogen.

Irritability and tetany are related to K^+ and other ions.

$$\text{Irritability} = \frac{(Na)^+ + (K)^+}{(Ca^{++}) + (Mg^{++}) + (H^+)}$$

$$\text{Tetany} = \frac{(K)^+ (HPO_4)^{--} (HCO_3)^-}{(Ca^{++})(Mg^{++})(H^+)}$$

K^+ and Na^+ have reciprocal relationship. K^+ shifts under different conditions given below.

Condition	Extracellular	Intracellular
Growth		→
dehydratin	←	
rehydration		→
acidosis		→
alkalosis	←	
diarrhoea	←	
vomiting	←	
glycogenesis		→

K^+ is not only filtered but actively secreted by the renal tubules. In fact, kidneys cannot conserve K^+ as they can Na^+. If Na^+ is to be retained, there should be an obligatory loss of K^+. There is a daily excretion of about 160 mg K^+ in urine. This is aggravated in renal failure and diuretics therapy.

The kidneys handle potassium in a remarkable manner. The excretion of potassium by the kidneys is dependent upon the acid-base status of the individual as well as the adrenal cortical activity. If the kidneys are normal, their ability to excrete potassium in the urine supplemented by the addition of potassium to the urine by the renal secretion is such that hyperkalemia is not ordinarily encountered even after intravenous infusion of potassium. It is a fact, therefore, that hyperkalemia is not commensurate with the degree of uremia unless it is a case of chronic renal failure with terminal, severe uremia. In the light of these facts, in the face of inadequate renal function, it is unwise to administer potassium, particularly parenterally.

In K^+ deficiency, the cells take up H^+ for K^+ leading to intracellular acidosis and a consequent extracellular alkalosis.

Deficiency of K^+ is associated with paralysis of the skeletal muscle and aberrant conduction as well as abnormal activity of cardiac muscle. Other manifestations of potassium deficiency are retarded growth, bone fragility, sterility and kidney hypertrophy.

Hypokalemia is encountered in post-operative conditions, chronic wasting diseases, gastrointestinal losses and in metabolic alkalosis. Large doses of insulin administered in the management of diabetes mellitus drive K^+ along with glucose into the cells. Therefore, in such a treatment, the risk of hypokalemia should be wisely anticipated.

Hyperkalemia is met with in renal failure, severe dehydration, shock and in Addison's disease. It can also be precipitated by unduly excessive intravenous administration of K^+ salts. Weakness of respiratory muscles and flaccid paralysis of the limbs are seen in severe cases of hyperkalemia.

Sources of K^+ are foodstuffs like almonds, apple, asparagus, banana, beans, beef, dates, cabbage, potato, etc. A dietary deficiency is extraordinarily rare as many common foods contain K.

METABOLISM OF MINERALS

CHLORIDE

Chloride, as sodium chloride, plays a vital role in acid-base balance by way of chloride shift or Hamburger's phenomenon, as well as in fluid homeostasis. Chlorides also contribute to the formation of gastric HCl. Deficiency of chloride impairs growth and may also cause hypochlorhydria. Most often, sodium and chloride are associated with each other in their functions. Occasionally, there could be loss of more chloride than sodium as in profuse vomiting in pyloric obstruction precipitating a hypochloremic type of alkalosis that may even bring about tetany. Chloride deficiency may be encountered in diarrhoea, excessive perspiration and the rare fibrocystic disease of the pancreas. Electrolytes, Na^+ and Cl^- are elevated in sweat, in the above-mentioned disease of the pancreas which could cause muco-viscidosis particularly in the bronchial system.

Plasma chloride is around 100 mEq/L while the CSF chloride is higher, i.e., around 125 mEq/L.

Fluid and electrolyte homeostasis

The regulation of fluid and electrolyte balance is one of the finest homeostatic mechanisms obtaining in the system. A consideration of fluid balance cannot be divorced from that of electrolyte balance since in most of the altered states, fluid imbalance is secondary to electrolyte imbalance.

Body water constitutes 60–70 per cent of the body weight, irrespective of sex, provided that the lean body mass is taken into account. It is distributed between two main compartments, namely, the homogeneous intracellular compartment and the heterogeneous extracellular compartment composed of plasma, interstitial fluid, lymph, dense connective tissue, cartilages, bone and trans-cellular fluids.

Determination of distribution of body water

The distribution of heavy water or tritium oxide or antipyrine is determined by making a due correction for the fat content of the subject. The specific gravity of the subject in air and under water has been used to arrive at the lean body mass. Plasma volume can be determined by the classical Evans blue method or by the more recent and sophisticated tracer techniques employing intravenous administration of ^{32}P labelled erythrocytes or ^{131}I labelled human serum albumin. After allowing 10 minutes or more for mixing, their volume of distribution may be obtained from their concentration in an aliquot of blood or plasma. The determination of the extracellular fluid volume requires the use of a substance that does not enter the cell and it gets distributed quickly and uniformly in the entire plasma and the remainder of the extracellular fluid. Inulin or mannitol may suit the purpose though either substance is not ideal. The intracellular fluid volume is obtained by subtracting the extracellular fluid volume from the total body water.

Electrolyte composition of body fluids

While sodium is the predominant cation in the plasma as well as the interstitial fluid, the chloride of interstitial fluid and the protein of plasma are the principal anions. In the intra-cellular fluid, potassium is the predominant cation and phosphate is the chief anion. Since one chemical equivalent of any substance is equal in chemical reactivity to one equivalent of any other, changes in the chemistry of body fluids by compensatory shifts of one ion for the other can best be appreciated only if expressed in identical concentration units. This is obtained when the electrolyte levels are expressed as mEq/L obtained from mg per cent by multiplying by 10 and dividing by the equivalent weight of the substance concerned.

To strike the fluid balance obtaining in the body, the fluid intake has to be computed with the elimination of fluid by various channels. The intake of fluid includes preformed fluid taken as such and also contained in foodstuffs as well as the water from oxidation of lipids and carbohydrates. The loss of fluid from the body is mainly through the renal, intestinal, cutaneous and pulmonary routes. Insensible losses refer to pulmonary losses as well as imperceptible perspiration. Fluid losses can vary even physiologically, depending on dietary intake and climatic conditions. A high protein diet due to a high formation of urea, which is a diuretic, causes physiological polyuria. Extra-renal losses and renal elimination of fluids are reciprocally related. Insensible losses are increased post-operatively, during fever and due to debility. When the kidneys do not possess the normal concentrating capacity, renal loss is enhanced. Diarrhoea is a common cause of pronounced extra-renal loss and reciprocally a diminished excretion of urine. Acute and chronic kidney diseases, diabetes mellitus, diabetes insipidus, Addison's disease and Cushing's syndrome are well known examples of fluid imbalance.

Homeostatic mechanisms

The compartmental distribution of body fluids is constantly subject to fluctuations, osmosis being the cardinal factor apart from the ductless glands and the central regulator, namely, the hypothalamus.

Osmosis being a colligative property depending upon the number of the particles of the solute, the different solutes present in our system are of great moment not only in acid-base regulation but also in fluid balance. There are three categories of solutes that operate. Glucose, urea, amino acids, etc. provide examples of organic substances of small molecular size which diffuse freely across cell membranes and hence count only when they are present in high concentrations influencing the degrees of total hydration.

Plasma proteins are macro-molecular and by virtue of their hydrophilic property, are more closely concerned with the compartmental distribution of body water. The oncotic pressure of plasma proteins is 25–30 mm of Hg., mostly due to albumin, and it serves to imbibe fluid from the tissue spaces into the capillaries opposing the effective hydrostatic pressure. Hypoproteinemia is one of the well known causes of nutritional edema which can also be caused by other factors like accumulation of pyruvate as in thiamine deficiency. For the formation of urine, there has to be an effective filtration pressure due to the predominance of the effective hydrostatic pressure, over the opposing effective osmotic pressure. It accounts for oliguria or anuria in shock due to fall of blood pressure depending on the severity of the condition.

By far, the most important solutes that maintain fluid balance are the inorganic electrolytes chiefly sodium chloride of the extra-cellular fluid which has an important bearing not only on the extent of total hydration in the body but also in the compartmental distribution of body water.

Because of the free diffusion of water across the cell barrier, the changes in the concentration of extracellular sodium and intracellular potassium on either side of the cell membrane influence the migration of fluid from one compartment to the other. Sodium is appropriately considered the backbone of the extracellular fluid, and therefore, sodium chloride has to be necessarily restricted in the management of over-hydration irrespective of the causation.

The importance of potassium is illustrated by the fact that on certain occasions potassium leaves the cell, as in vomiting, diarrhoea and prolonged gastric suction. Under these circumstances, it would not suffice to give saline infusion alone which, if done inadvertently, leads to entry of sodium into the cells through the membrane in an attempt to replenish the intracellular potassium deficit, thereby precipitating a persistent intracellular alkalosis.

Such an undesirable state is avoided by institution of prompt and concomitant potassium therapy.

The intimacy between fluid and electrolyte balance is seen in a good number of conditions. Most of them show a fluid imbalance caused by electrolyte imbalance. It is exemplified well by the following conditions: When there is a restriction of water or when there is a considerable loss of fluid from the body, it promotes extracellular hypertonicity. This results in the migration of fluid from the cells resulting in intracellular dehydration. A prompt response is seen on the part of normal kidneys by way of oliguria with a raised specific gravity.

In Addison's disease, there is primary loss of electrolyte followed by that of fluid with subsequent dehydration. To combat the dehydration, saline has to be given. If water alone is ingested, it would lead to extracellular hypotonicity. Water will now flow into the cells due to osmotic necessity and there will be intracellular edema and extracellular dehydration. There is lowering of the extracellular fluid volume, blood volume and blood pressure, and in severe cases the renal circulation as a part of the peripheral circulation may fail and consequently the kidneys themselves are unable to compensate. In such a condition the urinary findings do not indicate the true state of affairs. The diagnostic features are an elevated hematocrit and a fall in the levels of plasma sodium chloride.

Hormonal controls of fluid and electrolyte homeostasis

Aldosterone of adrenal cortex is concerned with the obligatory reabsorption of water to the tune of 80 per cent in the proximal convoluted tubules of the kidneys. This is secondary to the reabsorption of sodium chloride, warranted by the necessity for maintaining the local osmotic environment, and not dictated by the body needs for water. The role of aldosterone is very well illustrated by Addison's disease and also Cushing's syndrome. In the former condition, dehydration is incident to loss of electrolytes namely sodium chloride from the body. In the latter condition, retention of sodium chloride is the cause of edema. Vasopressin or the anti-diuretic hormone of the neuro-hypophysis influences the facultative reabsorption of water to the extent of 20 per cent or less in the distal convoluted tubules of the kidney. This is dictated by the body needs for water. Diabetes insipidus is caused by the deficiency of this hormone resulting in polyuria without any glycosuria. The central regulator of fluid homeostasis is the hypothalamus which responds to the mediation brought about by osmo-receptors that are influenced by the fluctuations in crystalloid osmotic pressure, chiefly of sodium chloride of the extracellular fluid. Water diuresis is because of dilution of blood and lowering of the crystalloid osmotic pressure which, through the osmo receptors, brings about a suppression of the anti-diuretic hormone leading to diuresis. The converse effect is seen on restriction of fluids.

The rectification of dehydration and the type of fluids to be infused depend upon the cause as well as the anatomical situation from which fluid loss has occurred. For example, when the fluid loss is from the stomach or the upper intestinal tract, as in pyloric stenosis or obstruction or high intestinal obstruction or prolonged gastric suction, the loss of chloride is more than that of sodium relatively. Therefore, saline infusion (isotonic sodium chloride infusion) is indicated, as well as concomitant potassium infusion, provided that the renal function is all right. On the other hand, when the fluid losses originate from the lower intestinal tract, as in diarrhoea, pancreatic and biliary fistulas, there is comparatively more loss of sodium bicarbonate resulting in a relative chloride excess. Such a situation calls for infusion of a judicious mixture of 2/3 isotonic saline and 1/3 sodium lactate (1/6 molar). Sodium lactate is by far preferable to sodium bicarbonate as lactate is easily metabolised by oxidation and, unlike the latter, does not precipitate a metabolic alkalosis. Dehydration need not necessarily become manifest. It can be occult and recognised by changes in body weight. A loss of 8–12 per cent in body weight shows a severe degree of dehydration.

CALCIUM AND PHOSPHORUS

Calcium and phosphorus are associated with each other not only in the sources of their occurrence but also in metabolic processes. Eggs, milk and dairy products form some of the most important dietary sources of calcium. Many sea foods are rich in calcium. Green tops are excellent vegetable sources.

Proteins supply considerable phosphorous. Milk and cheese are important sources of phosphorus.

The daily requirement of calcium for an adult is 0.8 g while that of phosphorus is 1.0 g. The requirement is increased during pregnancy and lactation in women. The ratio between calcium and phosphorus in the diet should be between 1:2 and 2:1.

The human body contains 1–2 kg of calcium in a 70 kg individual. About 99 percent of this is in the bones and teeth. Though skeletal calcium is present as hydroxy apatite and non-crystalline calcium phosphate and carbonate, the bone is constantly being remodelled. As much as 700 mg of calcium may enter and leave the bone everyday. A small per centage of calcium is present in the cells and body fluids. Though low in concentration, i.e., about 20 mg/100 gm of tissue, its role in cellular activities is vital. Calcium as calcium phosphate is the chief salt of the teeth and bones and is required for the growth of the bone and thereby of the body.

Calcium ions are necessary for nerve and muscle functions and those of the heart. In fact, the trigger for muscle contraction is the interaction of calcium with troponin C. Many hormones use calcium as their 'second messenger'. Sometimes a hormone does not use cyclic AMP as its second messenger but uses calcium, e.g., epinephrine in the glycogenolysis of liver. In some cases, calcium becomes the third messenger, e.g., the para-thyroid hormone acts through cAMP and calcium. Calcium ions are necessary for the coagulation of blood. Calcium has a profound influence on membrane and capillary permeabilities and membrane formation, cellular motility and the function of the retina. It is a neuro-sedative, and in conditions like hypoparathyroidism, due to a deficiency of ionic calcium, tetany may result. In conditions of hypo-proteinemia, as in starvation and protein deprivation, total blood calcium may be lowered secondary to the lowered levels of proteins. This is because in the blood, a major amount of calcium is bound to proteins. Calcium is also needed for the activation of enzymes like lipase, adenosine triphosphatase, succinate dehydrogenase, etc.

In addition to the above, calcium is required for the curdling of milk and for certain types of bioluminescence.

The precise mechanism of function of the calcium ions in metabolism is their binding to an intracellular receptor protein known as calmodulin present in every nucleated cell. This Ca^{2+} – calmodulin complex modulates the activities of a great variety of enzymes including those involved in cyclic nucleotide metabolism, protein phosphorylation, secretory function (insulin secretion from pancreas is controlled by Ca), micro-tubule assembly, glycogen metabolism and calcium flux.

The serum calcium level ranges normally from 9–11 mg per 100 ml, half of which is ionic and the rest bound to plasma albumin. It is the ionic calcium which has tremendous physiological importance.

Factors that influence calcium and phosphorus absorption

Apart from the absolute levels of calcium and phosphorus in the diet, if the desirable range of ratio between the minerals is not there and if there is an excess of either, insoluble calcium soaps that hinder absorption are formed resulting in enhanced fecal excretion of calcium and phosphorus and negative calcium and phosphorus balance. The proportion of Ca and P in the diet, between 1:2 and 2:1 is an important factor.

The homeostasis of blood calcium is maintained by suitable absorption and management by vitamin D_3 as 1, 25-dihydroxy vitamin D_3, para-thyroid hormone and calcitonin. The absorption of calcium takes place mostly in the duodenum and proximal jejunum, and vitamin D is necessary for the same. Vitamin D gets converted to 1,25-dihydroxycholecalciferol (calcitriol), this conversion being effected in two steps in the liver and kidney, respectively. The function of this hormone in the intestinal mucosal cells is to induce the synthesis of a calcium binding protein which helps the absorption of calcium.

Digestion and absorption of lipids and their impact: When the absorption of fats is impaired, calcium soaps are formed to a greater extent retarding the absorption of calcium. On the other hand, the loss of unhydrolysed fats in feces due to imperfect digestion does not bring about calcium loss as unless the lipids are digested there are not sufficient fatty acids to bind the calcium. The extent of loss of calcium depends upon the severity of diarrhoea and particularly steatorrhoea.

Phytic acid, iron and oxalates: Massive iron therapy, particularly by the parenteral route, may bind phosphate and induce a low phosphate type of rickets. Oxalates precipitate calcium and foster a negative balance. The anti-calcifying action of cereals like oats is due to the high phytic acid which combines with calcium and magnesium to form calcium and magnesium phytates. However, such an action is mitigated by the presence of phytase.

pH and absorption of calcium and phosphorus: An increase in the acidophilic flora promotes the solubilisation of the calcium salts and their absorption. Administration of lactose or lactate along with calcium medication is beneficial. Since milk contains calcium and phosphate along with lactose, it is an easily assimilable source of these elements.

The pH of the duodenum and jejunum is important in the absorption of calcium, a neutral or acidic pH favouring its absorption since calcium salts are more soluble in acidic and neutral than in the basic medium. Lactic acid as well as some of the acidic amino acids present may enhance the solubility of calcium and consequently its absorption. The protein status of the diet matters a lot because a high protein diet facilitates the solubilisation of the calcium salts in the aqueous solution of amino acids.

The renal threshold is important in the regulation of the blood calcium level though the exact mechanism by which the kidneys handle calcium is yet to be elucidated. The reciprocal relationship between calcium and phosphorus holds good so far as the renal status is normal. For instance, in hyper-parathyroidism, uncomplicated by any renal involvement, there is a hypercalcemia with a reciprocal hypophosphatemia. But when the condition is complicated by the development of renal calculi, due to retention, there is hypercalcemia as well as hyperphosphatemia.

The product of ionic calcium and inorganic phosphate in serum is normally 30–40 in adults and 50–60 in growing children. If it is less than 30 on any account, due to lowering of calcium or serum phosphate, it predisposes to development of rickets or osteomalacia.

Normal kidney excretes little Ca^{2+} as 99 per cent of calcium filtered is reabsorbed. The normal urinary excretion of calcium is about 200 mg/24 hrs.

Role of vitamin D_3: It is now known that vitamin D_3 is inert in the body and the hormonal form is 1, 25-dihydroxy vitamin D_3. Hydroxylation at the 25th position takes place in the liver and a second hydroxylation at the first postition takes place in the kidneys. This derivative causes increased absorption of calcium in the small intestines and mineralisation and accretion of bone. Intestinal absorption is brought about by augmenting the transcription of messenger RNA and the formation of a protein known as the calcium binding protein (CaBP) which binds calcium and increases the transport of calcium across the intestinal mucosa. The enzyme Ca^{2+} ATPase has a permissive role in calcium transport.

The para-thyroid hormone is the chief hormone which helps in maintaining serum calcium. The hormone is secreted in response to lowered serum calcium. It causes demineralisation of bone calcium by its direct action and indirect stimulation of the formation of 1,25-dihydroxy cholecalciferol. In its direct effect, PTH activates the membrane-bound adenylate cyclase, increases the concentration of cAMP and thereby that of intra-cellular calcium which modulates the enzymes, iso-citrate dehydrogenase and pyro-phosphatase, and controls the mobilisation of bone calcium.

In hypercalcemia, another hormone, calcitonin, is elaborated by the 'C' cells of the thyroid, and, by acting on the bone through intracellular calcium, it brings down demineralisation and serum calcium.

Miscellaneous: It has been shown that each gram of the serum proteins, through albumin, binds 0.84 mg of calcium.

Total calcium = 0.84 × per cent of total serum proteins + diffusible calcium (ionic form).

This formula enables the indirect determination of ionic calcium which alone is metabolically active.

Sex hormones and calcium: Sex hormones seem to have an effect on calcium and phosphorus balance. It is known that after menopause women are prone to fractures due to osteoporosis.

Tetany

Though one of the well known causes of tetany is the lowering of ionic alcium, it is also decided by the level of some other ions as shown by the following equation:

$$\text{Tetany} = \frac{(K^+) \quad (HCO_3)^- \quad (HPO_4)^{2-}}{(Ca^{2+}) \quad (Mg^{2+}) \quad H^+} = \text{a constant to be maintained to avert tetany}$$

Irritability is also related to the concentration of Ca_{2+}

$$\text{Irritability} = \frac{(K^+) + (Na^+)}{(Ca^{2+}) + (Mg^{2+}) + (H^+)}$$

Disorders of calcium metabolism

Hyper-parathyroidism: In this condition, there is hyper calcemia as well as a reciprocal hypophosphatemia. In urine, there is hypercalcuria and hyperphosphaturia. Serum alkaline phosphatase may be elevated. If not controlled, there will be formation of renal calculi.

In hypo-parathyroidism, there is pronounced hypocalcemia, and tetany may follow. In rickets in children, serum calcium may be normal or lowered, but serum phosphorus is definitely lowered. The product of concentrations of Ca and phosphate which is 30 in the normal would be decreased. But in osteomalacia (adult rickets), serum calcium is usually lowered and tetany may occur.

Renal rickets is a familial condition as a sex-linked dominant trait. The features are hypophosphatemia, hyperphosphaturia and reduced intestinal absorption of Ca and P. It is vitamin D_3 resistant. The 1-hydroxylation of vitamin D_3 is affected in the renal tubules because of congenital absence of the hydroxylase enzyme even though the levels of the para-thyroid hormone may be normal.

The serum calcium level may decrease in severe renal disease due to enhanced renal losses.

PHOSPHORUS (PHOSPHATE)

Phosphate exists in cells as a free ion and also circulates in blood. The normal levels of serum phosphorum are 3–4.6 mg/100 ml in adults and 4–6mg/100 ml in children.

Phosphates are constituents of bone and teeth. They are required for the growth of bone and growth of body. In bone, PO_4 is present as hydroxy apatite. An adult has about 1 kg of phosphorus, 85 per cent of which is in the skeleton. Phosphates have an important role as high energy compounds like ATP and creatine phosphate. The phosphate group is present in some carbohydrates and vitamins. Phospholipids and proteins like casein of milk and vitellin of egg contain phosphate. Phosphates are constituents of cell membranes and nerve tissue. DNA, RNA, nucleotides like GTP, cAMP and a host of coenzymes like FMN, FAD, NAD^+, $NADP^+$ and pyridoxal phosphate contain phosphorus. Some enzymes are active only if phosphorylated, e.g., phosphorylase. Hormones like insulin act in some of their reactions through the phosphorylation of amino acids like tyrosine, serine and threonine in proteins.

Phosphates constitute one of the buffers of plasma.

Intestinal absorption of phosphate is similar to that of calcium, 1, 25-dihydroxy vitamin D_3 promotes intestinal absorption of phosphate in addition to that of calcium. It stimulates reabsorption of phosphate along with calcium in the proximal tubules of the kidneys. However, the para-thyroid hormone diminishes the renal tubular reabsorption of phosphate and thereby overrides the effect of 1, 25-dihydroxy vitamin D_3 on phosphate excretion. 1,25-dihydroxy vitamin D_3 plays a permissive role in the para-thyroid hormone mediated mobilisation of calcium and phosphate from bone.

Phosphate concentration has a part to play in tetany. As much as calcium is affected in hyper-parathyroidism, hypo-parathyroidism, rickets, osteomalacia and renal rickets, phosphate is also affected. In hyper-parathyroidism there is hypo-phosphatemia and hyper-phosphaturia, while in hypo-parathyroidism there is hyperphosphatemia and hypo-phosphaturia. In rickets, serum phosphate is lowered and the product of calcium and phosphate concentrations is less than 30. In renal rickets (vitamin D_3 resistant rickets) there is reduced intestinal absorption of calcium and phosphorus with hypo-phosphatemia and hyper-phosphaturia. In De Toni Fanconi's syndrome caused by a defect in the tubular reabsorption, the diagnostic features are phosphaturia, glycosuria, amino aciduria and ketonuria. The daily requirement of phosphorus for adults is about 1 g.

The urinary excretion of phosphate has a wide range of 700–1200 mg per 24 hrs depending on the phosphate intake.

MAGNESIUM

Most of the magnesium in the body is in the bones along with calcium and phosphorus. It is also one of the important cations of the soft tissues. The Mg ions are present in all the cells. Unlike calcium and sodium, magnesium and potassium are normally concentrated within the cells.

The blood level of Mg ranges between 1.7–3.4 m Eq/L. About 80 per cent of this is ionised and diffusible.

The absorption of Mg^{2+} is not an active process. High levels of calcium, proteins and phosphate diminish the absorption of magnesium from the intestines. Malabsorption in chronic diarrhoea, protein-calorie malnutrition and adult starvation in the form of alcoholism can all result in magnesium deficiency.

Magnesium is required in many reactions in which ATP participates as it is the ATP-Mg^{2+} complex that acts as a substrate. Mg^{2+} is chelated between the beta and gamma phosphates, and diminishes the dense ionic character of ATP so that it can easily approach and bind to specific enzyme sites. Mg is an activator of many glycolytic enzymes, particularly in the muscle. There is a clear antagonism between Mg^{2+} and Ca^{2+}. The metabolism of Mg simulates that of Ca and P.

In plasma, most of the Mg^{2+} exists in a form that can be filtered by the glomerulus. But the kidney has an extraordinary ability to conserve Mg^{2+} the daily loss of Mg is therefore insignificant. Aldosterone augments the renal clearance of Mg. In renal tubular acidosis or diabetes mellitus there is magnesium wasting by the kidneys, which increases the dietary requirement of Mg. A number of drugs including diuretics promote Mg wastage.

A low magnesium diet can induce tetany experimentally, the calcium in the diet being normal. Prolonged hyper-parathyroidism can deplete the body stores of magnesium. Muscle tremour, twitchings, convulsions and delirium can be caused by severe magnesium deficiency.

Magnesium toxicity is rare in the presence of normal renal function. In renal failure along with potassium retention, there is commensurate rise in serum magnesium levels. This produces depressant effects on nervous system.

Erythrocyte levels are more reliable indicators of the magnesium status of the individual.

Leafy vegetables, cocoa, nuts, soya beans and sea foods are rich in magnesium. The normal daily adult requirement is about 350 mg. During lactation, magnesium requirement is increased.

SULPHUR

Sulphur is present in the proteins of the biomembrane and the cells. It is present in compounds like acetyl CoA and succinyl CoA, which are needed for various reactions. In these and many enzymes, and coenzyme A, sulphur is present as sulph-hydryl groups. The acyl carrier protein also contains sulphur. Glycosaminoglycans like heparin, chondroitin sulphates, the enzyme 3-phosphoglyceraldehyde dehydrogenase, the vitamins, thiamine, biotin and lipoic acid and the hormone insulin contain sulphur. Bile acids and keratin of hair also contain sulphur. Sulphur metabolism is intimately linked with that of sulphur containing amino acids of proteins, namely cystine, cysteine and methionine. Glutathione is a valuable sulphur containing tripeptide needed for the detoxication of H_2O_2. Sulphur takes part not only in detoxications but also in tissue respiration.

If the sulphuric acid group has to be introduced in 1) glycosaminoglycans like chondroitin sulphates, and 2) phenols for detoxication as ethereal sulphates, it can be done only by PAPS, i.e., phospho-adenosine phospho-sulphate.

The major form of excretion of sulphur is by way of inorganic sulphate (about 80 per cent) derived by oxidation. A small percentage (8–10 per cent) is accounted for by ethereal sulphates which represent the products of detoxication of phenolic compounds obtained from tyrosine, and tryptophan. The phenol sulphuric acids are phenyl, paracresyl, indoxyl and skatoxyl sulphates. Urinary sulphur represents inorganic ionic sulphates, ethereal sulphates and neutral sulphur (about 10 per cent). The neutral sulphur includes S-containing amino acids which have escaped metabolism, thiocyanates, thiosulphate, etc.

TRACE ELEMENTS

Iron

The physiological role of iron is out of proportion to its body content. A normal adult of 70 kg possesses just 3–4g of iron. This is contained in hemoglobin, myoglobin, cytochromes, catalase and peroxidase. Special mention is to be made of the role of iron in the transport of oxygen by hemoglobin.

Legumes, molasses, nuts, amaranth and moringa leaves, spinach, dates, yolk of egg, fish, organ meats like liver, etc. are good sources of iron. Milk is a poor source.

Iron is considered a one-way substance because very little of it is excreted. The iron derived during the degradation of hemoglobin is conserved and recycled for the resynthesis of new heme. There is no loss of body iron except in women during menstruation. The normal daily requirement for an adult is 10–15 mg on an average and the requirements increase during pregnancy and lactation. Women require relatively greater amounts than men due to physiological losses during menstruation.

As there is no loss of body iron under normal circumstances, only very little of dietary iron is absorbed. Moore has shown that absorption of iron is about 10 per cent or less, i.e., just one mg. There is greater absorption in infants and children and in iron deficiency anemia. Increased phosphate in the diet diminishes iron absorption.

The gastric acidity and organic acids in the diet convert organic ferric compounds of the diet into free ferric ions (Fe^{3+}) which are reduced with ascorbic acid, reduced glutathione, etc. to the ferrous (Fe^{2+}) form. This is a necessity in the absorption of iron as it cannot be absorbed as Fe^{3+}. Not all the iron present in foodstuffs is assimilable and nutritionally available.

Heme absorbed by the intestinal mucosa as such is subsequently broken down and iron is released within the cell. Non-heme iron is absorbed in the ferrous state. The Fe^{2+} is absorbed into the mucosal cell of the duodenum and proximal jejunum and promptly oxidised to Fe^{3+}. Ferric ion is bound by an intra-cellular carrier molecule. Within the cell, the carrier molecule delivers Fe^{3+} to the mitochodria. Then depending upon the state of iron metabolism in the individual, the carrier molecule distributes Fe^{3+} in specific proportions to apoferritin and apotransferrin.

The only mechanism by which the total body stores of iron can be regulated is at the level of iron absorption. Physiological control of iron status exists at the level of intestinal absorption. More than 10 per cent of dietary iron would be absorbed in the intestines, only if there is greater metabolic need.

Apoferritin is a molecule with molecular weight of approximately 5,00,000. It assimilates upto 4,300 iron atoms into a single molecule to form ferritin which is the primary and most easily available iron-storage protein.

Apotransferrin is a protein of MW 90,000 that can bind just two atoms of iron to form transferrin (or siderophilin). Transferrin is the true carrier of iron. It is ordinarily 20–33 per cent saturated with iron. It occurs as micelles or colloidal particles. Ferric hydroxide and ferric phosphate are present as $(FeOOH)_3$ $(FeOOPO_3H_2)$. It is also suggested that apoferritin itself acts as a ferro-oxidase and oxidises bound ferrous iron to ferric iron which is then tightly bound to ferritin.

Under normal conditions, when about 1 mg of iron is absorbed daily by an adult, the intracellular iron carrier of the cell is nearly saturated. It transfers significant quantities of iron to apoferritin to form ferritin, and transfers the usual quantity of iron to the mitochondria. The remainder is transported across the serosal surface to apotransferrin.

In the *iron-deficient state*, the capacity of the intracellular iron carrier is expanded and more dietary iron is absorbed. Although the mitochondria receive their usual supply of iron, ferritin is not formed in the cell and the majority of iron goes to apotransferrin.

In the case of *iron-overload*, the carrier is diminished in capacity. A significant quantity of ferritin is formed within the mucosal cell and less iron is transferred to apotransferrin.

The transfer of iron from the storage ferritin as Fe^{3+} requires its reduction to Fe^{2+}. In plasma, ferrous iron is rapidly oxidised to the ferric form and is fixed to apotransferrin.

$$2Fe^{2+} + O_2 + 2CO_2 + \text{apotransferrin} \rightarrow 2Fe^{3+}CO_2 \text{ transferrin.}$$

Ceruloplasmin is a ferro-oxidase and oxidises plasma Fe^{2+} to Fe^{3+}. Transferrin is a beta-1 globulin and a glyco-protein. In addition to the mucosal cells, the reticulo-endothelial system of the liver, spleen and bone marrow also constitutes a second available storage form for iron. The Fe^{3+} of plasma transferrin is transferred to apoferritin of the above system to form ferritin.

Under physiological conditions, most of the transferrin iron is quickly taken up by the bone marrow. Possibly, only the reticulocytes can utilise bound Fe^{3+} while the mature erythrocyte can take up unbound ferric iron.

The plasma transferrin iron pool is in equilibrium with the iron in storage forms in GIT and RES. In a deficiency of blood iron, iron is mobilised from the ferritin stores in the following sequences:

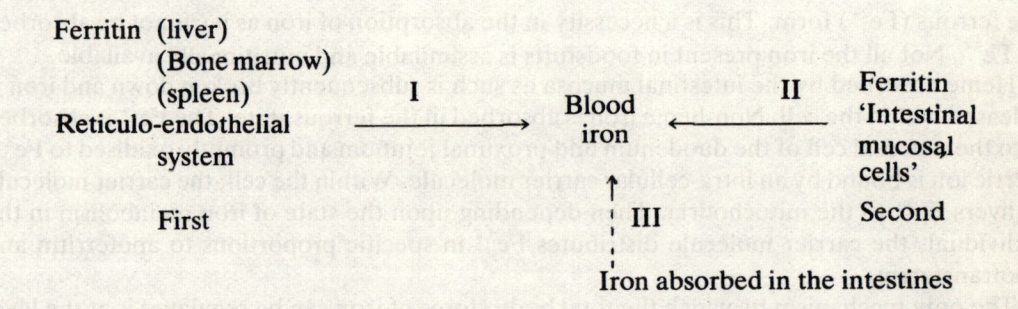

```
Ferritin  (liver)
          (Bone marrow)
          (spleen)            I                        II    Ferritin
Reticulo-endothelial    ─────────→   Blood    ←─────────    'Intestinal
system                                iron                   mucosal
                                       ↑                     cells'
First                                  ¦ III                 Second
                                       ¦
                          Iron absorbed in the intestines
```

But the Fe^{3+} of ferritin should first be reduced to Fe^{2+} to be released from ferritin.

Ferritin of the RES can become denatured, losing the apoferritin sub-units and subsequently aggregating into micelles of hemosiderin. While ferritin contains about 23 per cent of iron, hemosiderin contains more of iron and exists as microscopically visible iron staining particles. It is usually seen in states of iron over-load. The iron in hemosiderin is available for the formation of hemoglobin but the mobilisation of iron is much slower from hemosiderin than from ferritin.

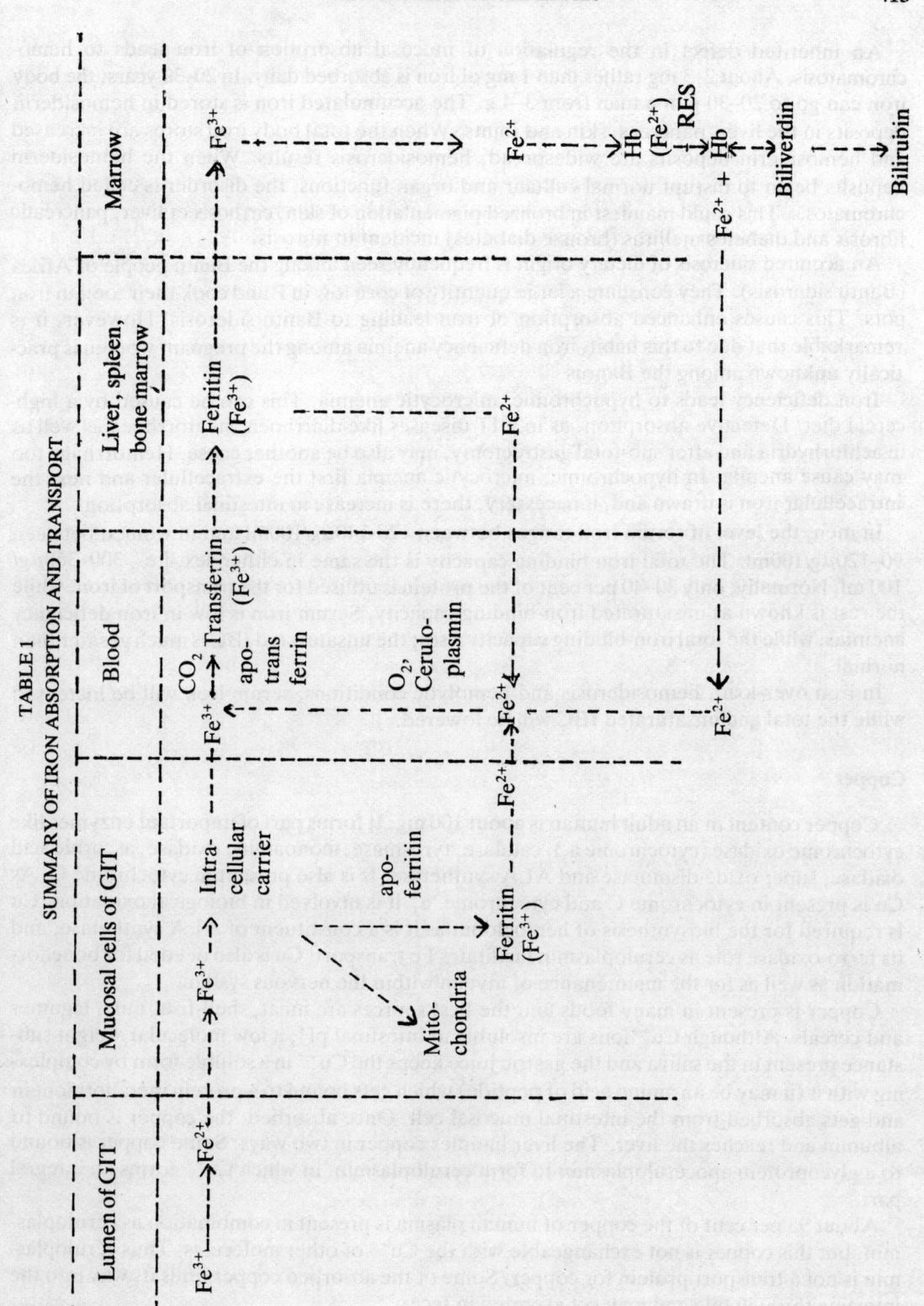

TABLE 1
SUMMARY OF IRON ABSORPTION AND TRANSPORT

Lumen of GIT	Mucosal cells of GIT	Blood	Liver, spleen, bone marrow	Marrow	RES
Fe^{3+} → Fe^{2+} →	Fe^{3+} → Intracellular carrier → apo-ferritin → Ferritin (Fe^{3+}) → Mitochondria	CO_2, Fe^{3+} → Transferrin (Fe^{3+}) ← apo-transferrin ← O_2, Ceruloplasmin ← Fe^{2+} ←	Ferritin (Fe^{3+}) ← Fe^{2+}	Fe^{2+} → Hb (Fe^{2+}) →	Hb → Biliverdin → Bilirubin; Fe^{2+} +

An inherited defect in the regulation of mucosal absorption of iron leads to hemochromatosis. About 2–3 mg rather than 1 mg of iron is absorbed daily. In 20–30 years, the body iron can go to 20–30 g in a man from 3–4 g. The accumulated iron is stored in hemosiderin deposits in the liver, pancreas, skin and joints. When the total body iron stores are increased and hemosiderin deposits are widespread, hemosiderosis results. When the hemosiderin deposits begin to disrupt normal cellular and organ functions, the disorder is called hemochromatosis. This could manifest in bronzed pigmentation of skin, cirrhosis of liver, pancreatic fibrosis and diabetes mellitus (bronze diabetes) incident to fibrosis.

An acquired siderosis of dietary origin is frequently seen among the Bantu people of Africa (Bantu siderosis). They consume a large quantity of corn low in P and cook their food in iron pots. This causes enhanced absorption of iron leading to Bantu siderosis. However, it is remarkable that due to this habit, iron deficiency anemia among the pregnant women is practically unknown among the Bantus.

Iron deficiency leads to hypochromic, microcytic anemia. This can be caused by a high-cereal diet. Defective absorption, as in GIT diseases like diarrhoea, steatorrhoea as well as in achlorhydria and after sub-total gastrectomy, may also be another cause. Hemorrhage too may cause anemia. In hypochromic, microcytic anemia first the extracellular and next the intracellular iron is drawn and, if necessary, there is increase in intestinal absorption.

In men, the level of serum iron ranges between 120-140 μg/100ml and in women between 90–120 μg/100ml. The total iron binding capacity is the same in either sex, i.e., 300–360 μg/100 ml. Normally, only 30–40 per cent of the protein is utilised for the transport of iron, while the rest is known as unsaturated iron-binding capacity. Serum iron is low in iron deficiency anemias, while the total iron-binding capacity rises; the unsaturated IBC is much greater than normal.

In iron over-load, hemosiderosis and hemolytic conditions, serum iron will be increased while the total and unsaturated IBC will be lowered.

Copper

Copper content in an adult human is about 100 mg. It forms part of important enzymes like cytochrome oxidase (cytochrome a_3), catalase, tyrosinase, monoamine oxidase, ascorbic acid oxidase, super oxide dismutase and ALA synthetase. It is also present in cytochrome C. As Cu is present in cytochrome C and cytochrome 'a_3' it is involved in biological oxidation. Cu is required for the biosynthesis of hemoglobin as it is a constituent of ALA synthetase, and its ferro-oxidase role as ceruloplasmin facilitates Fe transport. Cu is also needed for bone formation as well as for the maintenance of myelin within the nervous system.

Copper is present in many foods and the best sources are meat, shell-fish, nuts, legumes and cereals. Although Cu^{++} ions are insoluble at intestinal pH, a low molecular weight substance present in the saliva and the gastric juice keeps the Cu^{++} in a soluble form by complexing with it (it may be an amino acid or peptide) which gets bound to a protein metallothionein and gets absorbed from the intestinal mucosal cell. Once absorbed, the copper is bound to albumin and reaches the liver. The liver handles copper in two ways: Some copper is bound to a glycoprotein apoceruloplasmin to form ceruloplasmin, in which Cu^{++} forms an integral part.

About 95 per cent of the copper of human plasma is present in combination as ceruloplasmin, but this copper is not exchangeable with the Cu^{++} of other molecules. Thus ceruloplasmin is not a transport protein for copper. Some of the absorbed copper finds its way into the intestine through bile and may get excreted in feces.

Ceruloplasmin represents the blue, copper-containing protein of plasma exclusively synthesised in the liver, while erythrocuprein refers to the almost colourless, copper- containing protein in the erythrocytes. The brain contains cerebrocuprein. Ceruloplasmin is not a copper transport protein but has copper as an integral part of the ferro-oxidase and is a glycoprotein. It contains 0.34 per cent copper (6–8 atoms of copper per molecule, half as cuprous and the other half as cupric). Normal serum contains about 30 mgs per cent of ceruloplasmia.

In addition, there is a smaller fraction of serum copper loosely bound to the albumin and amino acids, particularly histidine. While this copper reacts with diethyl dithio carbamate giving yellow colour, ceurloplasmin-Cu does not react. The direct reacting copper (i.e., amino acids and albumin-bound) is present to the extent of about 90 micro-grams/100ml.

Copper deficiency has been observed in infants receiving only milk. The prime manifestation is microcytic normochromic anemia. As copper favours iron transport, hypoferremia may occur due to Cu deprivation. In continued severe deficiency, rats and pigs exhibit rickets-like syndrome as well as neurological disturbances. Chronic deficiency causes anestrus in female rats. In skin diseases like vitiligo, low levels of Cu have been reported.

One of the well known though rare diseases of copper metabolism is Wilson's disease, characterised by hepato-lenticular degeneration and, in advanced stage, cirrhosis of liver. It is an inherited disease. The salient features are accumulation of large amounts of copper in the liver and kidneys as well as in the lenticular nucleus of the corpus striatum of the brain and low levels of total copper (i.e., ceruloplasmin plus albumin and amino acids-bound Cu) in serum. Though the total copper is low as ceruloplasmin is low, the direct reacting copper may be comparatively high. There is an increased urinary excretion of copper and amino acids. The clinical picture is dementia and liver failure.

Various theories have been put forward to account for Wilson's disease. One of the postulates is that there is a pronounced increase in the intestinal absorption of copper resulting in accumulation of copper in tissues and high excretion in urine. The renal tubules loaded with copper may produce amino aciduria and at times glycosuria. Another view is that ceruloplasmin synthesis is at fault, or there is a deficient incorporation of copper into ceruloplasmin, leaving the copper unattached and combining in an abnormal way with albumin in large amounts. Consequently, the direct reacting copper in serum may be comparatively high in Wilson's disease though serum total copper and ceruloplasmin are low or normal. Yet another view is that the condition may be due to subnormal intestinal elimination of copper. Recently, a new protein in the liver has been isolated which binds copper in Wilson's disease to a greater extent than in normal persons. It is named cuprothionein.

Copper toxicity leads to blue-green diarrhoeal stools and saliva.

Zinc

Zinc is an essential and integral part of the insulin molecule during its storage in the ß-islet cells. But once released as a hormone, it need not have zinc bound to it. During the preparation of insulin from pancreatic extracts, the addition of zinc has been found helpful to purify and crystallise insulin. Long-acting insulin preparations like protamine Zn-insulin have zinc contained in it.

Zinc is secreted by the pancreatic juice and is found in the eye, tooth and testes. Significant quantities of zinc are excreted in sweat. Zinc is a part of the enzymes, carbonic anhydrase, carboxy peptidase, alcohol dehydrogenase, lactate dehydrogenase, glutamate dehydrogenase and alkaline phosphatase. The zinc-protein 'gusten' of saliva plays a role in taste. In the intestinal lumen, there is a zinc binding protein secreted by the pancreas. After absorption in the lumen, the binding protein transfers zinc to albumin on the serosal side of the mucosal cell membrane. Copper can interfere with zinc absorption.

High phosphate, calcium and phytates in the diet decrease zinc absorption.

In the liver, zinc is bound to a specific protein, metallothionein, as for Cu. In the erythrocytes, it is chiefly present in carbonic anhydrase.

Dietary deficiency of zinc is not likely due to its wide distribution in sea foods, liver, wheat germ, yeast and lettuce. However, zinc deficiency in man can occur due to consumption of cereals rich in phytic acid. This causes poor growth and hypogonadism during adolescence. Zinc deficiency is unlikely in those taking a diet not unduly rich in cereals. In zinc deficiency, there is poor wound healing. Zinc deficiency may occur in *acrodermatitis enteropathica*, a condition associated with dermatologic, ophthalmologic and intestinal disturbances and also hypogonadism. Growth retardation and decreased size and functions of the male genitals also occur. Zinc deficiency occurs in some parts of Iran, and hypogonadism and impotence are common there. Secondary zinc deficiency may occur in untreated diabetes mellitus and alcoholic cirrhosis. In the latter condition, there is zincemia and urinary excretion is very high. Zinc gets drained from the body.

In zinc deficiency, the erythrocyte carbonic anhydrase levels are low.

Deleterious effects on growth due to excess of calcium (para-keratosis) are alleviated by zinc. In this condition, there is anorexia, nausea and vomitting.

The daily requirement of zinc is less than 5 mg.

Manganese

A specific action of manganese in bone formation has been postulated. It is required for growth. Mn deficiency causes sterility in rats. Mn activates a host of enzymes like arginase, phosphoglucomutase, hexokinase, isocitrate dehydrogenase, superoxide dismutase and decarboxylases. The Mn^{2+} containing enzymes are hydrolases, kinases, decarboxylases and transferases. Mn is specific to arginase, peptidase and succinate dehydrogenase. The serum phosphatase activity is elevated in Mn deficiency. Mn has also a role in hemoglobin formation. The mammalian erythrocytes contain Mn in one or more porphyrins. Mn occurs in high concentrations in the mitochondria.

Mn is a necessary factor for the action of glycosyl transferases responsible for the synthesis of oligosaccharides, glycoproteins, etc. A deficiency of Mn reduces appreciably the synthesis of oligosaccharides and thereby that of glycoproteins. Perosis or slipped tendon disease develops in the chick due to the reduction of the mucopolysaccharide contents of epiphyseal cartilage.

A $beta_1$ globulin of plasma binds Mn as transmanganin. Mn absorption is inhibited by iron. Mn is universally distributed in plant and animal tissues, nuts, cereals and vegetables. Tea is exceptionally rich in Mn. The daily requirement of Mn is 3–9 mg.

Iodine

Iodine is required for the biosynthesis of the thyroid hormones, thyroxine (tetra-iodo thyronine. T_4) and tri-iodo thyronine (T_3).

Though iodine is needed for metabolism, only iodides are available in human diet. The body can however oxidise iodide to iodine. The thyroid gland is chiefly concerned with the uptake of iodine for the synthesis of the thyroid hormones.

The small intestines absorb the dietary inorganic iodide which is carried in the plasma in loose combination with proteins. About one third of the ingested iodide is taken up by the thyroid gland while the rest is eliminated by the kidneys. The thyroid stimulating hormone (TSH) of the anterior pituitary and the TSH regulatory factor (TRH) from the hypothalamus control the uptake of iodide by the thyroid gland which is an active transport. In the gland, iodide is oxidised by the thyro-peroxidase system to 1^+. This step requires H_2O_2 and is also controlled by TSH.

The protein of the thyroid, i.e., thyroglobulin, has 115 tyrosine residues per molecule. A step-wise iodination occurs, first in position 3 of the benzene nucleus of tyrosine forming mono-iodo-tyrosine, and then in position 5 forming di-iodo-tyrosine.

Two molecules of di-iodo-tyrosine get coupled to give tetra-iodo-thyronine (T_4) or thyroxine. One molecule of di-iodo-tyrosine can also couple with one molecule of mono-iodo-tyrosine to give tri-iodo-thyronine (T_3). But this takes place only to a limited extent. T_3 can form from T_4 also by de-iodination. About 30 per cent of tyrosine residues in thyro-globulin are iodinated. Thyroglobulin, after the formation of T_4 and T_3, is enzymatically hydrolysed and releases T_4 and T_3. TSH controls this process, too.

In the plasma, T_4 and T_3 are carried bound to the protein albumin, thyroxine binding globulin (TBG) and thyroxine binding pre-albumin (TBPA). About 0.05 per cent of the circulating thyroxine is unbound. It is only the free T_4 and T_3 which are metabolically most active but there is an equilibrium between the free and the bound forms. T_4 accounts for the major form of organic iodine and about 15 per cent is probably T_3. Nevertheless, T_3 is possibly the major thyroid hormone being more loosely bound to serum proteins, and hence penetrating the cells much more quickly than T_4. The onset of the action of T_3 as well as its degradation are very rapid.

Only about one-third of the maximum thyroxine binding capacity is utilised by a normal subject but it is much more in a hyperthyroid patient. The thyroxine bound globulin levels are elevated in pregnancy and after administration of estrogens. They are lowered in the nephrotic syndrome due to proteinuria and also after administration of androgens or anabolic steroids. There is a reciprocal relationship between TBG and TBPA. TBG is, however, the stable reservoir for T_3 and T_4. T_3 and T_4 undergo deamination and decarboxylation respectively to tri-iodo-thyro-acetic acid and tetra-iodo-thyro acetic acid, which possess a potent hypocholesterolemic effect without any side effects. De-iodination takes place in the peripheral tissues and saliva. T_4 is also converted to reverse T_3 in the peripheral tissues.

The iodide set free by de-iodination in the peripheral tissues and saliva is eliminated in the urine. The liver partly conjugates thyroxine, mostly with glucuronic acid, before excretion via, the bile. (For thyroid disorders and their clinical evaluation refer 'Thyroid Function Tests'.)

Cobalt

Cobalt is a component of vitamin B_{12} which contains about four per cent of the element. Elemental cobalt of the diet can be converted to cobalamin by the intestinal bacteria. In experimental rats, excessive administration of cobalt can induce polycythemia. It has also been demonstrated that cobalt stimulates erythropoietin production in animals.

Cobamide coenzymes are of considerable importance as is seen in the role of DBC cobamide in the prokaryotes in the reduction of ribo-nucleotides into deoxyribonucleotides. Methyl B_{12} converts homo-cysteine to methionine and relieves the 'folate trap' from N^5 methyl FH_4. Otherwise, part of the folate may not be available for one-carbon transfers.

Cobalt is an activator of the enzyme phospho gluco-mutase and glycyl glycine peptidase. Cobalt deficiency has not been demonstrated in non-ruminants.

Selenium

Selenium is an integral component of glutathione peroxidase, an enzyme which is an intracellular anti-oxidant. Selenium, thus, constitutes a second line of defence against peroxidation. It is known that vitamin E and selenium prevent peroxidative damage to cellular and sub-cellular elements and chiefly the membrane.

Selenium spares vitamin E requirements in three ways: 1) Selenium is required for normal pancreatic function and thereby digestion and absorption of lipids including vitamin E; 2) it is a component of glutathione peroxidase; and 3) in some unknown way, it aids retention of vitamin E in blood plasma lipoproteins. Conversely, vitamin E appears to reduce the requirement of glutathione peroxidase and thereby selenium. A selenium containing organic compound (Factor 3), isolated from yeast, is synergistic in action with vitamin E as well as cystine in protection against hepatic necrosis. In this process, vitamin E and the Se-containing glutathione peroxidase are anti-oxidants in cells and cysteine is a component of glutathione.

Selenium intake depends on the nature of the soil in which foodstuffs are grown. In China, some rural areas are reported as selenium deficient zones. Deficiency of selenium causes dilatation of the heart and congestive heart failure.

Human beings living in selenium-rich soil are prone to selenium toxicity-selenosis, i.e., selenium-poisoning, due to the replacement of sulphur of cysteine by selenium. An early symptom of selenium toxicity is garlicky breath due to dimethyl selenide in expired air. Administration of halogenated aromatic hydrocarbons is believed to disengage selenium from its association with cysteine and thereby ameliorate the symptoms of such toxicity.

Molybdenum

Molybdenum is required for the function of the metallo-enzymes xanthine dehydrogenase, aldehyde oxidase and sulphite oxidase. Hexavalent soluble forms of Mo are absorbed well across the intestines. Urine is the major route of molybdenum excretion. There is some evidence that Mo can interfere with copper metabolism by diminishing the efficiency of copper utilisation and perhaps even copper mobilisation from tissues.

The food content of Mo is highly dependent upon the soil in which the food-stuffs are grown. Mo deficiency has not been observed in humans or in any other species.

Fluorine

Bones and teeth contain this element. It is required in traces for the development of bones and teeth and for the prevention of dental caries, though large amounts beyond one part per million in water can cause fluorosis involving mottling of the enamel. In this condition, the enamel becomes stratified, has dull white patches and brown stains and shows pits. Mottled teeth have increased levels of fluoride in both the dentine and enamel. Fluoride is a well known inhibitor of enolase and thereby prevents glycolysis. It is used as an anti-coagulant during collection of blood for the determination of blood sugar.

Chromium

Chromium is absorbed in the small intestines perhaps along with zinc. It is transported bound to transferrin, and is excreted chiefly in the urine.

Chromium is found to play a functional role in the regulation of glucose metabolism, as a potentiator of Insulin action. Traces of Cr are shown to bring out efficient binding of insulin to its receptor sites. The trivalent form of chromium (Cr^{3+}) can improve the glucose tolerance of individuals suffering from protein-calorie malnutrition. Chromium is also said to be important in the metabolism of plasma lipoproteins.

Excess of Cr^{3+} is toxic and the hexavalent form is more toxic than the trivalent form. Chronic occupational exposure to chromate dust carries an increased risk of lung cancer.

Brewer's yeast is rich in chromium and most grain and cereal products contain significant quantities. Appreciable amounts of chromium are contributed to the diet by cooking in stainless steel containers.

20 Methods of Study of Metabolisms

Various methods have been employed for the study of metabolic reactions during the past 50 years, But since the advent of isotopic tracers, there has been a tremendous advancement in this field, as the label can be traced in the intermediate metabolites and the final products in the intact animal and isolated tissues.

In one of the methods, the constituents of blood, urine and tissue are investigated and the metabolism is studied. For example, glucose level in blood will be high about one hr after taking about 100 g of glucose. But if glucose is estimated in blood after taking an injection of insulin, there will be a decrease. If amino acids in blood are estimated immediately after a protein-rich meal they will record an increase. A high-protein diet will cause excretion of higher amounts of urea in urine showing that urea is a metabolic waste of proteins in man. By analysing tissue constituents, one can demonstrate that glucose is converted into glycogen. If, after giving plenty of glucose to an experimental animal, the liver is removed and analysed for glycogen, there will be high glycogen content. But on fasting, liver glycogen values will be negligible showing that liver glycogen is used up during fasting and is labile.

The analysis of the constituents can be done using colorimetric procedures, chromatography, electrophoresis, counter current distribution, flame photometry, ultra-centrifugation, tracer technique and radio autography. In addition, microscopy including phase contrast and electron microscopy and biochemical cytology are also very popular in the study of the distribution of substances including enzymes in various cellular compartments like the nucleus, cytosol, mitochondria, lysosomes, etc.

Metabolism can be studied at six levels:

1) The highest level is the whole animal in which all the cells of tissues are intact, operate in an integrated manner and are subject to hormonal influence and feed-back mechanisms.

2) The second level, slightly lower than the first, is the perfusion experiments while the organ is still intact. By giving injections of drugs, vitamins, hormones, etc., the metabolism is studied.

3) The third level is to investigate on isolated organs like the liver and small intestines. Here the nervous and hormonal controls are cut off.

4) The fourth level is to use tissue slices in which most of the cells of the concerned tissue are intact.

5) The fifth level is to prepare the homogenate of the tissue in a homogeniser. In this, there is complete disintegration of the cells. Enzyme activities are investigated in many cases in homogenates.

6) Lastly in the sixth level, the metabolite or enzyme can be isolated, purified and investigated.

Each level of study has its own advantages and limitations. The tracer technique can be employed for all the levels with precision and advantage.

Methodologies

Different methodologies are employed in the study of metabolism, e.g., in an intact animal. They are:

1) Estimation of the constituents in blood, urine and other biofluids.

2) Arterio-venous difference in blood constituents for the assessment of the function of an organ in respect of utilisation of oxygen, glucose, free fatty acids, etc.

3) Balance studies by the intake and excretion. For example, if the amounts of nitrogen in food and nitrogen excreted are known, it is possible to find out whether there is positive N-balance (growth), nitrogen equilibrium, or negative N-balance (tissue waste).

4) The respiratory quotient, i.e., ratio of CO_2/O_2, gives information as to which of the dietary constituents is metabolised. The brain has the RQ of 1 indicating that it uses glucose as its fuel, as the RQ for carbohydrates is 1.

5) By removal of the endocrines like the pituitary, adrenals and thyroid, the effect of hormones could be studied, or the hormones can be injected in different doses to study their effects in the animal.

6) Organs like the liver and pancreas can be removed to study the site of metabolism or the deficiency effects of these organs.

7) Destroying the cells of an organ, e.g., by cirrhotic diet to induce cirrhosis of liver, giving toxic substances like CCl_4 for causing necrosis of liver cells, or injecting alloxan to destroy the beta cells of the islets of Langerhans of pancreas, the role of liver and pancreas in metabolism can be investigated. Similar studies can be extended to other tissues.

If tissue slices of homogenates have to be investigated, Warburg's apparatus or Thunberg's tubes can be used.

8) The effect of vitamins and coenzymes in metabolism can be followed by cutting of the vitamins from the diet or by giving anti-vitamins. If enzymic influence on metabolism is to be studied, it has to be done at suitable dilution, temperature and pH, as these factors decide enzyme catalysis.

9) Tracer technique using radio-active as well as stable isotopes is very popular, since the living organism uses the isotopic atoms (label) as it does the metabolites and as complex elements always exhibit a constant isotopic composition in their natural state. The nutrient or the metabolite is diluted with a suitable radio-active compound with a label or stable isotope and traced during and after reactions. Depending on the presence of the tracer and its quantity, the qualitative and quantitative changes during metabolism can be followed.

The common radio-isotopes are ^{14}C, ^{32}P, ^{131}I, ^{60}Co, ^{59}Fe, ^{3}H, while stable isotopes are ^{13}C, ^{13}O, ^{15}N and ^{2}H. The instruments used to trace the radio-active isotope are electronic counters like the Geiger Muller counter and Scintillation counters while the stable isotopes are traced by the Mass Spectrograph.

In studies on carbohydrate metabolism, ^{14}C glucose, ^{14}C acetate, ^{14}C Na HCO_3 have been extensively used. It has been shown that glucose is converted to CO_2, glycogen, fat and even cholesterol, *in vitro* and *in vivo* using tissues like the liver. Even CO_2 normally considered a waste, can be shown to be utilised in the body. The extent of the Embden Meyerhof-TCA and HMP pathways in different tissues like the liver and kidneys, RBC and mammary glands has been followed using 6-^{14}C glucose and 1-^{14}C glucose. Again, though citric acid is a symmetrical molecule, the 2-CH_2 COOH groups are treated separately by the enzyme and CO_2 is eliminated from the —CH_2COOH which comes from oxalo-acetate while -CH_2-COOH from acetyl CoA remains intact. Fixation of CO_2 in photosynthesis has been proved by using $^{14}CO_2$.

In lipid metabolism, use of ^{14}C acetate has shown that it is the precursor for cholesterol biosynthesis and also of fat. The utilisation of CO_2 in extra-mitochondrial lipogenesis in the formation of malonyl CoA from acetyl CoA, and the subsequent release of the same (CO_2) from the acyl malonyl enzyme, has been established. Using ^{14}C and ^{15}N leucine, studies have been made on the synthesis and fate of lipoproteins. ^{14}C palmitate and ^{14}C essential fatty acids have been used to find out the source of fat in fatty livers.

^{14}C and ^{15}N leucine and other amino acids and deuterium labelled compounds have been used extensively in protein metabolism. It has been shown that there is a dynamic equilibrium between plasma proteins and tissue proteins and that the body proteins are in a state of flux. The extent of synthesis of proteins can be followed using labelled leucine. Fixation of ammonia into liver glutamate and many amino acids (except lysine) has been shown using ^{15}N ammonium acetate. Studies on transmethylation have shown that $-CH_3$ is transferred intact in such reactions. Again, in proteins like collagen, proline and lysine have been shown to undergo hydroxylation only after their incorporation into the protein and to form hydroxy proline and hydroxy lysine.

In nucleoprotein metabolism using radio-active thymidine (tritiated (^3H) thymidine), it has been shown that DNA replicates by the semi-conservative process. ^{32}P-labelled nucleotides have shown the nature of DNA in the sequence studies of Kornberg (known as Nearest Neighbour analysis) and in the recent studies on the sequence of DNA by synthesis under different conditions, employing poly acrylamide gel disc electrophoresis and auto-radiography. That DNA is the template for the formation of RNAs like messenger RNA, has been proved.

Under purine metabolism, the source of each atom in *de novo* synthesis of purine in the body is known. For instance, the C atom No. 6 comes from respiratory CO_2. Dietary adenine and not guanine may be used for the synthesis of nucleic acid in the body.

Glycine has been shown to be used in toto for the synthesis of heme.

^{45}Ca, ^{65}Zn, ^{60}Co, ^{59}Fe, ^{131}I, ^{35}S, ^{32}P, ^{51}Cr and ^{24}Na have all been used in the study of mineral metabolism. The influence of amino acids on the absorption of dietary Ca and P has been studied by using ^{45}Ca. Effects of the para-thyroid hormone and calcitonin, on bones have been studied using radio-active calcium. Considerable significance is attached to ^{131}I uptake by thyroid for thyroxine synthesis, as this has helped in the diagnosis of thyroid disorders.

Use of ^{59}Fe has shown that Fe is stored in the liver, bone-marrow and spleen. It has helped considerably in studying different kinds of anemias.

Another field in which labels have been used is enzymology. Elucidation of active sites of enzymes like chymotrypsin and choline esterase has been made using ^{32}P compounds. Serine was found to be the active centre. The existence of two types of dehydrogenases, alpha and beta, was found by using ^2H (deuterium)-labelled compounds. While alcohol dehydrogenase is an alpha dehydrogenase, beta hydroxy steroid dehydrogenases are beta. The handling of a symmetrical molecule like citric acid or glycerol in an unsymmetrical way by enzymes and the three pronged attack of the substrate on the enzyme have been indicated by tracer technique.

INHERITED DISEASES OF METABOLISM

The term *inborn errors of metabolism* was coined by Garrod in 1908, for four rare and relatively harmless diseases – albinism, alkaptonuria, cystinuria and pentosuria. Subsequently, the list has been enlarged and consists of more than thirty diseases. As the name itself indicates, such diseases occur even during birth are thus inherited. Hence the term 'inherited errors of metabolism' is employed at present. Inherited errors of metabolism lead to inherited diseases.

Metabolism comprises the anabolic and catabolic changes in the body. Whether the metabolic changes are exergonic or endergonic, most of them have to be catalysed by the enzymes. If the metabolic process including the transport of substances across cellular membranes are to take place as they should, the proper enzymes must be produced in the body without any errors of omission or commission and they should also be available in adequate amounts.

Due to hereditary absence of deficit of a specific enzyme involved in a definite reaction, there is a block in the course of biochemical reactions leading to metabolic abnormalities which are present throughout life and handed over to the progeny. The absence or deficiency of an enzyme will cause an abnormal accumulation of the intermediate products of metabolism in the body and increased excretion in urine as such or their degradation products. Some of the intermediates could be toxic, which explains the clinical mainfestations of the disease.

One gene-one enzyme (one polypeptide) hypothesis

How is that the diseases due to the absence or deficiency of enzymes are inborn and inherited? This is because an enzyme usually controls one step in a sequence of reactions. Beadle and Tatum put forth their theory 'one gene-one enzyme hypothesis', that one gene controls the synthesis of a single enzyme. It is now known that enzymes being proteins, whose synthesis is governed by DNA, the aberration of the enzyme proteins will be definitely brought about by mutations in the cistrons of DNA (refer 'Biochemical Genetics'). Thus diseases due to the absence or deficiency of enzymes are due to defective genes (DNA), hence cogenital diseases cannot be cured. The patient may be a heterozygote (one gene in the allele affected) of homozygote (both the genes in the allele affected). The defective gene may be present in the autosomal chromosomes (22 pairs) or in the sex chromosome, and be recessive or dominant. Phenylketonuria is inherited as an autosomal recessive disease whereas acute intermittent porphyria and hereditary spherocytosis are autosomal dominant conditions. Lesch-Nyhan syndrome and gout are sex-linked recessive conditions while vitamin D-resistant renal rickets and glucose-6-phosphate dehydrogenase deficiency are sex-linked dominant conditions. Luckily many disorders are due to a recessive gene and a heterozygote, autosomal recessive will apparently be normal.

Recently, the 'one gene-one enzyme hypothesis' has been replaced by the 'one gene-one polypeptide theory' as it has been proved that it need not be the whole protein that is controlled by one gene but only a polypeptide portion of protein is controlled by the gene. For example, in hemoglobin S and hemoglobin A, the protein is globin containing two alpha and two beta chains. Two genes are controlling the production of the alpha and beta chains proving the one gene-one polypeptide theory. Very recently, it has been shown in $\emptyset \times 174$ bacteriophage that one gene can control the synthesis of more than one polypeptide.

In addition to defect in the synthesis of enzymes, the proper proteins themselves cannot be synthesised by defective DNA through suitable transcription and translation. Due to alteration in the *quality* of *enzymes* or *proteins* there could be some inherited errors. In other cases, it could be on the *quantity* and not the quality of protein synthesised. This might be due to defective regulation of protein synthesis by the regulator and operator genes. Thirdly, there are some diseases which are due to impaired transport, in which case it is likely that some permeases are involved.

Disorders affecting amino acid metabolism

The metabolism of phenyl alanine is of considerable historic, academic and clinical importance because it offers the site for four inherited errors of metabolism.

1. *Phenyl alanine metabolism,* a) Phenylketonuria (hyper phenyl alaninemia, Type I) is due to the absence of the enzyme phenyl alanine oxidase (hydroxylase). As a result, there is accumulation of large amounts of phenyl alanine and phenyl pyruvic acid in blood and excretion of the same in urine. Such patients have mental retardation. The urine gives a green colour with $FeCl_3$. This is an autosomal recessive disease.

b) Alkaptonuria: This is due to the absence of homogentisic oxidase leading to continuous excretion of several grams of homogentisic acid in urine. The sample of urine becomes dark on standing and finally turns black. Alkaptonuria is inherited as an autosomal recessive trait.

c) Tyrosinosis is a rare disease due to the deficiency of parahydroxy phenyl pyruvic oxidase or tyrosine transaminase, or fumaryl aceto – acetate hydrolase.

d) Albinism is caused by the absence of tyrosinase in the melanocytes and hence the individual appears 'bleached'. In this condition, melanin is not synthesised in the melanocytes and affects the skin, hair, sclera, choroid, etc.

2. *Branched chain amino acid metabolism (Maple syrup urine disease):* This is due to the lack of the enzyme that decarboxylates the keto acids of branched chain amino acids, valine, leucine and isoleucine. The urine has the odour of maple sugar. The patients present feeding problems and may not live long. When milk is fed to such children, all of it is vomited. Brain damage may occur during the first week of extra-uterine life.

3. *Tryptophan metabolism (Hartnup's disease):* This is an inborn error associated with the metabolism of tryptophan. One of the clinical symptoms is mental retardation. The urine of patients with Hartnup's disease contains increased amounts of indole acetic acid and tryptophan. The disease is named after the family in which it was discovered first and is due to a deficiency of tryptophan dioxygenase in the liver. It is also suggested that it could be due to defective indole transport and a consequent impairment of tryptophan metabolism.

Disorders of protein metabolism

1) Clotting diseases: There will be abnormal bleeding due to the absence of one or more of clotting factors.

(i) Hemophilia A occurs due to deficiency of Factor VIII or the anti-hemophilic globulin (AHG). It is also known as Royal hemophilia as it was found in the Royal family in Great Britain for some time. This is transmitted by a sex-linked recessive gene.

(ii) Hemophilia B or Christmas disease is due to the absence of Factor IX or the Christmas factor.

(iii) von Willebrands' disease is due to the absence of the Willebrand factor along with the deficiency of Factor VIII. Here, the defect is platelet adherence.

(iv) Afibrinogenemia is due to the absence of fibrinogen in plasma.

2) Wilson's disease or hepato-lenticular degeneration, which is autosomal recessive, is characterised by decreased synthesis of ceruloplasmin or defective incorporation of Cu in ceruloplasmin or both. There is also a decrease of cytochrome oxidase. Copper gets deposited in tissues, chiefly in the liver and brain.

3) Agamma globulinemia (sex-linked recessive) is a disease with absence of gamma globulin in plasma and failure to develop antibodies on antigenic stimulation.

4) Analbuminemia is another disease, though not a very serious one.

5) Abnormal hemoglobins (hemoglobinopathies): a) Sickle-cell anemia: In this, the normal adult Hb (Hb A) is replaced by Hb S. Sickling of the red cells is caused by crytallisation of Hb S when oxygen tension is low into tactoids which distort the cell into a sickle-cell shape. As a result, there is increased fragility of RBCs causing hemolytic anemia. Hb S and Hb A are different in that valine is present in the place of glutamic acid in the beta chain of globin in Hb S. The disease is autosomal recessive.

Sickle-cell anemia is a public health problem in the black African population. Patients with sickle cell anemia are homozygous for an abnormal gene located on one of the autosomal chromosomes. Hence, offsprings who receive the abnormal gene from one parent and its normal allele from the other have the sickle-cell trait only and are asymptomatic, unlike the homozygotes who receive the defective genes from both parents. In certain parts of Africa the frequency of the sickle-cell gene is as high as 40 per cent. The sickle-cell trait is however not manifested clinically as those having it are heterozygotes. They will have both Hb S and Hb A.

b) Thalassemias are a group of anemias in which the rate of synthesis of Hemoglobin A is diminished. There would be Hb F. This is also a hereditary disorder transmitted as autosomal recessive. Thalassemia major and thalassemia minor refer to the homozygous and heterozygous states, respectively.

c) Glucose-6-Phosphate dehydrogenase deficiency: Coming under the group of diseases concerning hemoglobin and red cells is the deficiency of glucose-6-phosphate dehydrogenase of the red bood cells. This may be present clinically as primaquine sensitivity or favism. Primaquine which is an anti-malarial drug, when given to persons with glucose-6-phosphate dehydrogenase deficiency, induces massive hemolysis involving a fall in the Hb level by about one-third in one week. Favism is an extensive hemolysis associated with the ingestion of the Vicia fava bean during the spring months.

Due to the primary defect of glucose-6-phosphate dehydrogenase enzyme deficiency in red cells, the pentose phosphate pathway does not operate efficiently and there is not sufficient NADPH available for the reduction of G-S-S-G to GSH and the reduction of methemoglobin to normal hemoglobin. This results in damage to the red cell membrane and hemolytic anemia. G-6-PD deficiency is inherited as a sex-linked trait. Female heterozygotes have two types of RBCs, one with normal G-6-PD and the other with deficiency. Hence the deficiency is not expressed and the females will only be carriers. But in male homozygotes the deficient red cells will be more than 50 per cent and hence the defect is manifested only in males. G-6-PD deficiency among black Americans is nearly 11 per cent. Higher frequencies occur in populations living in malaria-infested areas and in communities where consanguineous marriages have been in practice for several generations.

d) Methemoglobinemia is associated with increased amounts of methemoglobin in the RBCs due the deficiency of the enzyme required for converting methemoglobin to reduced hemoglobin.

Disorders carbohydrate metabolism

1. Galactosemia: Galactose is not epimerised to glucose in liver for want of galactose-1-phosphate uridyl transferase. Hence galactose mixes with systemic blood and is excreted in urine. This disease is autosomal recessive. Mental retardation, wipdespread tissue damage (cirrhosis of liver) and cataract of the eye are the manifestations.

2. Glycogen storage diseases (Glycogenoses): The term 'glycogen storage diseases' describes a group of inherited disorders characterised by the deposition of either abnormal glycogen or huge quantities of glycogen in tissues. There are many types.

Type I glycogen storage disease — Glycogenosis — is also known as Von Gierkes' disease. In this, the enzyme which is deficient is glucose-6 phosphatase. Both the liver cells and the cells of renal convoluted tubules are loaded with glycogen. Ketosis and hyperlipemia are also present.

Very recently, Type I has been divided into Ia, Ib and Ic due to the congenital deficiencies of glucose-6-phosphatase, permease and translocase, respectively.

Type II is known as Pompe's disease. It is due to a deficiency of lysosomal 1,4-and 1,6-glucosidase (acid maltase) which is needed to degrade glycogen.

Type III is the limit dextrinosis, Forbe's or Cori's disease manifest in the accumulation of limit dextrins due to the deficiency of the de-branching enzyme.

Type IV is also known as amylopectinosis and Andersen's disease with the hereditary deficiency of the branching enzyme. The result is an accumulation of glycogen having a few branch points only, i.e., qualitatively different glycogen.

The absence of muscle phosphorylase (myophosphorylase) results in Type V glycogenosis popularly known as McArdle's syndrome. Patients exhibit a marked diminished tolerance to exercise, as little or no lactate is detectable in their blood after the exercise.

Type VI is due to a deficiency of liver phosphorylase and, Type VII (Tarui's disease) with a deficiency of phospho-fructokinase in the muscle and erythrocytes.

3) Essential pentosuria and fructosuria: Pentosuria is harmless excretion of excess of L.xylulose in urine due to the deficiency of L.xylulose reductase. Fructosuria is due to deficiency of liver fructokinase.

4) Diabetes mellitus: This is an autosomal recessive state caused by the production of decreased quantity of insulin. This disease is an example of an inborn error brought about by the deficit in quantity of protein synthesised.

5) Gargoylism: A well known type of acid muco-polysaccharidosis is gargoylism. Different sub-types are encountered like Hurler's syndrome and Hunter's disease. The causative enzyme deficiencies pertain to α-L-iduronidase, sulphoiduronate sulphatase, heparan sulphate sulphamidase and lysosomal α-glucosidase. The manifestations include grotesque facial features, bone changes like deformed extremities, stiff joints as well as clouding of the cornea and mental retardation. The salient biochemical finding is enhanced excretion of glycosaminoglycans in urine.

Disorders of lipid metabolism

1) Familial hyper-lipoproteinemias: There are as many as five types of familial hyperlipoproteinemias with two sub-types of Type II. In these diseases, there is an elevation of one or more of lipoproteins in blood like chylomicrons, LDL, VLDL and VLDL remnants (IDL),due to the hereditary deficiency of enzymes. Some of these types could cause atherosclerosis.

2) Gaucher's disease: In this disease, the cerebroside content of the cells of the enlarged spleen, liver and lymph nodes is increased. Glucose replaces galactose in the cerebroside.

3) Niemann Pick's disease: Cells of the liver and spleen become foamy and there is swelling of the ganglion cells. Sphingomyelin gets accumulated in these organs.

4) *Tay-Sach's disease* is an inherited disorder of ganglioside breakdown. Gangliosides are lipids present in the grey matter in the brain and CNS and are continuously broken down by sequential removal of terminal sugar residues by the ß N-acetyl hexosaminidase A enzyme. The congenital absence of this enzyme in Tay-Sach's disease results in the accumulation of the GM_2 ganglioside in the brain and CNS. Tay-Sach's disease is inherited as an autosomal recessive trait and occurs only in infants. It is usually fatal during infancy itself, mental retardation, blindness, etc. preceding the fatal episode.

5) *Refsum's disease:* An abnormal fatty acid called phytanic acid, present in vegetable oils is normally metabolised in the liver by $\propto$-oxidation by the enzyme, phytanic acid $\propto$-hydroxylase. This enzyme is congenitally absent in patients with Refsum's disease, and therefore phytanic acid accumulates in the liver and in other organs including the retina and also the CNS. If it accumulates in the retina, it causes retinitis pigmentosa, when retinal metabolism is impaired by inhibition by this fatty acid. Refsum's disease is an autosomal recessive condition.

PORPHYRIAS

Porphyrias are diseases of heme metabolism. In this type of disease, the intermediates in heme synthesis might be produced in abundance and not utilised. The porphyrins and/or porphyrin intermediates excessively produced are excreted in large amounts in urine and feces and may also be deposited in the teeth and bones. As porphyrins absorb light at definite wavelengths, there may be cutaneous manifestations due to photosensitivity.

Hereditary porphyrias: Hereditary porphyrias are broadly classified as follows:
1) Congenital erythropoietic porphyria
2) Hepatic porphyria
3) Others like porphyria cutanea tarda, proto-porphyria

1) Congenital erythropoietic porphyria: In this disease, there is excessive porphyrin formation in the developing red cells of bone marrow, increased erythropoiesis and splenomegaly. The porphyrins so produced get deposited in the liver, spleen, bones, teeth and skin. As porphyrins absorb light, there is cutaneous photosensitivity early in life with a tendency for cutaneous fragility and hemolysis. Continued exposure causes vesicles; nose and ears are deformed; there is mutilation of hands, face, etc. Bones and teeth have a red tinge, urine has large amounts of uroporphyrin I and coproporphyrin I, concomitant with increased production of Type I porphyrins; relatively larger amounts of porphobilinogen and amino levulinic acid (AmLev) than normal are also excreted. However, in the hepatic type there is a much greater amount of porphobilinogen and AmLev.

The enzymes deficient are uroporphyrinogen I synthase and/or uroporphyrinogen III cosynthase. AmLev synthase is induced.

The disease is autosomal recessive and very rare. Patients with congenital erythropoietic porphyria should be protected from exposure to light.

2) Hepatic porphyria: This type is due to liver dysfunction, and is also hereditary. An important type of hepatic porphyria is Intermittent Acute Porphyria (IAP).

The disease is associated with acute attacks of abdominal pain, usually colicky, peripheral neuropathy, convulsions, vomiting, constipation and cardio-vascular disturbances. There is no cutaneous manifestation. Type I and Type III porphyrins are excreted in urine. The amounts of porphobilinogen and AmLev excreted are quite high compared to the erythropoietic type. Urine darkens on standing due to excessive porphobilinogen.

The enzyme deficient is uroporphyrinogen 1 synthase. AmLev synthase is induced. Liver function tests are abnormal in IAP. Drugs like barbiturates and steroid hormones precipitate relapse and hence are contra-indicated. The disease is autosomal dominant.

3) Others: a) Porphyria cutanea tarda is hepatic and cutaneous and is more common than the other two. Though the disease is hepatic, there is cutaneous photo-sensitivity and is hence called 'cutanea'. Again, it appears late in life, hence 'tarda' (means late). There could be exceptions and a case of a five year old boy is reported. Porphyria cutanea tarda is associated with some form of liver injury, particularly due to alcohol or iron over-load. Not all alcoholics exhibit this condition. There is increased urinary excretion of Type I and III uroporphyrins. There is no increase in AmLev and porphobilinogen. The disease is not identical with erythropoietic porphyria as there is no splenomegaly or red bones and teeth, no increase of AmLev and porphobilinogen and there is excretion of Type III. At the same time, it is associated with liver injury and, strangely with cutaneous photo-sensitivity. It is different from IAP as IAP has no cutaneous sensitivity. There is also no increased excretion of porphobilinogen and AmLev.

Porphyria cutanea tarda is due to a partial deficiency of uroporphyrinogen decarboxylase and is autosomal dominant.

b) Protoporphyria or erythropoietic protoporphyria: In this condition the enzyme deficient (partial) is ferrochelatase. Hence there is increased protoporphyrin IX in erythrocytes. Urine does not have any increased amounts of porphyrins and their precursors.

On exposure to light patients exhibit acute urticaria. The disease is autosomal dominant.

Acquired porphyrias: Porphyrias could also be acquired (toxic). Some authors use the expression porphyrinuria for the acquired (toxic) type. It is due to the toxic effects of lead and heavy metal salts, gammexane, chloroform, etc. which poison the enzyme systems. It is also encountered in liver diseases like alcoholic cirrhosis, in leukemia and pernicious anemia. In the acquired (toxic) type, only about 1 – 4 mg of porphyrins are excreted in urine and they are only coproporphyrins. In the hereditary type, 10 – 100 mg of porphyrins may be excreted in the urine. They are both copro-porphyrins and uroporphyrins in hereditary porphyria.

Acquired (toxic), porphyria (porphyrinuria)	Hereditary porphyria
1) Acquired	Inherited
2) Caused by lead, gammexane, CCl_4, etc., Inhibition of enzymes.	Congenital deficiency of enzymes.
3) Only coproporphyrins	Both copro- and uroporphyrins.
4) 1 – 4 mgs.	10 – 100 mgs.

Gout

Gout is an inborn error of metabolism of nucleic acids and purines characterised by purine over-production and over-excretion. It is inherited as a sex-linked recessive disorder.

The disease is associated with recurrent attacks of acute pain and swelling at the joints. It initially affects only one joint, usually the metatarso-phalangeal joint of the big toe. Later, it become polyarticular. It is infrequent before the age of 40 and very rare in women.

The pathology is due to the deposition of urates, chiefly monosodium urate, in the articular and peri-articular tissues, and in the extra-articular cartilages especially in the ears and the kidneys. There is inflammatory reaction in the synovial membrane. Urates are deposited as tophi in the subchondral bone and synovial membrane. Joints become red, swollen, stiff and painful due to the deposition of sodium urate precipitating the formation of gouty tophi with hyperuricemia and increased excretion of uric acid in urine. Finally, the characteristic changes of osteo-arthritis appear. Between attacks, the patient is free from symptoms.

The deposition of urates is a consequence of hyper-uricemia, i.e., increased levels of urates in the blood. Normal level are 2 – 6 mg per dl. The physical property responsible for deposition is the insoluble nature of urate at the pH of the tissue cellular contents. In addition to elevation of uric acid in the blood, the total amount of uric acid in the body, i.e., miscible pool of uric acid, which is about 1200 mg increases to 2,000 – 4,000 mg in gout without tophi, and to 31,000 mg in gout with tophi. Apart from hyper-uricemia, there may be moderate polymorpho-leucocytosis in the blood.

Hyper-uricemia may be due to various causes. It may be hereditary or acquired. If hereditary, it is due to abnormal enzyme function in purine metabolism. This is referred to as primary metabolic gout. If there is derangement of some other metabolism, causing hyperuricemia, it is referred to as secondary metabolic gout.

Hyper-uricemia may also be due to impaired kidney function with no disturbance in metabolism.

Primary metabolic gout: Primary metabolic gout may occur because of increased biosynthesis of urate due to one of the following causes. It may be noted that the enzyme affected is involved in purine metabolism.

1) Increased activity of PRPP synthetase: This could be due to a) increased Vm of the enzyme, b) decreased Km of the enzyme for ribose-5-phosphate, or c) the enzyme not susceptible to feed-back inhibition by purine nucleotides, AMP ADP, GMP and GDP.

2) A different enzyme, i.e., PRPP glutamine amido transferase, may be affected. Its activity is uncontrolled due to diminished feed-back inhibition from AMP and GMP.

The diminished feed-back inhibition is due to decreased production of AMP and GMP and their di nucleotides in the salvage pathway. For the formation of GMP by the salvage pathway, the enzyme HGPRTase is needed. If the enzyme is deficient, GMP production and thereby the feed-back inhibition on PRPP glutamine amido transferase will decrease. In addition, there is greater availability of PRPP inside the cell. This is because PRPP is a substrate for HGPRTase. In the deficiency of HGPRTase, the concentration of this substrate PRPP increases due to poor utilisation. This PRPP is utilised for increased urate synthesis.

The deficiency of HGPRTase may be partial or total. If it is total, the condition is referred to as the Lesch-Nyhan syndrome.

3) A third cause of primary metabolic gout is a total absence of HGPRTase. This is referred to as the Lesch-Nyhan Syndrome. As described earlier, in a total deficiency of HGPRTase, there will be an increased concentration of intracellular PRPP. This is utilised for increased urate synthesis. In addition, there will be decreased production of GMP and GDP. Hence the feed-back inhibition of PRPP synthetase by GMP and GDP and that due to GMP on PRPP glutamine amido transferase is diminished, with a consequent overproduction of uric acid. Patients with the Lesch-Nyhan syndrome are usually children with manifestations of gout and exhibit cerebral palsy, mental retardation, tendency for self-mutilation and aggressive behaviour. The disease is inherited, sex-linked and is associated with renal calculi and renal failure.

4) Another example is glycogen storage disease, Type I (Von Gierke). In this, due to increased availability of ribose-5-phosphate, there is uric acid over-production. All the above four conditions are hereditary.

Secondary metabolic gout: In some instances, the synthesis may be normal but there may be hyper-uricemia. This is due to extensive degradation of the cells and their nucleic acids. Examples are polycythemia, leukemia, psoriasis, cancer etc. In other diseases also like alcoholism, toxemia pregnancy, diabetic ketoacidosis and hypothyroidism there could be hyper-uricemia. These are acquired.

Gout could also be due to renal dysfunction. In this case, the synthesis or production by degradation of cells might be normal. If there is renal retention, there may be hyper-uricemia. If it concerns only the excretion of uric acid, it is primary renal. This may be due to an elevated renal threshold. The level of serum urate needs elevation to 'flow over the dam' so to speak. If the block is common for many metabolities, it is secondary renal. In glomerulo-nephritis and essential hypertension, there is hyper-uricemia. Primary metabolic gout could also lead to secondary renal involvement.

It has to be understood that all hyper-uricemic conditions need not lead to gout and all cases of gout to gouty arthritis. If it does, it may be acute or chronic. Gouty arthritis is differentiated from the more common rheumatoid arthritis which also involves joints but is not associated with hyper-uricemia.

The treatment of gout is based on two principles: 1) to dislodge the crystals of urate from the joints and to prevent further deposition, 2) to remove urate from the body.

The first, i.e., prevention of progressive deposition is effected by colchicine. This drug prevents the degranulation of phagolysosome. The removal of urate from the body is effected by the so-called uricosuric drugs. These drugs inhibit tubular secretion and reabsorption, the predominant effect being on reabsorption. By inhibiting reabsorption of urate, large amounts of urate are made excretable by uricosuric drugs. Examples are probenecid (benemid), sulfin pyrazole (anturan), salicylates in large doses, etc.

As both the components are needed, i.e., prevention of deposition of urate and removal from the body, both colchicine and one uricosuric drug are given. Different medications are prescribed for acute and chronic conditions.

Another method of treatment is with allopurinol which inhibits the enzyme xanthine dehydrogenase and hence the formation of uric acid. Allopurinol also reacts with PRPP and depletes it. Administration of lithium salts or lithia water from springs is reported to have desirable effects against the onset of gout as lithium urates are soluble and are excreted easily from the body.

Orotic aciduria

Orotic aciduria which is a rare inherited metabolic error is characterised by an enhanced excretion of orotic acid in urine. The condition is accompanied by a severe type of anemia which is refractory to medication with iron, pyridoxine, vitamin B_{12}, folic acid and ascorbic acid. The disorder is due to the diminished activity of orotidylic pyrophosphorylase and orotidylic acid decarboxylase. As uridine and cytidine nucleotides can bring about a feed-back inhibition of the formation of orotic acid, a mixture of UMP and CMP in the form of yeast extract brings about a dramatic cure of the condition.

Disorders of liver and renal transport systems

Inborn errors may result due to disorders in liver transport and renal transport systems. The Crigler Najjar syndrome is the result of an absence of the enzyme for the conjugation of bilirubin as bilirubin glucuronide. In Gilbert's disease there is a congenital inadequacy of the UDP glucuronyl transferase system. In Dubin Johnson disease there is a defect in the liver transport system while the conjugation of bilirubin is normal.

Among renal transport disorders are diabetes insipidus due to absence or absence of response to ADH (anti-diuretic hormone), and renal glycosuria in which the renal threshold for glucose is low with glycosuria and no hyperglycemia.

Cystinuria is associated with abnormal urinary excretion of cystine together with other basic amino acids like lysine, arginine and ornithine. There might be urinary cystine stones. (or details refer 'Protein Metabolism'.) Cystinosis is another inborn error characterised by a distribution of cystine crystals in many tissues throughout the body with generalised amino aciduria. This is more serious than cystinuria. All the clinical features of cystinuria are attributable to the recurrent formation of urinary stones. Fanconi syndrome is another genetically determined abnormality in which there is impairment in the reabsorption of glucose, calcium, phosphate, potassium and amino acids. Large quantities of amino acids are excreted in the urine.

Congenital diseases like adrenal virilism, adrenal hyperplasia with electrolyte disturbances are also known.

All inherited disorders are not equally dangerous. No doubt some inherited disorders are dangerous only in the early days or months of the patient's life. If precautionary treatment is instituted in that period, the patient will survive. For example, low phenyl alanine in the first 3 – 4 years of life in the diet of patients with phenyl ketonuria and low galactose diet in the initial periods for galactosemic patients have been used with success. Relatively satisfactory methods of prolonged treatment are available for other inherited metabolic diseases like gout, diabetes mellitus and Wilson's disease.

As inborn errors are due to mutation in genes, mutation at various genetic loci can produce a large number of diseases. This however does not happen as mutations are usually single point mutations, i.e., around only one locus. As mutation tends to cluster about particular regions of a gene, the number of inborn errors is not alarmingly high.

Recent synthesis of gene (DNA) by genetic engineers like Kornberg, Khorana and Sanger has raised hopes that in some distant future, inherited errors of metabolism could be corrected by synthesising DNAs required for the production of the missing or deficient enzymes.

MOLECULAR AND GENETIC BASIS OF DISEASES

Molecular disease refers to the condition fostered by the abnormality of a single component at the molecular level, usually a protein including a protein enzyme. The aberrant coding leads to an abnormal structure of the protein enzyme which accounts for its less effective functioning that proves harmful to the organism. Pauling first applied the term 'molecular disease' to sickle-cell anemia where the mutant hemoglobin differed from the normal hemoglobin in its rate of electrophoretic mobility. Ingram demonstrated that a single amino acid substitution made all the difference between hemoglobin S and hemoglobin A and explained the highly reduced solubility of hemoglobin S. Since Pauling's earliest discovery, a host of molecular diseases have been recognised, thanks to sophisticated laboratory procedures, and many abnormal hemoglobins have been investigated. All molecular diseases are inherited diseases while all inherited diseases need not be molecular diseases e.g., (cystinuria is an inborn error but not a molecular disease).

Patterns of inheritance

The words 'dominant' and 'recessive' pertain to patterns of inheritance of specific traits in human mutations. They provide valuable information pertaining to the mechanism of inheritance of many of the inborn errors of metabolism. In fact, the mode of inheritance is the criterion for the proper genetic counselling of families with specific disorders. Hence, the genetic consultant derives significant information from chance ratings which may result in such metabolic errors. Consanguineous alliances pose a problem in this respect which, however, must be wisely anticipated so that it can ensure prophylaxis of mutant homozygotes appearing in greater than expected proportions. As most of the inborn errors of metabolism are inherited as rare recessive traits, it is quite incumbent to discourage inter-marriages in families known to carry a rare recessive trait.

If one has to have a proper biological perspective regarding the genesis of the various types of metabolic errors, one has to be aware of the transmission of genetic characters by one of four modes of inheritance: 1) autosomal dominant, 2) autosomal recessive, 3) sex-linked recessive, and 4) occasionally sex-linked dominant.

The autosomal dominant type is seen in conditions like renal glycosuria, and hereditary spherocytosis. If the father happens to be a heterozygote with one abnormal gene for the said condition, he shows detectable glycosuria. If the mother has both genes normal, half her children will have one normal and one abnormal gene and will be heterozygotes like the father with glycosuria. The other half will not have glucose in urine, as they will have normal genes. If rarely both the parents are affected, one of the offsprings will be normal, two will be heterozygotes and one a homozygote with two aberrant genes resulting in a severe form of inheritance.

A typical autosomal dominant pedigree is given in the figure below. Normally there would be no skipped generations.

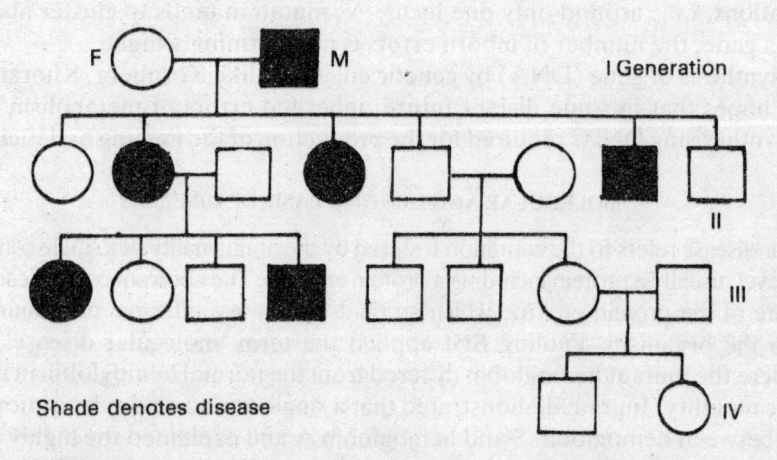

A typical autosomal dominant pedigree with no skipped generations.

Autosomal recessive inheritance is encountered in alkaptonuria, phenylketonuria, sickle-cell anemia, galactosemia, etc. The parents appear to be apparently normal and the condition affects siblings in the family in a ratio of one affected to three unaffected. There is usually a history of consanguineous marriages accounting for the greater frequency of the homozygous recessive state.

Skipped generations are the rule in autosomal recessive pedigree. Many recessive genes cause pathological conditions and hence the undesirability of consanguinous marriages.

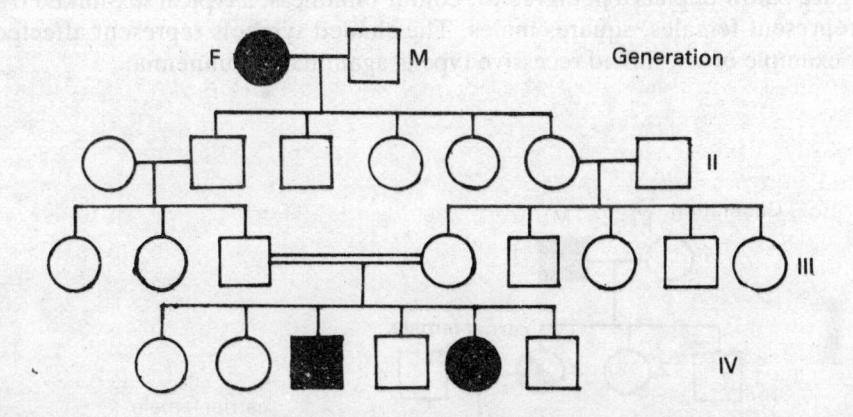

A typical autosomal recessive pedigree with skips in generations
(shade denotes disease)

Sex-linked recessive inheritance: Hemophilia provides an example of a sex-linked recessive state not manifest in the heterozygous female with only one abnormal gene. However, when she marries a normal male, out of the offspring one female and one male are quite normal while one female may be heterozygous like the mother and one male child be a bleeder (see

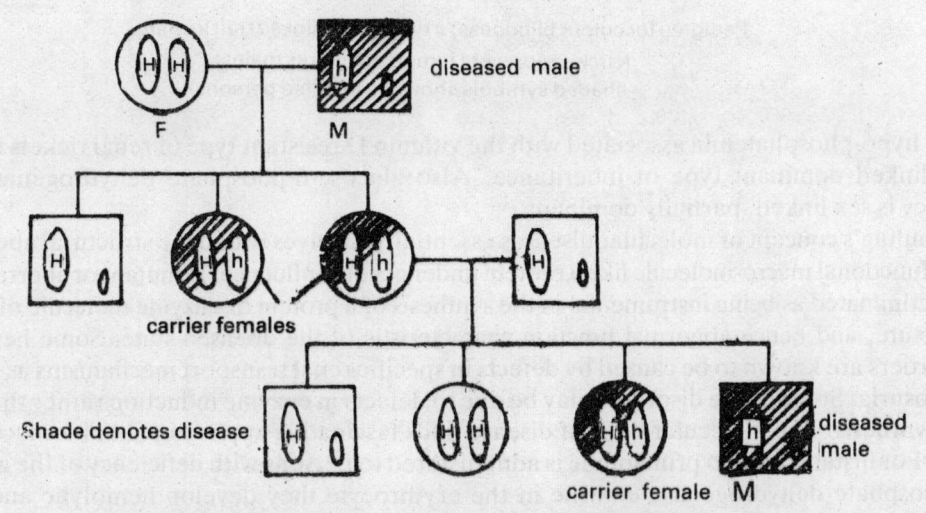

A pedigree for hemophilia in man (sex linked recessive chromosome carrying the defective gene is indicated by 'h')

the last row in the pedigree). It is a sex chromosome-linked recessive type of inheritance. Sex-linked recessive characteristics are not invariably associated with males, but tend to be so.

In the pedigree, a male with hemophilia produces normal male children, but heterozygous female carriers. When the female carrier marries a normal male, one male out of four children will be a bleeder.

The figure below depicts a pedigree for colour blindness, a typical sex-linked trait in man. Circles represent females, squares males. The shaded symbols represent affected persons. Another example of sex-linked recessive type is agamma globulinemia.

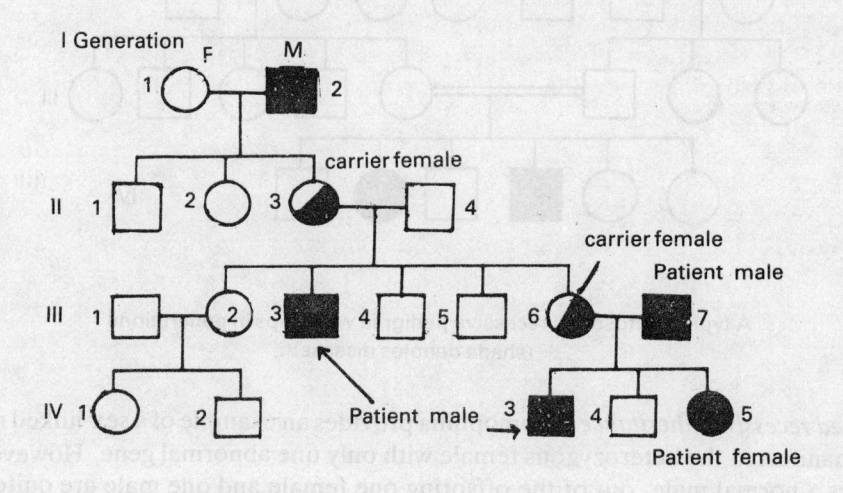

Pedigree for colour blindness; a typical sex-linked trait in man
(circles represent females, squares males;
shaded symbols show the affected persons)

A hypo-phosphatemia associated with the vitamin D resistant type of renal rickets is a rare sex-linked dominant type of inheritance. Also glucose-6-phosphate dehydrogenase deficiency is sex linked, partially dominant.

Pauling's concept of molecular diseases essentially revolves round the structural aberration of a functional macro-molecule like a protein under genetic influence. A mutant or aberrant gene is incriminated as being instrumental in the synthesis of a protein or enzyme molecule of altered structure, and hence abnormal function characteristic of the diseased state. Some hereditary disorders are known to be caused by defects in specific renal transport mechanisms as in renal glycosuria. Some of the disorders may be due to defects in enzyme induction rather than their biosynthesis. The molecular basis of disease finds fascinating application in pharmaco-genetics. For instance, when primaquine is administered to persons with deficiency of the glucose-6-phosphate dehydrogenase enzyme in the erythrocyte, they develop hemolytic anemia, a familial state encountered occasionally in malarial areas. Such alterations exemplify the widening horizons of molecular concepts of hereditary diseases.

21 Body Fluids

Water is essential to life. In the living cells, most of the biochemical reactions occur in an aqueous environment. Water forms an essential part of all body cells and body fluids and comprises 50–90 per cent of the human body mass. It acts as a solvent for various molecules, helps ionization of salts and is a medium for the excretion of nitrogenous waste products, metabolites and products of detoxication. It takes part in biochemical reactions and is produced in the body during oxidations as well.

Body water is distributed between two compartments, the extracellular fluids and intracellular fluids. The extracellular fluid (ECF), which comprises 40–50 per cent of the total body water, consists of blood plasma, lymph, interstitial fluid, connective tissue, cartilage, skin, bone and the various secretions including digestive juices, bile, tears, CSF, milk, seminal fluid, etc. The intracellular fluid (ICF) is the fluid present within the cells in the cytoplasm and nucleus and comprises 50–70 per cent of the total body water.

The typical ECF is the blood plasma. Plasma and interstitial fluid are both considered as ECF but they differ in the fact that plasma contains plasma proteins which are replaced by chloride as anions in the interstitial fluid. The interstitial fluid supplies nutrients from plasma into the cells and its volume changes according to physiological conditions.

The extracellular fluid (ECF) differs from the intracellular fluid (ICF) mainly in three respects: 1) ECF has Na^+ while ICF has K^+ as the predominant cation; 2) the principal anion of ECF is chloride while it is phosphate in ICF; and 3) the protein concentration of ECF is always lower than that of ICF. The composition of the interstitial fluid differs from blood plasma and the intracellular fluid and a comparison of the distribution of various components in these fluids can be made from the following table:

TABLE 1
COMPOSITION OF ECF & ICF

Distribution of proteins and other components (mEq/L)	Blood plasma	Interstitial fluid	Intercellular fluid (e.g., muscle cell)
Proteins	16	1 (traces)	40
Organic acids (Lactic acid etc.)	6	6	5
Chloride	103	114	2
Bicarbonate	27	30	10
Phosphate (HPO_4)	2	2	140
Na^+	142	145	10
K^+	5	4	150
Ca^{++}	5	3	2
Mg^{++}	3	2	40

The chemical composition of various body fluids is maintained more or less at constant levels by homeostatic control mechanisms. Dehydration which occurs due to water loss from the body can be due to excessive excretion resulting from exercise, sweating, diarrhoea, vomiting, exposure to hot environment, and also because of decreased intake of water which does not compensate for its loss. When water is lost from the body, electrolytes too are lost. Rehydration requires supplementation of electrolytes as well along with water for the maintainance of electrolyte balance. If dehydration occurs due to mild water loss only without any loss of electrolytes, then the body adjusts to it easily by a shrinkage of the interstitial fluid without any appreciable harm to the body. This shrinkage disappears and the normal state is restored following water intake.

Blood plasma

The total blood volume of an adult human body is about six litres, of which 3.5 litres are due to plasma. The protein and ionic composition of plasma is given in Table 1. Human plasma is straw yellow in colour and has a specific gravity of 1.015 – 1.035. The specific gravity of plasma is related to its protein content. Plasma contains 90 – 92 per cent water and the rest are solids. The proteins of plasma are present in the colloidal form and the lipids are present in combination with proteins as lipoprotein complexes. Several compounds are present in plasma in soluble form, and these include glucose, amino acids, urea and other non-protein nitrogen compounds, and various cations and anions. The chief cation of plasma is Na^+ while the chief anion is chloride. The levels of the various components in plasma that alter during pathological conditions are useful indices for diagnostic purposes. Clinical investigations using plasma are carried out in all hospital laboratories as an aid to diagnosis and treatment.

For discussion of plasma proteins refer 'Chemistry of Blood' (Chapter 9).

Lymph

Lymph resembles the interstitial fluid in many respects, since both these fluids have more or less similar compositions. The lymph capillaries drain the interstitial spaces and carry with them many nutrients including chyle (colloidal fat). Lymph has a crystalloid composition more or less like that of plasma, and thus contributes to the same amount of crystalloid osmotic pressure as plasma. However, the protein composition of lymph is very low and it consists mostly of albumin, since globulin cannot readily diffuse into it from plasma. Hence, lymph does not contribute much to the colloidal osmotic pressure (oncotic pressure) as does plasma. Unlike plasma, lymph clots very slowly because of the presence of the smaller amount of fibrinogen in this fluid as compared to plasma. Lymph is the main route for the absorption of long-chain fatty acids, partially digested fats and diacyl glycerols, cholesterol from the intestine and their transport via the thoracic duct. Lymph also helps to keep the tissues from 'drying up' by maintaining contact with the interstitial fluids.

Interstitial fluid

The fluid present in the interstitial spaces of tissues has a composition that resembles lymph and blood plasma in most respects except that the protein content is very low (only traces). Its function is to act as a 'middle man' to supply nutrients from plasma to the cells. It is also involved in the homeostatic control of the water content of tissues and thus it prevents dehydration. In conditions of drought or starvation, during which there is decreased water intake or increased water loss, the interstitial fluid shrinks in volume, while it expands after excess intake of fluid thus adapting very well to physiological needs. Lymph and the interstitial fluid together comprise about seven litres in a healthy human body while plasma comprises only 3.5 litres.

Cerebrospinal fluid (CSF)

The CSF is a clear colourless fluid that fills the non-tissue spaces of the brain and spinal cord and maintains the intracranial pressure constant. It also acts as a protective jacket for the nerve tissue. The CSF is formed by the choroid plexus and is secreted into the cerebral ventricles from where it passes into the sub-arachnoid space. The total volume of this fluid at any time is not more than 150 ml. The CSF is an ultra-filtrate of plasma but with a varied composition. This ultra-filtration involves an active secretory process.

For biochemical analysis, the CSF is usually collected by lumbar puncture, and the protein, glucose and chloride contents are determined.

A normal sample of CSF is a clear, colourless fluid of low viscosity. It does not coagulate unlike lymph and blood plasma, because it does not contain fibrinogen. The protein content of the CSF is only 15 – 20 mg per 100 ml of which 80 per cent is due to albumin. The protein content of serum is 200 – 400 times that of CSF and therefore any contamination with blood during its collection may raise its value. Usually, the total proteins in the CSF are determined. The total protein of the CSF is elevated in multiple sclerosis, encephalitis, tuberculous meningitis, purulent meningitis, neuro-syphilis, spinal cord tumour and brain abscess. In many of the above pathological conditions, the increase in protein content is due to the globulin (which is almost absent in normal CSF) and its assessment is useful for clinical diagnosis.

The glucose level of CSF is proportional to that of the blood plasma but uniformly less. At a normal plasma glucose level, the CSF glucose level varies from 45 – 80 mg per 100 ml while in diabetes mellitus and other hyperglycemic conditions it may rise to 100 mg per 100 ml or above.

CSF chloride is normally in the range of 700 – 750 mg per 100 ml (120 – 130 mEq per litre) and is higher than that of plasma. A decrease of chloride level in the CSF occurs in tuberculous meningitis. The level of HCO_3^- in the CSF, however, is the same as in plasma, i.e., 24 – 29 mEq per litre. The Ca^{++} content of the CSF is about half that of plasma. Normal CSF calcium levels are 4.2 – 5.8 mg per 100 ml which is almost the same as the ionised Ca^{++} levels of serum.

Synovial fluid

The synovial fluid is formed in the synovium cells. It is a highly viscous fluid, its high viscosity being due to the presence of the non-sulphated muco-polysaccharide hyaluronic acid. Because of its high viscosity, it helps the lubrication of joints and the moving parts of the body thereby allowing minimum damage or no damage during movements. The synovial fluid has the same pH as that of plasma and the same composition regarding glucose, electrolytes and NPN as in plasma. However, the protein composition is different from that of plasma. The synovial fluid has a protein content of only one g/100 ml; the ratio of albumin to globulin is relatively higher than in plasma and fibrinogen is absent. The synovial fluid also does not have any lipids present. The electrolytes of the synovial fluid can exchange readily with plasma.

Bile

(See chapter on 'Biochemistry of digestion and absorption'.)

Seminal fluid (semen)

Semen or the seminal fluid has a fluid part called seminal plasma in which innumberable spermatozoa are present. The composition of the seminal fluid is most suitable for the maintenance and survival of the spermatozoa, and any abnormality in its composition may affect it causing infertility in the individual. Analysis of the semen in the clinical laboratory has therefore assumed great significance in fertility clinics.

The seminal fluid or semen is the net fluid formed by the mixing of the testicular fluid, prostatic fluid and the secretion from the seminal vesicles. The testicular fluid brings the spermatozoa, and the seminal vesicle synthesises fructose which is contributed as a nutrient for the spermatozoa during its storage in the seminal vesicle. The prostate gland contributes to the production of citrate which is present in considerable amounts in the seminal fluid. The spermatozoa are constituted almost fully of nucleoproteins.

Human seminal plasma has the same pH as blood plasma. The content of phosphate, citrate and lactate in semen is higher than that in blood plasma, while chloride and cholesterol are lower. The sugar content of semen is high and most of it is due to fructose which is synthesised in the seminar vesicle. If the fructose content is low (as in many infertile men), the spermatozoa cannot survive.

Saliva

Saliva is the mixed secretion from four different glands, the parotid, sub-maxillary, sublingual and buccal glands. It contains 99.3 – 99.7 per cent water and is a colourless viscid fluid with neutral reaction (pH range 6.9-7.1). It is secreted from the mouth by psychic and reflex stimulation. Thus sight, smell and thought of food or mechanical acts like chewing stimulates salivary secretion. Some chemical compounds, salts and acids like acetic acid can also stimulate the taste buds and initiate salivary secretion.

The constituents of saliva are proteins, inorganic salts and non-protein nitrogenous compounds. The proteins comprise albumin, globulin, mucins and the enzyme ptyalin (salivary amylase). Salivary mucin is a mucoprotein containing 45 per cent carbohydrate. Mucin because of its high viscosity acts as a lubricant in the mouth and helps the processes of chewing (mastication) and swallowing (degluttition). The zinc-containing protein 'gusten' of saliva is implicated in getting the taste. The NPN compounds present in the saliva are urea and uric acid and small quantities of amino acids which arise from the blood. K^+, phosphate and chloride are present in saliva in greater amounts than Na^+, Ca^{++} and sulphate. In certain conditions, calcium phosphate and calcium carbonate together may form insoluble tartar on teeth. Increased acidity and the presence of oxalate may give rise to salivary calculi which is due to calcium oxalate.

Saliva has many functions. It moistens and lubricates the food and helps in chewing and swallowing. It keeps the mouth moist, cleanses it, and keeps the taste producing food materials in a soluble form so that the taste buds can sense them. Apart from these, saliva starts the digestion of starchy foods by virtue of the action of the $\propto$-amylase enzyme which converts starch to dextrins and small amounts of maltose. Na^+ and Cl^- ions enhance the salivary amylase activity and solubilisation of starchy foods.

Gastric juice

Gastric juice is the mixed secretion arising from three different types of cells on the walls of the stomach. The parietal cells secrete hydrochloric acid, while the chief cells produce and secrete pepsin and the mucous cells mucin. To this is added the mucoid secreted by the surface epithelium and the glandular mucoprotein secreted by the neck mucous cells of the gastric glands. The latter carries the 'intrinsic factor' needed for the absorption of vitamin B_{12}.

(For detailed discussion on gastric digestion, role of HCl and pepsin, etc., see Chapter 11 on 'Biochemistry of Digestion and Absorption'.)

Pancreatic juice

(See Chapter 11 on 'Biochemistry of Digestion and Absorption'.)

Tears

The lachrymal glands secrete tears which are stimulated during emotional states like grief, intense affection and happiness, and also when the eye is exposed to irritants like tear gas, harmful chemical or oil vapours, ingredients from raw onion or garlic and also glycerol and acrolein. Tears keep the eye moist, clean and healthy and also free from bacterial infection. The enzyme known as lysozyme present in the tears has the ability to hydrolyse and destroy the mucopeptide of the bacterial cell wall. Thereby it protects the cornea from infection by such organisms.

Tears have a chemical composition more or less resembling that of the interstitial fluid. The protein content of tears is very low. The ionic composition of tears resembles that of plasma, especially in its Na^+ and chloride content.

Milk

Milk is the secretion of the lactating mammary gland in women and is a fluid derived from blood plasma, to which is added some constituents like lactoglobulin, casein and lactose, all of which are synthesised in the mammary gland. Milk is secreted as a body fluid by the alveoli of the mammary gland of the lactating mother. It may also be considered as excretion from the mother but used as food for the infant for supplying all essential nutrients for its growth.

In this chapter, it is discussed only as one of the body fluids, and its nutritional importance as a food for the infant will be discussed later under 'Nutrition and balanced diet' in Chapter 23.

During pregnancy in women, the estrogens and progesterone stimulate an increased growth of the breast and proliferation of the functioning cells. However, milk secretion does not start till the end of pregnancy because the lactogenic hormone prolactin is opposed in its action by estrogens and progesterone. However, immediately following parturition (delivery and removal of the influence of the placenta), prolactin can act without antagonism and milk production starts. The secretion and milk ejection are stimulated by nervous stimulation induced by suckling and mediated through the posterior pituitary hormone, oxytocin.

Human milk has a slightly different composition from the milk of the other species of the animal kingdom, like the buffalo, cow or goat, especially with respect to the lactose content and the protein composition. Human milk has less casein and more of the lactalbumin and lactoglobulin as compared to bovine milk which has a predominance of casein. The lactose content of human milk is also high (almost 7 g/100 ml) as compared to bovine milk which is 4.8 per cent. The ash content of human milk is also lower than in bovine milk. Milk is rich in calcium and phosphate and it supplies all the calcium and phosphate requirement of the growing infant for several months, this calcium and phosphate being present as inorganic form in milk fluid as well as in organic combination in casein (as calcium caseinate). The lipids of milk are present in the form of fat globules suspended in an aqueous medium in the form of an emulsion stabilised by the presence of milk proteins. The fats consist of triacyl glycerols containing palmitic, oleic, stearic, myrsistic and other higher fatty acids, along with some short-chain fatty acids like butyric, caproic and caprylic acids. There is very little or only negligible amount of cholesterol in milk. The fat part of the milk is synthesised in the lactating mammary gland, the glycerol being formed from glucose.

Among the fat soluble vitamins, only vitamin A is present in milk in appreciable amounts needed for the body. Vitamin D is present only in meagre amounts. Vitamin C is almost absent in milk. However, fresh human milk has a higher content of vitamin C than cow's milk, on a comparative basis. Among the vitamins of the B-group, riboflavin is present in adequate amounts needed for the body, while most of the other vitamins are present only in trace quantities. Most of the minerals needed by the body are present in very small amounts except calcium and magnesium which are adequately available in milk. However, iron is one mineral which is present only in traces in milk, and thereby milk is considered as inadequate in iron as a food source.

Urine

The physical characteristics of urine vary even under normal conditions due to various physiological factors. The normal adult excretes 800 – 2500 ml of urine daily according to the quantity of fluids ingested, quality of foods taken, environmental and physical states. A high-protein diet produces physiological polyuria due to the diuretic effect of urea. Coffee and tea are known to possess diuretic action. The diurnal output of urine is by far greater being related to activity. The extra-renal losses of fluid, for example in diarrhoea, have a reciprocal effect on the urinary output.

The specific gravity of normal urine ranges between 1010 – 1025 being almost inversely related to the volume, except in diabetes mellitus. The last two digits of the specific gravity multiplied by 2.66 (Long's coefficient) give the approximate total amount of solids in grams per litre at 25°C. The reaction of normal urine is usually acidic but it may range between 5 – 7.5 depending upon physiological as well as pathological circumstances. A high-protein diet produces a much more acidic urine due to acid metabolites like phosphates and sulphates that arise during protein catabolism. Even normally, urine becomes alkaline on standing on account of bacterial decomposition of urea into ammonium carbonate. Toluene as well as thymol are used as preservatives to prevent bacterial decomposition. The alkaline urine tends to become turbid due to the precipitation of phosphates of alkaline earth elements and a distinct ammoniacal smell results. There is a transient post-prandial alkaline tide in urine which is explained by the compensatory role of the kidney which combats the alkalemic tendency due to copious secretion of hydrochloric acid.

The colour of normal urine is straw yellow and it varies with the concentration of urine, the chief pigment responsible being urochrome. The colour darkens when urine is allowed to stand. In fevers, the colour is more intense due to concentration. Bile pigments colour the urine in obstructive jaundice. Whole blood or hemoglobin imparts a smoky to red colour. In alkaptonuria the urine looks normal as soon as it is voided but when exposed to the atmosphere in a container, there is blackening from above downwards due to atmospheric oxidation of unmetabolised homogentistic acid. Various drugs may colour the urine as well as uroflavin.

Usually, urine is transparent unless allowed to stand without any preservative. The salts of uric acid are precipitated in strongly acidic urine. Pus, mucus, micro-organisms and excess of nucleo-proteins may make the urine turbid.

Normally, the odour of urine is faintly aromatic but in ketosis there is a fruity odour. Drugs may impart characteristic odours.

Normal components of urine: Urea accounts for half the total solids present, while sodium chloride comes next, dependent on the salt intake. There are inorganic as well as organic substances present in urine and they may be nitrogenous or non-nitrogenous. The non-protein nitrogenous substances of biochemical importance excreted in urine are urea, uric acid, creatinine, ammonia, amino acids and hippuric acid.

Urea is essentially exogenous in origin, its amount being proportional to the amount of protein ingested. On a mixed diet 80 – 90 per cent of urinary nitrogen is due to urea. The non-protein nitrogenous substances in urine can be increased in blood and in urine as well in the hypercatabolic state, but when there is gross renal dysfunction the blood levels go up while the urinary levels are decreased. Hemo-concentration due to dehydration can cause an enhancement of NPN in blood. The normal daily output of ammonia in urine ranges between 500 – 900 mg per 24 hr and is related to the acid-base status of the person. Unless renal damage is responsible for the acid-base imbalance, ammonia production by the kidneys is increased in acidosis and decreased in alkalosis.

Creatinine and creatine excretion in urine are reciprocally related. Women and children excrete by far less of performed creatinine than men due to their having less effective functioning muscle mass. Even on a creatine-free diet, the normal adult excretes a constant amount of creatinine in urine due to the endogenous synthesis of creatine. The average output for a man is 1.5 g per day while it is 1.3 g for a woman due to the less effective functioning muscle mass.

The uric acid output in urine ranges between 500 – 800 mg per day and even on a purine-free diet there is a constant urinary output indicating the biosynthesis of purines. Uric acid is the end product of purine metabolism in human beings and the output is increased in conditions like leukemias, polycythemia and gout.

Amino acids are excreted by the normal adult to a level not exceeding 200 mg for 24 hr. However, enhanced amino-aciduria may be generalised or renal in origin. At times, there may be selective amino-aciduria. It may be an inherited tubular defect in reabsorption, or acquired and may be associated with glycosuria and phosphaturia due to the renal defect.

Hippuric acid is normally excreted with a mean value of 700 mg daily. It represents the detoxication by the liver of benzoates by conjugation with glycine.

Chlorides are excreted in the urine mainly as sodium chloride to the extent of 10 – 12 g daily but the excretion varies according to the salt intake. Starvation is one of the causes of loss of salt. Sweating is another cause, particularly when it is profuse. Addison's disease shows an enhanced excretion of sodium chloride in urine while Cushing's syndrome shows a retention.

The sulphates in urine are chiefly derived from sulphur-containing amino acids. Inorganic sulphates represent the major form accounting for about 80 per cent due to the oxidation of sulphur. Along with urinary nitrogen, sulphur excretion is an index of protein catabolism. Ethereal or organic sulphur denotes the products of detoxication of phenol, paracresol, indol and skatol by oxidation and conjugation by the liver with PAPS. The toxic phenolic substances are obtained by intestinal bacterial enzymatic action on unabsorbed amino acids namely phenyl alanine, tyrosine and tryptophan. Neutral sulphur is normally excreted in very small amounts in urine but is increased in certain types of poisoning due to the defective oxidation of sulphur.

The phosphates in urine are those of sodium, potassium, calcium as well as magnesium. A protein-rich diet enhances phosphate excretion. Cellular breakdown can also increase the excretion of phosphate. There is an increased excretion in osteomalacia, renal rickets and hyper-parathyroidism. Hypo-parathyroidism and retention due to renal disease may result in diminished excretion of phosphate in urine.

Oxalates are normally excreted in negligible amounts in urine but a primary hyper-oxaluria may occur due to disorder of glyoxylate metabolism which is pyridoxine-dependent.

Minerals in urine: They are sodium, potassium, calcium, magnesium, etc. Sodium and potassium excretion are greatly influenced by the activity of the adrenal cortex. Excessive tissue catabolism results in enhanced potassium excretion due to cellular breakdown. Alkalosis increases potassium excretion. Calcium and magnesium are mostly excreted by the intestine. Bone involvement produces variations in calcium and magnesium excretion.

Vitamins, enzymes and hormones are excreted in negligible amounts in urine. Significant alterations are met with in pathological states.

Common abnormalities in urine: Proteinuria may sometimes be functional or physiological. It may be met with after strenuous exercise or it may be orthostatic when a person is in erect posture. A high-protein meal may at times cause transient proteinuria. It is not uncommon in normal pregnancy, though it is transient and does not indicate a pre-toxemic state.

Organic or pathological proteinurias may be pre-renal, renal or post-renal. Congestive cardiac failure is a pre-renal cause. The nephrotic syndrome is characterised by massive proteinuria and consequent hypo-albuminemia. Hyper-cholesterolemia is an additional feature. The globulins are less common than albumins due to their higher molecular weight.

Bence-Jones proteins are the globulins found in urine at times in multiple myeloma and myeloid leukemias, and they are recognised by their heat coagulability at $50°-60°C$, solubility at $100°C$ and reappearance on cooling.

Detectable glycosuria may be alimentary or physiological. It may be renal in origin but more usually it is caused by diabetes mellitus. Other causes of hyperglycemia resulting in glycosuria are hyperpituitarism, hyperthyroidism and hyperadrenalism of the medulla or the cortex. Emotional stress results in the elaboration of epinephrine causing transient hyperglycemia and glycosuria. When the renal threshold is lowered or increased or when the filtration is defective, urinary findings alone may be misleading, in which case blood sugar should necessarily be determined for correct diagnosis.

Fructosuria may be alimentary or it can be essential due to an inborn error of metabolism. Congenital galactosemia followed by galactosuria is a rare condition due to the deficiency of galactose-1-phosphate uridyl transferase. Pentosuria may be due to an inborn error caused by the deficiency of L-xylulose reductase.

Ketone bodies in urine are normally negligible but become detectable in advanced starvation and severe diabetic state. The ferric chloride test gives a wine red colour but it is not as sensitive as Rothera's test and hence if it is positive it shows the severity of ketosis.

Bilirubin in urine, if present, shows either intra-hepatic or post-hepatic obstruction. In hemolytic jaundice it is absent, but the urine shows excess of urobilinogen. In hepato-cellular jaundice, before obstruction supervenes, there is excess of urobilinogen in urine. Complete absence of urobilinogen in urine but with bilirubin shows established obstruction, most often post-hepatic.

Blood in urine may be in the form of whole blood when it is called hematuria. Free hemoglobin when present in urine points to a hemolytic process caused by bacteria, viruses or toxins, or incompatible blood transfusion. Pus should be distinguished from blood in urine by boiling the urine and destroying the peroxidase if any. Porphyrins in urine are normally eliminated in small traces, but increased excretion of uroporphyrins and coproporphyrins is seen in hepatic conditions as well as in erythropoietic conditions. Lead poisoning is one of the known causes of porphyrinuria.

Sweat:

The volume of sweat ranges between 500 – 400 ml/24hr depending on the environmental temperature, degree of exercise done and the type of work engaged in. There are pathological states showing alterations in the volume as well as the composition of sweat. The sodium level varies from 30 – 70 mEq/L, K^+ is 0 – 5 mEq/L, and Cl^- 30 – 70 mEq/L. Excessive sweating may produce muscular cramps, as in miners, due to heavy loss of NaCl. Muco-viscidosis is a rare condition which is a fibrocystic disease of the pancreas. It is associated with recurrent pulmonary infection in children, due to a congenital exocrine deficiency of the pancreas. The mucus secreted in the pulmonary and biliary systems is unduly viscous and the salient diagnostic biochemical findings are enhanced levels of sodium and chloride in sweat.

The collection of sweat is to be done carefully. Stimulation by electrodes is one of the procedures. The sweat has to be collected in weighed ash-free filter papers, as otherwise results would be vitiated.

Feces

The products of digestion as well as compounds like vitamins and minerals are absorbed mostly in the small intestine. In the large intestine, with the exception of water, the absorption process is normally completed. The unabsorbed material is eliminated as feces. The consistency of the feces depends largely on the water content. Gastro-intestinal motility and the type of diet alters the consistency of the stools.

The water content of feces is normally 60 – 70 per cent by weight. The undigested dietary material comprises cellulose, fatty material, minerals and bacteria. Most of the nitrogen in the feces is of bacterial origin.

The chief pigment responsible for the dark brown colour of the feces is stercobilin derived from stercobilinogen by oxidation. Complete absence of stercobilinogen indicates an established obstruction of the biliary tract, particularly post-hepatic obstruction when the feces are colourless or clay white.

22 Detoxication (Bio-Transformation)

Detoxication refers to the protective synthesis performed chiefly by the liver, whereby the toxic substances in diet, drugs as well as toxic metabolities are rendered non-toxic and readily excretable.

The liver is the major site of detoxication in the body. Hepatectomised animals do not have the capacity to detoxity. Administration of a test dose of sodium benzoate and determination of the amount of hippuric acid excreted in urine in a given period is one of the liver function tests.

Toxic substances are of three categories. They may be present in foods or in drugs or may be formed during metabolism. The most important detoxication is that of ammonia formed from the metabolism of proteins. The toxic ammonia is converted to the innocuous, highly water soluble urea by the Krebs-Henseleit urea cycle in the human liver. Most of the toxic metabolites arise due to bacterial enzymatic action on normal digestive products which are unabsorbed from the intestine. Sometimes after absorption, they are metabolised in such a way that they are made readily excretable. Carbohydrate and lipid metabolites do not pose a problem as far from being toxic, they are even beneficial to the organism. It is the unabsorbed amino acids that are subjected to bacterial enzymatic action in the large intestines giving toxic products that must be detoxified.

Decarboxylation of unabsorbed amino acids by bacterial action yields primary amines. Tyramine is thereby obtained from tyrosine, indol ethylamine from tryptophan and his-

HO—〈 〉—CH_2—$CH(NH_2)$—COOH
Tyrosine

↓ $-CO_2$

HO—〈 〉—CH_2—CH_2—NH_2
Tyramine

$CH_2 CH(NH_2) COOH$ (imidazole) —$-CO_2$→ $CH_2 CH_2 NH_2$ (imidazole)
Histidine　　　　　　　　　　　　　**Histamine**

(indole)—CH_2—$\overset{H}{\underset{COOH}{C}}$—$NH_2$ —$-CO_2$→ (indole)—CH_2—CH_2—NH_2
Tryptophan　　　　　　　　　　　　**indol ethylamine**

DETOXICATION (BIO-TRANSFORMATION)

tamine from histidine. However, histamine when taken orally is destroyed by histaminase of the intestinal wall.

Deamination of certain amino acids by bacterial enzymes in the large intestine yields corresponding acids. Aromatic amino acids, in particular, give rise to toxic metabolities which have to be rendered readily excretable by one of the various detoxifying mechanisms obtaining in the liver. Hence, deamination is often followed by simple or oxidative decarboxylation, or oxidation may follow decarboxylation. Some of the well known reactions are represented in the chapter on 'Amino Acid Metabolism'.

Broadly, the transformations involved in detoxication fall into four categories. They are oxidation, hydrolysis, reduction and conjugation. Conjugation however can take place straightaway but most often it is preceded by oxidation, reduction, or hydrolysis which pave the way for ready conjugation.

Oxidation plays an important role in protective synthesis, particularly in respect of alcohols which are oxidised to aldehydes and acids. Sulphur compounds are also oxidised. Sometimes ring cleavage is brought about by oxidation. Hydroxylation comes under the oxidative processes. Alcohols can be oxidised eventually to carbon dioxide and water, or they may be conjugated without preliminary oxidation. Conjugation often follows oxidative processes, the conjugating substance being mostly glucuronic acid. Epinephrine is known to be detoxified by mono amine oxidase (MAO), Chloral, used as a hypnotic, is partly oxidised to trichloracetic acid and partly reduced to trichloro-ethanol. The sulphur of several aromatic sulphur-containing compounds is oxidised to sulphate. Neutral sulphur refers to the unoxidised sulphur from some sulphur-containing compounds. Hydrolysis provides an alternate pathway for the alteration of foreign compounds rendering them suitable for conjugation. Digitalis, for example, which is a cardiac glycoside is hydrolysed yielding the sugar fraction as well as the aglycone fraction. Atropine is hydrolysed in the liver to tropine and tropic acid by a specific enzyme. Acetyl-salicylic acid, i.e., aspirin, is partly split up by hydrolysis into acetic acid and salicylic acid.

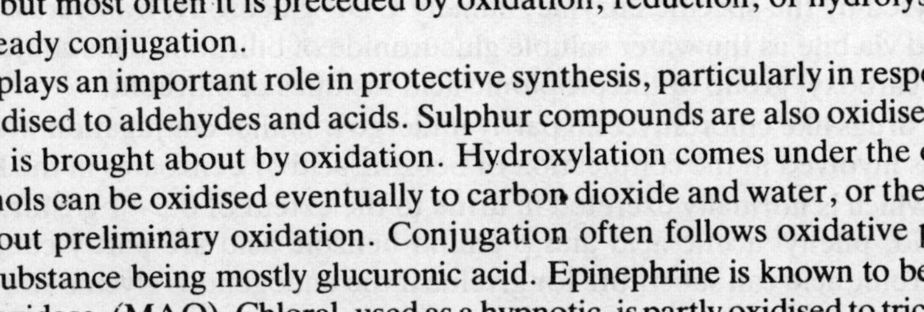

Acetyl Salicylic acid + H_2O → Salicylic acid + CH_3-COOH (acetic acid)

Nitro compounds like picric acid are reduced.

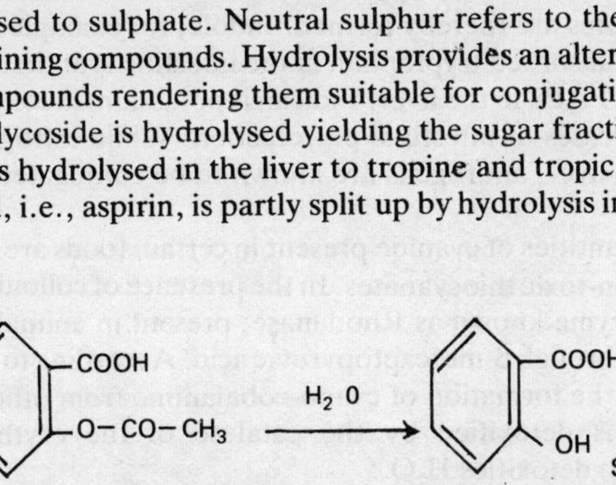

Picric acid → Picramic acid

Other important reduction reactions include conversion of disulphide linkages to the sulphydryl groups.

Mostly, the products formed as a result of oxidation, hydrolysis or reduction are not by themselves ready for excretion but they must be suitably conjugated so that they are made readily excretable. Conjugation therefore is indicated for the purpose of protective synthesis or for enhancing the excretability of toxic compounds following one of the above-mentioned preliminary processes. Conjugation either involves the formation of an ether type of linkage or an ester type of linkage, with the hydroxyl group or carboxyl group, respectively, of the foreign compound. Glucuronic acid is readily obtained by the oxidation of glucose. From uridine diphosphate glucuronic acid, the glucuronyl moiety is transferred to compounds like bilirubin catalysed by the specific enzyme, namely UDP-glucuronyl transferase. The conjugate is excreted via bile as the water soluble glucuronide of bilirubin. This conjugation takes place with the carboxyl group of the propionic acid residues of bilirubin.

Well known drugs like chloromycetin partly undergo a similar conjugation with glucuronic acid. Glycine is involved in the conjugation of benzoic acid or benzoates in the liver forming hippuric acid which is normally excreted in urine to the extent of 0.5 – 1 g/24hr.

Nicotinic acid, phenyl acetic acid and p-amino benzoic acid are partly conjugated with glycine. Glucuronic acid can substitute for glycine if the endogenous glycine stores happen to be exhausted.

Sulphuric acid in the form of PAPS is employed by the liver for the detoxication of phenolic compounds. Etheral sulphates are thereby formed. Indole, for example, obtained by bacterial enzymatic action on unabsorbed tryptophan in the intestine is at first oxidised to indoxyl before it is conjugated with PAPS in the liver. Indican is potassium indoxyl sulphate excreted in appreciable amounts in cases of intestinal putrefaction. While estrogens are conjugated with glucuronic acid in the liver, androgens are known to be conjugated with PAPS before they are excreted in urine.

Very small sub-lethal quantities of cyanide present in certain foods are detoxified by being oxidised to the relatively non-toxic thiocyanates. In the presence of colloidal sulphur the reaction is catalysed by the enzyme known as Rhodanase, present in animal tissue. Cyanide is also detoxified by cysteine through ß-mercaptopyruvic acid. According to one view, traces of CN could be detoxified by the formation of cyano-cobalamine from other forms of vitamin B_{12}. Hydrogen peroxide is detoxified by the catalase of the erythrocytes and liver. Glutathione peroxidase also detoxifies H_2O_2.

Methylation is one of the ways of protective synthesis which is exemplified by the disposal of excess of nicotinic acid by the liver. The nitrogen atom of the pyridine ring of nicotinic acid gets methylated resulting in the excretion of trigonelline. Epinephrine escaping oxidation by MAO and dihydroxymandelic acid undergo O-methylation and are excreted as methoxy derivative.

Acetylation plays an equally important role in certain detoxication mechanisms like the disposal of excess of sulphonamides, para-amino benzoic acid and aromatic halogenated hydrocarbons. Cysteine gets acetylated first, whereby acetyl cysteine is formed, which gets conjugated with toxic substances like bromo benzene to be excreted as p-bromophenyl mercapturic acid. In selenium poisoning, there is a reversal of mercapturic acid synthesis, and administration of halogenated aromatic hydrocarbons is ameliorative.

As much as half the excreted sulphonamides may be acetylated similarly.

Detoxication of certain poisons like heavy metals is effected by British Anti-Lewisite (BAL) which is 2,3-mercapto propanol. Heavy metals like arsenic, cadmium and mercurry

DETOXICATION (BIO-TRANSFORMATION)

Nicotinic acid —active methionine→ Trigonellin

Acetyl Cysteine + Bromo benzene → Para bromo phenyl mercapturic acid

are believed to bind the –SH groups of enzymes thereby inactivating them and British Anti-Lewisite acts as an antidote by virtue of pulling away the –SH groups from the heavy metals. Such a liberation helps in the detoxifying mechanism.

SECTION III
ENERGY METABOLISM, B.M.R., S.D.A., NUTRITION COMPOSITION OF FOODS AND BALANCED DIET

23 Energy Metabolism, Nutrition, Composition of Foods and Balanced Diet

Nutrition is the first need of man and his general health and well being are much dependent on his nutritional status. The availability of food is not uniform throughout the world and due to this unequal distribution, malnutrition and under-nutrition prevail in certain regions of the world, whereas excess food is available in certain countries where there are diseases due to overeating and over-nutrition.

The nutritional requirements of man have been assessed by various workers in their studies over long periods and the daily needs of proteins, carbohydrates, fats, minerals and vitamins have now been defined. Methods of classifying food in order to meet the daily requirements have also been adopted to solve nutritional problems.

Calorimetry: The measurement of energy requirements of the body under various physiological conditions and the determination of energy values of foods is called *calorimetry*. Energy value of foods and energy requirement for the body are measured usually in calories. In nutrition, the term *large calorie* is used. A large calorie or kilo-calorie (usually abbreviated as kcal) is defined as the heat required to raise the temperature of 1000 g of water through 1°C (say, from 15°C to 16°C) and is 1000 times the small calorie.

Energy value (caloric value) of foodstuffs

The total heat produced by the combustion of any foodstuff in the presence of oxygen can be measured in a bomb calorimeter. By this method, the energy values (caloric values) of carbohydrate, protein and fat have been measured. All carbohydrates, do not give the same value and different proteins and fats give values slightly different from each other. The average values are given in the table on the next page.

ENERGY VALUE OF FOODS

Foodstuff	Energy value in kilo-calories/g	
	in Bomb calorimeter	in the body
Carbohydrate	4.1	4.0
Protein	5.6	4.0
Fat	9.4	9.0

When foodstuffs are oxidised in the body, the oxidation is almost complete as far as carbohydrates and fats are concerned. However, proteins are not completely oxidised in the body. Proteins yield nitrogenous excretory products, like urea, creatinine, uric acid and ammonium salts after their overall metabolism, which still contain oxidisable carbon and/or hydrogen. Hence, their caloric value in the body is only 4.0 kcal/g as compared to the bomb calorimeter value of 5.6 kcal/g. It has also to be taken into consideration that when foodstuffs are ingested, the proteins, fats or carbohydrates are not absorbed totally. Thus the caloric value of foods calculated in this way has to be corrected for nutritional data.

RQ of foods

The respiratory quotient of foodstuffs depends upon the type of components and their varying proportions. The RQ of a mixed diet containing carbohydrate, fat and protein is around 0.85. If carbohydrate alone is considered, its RQ is unity, as shown by this equation:

$$C_6H_{12}O_6 + 6 O_2 \rightarrow 6 CO_2 + 6 H_2O$$

$$RQ = \frac{CO_2}{O_2} = \frac{6}{6} = 1$$

Fats have a much lower RQ since the ratio of oxygen to carbon in the molecule of fat is low when compared to that of carbohydrate. Thus

$$2\ C_{57}H_{110}O_6 + 163\ O_2 \longrightarrow 114\ CO_2 + 110\ H_2O$$
(Tristearin)

$$RQ = \frac{CO_2}{O_2} = \frac{114}{163} = 0.70$$

Proteins vary in their chemical structure and hence their RQ is not definitely and readily calculated unlike in the case of fats and carbohydrates. However, in some proteins whose molecular structure is known, the same method of calculation gives an RQ intermediate to that of fats and carbohydrates. Indirect methods of calculation give the RQ of 0.80 to the most common dietary proteins.

Basal metabolic rate (BMR)

The energy expenditure of the body is increased when the body and its organs are doing active work. However, a certain minimum amount of energy is always needed for the maintenance of life, even during the period of rest and sleep. The working of the heart and the blood circulation, the process of respiration by the lungs, the muscular twitchings and reflexes, the intestinal peristalsis, all go on even when a person is at rest physically and mentally. The metabolism under such a condition is referred to as 'basal metabolism'. The minimum energy production needed for the maintenance of cellular metabolism when the body is in the basal condition is called the basal metabolic rate (BMR).

For measurements of BMR, the subject should be prepared as follows:
1. The subject should be in the post-absorptive state, i.e., he/she should not have taken any food in the previous 12–14 hr period.
2. The subject should be physically relaxed and stay in bed or a reclining chair for at least an hour before the test.
3. He/she should be mentally at ease and relaxed.
4. He/she should be awake
5. It is preferable to have the room temperature around 25°C.

The measurement of BMR is done either directly from the oxygen consumption for 2–6 min from the graph obtained with Benedict-Roth apparatus or the Du Bois apparatus, or by any of the indirect methods by analysing the expired air and determining the oxygen consumption and CO_2 output. The total heat production is determined and is then calculated per square metre of the body surface per hour.

$$BMR = \frac{\text{Total heat production in kcal per hour}}{\text{Body surface area in square metres}}$$

In clinical practice, the oxygen consumption of a subject or patient is measured for six minutes and then corrected to 0°C and 760mm pressure. The resulting value is multiplied by 10 to obtain the oxygen consumption for one hour. This value is multiplied by 4.825 kcal in order to get the heat production for one hour (one litre of oxygen at NTP=4.825 kcal of heat). The correction for body surface area is done using the Du Bois formula.

Surface area $A = H^{0.725} \times W^{0.425} \times 71.84$, where H = height in cm and W = weight in kg. The value of A is obtained in square cm but has to be divided by 10,000 in order to get the surface area in square metres.

For convenience, a nomogram for the calculation of the surface area from height and weight is available.

A man of 35 years, 170 cm high and 70 kg weight consumes 1.2 litres oxygen (corrected to 0°C and 760mm pressure) for six minutes. His BMR is calculated as follows:

Consumption of oxygen for 1 hour = $1.2 \times 10 = 12$ litres.

1 litre of oxygen = 4.825 kcal

12 litres of oxygen = 58 kcal/hr

Surface area = 1.8 sq.m

BMR = 58/1.8 = 32 kcal/sq. m/hr.

For a man of 35 years, the normal BMR should be 39.5/kcal/sq.m/hr.

Difference is $\frac{(39.5 - 32)}{39.5} \times 100 = 18.5$ per cent

In this case it is minus 18.5 per cent.

If the difference is minus 15 to plus 20, it can be considered normal.

In hyperthyroidism it can be plus 50 or above. In hypothyroidism it can be minus 30–60. The young normal adult male has a BMR of 40 kcal/sq m/hr while women have 8–10 per cent less, i.e., about 36 kcal.

The international unit for energy measurement is the KJ (kilojoule) defined as the energy required to lift one kg up by one metre. The conversion of kcal (kilocalorie) to kilojoule can be done by multiplying by the factor 4.2. Thus a BMR of 32 kcal/sq.m/hr is equal to $32 \times 4.2 = 134.4$ KJ.

Factors influencing BMR

1. **Surface area:** Although the BMR is expressed in relation to surface area and therefore has to be constant with different individuals, it is observed that subjects with smaller body size have a higher BMR as compared to those with larger body size.

2. The BMR is low in the newborn, but rises to higher levels as the baby grows. The BMR is highest at the age of 5–6 years, and thereafter it remains constant or declines slowly till puberty. Then it decreases gradually with age.

3. The BMR of men is always higher than that of women in the same age group. However, in the winter season, the BMR of women rises so high and may reach almost the same level as that of men.

4. The BMR is decreased in summer months but increased in winter. A cold environment increases heat production in the body and hence also the BMR.

5. Racial variations in the BMR have been reported.

6. The BMR is lowered in undernutrition, semi-starvation and starvation.

7. Pathological states affect the BMR. Thus, febrile diseases, infection, pyrexia and fevers tend to increase the BMR. A raised BMR is encountered in the increased activity of cells, as in leukemia, polycythemia, etc. In hyperthyroidism, there is increased cellular activity and the BMR is raised by 30 per cent or more above the normal levels. In Addison's disease the BMR is subnormal while it is slightly increased in Cushing's syndrome. Epinephrine causes an increase at the outset. A hole in the eardrum will give a false-high BMR.

Resting metabolic expenditure (RME)

The normal method of measurement of the BMR under a defined set of conditions is sometimes found inconvenient and cumbersome, especially for the regular determination of the BMR for a number of patients in the hospital wards. Therefore, another index for energy measurement during resting condition can be adopted for an approximate measurement of the BMR. This is known as the RME or 'resting metabolic expenditure' of a patient on his hospital bed and can be taken 12 hours after his food intake. The value of the RME may not be different from the BMR beyond a 3–5 per cent variation, and is useful for quicker diagnostic purposes. However, the RME sometimes varies according to climatic changes and this fact should be taken into account. The SDA of foods also affects this value and hence the RME measurement must always be taken at least after 12 hours interval after the meal.

Specific dynamic action (SDA) of food (Thermogenic effect)

The extra heat production when food is used by the body over and above the calculated caloric value is called the specific dynamic action (SDA) or calorigenic action of the food. This is also known as the thermogenic effect of foods. When 25g of protein undergoes metabolism in the body, the theoretical value for heat production is 25×4 kcal = 100 kcal. However, it is found that the actual heat production may be 130 kcal. This extra 30 kcal is due to the specific dynamic action of protein in the body. Similarly, it is found that 11.1 g of fat instead of producing $11.1 \times 9 = 100$ kcal of heat, actually produces 113 kcal, whereas 25 g of carbohydrate actually produces 105 kcal instead of 100 kcal. Thus, proteins have the highest SDA value whereas carbohydrates have the lowest. It is observed that the SDA due to food components is greatest when they are fed individually, but the SDA is lowered when the components are mixed. The presence of fat or carbohydrate can thus decrease the SDA due to protein. Lard fat has great influence in reducing the SDA.

In stipulating the required calories for any one, suitable additions have to be made to compensate the loss through the SDA, i.e., expenditure during the utilisation of foods consumed. Usually 10 per cent is added to the total caloric need.

The following example will illustrate the effect of the SDA:

An individual has a BMR of 60 kcal/hr. He takes food equivalent to 200 kcal. During the next five hours, he expends 320 kcal while he should have spent $60 \times 5 = 300$ kcal only. Thus, an extra 10 per cent $\left(\dfrac{20}{200} \times 100\right)$ of calories taken is wasted. At this time the BMR comes to 60 kcal/hr and the SDA of food is over. The basal heat production of this individual is $(60 \times 24 =)$ 1440 kcal/day. As 10 per cent of this is lost as the SDA, it is not enough if he is given his needed 1440 kcal. He should be actually given 1600 kcal of which 10 per cent (160) may be wasted as SDA and the remaining 1600-160 = 1440 kcal will be available to him.

A person consuming 2500 kcal will have 810 kcal available for activity after subtracting (10 per cent) 250 for the SDA and 1440 for BMR.

The explanation regarding the exact mechanism for SDA is not clear. However, it is believed to be associated with cellular activity concerned with the metabolism of food components. In respect of amino acids, as both oral and intravenous administration of amino acids causes the same SDA, work involved in digestion and absorption has no influence on SDA. It is suggested that the high SDA of proteins could be due to deamination, oxidative deamination and transamination of amino acids and urea formation in the liver.

It is also known that hepatectomy abolishes the SDA effect, thereby indicating that the SDA is related to metabolic function in the liver.

Another observation reported has been that when glucose and thiamine were given, the SDA was 8 per cent while for glucose alone it was 4.2 per cent. The extra energy output represents that required to prepare glucose for deposition in the form of fat.

The SDA has a role in the regulation of body temperature. In cold climates the SDA of proteins makes us feel warm.

Energy requirement of individuals

The total energy requirement should be sufficient to cover the needs of 1) the basal metabolism, including the work involved in the action of organs like the heart, lungs and kidneys as well as the maintenance of body temperature, 2) muscular activity, 3) SDA, and 4) growth.

About 50 per cent of the total energy of basal metabolism is used for the active transport of Na^+, K^+, and Ca^{++} across the cell membrane, i.e., in or out of cells, and the remaining 50 per cent is used for all other purposes including muscle tone, heart beat, respiration, kidney function, brain activity and maintenance of body temperature.

A caloric intake of 1300 to 2000 kcal/24 hrs would cover the basal needs of most individuals. It has been shown that for a 70 kg man, 65 kcal/hr are expended in sleeping, 100 kcal/hr in sitting at rest, 200 kcal/hr in walking slowly, 570 kcal/hr in running and as much as 1100 kcal/hr in walking up a flight of stairs. The energy thus spent depends on the type of activity and, in addition, on age, sex and conditions like pregnancy.

The expenditure of energy (kcal/day) for men and women of different professions is given below:

Men	Average	Women	Average
Elderly, retired	2330	Elderly	1990
Office workers	2520	Middle aged	2090
Laboratory Technicians	2840	Laboratory Technicians	2130
University students	2930	University students	2290
Building workers	3000	Factory workers	2320
Steel workers	3280		
Coal miners	3660		

For guidance, the following is the approximate daily caloric requirement for men, women and children:

Age	Wt.	Nature of work	Energy required
Men (35-65)	65 kg	Sedentary	2600 kcal/day (2400)
Men (35-65)	65 kg	Moderately active	2900 kcal/ " (2800)
Men (35-65)	65 kg	Very active	3600 kcal/ " (3900)
Women (18-65)	55 kg	Most occupations	2200 kcal/ " (1900 - 2000)
Women (18-65)	55 kg	Very active	2500 kcal/ " (3000)
Women Pregnancy (+ 300)			2400 kcal/ " (2300)
Boys (12-15)	45 kg		2800 kcal/ " (2500)
Girls (12-15)	48 kg		2300 kcal/ " (2100)

(For other age groups the caloric needs are slightly different Data given in the brackets are for Indians.)

Diet and its components

The normal diet must contain six important components. They are: 1) carbohydrates, 2) fats, 3) proteins, 4) vitamins, 5) minerals, and 6) water.

The first two principles are for yielding energy while the third is to facilitate growth, maintenance and repair of tissues. Vitamins act as activators for cellular reactions and are essential parts in many of the chemical processes pertaining to the production and utilisation of energy and for the synthesis of many physiological compounds in the body. Minerals are needed for electrolyte balance, ionic equilibrium, activation of chemical processes and acid-base equilibrium, apart from many other functions.

There should also be some fibre in the diet (mostly carbohydrates like cellulose, lignin, pectin, etc) which is necessary for facilitating bowel movement and preventing constipation. Nutritionists now consider this as a very important factor for the present day diets.

Proteins, being the basis of protoplasm are needed in all living organisms for their normal functioning. The body needs dietary proteins for growth, tissue formation, maintenance and tissue repair. Dietary proteins supply all the nitrogen and sulphur needed by man. The amino acids derived from the dietary proteins of plant or animal origin, by digestion in the intestine, are absorbed and then utilised for the synthesis of tissue proteins and blood proteins. Only the nutritionally non-essential amino acids can be synthesised from the intermediates of glucose metabolism, but the nitrogen (as amino group) for this purpose is derived from other amino acids present in the body. The nutritionally essential amino acids which canot be synthesised by the human body have to come from dietary proteins only. Proteins also supply 15 per cent total energy required for the body, this being produced by the oxidation of the carbon skeleton of some of the amino acids that enter the TCA cycle. About 400g of protein are turned over by an adult man (70 kg body weight), 25 per cent of this is being oxidized for energy and the remaining 75 per cent recycled for synthesis of tissue proteins, blood proteins, etc.

The principal dietary sources of proteins are eggs, milk, milk products (like cheese), fish, poultry, meat, cereals, legumes and soya beans. These proteins differ from each other in their nutritional value and often a mixture of different proteins is needed for a full complementary effect, to all the nutritionally essential amino acids for the maintenance of nitrogen balance.

The protein requirement

A minimum amount of protein is always required in the diet as a provision for the replacement of tissue protein which undergoes constant destruction and has to be resynthesised. Some of the amino acids arising from the degradation of tissue proteins enter the amino acid pool, which are again used for the synthesis of new proteins, while a certain amount of amino acids undergo metabolic degradation and are excreted as urinary NPN. The daily requirement of protein to replace this loss of nitrogen and to maintain the N-balance is called 'the wear and tear quota'.

The value of 0.45g protein per Kg body weight is accepted as the minimum requirement of protein per day. Thus, a man 70kg body weight requires a minimum intake of $70 \times 0.45 = 31.5$ g protein per day. The value, however, does not hold good for children and adolescents whose requirement is higher. For children, the minimum requirement is 2.2 g per kg body weight. This requirement continues till adolescence, but decreases thereafter in middle age and old age. For the adult, the Food and Nutrition Board of the National Research Council (USA) has recommended a daily requirement of protein with an allowance of a safe margin of 0.8 g per kg body weight, taking into consideration defects in digestion, absorption and utilisation of amino acids for other purposes, slight changes in climate and body's changes in heat production, presence of other constituents like fat and carbohydrate in the diet, etc. Accordingly, a man needs $70 \times 0.8 = 56$ g and woman needs $55 = 0.8 = 44$ g protein per day. The protein can be even less, provided that adequate energy in the form of carbohydrate and lipid is already available. If there is an increased loss of nitrogen in urine (e.g., during stress, fever, infections, burns, etc.) an increased protein intake is needed to compensate for this nitrogen loss. Protein intake is also to be increased during pregnancy and lactation in women for the growth of the fetus and for milk production, respectively.

The quality of protein in the diet depends upon its easy digestibility as well as the content of amino acids and the efficiency with which they are absorbed. In general, proteins containing most of the nutritionally essential amino acids in the right proportions are considered as first class proteins. If the nutritionally essential amino acids are present in less amounts or if some of them are missing, the proteins are considered as inferior. Children require nutritionally essential amino acids much more than adults because the amino acids are required for their growth and maintenance, whereas they are required only for maintenance in adults. Infants and children thus require 35 per cent of food protein as nutritionally essential amino acids and 65 per cent as nutritionally non-essential amino acids. Adults need to take only 20 per cent of food protein as nutritionally essential amino acids and the rest as nutritionally non-essential amino acids. The nutritionally essential, semi-essential and non-essential amino acids for humans are given below.

Nutritionally Essential (Indispensable)	Semi-essential (Semi-indispensable)	Nutritionally Non-essential (Dispensable)
Lysine	Arginine *	Glutamic acid
Tryptophan	Histidine	Aspartic acid
Phenyl alanine		Alanine
Methionine		Proline
Threonine		Hydroxy proline
Leucine		Glycine *
Isoleucine		Serine
Valine		Cystine
		Tyrosine

(* Arginine and Glycine are nutritionally essential for chicks and turkeys.)

Tyrosine will spare but not completely replace phenyl alanine. Cystine will spare but not completely replace methionine. Nicotinic acid will spare but not completely replace tryptophan. (See also dispensable and indispensable amino acids under Protein Metabolism.)

Balance of dietary amino acids

Certain amino acids influence the requirement of some other amino acids, and this is an important fact to be remembered in nutrition. Thus the addition of leucine to an otherwise adequate diet of fat, increases the requirement of iso-leucine. Supply of methionine to a low protein diet deficient in this amino acid may cause threonine-deficiency. The admixture of lysine with a diet low in histidine increases the requirement of histidine and the urinary excretion of cystine.

It is thus observed that the addition of an excess of one amino acid to a previously adequate diet may reduce the utilisation of another amino acid to such an extent that a deficiency state occurs. The intake of certain amino acids also increases the requirement of certain vitamins, e.g., the addition of glycine and leucine to a sucrose diet increases the niacin requirement in rats. Excess of methionine added to the diet increases the requirement of pyridoxine.

Biological value of proteins

Food proteins vary considerably in their efficiency of utilisation in the body for the synthesis of hemoglobin, plasma proteins and tissue proteins. The evaluation of the quality of dietary proteins or biological value is thus very important and is related to their utilisation in the body. The biological value of a protein is its effectiveness in supporting growth in the young organism, and in tissue maintenance and repair (nitrogen balance) in the adult. It depends on; 1) the quantity (amount) and quality (amino acid composition) of the protein, 2) its digestibility, and 3) the efficient absorption of its constituent amino acids. The biological value of a protein is expressed in comparison with that of a standard, high-quality protein like ovalbumin (also lactalbumin).

The biological value of a protein is determined by various methods.

1. Measurement of its influence on the weight-increase in weanling animals: Weanling rats are fed on a ration containing the protein for a certain period and their weight increase is measured. If a rat given a ration containing 15 per cent protein for a certain period (say, one month) showed an increase of 54 g and consumed 240 g of food, then, the total protein content of the food consumed $= \dfrac{240 \times 15}{100} = 36$ g.

Therefore, protein efficiency ratio $= \dfrac{\text{weight increase (g)}}{\text{weight of protein consumed (g)}} = \dfrac{54}{36} = 1.8$

2) Determination of the biological value in terms of the percentage of N retained by the organism: This is one of the earliest methods of estimating the biological value. This method is time - consuming and difficult to carry out. The method makes use of young rats, and the percentage of absorbed N retained by the animals is determined.

3) Measurement of the influence of the protein on the rate of regain of body weight or of liver proteins of previously depleted animals (Osborne and Mendel).

4) Measurement of its effect on the rate of restoration of plasma proteins or hemoglobin in animals previously depleted of these specific proteins.

5) Assay of some liver enzymes, e.g., xanthine dehydrogenase:

It has been found that, evaluated on the above criteria, animal proteins in general are superior in biological value to vegetable proteins. Whole egg proteins and milk proteins rank the highest, with meat, kidney and other proteins ranking next. Gelatin, though derived from an animal protein collagen, lacks tryptophan and cysteine and has only a very low content of lysine and methionine. Hence it is considered an inadequate protein. Casein, on the other hand, contains little glycine and no cystine but is good for growth and is considered a good quality protein. It contains all the essential amino acids including leucine and methionine in high amounts, and is also rich in the non-essential amino acid, glutamic acid.

Among the vegetable proteins, soya bean proteins are considered the best, as they are rich in lysine, phenyl alanine, arginine, leucine and iso-leucine and may be as good as animal proteins. Gliadin of wheat is low in lysine and tryptophan, while zein of corn is low in glycine and contains no lysine and tryptophan. Zein is, however, rich in leucine and glutamic acid.

The biological value of proteins of some of the common food-stuffs along with their protein efficiency ratio is given in the following table:

Foodstuff	Biological value	Protein-efficiency ratio
Rice	68	2.2
Wheat	65	1.5
Maize	59	1.2
Bengal gram	68	1.7
Ground-nut	55	1.7
Egg	94	3.9
Milk	84	3.1
Meat	74	2.3
Fish	76	3.5

Supplementary value of proteins

If a dietary protein is seriously deficient in one or more of the essential amino acids, the N-equilibrium is upset and the N-balance cannot be maintained. Tissue proteins now come to the rescue by getting degraded in order to supply the missing amino acid. This results in an increased loss of NPN in urine. If, however, another protein containing the missing amino acid is added to the diet, the N-equilibrium and normal nutrition can be established. The capacity of one protein to make good the deficiency of another protein is known as the supplementary value of the protein. Most cereals and pulses have a mutual supplementary effect.

Effects of protein-deficiency

A lack of adequate amount of proteins in the diet results in insufficient availability of amino acids to the cells for their synthetic activities. The first effect of protein deficiency in a child is the slowing down of growth and loss of weight. Dietary proteins are needed for the formation of intestinal mucosal proteins and the effective functioning of the digestive glands. The intestinal mucosa and digestive glands normally have a rapid turnover of proteins. This, however, is affected in protein-deficient states, resulting in failure of digestion and absorption of food leading to diarrhoea, loss of water in stools and alteration of electrolyte balance. The liver also fails to perform its normal functions. Fat accumulates in the liver cells due to the lack of lipotropic factors, which have to come from proteins and amino acids, and also due to the decreased formation of lipoproteins. The liver fails to synthesise albumin which leads to the reversal of the A/G ratio in plasma resulting in edema of tissues. Protein deficiency leads to a failure in the maintenance of the integrity of the skeletal muscle and may lead to muscle wasting and atrophy. It manifests immediately in the decreased production of hemoglobin and erythrocytes and a nutritional anemia ensues. In severe protein deficiency, the glandular tissues are also affected and their function slackens due to depletion of their proteins. The proteins of the heart muscle, C.N.S. and brain, however remain, intact even in moderately severe protein deficiency, but decreased mental function, loss of memory, etc. may ensue in such conditions.

Secondary protein deficiency may occur in several conditions, e.g., on a low carbohydrate diet, protein may be utilised as a source of energy resulting in its deficiency later on. Proteins are lost in urine in certain kidney diseases and this may lead to secondary protein deficiency states. Hemorrhage, wounds and exudations also result in loss of protein reserves from the body. In liver diseases, the synthesis of proteins is depressed and the A/G ratio may be altered. Malabsorption syndrome and diarrhoea result in decreased absorption of amino acids from the intestine even if proteins are supplied in diet. The need for proteins is increased in burns, trauma and injuries to organs.

Diseases associated with protein malnutrition and protein calorie malnutrition

According to the WHO, protein-calorie malnutrition is the most important public health problem in developing countries. In these countries, the rates of infant mortality are very high. About 50 per cent of children born do not survive above six years, the death rate in these children being 20-50 times higher than that in USA, Soviet Union and other developed countries.

KWASHIORKOR

Kwashiorkor is an extreme condition of protein-calorie malnutrition (PCM) or protein-energy malnutrition. The term PCM covers a whole range of deficiency states, from mild to severe, and has been defined as 'a range of pathological conditions arising from coincidental lack in varying proportions of protein and calories, occuring most frequently in infants and young children and commonly associated with infection.

In kwashiorkor, the contribution of protein malnutrition is quite high and there is edema; in marasmus, the contribution of energy malnutrition is higher than that of protein and there is no edema.

Marasmus and kwashiorkor are two facets of the same disease entity, the final outcome being determined by the individual's capacity to adapt to nutritional stress. It has been proposed that an adequate adrenocortical response and adaptation to the nutritional deprivation leads to marasmus, whereas failure to adapt gives rise to kwashiorkor. Hence, in kwashiorkor, the level of corticoids in the blood will be low. Because of a decrease in adrenal corticoids, muscle protein is not changed to visceral and blood proteins.

Kwashiorkor is the disease the first child may get when the second is on the way. Growth is stunted and wasting of skeletal muscle is apparent, but the subcutaneous fat is maintained. Kwashiorkor in children frequently leads to death. Terminal infections are invariably present. Adult kwashiorkor may develop rather suddenly in response to protein deprivation and, in general, is characterised by low serum proteins and sub-normal immuno-competence.

Kwashiorkor is also encountered in children after weaning. Once the mother stops breast feeding the baby, a few months after its birth (perhaps due to the onset of a second pregnancy), the child gets displaced from the mother's attention. Instead of milk, it is fed with a low protein diet like gruel or porridge prepared from rice, maize or potatoes, and the prolonged feeding results in protein deficiency. The children develop enlarged abdomen, dermatosis, diarrhoea, glossitis, conjunctivitis, edema, fatty liver and anemia. Gradual feeding with good quality protein foods like milk and eggs or legumes may improve the condition, but the mental retardation occurring in childhood may not be corrected satisfactorily.

MARASMUS

Marasmus is a condition in which the infant is virtually starving due to lack of calories and proteins. Marasmus occurs in certain rural areas in the developing countries, where there is severe starvation and famine.

According to the Wellcome classification (British), there are four types of malnutrition in children, These are:

1) Marasmus (meaning 'neglected child'): Most common in infants under the age of six months. Occurs between one month and one year.

2) Marasmic kwashiorkor: An intermediate form of kwashiorkor and marasmus. Occurs in children between six months and two years.

3) Kwashiorkor (meaning 'displaced child'): Occurs between 1–4 years.

4) Underweight child: Usually occurs in children above five years but can occur from one year to 14 years, leading to nutritional dwarfing.

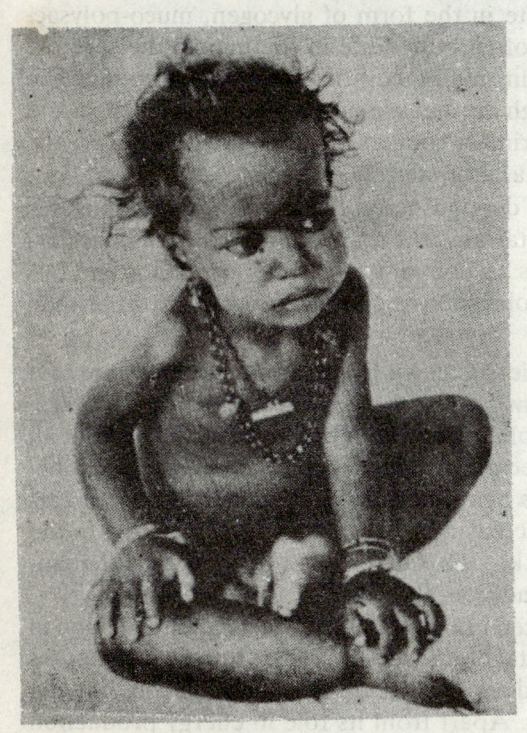

Kwashiorkor
(Note the edema.)

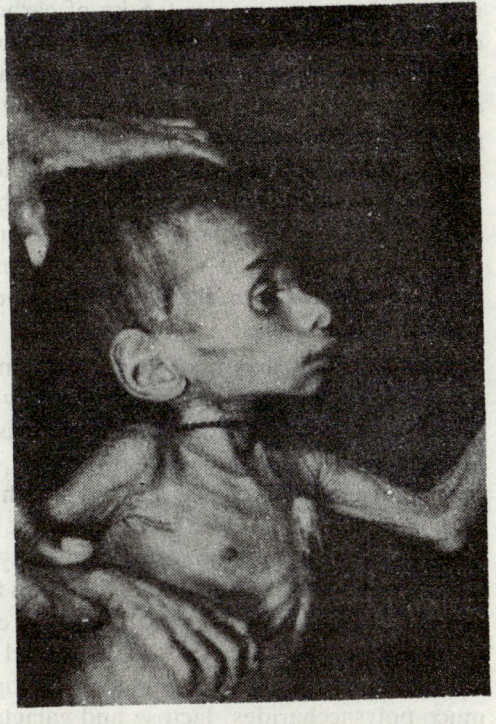

Marasmus

Courtesy: Department of Paediatrics
Jawaharlal Institute
Pondicherry 605 006

The distinguishing features of marasmus and kwashiorkor are many but the most important are edema and dermatosis seen almost always in kwashiorkor. Marasmic children are alert, have a ravenous appetite and will take food, if given. On the other hand children with kwashiorkor are lethargic, apathetic and irritable, and loss of appetite (anorexia) is always present. The underweight child is also restless, irritable and lethargic and refuses food due to different degrees of anorexia. Hepatomegaly (enlargment of liver due to accumulation of fat) leading to fatty liver occurs in kwashiorkor but seldom in marasmus. In children with kwashiorkor, plasma albumin and the A/G ratio are decreased leading to accumulation of the excess body fluid in the extracellular spaces which results in pitting edema of the trunk, limbs, eyelids, etc. This is not seen in marasmus. The underweight child looks apparently normal, but may have all the features of kwashiorkor in many respects. They have flabby muscles, protuberant abdomen, hypochromic anemia, mental retardation, low IQ, chronic illness thoroughout childhood, susceptibility to infection, digestive disturbances, etc.

PCM, shortly after birth, affects the human brain. Cell division is retarded resulting in permanent reduction in number of brain cells. In the post-weaning period, the reduction of DNA and deficit cell number is permanent. As the amino acid tryptophan and tyrosine are required to give neuro-transmitter substances like serotonin and dopamine, the synthesis of neuro-transmitters may be affected in protein deficiency.

Nutritional value of carbohydrates

Carbohydrates act as the chief energy suppliers for the body. The human body in the normal steady state contains 0.5 kg of carbohydrate in the form of glycogen, muco-polysaccharides and free sugar. The principal sources of carbohydrate in the human diet are the starches and sugars. Glycogen is also available in small quantities from animal foods. The most common sugars available to the body through the diet are sucrose, glucose, fructose and galactose with small amounts of pentoses. The primary sources of starch are grains, cereals, tubers and vegetables. Molasses, fruits and milk are the sources of sugars.

The amount of carbohydrate ingested by man depends upon the availability of foods, his food habits and economic status. In general, vegetarians take more of carbohydrates and starchy foods whereas non-vegetarians consume less carbohydrates but more proteins and fats than vegetarians. In India, from the economic point, non-vegetarian foods are costlier than vegetarian foodstuffs.

Is pre-formed carbohydrate required in our diet? Can we live without carbohydrate? Of course, our body does possess the ability to form glucose from non-carbohydrate sources. However, the extent of gluconeogenesis is not able to cope with the needs of the body at all times. Apart from adequate energy needs, we need pre-formed glucose for the effective functioning of the brain, CNS, muscles, erythrocytes and many of the glandular organs. If carbohydrate is not included in the diet, fat gets increasingly oxidised as fuel, resulting in the over-production of ketone bodies and may cause ketosis. On the other hand, if a small quantity of glucose or carbohydrate is given in this condition, the ketosis can be prevented.

The starch and glycogen in the diet are converted into glucose ultimately after digestion in the intestine and are absorbed and carried by portal blood into the liver. Glucose contributes to the blood sugar, muscle and liver glycogen and many other derivatives of glucose like the muco-polysaccharides, lactose and galactolipids. Apart from its role in energy production and synthesis of a wide variety of essential physiological compounds, glucose has also a sparing action on proteins and lipids. The metabolic intermediates of glucose breakdown may be transformed into non-essential amino acids, fatty acids and triacyl glycerols. The presence of glucose decreases the breakdown of proteins and amino acids for gluconeogenesis. During carbohydrate deprivation as in starvation, there is increased breakdown of protein resulting in increased NPN excretion in urine. This loss of NPN is prevented by ingesting carbohydrates. Glucose can also prevent the breakdown of fats and spare their loss from the body.

Fibre in the diet

Cellulose, pectin, lignin and gums are the carbohydrates present in our common foods. They contribute to the fibre content of food and help bowel movement (intestinal peristalsis) thus preventing constipation. Nutritionists are of the opinion that the deficiency of fibre in the diet may be one of the causative factors in the etiology of diverticulosis of the colon, appendicitis, colon cancer, and even coronary heart disease. Fibre has a triple beneficial role:

1) It helps to avoid hyperglycemia by reducing glucose absorption in the intestines. This is because the transit speed of digested food is increased in the intestines.

2) It helps to correct hypercholesterolemia by binding the bile salts and sequestering them. As a result, the concentration of bile salts reaching the liver by entero-hepatic circulation is decreased. The feed-back inhibition by the bile salts on the catabolism of hepatic cholesterol to bile acids is decreased. There will therefore be greater degradation of cholesterol to bile acids and disposal from the body.

Fibre, in addition, forms an insoluble complex with cholesterol and bile acids in the intestines and favours their fecal excretion.

3) Fibre prevents colon cancer. One possible factor for colon cancer is through toxins elaborated by some intestinal microbes. These toxins are adsorbed by dietary fibre and eliminated from the body.

These effects have been recognized in recent years, and more and more fibre-containing foods are being sought after in the Western dietary which mostly comprises refined foods containing sugar, protein and too much fat in the usual course. In this respect vegetarian food is always superior to the non-vegetarian one. Good sources of fibre are rice bran, wheat bran, legumes, fruits, vegetables and leafy greens.

Dietary sugars

Sucrose is the most abundant sugar present in the diet of civilised people. Refined cane sugar as well as jaggery are consumed by most people. Sucrose is hydrolysed to glucose and fructose by invertase in the small intestine. Fructose can be converted to glucose by the intestinal mucosa as well as the liver. Rarely, in some infants, sucrose is not tolerated due to the absence of the enzyme invertase in the jejunal mucosa, and diarrhoea of a sort ensues. It can be corrected by substituting glucose for sucrose.

Lactose is converted by the enzyme lactase in the intestine to glucose and galactose. The galactose absorbed from the intestine is converted to glucose by liver. In infants, most of the carbohydrate taken in is contributed by the lactose of milk. Lactose and galactose are not tolerated by children with galactosemia, a congenital condition, due to the absence of the liver enzyme galactose-1-phosphate uridyl transferase. In this condition, galactose-1-phosphate accumulates in the erythrocytes and other tissues causing damage to the liver, brain and lens of the eye. There is also lactose intolerance in some intestinal disorders.

Experimental feeding of high amounts of lactose in rats has been reported to elevate the blood sugar level, decrease the rate of growth and lead to cataract in about a month's time.

However, feeding of high amounts of glucose or starch is found to keep the rats healthy and normal. Thus, from a nutritional point, lactose and galactose are inferior to glucose and starch.

Nutritional aspects of lipids

Lipids are the necessary constituents of all body-tissues, contributing to about 14–18 per cent body weight. In some types of obesity, the percentage of body fat may be as high as 25 per cent. One gram of fat on complete oxidation in the body gives rise to 9.0 kcal, more than twice that given by carbohydrates and proteins. A considerable amount of fat is stored in the adipose tissue of the body and is available as a source of energy during periods of shortage. While fats can be synthesised in the body from carbohydrates, essential fatty acids (EFA) are necessary in the diet of man and animals.

Requirement of fats in diet

The total fat intake (especially in a diet of 2000 kcal per day) should not be more than 35 per cent of energy. Of this, 10 per cent energy as polyunsaturated fatty acids (EFA) is advisable. Since carbohydrate (glucose) is converted to fat in our body and that is also oxidised, the amount of fat in the diet can be reduced if carbohydrate intake can be increased. At least 15–25g of fat (containing also three per cent of polyunsaturated fatty acids) is needed by an average adult. The rest of the requirement is met by the conversion of glucose to fat (lipogenesis). The type of fat in the diet has to be so chosen as not to result in any deficiency of fat-soluble vitamins as well as EFA.

Effect of EFA deficiency in man

Biochemical determination of the iodine value of serum lipids has shown that it is decreased markedly in EFA-deficient patients. EFA deficiency in man also leads to phrenoderma, characterised by papular eruptions of the pilosebaceous follicles on the arm, thighs, buttocks and other regions. Eczema of various degrees and severity is also observed. The essential fatty acids for man and animals are linoleic acid, linolenic acid and arachidonic acid. Strictly speaking, only linoleic acid is essential for man. It cannot be synthesised, but other types of EFA can be produced, provided there is linoleic acid. The best sources for essential fatty acids are vegetable oils like cotton seed oil, soya bean oil and sun flower oil.

EFA plays certain important fucntions in the body. The phospholipids of cell membranes and of the mitochondrial membranes contain EFA. In EFA deficiency, the changes in the integrity of these membranes affect many a metabolic process. Red cell membranes become fragile and may easily get hemolysed. The skin becomes abnormally permeable to sweat and loss of weight from the body may occur. EFA is an important constituent of lipoproteins.

Disposal and oxidation of cholesterol require EFA and it is believed that the efficacy of many vegetable oils in diminishing the plasma cholesterol is by virtue of the EFA content. Arachidonic acid has been now recognised as a direct precursor of prostaglandins in the body. The function of prostaglandins is discussed elsewhere.

(For atherosclerosis, obesity and fatty livers, refer chapter on 'Lipid Metabolism'.)

Vitamins and nutrition

The importance of vitamins as accessory food factors, their sources and availability, their physiological role, biochemical functions, and symptoms of deficiency have all been discussed in the chapter on 'Vitamins'.

Minerals

Minerals too are to be obtained from the diet. NaCl is added to food at the time of cooking and is taken in the diet. Calcium, iron and magnesium are present in most foods. Their assimilation into the body requires efficient digestion of foods in stomach in the presence of HCl and factors concerned with their efficient absorption. Fe absorption is facilitated by acidity in the gastro-intestinal tract, presence of ascorbic acid, protein digestion products, etc. Ca^{++}, Mg^{++} and phosphate absorptions require vitamin D, fats and bile salts, the presence of certain acidic amino acids, and organic acids like lactic acid in the intestine. The trace elements I, Cu, Mo, Zn etc., are adequately available in foods and unless they are absent, normally no deficiency results. In regions where iodides are absent in the soil, people develop goitre due to their lack in the food.

Milk

Milk is the secretion of the mammary gland and forms the first food which the infants get during suckling. Though milk contains many useful nutrients it cannot be called a perfect food since it lacks iron and copper among minerals and contains practically very little ascorbic acid, thiamine and vitamin D. The proteins of milk are lactalbumin, lactoglobulin and casein.

Casein is present to the extent of 80 per cent in cow's milk and is of high biological value. The composition of the milk of a few species is given in the following table for comparision:

COMPOSITION OF MILK FROM A FEW SPECIES

Species	Water g per cent	Protein g per cent	Fat g per cent	Lactose g per cent	Minerals (ash) g per cent
Human	87.4	1.4	4.0	7.0	0.2
Cow	87.1	3.4	3.9	4.9	0.7
Goat	87.0	3.3	4.2	4.8	0.7
Buffalo	82.2	4.7	7.5	4.8	0.8

Though milk appears to be homogeneous to the naked eye, it is in fact an emulsion of fat in an aqueous medium, containing proteins, which help in stabilising it.

The fat of milk consists mostly of triacyl glycerols and only very small quantities of cholesterol and phospholipids. The fatty acids present in triacyl glycerols are butyric, caproic, caprylic, capric, lauric, myristic, palmitic, stearic and oleic acids, with very little amounts of polyunsaturated fatty acids.

Lactose is the principal sugar of milk and forms the carbohydrate consumed by infants and babies.

Ca, P, Mg, Na, K and Cl are present in the ash as minerals of milk with very small quantities of S, Fe, Cu, Zn and Mn. The Fe and Cu content of milk is so low that milk as the only food may lead to dietary anemia. The Ca and P of milk are nutritionally very important as they are much more easily assimilated than from other foods. Milk contains adequate amounts of vitamin A, riboflavin and pantothenic acid. However, most other vitamins are present in inadequate amounts. The ascorbic acid content of human milk is higher than cow's milk but it still is inadequate from the point of nutrition, and this vitamin has to be provided to the baby either through orange or tomato juice.

The vitamin D content of breast milk is also very low and its content can be increased if the mother takes food enriched with vitamin D.

Milk is the ideal food for the infant and for the growing child. As the child grows, the increased requirement of calories has to be met with by supplementing milk with carbohydrate and starchy foods. For the adult, it serves as a food supplement and is consumed by both vegetarians and non-vegetarians in increased amounts.

Egg

Like milk, egg is outstanding in its nutritional value as food for the young and for adults of all age-groups. Eggs contain protein (12.5 per cent) of very high biological value, i.e., ovalbumin and ovoglobulin in the egg white and ovovitellin in the yolk. The fat of egg yolk is very easily digested and assimilated in the body. It is a good source of lecithin, cephalin and linoleic acid. Egg contains very little carbohydrate (only 0.5 per cent). Egg is rich in calcium and phosphorus and is a good source of available iron. It contains vitamins, riboflavin and some other members of B-group as well as vitamins A and D. Raw egg white contains avidin, an antagonist to biotin, but cooking destroys it.

Importance of nutrition

Adequate knowledge about nutrition is necessary for man to plan his diet, to maintain good health and keep his body efficient for all activities. Ignorance and wrong food habits are mainly responsible for most illnesses. Adhering to the principles of nutrition and putting them into practice is the best way of preventing diseases. Prevention is better than cure, and nutritive food is the best way to achieve this.

SECTION – IV
MODERN TECHNIQUES IN BIOCHEMISTRY

24 Modern Techniques in Biochemistry

CHROMATOGRAPHY

The technique of chromatography is one of the popular techniques in the field of Biochemistry and can be employed even in ordinary laboratories. Its popularity is due to the following reasons:

1. Very small quantities of substances can be analysed qualitatively and quantitatively.
2. The equipment is very simple.
3. The actual operation is also fairly simple and no special skill is required.
4. The results are remarkably reproducible.

Important contributors to this technique are Tswett (1872), Tiselius, Lederer, Kuhn, Martin, Synge, Moore, Stein, Paul Karrer, Stahl and a host of others.

Definition

Chromatography is the technical procedure of analysis by the percolation of a fluid through a body of comminuted or porous rigid material, irrespective of physico-chemical processes that lead to the separation of substances.

The various physico-chemical factors involved in chromatography are adsorption, partition, ion exchange and molecular sieving.

There are different types of chromatography, which are listed in this table.

CHROMATOGRAPHY			
Column	Paper	Thin layer	Gas liquid

1) Adsorption	1) Single dimension
	   a) ascending
	   b) descending
		1) Use of ordinary adsorbents	

2) Partition			
3) Ion exchange	2) Two dimensions	2) Ion exchange resins for thin layer	

4) Gel filtration	a) ascending
	b) descending
5) Affinity chromatography	3) Reversed phase paper chromatography
	4) Circular paper and
6) HPLC	5) Use of ion exchange paper

(The term chromatography is a misnomer as it is no longer confined to coloured substances.)

In all types of chromatography there should be two phases, a stationary phase and a mobile phase. When the mobile phase, i.e., a liquid moves along stationary phase, e.g., a column of solid or a paper, the separation of substances takes place due to the operation of one or more physico-chemical factors. The movement of the mobile phase may be due to the capillary force, in addition, in certain cases, to gravity.

In column chromatography, a uniform column of a substance like alumina (2–150 cm long) is prepared in a long tube fitted with a separating funnel as shown in the diagram.

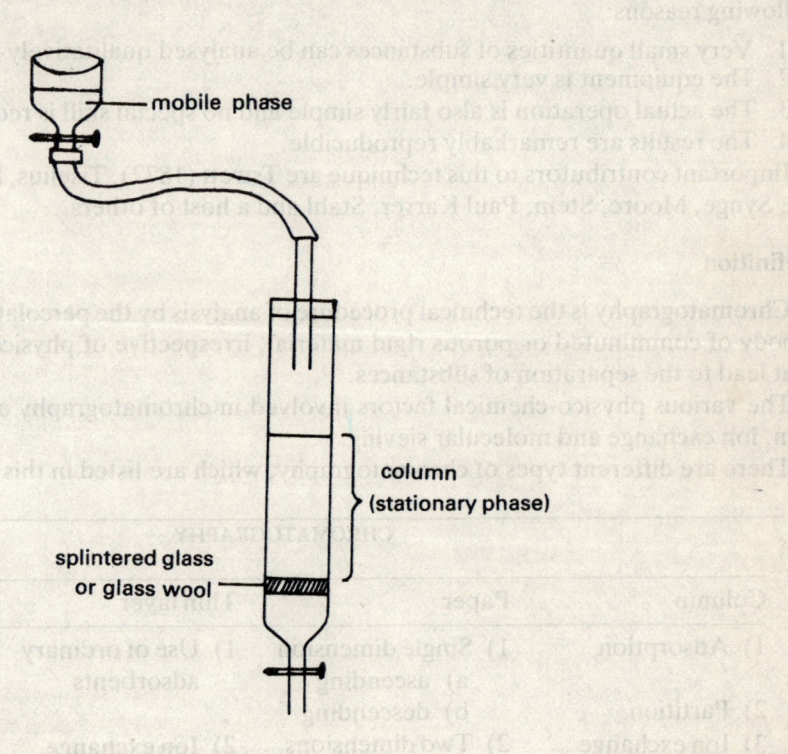

If the separating funnel is fitted with the help of a rubber tube, the rate of flow could be adjusted by increasing or decreasing pressure. While alumina is a general adsorbent, special adsorbents like bauxite for sugar, magnesia for carotenoids, calcium phosphate for proteins and silica gel for sterols and fatty acids could be used. The substances to be separated are dissolved in the smallest possible volume of a suitable solvent and applied on the top of the stationary phase with a pipette bent at right angles. As soon as the solution containing the sample has soaked into the adsorbent, it is followed by a small volume of wash liquid from the same vessel and finally by pure solvent.

In the case of coloured substances like chlorophylls and carotenoids, the separation of bands could be seen with the naked eye. Once individual bands are formed, they can be further processed by eluting with a suitable solvent and collecting individual fractions in different containers, or by removing the zones with the adsorbent (extruding) and extracting the material preferentially from the adsorbent. In other cases, fractions may be collected either at various time intervals or flow volumes and the individual fractions analysed. Automatic fraction collectors are available. By doing various tests, i.e., chemical, fluorescence, autoradiography for radioactive materials and counting the radiations, the contents of different fractions can be identified.

Column chromatography, using alumina, works on the basis of the physical factor of adsorption only and hence is called *adsorption chromatography*.

Partition can be the factor, e.g., in a column of gel or modified celluloses where the partition of the substances takes place between the gel (stationary phase) and the mobile phase.

If the column is an ion-exchange resin like Amberlite or modified celluloses, the chief factor is ion exchange. The chromatography in such cases is called 'ion exchange column chromatography'. The ion exchange resins may have a hydrophobic (water-hating) core as in the case of Amberlite or Dowex, and these in general should not be used for the separation of proteins or nucleic acids, which are hydrophilic in nature. For them, resins of hydrophilic nature, i.e., modified celluloses like diethyl amino ethyl cellulose (DEAE cellulose) resins, should be used. Again, there are cation exchangers and anion exchangers. The cation exchangers may be strongly acidic having, say, the sulphonic acid group or the weakly acidic (carboxylic acid) group, and the anion exchangers could be strongly basic (quaternary ammonium group) and weakly basic (tertiary and other types of amines). The substances separated should be ionisable and the ions of the substances are held together by the ionisable groups of the resin by electrostatic attraction. For instance, if a mixture of amino acids (at a pH of about 3) is passed through a cation exchanger with the sulphonic acid groups, the following interaction takes place:

$$\text{Resin } C_6H_4 - SO_3^- \cdots\cdots\cdots\cdots H_3N^+ - CH(R) - COOH$$

(Resin surface having ionisable groups) Cation exchanger   (amino acid, cationic form)

After the ionisable substance is thus held by the resin, using buffers of different pH or even HCl, the amino acids held can be released and eluted. If buffers of different pH are employed, as the dissociation of amino acids is influenced by pH, the acidic, basic and neutral amino acids can be conveniently separated, say, from a protein hydrolysate, and this is the principle of the auto analyser for amino acids introduced by Moore and Stein.

In column chromatography, in which the factor of molecular sieving is involved, gels of starch or dextran (Sephadex), with pores on the gel matrix are employed. When a heterogeneous mixture of, say, proteins and sodium chloride is passed through, the small ions of sodium and chloride enter the gel granules and come out to the mobile phase while the bigger protein particles cannot enter the gels in view of their size. They move only with the mobile phase along the interstices of the gel granules. Thus a sort of sieving takes place at the molecular level and hence this is called molecular sieving or molecular exclusion. As gels are involved in this, it is also called gel filtration. The volume between the gel granules, is referred to as the 'void volume'. As the bigger molecules cannot enter the gel, and are taken through the void volume, they come out first. This can be conveniently applied to separate the proteins from salts, urea, etc. in the urine sample from a patient with nephrotic syndrome. Choosing different gels with varying pore sizes, bigger molecules can also be fractionated. Sephadex gels G 25, G 50, G 100, G 200 etc, are commercially available for the same. The technique is employed for separation and purification of proteins and enzymes and determination of their molecular weights.

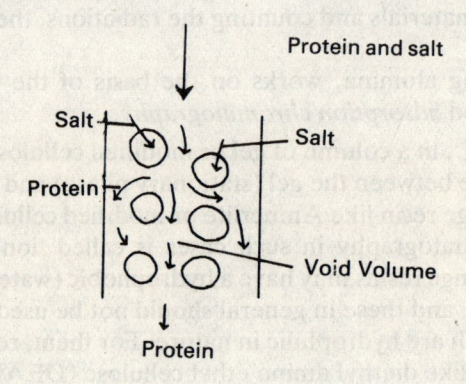

Affinity chromatography

Affinity chromatography is a type of column chromatography which is useful for the purification of enzymes, nucleic acids, antibodies, proteins, vitamins, drugs and hormone receptors. It can also be used for the concentration of dilute protein solutions.

In this type of chromatography, the column is a water-insoluble carrier. It can be inorganic like porous glass beads or organic like the matrix of cellulose derivatives, beaded agarose or poly-acrylamide gel.

The protein, enzyme or other substances to be purified may not combine directly with the water-insoluble column. For making them to selectively combine with the column material, a different substance known as the ligand is used. The ligand by certain groups can be made to unite (in most cases covalently) with the column matrix (say, beaded agarose) and by the other groups with the macromolecules like enzymes, nucleic acids, etc. (also by covalent linkage) as given below:

Beaded agarose	—	ligand	—	enzyme
polyacrylamide	—	ligand	—	nucleic acid

Thus, the macromolecule is made to bind preferentially while the other substances in the mixture like inhibitors and contaminants pass through the column freely. After suitable washing, the macromolecule-ligand bond is broken by suitable eluants and the macromolecule isolated.

Typical examples are given below:

1) $\text{Agarose} + H_2N(CH_2)_x-NH_2 \xrightarrow{CNBr}$
   ligand

   $\text{agarose - NH}(CH_2)_x - NH_2$
   $\downarrow \text{Protein - COOH + contaminants}$

   $\text{agarose - NH - }(CH_2)_x - NH - CO - \text{Protein}$
   $+ \text{contaminants freed}$

   $\text{agarose - NH}(CH_2)_x - NH - CO \text{ Protein} \xrightarrow{\text{eluant}}$

   $\text{agarose - ligand + Protein (purified)}$

2) $\text{Polyacrylamide-}\overset{O}{\underset{\|}{C}}\text{-NH}_2 + H_2N - NH_2 \longrightarrow$
   (Resin)　　　　　　　　　ligand

   $\rightarrow \text{Resin - }\overset{O}{\underset{\|}{C}}\text{ N}_3 \xrightarrow{H_2N\text{-protein}} \text{Resin-}\overset{O}{\underset{\|}{C}}\text{-NH-Protein}$
   azide

The azide group can combine with the amino groups of proteins to form the complex.

A list of some enzymes purified by affinity chromatography with the ligand and insoluble matrix is given below:

Matrix (column)	Ligand	Enzyme
Cellulose	Aminophenol	Tyrosinase
Agarose	ADP	Pyridoxal kinase
Polyacrylamide	P-amino phenyl thio galactoside	Beta galactosidase
Porous glass	NAD^+	Alcohol dehydrogenase

Affinity chromatography is very helpful in determining the existence and role of the multi-enzyme complex in lipogenesis.

Paper chromatography

The chief factor responsible for the separation of ingredients by paper chromatography is partition. Any paper used should have space in the interstices, because it is in the interstices, that the molecules of one phase are embedded. Hence, papers like Whatman No.1 and 3 are used. The paper functions as an inert support. When the mobile phase moves along the paper, the substances applied as a spot on the paper are partitioned between the mobile solvent and the molecules of the static solvent embedded in the interstices of the paper. As the partition coefficients may be different for different substances, the rates of flow of substances along the paper are also different and thus separation takes place.

Different types of substances like amino acids, sugars and pigments can be conveniently separated.

In the actual method, the required amount of the substance, usually a few micrograms (about 25µg for amino acids), in a suitable solvent (aqueous iso-propyl alcohol for amino acids) is applied as a spot to a diameter of about 0.5 cm with the help of a capillary tube or pipette and hot air blower. The spots should be about five cm from one end of the paper. The length and breadth of the paper will be dependent on the type of chromatography (single dimension or two dimensions), the number of spots to be applied and the size of the cabinet used. For a single dimension and for 3-4 spots, a paper of 15 cm × 50 cm can be used conveniently. Spots of known substances could be applied side by side if necessary. The chamber should have both the mobile phase and stationary phase and should be saturated with the concerned vapours.

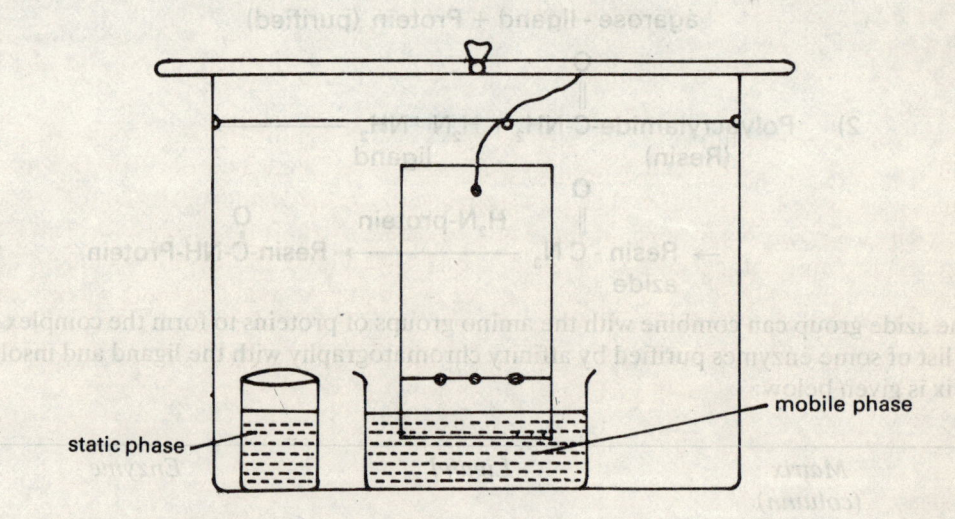

For example, for the separation of amino acids, the solvent is n-butanol, acetic acid and water 4:1:5. These are mixed in a separating funnel and shaken. The upper layer is butanol saturated with water which is the mobile phase and the lower layer is water saturated with butanol which is the stationary phase. The butanol layer is taken in the trough while the water layer is taken in a small beaker kept at one side of the chamber. The chamber having the two liquid systems and the paper with spots should be kept closed for about 30 minutes for saturation with the molecules of the system. It should however be noted that at this stage the paper should not touch any liquid system. Then, the end of the paper nearer to the spots is dipped in the mobile phase (butanol saturated with water) in the trough as shown in the diagram. The mobile phase moves up along the paper uniformly and pushes the substances in the spot. These substances get partitioned between the mobile phase and the static phase (in the paper). The liquid in the beaker (static phase) is only to have the paper always saturated with the static phase.

The paper is kept in such a position for about 12 – 16 hours, then taken out carefully, the maximum distance reached by the solvent is noted (solvent front) and the paper is dried in the air. Nothing will be seen in the paper. The paper is now sprayed with a solution of 100 mg per cent ninhydrin in acetone and dried in an oven at about 100°C for about five minutes. The amino acids separated react with ninhydrin and purple spots are produced. Proline and hydroxy proline give yellow spots. Identification can be made by having known amino acids side by side or by calculating Rf (ratio of fronts) and reference to literature, i.e., the centre of the spots is marked, and the distance of the centre of the spots from the starting point is measured. This distance divided by the solvent front gives the Rf value. Nowadays, instead of the solvent front, the Rf leucine is taken as '1', i.e., the distance travelled by leucine (in butanol system) is taken, instead of the solvent front, and the Rf for other amino acids is calculated. An advantage of this is that even if the solvent flow exceeds the periphery of the paper, the leucine spot will be in the paper itself and hence the Rf for all amino acids can be calculated. If estimation is to be done, the paper containing the coloured spots is cut suitably and the coloured material is eluted with aqueous alcohol and the intensity of the colour measured in a colorimeter. Or an instrument known as the Densitometer can be used to measure directly the intensity of the colours on the paper itself without elution.

Sugars like glucose, fructose and lactose can be separated by paper chromatography and the spraying reagent in their case is aniline oxalate, aniline phthalate, etc.

If the mobile phase moves, upwards as shown in the figure, it is 'ascending chromatography'. If it is made to move downwards, in which case the trough will be fitted above, it is called 'descending chromatography'.

If only one solvent is made to move as discussed above, it is likely that two or more substances will have the same Rf and will form a single spot. In such cases, 'two-dimensional chromatography' is employed, using a separate solvent for the second dimension. After drying the paper through which one solvent (say, butanol) has moved, the paper is rotated by 90° and dipped in another solvent (say, pyridine) system and chromatography continued. What was not possible by one solvent will be effected by the second and there will be lateral separation of the amino acids which have the same Rf in one solvent but different Rf in the second solvent. As in the case of single dimension chromatography, there could be ascending or descending techniques employed for two-dimension paper chromatography also.

Whatman filter paper discs have also been used, in which the liquid reaches the centre of the circular disc through a wick and spreads, in a circular fashion. This is called 'circular chromatography' which has only limited application now.

Instead of the water phase in the paper and a mobile organic phase as is usually employed, the organic liquid phase such as silicone grease is taken in the paper by dipping the paper at the beginning in the grease and if water is the mobile phase, such paper chromatography is referred to as 'reversed phase chromatography'. This technique is employed for the separation of esters of fatty acids, dinitrophenyl hydrazones, etc. and in HPLC.

Recently, papers made of ion exchange resin material as well as papers impregnated with ion exchange resins are used with advantage. One example is the cellulose acetate paper. Resolution in the ion exchange resin papers is better in view of an additional physico-chemical factor, i.e., ion exchange.

Thin layer chromatography (TLC)

This is an improvement over paper chromatography and uses thin layers of slurried adsorbents like silica gel, alumina, cellulose, kieselguhr, modified celluloses and polyacrylamide, on suitable glass plates. To have the thin layers firmly bound to the glass plates, one of the binders, like Plaster of Paris, gypsum, starch etc., is added to the substance used to give the thin layer. The method of running the chromatography is almost the same as that for paper. The advantage of TLC over paper chromatography is that the time required for development is considerably less, i.e., 1–2 hrs in TLC, and aggressive reagents like 50 per cent H_2SO_4 can be used for spraying (provided the thin layer is made of inorganic material like silica gel) which is not possible with paper. About 50 per cent H_2SO_4 is an universal spraying agent as all organic substances are reduced to carbon and thereby reveal themselves as black spots in TLC. The amino acids to be applied could be even as low as 2μg. TLC has more sensitivity than paper chromatography.

Gas chromatography (GLC)

This is actually gas liquid chromatography. The substances to be separated are carried as vapours in an inert gas like argon, nitrogen or helium over liquids when there is partitioning of the substances between the gas and the liquid. The liquids used are silicone oils, lubricating greases, etc. held in inert solids like diatomaceous earth or ground firebrick. Glass or metal tube 1–2 metres long and of 0.2–2 cm diameter could be used. If fatty acids are to be separated, they are first converted to methyl esters which are easily evaporated. The vapours are swept constantly by nitrogen through the tube containing the liquid phase kept at a temperature of 170–225°C so that the vapours of the esters could remain as vapours. Due to the partitioning of the esters between the gas and the liquid, separation takes place. The ingredients can be identified by physical or chemical means or flame ionisation.

Identification by thermal conductivity cell or hydrogen flame ionisation: Two fine coils of wire with a high temperature coefficient of resistance are placed in parts of metal C_1 and C_2 (see figure). Electrical resistors are inserted in the circuit of C_1 and C_2 to form Wheatston's bridge. Final equilibrium temperature of the wire depends on the thermal conductivity of the gas passing over C_1 and C_2. If carrier gas passes through C_1 and the carrier gas with the sample vapour passes through C_2, the wire temperature differs. Electrical resistance will therefore change and the bridge is unbalanced. Different samples will cause different resistances. The extent of imbalance is measured in a potentiometer.

In the hydrogen flame ionisation, the organic material is burnt in a hydrogen flame, and electrons and ions are formed. The negative ions and electrons move in a high-voltage field to anode and produce a very small current which is amplified. The electrical current is directly proportional to the amount of material burnt.

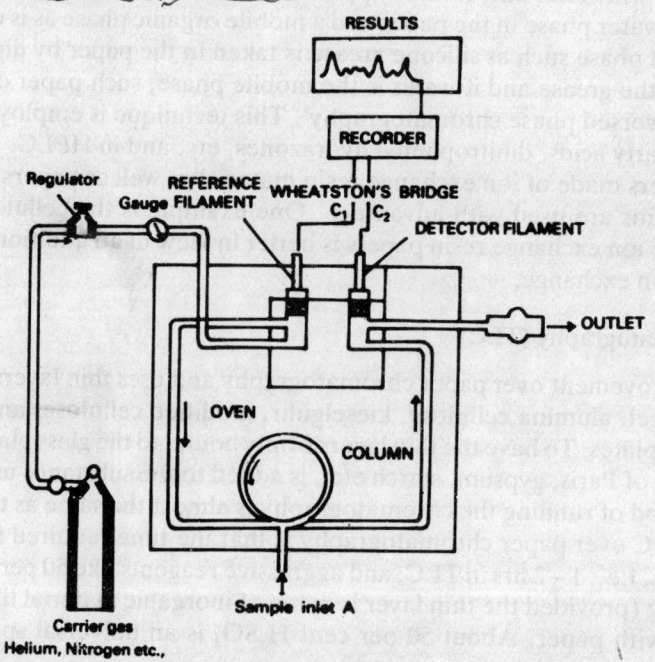

High performance liquid chromatography

One of the very recent and popular techniques is High Performance Liquid Chromatography (HPLC). This can be used for the resolution of a wide variety of compounds and drugs available in very small quantities and can be done in a short time. HPLC is indispensable in laboratories which do advanced research and sophisticated analysis of hormones, proteins, drugs etc.

HPLC has the following special features: The particles in the column matrix are extremely small, and increased pressure is applied for the movement. The column is narrow and pressurised with high resistance to the flow of mobile phase.

The HPLC equipment consists of 1) a solvent delivery system, 2) a sample injector, 3) a column, and 4) a detector.

The solvent delivery is effected by solvent pumps which generate pressure at several thousand psi and direct a pulse-free delivery of the solvent. Constant Pressure, Gas Displacement and Constant Volume pumps are used for the purpose.

The sample injection is done with a syringe. A valve with attached sample loop is an adjunct.

The column is usually of stainless steel of 2–5 mm internal diameter and 10–50 cm long.

The detectors may be 1) ultra-violet with fixed wavelength Hg lamp 254 nm or variable wavelength with dueterium lamp, 2) refractive index detectors of deflection type or Fresnel type, and 3) fluorescence detectors.

Usually, the technique applies a reverse phase system with non-polar stationary phase. The bonded phase is of stable silicones say Octa decyl silane (ODS).

The method is applicable forthe separation of carbohydrates, proteins, peptides, amino acids, vitamins, steroids, amines like the biogenic amines and the polyamines, neuro-peptides, hormones and drugs. Separations can be done in minutes, and derivatizations are not usually required if suitable columns are employed.

ELECTROPHORESIS

A technique which is as popular as chromatography is electrophoresis. In fact, electrophoresis is employed to a much greater extent in clinical laboratories than chromatography. While chromatography is done for the detection of amino acids in urine in diseases like phenylketonuria, cystinuria, Fanconi syndrome and collagen diseases which are not very common, electrophoresis is done almost daily for the separation of serum proteins in various diseases.

Electrophoresis is also popular for the same reasons as chromatography is i.e., only 'micro' quantities of biological materials are required, the equipment required is quite simple and so is the method of operation. The results are reproducible with accuracy.

Electrophoresis can be called electro-chromatography in that finally there is colour development, if suitable stains are used. One main difference between the two techniques is that the substances are made to move with the help of electric current in electrophoresis while electricity is not required for chromatography.

Some of the main workers in the field are Tiselius, Lederer, Durrum, Kunkel and Stein.

Definition

Electrophoresis is the movement of particles or molecules towards one of the electrodes under the influence of an applied electric current.

As the particles or molecules have to move under the influence of electric current, the essential pre-requisite is that they should be electrically charged. Sugar molecules, which are not charged, will not exhibit electrophoresis while proteins which are charge move easily. Enzymes being proteins could be subjected to electrophoresis. The charge on the protein is due to partial ionisation of groups like -COOH and -NH$_2$ in the molecule and the electrokinetic potential of the colloidal protein particle. As all the particles move towards one electrode, it is clear that all have similar charges.

Depending upon the mode of operation and separation, electrophoresis also is classified into various types as shown below.

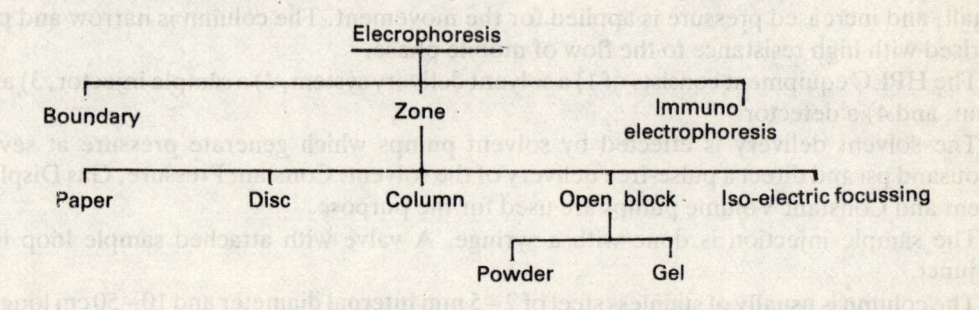

```
                        Elecrophoresis
        ┌───────────────────┼───────────────────┐
     Boundary              Zone              Immuno
                                           electrophoresis
    ┌──────┬──────┬──────┬──────┐
  Paper  Disc  Column  Open block  Iso-electric focussing
                        ┌────┴────┐
                     Powder      Gel
```

Of the three major types, one is the boundary electrophoresis which is at present seldom employed. All the components might not separate into distinct regions while, however, the fastest and the slowest moving fractions can be realised clearly. There may be considerable diffusion.

In zone electrophoresis, which is very popular, there are well-defined and permanent zones between the different fractions separated. In view of this, preservation, interpretation and quantitative estimation are done conveniently with zone electrophoregram. Paper and gel electrophoresis are important types of zone electrophoresis. If human blood serum is subjected to electrophoresis on paper or agar-gel plate, the proteins are separated into five bands, i.e., albumin, alpha$_1$ globulin, alpha$_2$ globulin, beta globulin and gamma globulin.

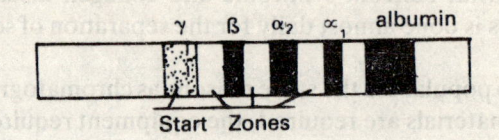

$\gamma \quad \beta \quad \alpha_2 \quad \alpha_1 \quad$ albumin

Start Zones

The electrophoretic migration depends on various factors, i.e., the intrinsic charge on the particle and its diameter, pH of the buffer used, temperature of operation, intensity of current strength and dilution of the sample.

The papers commonly used are Whatman No.1,2,3, Schleicher and Schull and even blotting paper. For routine investigations, veronal buffer (sodium diethyl barbiturate, pH 8.6) is employed, though for a better resolution, Tris buffer (pH 8.9) which separates serum proteins into nine fractions can be used. Two types of arrangement of paper are possible depending upon the apparatus. The paper can be kept horizontal in one case and vertical in the other.

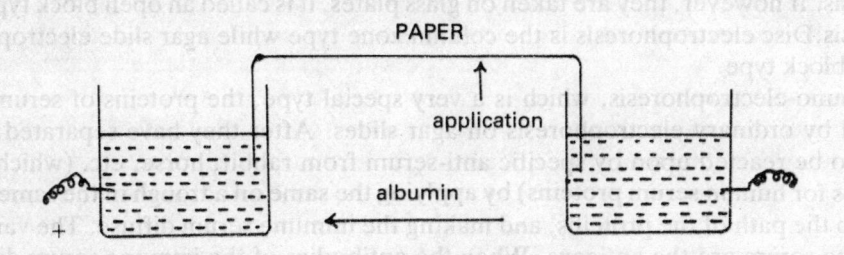

In the actual method, the paper might be about 32 cm long and 6 – 8 cm broad. It is moistened with the veronal buffer, with its free ends dipping in the buffer on either side. About 15 – 40 µl of serum is applied carefully as a thin line at the side of the paper near the cathode as shown in the diagram. This is to provide good distance for the fastest moving negatively charged albumin to migrate fully and still be within the limits of the paper.

A direct current of 0.5 – 7 mA with the help of a suitable powerpack is applied, with a voltage of operation of about 120 volts. The apparatus is kept closed with the lid provided and electrophoresis conducted for about 16 hours. The paper is removed, dried and stained with bromophenol blue (0.1 per cent solution in alcohol saturated with mercuric chloride) with or without acetic acid, when the proteins separated assume a blue colour. Other stains for proteins are amido black, azocarmine B etc. If lipoproteins are to be revealed, the technique is the same while the stains are different. Oil Red O and Sudan III are the usual stains for lipoproteins. Hemoglobins also can be separated by electrophoresis and in their case they need not even be stained as they themselves are coloured. Enzyme amylase can be revealed by starch iodide while esterase is made out by p-nitrophenyl butyrate.

If sugars have to be separated by electrophoresis, they should be converted to borate complexes (with boric acid) which are electrically charged.

The electrophoregrams can be interpreted by visual comparison with known patterns or by cutting the bands, eluting and using colorimeters or scanning by densitometers.

Recently, ion exchange papers like cellulose acetate paper are employed. In this, the adsorption which happens to some extent in ordinary papers is avoided and there is an additional factor of molecular sieving.

Gel electrophoresis

Instead of paper, microscopic glass slides coated with, gels of agar, starch and polyacrylamide could be used. There are a number of advantages in gel electrophoresis. There is considerable saving of time, as electrophoresis is done in about 30 minutes. There are no adsorption problems while the molecular sieving factor helps in better resolution. In view of the transparency of the medium, optical measurements could be done with better accuracy. Gels could also be had as columns. In Smithies' starch column electrophoresis, various fractions like prealbumin, post-albumin, haptoglobins, etc. normally not obtained in slides are separated. In disc electrophoresis using polyacrylamide gel also, many additional fractions of serum proteins are separated. In disc electrophoresis, the polyacrylamide gel is taken as a vertical cylindrical column about 10 cm long and the serum is applied on the top and electrophoresis conducted using suitable buffers.

When the powder or gel is taken as a column, it constitutes the column zone electrophoresis. If however, they are taken on glass plates, it is called an open block type of electrophoresis. Disc electrophoresis is the column zone type while agar slide electrophoresis is the open block type.

In immuno-electrophoresis, which is a very special type, the proteins of serum are first separated by ordinary electrophoresis on agar slides. After they have separated, they are allowed to be reacted upon by specific anti-serum from rabbit, horse, etc. (which contains antibodies for human serum proteins) by applying the same on a trough in the same agar side parallel to the path of the proteins, and making the immune serum diffuse. The various proteins of the serum are the antigens. When the antibodies of the immune serum diffuse and meet the antigenic proteins there is antigen-antibody reaction with formation of opaque lines in the form of ares (precipitins). As many as 30 fractions of serum proteins are known to be present and this has been revealed only by immuno-electrophoresis.

Iso-electric focussing

This is a very recent improvement in the technique of electrophoresis to reveal the heterogeneity and micro-heterogeneity of proteins and enzymes. Proteins move under the influence of applied electric current which is the primary gradient. A secondary gradient namely pH is super-imposed and this causes the molecular species to resolve further at equilibrium positions. In the usual electrophoresis, the pH of the buffer will not vary during the process.

Proteins or enzymes are put into a system of suitable carrier ampholytes (e.g., polyamino polycarboxylic acids) and electrolysis is done. The ampholytes cause a pH gradient, the more acid ampholytes with a lower iso-electric point will be focussed near the anode and the more alkaline ones with a higher iso-electric point near the cathode. Thus in the electrophoresis cell, the pH will show increasing values from anode to cathode. Each protein or enzyme will focus sharply at its respective iso-electric pH in the resulting pH gradient.

Using iso-electric focussing in polyacrylamide gels, the micro-heterogeneity has been shown for gamma globulins, ovalbumin, etc. That of myosin, globin chains of hemoglobin, myeloma proteins, $\propto$fetoprotein, lipoproteins, etc., has been exposed elegantly by this technique. Aldolase separated into five components, while the crystalline enzyme L amino acid dehydrogenase separated into as many as 18 components with iso-electric pH values ranging from 5.2 to 8.4. Enzyme, enzyme-substrate complex, enzyme-NAD^+ combination, etc. are shown by this technique.

By coupling iso-electric focussing with immuno-diffusion, the micro-heterogeneity of protein antigens can be analysed with precision. This technique is known as immuno-isoelectric focussing.

PHOTOMETRY

Photometry is the technique used to study the chemical nature, or more frequently the concentration of substances by employing the property of absorption of light of a definite wavelength by molecules or particles. This analytical procedure also requires small quantities of substances for simple and rapid estimation which offers a high order of accuracy. Unlike chromatography and electrophoresis, substances can not be separated by photometry but can only be estimated.

When light of a particular wavelength, say monochromatic light, is passed through the solution of a substance, it comes out with diminished intensity, as part of it is absorbed. For using this property for analytical work, the absorption of light by the substance should obey Beer-Lambert's laws.

Beer's law states that the intensity of a ray of monochromatic light decreases exponentially as the concentration of absorbing material increases (i.e., the proportion of light absorbed depends only on the total number of absorbing molecules through which the light passes).

$$\log \frac{I_0}{I} = K_1 C$$

where I_0 is the intensity of the light entering the solution and I the intensity of light emerging from the solution, C is the concentration of the solution in g/litre or mg/100 ml and K_1 is a constant.

Lambert's law states that the proportion of light absorbed by an absorbing substance is independent of the intensity of the incident light (i.e., when a ray of monochromatic light enters an absorbing solution, its intensity decreases exponentially with an increase in the length of the solution traversed).

$$\log \frac{I_0}{I} = K_2 l$$

Where l is the length of the solution traversed.

Combining the two laws,

$$\log_{10} \frac{I_0}{I} = KCl$$

The logarithmic ratio of I_0 and I is called the optical density (OD) or the extinction E or Absorbance. Substituting E in the above equation E = KCl. Hence, when light passes into a solution of an absorbing substance in a non-absorbing solvent, the extinction (E) is directly proportional to the depth of the solution (l) and the concentration (C) of the absorbing substance.

Transmission is the ratio of the transmitted light to that of the incident light.

$$T = \frac{I}{I_0}$$

Percentage transmission is the transmission when I_0 is equal to 100.

Relation between E and percentage transmission

$$E = KCl = \log_{10} \frac{I_0}{I}$$

But $I_0 = 100$

$$E = \log_{10} \frac{100}{I}$$

$$E = \log_{10} 100 - \log_{10} I$$

$(\log_{10} 100 = 2)$

$$E = 2 - \log_{10} I$$

Here I is the percentage transmission (T) as I_0 was 100

Hence $E = 2 - \log_{10} T$

In all the instruments used in photometry, the above principle is employed in estimating the substances. If the solute or the product of the reaction is coloured, the wavelength of light used for photometry is restricted to the visible region of the spectrum, i.e., 400–800 mμ (4000–8000Å). However, even colourless solutes can be estimated by photometry, for which a special instrument called spectro-photometer is required (the colourless purine bases of nucleic acids absorb light at about 260 mμ).

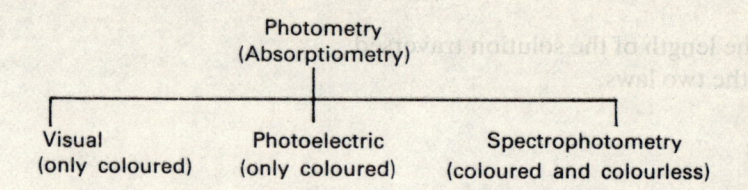

```
                    Photometry
                  (Absorptiometry)
                         |
     _____|_____
     |                   |                  |
   Visual           Photoelectric      Spectrophotometry
(only coloured)   (only coloured)   (coloured and colourless)
```

Colorimetry

Photometry becomes colorimetry if the solution is coloured. Light is absorbed by the coloured complex produced in the reaction. Two major types of colorimeter are employed: 1) the visual colorimeters, e.g., Hellige Dubosh, and 2) the photo-electric colorimeters, e.g., Eel colorimeter. In both cases, standard solutions of the substances are treated exactly like the test samples.

In the visual colorimeter, the length of the column through which light passes is changed with the help of plungers so that the intensity of transmitted light of both the standard and the test (I) as seen through the eye-piece is equal. The intensity of incident light (I_0) is also the same for both.

If C_1 the concentration of standard, l_1 the length of the solution column corresponding to standard and l_2 the length corresponding to the test solution are known, the only other unknown, i.e., C_2, the concentration of the test sample is calculated.

$$E_{std} = K\,C_1\,l_1$$
$$E_{test} = K\,C_2\,l_2$$

$$\text{But } E = \log_{10}\frac{I_0}{I}$$

As I_0 are1 are the same, E_{std} and E_{test} are the same. So

$$K\,C_1\,l_1 = K\,C_2\,l_2$$
$$\frac{C_1}{C_2} = \frac{l_2}{l_1}$$

In photo-electric colorimeters, the principle is different from that of visual colorimetry. Here, the length of the liquid column is kept constant for the standard and test as no matching of the colour intensity could be done.

The actual intensity of the transmitted light is measured by using a photocell and a galvanometer. Light energy is converted to electrical energy by the photo cell and this electricity is measured in a galvanometer. Either the percentage transmission or the optical density or both could be read from the galvanometer. Suitable filters are provided for admitting light of desired wavelength.

$$E_{std} = K\,C_1\,l_1$$
$$E_{test} = K\,C_2\,l_2$$
$$l_1 = l_2 \text{ (as lengths of solutions are constant)}$$

$$\text{Hence } \frac{E_{std}}{E_{test}} = \frac{C_1}{C_2}$$

If the extinction due to the known concentration (C_1) of the substance and the extinction of the test are known, the concentration of the unknown (C_2) can be calculated.

Instead of a single standard, a number of standards can be used and a graph drawn plotting the concentrations of the substance and extinctions. The use of a graph is desirable instead of a single concentration.

The extinction, if any, produced by the reagent blank should be subtracted from that due to the test and the standard.

There are colorimeters with a single photocell, e.g., eel, lumetron, etc., and the double photocell type, e.g., Klett Summerson colorimeter.

The photocell may be of two types, i.e., the barrier type and the high vacuum cell type. The barrier type consists of a photo-sensitive surface of selenium on a plate of copper or iron covered by a light transmitting layer of metal. The high vacuum cell type consists of a highly evacuated cell with a composite caesium cathode and requires application of a potential.

The spectro-photometer is an improvement over the photo-electric colorimeter as it works at various ranges of wavelength. Infra-red spectro-photometers work at the wave lengths above 800 mμ, while ultra-violet spectro-photometers work at wavelengths below 400 mμ. The latter is more useful in a biochemical laboratory in the estimation of many dehydrogenase-enzymes which require the extinction to be taken at about 340 mμ, as also purines and pyrimidines of nucleic acids with E at about 260 mμ. Many spectro-photometers work over a wide range of wavelengths, — Beckmann model DK_1 has a range of 185 to 3500 mμ while model DU works from 210 to 1000mμ.

The spectro-photometer is superior to photo-electric colorimeters in that it uses prisms or gratings as monochromator and the wave-length of incident light can be adjusted precisely. Hence, accuracy is greater with the spectro-photometer. Relatively smaller volumes of liquids and even samples of solids and gases which can not be used in colorimeters can be used.

Fluorimetry

Substances like thiochrome, riboflavin, etc. on irradiation emit light of a precise wavelength. Fluorescent light is used in the Fluorimeter for this purpose. Since the fluorescent light is emitted in all directions, the photocell is placed at right angles to the path of the incident beam. The rest of the procedure is analogous to photometry.

Flame photometry

In this technique, a definite colour is imparted to the non-luminous flame by the burning of, say, sodium ions, potassium ions, etc. The intensity of colour is proportional to the concentration of the ions. The different ions in serum will give the emission spectrum due to their burning. By using suitable filters, the light due to any one ion is allowed to fall on the photocell and measurements are taken in the galvanometer. Flame photometer is a versatile instrument used for the estimation of serum Na and K in the routine of a clinical biochemical laboratory.

Electron microscopy

While the nuclei and mitochondria of cells can be seen in a light microscope, smaller particles which have diameters of about 15 nm can be identified only by the electron microscope.

Electrons behave in many ways as waves of electro-magnetic radiation. A homogeneous beam of electrons will be absorbed or diffracted like the electro-magnetic waves of visible light by the components of a thin slice of material. Adjustable electro-magnets are used to guide the direction of the flow of electrons. The entire field of travel including the specimen should be in very high vacuum, as electrons will be stopped by colliding with gas molecules. The pattern of the emerging beam of electrons which is decided by the nature of the sub-cellular particles is visualised by using a fluorescent screen or is recorded photographically.

The material is embedded in plastic and extremely thin sections are cut. Osmic acid, which is an electron-dense substance, is often used to stain the material. Putting electron-dense phospho-tungstic acid in the surrounding medium (negative staining) is also practised. Phospho-tungstic acid does not penetrate the protein structure. Heat produced when the electrons strike the specimen is minimised by using very thin sections and by cooling in liquid nitrogen.

The elucidation of the ultra-structure of the cell has been made possible only by this technique of electron microscopy.

Respirometry (Manometry)

The study of oxygen uptake by tissue slices, homogenates or sub-cellular particles has to be made in many instances. Also, enzyme substrate interactions are effected and the level of enzyme is expressed in terms of oxygen uptake. For this, Warburg's manometer (Respirometer) is used. It actually measures the change in pressure at constant volume and constant temperature. A special flask with a side arm and a centre well is used. In the side arm, the homogenate or enzyme preparation or serum (source of enzyme) is taken, while the medium or substrate is taken in the flask. In the centre well, a strong alkali with a filter paper strip is taken to absorb the CO_2 produced during respiration or reaction. The flask is attached by means of a gas-tight joint to a capillary manometer containing a device for returning the manometric liquid to a fixed volume mark.

The homogenate and the medium or the enzyme and the substrate are equilibrated at 37°C in an atmosphere of oxygen or air as required. The substances are then mixed by tilting and the Warburg's flask is agitated at 37°C in air-tight condition. Due to respiration, oxygen will be consumed while the CO_2 formed is absorbed by KOH in the centre well. The contraction in volume due to CO_2 is measured at different time intervals and is equivalent to the oxygen consumed. It is the pressure difference that is noted in the manometer. But the volume is calculated by using the 'flask constant'. From the volumes of oxygen consumed for a definite time interval and the volume of the homogenate, weight of the tissue etc., the amount of oxygen respired by the tissue can be calculated for unit time. In enzymology, the quantity of enzyme in a sample can be expressed as micro-moles of oxygen consumed by unit weight of the tissue in unit time, when the enzyme of the tissue reacts with the required substrate in the flask.

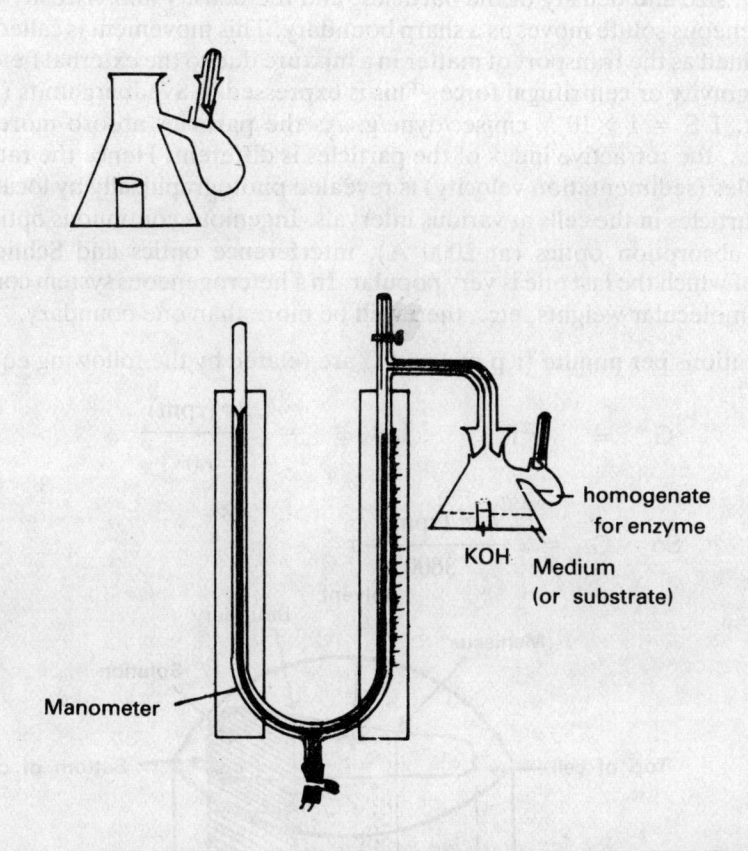

ULTRA-CENTRIFUGATION

Ultra-centrifugation is an important technique introduced by Svedberg and is responsible for the tremendous advancement in Biochemistry, especially in the field of proteins and nucleic acids. It is the technique of centrifugation of solutions at very high speeds, the centrifugal force being many thousands of times stronger than the force of gravity(g), and at very low temperatures and preferably in vacuum. The rotor revolves at high speeds, say 60,000 times per minute, giving a centrifugal force of the order of 500,000 times g. The centrifuge has refrigeration to offer low temperature. Glass or quartz cells can be used and volumes of solutions less than even 1 ml can used for analysis. There are different types of ultra-centrifuges like 1) the preparative, and 2) the analytical models. The analytical model incorporates the optical system and hence costlier than the preparative model.

The centrifugal force depends on the angular velocity of the centrifuge and the distance of the particles from the axis. The rate of movement of the particles depends on the centrifugal force, shape, size and density of the particles, and the density and viscosity of the medium. The homogeneous solute moves as a sharp boundary. This movement is called sedimentation which is defined as the transport of matter in a mixture due to the external field, in particular, that due to gravity or centrifugal force This is expressed in Svedberg units (S) in honour of the scientist. $1\ S = 1 \times 10^{-13}$ cm/sec/dyne/g. As the particles absorb more light than the medium does, the refractive index of the particles is different. Hence the rate of movement of the particles (sedimentation velocity) is revealed photographically by locating the boundary of the particles in the cells at various intervals. Ingenious continuous optical systems like ultra-violet absorption optics (at 2000 A), interference optics and Schlieren optics are employed, of which the last one is very popular. In a heterogeneous system containing solutes of different molecular weights, etc., there will be more than one boundary.

The revolutions per minute (r.p.m.) and G are related by the following equations:

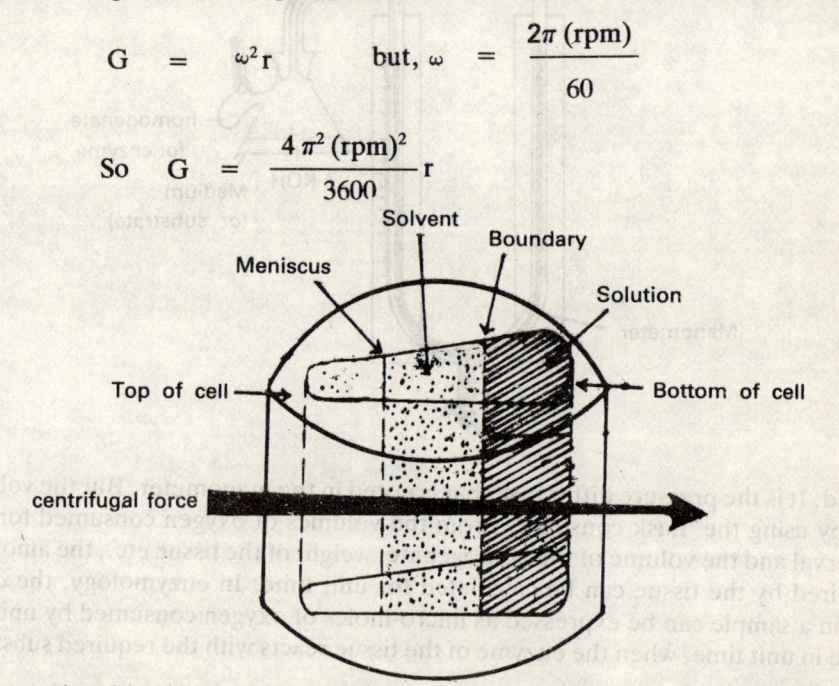

$$G = \omega^2 r \qquad \text{but,} \quad \omega = \frac{2\pi\ (\text{rpm})}{60}$$

$$\text{So} \quad G = \frac{4\pi^2\ (\text{rpm})^2}{3600} r$$

Centre section of the ultra centrefuge cell showing sector - shaped cavity containing sample material

Graph of solute concentration

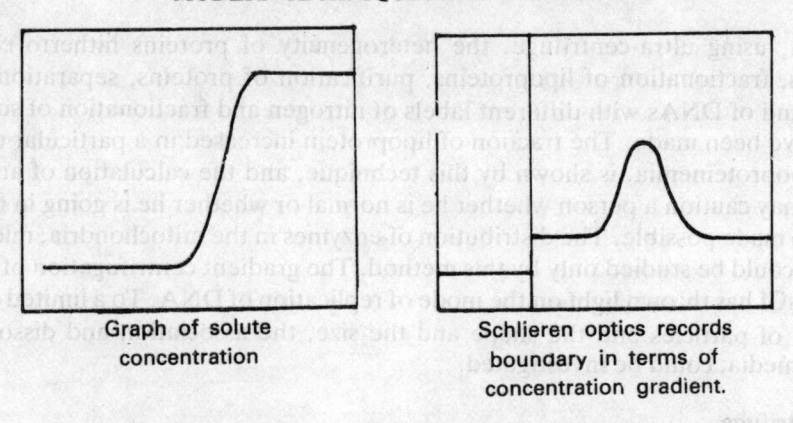

Schlieren optics records boundary in terms of concentration gradient.

Applications of ultra-centrifugation

One of the important uses of ultra-centrifuge is in the determination of the molecular weight of many proteins with some precision, as many other methods do not give accurate results.

The molecular weight of proteins, nucleic acids, enzymes, etc. can be determined either by the sedimentation velocity method or sedimentation equilibrium method. First, the solute is uniformly distributed. When it is centrifuged at, say, 55,000 r.p.m., a well defined boundary of the solute is formed. The movement of the boundary is measured in the sedimentation velocity method by Schlieren optics. If M is molecular weight of the protein, it is related to the sedimentation constant Z by the equation.

$$M = \frac{RTZ}{D(1-V\rho)}$$

where R is the gas constant, T the absolute temperature, D the diffusion coefficient, V the partial specific volume of the substance (for proteins it is about 0.74) and ρ the density of the medium. The sedimentation constant Z is obtained from the Sedimentation Velocity (velocity of the particles in a unit centimetre gm second field of force). Z is 4.4 for bovine serum albumin, 1.83 for cytochrome C and 185 for tobacco mosaic virus protein).

In the second method known as the sedimentation equilibrium method, the centrifuge is operated at a relatively low speed until there is no further movement of particles. The diffusion of the particles is balanced by their sedimentation velocity at which stage the sedimentation equilibrium is obtained. Using the following equation

$$M_E = \frac{2 RT \log_e \frac{C_2}{C_1}}{\omega^2 \{1-V\rho(x_2^2 - x_1^2)\}}$$

where C_1 and C_2 are the concentrations of substances at distances X_1 and X_2 from the axis of rotation, and ω is the angular velocity of the centrifuge. The molecular weight of many colloids has been determined. The analytical model of ultracentrifuge is used. The concentration of substances can also be determined if M is assumed.

In addition, using ultra-centrifuge, the heterogeneity of proteins hitherto considered homogeneous, fractionation of lipoproteins, purification of proteins, separation of DNA from RNAs and of DNAs with different labels of nitrogen and fractionation of sub-cellular organelles have been made. The fraction of lipoprotein increased in a particular condition, say, hyper-lipoproteinemia, is shown by this technique, and the calculation of atherogenic index which may caution a person whether he is normal or whether he is going in for a heart disease is also made possible. The distribution of enzymes in the mitochondria, microsomes, cytosol, etc., could be studied only by this method. The gradient centrifugation of the DNA of E.coli in CsCl has thrown light on the mode of replication of DNA. To a limited extent the configuration of particles and the shape and the size, the association and dissociation of molecules in media, could be investigated.

The infra-centrifuge

An infra-centrifuge has to produce sedimentation inwards instead of outwards. The rotor should be made to work in such a way that the centre becomes the periphery. The centrifugal force now acts inwards. An instrument working on this principle is the infra-centrifuge which has application in space ships.

AUTO-ANALYSERS

These instruments have introduced automation to the routine biochemical analysis of hospital specimens. Manual operation becomes tedious beyond a particular limit. With auto-analysers one can analyse nearly 60 samples of blood or serum per hour for as many as 18 constituents.

Auto-analyser is a train of interconnected modules that automate the use of time and step-by-step procedures of manual analysis. The main parts of an auto-analyser are: 1) sampler, 2) proportionating pump, 3) dialyser, 4) constant temperature bath, 5) flow-through colorimeter, and 6) recorder.

Working of the instrument

The flow of the analytical stream is directed through plastic tubing from one module to another. After the samples are loaded into the cup of the sampler, the channels of the proportionating pump aspirate them and dilute them. The diluted serum samples are led through one side of the dialyser unit. The pump also introduces suitable reagents through the other side of the dialyser. The two streams run side by side separated only by the dialysing membrane. Now a proportion of dialysable constituents of serum passes across the membrane to the reagent stream. Further treatment like incubation at suitable temperature is given in the constant temperature bath and the intensity of the colour developed is measured in a colorimeter. Suitable standards are treated in the same way for the purpose of estimation. If necessary, the colorimeter can be replaced by the spectro-photometer, flame photometer, fluorimeter or atomic absorption spectrometer.

Auto-analysers are very much in demand for the estimation of blood urea and sugar. For BUN (blood urea nitrogen), the samples of blood are diluted with physiological saline and dialysed against diacetyl monoxime. Urea diffuses into the reagent and moves in the stream. The mixture is acidified with H_2SO_4-HNO_3 mixture and heated to about $95°C$. The intensity of yellow colour is estimated at a wavelength of $480 m\mu$.

In the estimation by the auto-analyser, reactions are not carried to equilibrium or completion.

A constant proportion of sample is dialysed for the purpose of estimation. The same is extended to the standards. Air bubbles are introduced continuously into the flow to separate successive samples or standards and to segment the sample and the reagent streams.

Of late, Sequential Multiple Analysers have been introduced with multiple channels to estimate as many as 18 selected constituents including the non-dialysable proteins, enzymes like alkaline phosphatase, triacyl glycerols and electrolytes. Nearly 60 samples can be investigated per hour simultaneously for 18 parameters.

Auto-analysers are popular for routine biochemical analysis in clinical laboratories for their speed, flexibility of the methodology, increased quantum of clinical biochemical analysis, easy operation and accuracy of results. The disadvantages are the prohibitive cost, 'carry-over' errors and the usual hazards of any complicated equipment.

ENZYME IMMUNO-ASSAY

Until recently, substances such as hormones, proteins, etc. existing in the blood in minute concentrations, i.e., picograms or nanograms could be quantified only by using radio-active techniques. With the advent of the recent technique of 'enzyme immuno-assay, it has now become possible to do assays of hormones, etc. with all the advantages of radio immuno-assay and without some of its disadvantages.

The basic reaction is the interaction between an antibody and an antigen.

antigen + antibody → antigen-antibody complex

In enzyme immuno-assay, a highly specific antigen-antibody reaction is conducted.

Enzyme immuno-assay (EIA) may be classified into two broad types. They are;

1. homogeneous enzyme immuno-assay (HEIA), and
2. heterogeneous enzyme immuno-assay (HTEIA) or

Enzyme-linked immuno-sorbent assay (ELISA).

Homogeneous EIA

As the name suggests, there is only one phase, the liquid phase. Haptens like hormones, drugs, etc. are labelled with enzymes like lysozyme, malate dehydrogenase, etc. with no loss of enzyme activity. They will be made to react with a limited number of antibody molecules. There is competition between the enzyme-labelled hapten and the free hapten for the limited number of antibody molecules. There will not be any enzyme activity if only the enzyme-labelled haptens are present, as all of them unite with the antibody. But in a mixture of free haptens and enzyme-labelled haptens, depending on the concentration of free haptens, there will be no accommodation for some enzyme-labelled haptens. The latter will react with the substrate producing a colour. The intensity of colour is measured in a photometer. It may be seen that the reactions take place in one phase only i.e., the liquid phase. Hence, this is called homogeneous EIA and is applicable for T_3, T_4, estriol, cortisol, etc. The technique couples the three aspects: immunology, enzymology and photometry.

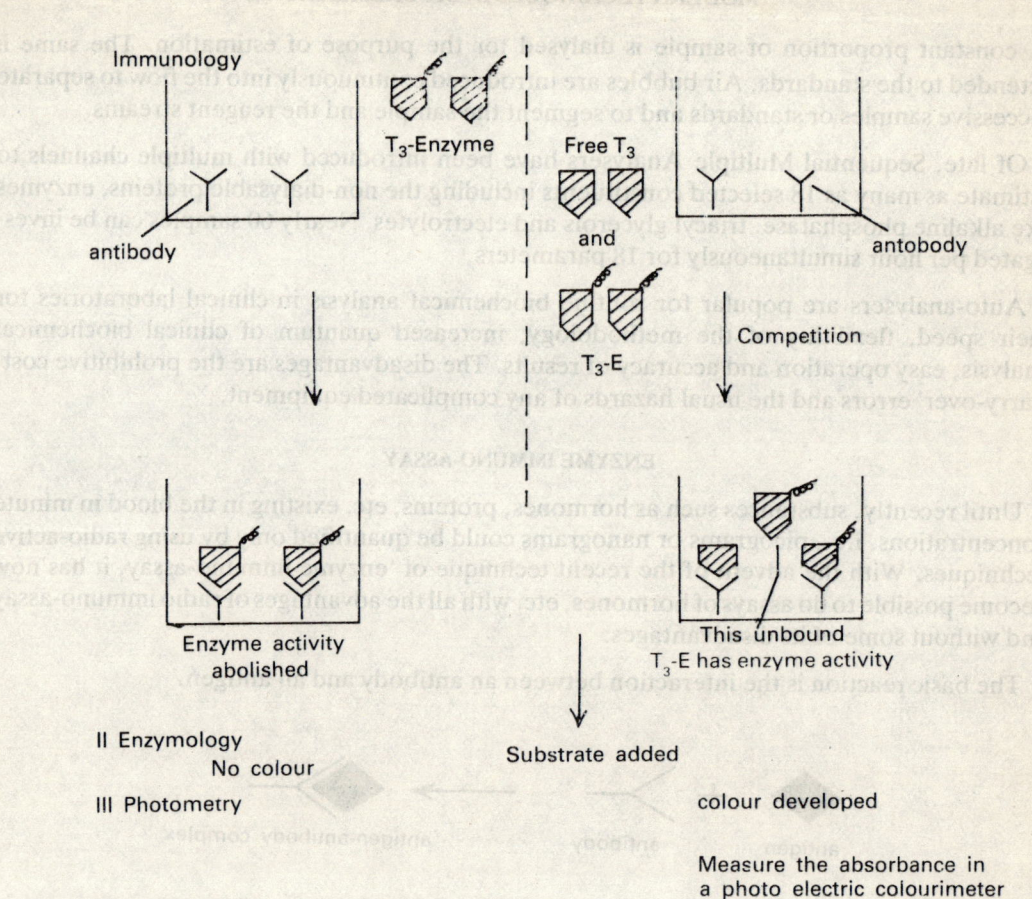

Heterogeneous EIA

In heterogeneous enzyme immuno-assay, the antibody is fixed to the wall of a small test tube. The test sample contains the antigen to be estimated. So, when the test sample is added, only the antigen is taken up by the antibody. Anything else contained in the medium is removed by aspiration and rinsing. This is the so-called bound-free separation.

As in homogeneous enzyme immuno-assay, the competition principle applies in this type also, i.e., to the container with antibodies fixd to the wall, the antigen of the sample and specifically labelled antigen are added. The label may be an enzyme, in most cases, peroxidase. There is now a competition for the antibody sites between the unlabelled antigen and labelled antigen. Some antigens which are bound may be labelled while some others may be unlabelled. The extent of the labelled antigen binding to the antibodies is governed by the concentration of unlabelled antigen in the specimen. The specimen liquid is then aspirated out and the tube is rinsed with a suitable liquid. After the bound-free separation, the test tube is treated with H_2O_2 and a chromogen. The peroxidase attached to the antigen and fixed to the wall via the antibody then reacts with the substrate to form a colour. The absorbance of the colour is measured in a photometer. The concentration of the enzyme-labelled bound antigen is decided by that of the unlabelled antigen. The latter can be assayed using a standard curve with known amounts of unlabelled antigen.

This diagram will illustrate the methodology.

Immunology

Antibody molecules have been fixed to the wall. Antigen of the specimen and enzyme-labelled antigen are added. They bind to the wall-fixed antibodies by competition.

Enzymology

Enzyme activity of the wall-fixed antibody-antigen-enzyme complex is measured with, say, H_2O_2 and a chromogen.

Photometry

Colour is developed due to enzyme activity. It is measured. Calculations follow using standards.

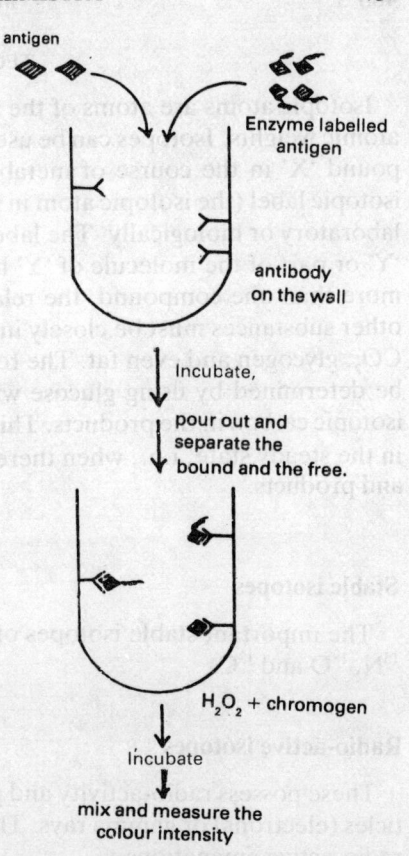

In some instances, instead of development of colour, turbidity can be produced, e.g., highly specific antibodies react with respective immunoglobulins in the serum. The increase in turbidity caused by the antigen-antibody reaction is measured in a cuvette.

In the titration principle, insulin labelled with enzyme is made to bind to antibodies in competition with unlabelled insulin. The sites are filled with enzyme-labelled insulin, in excess, in titrable quantities.

According to the sandwich principle, as applied to alpha feto protein, the antigen can bind to more than one site on the antibody. The first step involves filling the antibody sites fixed to the wall with say alpha feto proteins in the serum of a pregnant woman. The antigen though bound already has multiple sites for antibody binding. Now, the antibody labelled with the enzyme is added. This labelled antibody attaches to the site still free on the antigen. The contents are aspirated. The greater the amount of antigen, the greater the amounts of labelled antibody remaining on the wall. Colour reaction is performed for estimating the enzyme.

The above procedure involves sandwiching the antigen (alpha feto proteins) between antibody fixed to the wall and enzyme-labelled antibody which is added. Hence the name.

Apart from glass tube, polystyrene micro-titre plate, cellulose, iso-thiocyanate and polyacrylamide can be used for fixing the antibodies. The enzyme label for the antibody may be alkaline phosphatase, glucose oxidase, ß galactosidase, glucose 6(P) dehydrogenase and penicillinase. Apart from hormones, immunoglobulins, alpha feto proteins, etc., the antigens can be purified virus and virus components, bacterial antigens, tetanus or diphtheria toxoids, fungal antigens, etc.

TECHNIQUES USING ISOTOPES

Isotopic atoms are atoms of the same element with the same atomic number but different atomic weights. Isotopes can be used to find out whether a compound 'Y' is a product of compound 'X' in the course of metabolism. For this a pure specimen of 'X' should have the isotopic label (the isotopic atom in its molecule), and such compounds can be prepared in the laboratory or biologically. The label should be looked for in the compound 'Y' to prove that 'Y' or part of the molecule of 'Y' has come from 'X'. Again, if a compound is converted to more than one compound, the relative proportions of the parent compound in the various other substances must be closely investigated, e.g., glucose may be converted in the body to CO_2, glycogen and even fat. The relative quantities of glucose for the three metabolites may be determined by using glucose with isotopic carbon and analysing the distribution of the isotopic carbon in the products. Thirdly, isotopes can be used to measure the rates of reaction in the steady state, i.e., when there is no apparent change in the concentrations of reactants and products.

Stable isotopes

The important stable isotopes of biological interest are ^2H (deuterium, heavy hydrogen), ^{15}N, ^{18}O and ^{13}C.

Radio-active isotopes

These possess radio-activity and give out radio-active emanations like alpha rays, beta particles (electrons) or gamma rays. They are detected and estimated by taking advantage of the radio-active emanations.

A list of a few radio isotopes is given below together with the nature of the radioactive emanation and half life.

Element	Radiation	Half-life	
^3H	β	12.1	years
^{14}C	β	5,100	years
^{32}P	β	14.3	days
^{35}S	β	87.1	days
^{45}Ca	β	153	days
^{59}Fe	β & γ	45	days
^{60}Co	β & γ	5.3	days
^{131}I	β & γ	8.05	days
^{226}Ra	α & γ	1,622	years (not normally used as tracer)

Apart from naturally occurring rado-active elements like radium, most radio-active isotopes (from the otherwise non-radio-active elements) are made by bombarding a suitable target with the neutrons in an atomic pile.

There are different ways of expressing the distribution of isotopes. In case of isotopes not found naturally, the concentration can be expressed as the specific activity, i.e., number of isotopic atoms per unit of the substance (mg, μ mole etc.) of the compound under study. Secondly, the distribution of isotope in the precursor and product can be expressed by the dilution factor. The dilution factor is defined as,

$$\frac{\text{specific activity of precursor fed}}{\text{specific activity of product isolated}}$$

Thirdly, if the isotope is one that occurs naturally, e.g., ^{37}Cl which forms 24 per cent of all naturally occuring chlorine atoms, the concentration of the isotope is expressed in atoms per cent excess. This is the number of isotopic atoms per 100 total atoms of that element which is present in excess of the normal number occurring naturally. In experiments using radio-active isotopes, the incorporation of radio-activity is also expressed as counts per minute per g, or 100 mg etc. of the tissue for the comparison of metabolisms in the control and experimental animals.

Units of radio-activity

The amount of radioactivity present in any sample is defined as the number of atoms of each unstable isotope present, which in turn, will be given by the number of disintegrations in unit time. One curie is defined as 3.7×10^{10} disintegrations per second. Another unit is the roentgen (r). This is defined as the quantity of X radiation or γ radiation which, in passage through a volume of air weighing 0.001293 g (i.e., 1 cm^3 of air at 0°C and a pressure of 760 mm of Hg), produces ions of either sign equivalent to one electro-static unit of electricity. One (rhm) roentgen per hour at one meter of, say ^{131}I is the amount whose unshielded γ radiation produces one r/hr of ionisation in air one meter from the source.

The biological effects of radio-activity on tissues, etc. are expressed as rad (Roentgen absorbed dose) and rem (Roentgen equivalent man). The unit of radiation actually absorbed is defined in terms of energy absorbed: 1 rad = 100 ergs/g. The biological effectiveness of radiation is measured in rems, a unit which varies with the nature of tissues. Roughly one rem = rad for beta, gamma or X-rays, and 1 rem = 0.05 rad for alpha rays. As an example of the use of this unit, the maximum occupational exposure to radiation should not exceed five rems per year on an average.

Measurements

Except for deuterium, stable isotopes are estimated by means of a mass spectrometer. The samples are first converted to a gas (N_2 is used for ^{15}N; CO_2 for ^{13}C and O_2 for ^{18}O) which is introduced into the apparatus at a very low pressure. Deflecting the gases by magnetic fields, a spectrum of different masses of the compound of the same element can be produced and estimated.

The quantitative determination of radio-activity is based on the ionisation or excitation of matter by radiations emitted by radio-active materials.

For observataions of ionisations in gases, 1) Geiger-Muller counter, 2) proportional counter, and 3) ionisation chambers like electroscopes and electrometers are employed. Since 1950 a new detector, namely the scintillation counter, which measures the interaction of radiation with either solids or liquids has been popular. Between the alpha rays and ß particles, alpha causes a much higher ionisation of a gas through which it passes. X-rays and γ-rays (uncharged) ionise only through collision mechanisms and hence ionisations are only mild.

Geiger-Muller counter or a G-M tube: This consists of a large round outer electrode (metal cylinder, cathode) with a fine wire stretched in the centre as the second electrode. The tube is filled with an easily ionised gas such as argon at moderately low pressure together with a small quantity of an organic substance like ethanol as a 'quencher'. The central wire is maintained at a high potential (1000-2500 V) with respect to the outer electrode. The open end of the tube is covered with an extremely thin window of mica or synthetic plastic. Beneath this window (or a windowless tube with a positive pressure of argon and ethanol) the radio-active material is placed. The radiations enter the counter tube and ionise the gas molecules releasing a shower of electrons. The free electrons are accelerated to the positive wire because of the intense field around it. As they progress through the gas, they ionise the other molecules of the gas and produce secondary electrons. These secondary electrons ionise others producing what is termed the 'avalanche effect'. The end result is a substantial pulse (potential drop) at the anode) for each entering ray which causes ionisation. This is because the positive space charge diminishes the effective voltage gradient at the anode. The duration of the pulse is very short ~ 10^{-7} second. Before the G-M counter has to initiate another pulse, the positive ions must diffuse sufficiently towards the cathode so that the original voltage gradient at the wire is restored. This portion of the discharge cycle represents the 'dead' time of the counter and requires longer time from 10^{-3} to 10^{-4} sec. During the deadtime, any ray entering the tube will not be detected. The function of the quencher is to prevent the discharge from becoming continuous. (The positive argon ions collide with ethanol and develop positive alcohol ions. The high positive field of alcohol ions draws the electrons from the cathode so that excited neutral molecules, which dissociate before they can strike the wall, are created. This prevents the liberation of secondary electrons during the collision with the wall material. Thus, quenching is effected.)

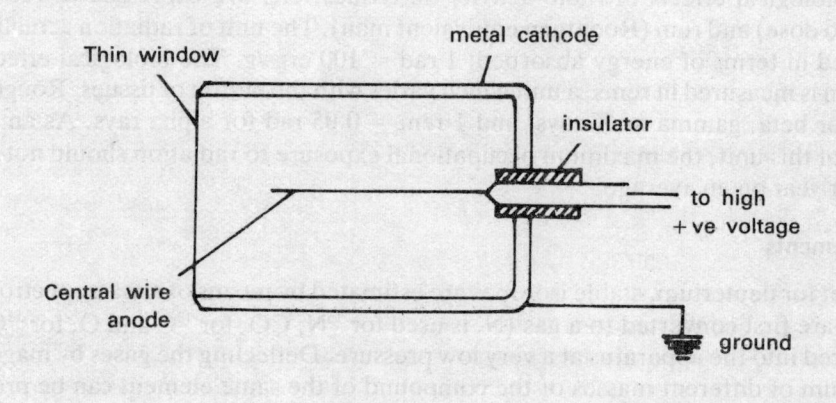

Geiger – Muller tube

Each tube has a certain voltage range over which the count rate is independent of voltage and this is called the 'plateau'. Beyond this, the tube discharges continuously and is rapidly spoiled.

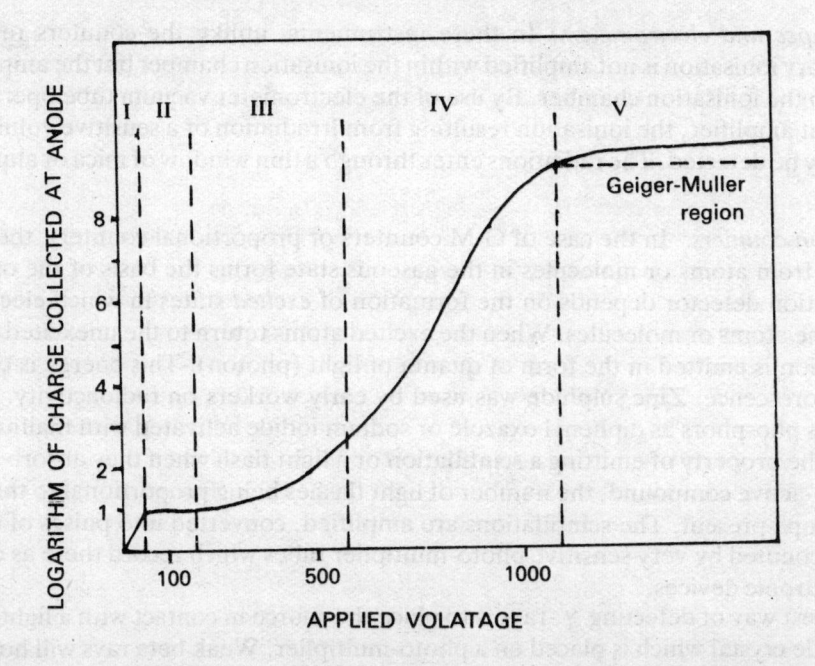

[Graph: Logarithm of charge collected at anode vs Applied Voltage, showing regions I, II, III, IV, V with the Geiger-Muller region marked in region V. X-axis values: 100, 500, 1000. Y-axis values: 1, 2, 4, 6, 8.]

G-M counters work best at relatively low count rates (<10,000 per min) since, then, the dead time is only a small fraction of the total counting time. The Geiger-Muller tubes are not efficient for gamma rays. They can be made to dip into liquid samples also but most samples are solid and are placed on a disc-planchette at a specified distance from the window. A problem called self-absorption arises with solid samples. Beta particles have only a short range in solids and frequently rays arising from the isotope at the bottom of the sample are absorbed by the sample itself before they can reach the window. Hence, the correction for self-absorption should be applied. However, if the solid samples are 'infinitely thin', the correction for self-absorption need not be made. Poor spreading of samples on the planchettes will not reflect actual counts. Hence, proper geometry is required. The radio-activity of the sample might be 3000-4000 cpm and need not be higher. Correction has also to be made for background counts, i.e., the counts recorded in the absence of any radio-active sample. The back-ground counts are subtracted from the counts registered for samples. The background count is due to ionisation caused either by cosmic rays or by radio-active contamination of the counting chamber striking the G-M tube.

Proportional counters: G-M tube is very efficient for ß-assay but is being replaced to some extent by the proportional counter. The proportional counter is so called because it operates in a region of applied voltage where the charge collected is proportional to the initial ionisation. The advantages of the proportional counter over the G-M counter are 1) it can be operated at reduced voltage, has greater stability and reproducibility, 2) the organic quenching gas is not consumed as rapidly as in the G-M tube, and 3) there is no dead time during each pulse so that very fast counting rates are possible. The most popular type of proportional counter is the 'flow' counter. In this instrument, a gas like argon, burshane or indane is continuously passed through an open-ended tube of the ionisation chamber at a pressure slightly higher than that of the atmosphere. As passage of weak ß-rays through a 'window' is difficult, the 'flow-counter' overcomes this pre-requisite. Hence, the proportional counter is more popular for studies with ^{14}C.

Electroscopes and electrometers: In these instruments, unlike the counters referred to above, primary ionisation is not amplified within the ionisation chamber but the amplification is external to the ionisation chamber. By use of the electrometer vacuum tube operating as a direct current amplifier, the ionisation resulting from irradiation of a sensitive volume of the chamber may be detected. The radiations enter through a thin window of mica or aluminium.

Scintillation counters: In the case of G-M counters or proportional counters, the *removal* of electrons from atoms or molecules in the gaseous state forms the basis of the operation. The scintillation detector depends on the formation of *excited* states in which electrons are *retained* in the atoms or molecules. When the excited atoms return to the unexcited (ground) state, radiation is emitted in the form of quanta of light (photon). This energy is converted into phosphorescence. Zinc sulphide was used by early workers on radioactivity. The present day uses phosphors as diphenyl oxazole or sodium iodide activated with thallium. Phosphors have the property of emitting a scintillation or a light flash when they absorb radiation from a radio-active compound, the number of light flashes being proportional to the amount of radio isotope present. The scintillations are amplified, converted into pulses of electrical energy and counted by very sensitive photo-multiplier tubes which record these as counts by suitable electronic devices.

The simplest way of detecting γ-rays is to place the source in contact with a light-shielded sodium iodide crystal which is placed on a photo-multiplier. Weak beta rays will however be absorbed by the container. For them the liquid scintillation counter is recommended. The phosphor and the sample are both dissolved in a non-aqueous solvent; the emission of photons takes place inside the transparent sample container which is placed in close contact with the photo-multiplier. This is the technique for counting ^{3}H and ^{14}C. (Strong ß-emitters like ^{32}P are just as easily counted with a G-M tube). Problems of geometry, coincidence and self-absorption are avoided in scintillation counters. However, there are certain difficulties with liquid scintillation counting; 1) Many chemicals quench the phosphorescence, i.e., they accept the energy from the excited atoms or molecules and abolish the emission of a photon. As water is a powerful quenching agent, a non-aqueous solvent (say, toluene) has to be used, 2) many coloured compounds, particularly yellow, if present in the samples, reduce the efficiency by absorbing the photons before they reach the photo-multiplier, and 3) there is difficulty in applying for polar components as they do not dissolve easily in non-polar liquids. Methods are now being developed to overcome these difficulties.

RADIO IMMUNO-ASSAY

Radio immuno-assay (RIA) is an elegant technique in analytical biochemistry and plays significant role in the diagnosis of diseases. If the substance to be analysed is in very low quantities, of the order of micrograms, nanograms or picograms, conventional methods like the gravimetric and colorimetric methods fail. Only the property of radioactivity comes to the rescue as even nanogram amounts of radio-labelled substance give appreciable counts detectable by electronic counters. These counts can be conveniently related to the concentration of the substance assayed. As such, the method finds extensive application in the assay of many substances which are present in trace amounts in blood. To quote just a few examples, RIA is employed for the estimation of vitamins like B_2 and folate, hormones like insulin, thyroxine, tri-iodothyronine, cortisol, testosterone, dihydrotestosterone and estrogens; tropic hormones like ACTH, FSH, LH; drugs like digoxin and digitoxin and antigens like the Australia antigen.

The principle of the method is as follows: A radio-labelled antigen (say, insulin labelled) with ^{125}I) is made to compete with an unlabelled antigen (insulin which has to be estimated, say, in the serum of a patient) for *a limited number of* binding sites of a specific antibody raised against insulin. The antigen binds to the antibody. In view of inadequate binding sites, some of the antigen will be free and will include radio-labelled antigen also.

$$\text{Antigen}^* + \text{Antibody} \rightleftharpoons \text{Antigen}^* \ldots\ldots \text{Antibody complex}$$
$$\text{(radioactive)}$$

$$\downarrow \text{Antigen (unlabelled)}$$

$$\text{Antigen} \ldots \text{Antibody complex}$$
$$\text{(non-radioactive)}$$

(*) = radioactive

After equilibrium, the antigen-antibody complex is precipitated by using suitable reagents. On centrifugation, the supernatant is separated from the precipitate. Both the precipitate (the bound antigen, B-form) and the supernatant (the free antigen, F) will have radio-activity as they have ^{125}I-insulin. If dextran-coated charcoal is used, it removes the free antigen. The supernatant will have, in this case, the bound form. The extent of radio-activity of the two forms is measured in a gamma ray well type scintillation counter. The magnitude of radio-activity of, say, the free form, may be related to the concentration of the unlabelled antigen. Alternately, the radio-activity of the bound form or the ratio of B/F could also be related to the concentration of the unlabelled antigen.

Different concentrations of the unlabelled insulin standrad are used separately with the same concentration of labelled insulin and hence the same extent of radio-activity and a graph is prepared with the radio-activity (counts) of, say, the free form, and the concentrations of the standard insulin. The serum, whose insulin content is to be determined, is treated in the same way. From the radio-activity of the free form and extrapolation, the concentration of insulin in serum is determined.

Labels used for RIA have high specific activity. Hence, the assay is very sensitive.

Normally, an antibody is raised for any antigen to be estimated. As the technique couples radio-activity and immune function (antigen binding to antibody), it is called radio immuno assay.

Competitive protein-binding assay

If, instead of an antibody, plasma protein is used to bind-the antigenic substance, the technique is called competitive protein-binding assay, as in this there is no immune function. For example, in the assay of cortisol, the physiological, protein transcortin is added to offer limited sites for binding. Likewise, the thyroxine binding globulin can be used for the competitive protein binding assay of T_4 and T_3.

In respect of many antigenic substances, ^{125}I is incorporated in the molecule. Thus we have ^{125}I-Insulin, ^{125}I-T_4, ^{125}I-ACTH, etc. For cortisol, it is ^{75}Se labelled cortisol.

While for the assay of insulin in serum, no extraction is required, T_3 and T_4 have to be extracted from serum with alcohol. ACTH is adsorbed on to glass while the serum is diluted and heated to denature transcortin and release cortisol in the assay of the latter. It is then processed suitably for subsequent estimation.

Importance of RIA

1) In addition to the estimation of hormones proper, even the receptor sites can be assayed. Such an analysis is possible only with RIA.   2) The patient need not take the radio-isotope as the label is used only in vitro. 3) RIA has value in diagnostic biochemistry. When it is necessary to estimate T_4 and T_3 separately, RIA comes in handy. In some thyroid disorders, T_4 may be normal but T_3 is elevated as in subclinical hyperthyroidism, ophthalmic Graves disease, autonomous thyroid nodules, T_3 toxicosis, etc. In pregnancy or in persons taking oral contraceptives, due to the increased level of the serum thyroxine-binding globulin, an apparent increase of the thyroid hormones is experienced. This should be discriminated from genuine hyperthyroidism.

It is only RIA that can help to differentiate the basic biochemical lesion in endocrinology — whether the increased level of a hormone is due to the production of the hormone as such or the tropic hormone, e.g., if there is an increased level of blood cortisol, it could be due to hyperfunction of either the adrenal cortex or anterior pituitary. This could be revealed by the estimation of ACTH in addition to cortisol by RIA.

In the use of drugs, if there is only a narrow margin between the therapeutic and toxic dosage, RIA offers safety to the patients. This is applied during digitalisation in the management of congestive heart failure. Low levels of serum digoxin mean under-digitalisation, while high levels are toxic. By RIA of digoxin in serum, the patient is given the optimum digitalisation.

RIA is also useful in diagnosing insulinomas, sex hormone-sensitive tumours, etc., and this facilitates proper treatment of the diseases.

SECTION – V

FUNCTION TESTS

25 Biochemical Evaluation and Function Tests

LIVER FUNCTION TESTS

While assessing liver function, one has to reckon with two factors, namely, the reserve capacity of the organ and the host of functions it performs. No single function test by itself would suffice, as the functions are manifold and diverse and a defect in one function need not necessarily be accompanied by a commensurate impairment of other functions. The reserve capacity confers the factor of safety due to which, unless a substantial degree of liver damage is there, very few of the tests are really sensitive. However, in spite of such limitations, the inherent merits of a good number of these tests lend themselves as useful biochemical parameters for diagnostic purposes. The various function tests are categorised according to the underlying group of functions.

The following are the diverse functions of the liver:
1) Detoxication of bilirubin by conjugating it as bilirubin diglucuronides and excreting through bile.
2) Epimerisation of galactose to glucose as UDP derivatives.
3) Synthesis of proteins like albumin.
4) Synthesis of prothrombin.
5) Handling of enzymes alkaline phosphatase, and release of transaminases, i.e., aspartate amino transferase and alanine amino transferase.

Detoxication of bilirubin

Bilirubin is formed daily from hemoglobin in the reticuloendothelial system chiefly of bone marrow and spleen. About 6.25 g of Hb are degraded every day. Bilirubin is transported in the plasma in combination with albumin to the liver where it is conjugated with UDP – glucuronic acid to form bilirubin glucuronide. Bilirubin is insoluble in water but bilirubin diglucuronide is soluble in water. This is an important difference in physical property. The glucuronide is excreted in the bile and through the bile goes to the intestines. There, it is reduced by bacterial enzymes to urobilinogen. Some urobilinogen is oxidised to urobilin. The urobilinogen not oxidised in the intestines is returned to the liver (entero-hepatic circulation) and is oxidised to bilirubin which is re-excreted into the bile. Normal urine therefore contains very little urobilinogen i.e., 1 to 4 mg per 24 hr.

Bilirubin metabolism is deranged in there important diseases. They are 1) hemolytic jaundice, 2) hepato-cellular jaundice, and 3) obstructive jaundice.

Hemolytic jaundice: What is the fate of bilirubin and urobilinogen in this condition? The liver is normal in this disease and can conjugate the usual amounts of bilirubin efficiently. But when there is hemolysis, there is extensive degradation of heme and over-production of bilirubin. There is subsequenty a rise of bilirubin in the blood as the liver cannot remove it efficiently. This bilirubin is not soluble in water. Hence it will not give a direct positive reaction with the Van den Bergh reagent. (The Van den Bergh reagent is a mixture of equal volumes of Van den Bergh solution A, i.e., sulphanilic acid in dil HCl, and Van den Bergh solution B, i.e., sodium nitrite.) If there is bilirubin diglucuronide which is water soluble, it will give a pink colour immediately. Such a reaction is called the Van den Bergh *direct* positive reaction. There may be no positive reaction but on adding methanol, the serum may answer the Van den Bergh test. This is because methanol dissolves bilirubin. Once a solution is obtained with methanol, the serum will give a pink colour with the Van den Bergh reagent. This is called the Van den Bergh *indirect* test. In hemolytic jaundice, the Van den Bergh test will be indirect positive. As the liver is normal, there is no regurgitation of whatever bilirubin that has been conjugated. Hence, there will be relatively greater amounts of bilirubin than normal going to the intestines. As a result, there will also be an increased excretion of urobilinogen in urine. There will not be any direct-reacting bilirubin (i.e., bilirubin glucuronide) in hemolytic jaundice as there is no regurgitation of bile into the blood. The jaundice therefore is due to increased water-insoluble bilirubin in blood.

In short, the following are the results of the blood and urine tests in hemolytic jaundice:
1) Bilirubin total – elevated (normal values are 0.1 – 0.8 mg/dl)
2) Bilirubin glucuronides – not detectable.
3) Van den Bergh test – indirect positive.
4) Urinary urobilinogen – increased.

Obstructive jaundice: The jaundice in this condition is due to the increased levels of both bilirubin and bilirubin diglucuronides in the blood. This is due to the regurgitation of bile. The regurgitation takes place due to obstruction in the bile duct. So, the bile instead of flowing into the gastro-intestinal tract regurgitates into the blood. As the obstruction is outside the liver, the condition is called extra-hepatic or post-hepatic obstruction. As the liver is normal except in chronic conditions, conjugation goes on, so that there is both water-insoluble bilirubin and water-soluble bilirubin glucuronides in the blood. The latter gives the Van den Bergh direct positive reaction. There is increased bilirubin glucuronides and total bilirubin. In view of the obstruction in the bile duct, there is no formation of urobilinogen. Hence the urine does not contain urobilinogen. Due to the regurgitation of bile, the activity of the enzyme alkaline phosphatase is quite high (greater than 30 – 35 King Armstrong Units) in the blood. The transaminases like alanine amino transferase may not be elevated significantly as there is no damage of the liver except in chronic conditions.

The following are the findings of the blood test:
1. Total bilirubin – elevated.
2. Bilirubin glucuronides (direct – reacting bilirubin) – increased .
3. Van den Bergh test – direct positive.
4. Alkaline phosphatase activity – above 30-35 KA.
5. Serum alanine amino transferase – slightly elevated.
6. Urinary urobilinogen – negligible

Hepato-cellular jaundice: In this condition the jaundice is due to the increased levels of bilirubin and bilirubin glucuronides in blood. This also is due to the regurgitation of bile but such a regurgitation is not due to post-hepatic obstruction but to intra-hepatic obstruction (obstruction from within the liver) because of inflammation caused by infection. The presence of bilirubin glucuronide in serum renders the Van den Bergh test direct positive. Total bilirubin is elevated. Alkaline phosphatase activity is also increased but is less than 30 - 35 KA units (normal values are 3 – 8 KA units).

So long as the obstruction is there, urinary urobilinogen is negligible. This is because the obstruction, intra-hepatic or post-hepatic, prevents bilirubin from reaching the intestines for reduction to urobilinogen. However, when the inflammation subsides during recovery there is no obstruction. Under such circumstances, urinary urobilinogen excretion may increase.

In view of infection, there is liver damage. Hence, the enzyme alanine amino transferase from the necrosed cells mixes with the blood. Hence, there is increase of this enzyme alanine amino transferase (normal values are 2 – 15 I.U/L).

The following are the diagnostic findings in hepato-cellular jaundice :
1) total bilirubin – increased.
2) Direct reacting bilirubin, i.e., bilirubin glucuronides – increased.
3) Van den Bergh test – direct positive.
4) Alkaline phosphatase activity – increased upto 30 – 35 KA.
5) Urinary urobilinogen – negligible.
6) Alanine amino transferase – increased.

The table below lists the blood and urine tests for the differential diagnosis of the three types of jaundice.

Condition	Total bilirubin	Bilirubin glucuronide	Van den Berg test	Alkaline phosphatase	Alanine amino transferase	Urinary urobilinogen
Hemolytic jaundice	Increased	Nil	Indirect positive	Normal	Normal	Increased
Obstructive jaundice	Increased	Increased	Direct positive	Increased, above 30 – 35 KA units	Slightly elevated	Negligible
Hepato-cellular jaundice	Increased	Increased	Direct positive	Increased, but below 30 – 35 KA units	Significantly elevated	Initial increase, but is lowered as the disease progresses

As hepato-cellular jaundice is due to infections like viral infection etc., it can be treated with medicines, and no surgery is needed. Hence, it is called medical jaundice. On the other hand, obstructive jaundice, if the obstruction is due to stone in the bile duct cannot be cured with medicines. Surgery is the only way and hence it is called surgical jaundice.

The van den Bergh's test is based upon the formation of purple colour due to azobilirubin obtained by the reaction between bilirubin and Ehrlich's diazo reagent. The reagent A is sulphanilic acid in hydrochloric acid while the reagent B is dilute sodium nitrite solution. The reagents are mixed fresh in the prescribed proportions just prior to the performance of the test. A purple colour within 30 seconds of mixing indicates direct reaction. Sometimes it may be delayed direct, if there is a partial conjugation defect. If the direct reaction is not positive, methanol is used as a solvent to dissolve the water insoluble bilirubin and the colour developed is measured after letting it stand for 30 minutes (indirect positive).

In hemolytic conditions the load of bilirubin is beyond the capacity of the conjugating ability of the liver, while in hepato-cellular jaundice, since the liver itself is diseased, it is not able to handle even the normal amount of bilirubin reaching it.

The normal serum bilirubin level ranges between 0.1 – 0.8 mg per 100 ml and the normal icteric index ranges between 4 – 6 units. False high icterus values may be caused by foodstuffs like carrots and certain drugs as well. The normal urinary urobilinogen output ranges between 1 – 4 mg per 24 hr. A collection of postprandial urine is necessary for the estimation of urobilinogen. The sterco-bilinogen level in feces may be normal or decreased in hepatic diseases. Presently icteric index is not estimated as the bilirubin levels are precisely analysed.

Serum alkaline phosphatase

The normal serum level of this enzyme ranges between 3 - 11 King Armstrong units. This enzyme is excreted by the liver via the bile and hence when the liver is out of order, the serum enzyme level goes up due to defective excretion but all the non-hepatic conditions which can cause a similar rise have to be scrupulously eliminated. Such conditions are rickets, osteomalacia, hyper-parathyroidism, post-hepatic obstruction, Paget's disease and bone tumours. It is belived that high flocculation with normal alkaline phosphatase level in serum shows hepatitis without obstruction whereas a high alkaline phosphatase level in serum (exceeding about 35 K.A. units) with normal flocculation points to post-hepatic obstrucion without hepatitis, though a prolonged post-hepatic obstruction may eventually, by back flow, cause liver damage in which case the vitamin K load test will be helpful. In infective hepatitis with intra-hepatic obstruction when Van den Bergh's reaction with serum is direct positive, the serum alkaline phosphatase levels are moderately raised, though not as high as in post-hepatic obstruction.

Serum transaminase assay

Jaundice is a contra-indication for the determination of serum transaminase levels as high values are obtained even in the absence of liver damage. Transamination is not the monopoly of any particular tissue in the body, so that in the absence of a second tissue damage, it is valuable in the diagnosis of liver dysfunction. Serum alanine amino transferase (originally called serum glutamic pyruvic transaminase) is raised early in liver disease even in the pre-icteric phase. Serum aspartate amino transferase (originally called serum glutamic oxalo-acetate transaminse) though diagnostic of myocardial infarction, can also be a useful liver function test, provided the liver involvement is not secondary to the involvement of the heart. Determination of serum iso-citrate dehydrogenase is belived to be a specific liver function test. Specificity is also seen in the estimation of the serum levels of the heat-labile hepatic isozymes of lactate dehydrogense, i.e., LD_4 and LD_5. The serum amylase levels are low in liver disease. The activity of pseudo-choline-esterase may also be lowered. High levels of serum leucine amino peptidase are seen in liver damage though it is not specific for liver dysfunction.

Plasma proteins levels and liver dysfunction

Hypo-albuminemia and hence hypo-proteinemia as well as a hypo-fibrinogenemia are met with in gross liver diseases, as the liver is concerned with the biosynthesis of these proteins. However, hypo-proteinemia can be seen in protein malnutrition as well as after hemorrhage so that a reversal of the ratio between albumin and globulin in serum is a more useful diagnostic aid, the normal ratio being 3:2. A reversal of the ratio, which is a compensatory effort on the part of the extra-hepatic situations where 20 per cent of globulins are synthesised normally, is the basis of the several flocculation tests for liver function, though it is said that the quality of albumin in liver diseases also differs from the albumin in normal serum. Electrophoretic separation of serum proteins can, for example, establish conditions like cirrhosis of the liver characterised by hypo-albuminemia and hyper-gammaglobulinemia. However, there are non-hepatic conditions where one finds a similar feature.

The uninhibited globulins due to inadequate albumin are responsible for the flocculation by different reagents in different tests. Maclagan's thymol turbidity test consists of treatment of 0,1 ml of serum with 6 ml of a saturated solution of thymol in barbital buffer (pH 8.6). The normal values range between 0 – 4 units and can increase in liver diseases and also lipemic conditions without any liver damage whatsoever. The zinc sulphate turbidity test is also done for liver function. A buffered zinc sulphate is used. The normal values range between 2 – 10 units. Hanger's cephalin cholesterol flocculation test involves treatment of serum with an emulsion of a mixture of cephalin and cholesterol and the degree of precipitation is read after 24 hr standing, which is a measure of the severity of liver involvement. Turbidity tests are losing their diagnostic importance in recent years, as many direct sensitive tests for liver function are available.

Bromsulphalein excretion test

This is the dye of choice for assessing liver function. It is non-toxic and is practically and almost exclusively excreted by the liver. Occasionally, allergy to this substance may be encountered. Extra-vasation of the dye outside the vein should be avoided as it is a bad irritant. The dye is injected slowly, intravenously, in a dose of 5 mg per kg body weight in a 5 per cent solution. Blood is drawn through the opposite median cubital vein, 30 and 45 minutes after the injection of the dye. With a normal liver, the retention of the dye is less than 10 per cent at 30 minutes and less than six per cent at 45 minutes. The presence of jaundice vitiates the results. Hence, this test is not to be done in the presence of jaundice but it is valuable in the screening of pre-icteric phases of conditions like infective hepatitis, and it also has prognostic value in the assessment of residual damage of the liver, if any, during the convalescent period even after the patient's bilirubin in serum reaches normal levels and the patient is apparently clinically normal. This test is superior to all other tests as it caters to specificity as well as to sensitivity.

Serum cholesterol

More than the absolute level of cholesterol in serum, the ratio between free cholesterol and ester cholesterol is diagnostic of liver damage as the liver is also a site of esterification of cholesterol. Normally, the ester fraction predominates in plasma but in liver damage it goes down. Non-hepatic cause of alterations in cholesterol level in blood have to be carefully excluded, e.g., in atherosclerosis, thyroid diseases, post hepatic obstruction, nephrotic syndrome and xanthomas.

Determination of prothrombin time

This is a very useful liver function test not only for diagnostic purposes but also as a measure of safety before one undertakes liver biopsy and operative procedures. Obviously, all the non-hepatic causes of prolongation of prothrombin time must be scrupulously excluded. The conditions are deficient intake of vitamin K, defective absorption of vitamin K, post-hepatic obstruction, steatorrhoea, administration of anti-coagulants and administration of antibiotics. Isotonic sodium citrate solution (3 8 per cent) is the anti-coagulant of choice for the collection of blood as it serves the purpose of converting ionic into the non-ionic form of calcium. Suitable thromboplastic material is used, and controls are done simultaneously along with every test. The normal values vary according to the type of the thromboplastic material used.

Tests based upon the role of liver in carbohydrate metabolism

Galactose is the sugar of choice and preferred to glucose because it imposes a strain on the diseased liver unlike glucose. Unless the patient has malabsorption, the test is done orally. The person is given 40 g of galactose dissolved in water and if he has a normal liver, he should not excrete more than 3 g of galactose in a five-hour-collection of urine after ingestion of the test dose. Galactose intolerance is also seen in congenital galactosemia. The fructose tolerance test is also a liver function test.

Hippuric acid test

The liver being the major site of detoxication in the body, a test dose of sodium benzoate, i.e., 5.99 g dissolved in water, is given one hour after breakfast and the amount of hippuric acid excreted by the person in a four-hour collection of urine is estimated. The first sample of urine excreted after the ingestion of a glass of water along with breakfast is to be necessarily discarded. The excretion of less than 3 g of hippuric acid in the four-hour collection of urine indicates hepatic dysfunction. Intravenous administration of benzoate is necessary in case of malabsorption. When there is hepato-renal involvement, para-amino benzoate is administered intravenously, and the para-amino hippurate found in the systemic blood in a stipulated period is estimated.

Blood ammonia

The normal level of blood ammonia does not exceed 40 μg as nitrogen per 100 ml of blood. In advanced liver damage bordering on coma or in hepatic coma, toxic levels are seen as ammonia accumulates due to the liver not being able to utilise it for the formation of urea.

In the light of the merits and limitations of the several liver function tests availabe nowadays, it is wise to perform a composite liver function test which involves the choice of three or four liver function tests (more correctly liver dysfunction tests!) which are fairly specific and sensitive and which also would eliminate the analysis of urine, since an involvement of the kidneys may be there as well which would vitiate the results.

GASTRIC ANALYSIS

Gastric analysis is valuable, despite its limitations, for screening peptic disorders as well as blood dyscrasias like pernicious anemia where due to lack of the gastric intrinsic factor, a macrocytic anemia is caused associated with a true and histamine-fast type of achlorhydria. Moreover even in simple achlorhydric-anemias, a dimorphic type of anemia is met with due to defective absorption of vitamin B_{12}, as well as iron.

The classical intubation procedure on account of the inconvenience caused to the patient, has been replaced by the tubeless test meals involving use of suitable cation exchange resins. In fact the Augmented histamine test meal is done more frequently, not only to differentiate true from spurious achlorydria but also to elicit the maximum parietal cell response.

Apart from quantitation particularly of free acidity making use of suitable indicators, and determination of pepsin, a qualitative analysis of the fasting contents of the stomach is useful

GASTRIC ANALYSIS

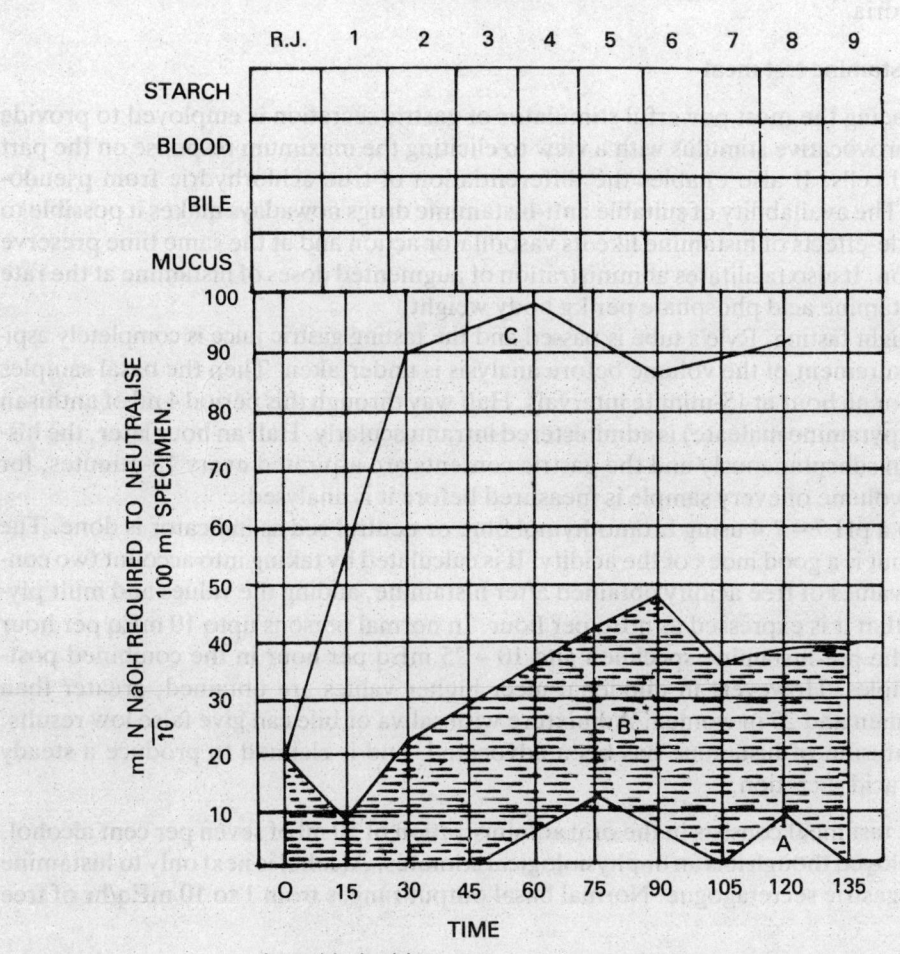

A – achlorhydria
B – Normal limits of free hydrochloric acid
C – hyperchlorhydria
R.J. Resting Juice

in ascertaining the presence of starch, bile, mucus and occult blood. Starch in the fasting juice is one of the evidences of stasis due to prolongation of the emptying time of the stomach, and bile is an evidence of regurgitation. Excess of mucus suggests gastritis. Blood is always pathological unless it is due to the trauma of intubation. An unduly large volume of the fasting gastric juice is yet another evidence of stasis. Lactic acid, if present, shows fermentation due to hypo-acidity.

The free acidity and total acidity are expressed as milliequivalents per litre (= clinical units). The normal range of free acidity is 15-35 mEq/L. It is high in duodenal ulcer, though not invariably. Chronic gastritis and carcinoma of the stomach are often associated with hypochlohydria or achlorhydria. A few normal persons, however show hypochlorhydria or even achlorhydria.

Augmented histamine test meal

Histamine being the most powerful stimulator of gastric secretion is employed to provide the requisite provocative stimulus with a view to eliciting the maximum response on the part of the parietal cells. It also enables the differentiation of true achlorhydria from pseudo-achlorhydria. The availability of suitable anti-histaminic drugs nowadays makes it possible to counter the side-effects of histamine like its vasodilator action and at the same time preserve its gastric action. It also facilitates administration of augmented doses of histamine at the rate of 0.04 mg histamine acid phosphate per kg body weight.

After overnight fasting, Ryle's tube is passed and the fasting gastric juice is completely aspirated for measurement of the volume before analysis is undertaken. Then the basal samples are collected for an hour at 15-minute intervals. Half way through this period 4 ml of anthisan (100 mg of mepyramine maleate) is administered intramuscularly. Half an hour later, the histamine is given subcutaneously and the gastric contents are aspirated every 15-minutes, for an hour. The volume of every sample is measured before it is analysed.

Titration to a pH 7 – 7.4 using bromothymol blue or neutral red as indicator is done. The peak acid output is a good index of the acidity. It is calculated by taking into account two consecutive high values of free acidity obtained after histamine, adding the values and multiplying by two so that it is expressed as mEq per hour. In normal persons upto 10 mEq per hour is present in the pre-histamine specimen and 10 – 25 mEq per hour in the combined post-histamine samples. However, in duodenal ulcer higher values are obtained, greater than 35 mEq/hr in men and 25 in women. Admixture with saliva or bile can give false low results. Continuous infusion of histamine has been advocated, and is claimed to produce a steady higher rate of acid secretion.

The alcohol test meal consists in the oral administration of 50 ml of seven per cent alcohol. It is often employed though it is an unphysiological stimulus. Alcohol is next only to histamine as a powerful gastric secretagogue. Normal basal output ranges from 1 to 10 mEq/hr of free acid.

Insulin test meal

The insulin test meal (Hollander's test) is designed to test the integrity of the vagi after vagotomy has been done to know whether the vagal resection has been successfully performed. The gastric secretion which results from the injection of insulin is due to the central stimulation of the vagus by the hypoglycemia induced by insulin. Hence, there is the necessity for bringing about a reduction in the blood sugar level to at least 40mg per 100ml, which can alone provide sufficient provocation for the stimulation of the vagus. Usually, the administration of

10 units of insulin intra-venously serves the purpose very well. Before vagotomy, there is a marked and prolonged output of acid in response to insulin, but after successful vagotomy there is no response to insulin which is foolproof evidence of complete resection of the vagal fibres.

Tubeless test meals

These are sophisticated procedures designed to obviate the inconvenience of swallowing the stomach tube. The principle consists in the administration of cation exchange resins, the cation of which gets exchanged for the hydrogen ions of the HCl of the gastric juice and gets quantitatively excreted in urine. The estimation of the exchanged cation in the urine is, therefore, indirectly a quantitative measure of the degree of free acidity in the gastric juice. Segal and his co-workers advocated the oral administration of quininium resin indicator whereby quinine hydrochloride is formed, absorbed in the intestine and excreted in urine 15 minutes after the resin is given. The quinine can be extracted from the urine and estimated by fluorimetry after ultra-violet exposure.

The procedure has been simplified in the Diagnex Blue test. Diagnex Blue is prepared by reacting a carbacrylic cation exchange resin with azure A, a safe indicator material. The hydrogen ions of the resin are exchanged with azure A ions, but this reaction is reversed in the stomach when free HCl is present in a concentration giving a pH of less than 3. The indicator which is then released is absorbed in the gastro-intestinal tract and is excreted in urine. Before the resin is administered on a fasting stomach and after emptying the bladder, caffein sodium benzoate tablets are given with a glass of water.

RENAL FUNCTION TESTS

Like the liver, kidneys also possess a remarkable factor of safety. Unilateral nephrectomy in cases of renal tumours, results in compensatory hyper activity of the remaining kidney. The renal function tests commonly done are given below

Total non-protein nitrogen in blood

The normal level ranges from 20 to 40 mg per 100 ml of whole blood. Azotemia refers to the elevation of non-protein nitrogen in blood. It can be caused either by over-production as in catabolic states and conditions of dehydration, or more often by retention due to renal dysfunction. Extra-renal conditions like starvation, fevers, thyrotoxicosis, Addison's disease, etc. must be excluded before the kidneys are incriminated. Uric acid in the blood rises first, followed by urea and creatinine simultaneously with deteriorating kidney function. Blood uric acid levels, however, can also be raised in conditions like leukemia, gout, etc.

Blood urea and urea clearance test

The normal blood urea level ranges from 20 to 40 mg per 100 ml (urea nitrogen ranges from 10 to 20 mg per 100 ml). Uremia denotes a high level of blood urea and is often due to retention caused by kidney damage, though as a part of general N.P.N. it can rise in catabolic states.

Normally, the plasma is cleared of urea only upto the extent of 10 per cent as it passes through the glomeruli of the kidneys. Plasma urea clearance, is therefore, defined as the volume of plasma cleared of urea per minute. When the volume of urine excreted per minute is equal to 2 ml or more than 2 ml, urea clearance is maximum. Hence, 2 ml is the augmentation limit for urea clearance. As long as the volume of urine excreted per minute is not less than 2 ml, there is a proportionality between urinary urea and blood urea levels, but when the volume of urine per minute is less than 2 ml, the proportionality is lost and urea clearance under this condition is called 'standard urea clearance', unlike maximum urea clearance in the former condition. The formula used for the calculation of maximum urea clearance is

$$\frac{U \times V}{B}$$

where U represents urinary urea nitrogen, V represents volume of urine excreted per minute and B stands for blood urea nitrogen. The normal maximum urea clearance is 75 ml per minute, though such a value is not seen in persons subsisting on a low protein diet. The formula used for calculation of standard urea clearance, is

$$\frac{U \times \sqrt{V}}{B}$$

The normal standard urea clearance is 54 ml per minute. The results are expressed better as percentage clearance of normal as follows. (Test value divided by normal value, multiplied by 100.) In the case of children, a suitable correction factor has to be applied for the surface area.

Procedure for urea clearance test: The patient need not fast overnight. A breakfast is allowed, and two glasses of water are given along with breakfast to ensure adequate urinary flow. The first sample of urine voided after the ingestion of water is to be discarded as false low urinary urea values can be caused by water diuresis. The time is noted immediately after the voiding of the first sample of urine and then two one-hourly samples of urine are collected and preserved for measurement of volume. At the end of the first hour, blood is drawn for the determination of urea.

Creatinine clearance test

It is an endogenous clearance test unlike the inulin clearance test. Creatinine like inulin is filtered by the glomerulus, but neither reabsorbed nor significantly secreted by the tubules so that whatever is filtered is excreted like inulin. Hence, it enables the assessment of the glomerular filtration rate (G.F.R.). Normal creatinine clearance values range from 95 to 105 ml per minute per 1.73 sq.m. body surface area.

Inulin clearance test

Inulin is filtered by the glomeruli but neither secreted nor reabsorbed by the tubules, so that inulin clearance measures the glomerular filtration rate. Mannitol can be used alternatively. When the test is done, a constant plasma inulin level is maintained by intravenous drip during the period of urine collections. The calculation is the same as for urea clearance. The normal inulin clearance value is 120 – 125 ml per minute per 1.73 sq.m. body surface area. Inulin clearance, however, cannot give any idea of the tubular status of kidneys.

Measurement of renal plasma flow (R.P.F.) (para-amino hippurate clearance)

Para-amino hippurate is filtered by the glomeruli but unlike inulin is secreted by the tubules and is removed completely during a single circulation of blood through the kidneys even at a low blood concentration of 2 mg per 100 ml. It is a measure of the plasma flow which is normally 574 ml per minute per 1.73 sq.m. body surface area.

Filtration fraction: GFR/RPF gives the fraction of plasma passing, through the kidney that is filtered by the glomeruli. Normally, it is 125/574 equal to about 21 per cent. In acute glomerulo-nephritis, the GFR is decreased considerably so that the filtration fraction is lowered; on the other hand, in essential hypertension, complicated by renal involvement, the RPF is decreased more than GFR so that the filtration fraction is increased.

Tm for PAH

By raising the para-amino hippuric acid concentration in the blood to 50 mg per 100 ml, maximum response can be elicited from the tubular secretory carriers of the kidneys. Diodrast is not used now, since it lyses the cells and also binds the plasma proteins. The normal Tm for the PAH is about 80 mg per minute per 1.73 sq.m. body surface area. The renal tubular integrity is assessed by this test.

Phenolsulphalein or phenol red excretion test for kidney function

The dye is non-toxic and exclusively excreted by the kidney and hence is the dye of choice for kidney function. After intravenous injection of the dye, the 15-minute sample of urine collected should contain 25 per cent or more of the injected dye. However, it is not a sensitive test. It is not uncommon to find on the other hand a temporary, increased elimination of the dye in urine possibly due to either a compensatory hyperactivity or due to the irritation caused by the inflammatory process in acute inflammatory states of the kidneys.

Concentration and dilution tests for kidney function

A constant and rigid specific gravity of various diurnal collections of urine is pathognomonic of renal function. Normally, the kidney tubules possess a remarkable osmotic adaptability according to the altered fluid content of the diet. The specific gravity and volume of the urine accordingly are reciprocally altered, the osmo-receptors of the hypothalamus being sensitive to fluctuations in the crystalloid osmotic pressure, chiefly to sodium chloride. The concentration test is seldom done as it entails fluid deprivation which is contra-indicated in obviously impaired renal function, in hot weather and also in adrenal insufficiency.

Procedure for dilution or water load test for kidney function

No water should be taken by the patient after midnight. The bladder is emptied at 7 a.m. the next morning and 1200 ml of water is given within the next half an hour. The bladder is emptied hourly for the next four hours. The voulme and specific gravity of each specimen are measured. The criteria for interpretation of results are that with normal kidney almost all the water drunk should be excreted within these four hours, and the specific gravity of at least one of the four samples should fall to 1.003 or below. However, in advanced kidney disease, the volumes may be less than 100 ml with specific gravities more than 1010.

THYROID FUNCTION TESTS

A basal metabolic rate below 30 per cent indicates hypothyroidism while elevated BMR is diagnostic of hyperthyroidism. However, it is difficult to enforce mental rest particularly in the hyperthyroid patient. Hence, one cannot entirely rely on BMR values in thyroid aberrations.

Protein-bound iodine in serum normally ranges from 3 to 7 μg/100 ml and less than $3\mu g$ shows a hypothyroid state while more than 7 μg/100 ml shows hyperthyroidism. A more reliable investigation is the determination of 'butanol-extractable iodine' (BEI) as otherwise inorganic iodine and iodo-tyrosines give a false high PBI value. The normal level of butanol-extractable iodine ranges from 2 to 6.5 µg/100 ml. It measures principally T_4 and T_3.

One of the latest diagnostic procedures is the measurement of ^{131}I uptake by the thyroid gland making use of suitable counters, chiefly the Gamma Ray Spectrometer. If the uptake is less than 15 per cent, it is characteristic of a hypothyroid state while an uptake of more than 40 per cent is seen in hyperthyroidism. The urinary excretion of ^{131}I shows a reciprocal relationship with the uptake in either condition.

Non-specific findings serve as accessory parameters. Serum cholesterol is elevated in hypothyroidism and markedly lowered in hyperthyroidism. The excretion of urinary neutral 17-ketosteroids is reduced in myxoedema usually. In hyperthyroid states, due to the wasting process, there is a significant alteration in the creatine-creatinine turnover.

An indirect measure of thyroxine-binding protein (TBG) can be had employing a test dependent on the uptake in vitro of ^{131}I labelled T_3 by the erythrocytes. This test is based upon the finding that when labelled T_3 is added to whole blood, the T_3 is bound by the serum protein as well as by the erythrocytes and the more it is taken up by the serum proteins, the less will be available for uptake by the red blood cells and vice versa. Hence, the amount of T_3 taken up by the red cell is an inverse measure of the degree of saturation of thyroxine-binding protein in the serum. In the hyperthyroid patient, the thyroxine binding capacity is more saturated than normal so that the residual serum binding capacity for T_3 is decreased while the T_3 uptake by the red cells is reciprocally increased. Diametrically opposite findings are obtained in the hypothyroid patient and the T_3 uptake by the erythrocytes is considerably less than normal. This test is of much help in investigations in children and elderly people who need not be hospitalised for the ^{131}I test. Very recently, radio-immuno assay is being used for the estimation of T_3 and T_4. From the precise levels of T_3 and T_4, the thyroid function is assessed.

Para-thyroid function tests

Absolute levels of calcium and phosphorus as well as determination of serum alkaline phosphatase are useful, but calcium balance studies are preferable. In a hyperparathyroid state, the salient laboratory features are hypercalcemia, hypophosphatemia, enhanced calcuria as well as phosphaturia. A calcium deprivation test to the extent of giving the patient only 100 mg of calcium in the diet per day is done and the excretion of more than 250 mg of calcium in urine, despite the deprivation, shows the severity of osteoporosis, characteristic of hyperparathyroidism though there are non-parathyroid causes which have to be eliminated. In chronic renal disease with secondary hyperactivity of the parathyroid gland, serum calcium is low due to the high serum phosphate characteristic of renal failure. Urinary calcium and phosphate are both lowered and there is resistance to the action of vitamin D. Hyperparathyroidism is associated with high levels of serum alkaline phosphatase.

In hypo-parathyroidism there is hypocalcemia, hyperphosphatemia, diminished excretion of calcium and phosphorus in the urine. The serum alkaline phosphatase is normal and serum magnesium levels are reduced as are hydroxy proline levels.

PANCREAS FUNCTION TESTS

Exocrine function tests of the pancreas relate to its acinar activity while the endocrine function tests pertain chiefly to the insulin status. Though the two groups are well demarcated, occasionaly one finds the development of a pancreatic diabetic state incident to pancreatic necrosis as well as recurrent attacks in chronic pancreatitis.

The pancreatic acinar activity can be directly studied by estimation of the output of fluid, bicarbonate and enzymes obtained by duodenal intubation under provocation by secretin and pancreozymin. Vagal stimulation or administration of mecholyl also increases the output of enzymes. Duodenal intubation is best done under screening.

More often, to obviate the inconvenience of duodenal intubation, the enzymes in serum and urine are determined as well as the fecal lipid and nitrogen.

Serum amylase, and serum lipase levels are considerably enhanced in acute pancreatitis or obstruction of the pancreatic duct, high intestinal obstruction, perforated peptic ulcer eroding into the pancreas, and mumps. The normal serum amylase levels range from 80 to 180 Somogyi units per 100 ml. The urinary amylase levels are correspondingly increased unless the kidney functions are deranged. In severe pancreatic deficiency, impaired digestion results in steatorrhoea and excess of nitrogen in the feces. The fats and nitrogen eliminated in the feces are determined on a standard diet containing known amounts of fat and protein. Intestinal steatorrhoea can be recognised and differentiated from a pancreatic lesion by the xylose absorption test which is diminished in the former condition. More than the total fats, the ratio between the split fat and unsplit one is a reliable parameter, since the split fraction is markedly lowered in sub-normal pancreatic activity.

Electrolytes in sweat, particularly sodium and chloride, are raised significantly in the fibrocystic disease of the pancreas which is a part of an inherited exocrine disorder causing muco-viscidosis, particularly in the bronchial and biliary systems.

Microscopic examination of feces may reveal the presence of undigested and unabsorbed foods as an evidence of pancreatic insufficiency and dyspepsia.

Endocrine function tests of the pancreas

These relate chiefly to the integrity of the beta cells of the Islets of Langerhans. Examination of urine for reducing sugars followed by quantitation of blood sugar levels are the usual procedures. Urine findings alone may not suffice, particularly when the renal threshold for sugar is either lowered or raised, or when glomerular filtration is diminished as in Kimmelsteil-Wilson's syndrome (inter-capillary glomerular fibrosis sometimes complicating diabetes mellitus).

The conventional glucose tolerance test is done after an oral ingestion of glucose except when there is malabsorption which necessitates parenteral administration of glucose. The test is done in the morning after twelve-hour overnight fasting.

The blood in the fasting state is drawn and the reducing sugar estimated, and the urine collected in the same state is tested for reducing sugar. The true glucose in blood can be determined, if necessary, by appropriate methods like the one employing glucose oxidase.

The usual amount of glucose given is 50 g though larger amounts have been used, i.e., 1.75 g per kg body weight. However, 1 g per kg body weight should suffice. For children, the test dose given is on the basis of the following formula.

$$50 \times \frac{(\text{Wt. in lb})^{2/3}}{140} \text{ g}$$

The person should be on an ordinary mixed diet with the normal requirement of carbohydrate prior to the test. Lowered levels in the diet are likely to vitiate the results. After the test dose of glucose dissolved in water is given, flavoured if necessary, to avoid retching, blood is drawn at half-hourly intervals for two and a half hours, and samples of urine are collected with every sample of blood.

GLUCOSE TOLERANCE TEST

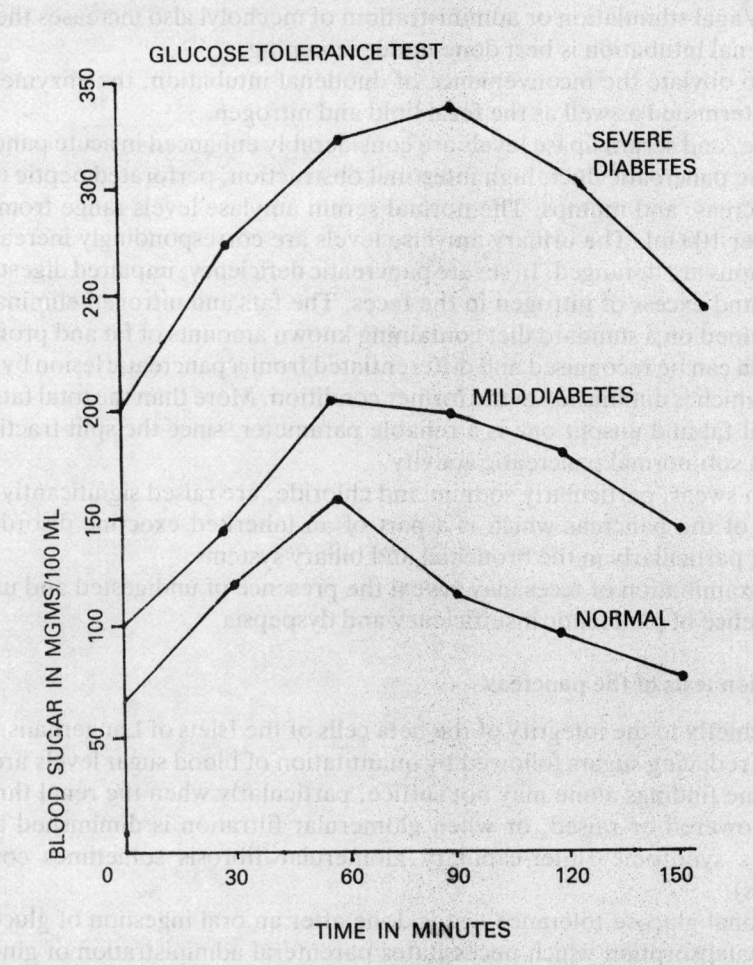

[Graph showing blood sugar in mgms/100 ml vs time in minutes (0 to 150) for three curves: Severe Diabetes (rising from ~200 to peak ~340 at 90 min, then declining to ~240), Mild Diabetes (rising from ~100 to peak ~205 at 60 min, declining to ~150), and Normal (rising from ~70 to peak ~160 at 60 min, declining to ~75).]

The results are expressed in the form of a graph with time intervals on the abscissa and the blood sugar levels on the ordinate. The nomal response, as shown in the diagram, shows the fasting level ranging from 70 to 90 mg per 100 ml. The maximum is reached in one hour, and the level reaches normal almost always in two hours. There is no detectable sugar in any or the urine specimens. Renal glycosuria is characterised by glycosuria due to lowered renal threshold, but there is no hyper-glycemia. The peak of the glucose tolerance curve is significantly lowered.

Glucose tolerance is lowered not only in the diabetic state but also in hyperpituitarism, hyperthyroidism, and hyperadrenalism. On the other hand, insulin sensitivity is increased in hyperinsulinism, hypopituitarism, hypoadrenalism and hypothyroidism. The insulin tolerance test is designed to screen insulin resistance in the former conditions mentioned, like hyperpituitarism and increased insulin sensitivity in conditions like Addison's disease, where insulin is administered along with glucose to minimise the risk of hypoglycemic coma.

There are some apparently normal persons showing a transient hyperglycemia and glycosuria but the blood sugar level returns to normal in the scheduled time. The curve is known as the 'lag curve' either due to a delay in insulin response or due to the enhanced rate of absorption of glucose following rapid emptying of the stomach. A similar finding is met with after sub-total gastrectomy.

A pre-diabetic state, often due to genetic factors, can be unmasked by a provocative administration of 11 oxo-steroids in the form of cortisone. Prednisone is used at the rate of 0.4 mg per kg body weight, half at midnight and half at 6 a.m. before carrying out the usual glucose tolerance test. Cortisone-induced hyperglycemia and glycosuria reveal a latent diabetic state. Tests for investigating hypoglycemia include the tolbutamide tolerance test, plasma insulin assay, the glucagon tolerance test as well as the leucine sensitivity test. Details can be obtained from practical clinical biochemistry books.

ADRENAL FUNCTION TESTS

Hypoadrenalism of the medulla is rarely encountered. Addison's disease is caused by the hypofunctioning adrenal cortex. The characteristic biochemical findings are the enhanced urinary output of sodium chloride, diminished excretion of potassium, hyponatremia, hyperkalemia and hypoglycemia. In women, 17-ketosteroids are very low or negligible in urine, but due to the testes, in men there is an excretion of 1-6 mg even in severe cases in men. Hypothyroidism as well as any chronic illness shows lowered values of excretion of 17-ketosteroids thereby rendering the result non-specific. A better index of adrenal cortical function, however, is assay of 17-ketogenic steroids and total 17-hydroxy cortico-steroids in urine.

Hyperactivity of the adrenal medulla is seen in the rare tumour known as pheochromocytoma. The malignant tissue in this condition abounds in nor-epinephrine. Apart from the determination of urinary catecholamines and their metabolites like vanillyl mandelic acid, a more specific test makes use of regitine or phentolamine, a structural antagonist of nor-epinephrine. A rapid intravenous injection of 5 mg of phentolamine should produce a sustained fall of blood pressure, at least 25 – 35 mm of mercury, within two to five minutes.

The hyperactivity of the adrenal cortex may be due to hyperplasia or tumours. Cushing's syndrome is one of the conditions recogised by diminished excretion of sodium chloride in urine, kaluresis, hypernatremia, hypokalemia, hyperglycemia, glycosuria and edema in severe cases. Cortical hyperactivity may be primary or secondary to hyperpituitarism. Hence, in addition to assay of urinary 17-ketogenic steroids and total 17-hydroxy cortico-steroids, their quantitation before and after stimulation with ACTH and suppression tests are invaluable in confirming the nature of the hypercortical activity. Very recently, cortical hormones and ACTH are estimated by radio-immuno assay, to assess the adrenal and pituitary functions, respectively.

Primary or essential aldosteronism is confined to excessive production of aldosterone. The important features are alkalosis, hypernatremia, persistent low serum K levels, hypertension, polyuria, polydipsia without edema. After sodium restriction, the tendency to alkalosis and hypokalemia is eliminated, and sodium disappears from the urine. Administration of spironolactone or aldactone which is an aldosterone antagonist restores serum potassium levels to normal. For detailed procedures and interpretations practical books have to be referred to.

SECTION VI
PHYSICO-CHEMICAL ASPECTS OF BIOCHEMISTRY

26 Physico-chemical Aspects of Biochemistry

HYDROGEN ION CONCENTRATION, pH AND BUFFERS

One of the important factors governing biochemical and biophysical changes is the concentration of various ions present in the blood, body fluids and glandular secretions. The important cations that are present are H^+, Ca^{++}, Na^+, K^+, etc., and the important anions are bicarbonate (HCO_3^-), chloride (Cl^-), phosphate (HPO_4^{--} and $H_2PO_4^-$), etc.

The origin of H^+ in living tissues is water, as all forms of life contain water and as water ionises feebly to liberate H^+ in addition to OH^-. Water contains 10^{-7}g of H^+/litre, and as it is source for H^+ in the body, the concentration of H^+ in biological fluids will be around 10^{-7}g of H^+/litre.

As a result of ionisation of water, there should be equivalent amount of OH^- for each H^+ according to the following equation:

$$H_2O \rightleftharpoons H^+ + OH^-$$

(It has to be remembered that the degree of ionisation is very slight, and it may be shown as just one molecule out of nearly 550 million molecules of water undergoes complete ionisation. Moreover, the H^+ ions gets hydrated and is present as hydroxonium ion, H_3O^+. In water, which is neutral, the H^+ concentration is 10^{-7} g ions/litre (equal to 10^{-7} g/litre). As there should be an equivalent number of OH^-, the concentration of OH^- is also 10^{-7}g ions/litre. But each gram ion of OH^- is equal to 17g. So 10^{-7}g ions of OH^- will be equal to 17×10^{-7}g.

The H^+ concentration is responsible for acidity, while it is OH^- for alkalinity. If H^+ and OH^- are of equal concentration in any system, the solution is neutral, e.g., water. If the H^+ concentration exceeds that of OH^-, the solution will become acidic. For example, if the concentration of H^+ in an aqueous medium is 10^{-2}, it is acidic. Alternately, if the H^+ is less than 10^{-7}, the solution will be basic, e.g., a solution with a concentration of 10^{-10} of H^+. In any aqueous solution the ionic product ($H^+ \times OH^-$) should not exceed 10^{-14}g ions/litre which is arrived at by the experimental findings that there are 10^{-7}g ions of H^+ and 10^{-7}g ions of OH^- ($10^{-7} \times 10^{-7} = 10^{-14}$) per litre of water.

In order to do away with the negative powers as 10^{-7} or 10^{-x} in representing H^+ concentration, especially of biological fluids, S.P.L. Sorensen suggested a convenient notation known as the pH notation (pH meaning Puissance Hydrogen, Potenz Hydrogen, or Hydrogen power). In this, the negative powers vanish and positive numbers result. At the same time one is able to appreciate the concentration of Hydrogen ions).

pH is defined as the negative logarithm to the base 10 of H^+ concentration (or better, H^+ activity).

$$pH = -\log_{10} H^+$$

pH of water: The H^+ of water is 10^{-7} g/litre

$$H^+ = 10^{-7} g$$
$$\log_{10} H^+ = \log_{10} 10^{-7} = -7$$
$$\text{Negative } \log_{10} H^+ = 7$$
$$(-\log_{10} H^+)$$

So the pH of water is 7

Thus the H^+ concentration of 10^{-7} involving negative power (-7) becomes 7, a positive number, in the pH notation. The different H^+ concentrations and their corresponding pH values are given below:

H^+	1	10^{-1}	10^{-2}	10^{-3}	10^{-4}	10^{-5}	10^{-6}	10^{-7}	10^{-8}	10^{-9}
pH	0	1	2	3	4	5	6	7	8	9
			acidic					neutral	basic	
		10^{-10}	10^{-11}	10^{-12}	10^{-13}	10^{-14}				
		10	11	12	13	14				
				basic						

As already indicated, the product of ionic concentrations of H^+ and OH^+ should be only 10^{-14} in any aqueous solution, and the highest pH could be only 14.

The pH of a solution containing one g of H^+ (say, 1 N HCl on complete ionisation) is 0 and that of 1 N alkali (on complete ionisation) is 14. The pH scale is thus between 0 and 14, pH 0 indicating almost pure acid solution and pH 14 almost pure alkali solution. pH 7 is that of water which is neutral as it contains equal number of H^+ and OH^-. An increase of H^+ concentration will mean decrease of negative power. The pH of acid solutions will be less than 7. If the pH is above 7, the solution is alkaline.

Neutral pH 7

Acid pH 0 upto 7

Alkaline pH above 7 to 14

The pH of blood is 7.35-7.45, i.e., slightly on the alkaline side. The pH of urine is on the acidic side, i.e., less than 7.

Buffer solutions

A buffer solution is one which is resistant to changes in pH on the addition of small amounts of acid or alkali. Such solutions usually consist of a mixture of a weak acid and its salt (e.g., acetic acid and sodium acetate), or a weakbase and its salt (ammonium hydroxide and ammonium chloride). A salt of a weak acid and a weak base (e.g., ammonium acetate) also has buffer action.

Mechanism of buffer action

The mechanism of buffer action can be explained by taking the buffer pair of acetic acid and sodium acetate.

Acetic acid ionises feebly to give acetate ion and hydrogen ion.

$$CH_3COOH \rightleftharpoons (CH_3COO)^- + H^+$$

Sodium acetate being a salt ionises considerably and gives a high concentration of acetate ions.

$$CH_3COONa \longrightarrow (CH_3COO)^- + Na^+$$

If a strongly ionising acid like HCl tending to decrease the pH is now added to the above buffer, the acetate ions available combine with the H^+ entering the buffer and convert it into the weakly ionising acetic acid. Thus, the influence of the added H^+ in decreasing pH is overcome by the salt of the buffer.

$$\underset{\text{of the acid added}}{H^+} + (CH_3-COO)^- \rightarrow \underset{\text{weakly ionising acid}}{CH_3-COOH}$$

On the other hand, if a strongly ionising base like NaOH tending to increase the pH enters the buffer, the H^+ of the acid combines with the hydroxyl ion and converts it into undissociated molecules of water. Thus, the influence of the added OH^- ion in increasing the pH is overcome by the acid part of the buffer.

$$\underset{\text{from the alkali added}}{OH^-} + \underset{\text{of the acid of buffer}}{H^+} \rightarrow H_2O$$

Thus a buffer solution resists any alteration in pH by the addition of small amounts of acids or alkalis.

Henderson and Hasselbalch's equation

In the buffer pair of acetic acid and sodium acetate, the following ionisations take place:

$$\text{Acid } CH_3-COOH \rightleftharpoons (CH_3COO)^- + H^+$$
$$\text{Salt } CH_3COONa \rightarrow (CH_3COO)^- + Na^+$$

It is the dissociation of acid that is responsible for the pH of the buffer.

Applying the Law of Mass Action to the ionisation of acid,

$$CH_3COOH \rightleftharpoons (CH_3COO)^- + H^+$$

$$K = \frac{[CH_3COO^-][H^+]}{[CH_3COOH]}$$

where K is dissociation constant of the acid.

Owing to the depressing effect of sodium acetate on the dissociation of acetic acid, the concentration of the undissociated acid (CH_3COOH) is practically the same as that of acetic acid itself. Also, the concentration of the acetate ions (CH_3COO^-) is practically the same as that of the salt, namely, sodium acetate.

$$(CH_3COOH) = (Acid)$$
$$(CH_3COO)^- = Salt$$

Substituting these in the above equation,

$$K = \frac{[Salt][H^+]}{[Acid]}$$

$$[H^+] = K \frac{[Acid]}{[Salt]}$$

$$\log_{10}[H^+] = \log_{10}K + \log_{10}\frac{[Acid]}{[Salt]}$$

$$-\log_{10}[H^+] = -\log_{10}K + \log_{10}\frac{[Salt]}{[Acid]}$$

$$\boxed{pH = pK + \log_{10}\frac{[Salt]}{[Acid]}}$$

This equation is called the Henderson-Hasselbalch equation and can be used to calculate the pH of any buffer from a knowledge of the dissociation constant of the acid and the concentration of the salt and acid.

The equation can be applied for the determination of pH of blood from a knowledge of the concentration of bicarbonate (salt) and carbonic acid (acid) and knowing the dissociation constant of carbonic acid.

Problem:

For example, in blood plasma, HCO_3^- is 0.25 moles/litre while H_2CO_3 is .00125 moles/litre. If the dissociation constant of carbonic acid is 7.943×10^{-7} find out the pH of blood.

First, pK is calculated from the dissociation constant K.

$pK = -\log_{10}K$

$pK = -\log_{10}7.943 \times 10^{-7}$

$\log_{10}7.943 \times 10^{-7} = \log_{10}7.943 + \log_{10}10^{-7} = 0.9 + (-7) = (-6.1)$

$pK = -\log_{10}K = -(-6.1) = +6.1$

The pH is now calculated using Henderson's equation.

$$pH = pK + \log_{10} \frac{[Salt]}{[Acid]}$$

Salt = .025 M
Acid = .00125 M

$$pH = 6.1 + \log_{10} \frac{.025}{.00125}$$

$$= 6.1 + \log_{10} \frac{20}{1} = 6.1 + 1.3010 = 7.401$$

Problem:

2) The pH of blood of a person is 7.46.
   What is the Hydrogen ion concentration?

$$pH = 7.46$$
$$-\log_{10}[H^+] = 7.46$$
$$\log_{10}[H^+] = -7.46$$
$$= -8 + 0.54$$
$$[H^+] = \text{antilog of } -8 \times \text{antilog of } 0.54$$
$$10^{-8} \times 3.467$$
$$[H^+] = 3.467 \times 10^{-8}$$

Indicators

Indicators are substances which change colour with change of pH of the solution. They are organic substances and are either weak acids or weak bases. They ionise partially. The undissociated molecules have one colour and the dissociated ions, another colour.

Indicator molecules ⇌ $H^+ + (Ind)^-$ ions
(one colour)   (different colour)

Just as acids have dissociation constant, the dissociation of indicator also has Indicator Constant K. The Henderson-Hasselbalch equation can be used for dissociation of indicator.

$$pH = pK + \log_{10} \frac{[\text{Indicator ions}]}{[\text{undissociated molecules}]}$$

The range of an indicator is that part of the pH scale (0-14) over which the eye can appreciate a change of colour of the indicator.

It is roughly two units in the pH scale.

For example, methyl orange range is pH 3.1–4.4; Phenolphthalein range is pH 8.3–10.

In addition to individual indicators, mixtures of indicators known as universal indicators have been found useful for covering the entire pH range and determining the approximate pH of an experimental solution.

A convenient and simple form of universal indicator covering a range of pH from 4 to 11 is made from methyl red, alpha naphtholphthalein, phenolphthalein and bromophenol blue. The colours and corresponding pH values for the above universal indicator are,

colour	red	orange red	yellow	greenish-yellow	green
pH	4	5	6	7	8

colour	blue green	blue	violet
pH	9	10	11

Determination of pH

pH can be determined either by colorimetric methods or EMF methods.

All colorimetric methods use indicators in one from or the other. Hence, they are also called indicator methods.

1) Using buffer solutions: The pH of the experimental solution is first determined approximately by means of a universal indicator, by matching the colour developed with the colour chart given for the various pH units. When a drop of universal indicator is added to about 2 ml of experimental solution and the colour developed is, say, red, the approximate pH is 4. Knowing the approximate pH, the individual indicator working at that range is chosen. In this case, it is methyl orange whose indicator range covers the pH of 4.

A fixed quantity of the individual indicator (methyl orange) is added to a certain volume of the experimental solution. The same quantity of the indicator is added to equal volumes of various buffer standards differing in pH by, say, 0.1 unit in the neighbourhood of 4, (say, 3, 3.1, 3.2, 3.3...3.9, 4, 4.1, 4.2...5). In the buffer standards, colours of different intensities will be produced. The experimental solution also would have developed a colour with the indicator. This colour obtained in the test solution is matched with the colours in the buffer standards. The pH of the buffer standard whose colour matches with the colour developed in the test solution is the pH of the test solution. If, for instance, the colour developed in the test solution matches with that in the buffer standard with pH 3.6, the pH of the test solution is 3.6.

2) Without using buffer solutions: i) By using universal indicator papers or individual indicator papers. The papers are supplied by companies. A drop of the solution is placed on the indicator paper. The colour developed is compared with the colour chart given for different pH units.

ii) By using Comparators: Various Comparators like Lovibond Comparator and Hellige Comparator are available. Discs with different shades of colours for different pH values are supplied. The test solution is treated with the prescribed quantity of indicator specified or supplied. The colour developed is matched with the colours in the disc. The pH for the colour in the disc matching with the colour of the experimental solution is the pH of the test solution.

iii) pH can also be determined without using buffer standards from a knowledge of the dissociation constant of the acid of the buffer and the concentrations of salt and acid in the buffer. For this the Henderson-Hasselbalch equation should be used.

In the EMF methods, different electrodes like the hydrogen electrode quinhydrone electrode or the glass electrode can be used. Direct reading pH meters working on this principle and using the glass electrode are available.

Determination of pH by glass electrode

When a glass surface is in contact with a solution, it acquires a potential difference, which has been found by experiment to be dependent upon the pH of the solution.

The glass electrode consists of a glass tube terminating in a thin walled bulb. A special glass of relatively low melting point and high electrical conductivity is used. The bulb contains a solution of constant pH and an electrode of definite potential. A platinum wire inserted in pH 4 buffer solution containing a small quantity of quinhydrone is usually employed. The glass bulb containing the above is inserted in the experimental solution whose pH is to be determined. This constitutes the glass electrode which is a half cell.

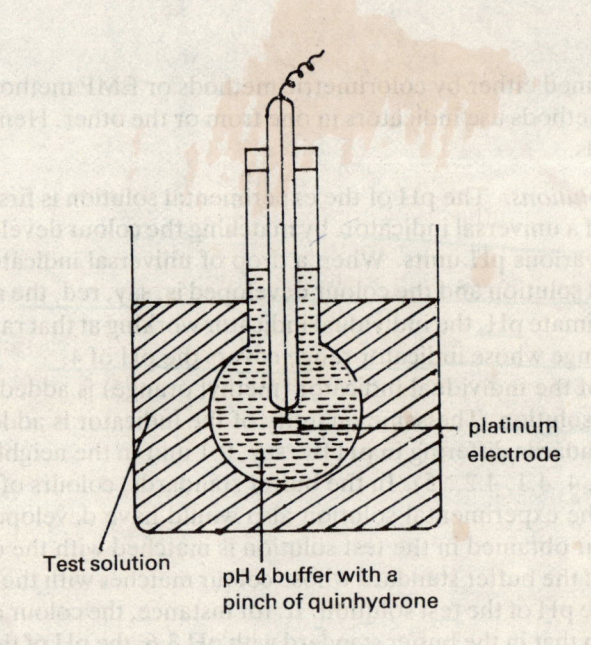

Test solution — *platinum electrode* — *pH 4 buffer with a pinch of quinhydrone*

In actual determination of pH, the glass electrode is coupled with the calomel electrode and the EMF of the cell is determined experimentally.

$$E_{cell} = E_{glass} + E_{calomel}$$
$$E_{glass} = E_{cell} - E_{calomel}$$

Assuming $E_{calomel}$ to be 0.3333, E_{glass} can be determined.

pH of the solution is related to the E_{glass} by the equation:

$$pH = \frac{E_{glass} - K}{.0002\,T}$$

where T is the absolute temperature and K is a constant for the glass electrode. (K for any glass electrode can be determined in a separate experiment using a solution of known pH).

The glass electrode can be used for almost any solution except that which is very acidic or very alkaline.

Direct reading pH meters incorporating the glass and calomel electrodes are available. Beckmann pH meter and Cambridge pH meter are such instruments which are routinely used in biochemical laboratories.

COLLOIDAL STATE

Graham, the father of colloid chemistry, classified substances into crystalloids and colloids. Substances like sodium chloride, sugar, etc. which, while present in solution, pass through parchment membrane were called *crystalloids,* while substances like albumin, gelatin etc. which were retained by the parchment membrane were called *colloids* (glue-like). This classification was later found to be untenable as the same substance can be a crystalloid or a colloid depending on the experimental conditions. Thus, sodium chloride which is a crystalloid according to Graham can be obtained as a colloid in benzene or alcohol. Soap behaves as a colloid in water but as a crystalloid in alcohol. Thus any substance can be obtained in colloidal form by employing suitable methods. The term 'colloid' therefore does not concern the nature of the substance but only the state of the subdivision of matter. Any substance is said to be in colloidal state if the diameter of its particles ranges from $1m\mu$ to $200m\mu$ ($1m\mu$ (millimicron) = 10^{-3} of μ (micron) or 10^{-6}mm). The colloidal state of matter is intermediate between molecules and suspensions: molecules having diameter below $1m\mu$ are not seen at all even with the help of a microscope, and suspensions having diameters above $200\ m\mu$ can be seen even with the naked eye. Colloidal particles can be seen with the aid of an ultra-microscope, each particle appearing as a small disc of light. Because of the very small size, the colloidal particles pass through ordinary filter paper, the pores of which are too big to retain the particles. However, colloidal particles are retained by parchment, collodion or animal membranes.

A colloidal system is a heterogeneous diphase system. The phase which constitutes the bulk is called the dispersion medium (external medium) which contains the dispersed or internal phase. Combination of any two phases can form a colloid of any of the following types–gas/liquid; gas/solid; liquid/gas; liquid/liquid; liquid/solid; solid/gas; solid/liquid; and solid/solid.

The phase written first forms the dispersed phase.

The solid/liquid colloid is termed as 'sol' while the liquid/liquid colloid is called an 'emulsion' which contains two immiscible liquids, the one in smaller amounts forming the dispersed phase.

Sols in which water is the dispersion medium are called hydrosols. If alcohol is the medium, these sols are called alcosols. Sols are divided into lyophobic (liquid-hating) and lyophilic (liquid-loving) types depending on the affinity between the solid and the liquid. If it is hydrosol, the terms hydrophobic (water-hating) and hydrophilic (water-loving) are employed. The terms 'suspensoid' for lyophobic sols and 'emulsoid' for lyophilic sols are used by biochemists, but not used by physical chemists.

Good examples of hydrophilic sols are those of proteins, gum, gelatin, starch and agar in water.

Properties

1) Dialysis: The retention of colloidal particles by a parchment membrane and the passage of the dispersion medium is called dialysis. This is employed to filter off the colloidal particles. If an electric field is applied in the vessel containing the dialysing bag with the colloid inside, the process is called electro-dialysis. Cerebrospinal fluid (CSF) is considered a dialysate of blood plasma. The technique of dialysis is employed in the 'artificial kidney' to remove urea and other unwanted small metabolic wastes from proteins of the blood.

2) *Osmotic pressure*: The osmotic pressure of the colloidal systems is usually small when compared with solutions of true molecules. This is to be expected because a colloidal particle is an aggregate of thousands of molecules. Osmotic pressure being a colligative property depends on the number of particles. The osmotic pressure of proteins of plasma is called 'oncotic' pressure which has a significant role in maintaining blood volume.

3) *Optical properties*: Colloidal particles scatter light. This property is called the Tyndall effect. It is the Tyndall effect which is applied in the ultra-microscope, in which the light scattered by the particles is seen.

4) *Brownian movement*: The colloidal particles are in a continual and haphazard motion. This is called the Brownian movement which is due to the bombardment of the colloidal particles by the molecules of the dispersion medium.

5) *Electrical properties of colloids*: The colloidal state of matter possesses a great surface area per unit weight in view of the presence of a large number of particles in it. Every particle possesses, about it, certain fields of force which are probably largely electro-magnetic. Such fields of force which are on the surface of each particle are relatively free when compared with those inside the particle. Inside the particles, the forces are neutralised by the atoms present, while forces on the surface are only partially neutralised. Hence, the surface of the colloidal particle has free fields of force or attraction which impart to it peculiar properties. One very important property arising out of surface forces is adsorption. Traces of ions are adsorbed on to the surface of colloidal particles. This adsorption together with primary ionisation of ionisable groups in the colloidal substance is responsible for the electrical properties of colloids.

Electrical potentials on the surface

There is a statically charged electrical system around each colloidal particle. The colloidal particle has a definite electrical charge. This charge, in the case of proteins, is due to the primary ionisation of groups like —COOH (of acidic amino acids) and —NH_2 of basic amino acids. In addition, there is adsorption of ions on the surface. Where there is no primary ionisation, the electrical charge is only due to the adsorption of ions. Adsorption results in the formation of two layers, one the immobile layer and the other mobile. This envelope of ion on the colloid is termed, the Helmholtz-Gouy electrical double layer. There are three types of potentials as a result of the double layer. The epsilon potential (ε), the electro-kinetic or zeta (ζ) potential and the Stern potential. The potential drop across all ionic layers from the surface into the solution is the epsilon or electro-chemical potential. The potential between the immobile layer of ions and the mobile layer is termed the zeta potential while the potential between immobile layer and the particle surface is the Stern potential.

Of the three potentials, the ζ potential is of primary importance in connection with the electrical properties of colloids. This is of the order of 30 millivolts for many colloids if water is the dispersion medium. All the colloidal particles thus have like electrical charges. When an electric current is passed through a sol, in view of like electrical charges, all the colloidal particles move towards the anode or cathode. Such a movement is called *electrophoresis*. There are colloids in which all the particles are positively charged, e.g., proteins in acid solution, hemoglobin, etc., while proteins in alkaline solution and metal sols are good examples of negatively charged colloids. The same colloid may be made positive or negative with suitable amounts of reagents used.

Immobile + ve ions adsorbed on the surface of colloid from the liquid

colloid particle (having, say, −ve charge)

mobile + ve ions around the particle

Helmoholtz Guoy double layer

Solid | Immobile layer | Mobile layer

Stern potential

ζ potential

ε Potential

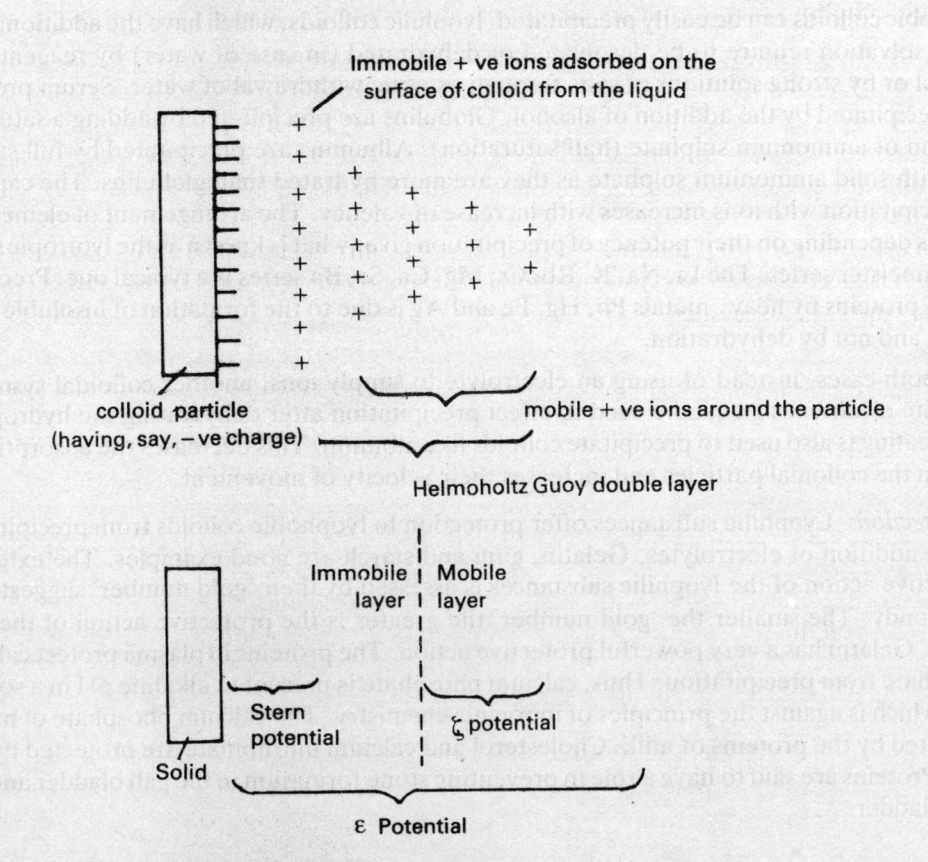

Electro-osmosis, endosmosis or electro-endosmosis

This is the opposite of electrophoresis. When an electric current is passed through a sol, if the colloidal particles are kept stationary, the dispersion medium will move towards an electrode. This is called electro-osmosis.

Stability of colloids

The stability of colloids, both lyophobic and lyophilic, is due to like charges brought about by the ζ potential and the force of repulsion between individual particles. In the case of lyophilic colloids, an additional factor operates for the stability. This is solvation or hydration (in case of water) of the particles which are associated with molecules of solvent or water.

Precipitation of colloids

For precipitation of both types of colloids, the force of repulsion should be overcome. For this, the charge on the particles should be neutralised by the addition of ions of opposite electrical sign. This brings about a decrease in the magnitude of the ζ potential. When it is lowered to less than 15 millivolts, the precipitation of colloids occurs. At this point at which the charges have been neutralised, the particles are said to be brought to the 'iso-electric point'. At the iso-electric point there is no electrophoresis. The iso-electric point of casein of milk is a pH of 4.6. Salt solutions of sodium chloride, ammonium sulphate, and acids like sulpho salicylic acid and trichlor acetic acid are employed for the precipitation of colloids. While the lyophobic colloids can be easily precipitated, lyophilic colloids, which have the additional factor of solvation require to be desolvated or dehydrated (in case of water) by reagents like alcohol or by strong solutions of salts through osmotic withdrawal of water. Serum proteins are precipitated by the addition of alcohol. Globulins are precipitated by adding a saturated solution of ammonium sulphate (half saturation). Albumins are precipitated by full saturation with solid ammonium sulphate as they are more hydrated than globulins. The capacity of precipitation with ions increases with increase of valency. The arrangement of elements in a series depending on their potency of precipitation gives what is known as the lyotropic series or Hofmeister series. The Li, Na, K, Rb, Cs, Mg, Ca, Sr, Ba series is a typical one. Precipitation of proteins by heavy metals Pb, Hg, Fe and Ag is due to the formation of insoluble complexes and not by dehydration.

In both cases, instead of using an electrolyte to supply ions, another colloidal system of opposite electrical sign can be used to effect precipitation after dehydrating the hydrophilic sol. Heating is also used to precipitate colloids like albumin. This decreases the adsorption of ions on the colloidal particles and increases their velocity of movement.

Protection: Lyophilic substances offer protection to lyophobic colloids from precipitation on the addition of electrolytes. Gelatin, gum and starch are good examples. The extent of protective action of the lyophilic substances is assessed by their 'gold number' suggested by Zsigmondy. The smaller the 'gold number' the greater is the protective action of the substance. Gelatin has a very powerful protective action. The proteins in plasma protect calcium phosphate from precipitation. Thus, calcium phosphate is present in alkaline pH in a soluble form which is against the principles of inorganic chemistry. The calcium phosphate of milk is protected by the proteins of milk. Cholesterol and calcium bilirubinate are protected by bile salts. Proteins are said to have a role in preventing stone formation in the gall bladder and urinary bladder.

Gels

Gel is a semi-rigid, jelly-like mass obtained by coagulating a sol, particularly a lyophilic sol, under certain conditions and it contains the whole of the liquid present in the sol. A gel can be considered a colloidal system of a liquid dispersed in a solid.

Gels are of two types: elastic and non-elastic. Partial dehydration of an elastic gel yields an elastic solid from which the original sol may be readily regenerated by the addition of water and warming, if necessary. Elastic gels when placed in water imbibe large quantities of water and swell in size. The imbibed gel, on keeping, exudes the water and shrinks. This is called 'syneresis' or 'weeping of the gel'. A non-elastic gel, on the other hand, becomes glassy or powdery and loses its elasticity on drying. It does not exhibit imbibition and syneresis. Agar and gelatin gels are good examples of elastic gels while silica gel is a non-elastic gel. The glandular secretions are said to be due to the imbibition of water by the gel-like colloids in the glands which give out the water of imbibition as glandular secretions by syneresis.

Emulsions

Emulsions are liquid-liquid colloids. The two liquids should be immiscible. Either liquid can be dispersed in the other. That which is in smaller amounts forms the disperse phase, the droplets of which have a diameter of 0.1–1 mμ. In general, two types of emulsions are recognised: 1) oil-in-water emulsions, and 2) water-in-oil emulsions, depending on whether there is excess water or the other liquid.

Emulsions are generally unstable unless a third substance known as an emulsifying agent is present. The emulsifier reduces the inter-facial tension between the oil (or organic liquid) and water and this imparts stability to the liquid/liquid colloid. Soaps, gelatin, gum, proteins and bile salts are good exmples of emulsifying substances.

Milk is a common example of an emulsion and contains droplets of liquid fat dispersed in an aqueous medium. The emulsifier is the protein casein.

Emulsion can also be separated by electrophoresis and can be coagulated, broken or de-emulsified by the additon of material which destroys the emulsifier and neutralises the electric charge. Boiling and freezing also bring about de-emulsification.

Membrane phenomena

The living cell membrane is quite different from artificial membranes made of non-living matter in composition, electrical poperties and permeability. The surface tension-lowering substances which concentrate at the interface according to the Gibbs Thomson effect, are found to predominate in the living membrane. The membrane is thus made up of phospholipids, cholesterol, fatty acids and protein, most of which are surface tension-lowering substances. Electron microscopy, X-ray diffraction and histo-chemical studies have shown the living membrane to be composed of two long thin monolayers of proteins between which occurs a layer of lipid, chiefly phospholipid and cholesterol two molecules thick (for details refer 'Cell Structure') with protein-invaginations.

The resting membrane has an electrical potential which is about –70mv. The electrical charge of the membrane is due to different factors like the ionisaton of the membrane protein, adsorption of ions and Helmholtz double layer and to a cosiderable extent diffusion potential due to concentration cells of Na^+, K^+, and Cl^-. The value of the electrical potential appears to depend on the K^+ion diffusion potential as a first approximation. The inside of a membrane is negative and the outside positive, while the pore walls also are negatively charged. The electrical nature of the membrane influences the permeability of substances across the membrane.

Many biochemical reactions take place inside the cells. Hence, the entry of substances across the cell membrane determines the metabolic reactions. Various factors operate in the transport of substances across the membrane. For small molecules and ions, it is simple diffusion. Urea and sugar are equally distributed between the cells and plasma. The differential distribution of Na^+ and K^+ between the cells and extra-cellular fluid and chiefly the accumulation of K^+ inside the cells is explained by relative decrease in the size of hydrated K^+ when compared with hydrated Na^+ (1:1.47), the increased motility of K^+ (1×10^{-6} mole/sq cm/sec of membrane while for Na^+ it is less, i.e. 2×10^{-8} moles only) and the Donnan effect.

Permeability is also concerned in some cases with the nature of the substances, structure and their configuration. Iron passes through the membrane of intestinal cells only if it is in ferrous form Fe^{++} and not Fe^{+++}, while in rats it is shown that Fe^{+++} can also be absorbed. Glucose is transported more easily than galactose; choleic acid with greater ease than fatty acids. Even the configuration of a substance in certain regions of the molecule is a factor in absorption. D glucose, D galactose and L arabinose have similar configurations around the 3-carbon atoms shown.

D-arabinose which has a different configuration is not transported easily as L arabinose.

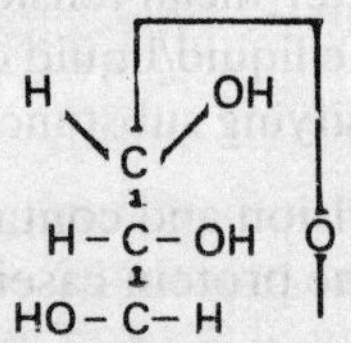

$$\begin{array}{c} HOH \\ \diagdown\diagup \\ C \\ | \\ H-C-OH \\ | \\ HO-C-H \\ | \end{array} \quad O$$

In some cases, when the substance crosses the membrane, chemical reaction can take place and this might hasten absorption. Sucrose when it passes through the membrane is hydrolysed to glucose and fructose by the enzyme effect. Cholesterol which does not pass through the membrane easily passes through when it is in the form of a complex in bile. The permeability of membrane is considerably influenced by cyclic AMP formed by membrane-bound adenylate cyclase.

The membrane phenomenon is of utmost importance in nerve induction. Acetyl choline causes an efflux of potassium and infux of sodium leading to a reversal of polarisation. A wave of negative electricity passes through the membrane until choline esterase hydrolyses acetyl choline and this restores the original condition in respect of potassium and sodium ions and the negative charge on the internal side of the membrane is restored.

Transport phenomena across the membranes

Transport of substances across the membrane is necessary both in absorption and metabolism. Firstly, at the level of the intestines, products of digestion like glucose have to pass through the membranes of the intestinal mucosal cells. Gases like oxygen, nutrients like glucose and water should pass into the cells through bio-membranes to cater to metabolic needs. Some metabolites like ATP, synthesised, have to come out through the mitochondrial membrane. Movement of sodium and potassium ions has to be under control for normal bio-electricity. Passage of Ca^{2+} ions through the membrane and release from intra-cellular stores are necessary for physiological functions like muscle contraction and hormone function.

There are different methods of transport. The simplest is physical diffusion of solutes and gases along their concentration gradient. The diffusing molecule is neither chemically modified nor associated with other molecular species in its passage through the membrane. There is no saturation or limit to diffusion. There is also no requirement of energy. Electrolytes diffuse slowly as the biomembrane has the hydrophobic core due to lipid bilayer. The greater the charge, the slower the diffusion. Movement of D amino acids into the muscosal cells is by passive diffusion. Glucose enters the hepatic cells by simple diffusion once a concentration gradient is set up. Pore-like structures constitute ion-conductive pathways. Some peptides provide ion-channels or ionophores, e.g, nonactin for K^+. A complex is formed and K^+ is transported. Uncouplers like dinitrophenol increase the transport of H^+. The diphtheria toxin also increases membrane permeability. cAMP is said to increase the permeability of membrane for Ca^{2+}

Facilitated transport, also known as carrier-mediated transport, is different from physical diffusion in the following aspects: 1) saturation kinetics, i.e., the transport system can become saturated with the solute transported, and 2) specificity for the substance transported. The solutes are transported more rapidly than expected from the size, charge and partition.

Facilitated transport is made possible by proteins capable of reversibly binding specific substrates; these transport molecules have been variously termed transport systems, carriers, porters and translocases. The compound moves across the membrane only in defined passages. There is no requirement of energy of ATP for facilitated transport. The transport of glucose, galactose, D xylose and L arabinose across the plasma membrane of muscle and adipose cells is carrier-mediated facilitated transport. Insulin increases the membrane transport of glucose by translocating the glucose (hexose) carriers from intra-cellular organelles like the Golgi bodies to the plasma membrane. For this insulin has to first bind to its receptors on the plasma membrane. Glucose enters the red blood cells by facilitated transport. The movement of phosphate, ADP, –OH ions, ATP, dicarboxylic acids, tricarboxylic acids, etc., across the mitochondrial membrane is by facilitated movement with the help of carriers.

The third method of movement across the membrane is by active transport. The solute moves against an electro-chemical or concentration gradient. For this purpose, continuous supply of energy through ATP is needed. The movement is unidirectional. Nearly 30–40 per cent of total energy input is used for active transport. Good examples are given by sodium and calcium pumps. There will always be, in resting state, high intra-cellular K^+, low intra-cellular Na^+ and high extra-cellular Na^+. During the development of action potential, there is a large influx of Na^+. This Na^+ has to be pumped out of the cell. The sodium pump does the job. Na^+, K^+ ATPase is an integral membrane protein. It interacts both with ATP and Na^+ on the cytosolic side and K^+ on the extra-cellular sites. By hydrolysing ATP, energy is made available for pumping out sodium. Ouabain inhibits Na^+, K^+ ATPase. The para-thyroid hormone increases intracellular calcium. Subsequently, calcitonin brings about a calcium pump mechanism and decreases intra-cellular calcium. Again, when muscle contraction has to be followed by relaxation, calcium ions are pumped into the sarcoplasmic reticulum. These processes derive energy through ATP hydrolysis with the help of the Ca^{2+} ATPase. The transport of L amino acids into the mucosal cells is an energy-dependent process — so also, the gamma glutamyl cycle operating in the transport of some amino acids needs ATP at certain steps.

Transport of glucose across the intestinal mucosal cell membrane, though not directly requiring energy, is indirectly dependent on the availability of energy. Sodium and glucose move together (co-transport) with the same carrier. For transport of Na^+, ATP is needed for operating the pump. While Na^+ and glucose have co-transport in the intestinal absorption, K^+ and glucose have the co-transport in the muscle cells. Active transport fails in conditions when oxygen supply is inadequate or under the influence of uncouplers of oxidative phosphorylation, for want of ATP.

Endocytosis is another method of membrane transport. The extracellular material is internalised as endocytic vesicles. Segments of plasma membrane invaginate, pinch off and fuse with the endocytic vesicle. The vesicle can fuse with other membrane structures, say, lysosomes. Phagocytosis occurs in specialised phagocytic cells like the macrophages and granulocytes in blood. Pinocytosis is a similar phenomenon occurring in all cells, e.g., pinocytosis of the LDL molecule and its fusion with lysosome. If lysosome is not involved, it is referred to as receptosome.

There is also pinocytosis and phagocytosis in the case of thyroglobulin, with thryoxine and tri-iodothyronine bound to it from the follicular space towards the intracellular site.

Donnan membrane equilibrium

The colloidal state is responsible for the so-called selective membrane permeability of certain ions. This role of colloids is explained by the Donnan membrane equilibrium.

If a membrane is freely permeable and if, on either side of the membrane, there are electrolytes (e.g. NaCl and KNO_3), all the four ions Na^+, Cl^-, K^+ and NO_3^- will be equally distributed on each side of the membrane at equilibrium. On the other hand, if the membrane is not permeable to all the ions, then there will be an unequal distribution of ions on either side of the membrane. According to Donnan, the non-diffusible ions on one side of the membrane influence the diffusion of the diffusible ions. Both the quality and quantity of diffusible ions will be influenced.

Sodium salt of protein (sodium proteinate) is a colloidal electrolyte as it ionises to give the sodium ion and proteinate (Pr^-) ion. The proteinate ion is colloidal and is not diffusible through a membrane. Its nature and concentration will influence the diffusion of electrolytes across the membrane and cause accumulation of certain ions on one side.

Let it be assumed that sodium proteinate is separated from NaCl by a membrane.

$$
\begin{array}{c|c}
Na^+ & Na^+ \\
Pr^- & Cl^- \\
\end{array}
$$

membrane
(A)   (B)

Diffusion occurs across the membrane of the easily diffusible ions Na^+ and Cl^-. After equilibrium is attained, the following will be the state of affairs.

$$
\begin{array}{c|c}
Na^+ & Na^+ \\
Pr^- & Cl^- \\
Cl^- & \\
\end{array}
$$

(A)   (B)

On side A, the Na^+ ion has to balance in addition to the existing Pr^- ion the newly entered Cl^- ion, while on side B it has to balance only the Cl^- ion. So the concentration of Na^+ on side A will be greater than that on side B.

$$Na^+ (A) > Na^+ (B)$$

The total ionic concentration on side A will be greater than on side B.

From thermo-dynamical considerations, after equilibrium is established, the concentrations (more correctly 'activity') of NaCl on both sides should be the same.

$$Na^+ Cl^- (A) = Na^+ Cl^- (B)$$
$$Na^+ (A) > Na^+ (B)$$

It follows therefore that the concentration of chloride ions on side A should be less than on side B.

$$Cl^-(A) < Cl^-(B)$$

or chloride concentration in B should be greater than that in A.

Thus, the Donnan equilibrium has brought the following effects on the distribution of Na^+ and Cl^-. 1) On the side in which the non-diffusible ion is present, there is accumulation of oppositely charged diffusible ion. (In the present case Pr^- ion has resulted in the accumulation of Na^+ on the same side). 2) On the other side of the membrane, the non-diffusible ion causes accumulation of the diffusible ion of the same charge. It is as if the diffusible ion of the same charge is excreted from its side by the non-diffusible ion (Pr^- ion causes the accumulation of Cl^- ion on the opposite side). 3) The total concentration of all the ions will be greater on the side in which the non-diffusible ion is present. This will lead to an osmotic imbalance between the two sides.

The amount of sodium chloride diffusing across the membrane can be related mathematically, as follows, with the initial concentration of sodium chloride.

If the initial concentration of sodium proteinate ($Na^+ Pr^-$) is 'a' moles on side I and that of sodium chloride on the other side (side II) is 'b' moles, and both of them are separated by a membrane,

$$\begin{array}{c|c} Na^+(a) & Na^+(b) \\ Pr^-(a) & Cl^-(b) \\ I & II \end{array}$$
$$x$$

'a' moles of Na^+Pr^- would mean 'a' moles of each ion while 'b' moles of sodium chloride would mean 'b' moles of each ion. If it is assumed that 'x' moles of sodium chloride diffuse through the membrane from side II to side I, then after equilibrium, the concentration of sodium ions and chloride ions on either side would be as follows:

$$\begin{array}{c|c} Na^+(a+x) & Na^+(b-x) \\ Pr^- & \\ Cl^-(x) & Cl^-(b-x) \\ I & II \end{array}$$

(After equilibrium also, there will be diffusion of sodium chloride both ways but the rates of diffusion from II to I and I to II will be equal. Such a dynamic equilibrium will not affect the distribution of ions after equilibrium is established.)

The amount of sodium chloride diffusing from either side is proportional to the products of their concentrations (more correctly their activities). So at equilibrium, the amount diffusing from II to I is proportional to $(b-x)(b-x)$; and the amount diffusing from I to II is proportional to $(a+x)(x)$.

As the rates of diffusion at equilibrium are equal,

$$(a + x)(x) = (b - x)(b - x)$$
$$ax + x^2 = b^2 - 2bx + x^2$$
$$ax + 2bx = b^2$$
$$x(a + 2b) = b^2$$
$$x = \frac{b^2}{(a + 2b)}$$

The above equation shows that the amount of sodium chloride diffusing i.e., x is dependent not only on the concentration of sodium chloride (b) but also on the concentration of non-diffusible ion, i.e., 'a'. Thus the quality and quantity of non-diffusible ion influences the extent of diffusion of easily diffusible ion across a membrane. For example, in the red blood cells the pH is relatively less than in plasma. This could be because the negatively charged protein hemoglobin of the RBC abstracts the oppositely charged H^+ ion by the Donnan effect tending to cause a decrease of pH inside the RBC.

By Donnan effect, there will be unequal distribution of ions on either side of the membrane and this will create a difference of osmotic pressure. The total number of ions on one side is greater than that on the other side. This can be shown qualitatively that on the side containing the non-diffusible ion (Pr^-) in the above example, in addition to the existing Na^+Pr^-, sodium chloride (Na^+ & Cl^-) also enters. Hence, the ionic concentration on such a side increases. See the equation below:

$$(a + x)x = (b - x)(b - x)$$
$$\text{I} \qquad\qquad \text{II}$$

On side II it is a square, say, 8×8. To get the product 64, for side I, if (a+x) could be greater than 8, x will be less than 8. But the total of (a+x) and x will always be mathematically greater than 16 of the right side (i.e. (b − x) + (b − x) = 8 + 8). (For e.g. 32 × 2 = 64; 32 + 2 = 34); 34 is greater than 16.(ii) 16 × 4 = 64; but 16 + 4 = 20; 20 is greater than 16. Thus mathematically also the ionic concentration could be shown to be greater on the side having the non-diffusible ion. As osmotic pressure depends on the number of ions, the side having the non-diffusible ion will exert great osmotic pressure than the other side. For example, if a = 1 mole and b = 2 moles, x, the amount diffused, will be

$$x = \frac{b^2}{a + 2b} = \frac{4}{1 + 4} = 0.8$$

Side I will have 1.8 (a+x) of Na^+; 1 of Pr^- (a) and 0.8 (x) of Cl^- with a total of 3.6 moles, while side II will have 1.2 (b−x) of Na^+ and 1.2 (b−x) of Cl^-, making a total of 2.4 moles.

Thus, side I has (3.6–2.4) = 1.2 moles more leading to greater osmotic pressure on that side. Imbibition of gels is said to be due to this type of osmotic imbalance. Inside the gel are large non-diffusible protein ions. At equilibrium, the total ionic concentration will be greater within the gel than in the solution in which it is placed. The osmotic flow of water will therefore occur and thus the gel imbibes.

Membrane hydrolysis

The Donnan effect causes membrane hydrolysis and makes one side relatively acidic, or alkaline depending on the charge of the non-diffusible ions. Since in this system one side is concerned with H^+ and OH^- of water, this shift of ions is called membrane hydrolysis.

The interaction between sodium proteinate and water (H^+ and OH^-) through a membrane results in the accumulation of H^+ (oppositely charged to Pr^-) on the side containing sodium proteinate. Hence the pH of that side drops and it becomes acidic.

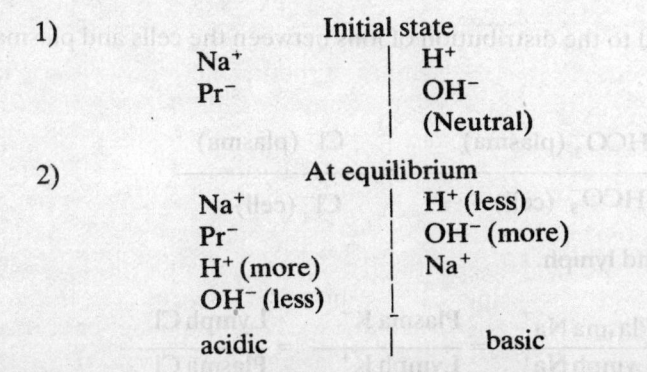

1)   **Initial state**

Na^+	H^+
Pr^-	OH^-
	(Neutral)

2)   **At equilibrium**

Na^+	H^+ (less)
Pr^-	OH^- (more)
H^+ (more)	Na^+
OH^- (less)	
acidic	basic

On the other hand, if protein salts as Pr^+Cl^- are separated from water by a membrane, the protein ion with positive charge will cause the accumulation of OH^- of water making the pH of the side containing water acidic.

1)   **Initial**

Pr^+	H^+
Cl^-	OH^-
	(Neutral)

2)   **At equilibrium**

Pr^+	H^+ (more)
Cl^-	
H^+ (less)	OH^- (less)
OH^- (more)	Cl^-
basic	acidic

Thus it could be stated that acid is excreted by Pr^+Cl^- across the membrane. Donnan points out that it is possible to attain in this way a concentration of H^+ ions as great as present in the gastric juice.

If a number of ions are separated by a membrane, a relationship could be arrived at for their relative concentrations. It is known that,

$$Na^+(1) \times Cl^-(1) = Na^+(2) \times Cl^-(2)$$

$$\text{or } \frac{Na^+(1)}{Na^+(2)} = \frac{Cl^-(2)}{Cl^-(1)}$$

If a system contains a number of diffusible ions like Na^+, K^+, Ca^{++}, Cl^-, SO_4^{2-} distributing on either side of a membrane, then,

$$\frac{Na^+ (1)}{Na^+ (2)} = \frac{K^+ (1)}{K^+ (2)} = \frac{Ca^{++} (1)}{Ca^{++} (2)} = \frac{Cl^- (2)}{Cl^- (1)} = \frac{SO_4^{2-} (2)}{SO_4^{2-} (1)}$$

This can be applied to the distribution of ions between the cells and plasma,

$$\frac{HCO_3^- \text{ (plasma)}}{HCO_3^- \text{ (cell)}} = \frac{Cl^- \text{ (plasma)}}{Cl^- \text{ (cell)}}$$

or between plasma and lymph.

$$\frac{\text{Plasma } Na^+}{\text{Lymph } Na^+} = \frac{\text{Plasma } K^+}{\text{Lymph } K^+} = \frac{\text{Lymph } Cl^-}{\text{Plasma } Cl^-}$$

As plasma contains more of negatively charged proteins, it has more of the positively charged ions Na^+, K^+ etc., and less of the negatively charged ion, i.e., Cl^-. However, the concentration of HCO_3^- is greater in plasma than in lymph and this is an exception. This is required for the transport of CO_2 produced by the cells during metabolism.

Physiological importance of colloids

1) Enzyme action due to the large surface area of colloids and adsorption: The small size of the colloidal particles gives them a great specific surface. The enzymes which are colloidal have thus a great specific surface. Because of great specific surface there is efficient adsorption which leads to efficient enzyme catalysis.

Adsorption is also important in drug action, detoxification, orientation of protoplasmic constituents, etc. Thus the colloidal state leads to good adsorption in the above cases.

2) Maintenance of blood pH due to electro-kinetic potential: Because of the electro-kinetic potential, proteins, phospholipids etc, serve as amphoteric colloids reacting either with acids or alkalies. Because of the amphoteric nature, they help in maintaining the pH of blood.

3) Fat absorption and protection of bile contents: Colloidal proteins and phospholipids hold the fatty substances of living cells in an invisible colloidal form and protect them. Again, colloidal lyophilic proteins keep the not easily soluble substances like cholesterol, calcium bilirubinate etc. in the dissolved state in bile and protect them from precipitation by electrolytes.

4) Water transport and formation of urine due to colloidal osmotic pressure: The colloidal systems exert osmotic pressure which is very important in water transport in the body and in the formation of urine.

5) Glandular secretions by imbibition and syneresis: Tissue fluid is imbibed by the colloids in the glandular cells. The imbibed fluid is squeezed out as glandular secretions by syneresis.

6) *Donnan membrane equilibrium:* The colloidal state is responsible for the so-called slective absorptions or selective membrane permeabilities of certain ions. This application of colloids is explained by Donnan membrane equilibrium, and membrane hydrolysis.

OSMOSIS

Osmosis is an important property of solutions. It is the spontaneous flow of a solvent into a solution when the two are separated by a semi-permeable membrane. It is also the flow of a solvent from a more dilute solution to a relatively concentrated solution through a semi-permeable membrane. Osmosis refers strictly to the flow of the solvent. It is a colligative property depending on the number of the particles of the solute.

Osmotic pressure

Osmotic pressure is defined as the equivalent of excess pressure which much be applied to the solution to prevent the passage of the solvent into it through a semi-permeable membrane separating the two. Osmotic pressure can also be defined as the excess pressure which must be applied to the solution in order to increase the vapour pressure until it becomes equal to that of the solvent.

Strictly speaking, even a semi-permeable membrane is not required for the osmotic flow of a solvent. If a solvent and a solution are kept side by side in a closed chamber there will be osmosis of the solvent to the solution though there is no semi-permeable membrane between the two. The osmosis of the solvent is due to the fact that the chemical potential or activity or partial free energy of the solvent is less on the solution side than on the solvent side. Hence the movement of the solvent to the solution is accompanied by a decrease of free energy and is, therefore, spontaneous.

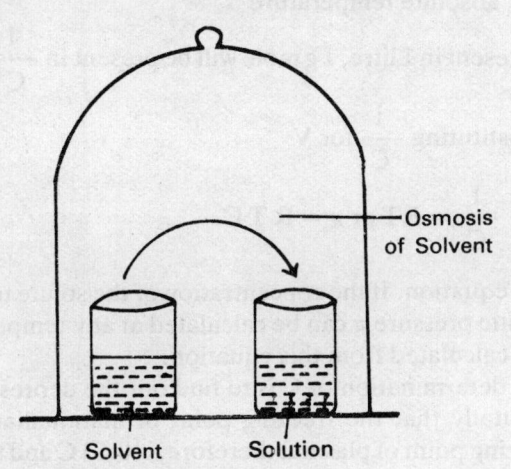

In the body there is no membrane which is typically semi-permeable, i.e., permeable to solvent and impermeable to the solute. Hence the term 'osmotic potency' is preferred to 'osmotic pressure' by some authors.

The osmotic pressure of a solution is directly proportional to the concentration of the solute and in turn depends on the number of molecules or ions contained in the solution. Though dependent on the number of molecules of the solute, osmotic pressure is independent of the nature of the solute. Hence it is a colligative property. A substance of lower molecular weight will have more molecules per unit volume than a substance of higher molecular weight and will therefore exert greater osmotic pressure. Albumin which has a lower molecular weight than globulin exerts greater osmotic pressure than globulin on a weight to weight basis.

If two solutions have the same osmotic pressure these are described as iso-osmotic solutions. Iso-osmotic solutions will have the same vapour pressure. When the osmotic pressure of one is greater than that of the other, it is called hypertonic. If it is less than that of the other, it is called hypotonic.

Vant Hoff's theory of dilute solutions

Vant Hoff observed a striking parallelism between the properties of gases and those of dilute solutions. He put forward the theory of solutions according to which a substance in a solution behaves exactly like a gas and the osmotic pressure of a dilute solution is equal to the pressure which the solute would exert if it were a gas at the same temperature and occupied the same volume as the solution. Laws analogous to gas laws are obeyed by substances in solution.

The gas law is, $PV = RT$

Vant Hoff's equation analogous to the above is

$\pi V = RT$

where π is the osmotic pressure in atmospheres
  V = volume in litres containing 1 mole
  R = solution constant
     (.082 litre atmosphere/degree/mole)
  T = absolute temperature

If C g moles are present in 1 litre, 1 g mole will be present in $\frac{1}{C}$ litres. Then $V = \frac{1}{C}$

Substituting $\frac{1}{C}$ for V

$\pi \times \frac{1}{C} = RT$ or $\pi = RTC$.

This is a very useful equation. If the concentration of the solute in moles/litre (C) is known, assuming R, the osmotic pressure π can be calculated at any temperature. The osmotic pressure of plasma can be calculated from this equation.

An easy method of determination of C is to find out the depression in the freezing point. It is found experimentally that the freezing point of mammalian plasma is –0.53°C. The **depression in the freezing point of plasma, therefore, is 0.53°C and this is brought about by the substances present in plasma.** The depression in freezing point in related to molar concentration by the following equation:

$$\text{Molar concentration (C)} = \frac{\text{depression in freezing point (d)}}{\text{molar depression constant for water (K)}}$$

  $d = 0.53$
  $K = 1.8$ for water
So $C = \frac{.53}{1.8} = 0.3$ (approximated)

This value of C is introduced in the osmotic equation of Vant Hoff.

$$\pi = RTC$$

$$(R = 0.82; T \text{ for } 37°C = 310, C = \frac{.53}{1.8}$$

$$\pi = 0.82 \times 310 \times \frac{.53}{1.8} = 7.6 \text{ atmospheres}$$

Thus the osmotic pressure of plasma is determined by the indirect method, i.e., from depression in freezing point studies. In terms of moles, osmotic pressure is .3 mole or 300 milli-moles (or 300 milli-osmols) and in terms of atmospheres it is 7.6 atmospheres. It is also equal to 7.6×76 cm of mercury or 7.6×760 mm of mercury.

Abnormal osmotic pressures:

The osmotic pressure of electrolytes gives results considerably higher than those calculated from their formulae. This is due to the ionisation of the electrolytes in solution. The number of particles will increase due to the ionisations of electrolytes in the aqueous medium. Therefore, the osmotic pressure of electrolytes determined experimentally is greater than the theoretically calculated value not assuming ionisation. This greater osmotic pressure is referred to as the abnormal osmotic pressure of electrolytes.

Vant Hoff's factor 'i'

The ratio between the observed osmotic pressure and theoretically calculated osmotic pressure gives the Vant Hoff's factor 'i':

$$\text{'i'} = \frac{\text{observed osmotic pressure}}{\text{theoretical osmotic pressure}}$$
$$\text{(not taking into account ionisation)}$$

As ionisation means increase in the number of particles, 'i' will be greater than 1 in the case of electrolytes. For example, for sodium chloride, if there is complete ionisation 'i' will be equal to 2 and for potassium sulphate it will be 3. If ionisation is not taken into account, and if the theoretical calculation for osmotic pressure considers 1 mole, in practice, 2 ions, i.e., two osmotically active particles are got by ionisation in the case of sodium chloride. So the denominator is 1 while the numerator becomes 2. So, 'i'=2 for NaCl. Thus molality becomes equal to 2 × molality or 2 osmolality for NaCl. For potassium sulphate, it is molality × 3. For non-ionisable substances like glucose and urea, 'i' will be equal to 1 and molality and osmolality will be the same. As the particles concerned with producing osmotic pressure are of real interest to the biochemists, the term osmolality is frequently used by them. Thus plasma has solutes whose concentration is 300 milli-osmoles/litre.

Role of osmosis in physiological processes

Osmosis has an important role in 1) the regulation of blood volume, 2) the excretion of urine, 3) absorption, etc. Osmosis also explains the hemolysis of red cells which is applied in the fragility test.

Regulation of blood volume: The concentration of electrolytes and of organic solutes in plasma and tissue fluids is substantially the same. So, the osmotic pressure due to these substances are practically identical (see figure).

The blood proteins are responsible for about 25 mm of total osmotic pressure of plasma (about 7.6×760 mm of Hg). The osmotic pressure of proteins is referred to as oncotic presure. This oncotic pressure is opposed by the oncotic pressure of tissue fluid protein which is about 10 mm. Hence, the effective osmotic pressure of blood over that of tissue fluid is $(25-10) = 15$mm. The net hydrostatic pressure on the arterial side is $(30-8)=22$mm. Hence there is excess hydrostatic pressure over osmotic pressure by $(22-15)=7$mm at the arterial end and this favours the forcing of fluid outward from the capillary to the tissue spaces. On the other hand, the net hydrostatic pressure at the venous end is $(15-8)=7$mm, while the effective osmotic pressure is 15mm. Hence the excess osmotic pressure over the hydrostatic pressure by $(15-7) = 8$ mm is in favour of reabsorption of water as intra-vascular osmotic pressure predominates.

STARLING'S HYPOTHESIS

ARTERIAL SUPPLY → Blood pressure (hydrostatic pressure): 30mm / 15 mmHg
← Tissue hydrostatic pressure: −8mm / −8mm
← Blood osmotic pressure: 25mm / 25mm
→ Tissue osmotic pressure: −10mm / −10mm

Net filtration pressure: 7mmHg
Net absorption pressure: 8mmHg

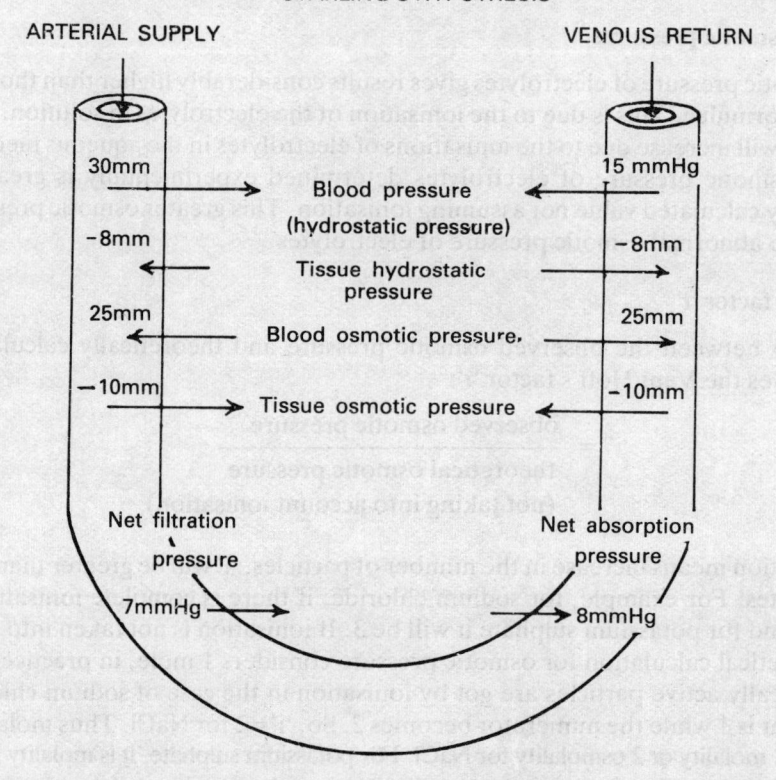

The difference in pressures explains the mechanism of exchange of fluids and dissolved materials between the blood and tissue space as well as regulation of blood volume. This is known as Starling's hypothesis.

Osmosis and excretion of urine: The cells in the body are isotonic (in osmotic equilibrium) with the tissue fluids and blood plasma. An appreciable osmotic dilution of plasma would create a dangerous hydrostatic pressure in the red cells and tissue cells which would take in water to achieve osmotic equilibrium. This does not occur in the body because water or salt (chiefly NaCl) are excreted by the kidneys so as to keep the blood isotonic with the cells.

Urine is formed by a process of filtration, the energy for which is derived from the hydrostatic pressure of blood. A gross filtration force of about 75 mm of Hg is the capillary pressure. This is opposed by the oncotic pressure of plasma proteins (about 30mm), renal interstitial pressure of 10 mm Hg and renal intra-tubular pressure of 10mm Hg. The total force opposing filtration is 50mm (30+10+10). The net filtration pressure of 25 mm and the amount of blood flowing through the kidneys decides the quantity of the glomerular filtrate. For the proper filtration of urine, it is thus clear that both the hydrostatic pressure of blood and the oncotic pressure of proteins have a part to play, and changes in their pressures will affect the extent of glomerular filtration. Osmosis is of great importance in the process of urine excretion.

3) Absorption   Intestinal | Blood plasma
                 contents | (Capillaries of
                 (Lumen) | the villi)
                         Mucosa

In absorption, the processes of osmosis and diffusion or active transport work hand in hand. Hence, whatever may be the osmotic pressure of the intestinal contents, hypotonic or hypertonic with blood plasma, water of course moves osmotically depending on the osmotic pressure but salts or sugars move by diffusion or active transport. (The mucosal membrane is permeable to salts or sugar in either direction.) This may be illustrated in the case of the absorption of, say, isotonic glucose which is easily absorbed into blood. First, the sugar moves into the blood and makes the isotonic solution hypotonic. Now, there is osmotic flow of water into blood and this makes the sugar solution isotonic again. Thus, by diffusion or active transport and osmosis, the absorption of sugar and salt solutions takes place.

4) Destruction of red cells by hemolysis occurs both *in vivo* and *in vitro*.

Since the red cell membrane is permeable to water, the volume of the cell changes according to its osmotic environment. When placed in a hypotonic solution, red cells swell owing to water passing in osmotically (endosmosis). If the solution is sufficiently hypotonic, the cells may even rupture, with the cell contents diffusing into the surrounding fluid. This is called hemolysis.

5) If the RBCs are placed in a hypertonic solution, water passes out of the cells (exosmosis) and the red cells shrink due to diminution in volume. This process is called crenation.

Clinical applications

1) The practice of giving five per cent glucose or 0.9 per cent saline for intravenous infusions has a tremendous bearing on osmotic pressure considerations and Vant Hoff's factor 'i'.

The osmotic pressure of plasma is about 280-300 milli-osmols/litre. Any intravenous infusion should have iso-osmolality. A 5 per cent glucose (50 g/litre) solution will give rise to 50/180 mole or $(50/180) \times 1000$ = about 280 milli-osmols/litre. In case of 0.9 per cent saline (9 g/litre), this amount equals 9/58.5 mole or $(9/58.5) \times 1000$ = about 150 milli moles/litre. But as NaCl (of saline) is an electrolyte unlike glucose, it undergoes ionisation. Each molecule of NaCl will give rise to two osmotically active ions (Vant Hoff's factor being 2). Hence 150 milli moles/litre of NaCl will be equal to $150 \times 2 = 300$ milli-osmols/litre. Thus five per cent glucose and 0.9 per cent saline are iso-osmotic with plasma and are hence used for intravenous infusions.

2) *Fragility test:* The extent of the hemolysis of RBCs is observed in a series of tubes with hypotonic NaCl solutions adding a drop of blood in each and observing after about two hours. If there is no hemolysis there will be a colourless supernatant layer over an opaque red suspension of cells. If hemolysis is complete, a transparent red solution is seen. In normal human blood, complete hemolysis occurs only below 0.35 per cent NaCl; the cells resist hemolysis above 0.45 per cent. In between, there is partial hemolysis. Diminished resistance (increased fragility) is seen in hemolytic jaundice (upto 0.7 to 0.8 per cent) and increased resistance in certain anemias.

3) *Cause of edema in cases of albuminuria:* In albuminuria, plasma proteins, mainly albumin, are excreted. When the concentration of plasma proteins is thus reduced, the oncotic pressure of plasma is lowered. This would naturally mean increase of hydrostatic pressure causing edema which is accumulation of excess fluid in tissue space.

4) Another clinical application of osmotic pressure is the injection of hypertonic solutions of salts such as sodium chloride to reduce cerebral edema. Water is withdrawn from the brain osmotically.

ADSORPTION

Adsorption is an important property of surfaces. It describes the existence of a higher concentration of any particular substance at the surface of a liquid or solid than is present in bulk.

Adsorption of gases by solids

Adsorption of gases by solids refers to an excess concentration of gases at the surface and is different from 'absorption' which means uniform penetration of solids by gas. Best known adsorbents are charcoal (by burning coconut shells in a limited supply of air), silica gel, alumina, metals like platinum, palladium, etc. Increase of pressure and decrease of temperature, increase the extent of the adsorption of a gas by a solid.

Adsorption isotherm

The term isotherm or isothermal is used to describe a curve which gives the variation of volume with the pressure of a gas at constant temperature.

The variation of gas adsorption with pressure at constant temperature can be represented by the equation $a = Kp^n$ where 'a' is the amount of gas adsorbed by the unit mass, e.g., 1 g of adsorbing material at the pressure p; k and n are constants for the given gas and adsorbent at the particular temperature. An equation of this type is known as an adsorption isotherm since it is applicable at constant temperature.

Gibbs adsorption equation

Gibbs adsorption equation is a significant equation in connection with adsorption at the surface of a solution. For a dilute solution of concentration 'C' and surface tension 'γ', the equation is,

$$S = -\frac{C}{RT} \frac{d\gamma}{dC}$$

where S is the excess concentration of solute per sq. cm. of surface. as compared with that in the bulk of the solution; $d\gamma/dC$ is the rate of increase of surface tension of the solution with the concentration of the solute, R is the gas constant and T the absolute temperature.

According to the equation, if the solute causes the surface tension of the solvent to decrease, $d\gamma/dC$ will become negative and S positive indicating that the solute will have a higher concentration in the surface than in the bulk of the solution. Thus, according to the Gibbs adsorption equation, a substance which decreases the surface tension (or interfacial tension) will be adsorbed at that interface. On the other hand, if any substance increases the surface tension of the solvent, $d\gamma/dc$ will be positive and S will be negative, that is, the concentration of the solute will be lower in the surface than in the bulk of the solution. In other words, surface tension increasing substances like some electrolytes are not adsorbed at the interface.

SURFACE TENSION

A molecule in the interior of a liquid is completely surrounded by other molecules and so, on an average, it is attracted equally on all directions and hence can move with equal freedom in any direction. On the molecule on the surface, however, there is a resultant attraction inwards because the number of molecules per unit volume is greater in the bulk of the liquid than in the vapour (which would always be present above the liquid column). Hence, the freedom of movement of the molecule at the surface is restricted. As a consequence, the surface of a liquid pulls itself together tending to occupy the least possible area. Because of this tendency to contract, a surface behaves as if it were in a state of tension. The surface film of a liquid has thus the properties of an elastic skin and is resistant to rupture as can be seen by the experiment of floating a needle on water.

The force with which the surface molecules are held is called the 'surface tension' of the liquid. It is defined as the force in dynes acting at right angles to any imaginary line of 1 cm length on the surface.

Other manifestations of surface tension are the formation of drops of liquids falling through air, the rise of liquid in a capillary tube and formation of a meniscus at the surface of liquids.

The surface tension occurring at the surface of separation of two immiscible phase (e.g., liquid-liquid, or liquid-solid) is called interfacial tension.

Role of surface tension in physiological processes

Gibbs-Thomson principle: Substances which lower the surface tension become concentrated in the surface layer whereas substances which increase surface tension are distributed in the interior of the liquid. Soaps, oils, proteins and bile salts reduce the surface tension of water while sodium chloride and most inorganic salts increase surface tension. Substances which reduce surface tension are used for emulsification.   1) Bile salts which reduce surface tension bring about a stable emulsion of fats and help in fat absorption.   2) Lipids and proteins which are both surface tension-lowering substances are found concentrated in the cell wall. This facilitates adsorption of these substances (taking up of substances from solutions by surfaces).

Surface tension leads to efficient adsorption. This is applied in a) enzymatic reactions, b) formation of complex compounds of proteins and lipids, of proteins and salts, etc. in protoplasm, and c) in the action of drugs and poisons.

A practical application of lowering of surface tension is Hay's test for bile salts in urine. The surface skin of normal urine is sufficiently dense to prevent fine particles of sulphur sprinkled on the surface from penetrating the skin and sinking to the bottom. The presence of bile salts as in the urine of certain types of jaundice lowers the surface tension so much that the surface-skin can no longer support the sulphur particles, which sink to the bottom of the tube. Thus bile salts are detected in urine.

VISCOSITY

The viscosity of a liquid is its resistance to flow. It is the manifestation of the frictional effect due to the passage of one layer of liquid over another.

Coefficient of viscosity: The coefficient of viscosity of a fluid is defined as the force required per unit area to maintain unit difference of velocity between two parallel planes in the fluid 1 cm apart. The smaller the coefficient of viscosity, the more rapidly the liquid flows. Oils, and liquids like glycerine have a high coefficient of viscosity while ether and the so-called mobile liquids possess a low coefficient of viscosity.

The unit of viscosity is a poise.

Determination of coefficient of viscosity

When a liquid flows through a capillary tube of length 1 and radius 'r' for a time 't', under a constant pressure head 'p', and if the volume of the liquid flowing out from the tube is 'v', then the coefficient of viscosity η is given by the equation.

$$\eta = \frac{p\pi r^4}{8 lv} \times t$$

Instead of finding the absolute coefficient of viscosity, it is easy to determine the relative coefficients of viscosity of liquids with respect to water. Ostwald's viscometer is used for this purpose.

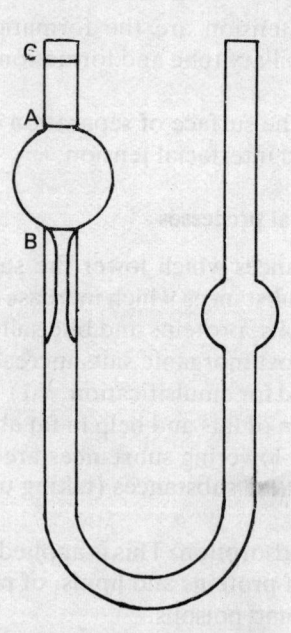

First, water is taken in the viscometer and is sucked up to level C. Time taken for the flow of water from mark A to B is noted using a stopwatch. The experiment is repeated with the liquid whose coefficient of viscosity is to be known. If the densities of water and the liquid are d_1 and d_2, their coefficients of viscosity η_1, and η_2, the times taken being t_1 and t_2, then

$$\frac{\eta_1}{\eta_2} = \frac{d_1 t_1}{d_2 t_2}$$

If η_1 is assumed, η_2 can be known.

Application of viscosity

a) To determine the approximate molecular weight of certain liquids and to know whether the liquids are normal or associated. According to Dunstan,

$$\frac{d}{M} \times \eta \times 10^6 = 40 \text{ to } 60$$

where d is the density of the liquid, η the coefficient of viscosity and M the molecular weight. If 'd' and η_1 are known, the approximate value of M can be calculated.

This relation holds good for non-associated liquids only. For associated liquids the value is considerably greater than 60. Benzene has the value of 73 while water 559.

b) Viscosity and chemical constitution: In the homologous series of organic compounds, increase of viscosity per CH_2 group is approximately constant.

From a knowledge of viscosity, molecular weight and density, the molecular viscosity of a liquid can be determined.

$$\text{Mol. viscosity} = \left(\frac{M}{d}\right)^{2/3} \times \eta$$

Thrope and Rodger found that molecular viscosity is an additive property at the boiling point. For H it is 80, O in OH 196. Like Parachor, molecular viscosity can be employed to solve structural problems.

Viscosity of protoplasm

This is determined by the gravity or centrifugation methods. In this, granules or inclusions are moved throught the protoplasm by gravity or centrifugal force. By applying Stokes's law ($F = 6\pi \eta a v$ where F is the force pushing a sphere through the liquid, 'a' the radius of the sphere and v its velocity, η can be determined.

In another method, the speed of Brownian movement is used as a measure of viscosity.

Heilbrunn has shown that the viscosity of protoplasm depends on the nature and concentration of granular suspension. The value is roughly 3–5 centi-poises. It is affected during cell division, or when a muscle or nerve is thrown into activity or when a cell is subjected to the influence of an anesthetic like ether. Usually, the viscosity of protoplasm at the cortical region is higher than that at the interior. The high protoplasmic viscosity in the cortical region seems to be due to the presence of increased amounts of calcium at the cortex and the influence of calcium in clotting. This is the case with amoeba. When the organism is exposed to stimulus as mechanical agitation, electric shock or ultra-violet irradiation, calcium ions move to the interior. This results in the sharp increase of viscosity of the protoplasm of the interior.

Change of protoplasmic viscosity in the cells under stimulus is thus primarily due to the movement of the calcium ions.

Anesthetics prevent the clotting reaction of protoplasm. They cause 'liquefaction' of cortex, tend to prevent the gelation of the interior and bring about alterations in viscosity.

Viscosity of blood

This is determined by the Hess viscometer. This consists of essentially two capillaries of equal bore and equal length, connected by T tube with suction bulb. Simultaneously, blood is sucked through one capillary and water through the other. The relative viscosity of blood as compared with that of water is determined from the volumes of water and blood that have flowed through the capillaries, the viscosities being inversely proportional to the volume of flow for a given time.

Blood is nearly 4.5 times as viscous as water. The viscosity of blood is lowered in anemia, nephritis, leukemia, malaria, diabetes mellitus, jaundice and pneumonia. Excessive sweating and traumatic shock lead to increase of blood viscosity. High viscosity of blood is encountered in Waldenstrom's macroglobulinemia, a congenital disease.

SECTION VII
BIOSTATISTICS

27 Fundamentals of Biostatistics

Statistics refers to a body of scientific methods used to acquire systematic information in almost every field of science. It does not rely on the statement that 'facts speak for themselves', as facts tend to be conditioned by the particular process used to obtain them; they may be distorted by the prejudices of the persons who collect, select and publish them. The interpretation of the facts depends upon the orientation of the reader towards the subject matter under consideration. The facts and observations may concern anything, e.g., physical science, engineering, industry, social science, business or government. In statistics, these are put together on a rigorously logical basis involving random sampling and the theory of probability. The making of inferences is therefore reduced to one of testing hypotheses which are formulated as a result of an analysis of the problem situation. Statistics is defined as *the science which applies the theory of probability to the making of estimates and inferences about the quantitative characteristics of a population of objects*. The numerical and quantitative data affected by multiple causes are analysed by statistics which combines mathematics and logic.

Statistics helps to summarise the results of investigations by way of charts, graphs or tables, and gives the mean, range, standard deviation, standard error, frequency distribution, probability, etc. Alternately, it can deal with the design of experiments and the interpretation of results to permit generalised inferences.

When the science of statistics is applied to biological experimentation, the body of methods used in the design of the experiments, analysis, summary and interpretation of the results, it is referred to as biostatistics.

Statistics has 12 basic concepts. They are 1) attribute, 2) variate, 3) estimate, 4) frequency distribution, 5) population, 6) sample, 7) randomness, 8) probability, 9) correlation, 10) error, 11) hypothesis and 12) inference.

Attribute (Qualitative data)

There are two types of numerical data which are fundamental in statistics. One of them refers to the characteristic of 'all-or-none' type which can be counted or enumerated but *not measured*. Such a characteristic is called the 'attribute'.

In defining an attribute, objects are classified first into two groups—those that have the characteristic and those that do not. Then, either one or both of the classes may be further sub-divided.

The attribute 'history' books in a library may be sub-divided into ancient history, medieval history and modern history. In some attributes such as 'nationality', the have-not class may be practically zero. In some attributes such as 'occupation' we may further sub-divide the series of objects by a characteristic such as 'weekly wages'. The individuals of a group may be classified as those who suffer from blindness and those who do not. Here, blindness is the 'attribute'.

Variate (quantitative data)

Variate means a characteristic which can take on a series of magnitudes, each with a specified frequency of occurrence. A variate may be a measurment or a count. Examples of variates are height, weight and volume in physical measurement; intelligence quotient, mental age and achievement scores in psychological measurement; prices, wages and costs in economics; electrical and mechanical characteristics in manufactured products; family size, crime rates and population density in sociology; accidents, fires and death in the field of insurance.

Estimate

Variate further includes more than a series of measurements or counts. It includes any characteristic which is derived from measurements or counts. An *estimate* is derived by calculation from measurements or counts, e.g., percentages, averages, totals, etc. (Full discussion of *estimate* is done later). An estimate is obtained on the basis of a sample from a population.

Frequency distribution

If a reasonably large number of observations are included in a sample, it is likely that some or many values occur more than once. In the case of adult male height as an illustration—a few may be as short as four feet, while a few could be as tall as seven feet. These extreme values have a relatively low frequency of occurence. Thousands of adult males must have heights between 5' and 6' with a bunching in the neighbourhood of 5' 8"; in other words, the frequency of occurence of adult male height varies over a limited range of values, with concentration in some range of height. The farther we go in either direction the fewer the individuals. If, say, 'n' observations of a variate are distributed over 'k' different values $x_1, x_2 \ldots x_n$ such that these occur $f_1, f_2 \ldots f_k$ times respectively, the sum of $f = n$. These data are conveniently represented by what is called a *frequency table*.

Table 1
Frequency distribution of variate x

Value of variate	Frequency
x_1	f_1
x_2	f_2
x_3	f_3
x_n	f_k
Total frequency	n

The data could also be given in an ordinary graph with values of variates measured horizontally and the corresponding frequencies of occurrence measured vertically. Thus, frequency distribution refers both to a table of data or bar diagram, a pie diagram, a histogram, a frequency polygon or a linear graph or line diagram.

For example, the frequency distribution of heights of 1784 males (n) is given below:

Different groups	Frequency	
55.5–56.4 (inches)	105	f_1
56.5–57.4 "	56	f_2
60.5–61.4 "	498	f_3
61.5–62.4 "	1125	f_4
n =	1784	

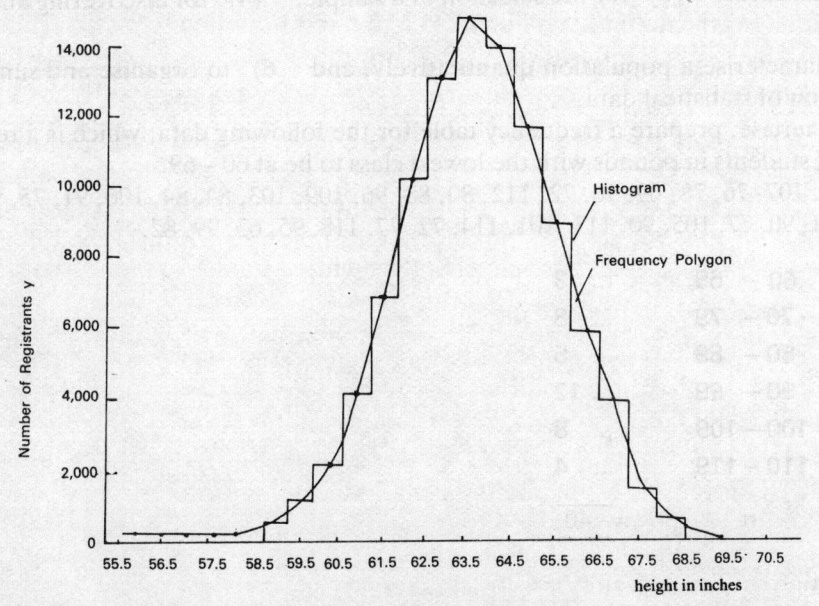

Frequency distribution of heights of males

As a further illustration, if one wants to study the death rates in a population, one has to study separately the death rates in
1) infants,
2) pre-school children,
3) school children,
4) boys, and
5) men.
In this case 'age' can be taken to grade them as follows:
1) less than 1 year
2) 1 – 4 years of age
3) 5 – 9 years
4) 10 – 14 years
5) 15 – 24 years etc.
Frequeney distribution has three important characteristics.

1) It shows the approximate range of values, over which the variate is found. These limit or bound the values with which one has to work.   2) It indicates any tendency of the values of the variate to be concentrated or bunched among certain values.   3) It indicates the most probable and the least probable values of the variate, the former referring to values with a high frequency of occurrence and the latter to those with a relatively low frequency of occurence. There are no hard and fast rules for the construction of a frequency distribution. Any frequency distribution should achieve condensation of the bulk of the data without sacrificing the essential infomation contained in the sample.

The proportion of Relative Frequency Distribution: For many types of problems, the important characteristic is not the number of persons or things falling within each interval, but the proportion of the total number of cases for each number.

Frequency has the following uses: It has the basis   (1) for making estimates,   (2) for making inferences, (3) for the selection of a sample, (4) for discovering and interpreting error

5) to characterise a population quantitatively, and 6) to organise and summarise the presentation of statistical data.

As an exercise, prepare a frequency table for the following data, which is a record of the weights of students in pounds with the lowest class to be at 60 – 69.
61, 73, 93, 107, 76, 78, 69, 96, 72, 112, 80, 88, 96, 109, 103, 84, 84, 106, 91, 75, 91, 92, 102, 91, 90, 101, 90, 77, 105, 90, 113, 101, 114, 72, 77, 118, 95, 63, 99, 82.

60 – 69	... 3
70 – 79	... 8
80 – 89	... 5
90 – 99	... 12
100 – 109	... 8
110 – 119	... 4
n	= 40

Population

Population is the hypothetical infinite data which one would have got from an infinite number of repetitions of an experiment. It refers to the statistical population not limited to people although human beings are one type of statistical population. In short, population consists of all the objects (infinite) about whom conclusions are to be drawn from a study of a few or a sample of them (finite). Statistical populaton is to give boundary to the measurements and enumerations and of the estimates and inferences based upon them. The population object or an element of the population has many characteristics but only a few are used at a time, and statistical principles are applied to the population characteristics. The following illustration will help to understand the population object and the population characteristics.

Population object	Population characteristics
Persons	Income, height, weight
Desks	Length
Light	Velocity, ref. index
Farms	Number of acres, number of cows

Sample

This has the same general meaning as in common usage; a sample is a part of a whole, total or aggregate. It is actually a part of a statistical population which is selected at random and is used as a basis for making estimates and inferences about the population. A sample has a number of important aspects or characteristics: 1) the size of the sample, i.e., the number of elements or groups of elements of the population, 2) how the sample is selected, (3) individual population element or groups of population elements, and (4) whether a sample selected continues through time to be representative of the population.

The sampling unit may be the individual or the family, a house or a city block, or the census enumeration data or some other area. For e.g., in sampling a population of high school students, the population element is the individual student while the sampling unit may be the individual, a class, a grade or an entire school. Likewise, we can have a sample of cities, a sample of stores within the cities and a sample of each commodity within each sample store. In the laboratory the result obtained from a single experiment or a small series of experiments gives a sample of data.

Random sampling

In biological experimentation, one works with samples. Any sample should be a representative of the population to permit a valid inference about the population. For this, statisticians do 'random sampling'. By this process each possible sample of a given size has an equal chance of being drawn.

For example, to select a random sample of 40 students from a class of 130 for the same investigation, first serial numbers are given to all. The numbers 1 to 130 are written in separate slips, shuffled and any 40 are taken. By this, equal chance is given to all the 130. If the sampling units chosen are affected by the procedure used to select them so that some are chosen more often or less often on the average than their proportion demands, then a biased sample results.

What has been illustrated above is an example of simple random sample. The definition of random selection as giving every population element an equal chance of being selected is adequate so long as one limits oneself to sampling simple populations. The classical approach to random selection has now been modified in order to include sample situations in which random selection is used but where all the population elements *do not have an equal chance* of being selected. Two such situations are 1) stratified random sampling, and 2) sampling with probabilities proportional to size.

1) *Stratified sampling* is another type of sampling procedure which is used when the population is not homogeneous. For example, the population of a country is not homogeneous with regard to spoken language or food habits. It can be divided into sub-groups which are homogeneous. The people who speak the same language will form a homogeneous group. The samples are selected from each group. The size of the sample selected is usually proportional to the size of the population in that sub-group. The sub-group is called a 'stratum'. This method of dividing the population into different strata and selecting the sample from each stratum is called the 'stratified sampling method'. If the samples from each stratum are selected by the random sampling method, such a sample is called the 'stratified random sample'. In respect of sampling with probabilities proportional to size, the important fact is that the probabilities in the sampling need not be equal but should be known. It is for this reason that the Sub-Commission on Statistical Sampling has defined randomisation as follows:

A process is properly described as random if to each unit has been initially assigned an independent and determinate probability of being selected.

Numbers: In representing numbers, one should take into account the digits. The numbers 40 and 40.00 do not have the same statistical significance. The number 40.00 has greater significance though it apparently looks the same as 40. This is because 40 can be anything between 39.5 and 40.5 with a difference of 1. On the other hand 40.00 is anything between 39.995 and 40.005 with a difference of 0.01. This is because of 'digits'. The number 40 has only one significant digit, while 40.00 has as many as four significant digits. This is arrived at by the following principles: 1) All digits are significant except zero, 2) zero is not significant when it is at the extreme left of a number and 3) zero on the left of a decimal point is also not significant. e.g.,

Numbers	Significant digits
125	3
1250	3
1205	4
0.0125	3
1250.00	6
40	1
40.00	4

Rounding of numbers: As an illustration, to round off the following numbers to two significant digits,

Numbers	Rounded off to 2 digits
1465	1500
1390	1400
1425	1400
1515	1500
1350 *	1400
1250 **	1200
1251	1300
1996 ***	2.0×10^3

In rounding off 1465, 1390, the last two digits are above 50, so they are rounded off to 1500 and 1400, respectively. In respect of 1425 and 1515 the last two digits are less than 50. So, the number is rounded off to 1400 and 1500, respectively. But 1350* can be rounded off to 1300 or 1400. it should be rounded off only to 1400 according to the statistical principle that if the second number is 'odd', it should be raised by 1. On the other hand, 1250** is rounded off to 1200, according to the principle that if the second number is 'even' it should be left as such. Regarding the number 1996***, though it is to be rounded off to 2000, it should be written as 2.0×10^3. This is because, as per instruction, there should be two significant digits. The number 2000 has only one digit. So, it should be rounded off to 2.0×10^3.

When the average is worked out, it should be confined to significant digits. e.g., the body weight of six rats (in grams) is: 1) 148.0, 2) 145.0, 3) 145.5, 4) 139.0, 5) 142.5 and 6) 151.5. The animal balance can weigh only upto 0.5 g. The averages calculated by the arithmetic method could be any one of the following: 144.416, 144.4, 144.42, or 144.417. But from the point of view of statistics, the average should be restricted to significant digits of four. As the balance can weigh only upto 0.5 g, the average is 144.0

Location of central tendency: Central tendency gives the whereabouts of some central values. The average, the mean or the arithmetic mean of a number of data gives the measure of central tendency of a sample of data. It is arrived at by summing up the individual values of observatious and dividing this by the total number of observations. If the values are $x_1, x_2, x_3, x_4, x_5.....x_n$ the sum is $x_1+x_2+x_3+x_4........x_n$. If n is the number of observations,

$$\bar{x} = \frac{x_1 + x_2 + x_3 + x_4 ..... x_n}{n} = \frac{\Sigma x}{n}$$

When $\bar{x}$ is the mean, Σx is the sum of x values, $\bar{x}$ is a statistic, and the corresponding parameter in the population is given by the symbol μ(Mu).

If the population is a finite one,

$$\mu = \frac{\Sigma x}{n}$$

Where Σx stands for the sum of all the observations in the population and n the number of observations in the population.

If the sample of data is divided into groups, the mean of each group is called the group mean, e.g., the pulse rate of 20 students is a) 77, 96, 70, 73; b) 74, 86, 75, 70, 84, 82; c) 78, 76, 60, 88, 76, 99, 75, 76, 68 and 82.

$$\bar{x}^1 \text{ for group a (4)} = 79$$
$$\bar{x}^{11} \text{ for group a (6)} = 78.5$$
$$\bar{x}^{111} \text{ for group a (10)} = 77.8$$

$$\text{The grand mean} = \frac{\bar{x}^1 + \bar{x}^{11} + \bar{x}^{111}}{\text{No. of groups}} = \frac{79 + 78.5 + 77.8}{3} = 78.1$$

Weighted mean: Weighted mean is a special type of average. In finding the arithmetic mean, it is assumed that all the observations have the same importance. But sometimes, each variable may have an assigned importance. For example, x_1 has importance of w_1 (called the weight of x_1) and x_2 has the weight of w_2 and so on.

So the weighted arithmetic mean (A.M.) is given by the formula,

$$\text{weighted A.M.} = \frac{x_1 w_1 + x_2 w_2 + x_3 w_3 + \ldots x_n w_n}{w_1 + w_2 + w_3 \ldots w_n}$$

In fact, the A.M. for the frequency distribution given by

$$\bar{x} = \frac{x_1 f_1 + x_2 f_2 + \ldots x_n f_n}{f_1 + f_2 + \ldots f_n}$$

is a form of weighted A.M. only

Supposing there are 500 samples, a) 40 samples with a mean of 47.5, b) 120 samples with a mean of 85, c) 250 with a mean of 22.4, and d) 90 with a mean of 51.5;

the grand mean is, $\dfrac{47.5 + 85 + 22.4 + 51.1}{4} = \dfrac{206.0}{4} = 51.5$

Weighted mean is $\dfrac{(47.5 \times 40) + (85 \times 120) + (22.4 \times 250) + (51.1 \times 90)}{40 + 120 + 250 + 90} = 44.6$

Though the weighted mean will be different from the grand mean it may be close to it but more accurate.

The arithmetic mean may sometimes be unacceptable if one or more values are very distant. Supposing there are individual values as 4, 1, 2, 4 and 7880, the mean is 716 which is neither near 4 nor near 7880.

Median: This is another way to locate central tendency. The items are arranged in the order of magnitude and the central item is picked out, – for e.g., 70, 79, 73, 96, 77. On arranging as 70, 73, 77, 79, 96, the median is 77.

If the total number of samples is even say, 70, 77, 79, 73, the average of the middle two is taken.
70 (73, 77) 79 — the median is 75. The median for many numbers e.g., 171,

$$\frac{n+1}{2} \quad \text{i.e.,} \quad \frac{171+1}{2} = 86$$

If it is 86.5, the values of the 86th and 87th items and their average are taken.

The median 50 values is used extensively in Pharmacology. LD_{50} refers to the median lethal dose.

Median is not distorted by extreme values, does not require the measurement of individual values, but is tedious to operate.

Geometric mean: When large numbers are not available, this geometric mean can be taken,

$$(GM = \sqrt[n]{x_1 \times x_2 \times x_3})$$

If x_1, x_2 and x_3 are 1, 3 and 9

$$GM = \sqrt[3]{1 \times 3 \times 9} = \sqrt[3]{27} = 3$$

For very large numbers, log GM is used,

$$\text{Log GM} = \frac{\log x_1 + \log x_2 + \log x_3 \ldots}{n}$$

Probability: An important method of measuring the degrees of uncertainty is that of *probability*.

Probability is simply the proportion or occurrence of a specified characteristic in a given population. The characteristic may be either an attribute or an interval of a variate. It may be the sex of a person, the make of an automobile or the presence of a defect in a manufactured product. It may be a specified range of height, income or size of a family. The characteristic may be an event, or a property of a physical object or an attribute of a human being. In short, it is the relative frequency of occurrence of an event in the long run. If P is a proportion or probability, Ni the number of objects or events with a specified characteristic and N the total number of objects or events included in the population under consideration, then

$$P = \frac{Ni}{N}$$

Thus, the definition of P is based upon population values which in many problems are not known. Hence, sample values drawn at random are used as a means of obtaining an estimate of P.

$$p = \frac{ni}{n}$$

Where p is the estimated proportion or probability, ni the number of elements of objects or events with the specified characteristic and n the total number of elements or objects or events in the sample.

Characteristics of probability: 1) Probability as defied, $P = \dfrac{N_1}{N}$ can very from 0 to 1. If $P = 0$, no element in the population has the characteristic or quality in question. On the other hand, if $P=1$ then every element has the characteristic and $N_i=N$. If $P=0.5$ half the elements have the specified characteristic.

2) Probabilty is a valid means of prediction, only to the extent that past relationship on which the probability is based remains stable.

3) A value of probability applies to the average for an aggregate. It does not predict exactly for a single event.

4) The estimation of probability from a sample is based upon the assumption that every element in the population has an equal chance of being selected in the sample.

In estimating P from direct frequencies of occurrence for high accuracy, it is often necessary to use larger random samples especially in those cases where the specific event or characteristic is rare.

Problem: What is the probability of drawing an ace from a regular deck of 52 cards?

Population = 52
Characteristic is the ace.
In a single deck of cards, ther are 4 aces

$$P = \text{probability (or chance)} = \dfrac{4}{52} = \dfrac{1}{13}$$

P-estimate or p

There are 180,125 youths in a city who are 16 to 24 years of age (inclusive). Of these 161,92 are 16 years. What is the probability of drawing 16 year olds in a sample selected from the youth population assuming that all youths in the population are given equal chance. $P = \dfrac{16{,}912}{180{,}125} = 0.094$. In practice, we do not know the total number in the population. In such a case we must estimate the proportion or probability for a larger sample down from the population. If every element in the population is given an equal chance of being selected, then, the estimate might be quite close to the true but unknown population value. Suppose in the last e.g., we observe 1755 youths, 16 years of age in a 10 per cent random sample of the larger group. A 10 per cent sample of 180125 would be 18013. The estimated probability (small p) of 16 year olds would be.

$$p = \dfrac{1755}{18013} = 0.097$$

If we do not know the probability of 0.094 for the entire population, we would use 0.097 as the best estimate of it. On the other hand, if we know both P and p as .094 and .097 respectively, we would infer from P estimate i.e., p, that the 16 years olds have been slightly but not seriously oversampled.

Probabilty is governed by 2 important laws. Addition law: This law states that the probability of any one of several possible mutually exclusive events occurring, is the sum of the probabilities of the separate events. For e.g.,.

While tossing a die (a cube whose 6 faces are marked with numbers 1, 2, 3, 4, 5 and 6 respectively), what is the probability of getting either 2 or 4 or 6?

The probability of getting a 2 on a single toss of the die = 1/6 by definition. Similarly, the probabilities of getting a 4 as well as a 6 are each 1/6. By the addition law of probability, the probability of getting either a 2 or a 4 or a 6 is

$$\frac{1}{6} + \frac{1}{6} + \frac{1}{6} = \frac{1}{2}$$

If two events are such that the occurrence of one prevents the occurrence of the other, then they are said to be mutually exclusive. The two sides of a coin are mutually exclusive.

Multiplication law: The Multiplication law states that the probability of the combined or simultaneous occurrence of two or more independent events is the product of the individual probabilities of the two events. For example in tossing a coin twice, the probability of getting a head on both tosses is given by this law as. P (Head, Head) = P (Head) × P (Head) = ½ × ½ Two events are said to be independent if the probability of occurrence of either is not affected by the occurrence or non-occurrence of the other. For example getting a two on each of the two dice thrown are independent events.

Probability distribution: If x is any variable that can take values $x_1, x_2, x_3 \ldots x_n$ with probabilities $p_1, p_2, p_3, \ldots p_n$, respectively, the probability distribution of x is the listing of each one of those possible values with corresponding probabilities. The sum of $p_1, p_2, p_3 \ldots p_n$ is p=1 in a probability distribution.

To sum up, the relative frequency of an event in a uniform series of trials tends to a finite limit as the number of trials in the series is indefinitely increased. The limiting value of the relative frequency is taken as the measure of the probability of occurence of the event in a single trial.

Measures of variation: the mean and median do not provide any indication of the variability of the individual values of observations in a sample. Two statistics that are most commonly used in biological experiments to measure variability are the variance and its square root, the standard deviation (S.D. or S).

The variance is defined as the sum of the squares of the deviations of the individual values from their mean divided by (n – 1) — one less than the number of observations.

$$S^2 = \frac{(x_1 - \bar{x})^2 + (x_2 - \bar{x})^2 + \ldots + (x_n - \bar{x})^2}{(n-1)}$$

$$= \frac{\Sigma(x - \bar{x})^2}{(n-1)}$$

For convenience, S^2 can be written as,

$$S^2 = \frac{\Sigma x^2 - x \Sigma x}{(n-1)}$$

S^2 is the variance

The numerator in the above formulae is referred as the sum of squares. The denominator (n – 1) is called the number of degrees of freedom associated with S^2.

The variance is measured in terms of the square of the actual units of measurement. A measure of variability in the same units of measurement as the original observations is given by the standard deviation S.D. or simply S. In a sample, the standard deviation is given by

$$S = \sqrt{\frac{\Sigma(x - \bar{x})^2}{(n-1)}}$$

or (ii) $\quad S = \sqrt{\dfrac{\Sigma x^2 - \bar{x}\,\Sigma x}{(n-1)}} \quad$ or $\quad S = \sqrt{\dfrac{\Sigma x^2 - \dfrac{(\Sigma x)^2}{n}}{(n-1)}}$

$$S = \sqrt{\frac{1}{n-1}} \ \sqrt{\Sigma x^2 - \frac{(\Sigma x)^2}{n}}$$

It is the measure of the variability of the observations around the mean.
S is the S.D., i.e., Standard Deviation.

Problem:
Calculate the standard deviation of the following values: 5, 9, 9, −2, 5, 2, −2, 8, 9, 6
First method:

Values x	Deviation $(x - \bar{x})$	Squared deviation $(x - \bar{x})^2$
5	0.1	.01
9	4.1	16.81
9	4.1	16.81
−2	−6.9	47.61
5	0.1	.01
2	−2.9	8.41
−2	−6.9	47.61
8	3.1	9.61
9	4.1	16.81
6	1.1	1.21
Sum Σx 49 Mean $\bar{x} = 4.9$	$\Sigma(x - \bar{x}) = 0$	$\Sigma(x - \bar{x})^2 = 164.9$

$\Sigma(x - \bar{x})^2 = 164.9$

$$S = \sqrt{\frac{\Sigma(x - \bar{x})^2}{(n-1)}} = \sqrt{\frac{164.9}{9}} = \sqrt{18.322} = 4.28$$

Second method:
In this, the term $\Sigma(x - \bar{x})^2$ is equated to $= \Sigma x^2 - \dfrac{(\Sigma x)^2}{n}$

Values	Squares (x^2)
5	25
9	81
9	81
−2	4
5	25
2	4
−2	4
8	64
9	81
6	36
$\Sigma x = 49$	$\Sigma x^2 = 405$

Calculate $\Sigma x^2 - \dfrac{(\Sigma x)^2}{n}$

$\Sigma x^2 = 405$
$\Sigma x = 49$
$(\Sigma x)^2 = 49 \times 49 = 2401$
$\dfrac{(\Sigma x)^2}{n} = \dfrac{2401}{10} = 240.1$

$\Sigma x^2 - \dfrac{(\Sigma x)^2}{n} = 405 - 240.1 = 164.9$

This is equal to $\Sigma (x - \bar{x})^2 = 164.9$

$S = \sqrt{\dfrac{\Sigma (x - \bar{x})^2}{(n-1)}} = \sqrt{\dfrac{164.9}{9}} = \sqrt{18.322} = 4.28$

S.D. = S.

Third method

In this method, the derivations of values from some arbitrary numbers are taken and the following formula is used:

Values x	Deviations from 4	(deviations)^2
−2	−6	36
−2	−6	36
2	−2	4
5	1	1
5	1	1
6	2	4
8	4	16
9	5	25
9	5	25
9	5	25
$\Sigma x = 49$ Mean 4.9	sum of deviations = 9	sum of squared deviations 173

$\Sigma(x - \bar{x})^2$ = Sum of squared deviations − n (mean of deviation)^2
Sum of squared deviation = 173
n = 10
Sum of deviation = 9
Mean of deviation = $\dfrac{9}{10}$ = 0.9
$\Sigma(x - \bar{x})^2$ = Sum of squared deviation − n (mean of deviation)^2
= 173 − 10 (0.9)^2 = 164.9

$$S = S.D. = \sqrt{\dfrac{164.9}{9}} = 4.28$$

Variance is the square of standard deviation
Variance = S.D.^{2} = S^2
Standard deviation S.D = S = $\sqrt{\text{variance}}$

Coefficient of variation: Another important measure of variation is the 'coefficient of variation'. This is defined as the standard deviation expressed as a percentage of the mean value.

$$\text{Coeff. variation} = \dfrac{S.D.}{\bar{x}} \times 100$$

This measure of variation is a pure number and is free of the units of measurements, as different from the variance and S.D. which are expressed in the same units as the observations. This measure helps to assess variation in relation to the size of the mean. A standard deviation of 6, around a mean of 60 has a lower variability, i.e., 6/60 × 100 = 10 per cent coefficient of variation compared to a S.D. of 6' around a mean of 30, i.e., 6/30 × 100 = 20 per cent (coefficient of variation)

Error: Error (e) is the deviation of an observed value of x from a true or expected value X. So e = x − X. An error has both magnitude and direction since x can be smaller or greater than X.

Standard error: Standard Error is a variability. Due to inherent variability in all biological material, any quantity computed from a sample, i.e., statistic, tends to vary from sample to sample. The standard deviation measures the variability of individual observations in a sample of data or in the population.

The standard error of the sample mean is given by S.E. of $\bar{X} = S_{\bar{x}} = \dfrac{S.D.}{\sqrt{n}}$

Similarly, the S.E. of a difference between two sample means is given by

$$\text{S.E. of } (\bar{x}_1 - \bar{x}_2) = S \sqrt{\dfrac{1}{n_1} - \dfrac{1}{n_2}}$$

where S is the S.D. of the two samples pooled together and n_1 and n_2 are the numbers in the two samples.

Normal distribution: Normal distribution is continuous distribution which can, in theory, take any value within the range of possible values. The number of values that a continuous variable can take within a given range is infinite. Because of this, it is not strictly correct to talk of the probability of the variable having a particular value since that would tend to be zero. Instead, the probability that the variable may fall within a specified interval is considered.

Normal distribution has only two parameters, the mean x̄ and the standard deviation. *Normal* is used in the usual sense since the distribution of a number of biological variables were found to approximate to this distribution.

Figures below shows a set of normal curves with differing values of mean x̄ and S.D. Figure 1 shows three normal distributions which have different S.D., but the same mean.

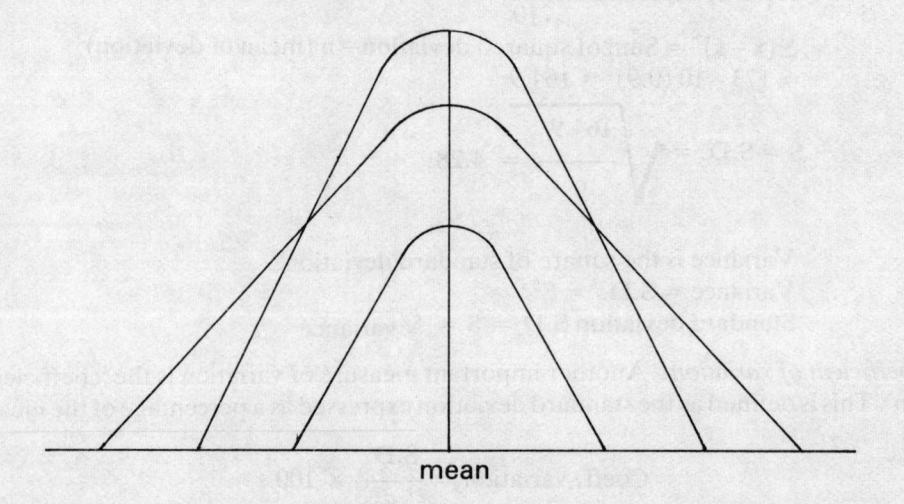

mean

Figure 2 below shows three normal distributions with different mean values, but the same standard-Deviation.

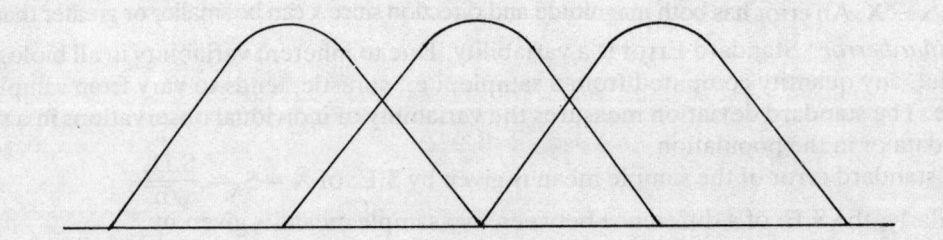

The important features of normal distribution are
1) The distribution is bell-shaped and symmetric around the mean value;
2) the mean and the median coincide; and
3) the distribution has no upper or lower limit.

Probability from normal distribution: Defined areas of the normal curve are enclosed within defined limits on either side of the mean.

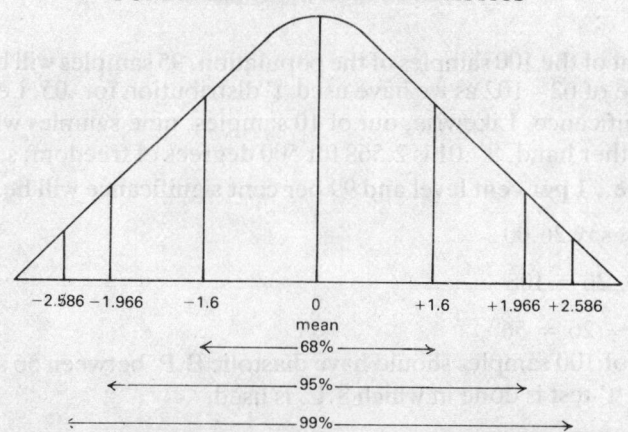

The range, Mean − S.D. (1.6) to Mean + S.D. (1.6) encloses 68 per cent of the area, and the range, Mean − 1.96 (S.D.) to the Mean + 1.96 (S.D.) encloses 95 per cent of the area of the curve, and the range, Mean − 2.58 (S.D.) to Mean + 2.58 (S.D.) encloses 99 per cent of the area of the curve.

THE DISTRIBUTION OF TEST STATISTICS

Statistical inference or the testing of hypothesis: The most common test statistics are identified with the following six distributions:
1) The 't' distribution
2) Wilcoxon's Rank Sum Test
3) The binomial distribution
4) The normal probability curve
5) The *Chi* square x^2 distribution and
6) The 'Z' and 'F' distribution

The first two i.e., the 't' distribution and the Rank Sum Test are often used for biological research and therefore will be discussed here.

The 't' distribution: In 1908, W.S. Gosset (better known to statisticians as the student) in a paper on the probable error of the mean, obtained the distribution of $\dfrac{\bar{x}}{S\bar{x}}$ so that an exact test could be obtained by using the estimated S.D., $S_{\cdot x}$. R.A. Fisher extended the work of 'the student' and showed that the 't' distribution could be applied not only to differences between means but to regression of coefficient as well. The 't' distribution table is available. The data can be noted for the particular value of (n − 1) which is technically known as the degree of freedom. Normally $\bar{x}\, t_{0.5}$ i.e., distribution of t for .05 i.e., 5 per cent level and 95 per cent significance, is used for testing statistical significance.

Illustration: (a) for large samples; from the table of values for 't' distribution for 500 degrees freedom, the probability of 't' for .05 is 1.965.

In a population of 500 men, the diastolic blood pressure ranged from 65 to 115 with a mean value of 82.
The S.D. was 10.
The mean diastolic B.P. of 500 men = 82 ± 10.
As the value of 't'$_{.05}$ is 1.965, the product of this value and the S.D. is 1.965 × 10 = 19.65, say, 20.00.

    82 − (S.D. × t value) = (82 − 20) = 62
    82 + (S.D. × t value) = (82 + 20) = 102

This means that out of the 100 samples of the population, 95 samples will be having diastolic B.P. within the range of 62 – 102 as we have used 't' distribution for .05, i.e., 5 per cent level with 95 per cent significance. Likewise, out of 10 samples, nine samples will have this range of 62 – 102. On the other hand, 't' .01 is 2.568 for 500 degrees of freedom; so, product of S.D. × 't' value for .01, i.e., 1 per cent level and 99 per cent significance will be,

$10 \times 2.568 = 25.68$, say 26.00

Mean + 26 = 82 + 26 = 108

Mean − 26 = 82 − 26 = 56

i.e., at $'t'_{0.01}$, 99 out of 100 samples should have diastolic B.P. between 56 and 108. For smaller samples, student 't' test is done in which S.E. is used.

Illustration:

Blood-serum sodium concentration in 18 cases —

$\bar{x} = 115 \pm$ S.E. i.e., 115 ± 2.83

The 't' distribution for 18 at .05, i.e., 95 per cent confidence limit from the table is 2.1.

't' value × S.E. i.e., $2.1 \times 2.83 = 5.94$

Mean + 5.94 = 115 + 5.94 = 120.94

Mean − 5.94 = 115 − 5.94 = 109.06

Out of 100 samples, 95 per cent will be within this range of 109.06 to 120.94, i.e., 95 per cent samples will have the range of mean + (S.E. × $'t'_{.01}$ values) to mean − (S.E. × $'t'_{.01}$ value).

Difference between two groups: The difference between two groups is assessed by the Null hypothesis. 'Null' means no difference between the two sets of observations. However, there is definitely an obvious difference. It has to be ascertained whether the difference is due to chance. Tests have to be applied to reject the hypothesis if the difference is not due to chance and is statistically significant. If we cannot reject the Null hypothesis, there is no difference.

If in a comparison of a characteristic between two groups, the Probability 'P' is less than 0.05 ($P < .05$) it means that in 95 per cent of the cases there is difference and only in 5 per cent there is no difference. Hence, the Null hypothesis is rejected, and the difference is said to be significant.

Consider the following schematic diagram.

The mean $\bar{x}_1$ is for values falling on OA with ± AC and AD as the standard error. The mean $\bar{x}_2$ is for values falling on OB with standard error ± BD and BE.

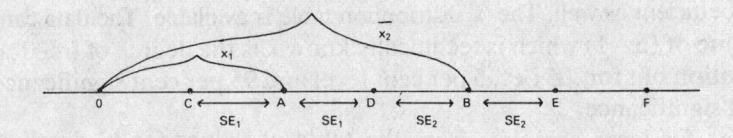

Point D is overlapping for the means $\bar{x}_1$ and $\bar{x}_2$

$(\bar{x}_1 - \bar{x}_2) = SE_1 + SE_2$

As there is overlap, the difference is not significant. On the other hand take the following example.

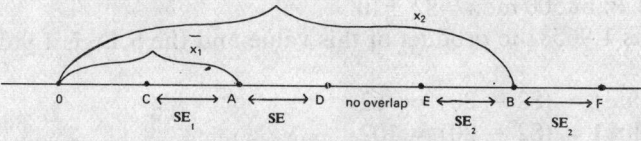

If $\bar{x}_1 - \bar{x}_2$ is more than twice $(SE_1 + SE_2)$, the difference is significant, i.e., when the observed difference is more than twice the standard error of the difference in the mean, the difference is said to be satistically at the 5 per cent level.

i.e., $$\frac{\bar{x}_1 - \bar{x}_2}{(SE_1 + SE_2)} > 2$$

This is given by the working formula,

$$t = \frac{\bar{x}_1 - x_2}{\sqrt{SE_1^2 + SE_2^2}}$$

Illustration: Treatment of 11 cases with A; mean blood sugar 83.40
Treatment of 11 cases with B; mean blood sugar 68.4

$$SE_1^2 + SE_2^2 = 43.29$$
$$\sqrt{SE_1^2 + SE_2^2} = 6.58$$

$$t = \frac{(\bar{x}_1 - \bar{x}_2)}{\sqrt{SE_1^2 + SE_2^2}} = \frac{83.40 - 68.40}{6.58} = 2.28$$

The degree of freedom is, n –1 = (11 – 1) = 10.
For each group it is, 10.
For two groups, 10 + 10 = 20.
't' .05 for 20 is, 2.09 from the table.
Value at hand is 2.28.
  If the value at hand is greater than the value found in the table, the finding is statistically significant.
So, P<.05, as 't' was taken for .05.

The 't' test for small samples: In finding out whether the difference in mean between two small samples is significant, the following method is used 1) The samples are pooled together to determine the common variability of the universe from which the samples are drawn:
Common variablity

$$(i) \quad S = \sqrt{\frac{\Sigma(x_1 - \bar{x}_1)^2 + \Sigma(x_2 - \bar{x}_2)^2}{(n_1 + n_2) - 2}}$$

Where $\bar{x}_1$ = mean of first sample
n_1 = No. in the first sample
$\bar{x}_2$ = mean of second sample
n_2 = No. in the second sample

(ii) $SE_{X1} = \dfrac{S}{\sqrt{n_1}}$

$SE_{X2} = \dfrac{S}{\sqrt{n_2}}$

$t = \dfrac{\bar{x}_1 - \bar{x}_2}{\sqrt{\dfrac{S^2}{n_1} + \dfrac{S^2}{n_2}}}$ OR $t = \dfrac{\bar{x}_1 - \bar{x}_2}{S}\sqrt{\dfrac{n_1 n_2}{n_1 + n_2}}$

3) Finding the number of degrees of freedom:
$(n_1 + n_2) - 2$. This is the number of independent differences used in the calculation of S.

4) Finding in 't' table the probability of such a value of 't' to occur by chance with reference to the degrees of freedom.

Paired 't' test: Some times, it is necessary to assess the effect of a drug or treatment on a set of inividuals, for which the values of a characteristic 'before' treatment and 'after' treatment are recorded. There will be a difference in these values. Is the difference statistically significant?

In this case, the two samples are individuals before and after treatment, and they are not independent samples. The following method is used:
1) Calculate the difference in each set of observations (d).
2) Find out the mean difference 'd'.

$$\bar{d} = \dfrac{\Sigma d}{n} \text{ where n is the number of individuals}$$

3) Find out S.D.
$$S.D. = \sqrt{\dfrac{(d - \bar{d})^2}{n - 1}}$$

4) Find out the S.E. of the difference.
$$S.E. = \dfrac{S.D.}{\sqrt{n}}$$

5) Find out 't'.
$$t = \dfrac{\bar{d}}{S.E.}$$

The Degree of freedom is $(n - 1)$

Refer to the 't' table for (n - 1) degree of freedom for probability. The 't' distribution test of significance is for normal distribution. If 't' is not normal distribution say, it is skew distribution. Skew distribution has to be converted to normal (normal does not mean that all measurements occurring in nature should conform to normal distribution but many characteristics do follow this distribution closely). 'Normal' here does not mean the opposite of abnormal. Normal or Gaucien distribution means that the curve has a symmetrical and bell-shaped character. It includes a family of curves, flatter or steeper. The important property of the normal curve is that the proportion of the area under the curve, lying between the mean and any given multiple of the S.D. away from it, is constant for every normal curve.

In case there is no normal distribution, non-parametric values can be used, i.e., parameters are changed to metameters like log values, square root values, etc. Non-parametric values can be used for normal distribution as well.

Wilcoxon's Rank Sum Test:

For normal and other distributions like the skew. This consists of two types, one using unpaired data and the other paired data.

1) Signing rank test (paired data)

Two sample tests (unpaired data)

Signing Rank Test: Ten pairs of patients with rheumatoid arthritis are treated with drug A first and later with drug B. Their serum globulins are investigated for following the lowering of globulin fraction.

Values are paired, and difference and ranks are noted:

	Treatment with A	Treatment with B	Difference		Rank (no sign)	
Ist pair	38	45	−7		6½	VI
2nd pair	26	28	−2	2nd + 3rd	2½	II
3rd pair	38	36	2	2	2½	III
4th pair	33	30	3			IV
5th pair	33	48	−15			X
	−	−	−4		5	V
	−	−	−11			IX
	−	−	−7		6½	VII
			−9			VIII
10th pair	−	−	1		1	

If the two ranks have the same difference, as −2 and 2 for the second and third ranks, the ranks are added up and divided by 2, i.e.,

$$\frac{2nd + 3rd}{2} = \frac{5}{2} = 2½$$

and 2½ is given for each. The next one should be the fourth rank. Similarly if the sixth rank is 7 and the seventh also is 7. Then it is

$$\frac{6 + 7}{2} = \frac{13}{2} = 6½$$

So, both the sixth and seventh ranks will have 6½. The next rank is the eighth. The signed ranks are as follows:

```
        − 6½
        − 2½
        −        + 2½
        −        + 4
        − 10
        − 5
        − 9
        − 6½
        − 8      + 1
        ─────────────
        − 47½    + 7½
```

The total rank is added up. It should be $\dfrac{n(n+1)}{2}$

$n = 10;\quad n + 1 = 11;\quad \dfrac{10 \times 11}{2} = \dfrac{110}{2} = 55\ (\text{Ranks})$

All + signs are added and, all (−) signs are also added: out of the two, the small rank is taken. e.g., of 55 ranks, if (+) is 7½ and (−) is 47½, 7½ is taken. The chart for 5 per cent level should then be referred to.

No. of pairs	*5 per cent*
7	2
8	2
9	6
10 --------------	⑧
11	11

If the values at hand are less, the finding is significant.

For n = 10, the values are 7½. The value from the chart is 8. So, the finding is statistically significant.

Unpaired test:

All the findings are combined and noted down according to the rank.

	Rank
26	*1*
27	*2*
28	*3*
29	*4*
30	*5*
31	
32	

Groups of the total $n \times \dfrac{(n+1)}{2} = \dfrac{20 \times 21}{2} = \dfrac{210^*}{2} = 105$

* Values will be double.

Ranks are underlined and totalled for A and B. The lower value for A and B is taken. If sample A is 81.5 and B is 128.5, 81.5 is taken.

n_1	= 1	2	3	...	10
n_2					
4					
5					
6					
–					
–					
10					78

For 20 samples, the figure against 10 (n_1) and 10 (n_2) is noted. It is 78. In this work, there is a higher value, i.e., 81.5. Hence, the finding is not significant. In student's 't' test if the derived value is greater than that in the table, the difference is significant. Thus, the last steps are opposite for the 't' distribution method and Wilcoxon's Rank method.

Value of 't'

n	0.10	.05	.01
1	6.314	12.706	63.657
2	2.920	4.303	9.925
3	2.353	3.182	5.841
4	2.132	2.776	4.604
5	2.015	2.571	4.032
6	1.943	2.447	3.707
7	1.895	2.365	3.499
8	1.860	2.306	3.355
9	1.833	2.262	3.250
10	1.812	2.228	3.169
11	1.796	2.201	3.106
12	1.782	2.179	3.055
13	1.771	2.160	3.012
14	1.761	2.145	2.977
15	1.753	2.131	2.947
16	1.746	2.12	2.921
17	1.74	2.11	2.898
18	1.734	2.101	2.878
19	1.729	2.093	2.861
20	1.725	2.086	2.845
21	1.721	2.080	2.831
22	1.717	2.074	2.819
23	1.714	2.069	2.807
24	1.711	2.064	2.797
25	1.708	2.060	2.787
26	1.706	2.056	2.779
27	1.703	2.052	2.771

28	1.701	2.048	2.763
29	1.699	2.045	2.756
30	1.697	2.042	2.750
60		2	2.66
100		1.984	2.626
200		1.972	2.601
500		1.965	2.588
1000		1.962	2.581
∞	1.64485	1.95986	2.57582

REFERENCES

1. Annino, J.S.
   Clinical Chemistry, Principles and Procedures, J & A Churchill Ltd., London, 3rd edn., 1964.
2. Astwood, E.B.
   Recent Progress in Hormone Research, Vol. 24, Academic Press Inc., New York, 1968.
3. Baldwin, E. and Bell, D.J.
   Cole's Practical Physiological Chemistry, Sagar Publications, New Delhi, 1967.
4. Bell, G.H., Davidson, J.N. and Scarborough
   Textbook of Physiology and Biochemistry, The English Language Book Society and E & S Livingstone Ltd., 6th edn., 1965.
5. Best, C.H. and Taylor, N.B.
   The Physiological Basis of Medical Practice, Williams & Wilkins Co., Baltimore, 8th edn., 1966.
6. Bier, M.
   Electrophoresis, Theory, Methods and Application, Academic Press Inc., New York, 1959 and 1966.
7. Bloemendal, H.
   Zone Electrophoresis in Blocks and Columns, Elsevier Pub. Co., 1st end., 1963.
8. Cantarow, A. and Schepartz, B.
   Biochemistry, W.B. Saunders Co., Philadelphia & London, 4th edn., 1967.
9. Cheyne, G.A.
   Techniques in Chemical Pathology, Blackwell Scientific Publications, Blackwell, Oxford, 1st edn., 1964.
10. Colowick, S.P. and Kaplan, N.O.
    Methods in Enzymology, Vol. 3, Academic Press Inc., New York, 1957.
11. Dixon, M. and Webb, E.C.
    Enzymes, Academic Press Inc., New York, 2nd edn., 1964.
12. *Duncan's Diseases of Metabolism,* Eds. Bondy, P.K. and Rosenberg, L.E., W.B. Saunders Co., Philadelphia, London, 6th edn., 1969.
13. Easthem, R.D.
    Biochemical Values in Clinical Medicine, John Wright & Sons Ltd., Bristol, England, 3rd edn., 1967.
14. *Enzyme Nomenclature Recommendations (1964) of the International Union of Biochemistry on the Nomenclature and Classification of Enzymes,* Elsevier Publishing Co., New York, 1965
15. Fieser, L.F. and Fieser, M.
    Organic Chemistry, Reinhold Pub. Co., New York, 3rd edn., 1960.
16. Fruton, J.S. and Simmonds, S.
    General Biochemistry, John Wiley & Sons, New York, 2nd edn., 1956.
17. Ganong, W.F.
    Review of Medical Physiology, Lange Medical Publications, Maruzen Co. Ltd., 6th edn., 1973.
18. Glasstone, S.
    Textbook of Physical Chemistry, Macmillan & Co., Ltd., London, 2nd edn., 1956.
19. Gordon, A.H. and Eastol, J.E.
    Practical Chromatographic Techniques, George Newnes Ltd., London, WC., 1964.
20. Guyton, A.C.
    Textbook of Medical Physiology, W.B. Saunders Co., Philadelphia & London, 3rd edn., 1966.

21. Martin, D.W. Jr., Mayes, P.A., Rodwell, V.W. and Granner D.K.
    Harper's Review of Biochemistry, Lange Medical Publications, 20th edn., 1985.
    Harper's Review of Bio Chemistry, Lange Medical Publications, 20th edn., 1985.

22. Harrison, A.
    Chemical Methods in Clinical Medicine, J & A Churchill Ltd., 4th edn., 1957.

23. *Hawk's Physiological Chemistry* (ed. Oser, B.L.), Blakiston Division, McGraw-Hill Book Co., New York, 14th edn, 1965.

24. Helfferich, F.
    Ion Exchange, McGraw-Hill Book Co., Inc., 1962.

25. Hoffman, W.S.
    The Biochemistry of Clinical Medicine, Year Book of Medical Publishers Inc., Chicago, 4th edn., 1960.

26. Homburger, F. and Bernfeld, P.
    The Lipoproteins, Methods and Clinical Significance, Basal, S. Karger, New York, 1958.

27. Kamen, M.D.
    Isotropic Tracers in Biology, Academic Press Inc., New York, 1957.

28. Keela, C.A.
    Samson Wright's Applied Physiology, The English Language Book Society and Oxford University Press, London, W.I., 12th edn., 1971.

29. King, J.
    Practical Clinical Enzymology, D. Van Nostrand Co., London, 1965.

30. Kleiner, I.S. and Orten, J.M.
    Biochemistry. The C.V. Mosby Co., Saint Louis, 7th edn., 1966.

31. Lederer, M.
    An Introduction to Paper Electrophoresis and Related Methods, Elsevier Publishing Co., 1957.

32. Mahler, H.R. and Cordes, E.H.
    Biological Chemistry, Harper and Row, New York and John Weatherhill Inc., Tokyo, 1966.

33. McGilvery, R.W.
    Biochemistry – A Functional Approach, W.B. Saunders Co., Philadelphia, 1970.

34. Mountcastle, V.B.
    Medical Physiology, Vol. 1, C.V. Mosby Co., St Louis, 12th edn., 1968.

35. *Nutritional Data*
    Heinz International Research Center and Heinz Research Fellowship of Mellon Institute, Pennsylvania, H.J. Heinz Co., Pennsylvania, 1962.

36. Paoletti, R. and Kritchevsky, D.
    Advance in Lipid Research, Vol. I, Academic Press Inc., New York, 1963.

37. Pritham, G.H.
    Anderson's Essentials of Biochemistry. The C.V. Mosby Co., 1968.

38. Ramachandran, G.N.
    Aspects of Protein Structure, Academic Press, London, 1963.

39. Rao, C.N.R.
    Ultraviolet and Visible Spectroscopy – Chemical Applications, Butterworths, London, 2nd edn., 1967.

40. Sakami, W.
    Handbook of Isotopic Tracer Methods, Western Reserve School of Medicine, U.S.A., 1965.

REFERENCES

41. Sunderman, F.W. and Sunderman, F.W., Jr.
    Serum Proteins and the Dysproteinemias, Pitman Medical Pub. Co. Ltd., Philadelphia, 1st edn., 1964.
42. Thompson, R.H.S. and King, E.J.
    Biochemical Disorders in Human Disease, J & A Churchill Ltd., London, 2nd edn., 1964.
43. Thompson, R.H.S. and Wootton, I.D.P.
    Biochemical Disorders in Human Disease, J & A Churchill Ltd., London, 3rd edn., 1970.
44. Umbriet, W.W., Burris, Z.H. and Stauffer, J.F.
    Manometric and Biochemical Techniques, Burgess Pub. Co., Minneapolis, 5th edn., 1972.
45. Varley, H.
    Practical Clinical Biochemistry, The English Language Book Society and William Heinemann Medical Books Ltd., London, 1969.
46. Von Euler, U.S. and Eliasson, R.
    Prostaglandins, Medical Chemistry, Vol. 8, Academic Press Inc., New York 1967.
47. Weber, G.
    Advances in Enzyme Regulation, Pergamon Press, New York, 1965.
48. West, E.S. Todd, W.R., Mason, H.S. and Van Bruggen, J.T.
    Textbook of Biochemistry, Macmillan Co., New York, 1966.
49. White, A., Handler, P. and Smith, E.L.
    Principles of Biochemistry, The Blakiston Division, McGraw-Hill Book Co., Inc., New York, 1968.
50. Wilkinson, R.
    Isoenzymes, Chapman & Hall Ltd., London, 2nd ed., 1970.
51. Williams, R.
    Textbook of Endocrinology W.B. Saunders Co., Philadelphia, 1968.
52. Wootton, I.D.P.
    Microanalysis in Medical Biochemistry, J & A Churchill Ltd., London, 1964.

Index

Abnormal hemoglobin 425
Abnormal sickle cell anemia 425
Abnormal thalassemia 425
Abnormal osmotic pressure 531
Absorbance
   (extinction) 477
Absorption 170
Absorption glucose and sodium 171
Absorption lipids 172
Absorption sugars 171
Acetaldehyde 271
Aceto acetate thiokinase 221
Aceto acetic acid 202, 221, 222, 272, 277
Aceto acetyl CoA 220, 234
Aceto acetyl succinic
   thiophorase 221
Acetone 202, 221, 222
Acetylase 301, 302
N-acetyl aspartate 274
Acetyl carnitine 219
Acetyl choline 355, 396
Acetyl CoA 188, 189, 218, 225
Acetyl CoA carboxylase 224, 225
N-acetyl glutamate 105, 255
Acetyl lipoamide 188
Acetyl malonyl enzyme 225
N-acetyl mannosamine
   6 phosphate 197
N-acetyl 5 methoxy serotonin
   (melatonin) 285
N-acetyl neuraminic acid 28, 197
Acetyl phosphate 134
3- acetyl pyridine 345
Acetyl salicylate (asperin) 445
Acetyl transacylase 225
Acetyl globulin 149
Achlorhydria 501
Achrodermatitis enteropathica 416
Achrodextrin 23
Acid-base balance 142
   assessment of 148, 149
   hemoglobin 144
   kidneys and 147
Acid hematin 93, 94
Acidification of urine, 147
Acid mucopolysaccharides
   (glycosamino glycans) 25
Acid number 36
Acidosis
   respiratory 148
   metabolic 148
Acid phosphatase 115
Aconitase 188, 189

Aconitate 188
ACP 223
Acromegaly 389
ACTH 214
   biological and metabolic effects 391
Actin 157
Actinin 158
Actinomycin D 317
Active cAMP-dependent protein
   kinase 178, 180
Active iodine, I 280, 364
Active methionine 259, 260, 265
Active serine 230, 239
Active succinate 85
Active sulphate 134
Active transport 523
Actomyosin 158
Acyl carrier protein ACP 225, 374
Acyl CoA 229, 230
Acyl CoA dehydrogenase 218
Acyl CoA synthetase 217, 219, 229, 230
Acyl enzyme 225
Addison's disease 203, 400, 405
Adenine 67
Adenine phosphoribosyl transferase
   (APR Tase) 326, 327
Adenosine 68, 69
Adenosine deaminase 322
Adenosine di phosphate (ADP) 70, 80
Adenosine 3' 5' diphosphate
   (di phospho adenosine) 80
Adenosine di phospho glucose 81
Adenosine tri phosphate (ATP) 71, 79
S- adenosyl ethionine 240
S- adenosyl methionine 83, 134, 259, 260,
   264, 280, 289
Adenylate cyclase 214, 379
Adenylate kinase 161
Adenylic acid (adenosine mono
   phosphate, AMP) 70
Adenylo succinase, 325, 326
Adenylo succinate 326
Adrenal cortical hormones 378
Adrenalectomy 380
Adrenal function tests 509
Adrenal hyperplasia 431
Adrenal virilism 431
Adreno corticotropic hormone
   (ACTH) 249, 388, 391
Adsorption 534
   chromatography 467
   Gibb's adsorption equation 534
   isotherm 534
Aerobic dehydrogenase 119

Affinity chromatography 468, 469
Afibrinogenemia 152, 424
Agammaglobulinemia 425
A/G reversal 138
Air hunger 202
Alanine 271
ß alanine 271, 286, 320, 321
Alanine amino transferase
   (ALT), 252, 271
Alanine tRNA gene 73
Albinism 279, 281, 422, 424
Albuminoids (sclero proteins) 55
Albumins 55, 137
Alcohol 243
   cholesterol accumulation 243
   cirrhosis of liver 244
   deleterious effects of 243
   fat accumulation 243

fall in liver 243
hepatitis 244
hypercholesterolemia 244
hypertriacyl glycerolemia 244
hyperuricemia 244
pulmonary surfactant 244
safe limits 243
Alcohol dehydrogenase 244
Alcoholism 242, 243
Aldactone 509
Aldehyde deleterious effects 243
Aldehyde dehydrogenase 130, 244
Aldehydism 243
Aldolase 115
Aldose reductase 384
Aldosterone 379, 380, 410
Aldosteronism 509
Alkali hematin 93, 94
Alkaline phosphatase 115, 498
Alkali reserve 143
Alkalosis
   metabolic 148
   respiratory 148
Alkaptonuria 278, 279, 281, 422, 424
Allantoin 320
Allopurinol 430
ALT 115
Amadori rearrangement 16
Amethropterin 349, 355
Amido black 475
Amination 252
Amino acetone 271
Amino acids 46
   aromatic 48

assay of 53
classification 46
configuration 50
definition 46
energy from 249
glucogenic 249
ketogenic 249
metabolism 245, 248
non-protein nitrogen compounds
nutritionally dispensable
  (non-essential) 248
nutritionally indispensable 248
reactions of 49
role in detoxication 261
separation 53
sulphur-containing 48
urinary 248
D amino acid dehydrogenase 251
L amino acid dehydrogenase 250
Amino acid nitrogen
  blood levels 246
Amino acid pools 246
Amino acid uria 268
Amino acyl adenylate 134
Amino acyl tRNA 307
  acceptor specificity 308
  recognition site 308
  transfer specificity 308
Amino acyl tRNA synthetase 307
α amino adipic acid 273
α amino adipic E semialdehyde 273
ρ amino benzoic acid 349, 353
γ amino butyric acid 49, 254
ρ amino hippurate clearance 504
Amino imidazole carboxamide-
  ribosyl 5 phosphate 325
Amino imidazole carboxylate-
  ribosyl 5 phosphate 324
Amino imidazole ribosyl 5 phosphate 324
Amino imidazole succinyl carboxamide-
  ribosyl 5 phosphate 325
β amino iso butyric acid 246, 320, 321
α amino β keto adipic acid 85
δ amino levulini acid 85, 262, 264, 427
Amino propanol link 350
Aminopterin 106, 349, 355
Ammonia
  blood ammonia 500
  detoxication of 255
  urinary 147
Ammonotelic 245
Amobarbitol (amytal) 125
5' AMP 70, 79, 214, 326
Amygdalin 20
α Amylase 22, 23, 115
β amylase 23
Amylo dextrin 23
Amylo pectin 22, 23

Amylum (starch) 22, 24
Amytonia congenita 366
Anaerobic dehydrogenase 119
Analbuminemia 425
Anaplerotic reaction 191
Andersen's disease 426
Androgens 382
  biochemical effects 383
  biosynthesis 383
Androstane 382
Androstene 3, 17, drone 383, 384
▲^{4}androstene 3, 17, drone 19 or 384
Androstene 3, 11, 17, trione 383
Androsterone 382, 383
Anemia
  folate deficiency 350
  pernicious anemia 353
  pyridoxine deficiency 88
Angular conjunctivitis 343
Angular stomatitis 343
Anorexia 341
Anserine 161, 286
Anterior pituitary hormones 388
Anthrione test 15
Antibody 140
Anti-coagulants 153
Anti-codon 300, 308
Anti-diuretic hormone 395
Anti-egg white injury factor 340
Antigen-antibody complex 491
Anti-grey factor 353
Anti-hemphilic globulin 150
Antimycin A 125
Anti-oxidants 37
Anti-pellagra 339
Anti-pernicious anemia factor 350
Antipyrin 403
Anti-rachitic vitamin 334
Anti-scorbutic vitamin 355
Anti-sterility vitamin 337
Anti-thrombin III, 153
Anti-thyroid agents 367
Anti-vitamins 355
Anti-xerophthalmic vitamin 331
Anturan 430
Apo B 48, 238
Apo B 100, 238
Apo C 238
Apo ceruloplasmin 414
Apo enzyme 97
Apo ferritin 411
Apo lipo proteins 211
Apo transferrin 411
Arabinose 5 phosphate 192
Arachidonic acid 228, 229, 397, 399
Arbutin 20
Argentaffinoma 284
Arginase 257, 258

Arginine, 256, 258, 287
Arginine glycine transamidinase 254
Arginine phosphate 134
Arginino succinase 257, 258
Arginino succinate, 256, 258
Arginino succinate synthetase 256, 258
Arginino succinic aciduria 259
Arsenate 189
Arterenol 280
Ascheim Zondek test 395
L asunbic acid (vitamin C) 197, 355
Ascorbic acid saturation test 357
Asparaginase 251
Asparagine 251
Aspartate 190
Aspartate amino transferase (AST) 252
Aspartate trans carbamoylase 327
Aspartate trans carbomoylase
  feed back inhibition by CTP 327
Aspartic acid 274
Aspergilles oryzae 231
Aspirin (acetyl salicylic acid) 398
AST 115
Atherosclerosis 241
  ageing 241
  diseases associated with 241
ATPase 127, 128, 171
ATPase vectorial 128
ATP citrate lyase (citrate cleavage
  enzyme) 227
ATP citrate lyase trans phosphorylase 160
ATPs from TCA cycle 189
Atractyloside 126
Atromide S$^{(R)}$ 237
Atropine 445
Attenuation
  (gene expression in *E.coli*) 303
Attribute 539
Augmented histamine test 502
Auto analysers 484
Autosomal, dominant, 423, 432
  acute intermittent porphyria 423
  hereditary spherocytosis 423
Autosomal recessive 423, 432
  phenylketonuria 423
Avidin 227, 348
Azide 126
Azo carmine B 475
Azure A 503

$\emptyset$ X 174 Bacteriophage 74, 296
Bantu siderosis 414
Barfoed's reagent 15
Basal metabolic rate (BMR) 449, 450, 505
  Addison's disease 451
  conditions for measurement 450

INDEX

Cushing's syndrome 451
  factors influence 451
  ear drum hole 451
  febrile diseases 451
  hyperthyroidism 451
  hypothyroidism 368, 451
  infection 451
  leukemia 451
  polycythemia 451
  pyrexia 451
Bearberry 20
Beer's law 477
Bence Jone's proteins 139
Benedict Roth's apparatus 450
Benedict's qualitative reagent 14
Benedict's quantitative reagent 15
Benemid 430
Benzimidozole 351
Benzoic acid 261, 446
Beri beri 342
Betaine 239, 261, 262, 263
Beta sheet 62
Bial's oncinol test 18
Biguanides 373, 377
Bile 168, 437
  cholagogues 168
  choleretic effect 168
  composition 168
  acids 169
  pigments 94
  reactions of 96
  salts 169
Bilk functions of 170
Biliary calculi 169
Bilirubin 94, 95, 96
  and jaundice 495
Biliverdin 94, 95, 96
Biochemical genetics 295
Biocytin 348
Bioenergetics 131
  TCA cycle 189
Bioflavonoids 355
Biological oxidation 117
  definition 117, 118
  chemi-osmotic hypothesis 128
  experimental evidences of
    chemi-osmotic hypothesis 128
Biological value of proteins 456
Biostatistics 539
Biotin 227, 263, 348
  biochemical function 348
  deficiency effects 348
  coenzymes 112
Biotransformation 444
, 3 bis phosphoglycerate 187
, 3 biso phospho glycerate phosphatase 187
Bitter almond 20
Biuret reaction 59

Blood
  buffers 144
  chemistry 135
  clotting 149
    biochemistry of 149
    factors 149
  constituents 143
    normal values 143
  group polysaccharides 28
  plasma 436, 530
    molar concentration 530
  pressure 242
  sugar, homeostasis 198
    role of kidneys 200
    role of liver 200
  tests for 94
  urea 503
  volume 532
    regulation 532
Body fluids 435
Body composition 403, 435
  extracellular fluid (ECF) 435
  Interstitial fluid 435
  Intracellular fluid (ICF) 435
Body water
  distribution of 403
  electrolytes 403
Bohr effect 91
Bradykinin 151
Biostatistics 539
Branching enzyme
  (amylo 1 4, 1 6 transglucoxidase 177
British anti-Lewasite, BAL
  (dimercaprol, 2,3 mercapto propanol)
  125, 446
Britten and Davidson's model (gene
  expression in eukaryotes) 304
Bromobenzene 261, 290, 446
Bromophenol blue 475
p-bromo phenyl mercapturic acid 446, 447
Bromosulphalein excretion 499
Bronchial asthma 148
Bronze diabetes 414
Brownian movement 518
Buffer 511
Buffer action 512
  mechanism 512
Butanol extractable iodine (BEI), 506

Caffeine 214
Calcitonin 249, 368
  mode of action 368
Calcitriol
  (1, 25 dihydroxy Vitamin D_3) 369, 397
Calcium 406

  amount in the body 406
  factors for intestinal absorption 407
  function of 406
  parathyroid hormone 408
  second messenger to hormones 406
  serum 408
  third messenger to hormones 406
  vitamin D3 408
Calcium ATPase 408
Calcium binding protein 369, 408
Calcium calmodulin dependent protein
  kinase 178
Calcium calmodulin phosphorylase
  kinase 178
Calcium deprivation test 506
Calcium ions
  regulation of hormone action 363
Calmodulin 406
Calmodulin-component of phosphorylase
  kinase 180
Calorimetry 448
cAMP 291
Canlicular wall 163
CAP-cAMP complex 302, 303
Carbamino hemoglobin 90
N-carbamoyl B alanine 321
N-carbamoyl B amino isobutyrate 321
Carbamoyl aspartate 328
Carbamoyl phosphate 255, 258, 328
Carbamoyl phosphate synthetase 255, 258
  327, 328
Carbohydrases 98
Carbohydrates 8
  classification 8
  disaccharides 9
  metabolism, role of
    hormones 200
  monosaccharides 8
  polysaccharides 9
Carbon dioxide combining power 143, 149
Carbon dioxide content 148, 149
Carbon dioxide fixation (dark reaction of
  photosynthesis) 207
Carbonic anhydrase 165
Carbon tetrachloride 240
N-carboxy biotin 348
V-carboxy glutamate (GLA residue)
  150, 151, 152, 339
  and vitamin K 339
Carboxy hemoglobin 92
Carboxy peptidase 166
Cardiac glycosides 20
Cardiolipins 40, 235
Carnitine 219
  acyl transferase I 219
  acyl transferase II 219
Carnitine acyl carnitine translocase 219
carnosine 286

carotenes 331
carotenoids 204
Carrier-mediated transport
  (facilitated transport) 522
Carr Price reaction 334
Casein 58
Catalase 130, 251
Catabolite activator protein (CAP) 80
Catecholamines 280, 371
Catechol O methyl transferase 372
CDP - choline 82, 229, 230
CDP - diacyl glycerol 230
CDP ethanolamine 82, 230
CDP glycerol 82
Cell 1
  biochemical constituents and function 6, 7
  centrosomes 5
  cytosol 5
  endoplasmic reticulum 4
  eukaryote 1
  fractionation 5, 6
  Golgi bodies 4
  isolation of components 5
  lysosomes 4
  marker enzymes for cytological structures 6, 7
  membrane 2
    fluid mosaic model 3
  mitochondria 3
  nucleus 3
  prokaryotic 1
  ribosomes 4
  structure 2
cellulose 207
cephalin 39, 229, 270
  (phosphatidyl serine and phosphatidyl ethamolamine)
Cerebrocuprin 415
Cerebron (phrenosin) 42
Cerebronic acid 238
Cerebroxides (glycolipids) 41, 238
Cerebrospinal fluid 437
Ceruloplasmin 141, 312, 414, 415
Ceruloplasmin feroxidase 412
Chargaff's rules 72, 73
Chaulmoogra oil 31
chemodeoxycholic acid 169
Chloride 403
Chloride deficiency effects 403
Chloride - bicarbonate shift 145, 146
Chloroform anesthesia 268
P-chloro mercury benzoate 105
Chloromycetin
  (chloramphenicol) 317
Cholorophyll a 204
Cholorophyll b 204
Cholecalceferol 237, 334
  (Vitamin D_3) 168

Cholecystokinin 168, 387, 393
Cholera 268
Cholesta polyene carbonium ion 45
Cholesta polyene sulphonic acid 45
Cholesterol 43
  biosynthesis 233
  androgens and estrogens 236
  effect of drugs 237
  fasting and diabetes 236
  feed back control
  hyperthyroidism 236
  hypophysectomy 236
  hypothyroidism 236
  insulin and 236
  polyunsaturated fatty acid 236
  regulation 236
  species variation 236
Cholesterol catabolism 237
Cholesterol, dietary 232
Cholesterol ester transport protein 238
Cholesterol metabolism 232
  tissue synthesis 233
Cholestyramine 237
Cholic acid 169
Cholilithiasis 243
Choline 230, 231, 248, 260, 354
  (trimethyl amino ethanol)
  biochemical function 354
  deficiency effects 354
Choline acetylase 396
Choline esterase 396
Choline phosphate 231
choloxin 237
Chondroitin sulphate A
  (chondroitin 4 sulphate) 26
Chondroitin sulphate B
  (dermatan sulphate) 26
Chondroitin sulphate C
  (chondroitin 6 sulphate) 26
Chorionepithelioma 395
Chorionic somato mammotropin
  (CS, placental lactogen) 394
Christmas disease 152
Christmas factor, 150, 152, 153
Chromatin 306
Chromatin fibres 74, 75
Chromatin fibrils 74, 75
Chromatography 465
  adsorption 467
  column 467
  gel filtration 468
  ion exchange 467
chromium 418
  and insulin receptors 418
  potentiator of insulin 418
Chromoproteins 56
Chromosomes 295, 423, 432

autosomal 395
gene relationship 295
sex-linked 395
Chylomicrons 173, 210, 212, 237, 238
Chymotrypsinogen 166
Chymotrypsin 166
Cincophan 320
Cis aconitate 185
Cistron 295, 299
Citrate 188
Citrate-cleavage enzyme
  (ATP citrate lyase) 227
Citrate synthetase 190
Citric acid cycle 188
Citrulline 255, 258
Citrullinemia 259
Clostridium welchii 231
Clotting diseases 424
CMP-NANA 239
Cobalamines 350
Cobalt 417
Cobaltous chloride 367
Cobamide coenzymes 112, 351, 417
  adenyl 351
  benzimidazol 351
  5, 6 dimethyl benzimidazole 351
Codons 299
  AAA 310
  AUG 310
  UAG 310
  UUU 310
Codons, chain terminating 299
  degenerate 299
  universal 299
Coefficient of viscosity 536
  determination 536
Coenzymes 108
  and B vitamins 108
Coenzyme A 81, 110, 111, 347
  reactions 347
Coenzyme Q 121, 232
Colchicine 430
Collagen 63, 273
Collagenase 166, 333
Collagen diseases 138
Colligative property 529
Colloidal state 517
Colloids
  physiological importance 538
Colorimetry 478
  photo-electric 479
  spectro-photometry 479
  visual 479
Colour vision 333
Compactin 237
Comparator 515
Competitive protein-binding assay 49?
Concentration and dilution

INDEX

test of urine 505
Condensing enzyme 226, 230
Congeners 244
Congestive cardiac failure 148
Conjugated proteins 55
Copolymerase III 296, 329
  (Copol III)
Copper 414
  body content 414
  deficiency effects 415
  functions 414
Coproporphyrin I 87
Coproporphyrin III 87
Coproporphyrinogen 86
Coprosterol 45
Corepressor 303
Cori's disease, (Forbe's disease, glycogen storage disease III) 426
Coris ester (glucose I phosphate) 19
Coris lactate - glucose interconversion 182
Corpus luteum 396
Corrin ring 350
Corrinoid coenzymes 351
Corticosterone 379, 380
  binding globulin 382
Corticotropin releasing hormones 387
  α 1 CRH 392
  α 2 CRH 392
  β CRH 392
Cortisol
  (17 hydroxy corticosterone) 379, 380
Cortisol binding globulin 141
Cortisone
  (17 hydroxy 11 deoxy corticosterone) 379
Creatine 248, 265
  urinary 266
Creatine kinase 114, 115
Creatine phosphate 260
Creatinine 265
  clearance 504
  excretion in urine 266, 267
  level in muscle 266
Cretinism 368
Crigler Najjar syndrome 431
CRO (control of repressor and other genes) 303
Crotony CoA 273
Crystalloids 517
Cuprothionein 415
Curie 489
Cushing's syndrome 401, 405
Cyanide 172
Cyanide poisoning 268
Cyanocobalamine
  (Vitamin B12) 350
  biochemical function 352
Cyanopism 333

Cyclic AMP (cAMP) 79, 80
Cyclic GMP (cGMP) 82
cyclo pentemo perahydro phenanthrene, 43
cystathionine 270, 288
Cystathionine reductase 288, 291
Cysteic acid 290
Cysteine 270, 288
Cysteine desulfhydrase 289
Cysteine sulfuric acid 289
Cystine 288
Cystine calculi 291
Cystinosis 291, 431
Cystinuria 268, 290, 291, 422, 431
Cytidine 68
Cytidine nucleotide 82, 112
Cytidylic acid (CMP) 70
Cytochrome 110
  a 121
  a3 (Oxidase) 121
  b 121
  b5 130
  b6 205
  b559 206
  C 121
  C reductase 343
  C1 121
  C552 206
  f 205
  P450 130
Cytosine 66
Cytosolic hydroxylation 130
Cytosolic oxidations 129

Dalmatian coach dog and uric acid 320
Dark reaction 206
Deacylase (thioesterase) 189, 221, 224, 225
Deamidation 251
Deamination 250
  non-oxidative 250
  oxidative 250
Debranching enzyme
  (amylo 1, 6 glucosidase) 179, 180
Decarboxylation 253
Degree of Freedom 555
Dehydrases 250
Dehydration 405
Dehydro ascorbic acid 356
7 dehydro cholesterol 45, 237, 334
11 dehydro corticosterone 379
Dehydro epi androsterone 360
α dehydrogenases 119
β dehydrogenases 119
Dental caries 418
Deoxy adenosyl B12 265
5 deoxy adenosyl cobalamine 352

Deoxy cholic acid 169
Deoxy corticosterone 379, 380
11 deoxy cortisol 380
Deoxyribo nucleic acid (DNA) 71
Deoxy antitemplate 298, 330
DNA directed RNA polymerase I 298
DNA directed RNA polymerase II (Transcriptase) 298, 329, 330
  rewindase activity 298
  unwindase activity 298
DNA directed RNA polymerase III 298
  extraction 71
  gene relationship 295
  in vitro synthesis 329
  in vivo synthesis 329
  ligase 297
  mitochondrial 297
  polymerase III (pol III) 296, 329
  polymerase L (maxi) 296
  primary structure 71, 72
  primer 329
  replication 296
  secondary structure 72, 73
  sequence studies 73, 74
  single-stranded 74
  template 298, 329, 330
  retiary structure 74
Deoxy ribose 18
Deoxy uridine di phosphate 328
Deoxy uridine mono phosphate 327, 328
Depression in freezing point 530
Derepression 302
Dermatan sulphate 26
Desmosterol 236
Desulphhydrases 254
De Toni Fanconi syndrome 409
Detoxication
  (bio-transformation) 444
  conjugation 445
  deamination 445
  decarboxylation 444
    histamine 444
    indole ethylamine 444
    tryptamine 444
  hydrolysis 445
  oxidation 445
Diabetes insipidus 395, 405
Diabetes mellitus 201
  cholesterol in 203
  dehydration 203
  fatty acids in blood 202
  hyponatremia 202
  IDDM (juvenile) Type I 201
  ketoacidosis 202
  ketone bodies 202
  lipid mobilisation 202
  lipid synthesis 202
  nephropathy 203

neuropathy 203
NIDDM
(Type II, maturity-onset) 201, 376, 377, 426
Non-protein nitrogen 202
  peripheral resistance 203
  pH of blood 202
  plasma bicarbonate 202
  receptors of insulin 203
  retinopathy 203
  triacyl glycerol 202
  VLDL 202
Diacyl glycerol 214, 227
Diacyl glycerol acyl transferase 228
∝ β diacyl glycerol phosphate 229, 230
Diagnex blue 503
Dialysis 517
∝ ε diamino S hydroxy caproate 273
Diaphorase 342
2, 6 dichloro phenol indophenol 357
Dicoumarol 152
Diet 454
  components 454
  protein requirements 454, 455
  sugars 461
Diethyl stilbesterol 385
Digestion 162
  gastric 162
  salivary 162
Digitogenin 20
Digitonin 20, 45
Dihydro biopterin 276, 279
Dihydro biopterin reductase 276, 281
Dihydro folate 349
Dihydro folate reductase 349
Dihydro lipoyl dehydrogenase 188
Dihydro lipoyl transacetylase 188
Dihydro orotase 328
Dihydro orotic acid 327, 328
Dihydro sphingosine 230
Dihydro sphingosine reductase 230
Dihydro testosterone 360, 383
Dihydro thymine 321
Dihydro uracil 321
  uracil loop in RNA 308
Dihydro uridine 298
Dihydroxy acetone phosphate 182, 185, 207
5, 6 Dihydroxy indole 279
Dihydroxy mandelic acid 446
2, 5 dihydroxy phenyl acetic acid
  (homogentisic acid) 276, 277
3, 4 dihydroxy phenyl alanine (DOPA) 279, 280
1, 25 Dihydroxy vitamin D3 (calcitrol) 335, 369
24, 25 Dihydroxy vitamin D3 335, 369
Diido thyronine 367

Dilodo tyrosine 280, 365
Di isopropyl fluoro phosphate 105
Di methyl allyl pyro phosphate 234
Di methyl amino enthanol 260, 265
  6 dimethyl benzimedazol 351
Di methyl glycine 263, 264
Di nitro fluoro benzene 52, 60
Di nitro phenol 126
Di nitro phenyl devivative 60
2, 4 dinitro phenyl hydrazine 357
Di oxygenase 118
Dipalmityl lecithine 39
Diphenyl oxazole 492
1. 3 diphospho glycerate 134, 183, 185, 207
3 diphospho glycerate 91
Disaccharides 20, 22
Diseases, molecular and genetic basis 431
  8 dithio octanoic acid
    (lipoic acid) 347
Donnan membrane equilibrium 524
  hydrolysis 527
  transport of diffusible ions 524, 525
Dopa decarboxylase 280
Dopamine 280
Dopamine hydroxylase 280
Dopa quinone 279
Dubin Johnson disease 431
Dunstan's equation 537

E. coli 296, 299, 327
Eczema 462
Edema 534
EDTA 105
Egg in nutrition 464
Eicosanoids 31
Eicosatrienoate 229
Elastase 166
Elastomucase 241
Elastoproteinase 241
Electrical potential on the
  suface of colloids 518
Electrodialysis 517
Electrometers 492
Electron microscope 480
Electro-osmosis
  (endosmosis) 519
Electrophoresis 473, 518
  definition 473
  different types 474
  gel 475
Electroscope 492
Elemente constant 212, 231
Elemente variable 213
Elements
  bulk 400
  trace 400
Elongation factors 307, 312, 314

Embden Meyerhof pathway
  glycolysis 182, 184
Emerson enhancement effect 205, 206
Emotional stress 242
Emphysema 148
Emulsion 520
  oil in water 520
  water in oil 520
Encephalitis 148
Endergonic reactions 131
Endocytosis 523
Endoperoxide 397, 399
Endorphisis 249, 393
  ∝ 393
  β 393
  γ 393
Endosmosis 519
Energy expended — men,
  women & children 453
Energy expenditure
  persons of different professions 453
Energy metabolism 448
Energy value of foods 448, 449
Enolase 183, 186, 189
Enol phosphate 134
Enol pyruvate 183, 185
Enoyl CoA hydratase 214
Enoyl hydrase 226
Enoyl reductase 224, 225
Enterocrinin 168
Entero-hepatic circulation 170
Enterokinase 166
Enthalpy 131
Entropy 131
Enzymes 97
  activation, allosteric 104, 105
    metals 104
    organic substances 105
  active sites of 106, 107
  classification 97
  definition 97
  diagnosis of diseases 113
  extraction 99
  factors influencing enzyme action 99
    concentration of enzyme 99, 100
    concentration of substrate 101
    pH 99, 100
    temperature 101
  inhibition 105
    competitive 105
    feed back 105
    metals 105
    organic substance 105
  mechanism of enzyme action 102
    enzyme-substrate complex 106
    Induced-fit theory 106
    lock and key 106
    michaelis menten constant 102

three pronged attack 107
plasma, non specific 113
plasma, specific 113
properties 99
specificity 99
Enzyme immunoassay 485
 heterogeneous 486
 homogeneous 485
 sandwich principle 487
Epinephrine 201, 214, 265, 280, 371
 biochemical effects 371, 372
 blood pressure 371
 fright, fight and flight 371
Epsilon potential 518, 519
Ergo calciferol (Vitamin D2) 334
Ergo sterol 43, 334
Ergo thionine 267, 286
Error 539, 551
 Standard error 551
Erythro cuprein 415
Erythro dextrin 23
Erythrose 4 phosphate 192, 193, 207
Essential fatty acids 228, 229, 462
Essential deficiency effects 228, 462
Estimate 539, 540
Estrogens 384
  biochemical effects 385
  estradiol 384
  estriol 384
  estrone 384
Ethanolamine 230, 261, 263, 265
Ethenyl estradiol 385
Ethereal sulphates 265, 268
Ethronine 240
Ethyl para chlorophenyl isobutyrate 237
Evans Blue method 403
Exergonic reactions 131
Exons 298, 309
Exophthalmos 368
Extrinsic factor of Castle 350

Facultative reabsorption 395
Facultative transport 522, 523
FAD 81, 110, 120, 121
Fanconi syndrome 268
Farnesyl pyrophosphate 234, 235
F1 ATPase 127
Fats 209
 metabolism 215
 physiological value 209
 transport in blood 210
Fats and oils 35
 rancidity 37
 separation of 37
 tests for purity 35
Fatty acids
 biosynthesis 223

chemistry 30, 31
properties 33, 34
spiral 216, 218
Fatty liver 239
 alcoholism 240
 causes 239, 240
 deficiency of vitamins and proteins 240
 experimental 240
 essential fatty acid deficiency 240
 high cholesterol diet 240
 high fat diet 240
 hormones 240
 lipotropic factors deficiency 240
 starvation 240
 uncontrolled diabetes 240
Favism 192, 425
Feces 174, 175, 443
Fehlings solution 14, 15
Ferredoxin 205
Ferritin 411, 412
f5 FH4 262, 286
f10 FH4 262, 286
Fibre in the diet 461
Fibre, triple beneficial effects 461
Fibrin monomer 152
Fibrinogen 137, 149
Fibrino peptides 152
Fibrous proteins 62
FIGLU 350
Filtration factor 505
Flame photometer 480
flask constant 481
Fluid and electrolyte homeostasis 403
Fluid losses 405
fluoride 183, 189
Fluorimetry 480
Fluorine 418
Fluo acetate 106, 189
Fluorosis 418
5 fluoro uracil 328
FMN 81, 110, 120, 121
Folate coenzymes 112
folate trap 353, 417
Folic acid (pteroyl glutamic acid) 349
 biochemical functions 349
 deficiency effects 350
Folin and Wu method 199
Folinic acid
 (N5 formyl FH4) 350
Folins reaction 53
Follicle stimulating hormone (FSH) 388, 394
Forbe's disease (Cori's disease, glycogen storage disease III) 426
Formaldehyde 262
Formate 262
Formic acid 269

N5 formimino FH4 286
N formimino glutamic acid (FIGLU) 285
Formol titration 50
Formyl glycinamide ribosyl 5 phosphate 323
Formyl glycinamidine ribosyl 5 phosphate 323
Formyl transferase 323
Fouchet's test 96
Foulger's test 17
Foxglove 20
Fractional test meal (FTM) 501
Fragility test 534
Free energy change 124
 determination 133
Free fatty acids (FFA) 211
Free fatty acids from carbohydrates 215
Frequency distribution 539, 540, 541
Fried Mann's test 395
Fructokinase 198
Fructose 16, 17
Fructose 2, 6 bisphosphate 183
Fructose 1, 6 diphosphatase 185, 194, 207
Fructose 1,6 diphosphate
 (Harden young ester) 182, 185
Fructose 1 phosphate 198
Fructoe 1 phosphate aldolase 198
Fructose 6 phosphate
 (Neuberg ester) 182, 184, 185, 193 207
Fructosuria 426
Fucose 28
Fuel stores in a normal individual 295
Fumarase 190
Fumarate 190
Fumaryl aceto acetate 277, 278
Fumaryl aceto acetate hydrolase 277, 278
Function tests 445
Furfural 14

GABA (gamma amino butyric acid) 254, 274, 345, 397
Galactoflavin 344
Galactokinase 198, 199
Galactose 17
Galactosemia 198, 199, 425
Galactose 1 phosphate 198, 199
Galactose 1 phosphate uridyl transferase 198, 199, 425
Galactose tolerance test 500
Gangliosides 42, 238, 239
Gargoylism 426
Gas chromatography 472
Gastric analysis 500
Gastric inhibitory peptide 387
Gastric juice 162, 468
 formation 162

function 163
   hydrochloric acid 162
   lipase 165
Gastrin 386, 393
Gaucher's disease 142, 243, 426
GDP 188
Geiger Muller counter (GM counter) 490
Geiger Muller tube 490
Gels 520
   elastic 520
   non-elastic 520
   electrophoresis 475
   Poly acylamide 475
   starch 475
   filtration 468
   (Molecular exclusion)
gene 67
   alanine tRNA 77
   expression
      during cell cycle 304
      gene amplification 305
      hestomes and 304
      in eukaryotes 304
      in Lambda bacteriophage 303
      regulation of 301
   types of 295, 296
      constitutive 301
      inducible 301, 302
      operator 296
      regulator 296
      structural 296
Genetic code 299
Genome, human 295
Geranyl pyrophosphate 234
Gibb's Adsorption Equation 534
Gibbs Thomson Principle 535
Gilbert's disease 431
Glass electrode 516
Globins 55
Globulins 55, 137
γ.Globulin 138
Glucagon 201, 249, 377
   biochemical effects 377
   mode of action 377, 378
1←4, 1←4 glucon transferase 179, 180
Glucaric acid 14
   (saccharic acid)
Gluco corticoids 381
Gluco anti-inflammatory response 381
Gluco metabolic effects 381
Gluco kinase 177, 182, 184, 199
Gluco difference from hexokinase 182, 200
Gluconeogenesis 194
Gluconeogenesis energy barriers in 194, 195
Gluco enzymes 194

Gluco glycine effect (protein effect) 269
Gluconic acid 14
Glucosamine 19
Glucosazone 16
Glucose 9
   amount in the body 177
   configuration 10
   conformation 11
   mutarotation 12
   reactons of 12
   structure 9
Glucose alanine cycle 196
   branched chain amino acids 196
Glucose glycerol cycle 196
Glucose oxidase 14
Glucose 6 phosphatase 180, 199
Glucose 1 phosphate 177, 184
Glucose 6 phosphate 177, 182, 184
Glucose 6 phosphate dehydrogenase 191, 193
Glucose tolerance test 508
Glucuronic acid 14, 19
Glutaconyl CoA 273
Glutamate decarboxylase 254, 274
Glutamate dehydrogenase 252, 253, 274
Glutamate oxalo acetate transaminase (GOT) 252
Glutamate pyruvate transaminase (GPT) 252, 271
Glutamate γ semi-aldehyde 287
Glutamate acid 274
Glutammase 274, 275
Glutamine 261, 274, 275
L Glutamine D fructose 6 phosphate transamidase 197
Glutamine synthetase 274, 275
Glutaryl CoA 273
Glutathrone 54, 246, 249, 277
Glutathrone peroxidase 130, 418
Glutathrone reductase 130
Glurelin 55
Glycans (polysaccharides) 22
Glyceraldehyde 3 phosphate 182, 193, 207
Glyceraldehyde 3 phosphate dehydrogenase 185, 189
D glycero D ido octulose 8 phosphate 192
D Glyceero D ido octulose 1, 8 diphosphate 192
Glycero phosphate acyl transferase 230
Glycerokinase 227, 228, 230
Glycerol 194
Glycerol 3 phosphate (glycero phosphate) 185, 227, 229, 231
Glycero phosphate acyl transferase 230
Glycero phosphoryl choline 231
Glycero phosphoryl choline esterase 231
Glycinamide ribosyl 5 phosphate 323

Glycine 269
Glycine oxidase 269, 343
Glycinura 270
Glyco cholic acid 269
Glycogen 24, 184
   amount in the body 176
Glycogenesis 176
   control 178
Glucogenolysis 178
   epinephrine 179
   glucagon 179
   phospharylases 179
Glycogen storage diseases (glycogenoses) 426
Glycogen synthase kinases 3, 4, 5...178
Glucogen synthetase (UDP glucose-glycogen) transglucosylase) 177
Glucogen synthetase a 105, 178
Glucogen synthetase b 105, 178
Glycolipids 41
   metabolism 238
Glycolysis 182
   erythrocytes 187
Glycophorin 28
Glyco proteins 56
Glycosamino glycans 25
   (acid mucopolysaccharides)
Glycosides 20
Glyucosuria 201
Glyoxylic acid 269
Gmelin's test 96
GmP 326
G/N ratio 250
Goitre 368
Gold Number 520
Gonadotrophic hormone (Gonadotrophins) 394
   biological effects 394
Gonadotropin releasing hormone GnRH, (HRH, FSHRH) 380
Gopalan's syndrome 347
Gout 320
   feed back inhibition of enzymes 428
   primary metabolic 428
   secondary metabolic 430
   renal 430
Grave's disease 368
Groundnut oil 242
Growth hormone 388, 389
   metabolic effects 389, 390
   regulation of secretion 389
Growth hormone release inhibitory hormone (somatostatin) (GHRIH) 388, 389
Growth hormone releasing hormone 387, 389
GTP 188
G^{m}TP Cap 309

Guanase 322
Guanido acetic acid
 (glycocyamine) 254, 260, 265
Guanine 67
Guanosine 68
Guanylate cyclase 82
Guanylic acid 70
L Gulonic acid 197
Gusten 415

Hageman factor 150
Haptoglobin 141
Harden Young ester 20
Hartnup's disease 285, 424
HDL cholesterol 242
HDL/LDL ratio 237, 242
Heavy chains 139
Heavy chain disease 138
Heavy water 403
α helix 62
Helmholtz Gouy electrical double
 layer 518, 519
Hematin 93, 94
Hematoporphyrin 93
Heme 87, 88, 94, 249
Hemoglobin 84, 88, 94
Hemochromogen 93, 94
Hemo globin 84, 88, 94
 adult 91
 degradation products 94
 fetal 91
 oxygen binding 91
  relaxed (R) 91
  taut, tense (T) 90, 91
 oxygen dissociation curve 90
 oxyhemoglobin 89
 HbS 92
 synthesis 85
Hemoglobinopathies 425
Hemolytic jaundice 496
Hemophilia A 152
Hemophilia B (Christmas disease) 152
Hemosiderin 414
Henderson-Hasselbalch's equation
 144, 512, 513
Heparan sulphate 238
Heparin 26
Hepatic coma 148
Hepatocellular jaundice 497
Hesperidin 355
Hess Viscometer 538
α, β Heterodimer 374
Heterogeneous nuclear RNA
 (hnRNA) 308, 309
Heteroglycans 25
Hexokinase 177, 182, 199
Hexosaminidase 42

High diversity lipoproteins 211, 237, 238
High energy compounds 134
High Performance Liquid
 Chromatography (HPLC) 473
Hippuric acid 269, 446
 test 500
Hill reaction 204
Histaminase 286
Histamine 444, 502
 allergic reaction 286
Histidase 284
Histidinura 286
Histone octamer 74
HMP shunt (Hexose Mono Phosphate)
 energetics 192
 eye lens 192
 fat synthesis 192
 F type 192
 galactosemia 192
 L type 192
 pentoses for nucleic acid
  synthesis 191
 significance 191
Holoenzyme 97
Homocupterine 261, 263, 265, 270, 288
Homocystine 288
Homocuptinura 291
Homogentisate dioxygenase 118
Homogentisic acid 276, 277
 (2,5 dihydroxy phenyl acetic acid)
Homogentisic oxidase 118, 277, 281
 (Homogentisate dioxygenase)
Homoserine 270, 288, 289
Hopken's-Cole test 53, 59
Hormones 358
 Calcium ions as "Second
  messenger" 362
 cAMP as 'second messenger' 361, 362
 classification 359
 definition 358
 feed back control 361
 general characteristics 359
 mode of action 361
 plasma proteins, carrying 360
 receptors 360
Human Chorionic Gonadotropin (hCG)
 394
Hyaluronic acid 26
Hydrocarpic acid 31
Hydratase 223, 225
Hydrogen flame ionisation 472
Hydrogen ion concentration 510
Hydrolases 98
Hydroperoxidases 119
Hydroquinone 20
β hydroxy acyl CoA 218
β hydroxy acyl CoA dehydrogenase
 218, 228

β hydroxy acyl enzyme 225
β hydroxy 5 β androstane 17 one 382
3 β hydroxy androst 5 β 17 one 382
17 α hydroxy androst 4 one 382
3 hydroxy kynurenic acid 282
β hydroxy butyric acid 202, 221, 222
Hydroxy cobalamine 350
18 hydroxy cortisone 380
18 hydroxy deoxy corticosterone 380
Hydroxy furfural 14
γ hydroxy glutamic semi aldehyde
 287
5 Hydroxy indole acetic acid (5 HIAA)
 284
β hydroxy isobutyric acid 272
β hydroxy isobutyryl CoA 272
3 hydroxy kynurenic acid 282
 (xanthurenic acid)
3 hydroxy kynurenine 283
1 hydroxylase 369
11 β hydroxylase 380
17 α hydroxylase 380
18 hydroxylase 380
21 hydroxylase 380
Hydroxyl free radical 130
Hydroxy lysine 273
Hydroxy methyl cytidylate 328
d Hydroxy methyl cytidylate synthetase
 328
β Hydroxy β methyl glutaryl CoA
 (HmG CoA) 220, 221, 233, 272
 reductase 233, 234, 237
 synthase 234
Hydroxy methyl glycine 262, 263, 264
Hydroxy methyl sarcosine 262, 263, 264
p.hydroxy phenyl lactic acid 278
p.hydroxy phenyl pyruvic acid 276, 281
p.hydroxy phenyl pyruvic oxidase 276,
 279, 281
17 hydroxy progesterone 380
19 hydroxy progesterone 384
Hydroxy progesterone 383, 384
Hydroxy proline 287
5 hydroxy tryptamine 283
5 hydroxy (serotonin) tryptophan 284
Hydroxy tryamine (Dopamine) 280
Hyperadrenalism 203
Hyperammonemia Type I 259
Hyperammonemia Type II 259
Hyperbilirubinemia 339
Hypercalcemia 408
Hyperchlorhydria 501
Hyperkalemia 402
Hyperlipoproteinemia 242, 426
Hyperparathyroidism 409
Hyperphenylalaminemia Type I 281
 (Phenyl ketonuria)
 types II & III 279, 281

IV & V 279
Hyperpituitarism 203
Hyperthyroidism 203, 368
Hyperuricemia 429
Hypokalemia 402
Hypoparathyroidism 409
Hypoprothrombinemia 152, 339
Hypotaurine 290
Hypothalamic hormones 387
Hypothesis 539
Hypothyroidism 368
Hypoxanthine 322
Hypoxanthine Guanine Phospho-
  ribosyl-transferase (HGPRT ase) 326,
  327
Hypoxanthine nucleotide 83
H zone 157

I^+ active iodine 364, 363
I C D, NAD^+ dependent 188
I C D, $NADP^+$ dependent 188
I D D M 201
I D L (Intermediate density lipoproteins)
  238, 242
I D P 188
L Iduronic acid 27
Imbibition of gels 520
Imidazole pyruvate 286
Imidazolone propionic acid 285
Immuno-electrophoresis 476
Immuno-globulins 138
  classes 138
  definition 138
  types 138
Immuno-globulin G 140
  action of papain 141
  action of pepsin 141
  structure 140
Immuno-globulin gene 305
  class switching 305
  VH, D, JH, CH segments 305
  VL, JL, CL segments 305
  V J joining 305
IMP (Inosinic acid) 326
Inborn errors of metabolism 422
Indican 267, 446
Indicators 514
  range 514
Indole 446
Indole acetic acid 282, 285
Indole ethyl amine 444
Indole lactic acid 282
Indole pyruvic acid 282
Indole 5, 6 quinone 279
Indoxyl 268
Inducers 301, 303

gratuitous 301
Inducible enzymes, names of 301
Induction 301
Inference 539
Infra centrifuge 484
INH (Inonictotinic acid hydrazide,
  Isoniazid) 346, 355
Inherited disorders of metabolism 422
Initiation factor $1F_1$, $1F_2$, $1F_3$
  307, 312, 313
Inosinic acid 325
Inositol 354
  biochemical function 354
Inositol hexa phosphate (phytic acid) 354
Inositol phosphatides 354
  Ca^{2+} release from mitochondria 354
  second messenger of hormones 354
Insulin 214, 373, 403
  antibodies 203
  biosynthesis 373
  C peptide 373
  diabetes mellitus 203
  effect on metabolism 373, 374
  enzymes induced 374, 375
  enzymes repressed 374, 375
  facilitated transport of glucose 375
  membrane permeability for glucose 374
  obesity 375
  physiological effects 376
  potentiation by chromium 375
  potentiation tolbutamide 375
  prepro-insulin 373
  pro-insulin 373
  protein metabolism 375
  receptors 374, 375
  secretion 373
  secretion mechanism 373
    secretion calcium ion 373
    cAMP 373
    secretion $NADPH/NADP^+$ 373
  structure 61
  test meal 502
Insulinase 376
Insulin-like growth factor (IGF_1,
  somatomedin c) 389, 390
  IGF_2 377
Integrator gene 304
Interferons 142
  mechanism of action 142
Inter-organ amino acid exchange 174
  Bram and branched chain amino acids 174
  muscle and branched chain amino
    acids 174
  post-absorptive 174
Interstitial cell stimulating hormone
  (ICSH, luteinizing hormone, LH)
  388, 394, 396
Interstitial fluid 436

Intestinal juice 168
Intestinal enzymes 168
Intravenous infusion 533
  basis for giving 5 per cent glucose 533
  basis for giving normal ralive 533
Intrinsic factor of Castle 164, 352
Introns 298, 309
Inulin 25
Inulin clearance test 504
Inversion 21
Invert sugar 21
Iodine 416
  active iodine 364, 365
  biochemical role 416
  number 35
Iodo acetamide 105
Iodo acetate 189
Iodopsin 333
Ion exchange chromatography 467
β ionone 331
Iron 411
  absorption 411
    iron deficiency 412
    iron overload 412
  binding capacity 414
  body stores 411
  serum levels 414
  sulphur proteins 206
  transport in blood 413
Islets of Langerhans 372
  A   372
  B   372
  D   372
  F   372
Isoalloxazine 120, 121
Isobutyryl CoA 272
Isocitrate dehydrogenase (ICD)
  115, 116, 188, 190
Iso-electric focussing 476
Iso-electric point 57, 58
Iso-hydric change 145
Iso-leucine 271
Isomerases 98
Isopentanol 244
Isopentenyl pyrophosphate 234
Isoprenoid units 233, 234
Isotopes 488
Isovaleryl CoA 272
Isozymes 113
  diagnosis of diseases 114, 115
ITP 188
131 $I-T_3$ uptake by erythrocytes 506
131 I- uptake by thyroid 506

Jaundice 496, 497
  differential diagnosis 497
  hemolytic 496

hepatocellular 497
obstructive 496

Kala azar 138
Kallikrein 150
Kerasin 42
Keratin 63
Keratomalacia 333
Kerato sulphate 27
Keto acids 202
β Keto acyl CoA 218
β Keto acyl reductase 223, 225
β Keto acid synthetase 223, 225
α Keto adipate 273, 283
α Keto E amino caproic acid 273
α Keto butyric acid 271
Ketogenesis 220
α keto glutonate 188, 190
α keto glutarate dehydrogenase complex 188, 190
keto L gulonic acid 197
α keto isocaproic acid 272
α keto isovaleric acid 272
α keto r methyl thio butyric acid 289
α keto ß methyl valeric acid 272
Ketone bodies 202, 222
blood levels 222
Ketonemia 202
Ketonuria 202
keto 6 phospho gluconate 192, 193
Ketosis 222
β keto thiolase 218
Kiliani's reaction 9
Kimmelsteil Wilson syndrome 200
Km valve 102
determination 102, 103
Krebs Henseleit urea cycle 255
Kussmaul breathing 202
Kwashiorkor 458
corticoid hormone levels in blood 458
Kynureninase 282, 283
Kynurenine 282, 283
Kynurenine formylase 282, 283
Kynurenine hydroxylase 283

Lac operon 301, 302, 303
E. coli 302
Lactate dehydrogenase 186
Lactobacillus arabinosis 345, 347
casei 271, 344, 349
lactus 353
Lactogenic hormone
(luteotropin, prolactin) 388
Lactosazone 16
Lactose 21

synthetase 199
Lactotropic effect 390
Lag GTT curve 501
Laki Lorand factor 150, 152
Lambda bacteriophage 303
gene expression 303
Lambert's law 477
Lanosterol 233, 235
LCAT 238, 242
LDH isozymes 113
LDL cholesterol 242
Lecithins (phosphatidyl choline) 38, 229
Lecithin cholesterol Acyl Transferase (LCAT) 39
Lesch Nyhan syndrome 429
Leucine 271
Leukotriene 32
Liebermann Burchard reaction 45
Ligases 98
Light chains 139
Light reaction 204
Lignoceric acid 238
Lineweaver Burk Plots 104
Linoleic acid 228
Linolenic acid 228
Lipase 115
Lipids 29
classification 29
compound 29
derived 29
simple 29
in adipose tissue 214
in blood 210, 211
fate in the body 213
metabolism 209
Lipogenesis 223
extra mitochondrial 223
mitochondrial 226
differences between 227
Lipoic acid (thioctic acid) 347
Lipoproteins 56, 212
blood 211, 212
body 212
Lipoprotein lipase 238
Lipositol (Inositol phosphatide) 40, 354
Lipo thiamide pyro phosphate 347
Lipotropic factors 293
ß lipotropin 249, 393
Liquid scintillation counter 492
Lithia water 430
Liver Function Tests 495, 496
Liver lipid metabolism 215
Lobry de Bruin Alberda van Ekenstein transformation 13
Location of Central Tendency 544
Lohmann's reaction 134
Long Acting Thyroid Stimulator (LATS) 392

Lorelco 237
Low Density lipoproteins (LDL) 210, 237, 238
Lung surfactant 39
Lupus erythematosis 138
Luteinizing hormone 396
Lyases 98
Lymph 436
Lymphogranuloma venereum 138
Lysine 273
Lysolecithin 231
Lysophospholipase (phospholipase) 38, 231
Lysophospholipid 231

Magnesium 410
role of 410
Malaria 138
Malate 190
Malate dehydrogenase 189, 190
Malena neonatorum 339
Maleyl aceto acetate 277, 278
cis-trans isomerase 277, 278
Malic enzyme 225
Malignant carcinoid syndrome 284
selected right-sided heart failure 284
Malonic acid 106, 189
Malonyl CoA 220, 224, 225
Malonyl trans acylase 223, 225
Maltosazone 16
Maltose 22
Mandelo nitrile 20
Manganese 416
biochemical function 416
Mannitol 403
D Mannosamine 28
Mannose 18
Maple syrup urine disease 274, 424
Marasmic kwashiorkor 459
Marasmus 458
McAndle's syndrome
(Type V, glycogen storage disease) 426
Mean 545
geometric 546
grand 545
weighted 545
Measures of variation 548
Median 546
Melanins 249, 279, 281
Melanocyte cells 279
Melanocyte stimulating hormone 249, 392
Melanotic sarcoma 268
Melatonin
(N acetyl 5 methoxy serotonin) 285
Membrane hydrolysis 527
hydrolysis gastric acidity 527
Membrane phenomena 521

Menopause 396
Menstruation 396
MEOS pathway
  (Microsomal Ethanol Oxidising System) 244
Mepyramine maleate (anthisan) 289
ß mercapto pyruvic acid 289, 446
Meromyosin 158
  light 158
  heavy 158
Metabolism
  methods of study 420, 421
  levels 420
  tracer techniques 421, 422
Metalloproteins 56
Metanephrine 372
Methanol 262
Methemoglobin 92, 93
  reductase 93
Methemoglobinemia 425
Methionyl adesonyl transferase 259, 269, 288
∝ methyl aceto acetyl CoA 273
Methyl acrylyl coA 272
Methyl amino ethanol 260, 265
Methylation of nitrogen bases 306
Methyl B^{12} 261, 264, 265, 352
∝ methyl butyryl CoA 272
∝ Methyl carnosine (anserine) 286
∝ methyl crotonyl CoA (tiglyl CoA) 273
ß methyl crotonyl CoA 272
5 methyl cytosine 321
2 methyl, 3 difarnesyl 1, 4 naphthoquinone 338
N5 N10 methylene tetra hydrofolate 261, 262, 265
Methyl glucoside 20
ß methyl glutaconyl CoA 272
N methyl guanido acetic acid (creative) 260
3 methyl histidine 286
∝ methyl ß hydroxy butyryl CoA 273
Methyl malonic aciduria 352
Methyl malonic semialdehyde 272
Methyl malonyl CoA 272, 352
L methyl malonyl CoA mutase 352
2 methyl 1, 4 naphithoquinone (menadione, vitamin K^3) 338
  biochemical functions 339
N- methyl nicotinamide 265, 289
w methyl pantothenic acid 347
2 methyl 3 phytyl 1, 4 naphtho quinone 338
N5 methyl retra hydro folate 261, 264, 265
Methyl transferase apo enzyme 352
Mevalonate 233, 234
  5 phosphate 234
  kinase 234

Micro-emulsion of lipids 173
Microsomal synthesis 227
Milk 439, 463
Millon's test 53, 59
Minerals in nutrition 463
  metabolism 400
Mitochondria 124
Mitomycin 317
Modern techniques 465
Molar depression constant 520
Molecular diseases 431
Molecular weight determination 483
  sedimentation equilibrium 483
  velocity 483
Molisch's test 15, 59
Molybdenum 418
1 mono acyl glycerol 173
2 mono acyl glycerol 173, 214
Mono amine oxidase (MAO) 283, 365
Monoclonal gammopathy 138
Mono-iodo tyrosine 280, 365
Mono-oxygenase 118
Monosodium urate 320
Moore's test 13
Morphia poisoning 148
Motilin 387
m RNA 308
∝ MSH 391, 392
ß MSH 392, 393
Mucin 28
Mucoproteins 56
Multi-enzyme complex 223, 224
Multiple myeloma 138, 140
Multiplication law 548
Muscle 155
  contraction 155
    energy for 160
    mechanism 159
    calcium ions 159
    relaxation 160
    structure 155
Mutarotation 12
Mutation 300
  frame shift 300
  transition 300
  transversion 300
Myasthenia gravis 269
Mycosterols 44
Myofibrils 155, 156
Myoglobin 63
Myo inositol (meso inositol) 354
Myosin 157
  action of papain 157
  action of trypsin 157
Myxedema 203, 368

NAD$^+$ 109, 119, 192, 193, 283
NADH

transport via glycero phosphate 187
  malate 188
NADP$^+$ 109, 119
NADPH 192, 193, 225, 244
NANA 239
Nascent chylomicrons 237
Nascent VLDL 238
Nearest Neighbour Analysis 73
Neonatal tyrosinosis 279
Nerve Growth Factor 377
Nervonic acid 238
Neuberg ester 19
Neumann's Test 59
Neurohypophysin I 395
Neurohypophysin II 395
Neutrial sulphur 268
Niacin (Nicotinic acid) 214.237.282 283, 344
Niacinamide (Nicotinamide) 265, 344
  biochemical function 344
  effects of deficiency 344
Nicotinic acid ribonucleotide (NMN) 282, 283
NIDDM (Type II diabetes mellitus) 2
Niemann Pick's disease 243, 420
Ninhydrin 51, 59
Nitro cobalamine 350
Nitrogen balance 249
  negative 249
  positive 249
Nitrogen equlibrium 245, 249
Nitroprusside 59
Nonactin 522
Non-protein nitrogen compounds 246
Nor-epinephrine 261, 265, 280, 371
  biochemical functions 371, 372
  blood pressure 371
Normal distribution 551
NPN in blood 503
Nucleases 166
Nucleic acid 71
  biological importance 78
  identification 83
Nucleic acids
  biosynthesis 329
Nucleoproteins 55, 65
  metabolism 319
  total hydrolysis 78
Nucleosides 68
Nucleosomes 74
Nucleotides 70
Null hypothesis 554
Numbers 543
  rounding of 54
Nutritionally essential (indispensable) amino acids 455
Nutritional non-essential

(dispensable) amino acids 455
Nutritional value of carbohydrates 460
   lipids 462
   proteins 457
Nyctalopia (night-blindness) 333

Obesity 241, 242
   causes 241
   environmental factors 241
   genetic 241
Obligatory reabsorption of water 395
Obstructive jaundice 496
Ochronosis 279
Oil Red O 475
Okasaki fragments 297, 329
Oleic acid 228
   synthesis 228
Oleo distearin 29
2', 5' oligo adenylate 142
Oncotic pressure 532
One carbon metabolism 350
One carbon transfer 350
One carbon units 261
One gene- one enzyme hypothesis 423
One gene- one polypeptide hypothesis 423
Operon 301
Opsin 332
Ornithine 256, 258, 287
ornithine trans carbamoylase
   (ornithine carbamoyl transferase) 256
Oroso mucoid 28
Orotic acid 327, 328
Orotic aciduria 430
Orotidine 5 phosphate
   (orotidylate) 328
Orotidylic acid decarboxylase 328
Orotidylic pyrophosphorylase 328
Osazones 15, 16
Osmosis 529
   and fluid balance 404
   role in physiological processes 531
      urine excretion 532
Osmotic potency 529
Osmotic pressure 529
   abnormal 531
Osteomalacia 335, 336, 407
Ouabain 20, 171, 523
Oxalic acid 269
Oxalo acetate 190
Oxalo succinate 188
Oxidases 119
ß oxidation of fatty acids 216
   energetics 220
Oxidative phosphorylation 124
   mechanism 126
   oligo mycin and 126
   sites of 125
   uncoupling 126
Oxido reductases 98
Oxy hemoglobin 90, 92
Oxynervon 42
Oxynervonic acid 240
Oxythiamine 342, 355
Oxytocin 249, 395

Palmitoleic acid 228
Palmityl CoA 230
Pancreatic function tests 506
   endocrine 507
   exocrine 507
Pancreatic juice 165
   amylase 167
   enzymes 165
   ionic composition 165
   lipase 167
   polypeptide 378
Pantothenic acid 346
   biochemical functions 347
   effects of deficiency 347
Pantothenyl – SH 226
Pantoyl taurine 347, 355
Paper chromatography 469, 470
PAPS 268
Para hemophilia 152
Para keratosis 416
Parathyroid function tests 506
Parathyroid hormone 249, 369
   biochemical effects 370
   control of secretion 370
   mechanism of action 370
Passive transport 522
Pasteur effect 182
p CO_2 148
Pellagra 284, 344
Pellagra preventive factor 339
Penicillium notatum 231
Pentagastrin 164
Pentasome 76
Pentosuria 422, 426
Pepsin 164
Pepsinogen 164
Peptones 56
Perchlorate 367
Permeability across biomembrane 521
Permease 301, 302
Pernicious anemia 353
Perosis 416
Peroxidase 280
Pestimate p 547
Petechial hemorrhage 357
pH 510, 511
   blood 144
   determination 515
   scale 511
Phagocytosis 365, 523
Phenethyl biguanide (phenformin) 377
Phenols 282
Phenol sulphaline
   (phenol red) excretion 505
Phenol sulphates 268
Phenyl acetic acid 261, 277, 278
phenyl alanine 275
Phenyl alanine hydroxylase 276, 278, 281
Phenyl alaninemia 278
Phenyl ethanolamine N methyl
   transferase 280
Phenyl ketonuria
   (hyper phenyl alaninemia, type I,
   phenyl pyruvic oligophrenia) 278, 424
Phenyl lactic acid 277, 278
Pheochromocytoma 372, 509
Phlorhizin 249
Phosphatidic acid 229, 230, 231
Phisphatidyl choline (lecithin) 38
Phosphatidyl ethanolamine 232
Phisphatidyl inositol (lipositol) 40
Phospho adenosine phospho
   sulphate (PAPS) 80
Phospho creatine 214
Phospho diestase 214
Phospho enol pyruvate 134, 186
Phospho enol pyruvate carboxy kinase
   (phospho pyruvate carboxylase) 190,
   194
Phospho fructo kinase 105, 182, 184, 185
Phospho gluco mutase 177, 180, 184, 199
6 phospho gluconate dehydrogenase 192,
193
3 phospho glyceraldehyde 185
3 phospho glyceraldehyde dehydrogenase
185
3 phospho glycerate 183, 185, 186
2 phospho glycerate 183, 186
Phospho glycerate kinase 183, 185, 189
Phosphoglycero mutase 183, 186
Phospho hexose isomerase 184
Phospholipases 38, 231
   A1 38, 231
   A2 38, 231
   B (lysophispholipase) 38, 231
   C 38, 231
   D 38, 231
Phospholipids 37, 231
   biosynthesis 230
   classification 38
   catabolism 231
   fractionation 43
   functions 231
   mevalonate kinase 234
   5 pyrophosphate 234
4 phospho pantotheine 223, 224, 225, 347
Phospho proteins 56

Phospho ribosyl pyrophosphate 286, 323, 327
Phospho ribosyl pyrophosphate glutamine-amido transferase 323, 325, 327
   feed back inhibition byproducts 327
Phosphoribosyl pyro phosphate synthetase 323
Phosphorus 240, 409
Phosphorus role in the body 409
Phosphorus serum 409
Phosphorus vitamin D3 409
Phosphorylase a 179, 184
   b 179
   b kinase 179
   cAMP 176
Phosphoryl choline cytidyl transferase 230
Phospho sphingosides (sphingo myelins) 41
Phospho triose isomerase 185
Photometry 476
   different types 478
Photo phosphorylation 204
   cyclic 205
   non-cyclic 205
   differences between 206
Photosynthesis 204
   carbon dioxide fixation 207
   proteins and fats 208
Phrenoderma 462
Phrenosin (cerebron) 42
Physical exercise 242
Phytic acid (inositol hexa phosphate) 354, 407
Phytosterols 44
Picramic acid 445
Picric acid 445
Piercidin 125
Pigment P^{680} 206
Pigment P^{700} 206
Pinocytosis 365, 523
Pipecolic acid 273
Piperidine 2 carboxylic acid 273
Placental hormones 394
Placental lactogen 394
Plasma, blood 436
Plasmalogen 39
   proteins 135, 499
   fractionation 136
   oncotic pressure 404
   thromboplastin component 150
Plasmin 153
Plasminogen 153
   activator 153
Plastocyanin 204
Plastoquinone 204
Pneumonia 148

Polyacrylamide gel disc electrophoresis 136
Poly A tail 298, 306
Poly cistronic m RNA 302
Polyclonal gammopathy 138
Polycythemia 417
Polynucleotide phosphorylase 3330
polyphenyl alanine 299
Polyribosome (Poly somes) 76, 307, 311
Polysaccharides (glycans) 22
Poly U 299, 330
Poly UG 299, 330
Poly uria 202, 395
Pompe's disease (Glycogen storage disease, Type II) 426
Population 539, 542
porphin 84
Porphobilinogen 85, 425
Porphyrias 427
   acquired 428
   congenital 427
      erythiopoietic 427
      hepatic 427
   cutanea tarda 427
   differences between hereditary and acquired 428
Porphyrin 84
   biosynthesis 84
   excretion 87
Porphyrinuria 428
Porphyropsin 333
Potassium 401
   deficiency effects 401
   hyper kalemia 402
   hypo kalemia 402
   plasma levels 401
   shifts between extracellular and intracellular compartments 402
Pre-allumin 141, 333
Pre-beta lipoprotein 212
Pre-diabetics 509
Pregnant mare serum gonadotrophin (PMSG) 395
Δ_5 pregnenolone 380, 383
Prekallikrein 150
Pre-mRNA 306, 308, 310
Pre-pro-hormone 370
Primaquine sensitivity 192, 425
Primary hyper oxalunia 270
Pro-accelerin 149
Probability 539, 546, 552
   characteristics 547
   estimation 547
Probability distribution 548
Probucol 237
Proconverin 150
Progesterone 380, 383, 386, 396
   and water retention 401

Pro-hormone 370
Prolactin 393, 396
Prolactin release inhibiting hormone (dopamine) 388, 393
Prolamine 55
Proline 287
Promoter site 302
Pro opio melano cortin 393, 360
Proportional Gas Flow Counter 491
Prostacyclins 31, 242, 397
   synthetase 399
Prostaglandins 31, 242, 397
   biosynthesis 397
   E^1 214
   levels in blood 398
   urinary levels 398
Prostaglandin synthetase (fatty acid cyclo oxygenase) 397
Protamines 55
Protamine zinc insulin 415
Protection of colloids 520
Proteins 46, 456
   biological value 456
   calorie malnutrition 458
   calorie cortical hormones in 458
   classification 55
   definition 46
   denaturation 58
   dynamic flux between tissue and plasma proteins 247
   efficiency ratio of common foods 456, 457
   flocculation 58
   hydrolysis 57
   integral 2
   isoelectric point 57
   isolation 59
   kinase 214
   labile 249
   malnutrition 458
   metabolism 245
   molecular weights 57
   peripheral 2
   phosphatase 178, 180
   precipitation 58
   reactions of 57
   storage 249
   and amino acids storage 249
   structure 60
      polar (salt) linkages 63
      primary 60
      quaternary 64
      secondary 61
      $\propto$ helix 61
      ß plait 61
      tertiary 63
   urine 248
Protein biosynthesis 306

INDEX

accommodation 312
active transport of amino acids 306
Bird's eye view 318
elongation 312
endocrine influence 306
energetics of 316
initiation 312
peptide synthetase
 (peptide transferase) 307, 314
termination 315
translocation 312
requirements 307
role of antibiotics 316
 actinomycin D 316, 317
 chloromycetin (chloramphenicol)
  316, 317
 mitomycin C 316, 317
 neomycin 316, 317
 puromycin 316, 317
  streptomycin 316, 317
  tetracyclins 316, 317
role of GTP 316
Protein–bound iodine 366, 505
Protein calorie malnutrition 458
 cortical hormones in 458
Proteoglycans 25
Proteoses 56
Prothrombin 149, 151, 153
 time 154, 500
Protoporphyrin 428
Protoporphyrin III, 86, 87
Protoporphyrinogen III 86
Pro-vitamin D³ 334
Pro-vitamin D² 334
Pseudo-choline esterase 115
Pseudo-uridine 69, 298
Psychosin (galactosyl sphingosine) 239
Pteridine 349
Pteroyl glutamic acid (Folic acid) 340
Pulmonary surfactant 244
Purines
 biosynthesis 323
  source of various atoms 323
Purine and pyrimidine metabolism 319
Purine nucleoside phosphorylase 322
Pyelo nephritis 391
Pyloric obstruction 148
Pyridine 3 sulphonic acid 345, 355
Pyridoxal 345
Pyridoxal phosphate 111, 112, 230, 253,
 280, 283, 345
 biochemical functions 345
 effects of deficiency 346
 reactions requiring pyridoxal phosphate 345, 346
Pyridoxamine 345
Pyudoxine (vitamine B⁶) 345
Pyrimidines 65, 327

biosynthesis 327
catabolism 320
Pyrithiamine 342, 355
Pyroglutamic acid 388
Pyrophosphatase 256
Pyrophosphate 256
Δ pyrioline 5 carboxylic acid 287
Pyruvate 186, 187
Pyruvate carboxylase 105, 190, 194
Pyruvate dehydrogenase complex 188,
 190
Pyruvate kinase 183, 186

Questran 237
Quick's method 154
Quinaldic acid 283
Quinolinic acid 283

Radio activity units 488
 Curie 489
 Rad 489
 Rem 489
 Roentgen 489
Radio immunoassay 491
 importance 494
Randomness 539
Random sampling 543
Rank Sum test 553, 557
Rapoport Luebering cycle 187
Red drop 206
Refsum's disease 427
Regitine 509
Reichert Meissl Number 36
Relaxin 377
Release Factor 307, 315
Renal function tests 503
Renal glycosuria 508
renal plasma flow 504
Renotropic 390
Replication bubble 297
Repression 301
 catabolic 301
Repressor protein 302
Respiratory chain 123
 inhibition of 125
Respiratory Quotient (RQ) 293
 foods 449
Respirometry (manometry) 480
Resting membrane potential 401, 521
Resting metabolic expenditure 451
Retinal 332
 all prans 332
 cis 332
 isomerase 332, 333
Retinoic acid 332, 333
Retinoids 332

Retinol 331
 all trans 332
 cis 332
Retinol-binding protein 141, 333
Reverse T3 364
Rh factors 28
Rhodanase 446
Rhodopsin 332
Rho factor 298
Riboflavin (Vitamin B2) 342
 biochemical functions 342
 effects of deficiency 343
 nucleotide reductase 325
 reduction 325
 reduction eukargotes 325
 reduction prokaryotes 325
Ribose 18
 5 phosphate 192, 193
Ribosome 76, 309
 ammo acyl site 311, 312
 peptidyl site 311, 312
Ribothymedine 68
Ribulose 1, 5 diphosphate cycle 206, 207
Ribulose 5 phosphate 191, 193, 207
 kinase 207
Richnar Hamhart syndrome
 (Tyrosinosis II) 279
Rich's scheme 310
Rickets
 renal nickets 336, 409
RNA 75
 common C-C-A termines 308
 isolation 70
 requence shidres 77
 structure
  primary 76
  secondary 76, 77
 synthesis in vitro 329
 types
  messenger RNA 76
  ribosomal RNA 76, 309
  transfer RNA 76
RNA dependent DNA polymerase 330
 (Reverse transcript)
RNA dependent RNA Polymerase 330
 (Replicase)
RNAse L 142
Robinson ester 19
Rotenone 125
Rothera's test 202, 223
R Proteins 352
Rutin 355

Saccharic acid (glucaric acid) 14
Saccharomyces Carlbergensis 346
Sakaguchi reaction 53

Salicylate poisoning 148, 149
Saliva 438
Salkowsky's test 45
Salvage pathway 326
Sample 539
   randomness 543
   stratified 543
Sapogenins 43
Saponification value 36
Saponin 20
Sarcolemma 156
Sarcomere 156
Sarcoplasm 156
Sarcosin 263, 264
Sarcotubular system 156
Scavenging of body cholesterol 242
Schardinger's enzyme 319
   (Xanthine dehydrogenase)
Schiff's base 253
Scintillatin counter 492
Scleroproteins (albuminoids) 55
Scurvy 356
SDA (Specific Dynamic Action,
   thermogenic effect) 452
   probable causes 452
Secretin 168, 387
Sedimentation equilibrium method
   (molecular weight determination) 483
Sedimentation velocity method
   (molecular weight determination) 483
Sedoheptilose 7 phosphate 192, 193, 207
Selenium 418
   garlicky breath 418
Selenosis 418
Selewanoff's test 17
Semi-indispensable amnio acids 455
Seminal fluid 438
Sensor gene 305
Serine 264, 270
   sulphhydrases 287
Serotonin 283, 284
   (enteramin, thrombocytin) 283, 284
Sex hormone binding globulin 383, 360
Sex-linked dominant (gene) 434
   vit D resistant rickets 434
   G 6 P D deficiency 434
Sex-linked recessive 423, 433
   Gout 423, 433
   Lesch Nyhan syndrome 423, 433
Sexual doublet concept 310
   (Wobble theory)
Sialic acid 27, 239
Sickle cell hemoglobin 92
Sickle cell anemia 425
   tactoids 425
Sigma factor 298
Signing Rank Test 557

paired data 557
unpaired data 558
Simmond's disease 203
Sitosterol 43
Smoking 242
Sodium 400
   deficiency manifrestratium
   hypernatremia 401, 400, 401
Sodium levels in plasma 400
   osmotic pressure
   retention and progesterone 401
Sodium Potassium ATPase 171, 523
Somatomedin C (Insulin-like)
   growth factor, IGF_1) 389
Somatostatin 372, 378, 388, 393
Sorbitol 14, 198, 384
Sorbitol dehydrogenase 198, 384
Soya bean 458
Specific activity 489
Spectacled-eye syndrome 348
Spectrin 2
Sphingol 270
Sphingomyelin (phosphosphingosides)
   41, 230
   biosynthesis 230
Sphingosine 230, 239
Sphingosyl phosphoryl choline 230
Spironolactone 509
Squalene 233, 235
   synthetase 235
Stability of colloids 519
Stable isotopes 488
Standard deviation 549
   methods of calculation 549, 550
Standard oxidation reduction potential
   125
Starch (amylus) 22, 27, 207, 222
Starlings hypothesis 533
   role in maintenance of blood
      volume 533
starvation 293
   blood pressure 294
   BMR 294
   brain demands 293
   enzymes affected 294
   β hydroxy butyrate 294
   insulin 294
   ketosis 293
   labile proteins 293
   loss of body water 294
   muscle proteins 294
   second messenger (cAMP) 293
Stearyl CoA 226
Stercobilin 94, 95
Stercobilinogen 94, 95
Stern potential 519
Steroidal hormones 362

mode of action 362
Steroids 43
Sticky end technique 329
Stigmasterol 44
Streptococcus lactus 350
Streptomycin 317
Strophanthus 20
Stuart Prower Factor (X) 150, 153
   activation 150
   extrinsic 150
   intrinsic 150
Student's t test 553
Substrate phosphorylation 185, 186, 188
Succinate dehydrogenase 189, 190
Succinate throkinase 188
Succinyl CoA 85, 188, 190
Succinyl CoA acetoacetate CoA
   transferase 221
Sucrose 21, 207
Sudan III 475
Sugar amines 19
Sugars
   definition 8
Sulfation factor 389
Sulfin pyrazone (anturan) 430
Sulpha drugs 355, 367
Sulphanilamide 106, 354
Sulphate
   excretion 267, 268
   ethereal 267, 268
   inorganic 267, 268
   neutral 267, 268
Sulphatides 42, 239
Sulphate of cysteine and homocysteine
   291
Sulpholipids 42
Sulphonyl ureas (tolbutamide) 203, 373, 377
Sulphur 410, 411
   metabolism 267
   physiological role 410
Super oxide dismutase 130
Super oxide ion 251
(Super oxide anion)
Surface tension 535
   and adsorption 535
Svedberg unit 482
Sweat 443
Swivelases 297
Syneresis (weeping of colloids) 520
Synovial fluid 437

Taurine 49, 290
Taurinuria 352
Taurocholic acid 290
Tay Sach's disease 427
't' distribution 553

INDEX

Tears 439
Terminal pathway 291
Testosterone 383
Testosterone—estrogen—binding globulin (TEBG) 383
Tetany 402, 408
Tetracyclins 317
Tetrahydrobiopterin 276, 279
Tetra hydrofolate 262, 349
Tetra hydro isoquinoline alkaloids 243
Tetra iodo thyro acetic acid 417
Thalassemia 425
Thermal conductivity cell 472 (GLC)
Thermogenic effect 452
Thiamine (aneurine), vitamin B, biochemical function 340
effects of deficiency 341
Thiamine pyrophosphate 111
Thin layer chromatography 471
Thiocarbamide 367
Thioctic acid (lipoic acid) 347
biochemical function 348
Thiocyanates 267, 268, 289, 367
Thioesterase 224, 225
Thiokinase 217, 219, 221
Thiolase 221, 234
Thiophorase 189, 221
Thioredoxin 325
Thioredoxin in reductase 325
Thiosulphates 268, 289
Thiourea 364
Threorine 271
dehydrase 271
Thrombin 151, 153
Thrombocytin (serotonin) 283
Thromboxanes 33
Thymedine 68
Thymedylate synthetase 327, 328
Thymidylic acid 70
Thymine 67, 263
Thyroglobulin 281, 364, 365
Thyroid function tests 505
Thyroid hormones 363
biosynthesis 364
mode of action 366
secretion 364
transport 366
Thyroid stimulating hormone 280, 388, 399
biochemical effects 392
Thyroperoxidase 364, 365
Thyrotropin-releasing hormone 387, 392
Thyroxine 281, 363, 364
Thyroxine-binding globulin 141, 360
Thyroxine-binding pre-albumin 141, 360
Thyroxine-deiodination 417

Tiglyl CoA ($\propto$ methyl crotomyl CoA 273
TmG (Tubular maximum for glucose) 200
Tm PAH (for para amino hippurate) 505
Tobacco mosaic-virus 296
Tocopherols 337
Tolleins naphtho resorcinol test 19
phloroglucinol test 19
Toller's reagent 15
Tophi 320
T ψ C loop 308
Trace elements 411
Transaldolase 192, 193, 207
Transamination 252, 275
mechanism 253
Transcobalamin I 352
Transcobalamin II 141, 352
Transcostin (control-binding protein) 141
Transcriptase 298
Transcription (mRNA synthesis from DNA) 297, 330
Transdeamination 253
Transferases 98
Transferrin 141, 412
Transformylase 325
Transglutaminase 152
Transhydroginase 385
Transketolase 192, 193, 207
Translation 300
role of hormones 306
Transmanganin 416
Transmission 477
Transport phenomena across the membrane 522
co-transport 522
Trans-sulphuration 261
Triacyl glyceroh 227
biosynthesis 228
Triacyl glocerol lipase 214
hormone sensitive 214
Tri carboxylic acid cycle 188
Trigonelline 446
Tri iodo thyro acetic acid 417
Tri iodo thyronine 281, 363, 364
Tri methyl ammonium ß hydroxy butyrate (carnitine) 219
Trimethyl hydroxy ethyl ammonium hydroxide 354
(tri methyl N ethanolamine hydroxide choline) 354
Tristearin 449
Tritium oxide 403
t RNA amino acids 307
Tropomyosin 157
Troponins 157
calcium binding 158
inhibitory 158

tropomyosin-binding 158
True sugar 201
Trypsin 166
Trypsinogen 166
Tryptophan 282
Tryptophan dioxygenase (11 pyrolase) 118, 282, 283, 285
Tryptophan hydroxylase 284
TSH 214
't' test 555
Tubeless test meal 503
Tuberculosis 138
't' values 559, 560
paired 't' test 556
two sample tests 556
Tyndall effect 518
Tyramine 282, 444
Tyrosine 275
Tyrosinemia Type I 278
Tyrosinosis 279, 281, 424
Tyrosinosis neonatal 279
Tyrosinosis Type II 279

UDP arabinose 207
UDP galactose 82
UDP galactose epimerase 199
UDP glucose 82
UDP glucose dehydrogenase 197
epimerase 239
pyrophosphorylase 177, 197, 199
UDP glucuronic acid 197
UDP glucuronyl transferase 446
UDP xylose 207
Ultracentrifuge 482
analytical 488
preparative 488
uses 483
Universal indicator 515
$\propto$, ß unsaturated acyl CoA 218
$\propto$, ß unsaturated acyl CoA reductase 215, 226
$\propto$, ß unsaturated acyl CoA thio esterase 272
$\propto$, ß unsaturated acyl enzyme 225
$\propto$, ß unsaturated stearyl CoA 226
Unwindase 298
Uracil 66, 263
Urea clearance 503
maximum 504
standard 504
Urea cycle 255, 258
disorders 259
energetics 258
Urea in urine 258
ureido isobutyric acid 321

uredo propionic acidf 321
ureido succinic acid 328
Ureotelic 245
uric acid 322
  miscible pool 320
Ureco oxidase (uricase) 320
uricosuric drugs 320
uricotelic 245
uridine 68, 69
uridine diphospho glucose 198, 207
uridine nucleotide 82, 112
uridylic acid 70
urine 440
urine
  abnormal 442, 443
  Bence Jones proteins 442
  bilirubin 442
  blood 442
  fructose 442
  galactose 442
  glucose 442
  ketones 442
  normal 441
  pentose 442
  porphyrins 442, 443
  proteins 442
Urobilin 94, 95
Urobilinogen 94, 95
Urocanase 285
urocanic acid 285
urokinase 153
uronic acid pathway 197
  significance 197
uronic uroporphyrinogen I 87
uronic uroporphyrinogen III 86
uroporphyrinogen I synthase 86, 427
uroporphyrinogen III cosynthase 427
UTP 239

Value 271
Vanadium salts 237
Van den Bergh's reaction 96
Van der Waal's forces 60, 63
Vanillyl mandelic acid VMA 372, 509
Van Slyke's method 53

Vant Hoff's factor 'i' 531
Vant Hoff's theory of dilute solution 530
Variance (s^2) 548
  coefficient of variance 551
  determination of variance 549, 550, 551
Variate (quantitative) 539, 540
Vaso-active intestinal peptide 393
Vasopressin 249, 395
  (anti-diuretic hormone ADH)
Very Low Density Lipoproteins 210, 237, 238
Viscosity 536
  blood 538
  chemical constitution 537
  molecular weight 537
  protoplasm 537
Vitamine 331
vitamin 331
  fat-soluble 331
  water-soluble 339
Vitamin A, anti-xerophthalmic vitamin 331
  biochemical action 332
  deficiency effects 333
  estimation 334
  hypervitaminosis 334
Vitamin B_1 (thiamine) 340
Vitamin B_2 (riboflavin) 342
Vitamin B_6 (pyridoxine) 345
Vitamin B_{12} 352
  DNA synthesis eukaryotes 353
  DNA synthesis prokaryotes 353
Vitamain C (L ascorbic acid) 355
  biochemical function 356
  hydroxylation of proline and lysine 356
  effects of deficiency 356
Vitamin D (anti-rachitic) 335
  biochemical function 335
  as a hormone 337
Vitamin E (anti-sterility) 337
  (tocopherols)
  biochemical function 338
  effects of deficiency 338
Vitamin K (k, K_1 K_2, K_3) 338
  K_3 (menadione) 338
  V carboxylation of glutamic

$^{3'}$
acid (GLA) 339
effects of deficiency 339
Vitamin P 355
  (Bioflavonoids)
Vitellin 56
VLDL remnants (IDL) 242
Void volume 468
Volhard's method 54
Von Gierke's disease 426
Von willibrand's disease 424

Waldenstrom's macroglobulinemia 138
Wald's visual cycle 332
Warburg's manometry 481
Water-soluble vitamins 339
Waxes 29
Weeping of gels (syneresis) 520
Wilcoxon's Rank Sum test 553, 557
Wilson's disease 141, 415
Wobble theory 310

Xanthine 282, 283
Xanthine dehydrogenase 130, 319, 322
Xanthoproteic reactors S_9
Xanthosine monophosphate 326
Xanthosine acid 346
Xerophthalmia 333
Xylan 207
Xylitol 197
Xylulose 5 phosphate 192, 197

Zak's method 45
Zein 457
Zeta potential 519
Zinc 415
  biochemical role 415
  intestinal absorption 416
Z line 156, 157
Zone electrophoresis 474
Zoosterols 44
Zymogen 165
Zymosterol 44, 236